Proceedings of

Dynamic Systems and Applications

Volume 3

Proceedings of

Dynamic Systems and Applications

Volume 3

Proceedings of the Third International Conference on Dynamic Systems and Applications held at Morehouse College, Atlanta, USA, May 26-29, 1999

Editors

G. S. Ladde
University of Texas at Atlington, TX, USA.,

N. G. Medhin
Clark Atlanta University, Atlanta, GA, USA., and

M. Sambandham
Morehouse College, Atlanta, GA, USA

Editorial Committee

R. P. Agarwal
Shair Ahmad
D. Bainov
R. Bozeman
C. Y. Chan
J. Graef
J. K. Hale
L. Hatvani
D. Kannan
J. de Klerk
V. B. Kolmanovskii
V. Lakshmikantham
V. M. Matrosov
R. E. Mickens
A. Msezane
M. Rama M. Rao
S. Sathanantham
R. Shivaji
S. Sumner
C. P. Tsokos
A. S. Vatsala
F. Zanolin

2001
Dynamic Publishers, Inc, U.S.A.

Dynamic Publishers, Inc.
P. O. Box 48654
Atlanta, Georgia 30362-0654

Library of Congress Catalog Card Number: 2001-132707

Ladde, G. S., N. G. Medhin, and Sambandham, M.
Proceedings of Dynamic Systems & Applications, Volume 3

ISBN 0-96-40398-34

* The third International Conference on Dynamic Systems and Application was funded by the Department of Navy Grant N00014-99-1-0629 issued by the office of Naval Research. The United States Government has a royalty-free license throughout the world in all copyright material contained in this publication.

Printed in the United States of America

CONTENTS

PREFACE

The **third International Conference on Dynamic Systems and Application** was held at Morehouse College, Atlanta, during May 26-29, 1999. This conference was held under the auspices of the International Federation for Nonlinear Analysts, and was sponsored by the Department of Mathematics, Morehouse College, Clark Atlanta University: Center for Theoretical Studies of Physical Systems (A National Science Foundation Research Center of Excellence), and the office of Naval Research.

This volume consists of the proceedings of the conference. It includes selected research reports and invited addresses as well as contributed papers. The aim of the conference was to discuss the trends in the related topics of dynamic systems and applications. There were over one hundred and fifty talks and the participants were from thirty countries. This proceedings attempts to put together the multidisciplinary work of researchers ranging from mathematicians, scientists and engineers.

The problems discussed are both deterministic and stochastic. This volume contains the research investigations on concerning wave equations, wavelets, cellular automata, Boolean networks, quantum gravity, various types of differential equations and difference equations, diffusion equations, heat equations, Integro-differential equations, Lyapunov's methods, mixed boundary value problems, numerical stability, partial differential equations, population models, quenching phenomena, singularly perturbed systems, stochastic approximations, stochastic controls, etc.

This proceedings will provide an upto date reference book for researchers working in the areas of mathematics, science and engineering who are interested in the recent works on all aspects of differential equations, integral equations, discrete analog of these equations, and applications.

We wish to express our sincere thanks to R. P. Agarwal, Shair Ahmad, C. Y. Chan, J. R. Graef, D. Kannan, V. Lakshmikantham, R. E. Mickens, A. Z. Msezane, M. Rama Mohana Rao, K. M. Ramachandran, A. S. Vatsala, who organized several sessions, and all participants for their spirit and support. We are extremely thankful to Walter E. Massey, President, and all faculties of Department of Mathematics, Morehouse College for their dedicated help to organize this meeting.

G. S. Ladde
N. G. Medhin
M. Sambandham
Editors

Proceedings of Dynamic Systems and Applications 3 (2001) 1-8

Regularity of Solutions to a Quasilinear Size-Structured Population Model

Azmy S. Ackleh
Department of Mathematics
University of Louisiana at Lafayette
Lafayette, Louisiana 70504

Abstract

We consider the following initial-boundary value problem that describes the dynamics of a size-structured population:

$$\begin{cases} u_t + (g(x, P(t))u)_x + m(x, P(t))u = 0, & (x,t) \in (0,l) \times (0,T) \\ g(0, P(t))u(0,t) = C(t) + \int_0^l \beta(x, P(t))u(x,t)\,dx, & t \in (0,T) \\ u(x,0) = u^0(x), \quad x \in [0,l], \end{cases}$$

where $P(t) = \int_0^l u(x,t)dx$. We establish regularity results for the unique bounded variation weak solution discussed in [2] under some smoothness and compatibility conditions on the initial data u^0.

Key words. Quasilinear size-structured population model, regularity of weak solutions

AMS subject classification. 35A40, 65M06, 92D25.

1 Introduction

In this short note, we consider the following initial-boundary value problem that describes the dynamics of a size-dependent population with nonlinear growth, reproduction and mortality rates:

$$\begin{cases} u_t + (g(x, P(t))u)_x + m(x, P(t))u = 0, & (x,t) \in (0,l) \times (0,T) \\ g(0, P(t))u(0,t) = C(t) + \int_0^l \beta(x, P(t))u(x,t)\,dx, & t \in (0,T) \\ u(x,0) = u^0(x), \quad x \in [0,l]. \end{cases} \tag{1}$$

Here $u(x,t)$ is the density of individuals of size x at time t, and $P(t) = \int_0^l u(x,t)\,dx$ is the total number of individuals in the population at time t. The parameters m and

β denote the mortality and reproduction rates of an individual of size x, respectively. The function g denotes the growth rate of an individual of size x and C represents the inflow of zero-size individuals from an external source.

Existence and uniqueness of solutions to the problem (1) have been discussed in [2, 3]. In [3] an argument which is based on the characteristic method and fixed point theory is used to establish the existence and uniqueness of solutions that are differentiable along characteristic curves. In [2] the authors developed a finite difference approximation to the problem (1). By passing to the limit (via a compactness argument) the existence of a bounded variation weak solution, which coincides with the solution defined in [3], is established. Uniqueness of this weak solution was then proved using the Holmgren uniqueness theorem. This approach has the advantage that it results in an approximation that can be easily computed and the convergence of which to the exact solution is established. See Section 2 for a review of the results presented in [2].

The goal of this paper is to prove additional regularity of the unique bounded variation weak solution to the problem (1) under some smoothness and compatibility conditions on the initial data u^0 (see Section 3).

2 Background Material

In this section we review the results established in [2]. To this end, we define a (weak) solution to the problem (1) to be a locally bounded and measurable function $u(x,t)$ which satisfies

$$\begin{aligned}\int_0^l u(x,t)\varphi(x,t)\,dx - \int_0^l u^0(x)\varphi(x,0)\,dx &= \int_0^t\int_0^l (u\,\varphi_s + gu\,\varphi_x - mu\,\varphi)\,dx\,ds \\ &+ \int_0^t \varphi(0,s)\left(C(s) + \int_0^l \beta(x,P(s))u(x,s)\,dx\right)ds.\end{aligned}$$

Furthermore, the following assumptions will be imposed on the parameters throughout the paper:

(H1) $u^0(x) \in BV(0,l)\cap L_\infty(0,l)$ and $u^0(x)\geq 0$.

(H2) $m(x,P)$ is a non-negative continuously differentiable function with respect to x and P.

(H3) $\beta(x,P)$ is a non-negative continuously differentiable function with respect to x and P.

(H4) $g(x,P)$ is twice continuously differentiable function with respect to x, and continuously differentiable function with respect to P, $g(x,P)>0$, $\quad x\in[0,l)$ and $g(l,P)=0$.

(H5) $C(t)$ is a continuously differentiable function and $C(t) \geq 0$.

(H6) $\sup_{(x,P)\in[0,l)\times[0,\infty)} \{\beta(x,P) - m(x,P)\} \leq \omega_1$.

(H7) For any sufficiently small $\delta > 0$

$$\sup_{(x,P)\in[0,l)\times[0,\infty)} \left| \frac{g(x+\delta,P) - g(x,P)}{\delta} + m(x,P) \right| \leq \omega_2.$$

Let $\Delta x = \frac{l}{N}$, and $\Delta t = \frac{T}{M}$ denote the spatial and time mesh size, respectively. Let the spatial and time mesh points be given by $x_j = j\Delta x$, $j = 0, 1, 2, \cdots, N$ and $t_k = k\Delta t$, $k = 0, 1, 2, \cdots, M$, respectively. We denote by u_j^k and P^k the difference approximations of $u(x_j, t_k)$ and $P(t_k)$ and we define

$$g_j^k = g(x_j, P^k), \quad \beta_j^k = \beta(x_j, P^k), \quad m_j^k = m(x_j, P^k) \text{ and } C^k = C(t_k).$$

We define the difference operator

$$D_h^-(u_j^k) = \frac{u_j^k - u_{j-1}^k}{\Delta x}, \quad 1 \leq j \leq N$$

and the ℓ^1 and ℓ^∞ norm by

$$\|u^k\|_1 = \textstyle\sum_{j=1}^N |u_j^k| \, \Delta x$$

$$\|u^k\|_\infty = \max_{j=0,1,2,\cdots,N} |u_j^k|.$$

In [2] the following implicit finite difference approximation was developed to approximate the solution of the problem (1):

$$\begin{cases} \dfrac{u_j^{k+1} - u_j^k}{\Delta t} + \dfrac{g_j^k u_j^{k+1} - g_{j-1}^k u_{j-1}^{k+1}}{\Delta x} + m_j^k u_j^{k+1} = 0, & 1 \leq j \leq N \\ g_0^k u_0^{k+1} = C^k + \sum_{i=1}^N \beta_i^k u_i^{k+1} \, \Delta x & \\ P^{k+1} = \sum_{i=1}^N u_i^{k+1} \, \Delta x & \\ u_j^0 = \dfrac{1}{\Delta x} \displaystyle\int_{(j-1)\Delta x}^{j\Delta x} u^0(x)dx, & 1 \leq j \leq N. \end{cases} \tag{2}$$

The following result concerning the existence and uniqueness of nonnegative solutions to the difference approximation (2) has been established in [2].

Lemma 1 *Assume that Δx and Δt are chosen so that*

$$\begin{cases} \Delta x \dfrac{\beta_j^k}{g_0^k} + \dfrac{\Delta t}{\Delta x} g_j^k (1 + \dfrac{\Delta t}{\Delta x} g_{j+1}^k + \Delta t \, m_{j+1}^k)^{-1} < 1 & 0 \leq j \leq N-1 \\ \Delta x \dfrac{\beta_N^k}{g_0^k} < 1. & \end{cases} \tag{3}$$

Then (2) has a unique nonnegative solution.

Notice, for example, that if

$$\Delta x \frac{\beta_j^k}{g_0^k} < \delta \quad \text{and} \quad \frac{\Delta t}{\Delta x} g_j^k \leq 1 - \delta$$

for $1 \leq j \leq N$ and some $\delta \in (0,1)$ then (3) holds. The following estimates on the difference approximations u_j^k were established in [2].

Lemma 2 *Assume that (3) holds and that* $\max(\omega_1, \omega_2)\Delta t < 1$. *Then there exist positive constants* $c_i, i = 1, \ldots, 4$, *such that*

$$||u^k||_1 \leq c_1, \quad ||u^k||_\infty \leq c_2, \quad ||D_h^-(u^k)||_1 \leq c_3, \quad \text{and} \quad \sum_{j=1}^{N} |\frac{u_j^m - u_j^p}{\Delta t}| \Delta x \leq c_4(m-p).$$

Using Lemma 2 and similar techniques as those in [4, 6] the authors in [2] show that the family of functions $\{U_{\Delta x, \Delta t}\}$ defined by

$$U_{\Delta x, \Delta t}(x,t) = u_j^k \quad \text{for } x \in [x_{j-1}, x_j), \ t \in [t_{k-1}, t_k), \ j = 1, ..., N, \ k = 1, ..., M$$

is compact in the space $L^1((0,T) \times (0,l))$ and hence there exists a sequence $\{U_{\Delta x_i, \Delta t_i}\} \subset \{U_{\Delta x, \Delta t}\}$ which converges to a $BV([0,T] \times [0,l])$ function $u(x,t)$ in the sense that for all $t > 0$

$$\int_0^l |U_{\Delta x_i, \Delta t_i}(x,t) - u(x,t)| \, dx \to 0,$$

and

$$\int_0^T \int_0^l |U_{\Delta x_i, \Delta t_i}(x,t) - u(x,t)| \, dx \, dt \to 0,$$

as $i \to \infty$. Furthermore, the limit function satisfies $||u||_{BV([0,T]\times[0,l])} \leq c_5(||u^0||_{BV}, ||C||_{C^1})$.

The next Lemma shows that the limit function $u(x,t)$ constructed via the difference scheme is actually a weak solution of the problem (1).

Theorem 3 *The limit function $u(x,t)$ constructed via the difference approximation (2) is a weak solution of (1).*

Finally using Holmgren uniqueness theorem the following uniqueness result concerning weak solution with bounded total variation was established in [2].

Theorem 4 *The bounded variation weak solution to the problem (1) is unique.*

3 Regularity of Weak Solutions

Our next goal is to prove that the bounded variation weak solution to the problem (1) is continuous. Such a result shows that *nonlocal,* quasilinear, hyperbolic initial-boundary value problems similar to (1) are mathematically different from *local,* quasilinear, first order hyperbolic equations (e.g., conservation laws). In particular, it is well known that solutions to the later ones may attain discontinuities in finite time. To this end, define $P^0 = \int_0^l u^0(x)dx$ and set $g^0(x) = g(x, P^0)$ and $\beta^0(x) = \beta(x, P^0)$. Now, assume that the initial data u^0 satisfies the following compatibility condition:

(H8) u^0 is a non-negative continuous function which satisfies

$$g^0(0)u^0(0) = C(0) + \int_0^l \beta^0(x)\, u^0(x)dx.$$

We now have the following theorem.

Theorem 5 *Under the additional assumption (H8) the bounded variation weak solution to the problem (1) is continuous.*

Proof. Let $\hat{\beta}$ and $\hat{m}$ be non-negative continuously differentiable functions. Assume also that $\hat{g}$ is twice continuously differentiable with respect to x and continuously differentiable with respect to t, $\hat{g}(x,t) > 0$, $x \in [0, l)$ and $\hat{g}(l, t) = 0$, $t \in [0, T]$. Now, consider the following linear initial-boundary value problem:

$$\begin{cases} v_t + (\hat{g}(x,t)v)_x + \hat{m}(x,t)v = 0, & (x,t) \in (0,l) \times (0,T) \\ \hat{g}(0,t)v(0,t) = C(t) + \int_0^l \hat{\beta}(x,t)v(x,t)\,dx, & t \in (0,T) \\ v(x,0) = u^0(x), \quad x \in [0,l]. \end{cases} \tag{4}$$

It follows from the results in [2] that the linear problem (4) has a unique bounded variation weak solution. Furthermore, arguing as in [1] we see that a characteristic curve passing through $(\hat{t}, \hat{x})$ is given by $(t, X(t; \hat{t}, \hat{x}))$, where X satisfies

$$\frac{d}{dt}X(t; \hat{t}, \hat{x}) = \hat{g}(X(t; \hat{t}, \hat{x}), t)$$

and $X(\hat{t}; \hat{t}, \hat{x}) = \hat{x}$. By the assumptions on $\hat{g}$ the function X is a strictly increasing function, and therefore a unique inverse function $\Gamma(x; \hat{t}, \hat{x})$ exists. Hence, if we define $G(x) = \Gamma(x; 0, 0)$, then $(G(x), x)$ represents the characteristic curve passing through $(0, 0)$ and this curve divides the (t, x)-plane into two parts. Furthermore, using the method of characteristics we see that the solution v to the linear problem has the following implicit representation:

$$v(x,t) = \begin{cases} u^0(X(0;t,x))\cdot \\ \exp\left\{-\int_0^t(\hat{g}_x(X(s;t,x),s)+\hat{m}(X(s;t,x),s))ds\right\} & t \leq G(x) \\ R(\Gamma(0;t,x))\cdot \\ \exp\left\{-\int_{\Gamma(0;t,x)}^t(\hat{g}_x(X(s;t,x),s)+\hat{m}(X(s;t,x),s))ds\right\} & t > G(x) \end{cases} \tag{5}$$

where $R(t) = (C(t) + \int_0^l \hat{\beta}(x,t)v(x,t)dx)/(\hat{g}(0,t))$.

Now, let $P(t) = \int_0^l u(x,t)dx$ where u is the unique bounded variation weak solution to the problem (1). Note that P is continuously differentiable (see [2]). Using (H2)-(H4) we see that $\hat{g}(x,t) \equiv g(x,P(t))$, $\hat{\beta}(x,t) \equiv \beta(x,P(t))$ and $\hat{m}(x,t) \equiv m(x,P(t))$ satisfy the above requirements. Hence, using (H8) and the results in [1] it follows that the solution v to the linear problem (4) is continuous for this choice of functions $\hat{g}$, $\hat{\beta}$, and $\hat{m}$ and by uniqueness of solutions it coincides with the solution u of the nonlinear problem (1). This establishes the result.□

Next we will show that the solution to the problem (1) is continuously differentiable on $(0,l)\times(0,T)\backslash\{(x,t) : t = G(x)\}$. To this end, we require that the initial data u^0 satisfies (H8) and the following condition:

(H9) The function u^0 is a non-negative continuously differentiable function.

We now have the following result.

Theorem 6 *Under the additional assumptions (H8)-(H9) the bounded variation weak solution u to the problem (1) is continuously differentiable on* $(0,l)\times(0,T)\backslash\{(x,t) : t = G(x)\}$.

Sketch of proof. Using the solution representation (5) and the assumption (H9) it follows from standard but lengthy computations (see, e.g., [5, 7]) that v the solution to the linear problem (4) is continuously differentiable on $(0,l)\times(0,T)\backslash\{(x,t) : t = G(x)\}$ for the choice of parameters $\hat{g}(x,t) \equiv g(x,P(t))$, $\hat{\beta}(x,t) \equiv \beta(x,P(t))$ and $\hat{m}(x,t) \equiv m(x,P(t))$. Since, for these parameters v coincides with the solution u of the quasilinear problem (1) we have the desired result. □

References

[1] A.S. Ackleh and K. Deng, *A Monotone Approximation for the Nonautonomous Size-Structured Population Model,* Quart. Appl. Math., **57** (1999), 261-267.

[2] A.S. Ackleh and K. Ito, *An Implicit Finite Difference Scheme for the Nonlinear Size-Structured Population Model*, Numer. Funct. Anal. Optim., **18** (1997), 865-884.

[3] A. Calsina and J. Saldana, *A Model of Physiologically Structured Population Dynamics with a Nonlinear Growth Rate*, J. Math. Biol., **33** (1995), 335-364.

[4] M. G. Crandall and Andrew Majda, *Monotone Difference Approximations for Scalar Conservation Laws,* J. Math. Comp. **34** (1980), 1-21.

[5] K. Ito, F. Kappel and G. Peichl, *A Fully Discretized Approximation Scheme for Size-Structured Population Models.* SIAM. J. Numer. Anal., **28** (1991), 923-954.

[6] J. Smoller, *Shock Waves and Reaction-Diffusion Equations*, Springer-Verlag, 1994.

[7] S.L. Tucker and S.O. Zimmerman, *A Nonlinear Model of Population Dynamics Containing Arbitrary number of Continuous Structure Variables*, SIAM J. Appl. Math., **48** (1988), 594-591.

Proceedings of Dynamic Systems and Applications 3 (2001) 9-14

A Global Existence Theorem for a Wave Equation with a Nonlinear Damping Term

Keith M. Agre[1] and Mohammad A. Rammaha[2]
[1,2]Department of Mathematics and Statistics
University of Nebraska-Lincoln
Lincoln, NE 68588-0323

ABSTRACT: We consider an initial-boundary value problem for a wave equation in high dimensions with a nonlinear damping term. We establish the existence and uniqueness of a global solution by using a compactness method and by exploiting the monotonicity property of the nonlinearity.
AMS (MOS) subject classification. 35L05, 35L20, 58G16

1. INTRODUCTION

In this article, we study the initial-boundary value problem:

$$u_{tt} - \Delta u + |u_t|^m \, sgn(u_t) = 0, \quad \text{in } \Omega \times (0,T), \tag{1.1}$$

$$u(x,0) = u^0, \quad u_t(x,0) = u^1, \quad \text{in } \Omega, \tag{1.2}$$

$$u(x,t) = 0, \quad \text{on } \Gamma \times (0,T). \tag{1.3}$$

Here, $0 < m < 1$ and Ω is an open, bounded, connected domain in $\mathbb{R}^n$ with a sufficiently smooth boundary $\partial\Omega = \Gamma$.

Equation (1.1) models a classical vibrating membrane with resistance that is proportional to the velocity u_t. Similar problems to (1.1)–(1.3) have been studied by many authors. In particular, we note the work of Lions & Strauss [**6**], Lasiecka & Triggiani [**3, 4**], and Ang & Dinh [**2**]. In Ang & Dinh [**2**], the authors studied a similar problem where they established the existence and uniqueness of a global solution by using a standard Galerkin approximation scheme. However, their proof relies heavily on the fact that the equation is in one-space dimension. In particular, having in hand explicit formulas for the eigenfunctions of the Laplacian $-d^2/dx^2$ on the unit interval $(0,1)$, namely $\{cos\frac{\pi}{2}(2n+1)x, \ n \in \mathbb{N}\}$, was a key ingredient in obtaining the needed estimates for their proof.

In this paper we deal with the initial-boundary value problem (1.1)–(1.3) in all space dimensions. However, due to the lack of smoothness of the nonlinearity in the equation, a standard fixed-point-theorem argument is not applicable. Our argument utilizes Aubin's compactness theorem and the monotonicity of the Nemytskii operator, $f(u')(x) := f(u'(x)) = |u'(x)|^m \, sgn(u'(x))$, to prove the convergence of the Galerkin scheme in the appropriate spaces.

2. PRELIMINARIES

In this section we introduce some notation and definitions that are necessary for the remaining sections of the paper. Throughout the paper, we assume

$$u^0 \in H_0^1(\Omega), \ u^1 \in L^2(\Omega).$$

Let X and Y be Banach spaces. We write $X \hookrightarrow Y$ if X is continuously imbedded in Y. More precisely, the natural injection $i : X \to Y$ is continuous and we identify X with the subspace $i(X)$ of Y. We denote by $\mathcal{L}(X,Y)$ the space of all continuous linear operators from X to Y. Let $A : L^2(\Omega) \to L^2(\Omega)$, where $A = -\Delta$ with its domain $\mathcal{D}(A) = H^2(\Omega) \cap H_0^1(\Omega)$. Since A is positive, self-adjoint, and has a compact inverse, then A has an infinite sequence of positive eigenvalues $\{\lambda_n : n = 1, 2, ...\}$ and a corresponding sequence of eigenfunctions $\{e_n : n = 1, 2, ...\}$ that forms an orthonormal basis for $L^2(\Omega)$. Namely, if $u \in L^2(\Omega)$, then $u = \sum_{n=1}^{\infty} u_n e_n$, where the convergence is in $L^2(\Omega)$, with $\|u\|_{L^2(\Omega)}^2 = \sum_{n=1}^{\infty} |u_n|^2$ and $u_n = \langle u, e_n \rangle_{L^2(\Omega)}$. The powers of A are defined as follows:
$A^s : \mathcal{D}(A^s) \subseteq L^2(\Omega) \to L^2(\Omega)$, $A^s u = \sum_{n=1}^{\infty} \lambda_n^s u_n e_n$, with the domain of A^s given by $\mathcal{D}(A^s) = \{u \in L^2(\Omega) : u = \sum_{n=1}^{\infty} u_n e_n,\ \sum_{n=1}^{\infty} \lambda_n^{2s} |u_n|^2 < \infty\}$. We remark here that

$$\mathcal{D}(A^s) = \begin{cases} H^{2s}(\Omega); 0 \le s < \frac{1}{4} \\ \{u \in H^{2s}(\Omega) : u|_\Gamma = 0\}\, ; s > \frac{1}{4}. \end{cases}$$

Moreover, $\mathcal{D}(A^{1/4}) \hookrightarrow H^{1/2}(\Omega)$, and the norm $\|u\|_{H^s(\Omega)}$ is equivalent to $\left(\sum_{n=1}^{\infty} \lambda_n^s |u_n|^2\right)^{1/2}$. Therefore, we set $\|u\|_{H^s(\Omega)}^2 = \sum_{n=1}^{\infty} \lambda_n^s |u_n|^2$.

Let $S(t)$ and $C(t)$ denote the sine and cosine operators associated with A, i.e., $S(t) = A^{-1/2} \sin(A^{1/2}t)$ and $C(t) = \cos(A^{1/2}t)$.

Finally, we rewrite the initial-boundary value problem (1.1)–(1.3) as follows:

$$u'' + Au + f(u') = 0, \;\; on\ (0,T), \tag{2.1}$$
$$u(0) = u^0, \;\; u'(0) = u^1, \tag{2.2}$$
$$u\,|_\Gamma = 0, \;\; on\ (0,T), \tag{2.3}$$

where here and later $f(u') := |u'|^m sgn(u')$ and $u'' = d^2u(t)/dt^2$.

Definition 2.1 Let $u^0 \in H_0^1(\Omega)$, $u^1 \in L^2(\Omega)$. We say that u is a regular solution to the initial-boundary value problem (2.1)–(2.3) on $[0,T]$ if $u \in L^2(0,T,H_0^1(\Omega))$, $u' \in L^2(0,T,L^2(\Omega))$ and u satisfies:

$$\langle u'(t), \phi\rangle_{L^2(\Omega)} - \langle u^1, \phi\rangle_{L^2(\Omega)} + \int_0^t \langle A^{1/2}u(s), A^{1/2}\phi\rangle_{L^2(\Omega)}\, ds$$
$$+ \int_0^t \langle f(u'(s)), \phi\rangle_{L^2(\Omega)}\, ds = 0,$$

for all $\phi \in H_0^1(\Omega)$ and almost every $t \in [0,T]$.

It should be noted here that if u is a solution of (2.1)–(2.3), then by the variation of constants formula, u satisfies the integral equations

$$u(t) = U_0(t) - \int_0^t S(t-\tau) f(u'(\tau))d\tau, \tag{2.4}$$

$$u'(t) = V_0(t) - \int_0^t C(t-\tau) f(u'(\tau))d\tau, \tag{2.5}$$

where $U_0(t) = C(t)u^0 + S(t)u^1$, and $V_0(t) = C(t)u^1 - AS(t)u^0$.

Remark 2.1 It is not too difficult to show that if $u \in L^2(0,T,H_0^1(\Omega))$, $u' \in L^2(0,T,L^2(\Omega))$, and u satisfies the integral equations (2.4)–(2.5), then u is a regular solution to (2.1)–(2.3) in the sense of definition 2.1. Moreover, the converse is also valid.

The following regularity result is well-known (for example, see Lasiecka & Triggiani [**3**, **4**]), and thus its proof is omitted.

Lemma 2.1 For $s \geq 0$, we have

$$\text{(i) } C(\cdot) \in \mathcal{L}\left(\mathcal{D}(A^s),\, C([0,T],\, \mathcal{D}(A^s))\right),$$

$$\text{(ii) } S(\cdot) \in \mathcal{L}\left(\mathcal{D}(A^s),\, C([0,T],\, \mathcal{D}(A^{s+1/2}))\right).$$

Our main result is the following.

Theorem 2.1 Let $u^0 \in H_0^1(\Omega)$, $u^1 \in L^2(\Omega)$. Then the initial-boundary value problem (1.1)–(1.3) has a unique regular global solution u such that for all $T > 0$,

$$u \in L^2\left(0, T, H_0^1(\Omega)\right) \text{ and } u' \in L^2\left(0, T, L^2(\Omega)\right).$$

The proof of theorem 2.1 is presented in sections 3 and 4.

3. APPROXIMATE SOLUTIONS

In this section, we construct the approximate solutions to the initial-boundary value problem (2.2)–(2.4) and obtain the necessary estimates for the passage to the limit. Let $\{e_k\}_{k=1}^{\infty}$ be the orthonormal basis for $L^2(\Omega)$, as described in section 2. Let $\mathcal{P}_N$ be the orthogonal projection of $L^2(\Omega)$ onto the linear span of $\{e_1, ..., e_N\}$. Let $u_N(t) = \sum_{k=1}^{N} u_{N,k}(t)e_k$ be a regular solution to the Galerkin system associated with the initial-boundary value problem (2.2)–(2.4), i.e., $u_N(t)$ satisfies the initial value problem:

$$\frac{d}{dt}\langle u_N'(t), e_k\rangle_{L^2(\Omega)} + \left\langle A^{1/2}u_N(t), A^{1/2}e_k\right\rangle_{L^2(\Omega)} + \langle \mathcal{P}_N f(u_N'(t)), e_k\rangle_{L^2(\Omega)} = 0, \tag{3.1}$$

$$u_{N,k}(0) = u_k^0, \quad u_{N,k}'(0) = u_k^1, \tag{3.2}$$

for $k = 1, 2..., N$, where $u_k^0 = \langle u^0, e_k\rangle_{L^2(\Omega)}$, and $u_k^1 = \langle u^1, e_k\rangle_{L^2(\Omega)}$. Now, (3.1) and (3.2) are equivalent to:

$$u_{N,k}''(t) + \lambda_k u_{N,k}(t) = -\left\langle \mathcal{P}_N f(u_{N,k}'(t)), e_k\right\rangle_{L^2(\Omega)} \tag{3.3}$$

$$u_{N,k}(0) = u_k^0, \quad u_{N,k}'(0) = u_k^1, \tag{3.4}$$

for $k = 1, 2..., N$. Since (3.3), (3.4) is an initial value problem for a second order $N \times N$ system of ordinary differential equations with continuous nonlinearities in the derivatives of the unknown functions $u_{N,k}$, it follows from the theory of ordinary differential equations that (3.3), (3.4) has a maximal solution on the interval $[0, T_N]$, for some $0 < T_N \leq T$. Moreover, $u_{N,k} \in C^2[0, T_N]$, and they satisfy the following integral equation on $[0, T_N]$:

$$u_{N,k}(t) = u_k^0 \cos \lambda_k^{1/2} t + u_k^1 \lambda_k^{-1/2} \sin \lambda_k^{1/2} t - \int_0^t \lambda_k^{-1/2} \sin \lambda_k^{1/2}(t-\tau) \left\langle \mathcal{P}_N f(u_{N,k}'(\tau)), e_k\right\rangle_{L^2(\Omega)} d\tau, \tag{3.5}$$

for $k = 1, 2..., N$. It follows easily from (3.5) and the definitions of the sine and cosine operators that u_N satisfies the following integral equations on $[0, T_N]$:

$$u_N(t) = U_{N,0}(t) - \int_0^t S(t-\tau)\mathcal{P}_N f(u'_N(\tau))d\tau, \tag{3.6}$$

$$u'_N(t) = V_{N,0}(t) - \int_0^t C(t-\tau)\mathcal{P}_N f(u'_N(\tau))d\tau, \tag{3.7}$$

where $U_{N,0}(t) = C(t)\mathcal{P}_N u^0 + S(t)\mathcal{P}_N u^1$ and $V_{N,0}(t) = C(t)\mathcal{P}_N u^1 - AS(t)\mathcal{P}_N u^0$.

Next, we show that $T_N = T$. If $T_N < T$, then it must be the case that $\|u_N(t)\|_{H_0^1(\Omega)} + \|u'_N(t)\|_{L^2(\Omega)} \to \infty$ as $t \to T_N$. However, in view of the a priori estimates in Lemma 3.1 below, this behavior is impossible. In the remaining parts of the paper, we shall refer to the following Hilbert spaces repeatedly:
$X = L^2(0, T, L^2(\Omega))$ and $Y = L^2(0, T, H_0^1(\Omega))$, where $T > 0$ is arbitrary.

The proof of the following lemma is contained in Agre & Rammaha **[1]**, and thus it is omitted.

Lemma 3.1 The sequence $\{u_N\}$ satisfies the following:
(i) $\{u_N\}$ is bounded in Y.
(ii) $\{u'_N\}$ and $\{\mathcal{P}_N f(u'_N)\}$ are bounded in X.

Now, by using Lemma 3.1 and Aubin's compactness theorem, we can extract a subsequence of $\{u_N\}$ (still denoted by $\{u_N\}$) and find functions $u \in Y$ and $w \in X$ such that:

$$\begin{cases} u_N \to u \text{ strongly in } X, \\ u_N \to u \text{ weakly in } Y, \\ u'_N \to u' \text{ weakly in } X, \\ \mathcal{P}_N f(u'_N) \to w \text{ weakly in } X. \end{cases} \tag{3.8}$$

Now, (3.6), (3.7), and (3.8) show that $u \in Y$, $u' \in X$, and u satisfies the integral equations:

$$u(t) = U_0(t) - \int_0^t S(t-\tau)w(\tau)d\tau, \tag{3.9}$$

$$u'(t) = V_0(t) - \int_0^t C(t-\tau)w(\tau)d\tau. \tag{3.10}$$

In view of Remark 2.1, we conclude that u is a regular solution on $[0, T]$ (in the sense of definition 2.1) to the initial-boundary value problem:

$$u'' + Au + w = 0, \quad on\ (0, T), \tag{3.11}$$

$$u(0) = u^0, \quad u'(0) = u^1, \tag{3.12}$$

$$u\,|_\Gamma = 0, \quad on\ (0, T), \tag{3.13}$$

where w is the weak limit of $\mathcal{P}_N f(u'_N)$ in X.

Remark 3.1 The following are immediate consequences of (3.8)–(3.10):
(i) There exists a subsequence of $\{u_N\}$, which is still denoted by $\{u_N\}$, such that for

almost all $t \in [0, T]$, we have

$$\begin{cases} u_N(t) \to u(t) \text{ strongly in } L^2(\Omega), \\ u_N(t) \to u(t) \text{ weakly in } H_0^1(\Omega), \\ u_N'(t) \to u'(t) \text{ weakly in } L^2(\Omega). \end{cases} \tag{3.14}$$

(ii) The pointwise convergence in (3.14) is indeed everywhere on $[0, T]$, and moreover

$$u \in C([0,T], H_0^1(\Omega)) \text{ and } u' \in C([0,T], L^2(\Omega)).$$

The proof of the following lemma is contained in Agre & Rammaha [**1**] and thus it is omitted.

Lemma 3.2 Let $u^0 \in H_0^1(\Omega)$, $u^1 \in L^2(\Omega)$, $w \in X$ and u be a regular solution to the initial-boundary value problem (3.11)–(3.13). Then $u'' \in L^2(0, T, H^{-1}(\Omega))$ with $\|u''\|^2_{L^2(0,T,H^{-1}(\Omega))} \leq C\left[\|u\|_Y^2 + \|w\|_X^2\right]$, for some constant $C > 0$.

4. UNIQUENESS AND THE FACT $\mathbf{w = f(u')}$

In order to be able to prove $w = f(u')$, we shall require several facts. First, the following energy identity for the approximate solutions $\{u_N\}$ can be easily obtained from (3.3). Specifically, we have

$$\begin{aligned} E_N(t) := \frac{1}{2}\left(\|u_N'(t)\|^2_{L^2(\Omega)} + \left\|A^{1/2}u_N(t)\right\|^2_{L^2(\Omega)}\right) \\ + \int_0^t \langle f(u_N'(\tau)), u_N'(\tau)\rangle_{L^2(\Omega)}\, d\tau = E_N(0). \end{aligned} \tag{4.1}$$

The following lemma can be proven by modifying the proof of Lemma 8.3 in Lions & Magenes [**5**]. Thus, its proof is omitted.

Lemma 4.1 Let $u \in C([0,T], H_0^1(\Omega))$ and $u' \in C([0,T], L^2(\Omega))$ such that u is a regular solution to the initial-boundary value problem (3.11)–(3.13). Then, u satisfies

$$\begin{aligned} E(t) := \frac{1}{2}\left(\|u'(t)\|^2_{L^2(\Omega)} + \left\|A^{1/2}u(t)\right\|^2_{L^2(\Omega)}\right) \\ + \int_0^t \langle w(\tau), u'(\tau)\rangle_{L^2(\Omega)}\, d\tau = E(0). \end{aligned}$$

Lemma 4.2 below is straightforward and follows from the continuity and monotonicity of the function $f(\phi) = |\phi|^m \, sgn(\phi)$.

Lemma 4.2 Let $f(\phi) = |\phi|^m \, sgn(\phi)$. Then f generates a monotone operator from $L^2(\Omega)$ into $L^2(\Omega)$, i.e.

$$\langle f(\phi) - f(\psi), \phi - \psi\rangle_{L^2(\Omega)} \geq 0, \text{ for all } \phi, \psi \in L^2(\Omega).$$

Moreover, the mapping: $\lambda \mapsto \langle f(\phi + \lambda\psi), \eta\rangle_{L^2(\Omega)}$ is continuous from $\mathbb{R}$ to $\mathbb{R}$ for every fixed $\phi, \psi, \eta \in L^2(\Omega)$.

Finally, we can utilize (4.1), Remark 3.1, Lemma 4.1 and Lemma 4.2 to prove that $w = f(u')$. See Agre & Rammaha[1] for the proof.

Proof of uniqueness:
Let v_1 and v_2 be regular solutions of the initial-boundary value problem (1.1)–(1.3). Let $v = v_1 - v_2$. Then, v satisfies

$$\begin{aligned}\langle v'(t), \phi\rangle_{L^2(\Omega)} + \int_0^t \langle A^{1/2}v(s), A^{1/2}\phi\rangle_{L^2(\Omega)}\, ds \\ + \int_0^t \langle f(v_1'(s)) - f(v_2'(s)), \phi\rangle_{L^2(\Omega)}\, ds = 0,\end{aligned}$$

for all $\phi \in H_0^1(\Omega)$ and almost every $t \in [0,T]$. Therefore, v is a regular solution to the initial-boundary value problem

$$\begin{aligned} v'' - \Delta v + f(v_1') - f(v_2') = 0, \quad &\text{in } \Omega \times (0,T), \\ v(x,0) = 0, \quad v'(x,0) = 0, \quad &\text{in } \Omega, \\ u = 0, \quad &\text{on } \Gamma \times (0,T). \end{aligned}$$

However, by applying Lemma 4.1, we have

$$\begin{aligned}&\frac{1}{2}\left(\|v'(t)\|^2_{L^2(\Omega)} + \left\|A^{1/2}v(t)\right\|^2_{L^2(\Omega)}\right) \\ &\qquad = -\int_0^t \langle f(v_1'(\tau)) - f(v_2'(\tau)), v_1'(\tau) - v_2'(\tau)\rangle_{L^2(\Omega)}\, d\tau.\end{aligned}$$

Therefore, by Lemma 4.2 it follows that $\left\|A^{1/2}v(t)\right\|^2_{L^2(\Omega)} = 0$ for almost every $t \in [0,T]$, which completes the proof of uniqueness.

REFERENCES

1. Agre, K.M., & Rammaha, M.A., Global solutions to boundary value problems for a nonlinear wave equation in high space dimensions, *Preprint.*

2. Ang, D.D., & Dinh, A.P.N., Mixed problem for some semi-linear wave equation with a non-homogeneous condition, *Nonlinear Analysis. Theory, Methods & Applications*, **12** (1988), 581-592.

3. Lasiecka, I., & Triggiani, R., A cosine operator approach to modelling $L_2(0,T; L_2(\Gamma))$ boundary input hyperbolic equations, *Appl. Math. Optim.* **7** (1981), 35-83.

4. Lasiecka, I., & Triggiani, R., Regularity theory of hyperbolic equations with non-homogeneous Neumann boundary conditions. II. General boundary data, *J. Diff. Equ.* **94** (1991), 112-164.

5. Lions, J.L., & Magenes, E., *Non-Homogeneous Boundary Value Problems and Applications I, II*, Springer-Verlag, New York-Heidelberg-Berlin, 1972.

6. Lions, J.L., & Strauss, W.A., Some non-linear evolution equations, *Bull. Soc. math. France*, **93** (1965), 43-96.

Proceedings of Dynamic Systems and Applications 3 (2001) 15-22

Periodic Solutions of Linear Impulsive Systems with Periodic Delays in the Critical Case

Haydar Akça[1] and Valéry Covachev[2]
[1] King Fahd University of Petroleum and Minerals, Dept. of Math.
Dhahran 31261, SAUDI ARABIA
[2]Institute of Mathematics, Bulgarian Academy of Sciences
Sofia, BULGARIA

ABSTRACT: A linear impulsive system with delay which differs from a constant by a small amplitude periodic perturbation is considered. If the corresponding system with constant delay has an r-parametric family of ω-periodic solutions, then the equation for the generating amplitudes is derived, and sufficient conditions are obtained for the existence of ω-periodic solutions of the perturbed system in the critical case of first order.
AMS (MOS) subject classification. 34A37, 34K10

1. INTRODUCTION

In the mathematical simulation of the evolution of real processes in physics, chemistry, population dynamics, radio engineering etc. which are subject to disturbances of negligible duration with respect to the total duration of the process, it is often convenient to assume that the disturbances are "momentary", in the form of impulses. This leads to the investigation of differential equations and systems with discontinuous trajectories, or with impulse effect, called for the sake of brevity impulsive differential equations and systems.

Impulsive differential equations with delay describe models of real processes and phenomena where both dependence on the past and momentary disturbances are observed. For instance, the size of a given population may be normally described by a delay differential equation and, at certain moments, the number of individuals can be abruptly changed. The interaction of the impulsive perturbation and the delay makes difficult the qualitative investigation of such equations. In particular, the solutions are not smooth at the moments of impulse effect shifted by the delay (Bainov et al [3]).

In the present paper we consider a linear impulsive system with delay which differs from a constant by a small amplitude periodic perturbation. Such delays are mentioned in the abstract (Schley & Gourley [11]). If the corresponding system with constant delay has an r-parametric family of ω-periodic solutions, then we shall obtain necessary and sufficient conditions for the existence of ω-periodic solutions of the perturbed system.

2. STATEMENT OF THE PROBLEM. PRELIMINARY ASSERTIONS

Throughout this paper we study a linear system with impulses at fixed moments and delay fluctuating around a constant value which may be assumed 1 without loss of generality:

$$\begin{cases} \dot{x}(t) &= A(t)x(t) + B(t)x(t-1-\varepsilon\varphi(t)) + f(t),\ t \neq t_i, \\ \Delta x(t_i) &= C_i x(t_i) + a_i, \qquad i \in \mathbf{Z}, \\ \Delta x(t) &= 0 \ \text{ if } \ t-1-\varepsilon\varphi(t) = t_i \ \text{ for some } \ i \in \mathbf{Z},\ t \notin \{t_i\}_{i\in\mathbf{Z}}, \end{cases} \tag{1}$$

where $x \in \mathbf{R}^n$; $f : \mathbf{R} \mapsto \mathbf{R}^n$, $A, B : \mathbf{R} \mapsto \mathbf{R}^{n\times n}$, $\varphi : \mathbf{R} \mapsto [-1, 1]$; $\mathbf{Z}$ is the set of all integers; $\Delta x(t_i) = x(t_i + 0) - x(t_i - 0)$ are the impulses at moments t_i and $\{t_i\}_{i\in\mathbf{Z}}$ is a strictly increasing sequence such that $\lim_{i\to\pm\infty} t_i = \pm\infty$; $a_i \in \mathbf{R}^n$, $C_i \in \mathbf{R}^{n\times n}$ $(i \in \mathbf{Z})$, $1 + \varepsilon\varphi(t)$ is the delay, $\varepsilon \in [0, \varepsilon_0]$ is a small parameter; ε_0 will be specified below.

As usual in the theory of the impulsive differential equations (Bainov & Covachev [2], Lakshmikantham et al. [9]), at the points of discontinuity t_i of the solution $x(t)$ we assume that $x(t_i) \equiv x(t_i - 0)$. It is clear that, in general, the derivatives $\dot{x}(t_i)$ do not exist. On the other hand, there do exist the limits $\dot{x}(t_i \pm 0)$. According to the above convention, we assume $\dot{x}(t_i) \equiv \dot{x}(t_i - 0)$.

Similarly, the derivative $\dot{x}(t)$ does not exist at the other points of discontinuity of the right-hand side $f(t, x(t), x(t - \varepsilon\varphi(t)))$, i.e. at points t which are solutions of the equations

$$t - 1 - \varepsilon\varphi(t) = t_i, \tag{2_i}$$

$i \in \mathbf{Z}$. We require the continuity of the solution $x(t)$ at such points if they are distinct from the moments of impulse effect t_i.

For the sake of brevity we shall use the notation $\bar{x}(t) = x(t-1)$, $x_i = x(t_i)$, $y^\varepsilon(t) = x(t - 1 - \varepsilon\varphi(t))$ (thus, for instance, $y_i^0 = x(t_i - 1) = \bar{x}_i$).

For a vector $x \in \mathbf{R}^n$ we denote by $|x|$ its Euclidean norm, and for an $(n \times n)$–matrix A we define the associated norm $|A| = \sup\{|Ax|/|x|;\quad x \in \mathbf{R}^n \setminus 0\}$.

Denote by $\tilde{C}_{\omega,n}\{t_i\}$ the space of all ω–periodic, piecewise continuous functions $w : \mathbf{R} \mapsto \mathbf{R}^n$ with discontinuities of the first kind at t_i $(i \in \mathbf{Z})$, equipped with the norm $\|w\| = \sup_{t\in\mathbf{R}} |w(t)|$.

Denote by $\tilde{c}_{m,n}$ the space of all sequences $\{a_i\}_{i\in\mathbf{Z}}$ such that $a_i \in \mathbf{R}^n$, $a_{i+m} = a_i$ $(i \in \mathbf{Z})$, equipped with the norm $\| \{a_i\}_{i\in\mathbf{Z}} \| = \max_{1\leq i\leq m} |a_i|$.

In the sequel we require the fulfilment of the following assumptions:

A1. $f \in \tilde{C}_{\omega,n}\{t_i\}$, $A, B \in \tilde{C}_{\omega,n\times n}\{t_i\}$.

A2. There exists a positive integer m such that $t_{i+m} = t_i + \omega$ $(i \in \mathbf{Z})$, $\{a_i\}_{i\in\mathbf{Z}} \in \tilde{c}_{m,n}$, $\{C_i\}_{i\in\mathbf{Z}} \in \tilde{c}_{m,n\times n}$.

A3. The matrices $E + C_i$ $(i \in \mathbf{Z})$ are nonsingular (E – the unit matrix).

A4. The function $\varphi(t)$ is ω–periodic and Lipschitz continuous:

$$|\varphi(t') - \varphi(t'')| \leq K|t' - t''|, \quad t', t'' \in \mathbf{R}.$$

If $\varepsilon_0 \leq \min\{1, 1/K\}$, then for $\varepsilon \in (0, \varepsilon_0]$ each equation (2_i) has a unique solution $t_i(\varepsilon)$. It obviously satisfies

$$|t_i(\varepsilon) - t_i - 1| \leq \varepsilon, \quad t_i(0) = t_i + 1.$$

It is natural to assume that the period ω is distinct from the unperturbed delay 1. For the sake of definiteness we assume that $\omega > 1$ and $t_i \neq 0 \ \forall i \in \mathbf{Z}$.

For $\varepsilon = 0$, from (1) we obtain

$$\left\{\begin{array}{rcl} \dot{x}(t) & = & A(t)x(t) + B(t)x(t-1) + f(t),\ t \neq t_i, \\ \Delta x(t_i) & = & C_i x_i + a_i, \qquad i \in \mathbf{Z}, \\ \Delta x(t_i+1) & = & 0 \ \text{ if } \ t_i + 1 \neq t_j \ \ \forall j, \end{array}\right. \tag{3}$$

so called *generating system.*

Together with (3) we consider the homogeneous system

$$\left\{\begin{array}{rcl} \dot{x}(t) & = & A(t)x(t) + B(t)x(t-1),\ t \neq t_i, \\ \Delta x(t_i) & = & C_i x_i, \qquad i \in \mathbf{Z}. \end{array}\right. \tag{4}$$

If the only ω–periodic solution of system (4) is $x \equiv 0$ (noncritical case), as in Boichuk et al [5], Samoilenko & Perestyuk [10] we can define *Green's function* $G(t,\tau)$ *of the periodic problem* for (3), i.e. for any $f \in \tilde{C}_{\omega,n}$ and $\{a_i\}_{i\in\mathbf{Z}} \in \tilde{c}_{m,n}$ system (3) has a unique ω–periodic solution given by the formula

$$x(t) = \int_0^\omega G(t,\tau)f(\tau)\,d\tau + \sum_{0<t_i<\omega} G(t,t_i+0)a_i.$$

Moreover, each ω–periodic solution of system (1) must satisfy

$$x(t) = \int_0^\omega G(t,\tau)[f(\tau)+B(\tau)\left(x(\tau-1-\varepsilon\varphi(\tau)) - x(\tau-1)\right)]\,d\tau + \sum_{0<t_i<\omega} G(t,t_i+0)a_i.$$

Using the Contraction Mapping Principle (or the Implicit Function Theorem), it can be shown that for $\varepsilon > 0$ small enough system (1) has a unique ω–periodic solution. In Akça & Covachev [1] we proved that for a more general nonlinear system.

Here we consider the *critical* case where

A5. System (4) has $r > 0$ linearly independent ω–periodic solutions.

Following Hale [8], we shall write down necessary and sufficient conditions for the nonhomogeneities $f(t)$ and $\{a_i\}_{i\in\mathbf{Z}}$ of system (3), under which (3) has an ω–periodic solution (more precisely, an r–parametric family of ω–periodic solutions).

Suppose, for the sake of definiteness, that $0 < t_1 < t_2 < \cdots < t_m < \omega$.

It is easily seen that for $f \in \tilde{C}_{\omega,n}$ and $\{a_i\}_{i\in\mathbf{Z}} \in \tilde{c}_{m,n}$ system (3) has a solution in $\tilde{C}_{\omega,n}$ if and only if

$$\int_0^\omega y(t)f(t)\,dt + \sum_{i=1}^m y(t_i+0)a_i = 0 \tag{5}$$

for any ω–periodic solution y of the adjoint system

$$\left\{\begin{array}{rcl} \dot{y}(t) & = & -A'(t)y(t) - B'(t+1)y(t+1), \quad t \neq t_i, \\ \Delta y(t_i) & = & -(E + C_i')^{-1}C_i' y_i, \qquad i \in \mathbf{Z}, \end{array}\right. \tag{6}$$

where A' is the transpose of the matrix A.

Henceforth we introduce some more notation. Let $NH_{\omega,m,n} = \tilde{C}_{\omega,n} \times \tilde{c}_{m,n}$ be the space of nonhomogeneities of system (3). We identify $\tilde{C}_{\omega,n}$ with the subspace $\tilde{C}_{\omega,n} \times \{0\}$ of $NH_{\omega,m,n}$. Let $\tilde{\pi} : NH_{\omega,m,n} \mapsto \tilde{C}_{\omega,n}$ be a continuous projector of $NH_{\omega,m,n}$ onto the space of ω–periodic solutions of system (4). It can be defined as follows: Let $\mathcal{U}$ be an $(n \times r)$ – matrix whose columns are linearly independent ω–periodic solutions of (4). Then we define

$$\tilde{\pi}\left(f, \{a_i\}_{i\in\mathbf{Z}}\right)(t) = \mathcal{U}(t)\left[\int_0^\omega \mathcal{U}'(\tau)\mathcal{U}(\tau)\,d\tau\right]^{-1}\left\{\int_0^\omega \mathcal{U}'(\tau)f(\tau)\,d\tau + \sum_{i=1}^m \mathcal{U}'(t_i+0)a_i\right\}.$$

There exists a continuous projection operator $J : NH_{\omega,m,n} \mapsto \tilde{C}_{\omega,n}$ such that the set of homogeneities satisfying the condition (5) coincides with $(\mathrm{Id} - J)NH_{\omega,m,n}$ (Id – the identity operator), and a continuous linear operator $\mathcal{K} : (\mathrm{Id} - J)NH_{\omega,m,n} \mapsto (\mathrm{Id} - \tilde{\pi})\tilde{C}_{\omega,n}$ such that $\mathcal{K}\left(f, \{a_i\}_{i\in\mathbf{Z}}\right)$ is an ω–periodic solution of (3) for each $\left(f, \{a_i\}_{i\in\mathbf{Z}}\right) \in (\mathrm{Id} - J)NH_{\omega,m,n}$.

Let $\mathcal{V}$ be an $(n \times r)$ – matrix whose columns are linearly independent ω–periodic solutions of system (6). Then we define

$$J\left(f, \{a_i\}_{i\in\mathbf{Z}}\right)(t) = \mathcal{V}(t)\left[\int_0^\omega \mathcal{V}'(\tau)\mathcal{V}(\tau)\,d\tau\right]^{-1}\left\{\int_0^\omega \mathcal{V}'(\tau)f(\tau)\,d\tau + \sum_{i=1}^m \mathcal{V}'(t_i+0)a_i\right\}.$$

Now $J\left(f, \{a_i\}_{i\in\mathbf{Z}}\right)(t) = 0$, i.e.

$$\int_0^\omega \mathcal{V}'(\tau)f(\tau)\,d\tau + \sum_{i=1}^m \mathcal{V}'(t_i+0)a_i = 0 \tag{7}$$

if and only if (5) holds.

For $\left(f, \{a_i\}_{i\in\mathbf{Z}}\right) \in (\mathrm{Id} - J)NH_{\omega,m,n}$ system (3) has an ω–periodic solution $x(t)$, and we can define

$$\mathcal{K}\left(f, \{a_i\}_{i\in\mathbf{Z}}\right) = (\mathrm{Id} - \tilde{\pi})x.$$

If condition (7) (i.e. (5)) is satisfied, then system (3) has an r–parametric family of ω–periodic solutions

$$x_0(t, c_r) = \mathcal{U}(t)c_r + \mathcal{K}(\mathrm{Id} - J)\left(f, \{a_i\}_{i\in\mathbf{Z}}\right)(t), \tag{8}$$

where $c_r \in \mathbf{R}^r$ is an arbitrary vector.

3. MAIN RESULT

3.1. Equation for the generating amplitudes

Let us find conditions for the existence of ω–periodic solutions $x(t,\varepsilon)$ of system (1) depending continuously on ε and such that, for some $c_r \in \mathbf{R}^r$, we have $x(t,0) = x_0(t,c_r)$. A necessary condition for the existence of such solutions is given by the following statement.

Theorem 1. *Let system (1) satisfying conditions A1–A5 and (7) have an ω-periodic solution $x(t,\varepsilon)$ which for $\varepsilon = 0$ turns into a generating solution $x_0(t,c_r)$. Then the vector $c_r \in \mathbf{R}^r$ satisfies the equation*

$$\begin{aligned}\mathcal{F}(c_r) \;\equiv\; & \int_0^\omega \mathcal{V}'(t+1)B(t+1)\varphi(t+1)[A(t)x_0(t,c_r) + B(t)x_0(t-1,c_r)]\,dt \\ + & \sum_{i=1}^m \mathcal{V}'(t_i+1)B(t_i+1)\varphi(t_i+1)C_i x_0(t_i,c_r) \;=\; 0.\end{aligned} \tag{9}$$

Proof: In system (1) we change the variables according to the formula

$$x = x_0(t, c_r) + z \tag{10}$$

and are led to the problem of finding ω–periodic solutions $z = z(t, \varepsilon)$ of the impulsive system of functional differential equations

$$\left\{\begin{array}{rcl} \dot{z}(t) & = & A(t)z(t) + B(t)z(t-1) + B(t)\left(x(t-1-\varepsilon\varphi(t)) - x(t-1)\right), \ t \neq t_i, \\ \Delta z(t_i) & = & C_i z_i, \quad i \in \mathbf{Z}, \end{array}\right. \tag{11}$$

such that $z(t, \varepsilon) \to 0$ as $\varepsilon \to 0$.

We can formally consider (11) as a nonhomogeneous system of the form (3) with nonhomogeneities $(B(\cdot)(y^\varepsilon(\cdot) - y^0(\cdot)), 0)$. Then the solvability condition (7) becomes

$$\int_1^{1+\omega} \mathcal{V}'(t)B(t)(y^\varepsilon(t) - y^0(t))\, dt = 0 \tag{12}$$

(the ω – periodicity of the integrand allows us to integrate over any interval of length ω). For the sake of brevity we still write x, respectively y^ε instead of $x_0 + z$ in the nonhomogeneity of (11), respectively in the integrand of (12).

Since the left-hand side of equality (12) tends to 0 as $\varepsilon \to 0$, we first divide it by ε and then study its behaviour as $\varepsilon \to 0$. For the sake of later convenience we denote by $\eta(\varepsilon, x)$ expressions tending to 0 as $\varepsilon \to 0$, and satisfying the Lipschitz condition with respect to x with a constant tending to 0 as $\varepsilon \to 0$ (or just independent of x). We shall sometimes write $\eta(\varepsilon)$ instead of $\eta(\varepsilon, x)$ if this leads to no misunderstanding.

For $a, b \in \mathbf{R}$ denote

$$]a, b[= \left\{\begin{array}{ll} (a, b) & \text{if} \quad a < b, \\ (b, a) & \text{if} \quad a > b, \\ \emptyset & \text{if} \quad a = b. \end{array}\right.$$

We may note that

$$\tau \in]t_i(\varepsilon), t_i + 1[\quad \Longleftrightarrow \quad t_i \in]\tau - 1, \tau - 1 - \varepsilon\varphi(\tau)[.$$

Define the "bad" set $\Delta_1^\varepsilon = \bigcup_{i=1}^{m}]t_i(\varepsilon), t_i + 1[$. If $\varepsilon > 0$ is small enough, then Δ_1^ε is a disjoint union of intervals contained in $(1, \omega + 1)$. We further define the "good" set $\Delta_2^\varepsilon = [1, \omega + 1] \setminus \Delta_1^\varepsilon$.

For the sake of convenience we assume that for $i = \overline{1, m}$ $\ t_i + 1 \neq t_j \ \forall j \in \mathbf{Z}$. Then for $\varepsilon > 0$ small enough the "bad" set Δ_1^ε contains none of the points t_i, $i \in \mathbf{Z}$.

Let $\varepsilon_0 > 0$ be so small that all the above assumptions are valid for $\varepsilon \in (0, \varepsilon_0]$.

If $t \in \Delta_2^\varepsilon$, then the interval $]t - 1 - \varepsilon\varphi(t), t - 1[$ contains none of the points t_i. Thus we have

$$\begin{array}{rcl} y^\varepsilon(t) - y^0(t) & = & x(t - 1 - \varepsilon\varphi(t)) - x(t - 1) \\ = \displaystyle\int_0^1 \frac{\partial}{\partial s} x(t - 1 - s\varepsilon\varphi(t))\, ds & = & -\varepsilon\varphi(t) \displaystyle\int_0^1 \dot{x}(t - 1 - s\varepsilon\varphi(t))\, ds \end{array}$$

$$
\begin{aligned}
&= -\varepsilon\varphi(t)\int_0^1 A(t-1-s\varepsilon\varphi(t))x(t-1-s\varepsilon\varphi(t))\,ds \\
&\quad -\varepsilon\varphi(t)\int_0^1 B(t-1-s\varepsilon\varphi(t))y^\varepsilon(t-1-s\varepsilon\varphi(\tau)))\,ds \\
&= -\varepsilon\varphi(t)\,[A(t-1)x(t-1)+B(t-1)x(t-2)] \;+\; \varepsilon\eta(\varepsilon)
\end{aligned}
$$

and

$$
\varepsilon^{-1}\int_{\Delta_2^\varepsilon} \mathcal{V}'(t)B(t)(y^\varepsilon(t)-y^0(t))\,dt \tag{13}
$$

$$
\begin{aligned}
&= -\int_{\Delta_2^\varepsilon} \mathcal{V}'(t)B(t)\varphi(t)\,[A(t-1)x(t-1)+B(t-1)x(t-2)]\,dt \;+\; \eta(\varepsilon) \\
&= -\int_0^\omega \mathcal{V}'(t+1)B(t+1)\varphi(t+1)\,[A(t)x(t)+B(t)x(t-1)]\,dt \;+\; \eta(\varepsilon).
\end{aligned}
$$

Further on we find

$$
\varepsilon^{-1}\int\limits_{]t_i(\varepsilon),t_i+1[} \mathcal{V}'(t)B(t)(y^\varepsilon(t)-y^0(t))\,dt \;=\; |\varphi(t_i(\varepsilon))|\mathcal{V}'(\tau)B(\tau)(y^\varepsilon(\tau)-y^0(\tau))
$$

for some point $\tau\in]t_i(\varepsilon),t_i+1[$. This means that the interval $]\tau-1,\tau-1-\varepsilon\varphi(\tau)[$ contains just one discontinuity point t_i. We have

$$
\varphi(t_i(\varepsilon))=\varphi(t_i+1)+\eta(\varepsilon),\quad \mathcal{V}'(\tau)=\mathcal{V}'(t_i+1)+\eta(\varepsilon),\quad B(\tau)=B(t_i+1)+\eta(\varepsilon).
$$

First let us suppose that $\varphi(t_i+1)\neq 0$ and that ε is so small that $\varphi(t)$ has a constant sign in the interval $]t_i(\varepsilon),t_i+1[$. Now

$$
\begin{aligned}
&y^\varepsilon(\tau) \;-\; y^0(\tau) \;=\; x(\tau-1-\varepsilon\varphi(\tau))-x(\tau-1) \\
&= (x(\tau-1-\varepsilon\varphi(\tau))-x(t_i-0\cdot\operatorname{sgn}\varphi(\tau))) \\
&+ (x(t_i+0\cdot\operatorname{sgn}\varphi(\tau))-x(\tau-1))-\Delta x(t_i)\cdot\operatorname{sgn}\varphi(\tau) \\
&= \int_0^1 [A(\tau_\sigma^1)x(\tau_\sigma^1) \;+\; B(\tau_\sigma^1)y^\varepsilon(\tau_\sigma^1)]\,d\sigma\cdot(\tau-1-\varepsilon\varphi(\tau)-t_i) \\
&+ \int_0^1 [A(\tau_\sigma^2)x(\tau_\sigma^2) \;+\; B(\tau_\sigma^2)y^\varepsilon(\tau_\sigma^2)]\,d\sigma\cdot(t_i-\tau+1) \;-\; C_ix_i\operatorname{sgn}\varphi(\tau),
\end{aligned}
$$

where $\tau_\sigma^1\equiv\sigma(\tau-1-\varepsilon\varphi(\tau))+(1-\sigma)t_i\in]\tau-1-\varepsilon\varphi(\tau),t_i[$ and $\tau_\sigma^2\equiv\sigma t_i+(1-\sigma)(\tau-1)\in\,]t_i,\tau-1[$.

Since $\tau-1-\varepsilon\varphi(\tau)-t_i=\eta(\varepsilon)$ and $t_i-\tau+1=\eta(\varepsilon)$, then we find

$$
y^\varepsilon(\tau) \;-\; y^0(\tau) \;=\; -C_ix_i\operatorname{sgn}\varphi(t_i+1) \;+\; \eta(\varepsilon)
$$

and

$$
\varepsilon^{-1}\int\limits_{]t_i(\varepsilon),t_i+1[} \mathcal{V}'(t)B(t)(y^\varepsilon(t)-y^0(t))\,dt=\mathcal{V}'(t_i+1)B(t_i+1)\varphi(t_i+1)C_ix_i+\eta(\varepsilon). \tag{14}
$$

It is easily seen that this equality still holds for $\varphi(t_i+1)=0$.

Adding together equality (13) and the equalities (14) for $i=\overline{1,m}$, we write down the solvability condition (12) in the form

$$
\begin{aligned}
&-\int_0^\omega \mathcal{V}'(t+1)B(t+1)\varphi(t+1)[A(t)x(t)+B(t)x(t-1)]\,dt \\
&-\sum_{i=1}^m \mathcal{V}'(t_i+1)B(t_i+1)\varphi(t_i+1)C_ix(t_i) \;=\; \eta(\varepsilon,x).
\end{aligned} \tag{15}
$$

Now we easily see that the equation (9) is derived from (15) by passing to the limit as $\varepsilon \to 0$. □

Equation (15) can be called *equation for the generating amplitudes* (see, for instance, Grebennikov & Ryabov [7] or Boichuk [4], Boichuk et al [6]) of the problem for finding ω–periodic solutions of the impulsive system with delay (1).

3.2. Reduction of the problem to an operator system in a suitable function space

In view of (10) and (9) the solvability condition (15) can be written down in the form

$$\begin{aligned} &\int_0^\omega \mathcal{V}'(t+1)B(t+1)\varphi(t+1)[A(t)z(t)+B(t)z(t-1)]\,dt \\ &+\sum_{i=1}^m \mathcal{V}'(t_i+1)B(t_i+1)\varphi(t_i+1)C_i z_i \;=\; \eta(\varepsilon,x). \end{aligned} \tag{16}$$

The ω–periodic solution $z(t,\varepsilon)$ of system (11) such that $z(t,0)\equiv 0$ can be represented in the form

$$z(t,\varepsilon)=\mathcal{U}(t)c+z^{(1)}(t,\varepsilon), \tag{17}$$

where the unknown constant vector $c=c(\varepsilon)\in \mathbf{R}^r$ must satisfy an equation derived below from (16), while the unknown ω – periodic vector–valued function $z^{(1)}(t,\varepsilon)$ can be represented as

$$z^{(1)}(t,\varepsilon)=\mathcal{K}(\mathrm{Id}-J)\left(B(\cdot)(y^\varepsilon(\cdot)-y^0(\cdot)),0\right)(t), \tag{18}$$

where as above $y^\varepsilon(t)=x(t-1-\varepsilon\varphi(t))$.

Next let us denote

$$\begin{aligned} \mathcal{B}_0 \;=\; &\int_0^\omega \mathcal{V}'(t+1)B(t+1)\varphi(t+1)[A(t)\mathcal{U}(t)+B(t)\mathcal{U}(t-1)]\,dt \\ &+\sum_{i=1}^m \mathcal{V}'(t_i+1)B(t_i+1)\varphi(t_i+1)C_i\mathcal{U}(t_i) \end{aligned}$$

– an $(r\times r)$ – matrix. Then we have

$$\begin{aligned} \mathcal{B}_0 c \;=\; &-\int_0^\omega \mathcal{V}'(t+1)B(t+1)\varphi(t+1)[A(t)z^{(1)}(t,\varepsilon)+B(t)z^{(1)}(t-1,\varepsilon)]\,dt \\ &+\sum_{i=1}^m \mathcal{V}'(t_i+1)B(t_i+1)\varphi(t_i+1)C_i z^{(1)}(t_i,\varepsilon) \;+\; \eta(\varepsilon,x). \end{aligned} \tag{19}$$

Thus we have reduced the periodic problem for system (1) to the equivalent operator system (10), (17), (19), (18).

3.3. Critical case of first order

Suppose that $\det\,\mathcal{B}_0\neq 0$. It is easy to see (Grebennikov & Ryabov [7]) that this condition is equivalent to the simplicity of the root c_r of the equation for the generating amplitudes:

$$\mathcal{F}(c_r)=0, \qquad \det\left(\frac{\partial\mathcal{F}(c_r)}{\partial c_r}\right)\neq 0.$$

This is the so called *critical case of first order.* Then equation (19) can be solved with respect to c and we obtain (Grebennikov & Ryabov [7]) a Fredholm operator system of the second type to which a convergent simple iteration method can be applied.

Theorem 2. *For system (1) let conditions A1–A5 and (7) hold. Then for any simple* (det $\mathcal{B}_0 \neq 0$) *root* $c_r \in \mathbf{R}^r$ *of equation (9) there exists a constant* $\varepsilon_1 \in (0, \varepsilon_0]$ *such that for* $\varepsilon \in [0, \varepsilon_1]$ *system (1) has a unique* ω*-periodic solution* $x(t, \varepsilon)$ *depending continuously on* ε *and such that* $x(t, 0) = x_0(t, c_r)$.

The convergence of the method can be proved by using Lyapunov's majorants technique as in Boichuk [4], Grebennikov & Ryabov [7].

4. ACKNOWLEDGMENT

The authors gratefully acknowledge the support provided by King Fahd University of Petroleum and Minerals, Dhahran, Saudi Arabia. Second author supported in part by the Bulgarian Ministry of Education and Science under Grant MM–706.

REFERENCES

1. Akça, H., & Covachev, V., Periodic solutions of impulsive systems with periodic delays. Proc. of the Internat. Conf. on Biomathematics – Bioinformatics and Applications of Functional Differential Difference Equations (July 14–19, 1999, Alanya, Turkey).

2. Bainov, D. D. & Covachev, V., Impulsive Differential Equations with a Small Parameter. Series on Advances in Mathematics for Applied Sciences, Vol. 24, World Scientific, Singapore, 1994.

3. Bainov, D., Covachev, V., & Stamova, I., Stability under persistent disturbances of impulsive differential–difference equations of neutral type. J. Math. Anal. Appl., Vol. 187 (1994) 799–808.

4. Boichuk, A. A., Constructive methods of the analysis of boundary value problems. Naukova Dumka, Kiev, 1990 (in Russian).

5. Boichuk, A. A., Juravliov, V. F., & Samoilenko, A. M., Linear Noether boundary value problems for impulsive differential systems with delay. Differential Equations, Vol. 30 (1994), No. 10, 1677–1682.

6. Boichuk, A. A., Juravliov, V. F., & Samoilenko, A. M., Generalized inverse operators and Fredholm boundary value problems. Inst. Math. Nat. Acad. Sci. Ukraine, Kiev, 1995 (in Russian); VSP, Utrecht, 1998.

7. Grebennikov, E. A., & Ryabov, Yu. A., Constructive methods of the analysis of nonlinear systems. Nauka, Moscow, 1979 (in Russian).

8. Hale, J., Theory of functional differential equations, Springer, New York – Heidelberg – Berlin, 1977.

9. Lakshmikantham, V., Bainov, D. D., & Simeonov, P. S., Theory of impulsive differential equations. Series in Modern Applied Mathematics, Vol. 6, World Scientific, Singapore, 1989.

10. Samoilenko, A. M., & Perestyuk, N. A., Impulsive differential equations, World Scientific Series on Nonlinear Science. Ser. A: Monographs and Treatises, Vol. 14, World Scientific, Singapore, 1995.

11. Schley, D., & Gourley, S. A., Asymptotic linear stability for population models with periodic time–dependent perturbed delays, in: Alcalá Ist International Conference on Mathematical Ecology (September 4–8, 1998, Alcalá de Henares, Spain), Abstracts, p. 146.

Proceedings of Dynamic Systems and Applications 3 (2001) 23-34

Block System of Parabolic Differential Inequalities and Comparison Theorems

M. J. Anabtawi[1] and G. S. Ladde[2]
[1]Center of Excellence in ISEM, Tennessee State University,
330 10th Ave. N. Suite 265G, Nashville, Tennessee 37203 – 3401
[2]Department of Mathematics, The University of Texas at Arlington
Arlington, Texas 76019

ABSTRACT : In this paper, we consider a system of partial differential equations of parabolic type. We employ the theory of parabolic differential inequalities to develop block system of parabolic differential inequalities and a very general block comparison theorem and its variants. These results provide a necessary background and tool to study the qualitative properties of systems of parabolic differential equations.

AMS(MOS) Subject classifications. 93E15, 93D20, 35K40.

1. INTRODUCTION

Parabolic differential inequalities[5] play a very important role in the study of the qualitative and quantitative behavior of systems of partial differential equations of parabolic type. The study of such inequalities leads to develop very general comparison principle which play a key role in obtaining various qualitative properties such systems[5].

In Section 2, we formulate the problem. In Section 3, we develop fundamental results in the theory of systems of parabolic differential inequalities and its variants. In Section 4, we employ the results developed in Section 3, to develop a very general block comparison theorem and its variants which will provide a very effective tool and necessary background to study systems of parabolic partial differential equations with random structural perturbations.

2. PROBLEM FORMULATION

Let H be a bounded domain in R^m, $Q_{t_0} = (t_0, \infty) \times H$ for $t_0 > 0$, $\Gamma_{t_0} = (t_0, \infty) \times \partial H$. A vector $\nu \in R^m$ is said to be an outernormal at $(t, x) \in \Gamma_{t_0}$ if $(t, x - h\nu) \in Q_{t_0}$ for small $h > 0$. The outernormal derivative is defined by

$$\frac{\partial u}{\partial \nu}(t, x) = \lim_{h \to 0^+} \left[\frac{u(t, x) - u(t, x - h\nu)}{h} \right].$$

We shall always assume that an outernormal exists on Γ_{t_0}.

Consider the initial boundary value problem (in short IBVP)

$$\begin{cases} \frac{\partial u_i^k}{\partial t} = W_i^k(t, x, u, u_{ix}^k, u_{ixx}^k), \\ u_i^k(t_0, x) = u_{i0}^k(x) & \text{for } x \in \overline{H}, \\ B_i u_i^k(t, x) = \phi_i^k(t, x) & \text{on } \Gamma_{t_0}, i = 1, ..., N,\ k = 1, ..., s, \end{cases} \tag{2.1}$$

where B_i is the boundary operator defined by

$$B_i u_i^k(t, x) = p_i(t, x) u_i^k + q_i(t, x) \frac{\partial u_i^k}{\partial \nu} \quad , 1 \le i \le N, \tag{2.2}$$

where p_i, $q_i \in C[\Gamma_{t_0}, R]$ are such that $p_i(t, x) \ge 0$, $q_i(t, x) \ge 0$, and $p_i(t, x)+ q_i(t, x) > 0$, $1 \le i \le N$; $\frac{\partial u_i}{\partial \nu}$ is the outernormal derivative as defined before; and $u_i \in R^s$; $u = (u_1^T, u_2^T, ..., u_i^T, ..., u_N^T)^T$ is an $N.s$-dimensional vector; $\phi \in C[\Gamma_{t_0}, R^{N.s}]$. We assume that f and the domain are smooth enough so that the IBVP (2.1) has a solution on Q_{t_0}.

For the IBVP (2.1), we consider the following two special cases:

$$\begin{cases} \frac{\partial u_i^k}{\partial t} = \sum_{j=1}^{m} \sum_{l=1}^{m} \frac{\partial}{\partial x_l} \left(A_{lj}^{ik}(t, x) \frac{\partial u_i^k}{\partial x} \right) + \sum_{j=1}^{N} B_{ij}^k(t, x,) \frac{\partial u_i^k}{\partial x_j} + F_i^k(t, x, u), \\ u_i^k(t_0, x) = u_{i0}^k(x) & \text{for } x \in \overline{H}, \\ B_i u_i^k(t, x) = \phi_i^k(t, x) & \text{on } \Gamma_{t_0},\ i = 1, ..., N, k = 1, ..., s, \end{cases} \tag{2.3}$$

where $A_{lj}^{ik}(t, x)$, and $B_{ij}^k(t, x)$ are smooth enough $1 \times m$ row matrix processes, and F is smooth enough.

$$\begin{cases} \frac{\partial u_i^k}{\partial t} = \sum_{j=1}^{m} \sum_{l=1}^{m} A_{lj}^{ik}(t, x) \frac{\partial^2 u_i^k}{\partial x_l \partial x_j} + \sum_{j=1}^{N} B_{ij}^k(t, x) \frac{\partial u_i^k}{\partial x_j} + F_i^k(t, x, u), \\ u_i^k(t_0, x) = u_{i0}^k(x) & \text{for } x \in \overline{H}, \\ B_i u_i^k(t, x) = \phi_i^k(t, x) & \text{on } \Gamma_{t_0},\ i = 1, ..., N, k = 1, ..., s, \end{cases} \tag{2.4}$$

where $A_{lj}^{ik}(t, x)$, and $B_{ij}^k(t, x)$ are smooth enough random functions defined on $\overline{Q}_{t_0}$ and the Quadratic form

$$\sum_{j=1}^{m} \sum_{l=1}^{m} A_{lj}^{ik}(t, x) \lambda_l \lambda_j \ge 0, \qquad \text{on } \overline{Q}_{t_0}$$

λ being an arbitrary vector in R^m. F as is defined in (2.3).

3. SYSTEM OF PARABOLIC DIFFERENTIAL INEQUALITIES

In this section we develop fundamental results in the theory of systems of parabolic differential inequalities and its important special cases. These results are very important and essential in developing block comparison theorems.

First we present a well-known lemma[5] that will be used frequently.

Lemma 3.1[5] Suppose that

(i) $v \in C[\overline{Q}_{t_0}, R^N]$, and v possesses the continuous partial derivatives v_t, v_x, and v_{xx} on Q_{t_0},

(ii) (a) $v(t_0, x) < 0$ for $x \in \overline{H}$;

(b) $Bv(t, x) < 0$ on Γ_{t_0} and the boundary operator B is as defined in (2.2),

(iii) for $t_1 > t_0$, $x_1 \in \overline{H}$, and an index i, $1 \leq i \leq N$, if $v_i(t_1, x_1) = 0$, $v_j(t_1, x_j) < 0$. $1 \leq j \leq N$ and $j \neq i$, $v_{ix}(t_1, x_1) = 0$, and the quadratic form

$$\sum_{l,k=1}^{m} \frac{\partial^2 v_i(t_1, x_1)}{\partial x_l \partial x_k} \lambda_l \lambda_k \leq 0, \tag{3.1}$$

λ being an arbitrary vector in R^m, then $v_{it}(t_1, x_1) < 0$.

Under these assumptions, we have

$$v(t, x) < 0 \qquad \text{on } \overline{Q}_{t_0}. \tag{3.2}$$

Corollary 3.1 Suppose that

(i) $v \in C[\overline{Q}_{t_0}, R^N]$, and v possesses the continuous partial derivatives v_t, v_x, and v_{xx} on Q_{t_0};

(ii) (a) $v(t_0, x) < 0.$ for $x \in \overline{H}$;

(b) $Bv(t, x) < 0$ on Γ_{t_0} and the boundary operator B is as defined in (2.2);

(iii) for $t_1 > t_0$, $x_1 \in \overline{H}$, and an index i, $1 \leq i \leq N$, if $v_i(t_1, x_1) = 0$, $v_j(t_1, x_j) < 0$. $1 \leq j \leq N$ and $j \neq i$, $v_{ix}(t_1, x_1) = 0$, then $v_{it}(t_1, x_1) < 0$;

Under these assumptions, we have

$$v(t, x) < 0, \qquad \text{on } \overline{Q}_{t_0}. \tag{3.3}$$

Proof Assume that the conclusion of the corollary is not true, then there exist an index i, $t_1 > t_0$, and $x_1 \in \overline{H}$ such that

$$\begin{cases} v_i(t, x) < 0, & \text{on } [t_0, t_1) \times \overline{H} \quad \text{and } v_i(t_1, x_1) = 0 \\ v_j(t, x) < 0, & \text{on } [t_0, t_1) \times \overline{H},\ j \neq i,\ 1 \leq j \leq N \ \text{and } 1 \leq i \leq N. \end{cases} \tag{3.4}$$

We can see that $v_i(t, x)$ has its maximum on $[t_0, t_1] \times \overline{H}$ at (t_1, x_1). Its maximum is equal to zero. We note that $(t_1, x_1) \notin \Gamma_{t_0}$. If this is false, then for sufficiently small $h > 0$ we would have

$$B_i v_i(t_1, x_1) = p_i(t_1, x_1) v_i(t_1, x_1) + q_i(t_1, x_1) \frac{\partial v_i}{\partial \nu}(t_1, x_1) \geq 0.$$

This contradicts with assumption (ii)(b). Hence $(t_1, x_1) \in Q_{t_0}$. From (3.4), we have $v_i(t_1 - h, x_1) < 0$ and $v_i(t_1, x_1) = 0$, for $h > 0$ sufficiently small, it follows that

$$v_{it}(t_1, x_1) \geq 0. \tag{3.5}$$

On the other hand $v_i(t, x)$ attains its maximum at $(t_1, x_1) \in Q_{t_0}$. From this we have

$$v_{ix}(t_1, x_1) = 0.$$

Now from condition (iii), we have

$$v_{it}(t_1, x_1) < 0.$$

This is incompatible with (3.5). Hence this establishes the fact that there do not exist such an index i, $t_1 > t_0$, and x_1. This proves the validity of the inequality (3.3) on $\overline{Q}_{t_0}$.

We are now in a position to state and prove a comparison theorem that plays a fundamental part in applications.

Theorem 3.1 Suppose that

(i) $v \in C[\overline{Q}_{t_0}, R_+^N]$, and v possesses the continuous partial derivatives v_t, v_x, v_{xx} in $\overline{Q}_{t_0}$;

(ii) $G \in C[\overline{Q}_{t_0} \times R_+^N \times R^m \times R^{m^2}, R^N]$,

$$G(t,x,v,v_x,v_{xx}) = (G_1(t,x,v,v_{1x},v_{1xx}), ..., G_i(t,x,v,v_{ix},v_{ixx}), \\ ...., G_N(t,x,v,v_{Nx},v_{Nxx}))^T, \qquad (3.6)$$

the differential operator T is parabolic and satisfies the inequality $T[v] \leq 0$ on Q_{t_0}, where $T[v]$ is defined by

$$T_i[v] = v_{it} - G_i(t,x,v,v_{ix},v_{ixx}) \qquad \text{for } 1 \leq i \leq N; \qquad (3.7)$$

(iii) $g \in C[J \times R_+^N, R^N]$, $g(t,y)$ is quasi -monotone non decreasing in y for each $t \in J$, $r(t,t_0,y_0)$ is the maximal solution of the differential system

$$y' = g(t,y), \qquad y(t_0) = y_0 \geq 0 \qquad (3.8)$$

existing for $t \geq t_0$, and for each $(t,x,z) \in \overline{Q}_{t_0} \times R_+^N$, G and g satisfies the inequality

$$G(t,x,z,0,0) \leq g(t,z); \qquad (3.9)$$

(iv) (a) $v(t_0,x) \leq y_0$ for $x \in \overline{H}$

(b) $Bv(t,x) \leq Br(t,t_0,y_0)$ on Γt_0.

Under these assumptions, we have

$$v(t,x) \leq r(t,t_0,y_0), \qquad \text{on } \overline{Q}_{t_0} \qquad (3.10)$$

Proof Consider the function

$$U(t,x,\epsilon) = v(t,x) - r(t,\epsilon) \qquad (3.11)$$

where for sufficiently small $\epsilon > 0$, $r(t,\epsilon) = r(t,t_0,y_0,\epsilon)$ is the maximal solution of the differential system

$$y' = g(t,y) + \epsilon, \qquad y(t_0) = y_0 + \epsilon, \qquad (3.12)$$

existing on any compact interval $[t_0, t_0 + \beta]$, $\beta > 0$. Also, we note that

$$U(t,x) = \lim_{\epsilon \to 0}[v(t,x) - r(t,\epsilon)] = v(t,x) - r(t,t_0,y_0). \qquad (3.13)$$

Now we will show that $U(t,x,\epsilon)$ satisfies the hypotheses of Lemma 3.1. Clearly conditions (i) and (ii) of the lemma hold immediately because of the assumptions(i) and (iv),(2.2), and the facts that $p_i(t,x) \geq 0$, $q_i(t,x) \geq 0$ and $p_i(t,x) + q_i(t,x) > 0$.

Now to verify condition (iii), let us assume for $t_1 > t_0$, $x_1 \epsilon \overline{H}$, an arbitrary $\lambda \in R^m$, and an index $i, 1 \leq i \leq N$, such that $U_i(t_1,x_1,\epsilon) = 0$, $U_j(t_1,x_1,\epsilon) < 0$, $j \neq i$, $U_{ix}(t_1,x_1,\epsilon) = 0$, and

$$\sum_{l,k=1}^{m} \frac{\partial^2 U_i(t_1, x_1, \epsilon)}{\partial x_l \partial x_k} \lambda_l \lambda_k \le 0.$$

This means that for $j \neq i$ and $i, j = 1, ..., N$,

$$\begin{cases} v_i(t_1, x_1) = r_i(t_1, \epsilon), \\ v_j(t_1, x_1) < r_j(t_1, \epsilon), \\ v_{ix}(t_1, x_1) = 0, \end{cases} \tag{3.14}$$

and

$$\sum_{l,k=1}^{m} \frac{\partial^2 v_i(t_1, x_1)}{\partial x_l \partial x_k} \lambda_l \lambda_k \le 0.$$

Since G_i is elliptic, we have

$$G_i(t_1, x_1, v(t_1, x_1), v_{ix}(t_1, x_1), v_{ixx}(t_1, x_1)) \le G_i(t_1, x_1, v(t_1, x_1), 0, 0).$$

From this and (3.9), (3.10), and also in view of assumption (ii) and the quasi montonicity of g, we have

$$U_{it}(t_1, x_1, \epsilon) = v_{it}(t_1, x_1) - r_i'(t_1, \epsilon)$$

$$\le G_i(t_1, x_1, v(t_1, x_1), v_{i_x}(t_1, x_1) v_{i_{xx}}(t_1, x_1)) - r_i'(t_1, \epsilon)$$

$$\le G_i(t_1, x_1, v(t_1, x_1), 0, 0) - g_i(t_1, r(t_1, \epsilon)) - \epsilon \ \le \ -\epsilon < 0$$

This verifies assumption $(iii)$of Lemma 3.1. Thus by the application of Lemma 3.1 we have

$$U(t, x, \epsilon) < 0 \quad \text{on } \overline{Q}_{t_0}.$$

This implies that

$$v(t, x) \le r(t, t_0, y_0) \ \text{ on } \overline{Q}_{t_0},$$

in view of $\lim_{\epsilon \to 0} r(t, \epsilon) = r(t, t_0, y_0)$. This completes the proof of the theorem.

In the context of Theorem 3.1, we present two similar results regarding the two special cases (2.3) and (2.4) of (2.1).

Corollary 3.2 Suppose that

(i) $v \in C[\overline{Q}_{t_0}, R_+^N]$, and v possesses the continuous partial derivatives v_t, v_x, and v_{xx} in Q_{t_0};

(ii) A linear partial differential operator L satisfies the inequality $L(v) \le 0$, where

$$L_i(v_i) = v_{it} - \left[\sum_{j=1}^{m} \sum_{l=1}^{m} \frac{\partial}{\partial x_l} \left(A_{lj}^i(t, x) \frac{\partial v_i}{\partial x_j} \right) \right.$$

$$\left. + \sum_{j=1}^{m} B_{ij}(t, x) \frac{\partial v_i}{\partial x_j} + \sum_{j=1}^{N} C_{ij}(t, x) v_j \right]$$

for $i = 1, ..., N$;
(iii) $g \in C[JxR_+^N, R^N]$, $g(t, y)$ is quasi-monotone non decreasing in y for each $t \epsilon J$, $r(t, t_0, y_0)$ is the maximal solution of the differential system (3.8) existing for $t \geq t_0$ and

$$\sum_{j=1}^{N} C_{ij} z_j \leq g(t, z) \,, \qquad 1 \leq i \leq N \quad , z \geq 0 \,;$$

(iv) (a) $v(t_0, x)\,, y_0$ for $x \epsilon \overline{H}$;
(b) $Bv(t, x) < B\, r\,(t, t_0, y_0)$ on $\Gamma\, t_0$,
where B is the boundary operator defined in (2.2).

Under these assumptions, we have

$$v(t, x) \leq r\,(t, t_0, y_0) \quad \text{on } \overline{Q}_{t_0}. \tag{3.15}$$

Corollary 3.3 Suppose that
(i) $v \in C[\overline{Q}_{t_0}, R_+^N\,]$, $v(t, x)$ possesses the continuous partial derivatives v_t, v_x, v_{xx} in Q_{t_0} ;
(ii) A linear parabolic partial differential operator L satisfies the inequality $L(v) \leq 0$, where

$$L_i(v_i) = v_{it} - \sum_{l,j=1}^{m} A_{lj}^i(t, x) \frac{\partial^2 v_i}{\partial x_l \partial x_j} + \sum_{j=1}^{N} B_{ij}(t, x) \frac{\partial v_i}{\partial x_j} + \sum_{j=1}^{N} C_{ij}(t, x) v_j \tag{3.16}$$

For $i = 1, ..., N$;
(iii) $g \in C[\,J \times R_+^N, R^N]$, $g(t, y)$ is a quasi-monotone non decreasing in y for each $t \in J$, $r(\,t, t_0, y_0)$ is the maximal solution of the differential system (3.8) existing for $t \geq t_0$, and

$$\sum_{j=1}^{N} C_{ij} z_j \leq g_i(t, z) \,, \qquad 1 \leq i \leq N \,, \; z \geq 0 \,;$$

(iv) (a) $v(t_0, x) \leq y_0$, for $x \in \overline{H}$;
(b) $Bv(t, x) < Br(t, t_0, y_0)$ on Γt_0;
where B is the boundary operator defined in (2.2) .

Under these assumptions, we have

$$v(t, x) \leq r(t, t_0, y_0) \qquad \text{on } \overline{Q}_{t_0} \tag{3.17}$$

Remark 3.1 To prove corollary 3.2, we need to verify the assumptions of Corollary 3.1. This can be achieved by using an argument similar to the argument used in the proof of Theorem 3.1. To avoid the repetition, the details are omitted. Similarly the proof of Corollary 3.3 follows from the application of Lemma 3.1.

4. BLOCK COMPARISON THEOREMS

In this section we state and prove a very general block comparison theorem and its variants, moreover, this result is a natural extension of Theorem 3.1.

Theorem 4.1 Suppose that

(i) $v^k \in C[\overline{Q}_{t_0}, R_+^N]$, v^k possesses the continuous partial derivatives v_t^k, v_x^k, v_{xx}^k for $(t,x) \in Q_{t_0}$, $k = 1, ..., s$, and

$$v(t,x) = (v^1(t,x)^T, v^2(t,x)^T, ...,v^s(t,x))^T; \tag{4.1}$$

(ii) $W \in C[\overline{Q}_{t_0} \times (R_+^N)^s \times (R^N)^m \times (R^N)^{m^2}, (R^N)^s]$, $G^k \in C[\overline{Q}_{t_0} \times (R_+^N)^s \times R^m \times R^{m^2}, R^N]$, such that

$$v_t(t,x) \leq W(t,x,v,v_x,v_{xx}) \tag{4.2}$$

where

$$W = (W^{1^T}, W^{2^T}, ..., W^{s^T})^T, \text{ and}$$

$$W^k(t,x,v,v_x,v_{xx}) = \Big(G_1^k(t,x,v,v_{1x}^k,v_{1x}^k) + \sum_{\substack{l=1\\l\neq k}}^{s} \pi_{kl}v_1^l - \pi_{kk}v_1^k, ...$$

$$.., G_N^k(t,x,v,v_{Nx}^k,v_{Nxx}^k) + \sum_{\substack{l=1\\l\neq k}}^{s} \pi_{kl}v_N^l - \pi_{kk}v_N^k\Big)^T,$$

and $G_i^k(t,x,v,P,Q)$ are elliptic for all $1 \leq i \leq N$, and $1 \leq k \leq s$;

(iii) $w \in C[J \times (R_+^N)^s, (R^N)^s]$,

$$w(t,y) = (w^1(t,y)^T, w^2(t,y)^T, ..., w^s(t,y)^T)^T,$$

$w^k \in C[J \times (R^N)^s, R^N]$ and it has the following structural representation

$$w^k(t,y) = \Big(g_1^k(t,y^k) + \sum_{\substack{l=1\\l\neq k}}^{s} \pi_{kl}y_1^l - \pi_{kk}y_1^k, ...$$

$$.........,g_N^k(t,y^k) + \sum_{\substack{l=1\\l\neq k}}^{s} \pi_{kl}y_N^l - \pi_{kk}y_N^k\Big)^T,$$

and $y^l, y^k \in R^N$, $l,k = 1, ..., s$, $r(t,t_0,y_0)$ is the maximal solution of the system of differential equations

$$y' = w(t,y), \qquad y(t_0) = y_0$$

(iv) (a) $v(t_0,x) < y_0$ for $x \in \overline{H}$;

(b) $Bv(t,x) < Br(t,t_0,y_0)$ on Γt_0;

where B is as defined in (2.2) ,$1 \leq k \leq s$, and $B^k v^k$ is an $N-$dimensional vector.

(v) For $(t,x,z) \in \overline{Q}_{t_0} \times R^N$,

$$W(t,x,z,0,0) \leq w(t,z)\ , \quad z \geq 0.$$

Then

$$v(t,x) \leq r(t,t_0,y_0) \qquad \text{on } \overline{Q}_{t_0}. \tag{4.3}$$

Proof The $s-$ dimensional block vector function whose components are $N-$ dimensional vector functions, can be represented as an $N.s-$ dimensional vector function $\overline{W}$ defined as follows

$$\overline{W}(t,x,v,v_x,v_{xx}) = \left(W_1^1(t,x,v,v_{1x}^1,v_{1xx}^1),\ldots\ldots, W_N^s(t,x,v,v_{Nx}^s,v_{Nxx}^s)\right)^T$$

$$= \left(G_1^1(t,x,v,v_{1x}^1,v_{1xx}^1) + \sum_{\substack{l=1\\ l\neq k}}^{s}\pi_{1l}v_1^l - \pi_{kk}v_1^1,\ldots\ldots\ldots\right.$$

$$\left.\ldots, G_N^s(t,x,v,v_{Nx}^s,v_{Nxx}^s) + \sum_{\substack{l=1\\ l\neq k}}^{s}\pi_{1l}v_N^l - \pi_{kk}v_N^s\right)^T. \tag{4.4}$$

Also the $s-$ dimensional Block vector function with $N-$ dimensional components can be described as $N.s-$ dimensional vector function $\overline{v}$ as follows

$$\overline{v}(t,x) = (v_1^1(t,x),\ldots., v_i^k(t,x),\ldots., v_N^s(t,x))^T. \tag{4.5}$$

From (4.4), for fixed $k, 1 \leq k \leq s,$ the kth component of (4.2) can be written as follows

$$\begin{bmatrix} v_{1t}^k(t,x) \\ \cdot \\ \cdot \\ v_{it}^k(t,x) \\ \cdot \\ \cdot \\ v_{Nt}(t,x) \end{bmatrix} \leq \begin{bmatrix} W_1^k(t,x,v,v_{1x}^k,v_{1xx}^k) \\ \cdot \\ \cdot \\ W_i^k(t,x,v,v_{ix}^k,v_{ixx}^k) \\ \cdot \\ \cdot \\ W_N^k(t,x,v,v_{Nx}^k,v_{Nxx}^k) \end{bmatrix} \tag{4.6}$$

From the respective representation of $W(t,x,v,v_x,v_{xx})$ in (4.4), and $v(t,x)$ in (4.5), (4.2) can be written as follows

$$\overline{v}_t(t,x) \leq \overline{W}(t,x,\overline{v},\overline{v}_x,\overline{v}_{xx})$$

which implies that

$$T[\overline{v}] \leq 0 \qquad \text{on } Q_{t_0}, \tag{4.7}$$

where

$$T[\overline{v}] = \overline{v}_t - \overline{W}(t,x,\overline{v},\overline{v}_x,\overline{v}_{xx}),$$

Here, our comparison system is

$$\overline{y}\,' = \overline{w}\,(t,\overline{y}\,), \qquad \overline{y}\,(t_0) = \overline{y}_0, \tag{4.8}$$

where $\overline{w}$, $\overline{y}$ $\in R^{N.s}$.
Using condition (iii), we can verify that $\overline{w}(t,\overline{y})$ is quasi-monotone non decreasing in $\overline{y}$, and conclude that the comparison system (4.8) has the maximal solution, $\overline{r}\ (t,t_0,y_0)$. From the representation of $w(t,y)$ and $v(t,x)$, condition (iv) can be represented as follows:

$$\begin{cases} \overline{v}(t_0,x) \leq \overline{y}_0 & x \in \overline{H}, \\ B\overline{v}(t,x) \leq B\overline{r}(t,t_0,y_0) & \text{on } \Gamma_{t_0}. \end{cases} \tag{4.9}$$

Moreover, from (4.4), (4.5), (4.7), and (4.8), hypothesis (v) can be rewritten as

$$\overline{W}(t,x,\overline{z},0,0) \leq \overline{w}(t,\overline{z}). \tag{4.10}$$

From (4.7), (4.9), (4.10), and condition (i) in the context of (4.6), all the conditions of Theorem 3.1 are satisfied. Therefore an application of the comparison Theorem 3.1 yields

$$\overline{v}(t,x) \leq \overline{r}(t,t_0,y_0) \qquad \text{on } \overline{Q}_{t_o}$$

From Definition 3.1, we get

$$v(t,x) \leq r(t,t_0,y_0) \qquad \text{on } \overline{Q}_{t_0}.$$

This completes the proof of the theorem.

Corollary 4.1 Suppose that
(i) $v^k \in C[\overline{Q}_{t_0}, R_+^N]$, and v^k possesses the continuous partial derivatives v_t^k, v_x^k, v_{xx}^k in Q_{t_0}, $k = 1, ..., S$;

$$v(t,x) = (v^1(t,x)^T, ..., v^s(t,x)^T)^T$$

(ii) For fixed k, $1 \leq k \leq s$, a linear partial differential operator L satisfies the inequality $L(v) \leq 0$, where

$$L_i^k(v_i^k) = v_{it}^k - \left[\sum_{j=1}^{m}\sum_{l=1}^{m}\frac{\partial}{\partial x_l}\left(A_{lj}^{ik}(t,x,k)\frac{\partial v_i^k}{\partial x_j}\right) + \sum_{j=1}^{m} B_{ij}^k(t,x,k)\frac{\partial v_i^k}{\partial x_j}\right.$$

$$\left. + \sum_{j=1}^{N} C_{ij}^k(t,x,k)v_j^k\right] - \sum_{\substack{l=1 \\ l\neq k}}^{s}\pi_{kl}v_i^l + \pi_{kk}v_i^k \tag{4.11}$$

$1 \leq i \leq N$;
(iii) $w \in C[J \times (R_+^N)^s, (R^N)^s]$,

$$w(t,y) = Cy, \tag{4.12}$$

$$w(t,y) = (w^1(t,y), ...w^k(t,y), ..., w^s(t,y))^T,$$

where C is $s \times s$ block matrix defined as follows

$$C = \begin{bmatrix} C^1 - \pi_{11}I & \pi_{12}I & . & . & . & \pi_{1s}I \\ \pi_{21}I & C^2 - \pi_{22}I & . & . & . & \pi_{2s}I \\ . & . & . & . & . & . \\ . & . & . & . & . & . \\ \pi_{k1}I & \pi_{k2}I & . & C^k - \pi_{kk}I & \pi_{kk+1}I & \pi_{ks}I \\ . & . & . & . & . & . \\ \pi_{s1}I & \pi_{s2}I & . & . & . & C^s - \pi_{ss}I \end{bmatrix} \tag{4.13}$$

where I is an $N \times N$ identity matrix, C^k is an $N \times N$ matrix with off diagonal elements being non-negative, $y = (y^{1^T}, y^{2^T}, \ldots\ldots, y^{k^T}, \ldots\ldots, y^{s^T})^T$, and $y^k \in R^N, 1 \le k \le s$

$r(t, t_0, y_0)$ is the maximal solution of the system of differential equations

$$y' = w(t, y), \quad y(t_0) = y_0; \tag{4.14}$$

(iv) (a) $v(t_0, x) \le y_0$ for $x \in \overline{H}$;
(b) $Bv(t, x) \le Br(t, t_0, y_0)$ on Γ_{t_0};

(v)
$$\sum_{j=1}^{N} C_{ij}^k z_j^k + \sum_{\substack{l=1 \\ l \neq k}}^{s} \pi_{kl}\, z_i^l - \pi_{kk} z_i^k = w_i^k(t, z) \quad , \; z \ge 0\,.$$

Then

$$v(t, x) \le r(t, t_0, y_0), \qquad t \ge t_0. \tag{4.15}$$

Corollary 4.2 Support that

(i) $v^k \in C[\overline{Q}_{t_0}, R_+^N]$, and v^k possesses continuous partial derivative ,v_t^k, v_x^k, v_{xx}^k in Q_{t_0}, $k = 1, \ldots, s$;

(ii) For fixed k, $1 \le k \le s$, a linear parabolic partial differential operator L, satisfies the inequality $L(v) \le 0$, where

$$L_i^k(v_i^k) = v_{it}^k - \left[\sum_{j=1}^{m}\sum_{l=1}^{m} A_{lj}^{ik}(t, x, k)\frac{\partial^2 v_i^k}{\partial x_l \partial x_j} + \sum_{j=1}^{m} B_{ij}^k(t, x, k)\frac{\partial v_i^k}{\partial x_j} + \sum_{j=1}^{N} C_{ij}^k(t, x, k) v_j^k\right] - \sum_{\substack{l=1 \\ l \neq k}}^{s} \pi_{kl} v_i^l + \pi_{kk} v_i^k$$

$i = 1, \ldots, N$;

(iii) $w \in C[J \times (R^N)^s, (R^N)^s]$,

$$w(t, y) = Cy,$$

where C is $s \times s$ block matrix as defined in (4.13), and $r(t, t_0, y_0)$ is the maximal solution of the system of differential equations (4.14);

(iv) (a) $v(t_0, x) \le y_0$ for $x \in \overline{H}$;

(b) $B\,v(t,x) \leq Br(t,t_0,y_0)$ on Γ_{t_0};

$$(v) \qquad \sum_{j=1}^{N} C_{ij}^{k}\, z_j^{k} + \sum_{\substack{l=1\\ l\neq k}}^{s} \pi_{kl}\, z_i^{l} - \pi_{kk} z_i{}^{k} = w_i^{k}(t,z) \quad , \qquad z \geq 0$$

Then

$$v(t,x) \leq r\,(t,t_0,y_0)$$

Remark 4.1 To prove the Corollary 4.1 and Corollary 4.2, we need to verity the assumptions of Corollary 3.3. This can be achieved by following the argument used in the proof of Theorem 4.1. To avoid the repetition, the details are omitted.

5. ACKNOWLEDGMENTS

This research is supported in part by NASA grant No. NCCW-0085 and NSF grant No. HRD-9706268 and the support is greatly appreciated.

REFERENCES

1. Ladde, G.S. and Lakshmikantham, V., "Random Differential Inequalities", Academic Press, New York, 1980.
2. Ladde, G.S., Lakshmikantham, V. and Liu, P.T., Differential Inequalities and Boundedness of Stochastic Differential Equations, Journal of Mathematical Analysis and Applications, 48, No. 2, 1974.
3. Ladde, G.S. and Lawrence, B.A., Stability and Convergence of Large-Scale Stochastic Approximation Procedures, Int. J. Systems Sci, Vol. 26, No. 3, 1995.
4. Ladde, G.S. and Šiljak, D.D., Multiplex Control Systems: Stochastic Stability and Dynamic Reliability, Int. J. Control, 38, No. 3, 1983.
5. Lakshmikantham, V. and Leela, S., "Differential and Integral Inequalities Volume II", Academic Press, New York, 1969.

Proceedings of Dynamic Systems and Applications 3 (2001) 35-42

Positive Solutions for Nonlinear Differential Equations with Periodic Boundary Conditions

F.MERDIVENCI ATICI and G. SH. GUSEINOV
Department of Mathematics,
Ege University,
35100 Bornova, Izmir-Turkey
atici@fenfak.ege.edu.tr; guseinov@fenfak.ege.edu.tr

Abstract

We prove the existence of positive solutions of second order nonlinear differential equations on a finite interval with periodic boundary conditions and give upper and lower bounds for these positive solutions. Obtained results yield positive periodic solutions of the equation on the whole real axis, provided that the coefficients are periodic.

AMS No. : 34B15.
Keywords: Periodic boundary conditions, positive periodic solution, fixed point theorem in cones.

1 INTRODUCTION

For fixed positive number ω we consider the following boundary value problem (BVP):

$$-[p(x)y']' + q(x)y = f(x,y), \quad 0 \le x \le \omega, \tag{1.1}$$

$$y(0) = y(\omega), \quad y^{[1]}(0) = y^{[1]}(\omega), \tag{1.2}$$

where $y = y(x)$ is a desired solution, and $y^{[1]}(x) = p(x)y'(x)$ denotes the quasi-derivative of $y(x)$. We will assume that the coefficients $p(x)$ and $q(x)$ of Eq. (1.1) are real valued measurable functions defined on $[0,\omega]$ and satisfy the following condition,

(H1) $\quad p(x) > 0, \quad q(x) \ge 0, \quad q(x) \not\equiv 0$ almost everywhere,

$$\int_0^\omega \frac{dx}{p(x)} < \infty, \qquad \int_0^\omega q(x)dx < \infty.$$

Boundary conditions (1.2) we call the periodic boundary conditions which are an important representative of nonseparated boundary conditions.

In Section 2 we find the Green's function for equation

$$-[p(x)y']' + q(x)y = h(x), \quad 0 \le x \le \omega \tag{1.3}$$

subject to the boundary conditions (1.2), and we prove its positiveness. This fact is very crucial to our arguments. The Green's function for the second order differential equation with periodic boundary conditions was constructed and used in [10]. The sign property of the Green's function for differential equations with separated boundary conditions is investigated in [20-22].

In Section 3 we prove the existence of a positive solution to the BVP (1.1), (1.2). Proof of the existence of positive solutions is based on an application of a fixed point theorem for the completely continuous operators in cones and uses the properties of the Green's function. This tecnique in the case of seperated boundary conditions was used in [12, 8, 9, 6, 7, 4, 5, 1, 2, 11, 14-18] and others articles.

In Section 4 the results obtained in Section 3 yield positive periodic solutions to the equation (1.1) considered for $x \in \mathbf{R} = (-\infty, \infty)$, provided that the coefficients are periodic with period ω. The problem of existence of positive periodic solutions for nonlinear differential equations comprehensively discussed by Krasnosel'skii in [13,12]. In the discrete case this problem was investigated by the authors in [3].

Finally, for easy reference we state here the fixed point theorem [12, p.148], [8, p.94] which is employed in this paper.

Theorem 1.1 *Let $\mathcal{B}$ be a Banach space, and let $\mathcal{P} \subset \mathcal{B}$ be a cone in $\mathcal{B}$. Assume Ω_1, Ω_2 are open subsets of $\mathcal{B}$ with $0 \in \Omega_1$, $\overline{\Omega}_1 \subset \Omega_2$ and let $A : \mathcal{P} \cap (\overline{\Omega}_2 \backslash \Omega_1) \longrightarrow \mathcal{P}$ be a completely continuous operator such that, either*

(i) $\|Ay\| \leq \|y\|$, $y \in \mathcal{P} \cap \partial\Omega_1$, and $\|Ay\| \geq \|y\|$, $y \in \mathcal{P} \cap \partial\Omega_2$; or

(ii) $\|Ay\| \geq \|y\|$, $y \in \mathcal{P} \cap \partial\Omega_1$, and $\|Ay\| \leq \|y\|$, $y \in \mathcal{P} \cap \partial\Omega_2$.

Then A has at least one fixed point in $\mathcal{P} \cap (\overline{\Omega}_2 \backslash \Omega_1)$.

2 GREEN'S FUNCTION AND ITS POSITIVENESS

Consider the linear nonhomogeneous equation (1.3) with boundary conditions (1.2). Denote by $u(x)$ and $v(x)$ the solutions of the corresponding homogeneous equation

$$-[p(x)y']' + q(x)y = 0, \quad 0 \leq x \leq \omega, \tag{2.1}$$

satisfying the initial conditions

$$u(0) = 1, \quad u^{[1]}(0) = 0; \quad v(0) = 0, \quad v^{[1]}(0) = 1. \tag{2.2}$$

Let us set

$$D = u(\omega) + v^{[1]}(\omega) - 2. \tag{2.3}$$

The condition $(H1)$ will be assumed throughout.

Lemma 2.1 *The number D defined by (2.3) is positive.*

Proof. Using the initial conditions (2.2), we can deduce from Eq. (2.1) for $u(x)$ and $v(x)$ the following equations

$$u(x) = 1 + \int_0^x \left[\int_s^x \frac{dt}{p(t)}\right] q(s)u(s)ds, \tag{2.4}$$

$$v(x) = \int_0^x \frac{dt}{p(t)} + \int_0^x \left[\int_s^x \frac{dt}{p(t)}\right] q(s)v(s)ds, \tag{2.5}$$

$$p(x)v'(x) = 1 + \int_0^x q(s)v(s)ds. \tag{2.6}$$

From (2.4) - (2.6), by the condition $(H1)$, it follows that

$$u(x) \geq 1, \quad v(x) \geq \int_0^x \frac{dt}{p(t)}, \quad x \in [0, \omega], \tag{2.7}$$

$$u(\omega) > 1, \quad v^{[1]}(\omega) > 1. \tag{2.8}$$

Now from (2.3) and (2.8), we get $D > 0$.

Theorem 2.1 *For the solution $y(x)$ of the BVP (1.3), (1.2) the formula*

$$y(x) = \int_0^\omega G(x, s)h(s)ds, \quad x \in [0, \omega] \tag{2.9}$$

holds, where

$$G(x,s) = \frac{v(\omega)}{D}u(x)u(s) - \frac{u^{[1]}(\omega)}{D}v(x)v(s)$$
$$+ \begin{cases} \dfrac{v^{[1]}(\omega)-1}{D}u(x)v(s) - \dfrac{u(\omega)-1}{D}u(s)v(x), & 0 \leq s \leq x \leq \omega, \\ \dfrac{v^{[1]}(\omega)-1}{D}u(s)v(x) - \dfrac{u(\omega)-1}{D}u(x)v(s), & 0 \leq x \leq s \leq \omega; \end{cases} \tag{2.10}$$

the number D is defined by (2.3).

Proof. It is easy to see that the general solution of Eq. (1.3) has the form

$$y(x) = c_1u(x) + c_2v(x) + \int_0^x [u(s)v(x) - u(x)v(s)]h(s)ds,$$

where c_1 and c_2 are arbitrary constants. Substituting this expression for $y(x)$ in the boundary conditions (1.2) we can evaluate c_1 and c_2. After not very complicated calculations we can get (2.9), (2.10).

The function $G(x, s)$ is called the Green's function of the BVP (1.3), (1.2).

Theorem 2.2 *Under the condition $(H1)$ the Green's function $G(x, s)$ of the BVP (1.3), (1.2) is positive:*

$$G(x,s) > 0 \quad \textit{for } x, s \in [0, \omega]. \tag{2.11}$$

Proof. Since $G(x,s) = G(s,x)$, it is enough to prove that $G(x,s) > 0$ for $x \in [0,\omega]$ and $0 \leq s \leq x$. Setting

$$E(x,s) = u(s)v(x) - u(x)v(s), \tag{2.12}$$

$$F(x,s) = [v(\omega)u(x) - u(\omega)v(x)]u(s) + [v^{[1]}(\omega)u(x) - u^{[1]}(\omega)v(x)]v(s), \tag{2.13}$$

we have, for $s \leq x$,

$$G(x,s) = \frac{1}{D}[E(x,s) + F(x,s)]. \tag{2.14}$$

Let us now show that

$$E(x,x) = 0 \text{ for } x \in [0,\omega]; \quad F(\omega,0) = 0; \tag{2.15}$$

$$E(x,s) > 0 \text{ for } s \in [0,\omega) \text{ and } x \in (s,\omega]; \tag{2.16}$$

$$F(x,s) > 0 \text{ for } s \in [0,\omega],\ x \in [s,\omega] \text{ and } (x,s) \neq (\omega,0). \tag{2.17}$$

Evidently (2.15) holds. To prove (2.16), we note that for fixed $s \in [0,\omega)$

$$\frac{\partial}{\partial x}\left[p(x)\frac{\partial E(x,s)}{\partial x}\right] = q(x)E(x,s), \quad x \in [s,\omega],$$

$$E(s,s) = 0, \qquad p(x)\frac{\partial E(x,s)}{\partial x}\,|_{x=s} = 1.$$

Hence it follows that, for all $x \in [s,\omega]$,

$$E(x,s) = \int_s^x \frac{dt}{p(t)} + \int_s^x \left[\int_\xi^x \frac{dt}{p(t)}\right] q(\xi)E(\xi,s)d\xi. \tag{2.18}$$

Using $(H1)$, we get from (2.18) that $E(x,s) > 0$ for all $x \in (s,\omega]$.

Further, passing to $F(x,s)$ we note that for fixed $s \in [0,\omega)$

$$\frac{\partial}{\partial x}\left[p(x)\frac{\partial F(x,s)}{\partial x}\right] = q(x)F(x,s), \quad x \in [s,\omega],$$

$$F(\omega,s) = v(s), \quad p(x)\frac{\partial F(x,s)}{\partial x}|_{x=\omega} = -u(s).$$

Hence it follows that, for all $x \in [s,\omega]$,

$$F(x,s) = v(s) + u(s)\int_x^\omega \frac{dt}{p(t)} + \int_x^\omega [\int_x^\xi \frac{dt}{p(t)}]q(\xi)F(\xi,s)d\xi. \tag{2.19}$$

Using $(H1)$, (2.7), we get from (2.19) that $F(x,s) > 0$, if $(x,s) \neq (\omega,0)$.

Now (2.11) follows from (2.14) by Lemma 2.1 and (2.15)-(2.17).

Remark 2.1 In the case of $p(x) \equiv 1$, $q(x) \equiv c^2 \quad (c > 0)$, where c is a constant, the Green's function $G(x,s)$ of the BVP (1.3), (1.2) has the form

$$G(x,s) = \frac{1}{2c(e^{c\omega} - 1)} \begin{cases} e^{c(x-s)} + e^{c(\omega+s-x)}, & 0 \le s \le x \le \omega, \\ e^{c(s-x)} + e^{c(\omega+x-s)}, & 0 \le x \le s \le \omega. \end{cases}$$

3 EXISTENCE OF A POSITIVE SOLUTION

Consider the nonlinear BVP (1.1), (1.2), where $p(x)$ and $q(x)$ satisfy the condition $(H1)$. We assume that the function $f(x,\xi)$ satisfies the following condition.

> **(H2)** $f : [0,\omega] \times \mathbf{R} \longrightarrow \mathbf{R}$ is continuous in (x,ξ) and $f(x,\xi) \ge 0$ for $\xi \in \mathbf{R}^+$, where $\mathbf{R}^+$ denotes the set of nonnegative real numbers.

Regarding Eq.(1.1), denote by $G(x,s)$ the Green's function of the problem (1.3), (1.2), by Theorem 2.2 the inequality (2.11) holds. Let us set

$$m = \min G(x,s), \quad M = \max G(x,s), \quad x,s \in [0,\omega]. \tag{3.1}$$

Consider the Banach space $\mathcal{B} = C[0,\omega]$ of real-valued continuous functions $y(x)$ defined on $[0,\omega]$ with the norm

$$\|y\| = \max |y(x)|, \quad x \in [0,\omega].$$

[We take the occasion to point out that on the p. 170 of [3] the absolute value sign in the definition of the norm is missing.]

Define the cones $\mathcal{P}$ and $\mathcal{P}_0$ in $\mathcal{B}$ by

$\mathcal{P} = \{y \in \mathcal{B} | y(x) \geq 0 \text{ for } x \in [0,\omega]\}$, $\mathcal{P}_0 = \{y \in \mathcal{P} | min_{x\in[0,\omega]} y(x) \geq \frac{m}{M}\|y\|\}$.

Throughout this paper, m and M will denote the numbers defined by (3.1).

By Theorem ?? solving the BVP (1.1), (1.2) is equivalent to solving the following integral equation in $\mathcal{B}$

$$y(x) = \int_0^\omega G(x,s) f(s, y(s)) ds, \qquad x \in [0,\omega], \tag{3.2}$$

and consequently, it is equivalent to finding fixed points of operator $A : \mathcal{B} \longrightarrow \mathcal{B}$ defined by

$$Ay(x) = \int_0^\omega G(x,s) f(s, y(s)) ds, \qquad x \in [0,\omega]. \tag{3.3}$$

Note that for each $y \in \mathcal{B}$ the function $Ay(x)$ satisfies the boundary conditions (1.2) by the definition of the Green's function $G(x,s)$.

An operator (nonlinear, in general) acting in a Banach space is said to be completely continuous if it is continuous and maps bounded sets into relatively compact sets. ¿From the continuity of $G(x,s)$ and $f(x,\xi)$ it follows that the operator A defined by (3.3) is completely continuous in $\mathcal{B}$.

Lemma 3.1 *$Ay \in \mathcal{P}_0$ for all $y \in \mathcal{P}$. In particular, the operator A leaves the cone $\mathcal{P}_0$ invariant, i.e., $A(\mathcal{P}_0) \subset \mathcal{P}_0$.*

Proof. For all $y \in \mathcal{P}$, by (2.11) and $(H2)$, we have from (3.3), $Ay(x) \geq 0$ for all $x \in [0,\omega]$;

$$\min_{x\in[0,\omega]} Ay(x) \geq m \int_0^\omega f(s,y(s))ds \geq \frac{m}{M} \int_0^\omega \{max_{x\in[0,\omega]} G(x,s)\} f(s,y(s))ds$$

$$\geq \frac{m}{M} \max_{x\in[0,\omega]} \int_0^\omega G(x,s) f(s,y(s))ds = \frac{m}{M}\|Ay\|.$$

Therefore, $Ay \in \mathcal{P}_0$.

In the next theorem we also assume the following condition on $f(x,\xi)$.

(H3) There exist numbers $0 < r < R < \infty$ such that for all $x \in [0,\omega]$:

$f(x,\xi) \leq \dfrac{1}{\omega M}\xi$, if $0 \leq \xi \leq r$; $f(x,\xi) \geq \dfrac{M}{\omega m^2}\xi$, if $R \leq \xi < \infty$.

Theorem 3.1 *Assume that conditions $(H1), (H2)$, and $(H3)$ are satisfied. Then the BVP (1.1), (1.2) has at least one solution $y(x)$ such that*

$$\frac{m}{M} r \leq y(x) \leq \frac{M}{m} R, \quad x \in [0,\omega]. \tag{3.4}$$

Proof. For $y \in \mathcal{P}_0$ with $\|y\| = r$ (hence $0 \le y(s) \le r$ for $s \in [0,\omega]$), we have for all $x \in [0,\omega]$,

$$Ay(x) \le M \int_0^\omega f(s, y(s))ds \le M \frac{1}{\omega M} \int_0^\omega y(s)ds \le \frac{1}{\omega}\|y\|\omega = \|y\|. \tag{3.5}$$

Now if we let $\Omega_1 = \{y \in B \mid \|y\| < r\}$, then (3.5) shows that $\|Ay\| \le \|y\|$, $y \in \mathcal{P}_0 \cap \partial\Omega_1$.

Further, let

$$R_1 = \frac{M}{m} R \text{ and } \Omega_2 = \{y \in B \mid \|y\| < R_1\}.$$

Then $y \in \mathcal{P}_0$ and $\|y\| = R_1$ implies

$$\min{}_{x\in[0,\omega]} y(x) \ge \frac{m}{M}\|y\| = \frac{m}{M} R_1 = R,$$

hence $y(s) \ge R$ for all $s \in [0,\omega]$. Therefore, for all $x \in [0,\omega]$,

$$Ay(x) \ge m \int_0^\omega f(s, y(s))ds \ge m \frac{M}{\omega m^2} \int_0^\omega y(s)ds \ge \frac{M}{\omega m} \int_0^\omega \frac{m}{M}\|y\|ds = \|y\|.$$

Hence $\|Ay\| \ge \|y\|$ for all $y \in \mathcal{P}_0 \cap \partial\Omega_2$.

Consequently, by the first part of Theorem ??, it follows that A has a fixed point y in $\mathcal{P}_0 \cap (\overline{\Omega}_2 \backslash \Omega_1)$. We have $r \le \|y\| \le R_1$. Hence, since for $y \in \mathcal{P}_0$ we have $y(x) \ge \frac{m}{M}\|y\|$, $x \in [0,\omega]$, it follows that (3.4) holds.

Remark 3.1 From (3.4) it follows that the solution $y(x)$ is positive : $y(x) > 0$ for $x \in [0,\omega]$.

Remark 3.2 If

$$lim_{\xi\to 0^+} \frac{f(x,\xi)}{\xi} = 0 \quad \text{and} \quad lim_{\xi\to\infty} \frac{f(x,\xi)}{\xi} = \infty$$

uniformly on $x \in [0,\omega]$, then the condition $(H3)$ will be satisfied for $r > 0$ sufficiently small and $R > 0$ sufficiently large.

In Theorem ?? we assume the following condition on $f(x,\xi)$.

(H4) There exist numbers $0 < r < R < \infty$ such that for all $x \in [0,\omega]$,

$$f(x,\xi) \ge \frac{M}{\omega m^2}\xi, \text{ if } 0 \le \xi \le r; \ f(x,\xi) \le \frac{1}{\omega M}\xi, \text{ if } R \le \xi < \infty.$$

Theorem 3.2 *Assume that conditions* $(H1), (H2)$, *and* $(H4)$ *are satisfied. Then the BVP* $(1.1), (1.2)$ *has at least one solution* $y(x)$ *with property (3.4).*

The proof is analogous to that of Theorem ?? and uses the second part of Theorem ??.

Remark 3.3 If

$$lim_{\xi\to 0^+} \frac{f(x,\xi)}{\xi} = \infty \quad \text{and} \quad lim_{\xi\to\infty} \frac{f(x,\xi)}{\xi} = 0$$

uniformly on $x \in [0,\omega]$, then the condition $(H4)$ will be satisfied for $r > 0$ sufficiently small and $R > 0$ sufficiently large.

4 EXISTENCE OF POSITIVE PERIODIC SOLUTIONS

Consider the equation (1.1) on the whole real axis:

$$-[p(x)y']' + q(x)y = f(x,y), \qquad x \in \mathbf{R}. \tag{4.1}$$

We assume that the coefficients of Eq. (4.1) are periodic, that is, for some positive real number ω

(H5) $p(x+\omega) = p(x)$, $q(x+\omega) = q(x)$, $x \in \mathbf{R}$;

(H6) $f(x+\omega,\xi) = f(x,\xi)$, $x,\xi \in \mathbf{R}$.

We are interested in existence of positive ω-periodic(i.e., periodic with period ω) solutions of Eq. (4.1).

Together with Eq. (4.1) we consider the BVP (1.1),(1.2). If conditions ($H5$) and ($H6$) hold, then every solution of the BVP (1.1),(1.2) extended from $[0,\omega]$ to $\mathbf{R}$ as an ω-periodic function will be a solution of Eq. (4.1).

Therefore, Theorems 3.1 and 3.2 yield the following results, respectively.

Theorem 4.1 *Assume ($H1$), ($H2$), ($H3$), ($H5$), and ($H6$) hold. Then Eq. (4.1) has at least one ω-periodic solution $y(x)$ such that*

$$\frac{m}{M}r \le y(x) \le \frac{M}{m}R \text{ for all } x \in \mathbf{R}. \tag{4.2}$$

Theorem 4.2 *Assume ($H1$), ($H2$), ($H4$), ($H5$), and ($H6$) hold. Then Eq. (4.1) has at least one ω-periodic solution $y(x)$ with property (4.2).*

References

[1] R.P. Agarwal and F.H. Wong, *Existence of positive solutions for non-positive higher-order BVPs*, J. Comput. Appl. Math. **88**(1998), 3-14.

[2] R.P. Agarwal and P.J. Wong, *On eigenvalue intervals and twin eigenfunctions of higher-order boundary value problems*, J. Comput. Appl. Math. **88**(1998), 15-43.

[3] F. Merdivenci Atici and G. Sh. Guseinov, *Positive periodic solutions for nonlinear difference equations with periodic coefficients*, J. Math. Anal. Appl. **232** (1999), 166-182.

[4] P.W. Eloe and J. Henderson, *Positive solutions for higher order ordinary differential equations*, Electronic J. Differential Equations **3** (1995), 1-8.

[5] P.W. Eloe, *Positive solutions of boundary value problems for disfocal ordinary differential equations*, J. Comput. Appl. Math., **88** (1998), 71-78.

[6] L. H. Erbe and H. Wang, *On the existence of positive solutions of ordinary differential equations*, Proc. Amer. Math. Soc., **120** (1994), 743 - 748.

[7] L.H. Erbe, S. Hu, and H. Wang, *Multiple positive solutions of some boundary value problems*, J. Math. Anal. Appl. **184** (1994), 640-648.

[8] D. Guo and V. Lakshmikantham, *Nonlinear Problems in Abstract Cones,* Academic Press , San Diego,1988.

[9] D. Guo and V. Lakshmikantham, *Multiple solutions of two-point boundary value problems of ordinary differential equations in Banach spaces,* J. Math. Anal. Appl. **129** (1988), 211-222.

[10] G. Sh. Guseinov, *Spectrum and eigenfunction expansions of a quadratic pencil of Sturm-Liouville operators with periodic coefficients,* Spectral Theory of Operators and Its Applications, Elm, Baku (Azerbaijan), **6** (1985), 56 - 97 (in Russian).

[11] J. Henderson and H. Wang, *Positive solutions for nonlinear eigenvalue problems,* J. Math. Anal. Appl. **208** (1997), 252-259.

[12] M. A. Krasnosel'skii, *Positive Solutions of Operator Equations*, Noordhoff, Groningen, 1964.

[13] M. A. Krasnosel'skii, *The Operator of Translation Along the Trajectories of Differential Equations,* Transl. Math. Monographs , Vol. 19, Amer. Math. Soc., Providence, RI, 1968.

[14] Z. Liu and F. Li, *Multiple positive solutions for nonlinear two-point boundary value problems,* J. Math. Anal. Appl. **203** (1996), 610-625.

[15] B. Lou, *Solutions of superlinear Sturm-Liouville problems in Banach spaces,* J. Math. Anal. Appl. **201** (1996),169-179.

[16] F. Merdivenci, *Two positive solutions of a boundary value problem for difference equations,* J. Difference Equations and Appl., **1** (1995), 262 - 270.

[17] F. Merdivenci, *Green's matrices and positive solutions of a discrete boundary value problem,* PanAmerican Math. J., **5** (1995), 25 - 42.

[18] F. Merdivenci, *Positive solutions for focal point problems for 2n-th order difference equation,* PanAmerican Math. J., **5** (1995), 71 - 82.

[19] M.A. Neumark, *"Lineare Differentialoperatoren",* Akademie - Verlag, Berlin, 1967.

[20] A. C. Peterson, *On the sign of Green's fuction beyond the interval of disconjugacy,* Rocky Mountain J. Math. **3** (1973), 41-51.

[21] A. C. Peterson, *On the sign of Green's functions*, J. Differential Equations, **21** (1976), 167-178.

[22] A. C. Peterson, *Green's functions for focal type boundary value problems,* Rocky Mountain J. Math. **9** (1979), 721-732.

Proceedings of Dynamic Systems and Applications 3 (2001) 43-48

Stability Criteria for Dynamic Equations on Time Scales with Periodic Coefficients

F. MERDİVENCİ ATICI *, G. SH. GUSEİNOV * and B. KAYMAKÇALAN **
* Department of Mathematics,
Ege University,
35100 Bornova, Izmir, Turkey
atici@fenfak.ege.edu.tr , guseinov@fenfak.ege.edu.tr
** Department of Mathematics,
Middle East Technical University,
06531 Ankara, Turkey
billur@metu.edu.tr

Abstract

In this paper we obtain sufficient conditions for instability and stability to hold for second order linear Δ-differential equations on time scales with periodic coefficients.
Keywords: Time scale, periodicity, stability, generalized zero
AMS Subject Classification: 39A10, 39A11

1. INTRODUCTION: FORMULATION OF THE MAIN RESULTS

Let $\mathbb{T}$ be a closed subset (so called a time scale or a measure chain) of the real numbers $\mathbb{R}$. If there exists a positive number $\Lambda \in \mathbb{R}$ such that $t + n\Lambda \in \mathbb{T}$ for all $t \in \mathbb{T}$ and $n \in \mathbb{Z}$, then we call $\mathbb{T}$ a periodic time scale with period Λ.Suppose $\mathbb{T}$ is a Λ-periodic time scale. For the sake of simplicity we will assume that $0 \in \mathbb{T}$.

Consider the second order linear Δ-differential equation given by

$$\left[p(t)y^{\Delta}(t)\right]^{\Delta} + q(t)y(\sigma(t)) = 0, \quad t \in \mathbb{T}, \tag{1.1}$$

where σ is the forward jump operator and the coefficients $p(t)$ and $q(t)$ are real valued Λ-periodic functions defined on $\mathbb{T}$,

$$p(t+\Lambda) = p(t), \qquad q(t+\Lambda) = q(t), \quad t \in \mathbb{T}. \tag{1.2}$$

Besides we assume that

$$p(t) > 0, \quad p(t) \in C^1_{rd}[0,\Lambda], \quad g(t) \in C_{rd}[0,\Lambda], \tag{1.3}$$

where $[0,\Lambda] = \{t \in \mathbb{T} : 0 \le t \le \Lambda\}$.

For the definition of the Δ-derivative and other concepts related with time scales we refer to [3, 10, 7, 12, 13].

Equation (1.1) is said to be unstable if all nontrivial solutions are unbounded on $\mathbb{T}$, conditionally stable if there exist a nontrivial solution which is bounded on $\mathbb{T}$, and stable if all solutions are bounded on $\mathbb{T}$.

Main results of this paper are the following two theorems;

Theorem 1.1. If $q(t) \leq 0$ and $q(t) \not\equiv 0$, then equation (1.1) is unstable. By definition, put

$$p_0 = \max_{t\in[0,\rho(\Lambda)]} \frac{\sigma(t)-t}{p(t)}, \quad q_+(t) = \max\{q(t), 0\}, \tag{1.4}$$

where ρ is the backward jump operator.

Theorem 1.2. If

$$\text{(i)} \qquad \int_0^\Lambda q(t)\Delta t \geq 0, \quad q(t) \not\equiv 0; \tag{1.5}$$

$$\text{(ii)} \quad \left[p_0 + \int_0^\Lambda \frac{\Delta t}{p(t)}\right] \cdot \int_0^\Lambda q_+(t)\Delta t \leq 4, \tag{1.6}$$

then equation (1.1) is stable.

We dwell on the three special cases as follows:

1. If $\mathbb{T} = \mathbb{R}$ we can take as Λ any $\omega \in \mathbb{R}, \omega > 0$. Equation (1.1) takes the form

$$[p(x)y'(x)]' + q(x)y(x) = 0, \qquad x \in \mathbb{R}, \tag{1.7}$$

and the periodicity condition (1.2) becomes

$$p(x+\omega) = p(x), \quad q(x+\omega) = q(x), \quad x \in \mathbb{R}.$$

The conditions (i) and (ii) of Theorem 1.2 transform into

$$\text{(i)}_1 \quad \int_0^\omega q(x)dx \geq 0, \quad q(x) \not\equiv 0; \qquad \text{(ii)}_1 \quad \left[\int_0^\omega \frac{dx}{p(x)}\right] \cdot \int_0^\omega q_+(x)dx \leq 4.$$

We note here that the last conditions lead to a well-known result. In [14] M.A. Lyapunov has proved that, if a real, continuous and periodic function $q(x)$ of period $\omega > 0$ satisfies the conditions

$$1) \quad q(x) \geq 0, \quad q(x) \not\equiv 0; \qquad 2) \quad \omega \int_0^\omega q(x)dx \leq 4,$$

then the equation

$$y''(x) + q(x)y(x) = 0, \qquad x \in \mathbb{R} \tag{1.8}$$

is stable. In [4] G. Borg extended Lyapunov's result to functions $q(x)$ of variable sign, showing that if

$$1') \quad \int_0^\omega q(x)dx \geq 0, \qquad q(x) \not\equiv 0; \qquad 2') \quad \omega \int_0^\omega |\, q(x) \,|\, dx \leq 4,$$

hold, then equation (1.8) is stable. Then in [16] M.G. Krein improved Borg's result replacing in condition 2') $|\, q(x) \,|$ by $q_+(x)$.

Notice that equation (1.7) can be transformed into an equation of the type (1.8) by the change of variable. For direct investigation of equation (1.7) we refer to [9].

2. If $\mathbb{T} = \mathbb{Z}$ we can take as Λ any integer $N > 0$. Equation (1.1) takes the form

$$\Delta\,[p(n)\Delta y(n)] + q(n)y(n+1) = 0, \qquad n \in \mathbb{Z}.$$

where Δ is the forward difference operator defined by $\Delta y(n) = y(n+1) - y(n)$. The periodicity condition (1.2) can be written as

$$p(n+N) = p(n), \quad q(n+N) = q(n), \quad n \in \mathbb{Z}.$$

The conditions (i) and (ii) of Theorem 1.2 become

$$\text{(i)}_2 \quad \sum_{n=0}^{N-1} q(n) \geq 0, \quad q(n) \not\equiv 0; \qquad \text{(ii)}_2 \quad \left[\frac{1}{p} + \sum_{n=0}^{N-1} \frac{1}{p(n)}\right] \cdot \sum_{n=0}^{N-1} q_+(n) \leq 4,$$

where $p = \min\{p(0), p(1), \cdots, p(N-1)\}$, $q_+(n) = \max\{q(n), 0\}$. This result was established in [2].

2. FLOQUET THEORY

Consider equation (1.1), where $\mathbf{T}$ is a Λ-periodic time scale containing zero, with the coefficients $p(t)$ and $q(t)$ being real valued and satisfying the conditions (1.2) and (1.3). Floquet theory applies for the equation (1.1). For the details of Floquet we may refer to, for example, [5,15] for differential equations, and [8, pp. 113-115], [2], [6, pp. 144-149] for difference equations.

Let us denote by $y^{[\Delta]}(t) = p(t)y^{\Delta}(t)$, the quasi Δ-derivative of $y(t)$. For arbitrary complex numbers c_0 and c_1, equation (1.1) has a unique solution $y(t)$ satisfying the initial conditions

$$y(0) = c_0, \quad y^{[\Delta]}(0) = c_1.$$

Denote by $\theta(t)$ and $\varphi(t)$ the solutions of equation (1.1) under the initial conditions

$$\theta(0) = 1, \theta^{[\Delta]}(0) = 0; \quad \varphi(0) = 0, \varphi^{[\Delta]}(0) = 1. \tag{2.1}$$

There exist a nonzero complex number β and a nontrivial solution $\psi(t)$ of equation (1.1) such that

$$\psi(t+\Lambda) = \beta\psi(t), \quad t \in \mathbf{T}. \tag{2.2}$$

The number β is a root of the quadratic equation

$$\beta^2 - D\beta + 1 = 0, \tag{2.3}$$

where

$$D = \theta(\Lambda) + \varphi^{[\Delta]}(\Lambda). \tag{2.4}$$

The roots of (2.3) are defined by

$$\beta_{1,2} = \frac{1}{2}\left(D \pm \sqrt{D^2 - 4}\right).$$

Since the coefficients of equation (1.1) and the initial conditions (2.1) are real, the solutions $\theta(t), \varphi(t)$ and hence the number D defined by (2.4) will be real.

Theorem 2.1. Equation (1.1) is unstable if $| D |> 2$, and stable if $| D |< 2$.

Proof. If the discriminant $D^2 - 4$ is nonzero, then (2.3) has two distinct roots β_1 and β_2, and hence there exist two nontrivial solutions $\psi_1(t)$ and $\psi_2(t)$ of equation (1.1) such that

$$\psi_1(t+\Lambda) = \beta_1\psi_1(t), \quad \psi_2(t+\Lambda) = \beta_2\psi_2(t), \quad t \in \mathbf{T}. \tag{2.5}$$

It is easy to see that $\psi_1(t)$ and $\psi_2(t)$ are linearly independent. From (2.5) it follows that, for all $k \subset \mathbb{Z}$,

$$\psi_1(t+k\Lambda) = \beta_1^k \psi_1(t), \quad \psi_2(t+k\Lambda) = \beta_2^k \psi_2(t), \quad t \in \mathbb{T}. \tag{2.6}$$

If $\mid D \mid > 2$, then the numbers β_1 and β_2 will be distinct and real. Therefore from the equality $\beta_1\beta_2 = 1$ it follows that $\mid \beta_1 \mid \neq 1$ and $\mid \beta_2 \mid \neq 1$, since if this were false, we would get $\beta_1 = \beta_2 = \pm 1$. Obviously, if $\mid \beta_1 \mid > 1$, then $\mid \beta_2 \mid < 1$, and if $\mid \beta_1 \mid < 1$, we have $\mid \beta_2 \mid > 1$. Consequently from (2.6) as $k \to \pm\infty$ it follows that every nontrivial linear combination of $\psi_1(t)$ and $\psi_2(t)$ will be unbounded on $\mathbb{T}$, that is equation (1.1) is unstable.

If $\mid D \mid < 2$, then the numbers β_1 and β_2 will be distinct, nonreal and such that $\mid \beta_1 \mid = \mid \beta_2 \mid = 1$. Therefore, from (2.5) we have

$$\mid \psi_1(t+\Lambda) \mid = \mid \psi_1(t) \mid, \quad \mid \psi_2(t+\Lambda) \mid = \mid \psi_2(t) \mid, \quad t \in \mathbb{T}.$$

Consequently, $\psi_1(t)$ and $\psi_2(t)$ and hence every solution of equation (1.1) which is a linear combination of $\psi_1(t)$ and $\psi_2(t)$ will be bounded on $\mathbb{T}$, that is equation (1.1) will be stable. This completes the proof of the theorem.

Remark 2.1. If $\mid D \mid = 2$, then equation (1.1) will be stable in the case $\theta^{[\Delta]}(\Lambda) = \varphi(\Lambda) = 0$; but conditionally stable and not stable otherwise.

3. PROOF OF THEOREM 1.1

Let $\theta(t)$ and $\varphi(t)$ be solutions of equation (1.1) satisfying the initial conditions (2.1). Our aim is to show that, under the hypotheses of the theorem, the inequalities

$$\theta(\Lambda) \geq 1, \quad \varphi^{[\Delta]}(\Lambda) > 1$$

hold. Then $D = \theta(\Lambda) + \varphi^{[\Delta]}(\Lambda) > 2$ will be obtained and therefore by Theorem 2.1, equation (1.1) will be unstable.

Firstly we show that;

$$\theta(t) \geq 1, \quad \theta^{[\Delta]}(t) \geq 0, \quad \varphi(t) \geq 0, \quad \varphi^{[\Delta]}(t) \geq 1, \quad \forall t \in \mathbb{T}, \ t \geq 0. \tag{3.1}$$

For this purpose we apply the induction principle developed for Time-Scales (see, for example, [5,8]), to the statement,

$$A(t) : \theta(t) \geq 1 \text{ and } \theta^{[\Delta]}(t) \geq 0 \text{ for all } t \in \mathbb{T} \text{ and } t \geq 0.$$

(I) The statement $A(0)$ is true, since $\theta(0) = 1$ and $\theta^{[\Delta]}(0) = 0$.

(II) Let t be right-scattered and $A(t)$ be true, i.e., $\theta(t) \geq 1$ and $\theta^{[\Delta]}(t) \geq 0$. We need to show that $\theta(\sigma(t)) \geq 1$ and $\theta^{[\Delta]}(\sigma(t)) \geq 0$. But, in view of the induction assumptions, the required results are immediate from the relations

$$\theta(\sigma(t)) = \theta(t) + \mu^*(t)\theta^{\Delta}(t), \quad \theta^{[\Delta]}(\sigma(t)) = \theta^{[\Delta]}(t) - \mu^*(t)q(t)\theta(\sigma(t)),$$

the first one following from the definition of Δ-derivative and the second from equation (1.1) for $\theta(t)$. Here $\mu^*(t) = \sigma(t) - t$.

(III) Let t_0 be right-dense, $A(t_0)$ be true and $t \in [t_0, t_1] = \{t \in \mathbb{T} : t_0 \leq t \leq t_1\}$, where $t_1 \in \mathbb{T}$ is such that $t_1 > t_0$ and is sufficiently closed to t_0. We need to prove that $A(t)$ is true for $t \in [t_0, t_1]$.

From equation (1.1) with $y(t) = \theta(t)$, the equations

$$\theta^{[\Delta]}(t) = \theta^{[\Delta]}(t_0) - \int_{t_0}^{t} q(s)\theta(\sigma(s))\Delta s, \tag{3.2}$$

$$\theta(t) = \theta(t_0) + \theta^{[\Delta]}(t_0) \cdot \int_{t_0}^{t} \frac{\Delta\tau}{p(\tau)} - \int_{t_0}^{t} \frac{1}{p(\tau)} \left[\int_{t_0}^{\tau} q(s)\theta(\sigma(s))\Delta s\right] \Delta\tau \tag{3.3}$$

follow.

It can be shown that the equation (3.3) for $\theta(t)$ can be solved by the method of successive approximations for $t \in [t_0, t_1]$. Hence it follows that $\theta(t) \geq \theta(t_0) \geq 1$ for $t \in [t_0, t_1]$. Then from (3.2) we get also that $\theta^{[\Delta]}(t) \geq 0$ for $t \in [t_0, t_1]$.

(IV) Let $t \in \mathbb{T}$, $t > 0$ be left-dense and such that $A(s)$ is true for all $s < t$, i.e., $\theta(s) \geq 1$, $\theta^{[\Delta]}(s) \geq 0$, $\forall s < t$. Passing here to the limit as $s \to t$ we get by the continuity of $\theta(s)$ and $\theta^{[\Delta]}(s)$ that $\theta(t) \geq 1$ and $\theta^{[\Delta]}(t) \geq 0$, thereby verifying the validity of $A(t)$.

Consequently, by the continuous type induction principle on a time-scale, (3.1) holds for $\theta(t)$ and $\theta^{[\Delta]}(t)$, $\forall t \in \mathbb{T}$, $t \geq 0$. Proof for $\varphi(t) \geq 0$ and $\varphi^{[\Delta]}(t) \geq 1$ is similar.

From (3.1), in particular, we have $\theta(\Lambda) \geq 1$, $\varphi^{[\Delta]}(\Lambda) \geq 1$. Actually $\varphi^{[\Delta]}(\Lambda) > 1$; Indeed, if $\varphi^{[\Delta]}(\Lambda) = 1$, then from the equation

$$\varphi^{[\Delta]}(t) = 1 - \int_{0}^{t} q(s)\varphi(\sigma(s))\Delta s,$$

we get $q(s) = 0$, $\forall s \in [0, \rho(\Lambda)]$. Hence employing the Λ-periodicity of $q(t)$ it follows that $q(t) \equiv 0$, $\forall t \in \mathbb{T}$, contradicting the hypothesis of the theorem. Hence $\varphi^{[\Delta]}(\Lambda) > 1$ and Theorem 1.1 is proven.

References

[1] R. P. Agarwal, Difference Equations and Inequalities, Marcel Dekker, New York, 1992

[2] F. Merdivenci Atici and G. Sh. Guseinov, Criteria for the stability of second order difference equations with periodic coefficients, Communications in Appl. Anal. 3 (1999), 503 - 516.

[3] B. Aulbach and S.Hilger, Linear dynamic processes with inhomogeneous time scale, in "Nonlinear Dynamics and Quantum Dynamical Systems", 9-20, Akademie Verlag, Berlin 1990.

[4] G. Borg, On a Liapounoff criterion of stability, Amer. J. Math. **71** (1949), 67-70.

[5] M.S.P. Eastham, The Spectral Theory of Periodic Differential Equations, Scotish Acad. Press, Edinburgh and London, (1973).

[6] S. N. Elaydi, An Introduction to Difference equations, Springer-Verlag, New York, 1996.

[7] L. Erbe and S. Hilger, Sturmian theory on measure chains, Differential Equations Dynamic Systems **1** (1993), 223 - 246.

[8] G. Sh. Guseinov, On the spectral theory of multiparameter difference equations of second order, Izvestiya Akad. Nauk SSSR. Ser. Mat. **51** (1987), 785 - 811 = Math. USSR Izvestiya **31** (1988), 95 - 120.

[9] G. Sh. Guseinov and E. Kurpinar, On the stability of second order differential equations with periodic coefficients, (Indian) Pure Appl. Math. Sci. **42** (1995), 11-17.

[10] S. Hilger, Analysis on Measure Chains - A unified approach to continuous and discrete calculus, Results Math. **18** (1990), 18-56.

[11] M. G. Krein, on certain problems on the maximum and minimum of characteristic values and on the Lyapunov zones of stability, Prikl. Math. Meh. **15** (1951), 323 - 348 = Amer. Math. Soc. Translations. Ser. 2, **1** (1955), 163-187.

[12] B. Kaymakalan and S. Leela, A survey of dynamic systems on time scales, Nonlinear Times and Digest **1** (1994), 37-60.

[13] V. Lakshmikantham, S. Sivasundaram, and B. Kaymakalan, Dynamic Systems on Measure Chains, Kluwer Academic Publishers, Boston, 1996.

[14] M. A. Liapounoff, Probléme géneral de la stabilité du mouvement, Ann. Fac. Sci. Univ. Toulouse, Ser. 2, **9** (1907), 203 - 475. Reprinted in the Ann. of Math. Studies No. 17, Princeton, (1949).

[15] W. Magnus and S. Winkler, Hill's Equation, Interscience, New York, (1966).

Proceedings of Dynamic Systems and Applications 3 (2001) 49-60

Multiple Positive Solutions of a Partial Difference Equation

Richard I. Avery
Dakota State University, College of Natural Sciences
Madison, South Dakota 57042

Abstract

We are concerned with the partial difference equation $\Delta_t u(t-1,x)+\Delta_x u(t,x-1) - f(u(t,x),t,x) = 0$ with $u(0,x) = 0$ for all $x \in [0,\infty)_D$ and $u(t,0) = 0$ for all $t \in [0,\infty)_D$. Under various assumptions on f we prove the existence of multiple positive solutions.

Key words: *partial difference equations, Green's function, fixed points.*
AMS Subject Classification: 39A10.

1 PRELIMINARIES

There are many papers on the existence of multiple positive solutions to boundary value problems, however there are very few on partial difference equations. The entire theory of partial difference equations has been described as being an underdeveloped research area [4]. A source of applications of partial difference equations arises from the study of mathematical physics problems [5]. In this paper we apply a generalization of the Leggett-Williams Fixed Point Theorem [9] to prove the existence of at least three positive solutions to a first order nonlinear partial difference equation. We will assume the reader has an understanding of Green's functions and their applications. Books on the subject include Kelley and Peterson [8] and Ahlbrandt and Peterson [1]. In this section we will state the fixed point theorem that we will use to prove our main result. First we make a few definitions.

Definition 1 *Let E be a real Banach space. A nonempty closed convex set $P \subset E$ is called a cone if it satisfies the following two conditions:*

(*i*) $x \in P, \lambda \geq 0$ *implies* $\lambda x \in P$;

(*ii*) $x \in P, -x \in P$ *implies* $x = 0$.

The cone $P \subset E$ induces an ordering on the Banach space E given by

$$x \leq y \text{ if and only if } y - x \in P.$$

Definition 2 *An operator is called completely continuous if it is continuous and maps bounded sets into relatively compact sets.*

Definition 3 *A map α is said to be a nonnegative continuous concave functional on a cone P of a real Banach space E if*

$$\alpha : P \to [0, \infty)$$

is continuous and

$$\alpha(tx + (1-t)y) \geq t\alpha(x) + (1-t)\alpha(y)$$

for all $x, y \in P$ and $t \in [0,1]$. Similarly we say the map β is a nonnegative continuous convex functional on a cone P of a real Banach space E if

$$\beta : P \to [0, \infty)$$

is continuous and

$$\beta(tx + (1-t)y) \leq t\beta(x) + (1-t)\beta(y)$$

for all $x, y \in P$ and $t \in [0,1]$.

Let P be a cone of a real Banach space E then for numbers a, b such that $0 < a < b$, α a nonnegative continuous concave functional on P, and β a nonnegative continuous convex functional on P we define the following convex sets:

$$P_a = \{x \in P : \|x\| < a\},$$

$$Q(\beta, a, b) = \{x \in P : \beta(x) < a, \|x\| \leq b\},$$

and

$$P(\alpha, a, b) = \{x \in P : a \leq \alpha(x), \|x\| \leq b\}.$$

The following fixed point theorem is a minor generalization of the Leggett-Williams Fixed Point Theorem. The proof of this theorem is nearly identical to the one given by Leggett-Williams in [7], [9] and can be found in [3].

Theorem 4 *Let P be a cone of a real Banach space E, $A : \overline{P_c} \to \overline{P_c}$ be completely continuous, α be a nonnegative continuous concave functional on P, β be a nonnegative continuous convex functional on P such that $\beta(0) = 0$ and $\alpha(x) \leq \beta(x)$ for all $x \in \overline{P_c}$. Suppose there exist $0 < d < a < b \leq c$ such that:*

(i) $\{x \in P(\alpha, a, b) : \alpha(x) > a\} \neq \emptyset$ and $\alpha(Ax) > a$ for $x \in P(\alpha, a, b)$;

(ii) $\beta(Ax) < d$ for $x \in \overline{Q(\beta, d, c)}$;

(iii) $\alpha(Ax) > a$ for $x \in P(\alpha, a, c)$ with $\|Ax\| > b$.

Then A has at least three fixed points $x_1, x_2, x_3 \in \overline{P_c}$ such that;

$$\beta(x_1) < d,$$

$$a < \alpha(x_2),$$

and

$$d < \beta(x_3) \text{ and } \alpha(x_3) < a.$$

2 INTRODUCTION TO A PARTIAL DIFFERENCE EQUATION

In this paper for $m_1 < m_2$ with m_1, m_2 integers we define the discrete intervals $[m_1, \infty)_D$ and $[m_1, m_2]_D$ by

$$[m_1, \infty)_D = \{m_1, m_1+1, m_1+2, \dots\}$$

and

$$[m_1, m_2]_D = \{m_1, m_1+1, m_1+2, \dots, m_2\}.$$

Define the real Banach space E (with the sup norm) by

$$E = \left\{ u \;\middle|\; \begin{array}{l} u : [0,\infty)_D \times [0,\infty)_D \to \mathbb{R}, \;\; u(0,x) = 0 \;\text{ for all }\; x \in [0,\infty)_D \\ \text{and }\; u(t,0) = 0 \;\text{ for all }\; t \in [0,\infty)_D \end{array} \right\}$$

and the cone P of E by

$$P = \{u \in E \mid u(t,x) \geq 0 \;\text{ for all }\; (t,x) \in [0,\infty)_D \times [0,\infty)_D\}.$$

Consider the partial difference equation,

$$\Delta_t u(t-1,x) + \Delta_x u(t,x-1) - f(u(t,x),t,x) = 0 \;\text{ for }\; (t,x) \in [1,\infty)_D \times [1,\infty)_D \quad (1)$$

$$u(t,0) = 0 \;\text{ for all }\; t \in [0,\infty)_D \quad (2)$$

$$u(0,x) = 0 \;\text{ for all }\; x \in [0,\infty)_D \quad (3)$$

where $f : [0,\infty) \times [0,\infty)_D \times [0,\infty)_D \to [0,\infty)$ is continuous. Define the operator L on E by

$$Lu(t,x) = \Delta_t u(t-1,x) + \Delta_x u(t,x-1)$$

for $(t,x) \in [1,\infty)_D \times [1,\infty)_D$ with boundary conditions

$$Lu(t,0) = 0$$

for all $t \in [0,\infty)_D$ and

$$Lu(0,x) = 0$$

for all $x \in [0,\infty)_D$. Define the function G for $(t,x,\tau,y) \in [0,\infty)_D \times [0,\infty)_D \times [1,\infty)_D \times [1,\infty)_D$ by

$$G(t,x,\tau,y) = \begin{cases} 0 & \text{if } t < \tau \text{ or } x < y \\ \dfrac{\binom{t+x-\tau-y}{t-\tau}}{2^{\,t+x-\tau-y+1}} & \text{otherwise} \end{cases}$$

In the following theorem we prove that G is the Green's function for the partial difference equation (1), (2), (3).

Theorem 5 *The function G satisfies the following properties:*

(1) *G is defined for* $(t, x, \tau, y) \in [0, \infty)_D \times [0, \infty)_D \times [1, \infty)_D \times [1, \infty)_D$;

(2) $G(t, 0, \tau, y) = 0$ *for all* $(t, \tau, y) \in [0, \infty)_D \times [1, \infty)_D \times [1, \infty)_D$;

(3) $G(0, x, \tau, y) = 0$ *for all* $(x, \tau, y) \in [0, \infty)_D \times [1, \infty)_D \times [1, \infty)_D$; *and*

(4) $LG(t, x, \tau, y) = \delta_{(t,x),(\tau,y)}$ *for* $(t, x), (\tau, y) \in [1, \infty)_D \times [1, \infty)_D$.

Proof: By the definition of G, properties one through three are clearly satisfied. We verify property four by exhaustion. Let $(t, x), (\tau, y) \in [1, \infty)_D \times [1, \infty)_D$.

Case 1: $t < \tau$ or $x < y$.

$$\begin{aligned} LG(t, x, \tau, y) &= -G(t-1, x, \tau, y) + 2G(t, x, \tau, y) - G(t, x-1, \tau, y) \\ &= -0 + 0 - 0 = \delta_{(t,x),(\tau,y)}. \end{aligned}$$

Case 2: $t = \tau$ and $x > y$.

$$\begin{aligned} LG(t, x, \tau, y) &= -G(t-1, x, \tau, y) + 2G(t, x, \tau, y) - G(t, x-1, \tau, y) \\ &= -0 + \frac{2\binom{x-y}{0}}{2^{x-y+1}} - \frac{\binom{x-y-1}{0}}{2^{x-y}} = \delta_{(t,x),(\tau,y)}. \end{aligned}$$

Case 3: $t > \tau$ and $x = y$.

$$\begin{aligned} LG(t, x, \tau, y) &= -G(t-1, x, \tau, y) + 2G(t, x, \tau, y) - G(t, x-1, \tau, y) \\ &= -\frac{\binom{t-\tau-1}{t-\tau-1}}{2^{t-\tau}} + \frac{2\binom{t-\tau}{t-\tau}}{2^{t-\tau+1}} - 0 = \delta_{(t,x),(\tau,y)}. \end{aligned}$$

Case 4: $t = \tau$ and $x = y$.

$$\begin{aligned} LG(t, x, \tau, y) &= -G(t-1, x, \tau, y) + 2G(t, x, \tau, y) - G(t, x-1, \tau, y) \\ &= -0 + \frac{2}{2} - 0 = \delta_{(t,x),(\tau,y)}. \end{aligned}$$

Case 5: $t > \tau$ and $x > y$.

$$LG(t, x, \tau, y) = -G(t-1, x, \tau, y) + 2G(t, x, \tau, y) - G(t, x-1, \tau, y)$$

$$= -\frac{\binom{t-\tau-1+x-y}{t-\tau-1}}{2^{t-\tau+x-y}} + \frac{2\binom{t-\tau+x-y}{t-\tau}}{2^{t-\tau+x-y+1}} - \frac{\binom{t-\tau+x-y-1}{t-\tau}}{2^{t-\tau+x-y}}$$

$$= \frac{-\binom{t-\tau-1+x-y}{t-\tau-1} + \binom{t-\tau+x-y}{t-\tau} - \binom{t-\tau+x-y-1}{t-\tau}}{2^{t-\tau+x-y}}$$

$$= \ 0 = \delta_{(t,x),(\tau,y)}.$$

Therefore,

$$LG(t,x,\tau,y) = \delta_{(t,x),(\tau,y)}.$$

□

Define the operator A on the Banach space E by,

$$Au(t,x) = \sum_{\tau=1}^{\infty}\sum_{y=1}^{\infty} G(t,x,\tau,y) f(u(\tau,y),\tau,y). \tag{4}$$

Note that G is zero for all but a finite number of terms in the summations on the right hand side of equation (4), hence the operator A is well defined. Since $f(z,\tau,y)$ is nonnegative for all $(z,\tau,y) \in [0,\infty) \times [1,\infty)_D \times [1,\infty)_D$ we also have that

$$A : P \to P$$

since G is nonnegative on its domain. In the following theorem we verify the relationship between fixed points of A and the solutions of the partial difference equation (1), (2), (3).

Theorem 6 *A vector $u \in E$ is a fixed point of A if and only if u is a solution of the partial difference equation* (1), (2), (3).

Proof:
($\Rightarrow$) Suppose $u \in E$ and $Au = u$. If $(t,x) \in [1,\infty)_D \times [1,\infty)_D$ then

$$\begin{aligned} Lu(t,x) &= LAu(t,x) \\ &= L\left(\sum_{\tau=1}^{\infty}\sum_{y=1}^{\infty} G(t,x,\tau,y) f(u(\tau,y),\tau,y)\right) \\ &= \sum_{\tau=1}^{\infty}\sum_{y=1}^{\infty} LG(t,x,\tau,y) f(u(\tau,y),\tau,y) \\ &= \sum_{\tau=1}^{\infty}\sum_{y=1}^{\infty} \delta_{(t,x),(\tau,y)} f(u(\tau,y),\tau,y) \\ &= f(u(t,x),t,x). \end{aligned}$$

Therefore u is a solution of the partial difference equation (1), (2), (3).

($\Leftarrow$) Suppose $u \in E$ is a solution of the partial difference equation (1), (2), (3). Then for $(t,x) \in [1,\infty)_D \times [1,\infty)_D$:

$$L(u(t,x) - Au(t,x)) = f(u(t,x),t,x) - f(u(t,x),t,x) = 0$$

hence

$$z(t,x) = u(t,x) - Au(t,x)$$

is a solution of the associated homogeneous partial difference equation

$$\Delta_t z(t-1,x) + \Delta_x z(t,x-1) = 0 \text{ for } (t,x) \in [1,\infty)_D \times [1,\infty)_D \tag{5}$$

$$z(t,0) = 0 \text{ for all } t \in [0,\infty)_D \tag{6}$$

$$z(0,x) = 0 \text{ for all } x \in [0,\infty)_D. \tag{7}$$

A simple inductive argument verifies that the only solution of (5), (6), (7) is the trivial solution; therefore,

$$Au(t,x) = u(t,x)$$

for all $(t,x) \in [0,\infty)_D \times [0,\infty)_D$.

□

3 EXISTENCE THEOREM

We now prove a Lemma which will be used in the proof or our main result at the end of this section. First some notation.

Definition 7 *For* $n \in [2,\infty)_D$ *and* $(t,x) \in [1,\infty)_D \times [1,\infty)_D$ *define,*

$$D_n(t,x) = \{(\tau,y) \in [1,t]_D \times [1,x]_D \;:\; \tau + y = n\}.$$

To invoke Theorem 4 we need A to be completely continuous and $A : \overline{P_c} \to \overline{P_c}$ for some $c > 0$. The following lemma provides sufficiency conditions for A to satisfy these hypotheses.

Lemma 8 *If f is continuous and there exists a real number $c > 0$ such that*

$$0 \leq f(z,\tau,y) \leq \frac{4c}{2^{\tau+y}}$$

for all $(z,\tau,y) \in [0,c] \times [1,\infty)_D \times [1,\infty)_D$ then

$$A : \overline{P_c} \to \overline{P_c}$$

is completely continuous.

Proof: Let $(t,x) \in [1,\infty)_D \times [1,\infty)_D$ and $k \in [2,\infty)_D$ then,

$$\begin{aligned}
\sum_{(\tau,y)\in D_k(t,x)} G(t,x,\tau,y) &= \sum_{(\tau,y)\in D_k(t,x)} \frac{\begin{pmatrix} t+x-\tau-y \\ t-\tau \end{pmatrix}}{2^{t+x-\tau-y+1}} \\
&= \sum_{(\tau,y)\in D_k(t,x)} \frac{\begin{pmatrix} t+x-k \\ t-\tau \end{pmatrix}}{2^{t+x-k+1}} \\
&\le \sum_{j=0}^{t+x-k} \frac{\begin{pmatrix} t+x-k \\ j \end{pmatrix}}{2^{t+x-k+1}} \\
&= \frac{(1+1)^{t+x-k}}{2^{t+x-k+1}} = \frac{1}{2}.
\end{aligned}$$

Therefore if $(t,x) \in [1,\infty)_D \times [1,\infty)_D$ and $v \in \overline{P_c}$ then,

$$\begin{aligned}
\sum_{\tau=1}^{\infty}\sum_{y=1}^{\infty} G(t,x,\tau,y)f(v(\tau,y),\tau,y) &= \sum_{\tau=1}^{t}\sum_{y=1}^{x} G(t,x,\tau,y)f(v(\tau,y),\tau,y) \\
&= \sum_{n=2}^{t+x}\sum_{(\tau,y)\in D_n(t,x)} G(t,x,\tau,y)f(v(\tau,y),\tau,y) \\
&\le \sum_{n=2}^{t+x}\sum_{(\tau,y)\in D_n(t,x)} \frac{4cG(t,x,\tau,y)}{2^{\tau+y}} \\
&= \sum_{n=2}^{t+x}\left(\frac{4c}{2^n}\sum_{(\tau,y)\in D_n(t,x)} G(t,x,\tau,y)\right) \\
&\le \sum_{n=2}^{t+x}\frac{4c}{2^{n+1}} \\
&< \sum_{n=2}^{\infty}\frac{4c}{2^{n+1}} = c.
\end{aligned}$$

Hence, if $v \in \overline{P_c}$ then for all $(t,x) \in [0,\infty)_D \times [0,\infty)_D$ we have

$$Av(t,x) < c$$

which implies

$$\|Av\| \le c.$$

Therefore $Av \in \overline{P_c}$ since by previous remarks we know that $Av \in P$. Also for any $k \in [2,\infty)_D$, $v \in \overline{P_c}$ and $(t,x) \in [1,\infty)_D \times [1,\infty)_D$,

$$\sum_{j=k}^{\infty}\sum_{(\tau,y)\subset D_j(t,x)} G(t,x,\tau,y)f(v(\tau,y),\tau,y) \le \sum_{j=k}^{\infty}\sum_{(\tau,y)\in D_j(t,x)} \frac{4cG(t,x,\tau,y)}{2^{\tau+y}}$$

$$\begin{aligned}
&= \sum_{j=k}^{\infty}\left(\frac{4c}{2^j}\sum_{(\tau,y)\in D_j(t,x)} G(t,x,\tau,y)\right)\\
&\leq \sum_{j=k}^{\infty}\frac{4c}{2^{j+1}}\\
&= \frac{4c}{2^k}.
\end{aligned}$$

Let $\{u_n\}_{n=1}^{\infty}$ be a sequence of functions in $\overline{P_c}$ and $\epsilon > 0$. Then there exists an $M \in \mathbb{N}$ such that

$$\frac{16c}{2^M} < \epsilon.$$

Without loss of generality we can assume that

$$u_n \to u \in \overline{P_c}$$

pointwise. Since f is continuous for $(z,\tau,y) \in [0,c] \times [0,\infty)_D \times [0,\infty)_D$ and a continuous function on a compact set is uniformly continuous there exists a $\delta > 0$ such that if $(w,\tau,y),(z,\tau,y) \in [0,c] \times [0,M-1]_D \times [0,M-1]_D$ with

$$|z-w| < \delta$$

then

$$|f(z,\tau,y) - f(w,\tau,y)| < \frac{\epsilon}{2(M-2)}.$$

Since $u_n \to u$ pointwise there exists an $N \in \mathbb{N}$ such that

$$|u_n(\tau,y) - u(\tau,y)| < \delta$$

for all $(\tau,y) \in \{(\tau,y) \in [1,\infty)_D \times [1,\infty)_D \ : \ \tau + y \leq M-1\}$ and $n \geq N$. Hence for all $n \geq N$ and $(t,x) \in [0,\infty)_D \times [0,\infty)_D$ we have,

$$\begin{aligned}
&\left|\sum_{j=2}^{M-1}\sum_{(\tau,y)\in D_j(t,x)} G(t,x,\tau,y)f(u_n(\tau,y),\tau,y) - \sum_{j=2}^{M-1}\sum_{(\tau,y)\in D_j(t,x)} G(t,x,\tau,y)f(u(\tau,y),\tau,y)\right|\\
&\leq \sum_{j=2}^{M-1}\sum_{(\tau,y)\in D_j(t,x)} G(t,x,\tau,y)\,|f(u_n(\tau,y),\tau,y) - f(u(\tau,y),\tau,y)|\\
&< \sum_{j=2}^{M-1}\sum_{(\tau,y)\in D_j(t,x)} \frac{\epsilon G(t,x,\tau,y)}{2(M-2)}\\
&\leq \sum_{j=2}^{M-1}\frac{\epsilon}{4(M-2)} = \frac{\epsilon}{4}
\end{aligned}$$

and if $v \in \overline{P_c}$ then

$$\begin{aligned}
&\left|Av(t,x) - \sum_{j=2}^{M-1} \sum_{(\tau,y)\in D_j(t,x)} G(t,x,\tau,y)f(v(\tau,y),\tau,y)\right| \\
= &\sum_{j=M}^{\infty} \sum_{(\tau,y)\in D_j(t,x)} G(t,x,\tau,y)f(v(\tau,y),\tau,y) \\
\leq &\frac{4c}{2^M} < \frac{\epsilon}{4}.
\end{aligned}$$

Therefore, for all $(t,x) \in [0,\infty)_D \times [0,\infty)_D$ and $n \geq N$,

$$\begin{aligned}
&|Au_n(t,x) - Au(t,x)| \\
\leq &\left|Au_n(t,x) - \sum_{j=2}^{M-1} \sum_{(\tau,y)\in D_j(t,x)} G(t,x,\tau,y)f(u_n(\tau,y),\tau,y)\right| \\
+ &\left|\sum_{j=2}^{M-1} \sum_{(\tau,y)\in D_j(t,x)} G(t,x,\tau,y)f(u_n(\tau,y),\tau,y) - \sum_{j=2}^{M-1} \sum_{(\tau,y)\in D_j(t,x)} G(t,x,\tau,y)f(u(\tau,y),\tau,y)\right| \\
+ &\left|\sum_{j=2}^{M-1} \sum_{(\tau,y)\in D_j(t,x)} G(t,x,\tau,y)f(u(\tau,y),\tau,y) - Au(t,x)\right| \\
< &\frac{\epsilon}{4} + \frac{\epsilon}{4} + \frac{\epsilon}{4} < \epsilon.
\end{aligned}$$

Hence if $n \geq N$ then

$$\|Au_n - Au\| < \epsilon$$

which implies that A is completely continuous.

□

We now prove our existence theorem for the partial difference equation (1), (2), (3).

Theorem 9 *Suppose there exist numbers a, d, c where $0 < 2d < 2a < c$ such that f satisfies the following properties :*

(i) if $z \in [0,c]$ and $\tau, y \in [1,\infty)_D$ then $0 \leq f(z,\tau,y) \leq \frac{4c}{2^{\tau+y}}$,

(ii) if $z \in [0,d]$ then $f(z,1,1) \leq 2d$, and

(iii) if $z \in [a,c]$ then $f(z,1,1) \geq 2a$.

Then, the partial difference equation (1), (2), (3) has at least three positive solutions.

Proof: For $u \in P$ define

$$\alpha(u) = \beta(u) = u(1,1).$$

Note that $\alpha : P \to [0,\infty)$ is a nonnegative continuous concave functional and $\beta : P \to [0,\infty)$ is a nonnegative continuous convex functional with $\alpha(u) = \beta(u)$ for all $u \in P$ and $\beta(0) = 0$. Let

$$u_0(t,x) := \begin{cases} 0 & t = 0 \text{ or } x = 0 \\ c & t \geq 1 \text{ and } x \geq 1 \end{cases}$$

which is an element of $\{u \in P(\alpha, a, c) : \alpha(x) > a\}$ and hence is nonempty.

Claim 1: If $u \in P(\alpha, a, c)$ then $\alpha(Au) > a$.

Suppose $u \in P(\alpha, a, c)$ then $u(1,1) \in [a, c]$ and using (iii) we have,

$$\begin{aligned} \alpha(Au) &= \sum_{\tau=1}^{\infty}\sum_{y=1}^{\infty} G(1,1,\tau,y) f(u(\tau,y),\tau,y) \\ &= G(1,1,1,1) f(u(1,1),1,1) \\ &= \frac{f(u(1,1),1,1)}{2} \\ &> \frac{2a}{2} = a. \end{aligned}$$

Thus $\alpha(Au) > a$ and Claim 1 has been shown.

Claim 2: If $u \in \overline{Q(\beta, d, c)}$ then $\beta(Au) < d$.

Suppose $u \in \overline{Q(\beta, d, c)}$ then $u(1,1) \in [0, d]$ and using (ii) we have,

$$\begin{aligned} \beta(Au) &= \sum_{\tau=1}^{\infty}\sum_{y=1}^{\infty} G(1,1,\tau,y) f(u(\tau,y),\tau,y) \\ &= G(1,1,1,1) f(u(1,1),1,1) \\ &= \frac{f(u(1,1),1,1)}{2} \\ &< \frac{2d}{2} = d. \end{aligned}$$

Therefore $\beta(Au) < d$ and Claim 2 has been shown.

By condition (i) and Lemma 8 we have that

$$A : \overline{P_c} \to \overline{P_c}$$

is completely continuous. Hence the hypotheses of Theorem 4 are satisfied, therefore the partial difference equation (1), (2), (3) has at least three positive solutions $u_1, u_2, u_3 \in \overline{P_c}$ such that:

$$u_1(1,1) < d,$$

$$u_2(1,1) > a,$$

and

$$d < u_3(1,1) < a.$$

□

REFERENCES

[1] Calvin D. Ahlbrandt and Allan C. Peterson, *Discrete Hamiltonian Systems: Difference Equations, Continued Fractions and Riccati Equations,* Kluwer Academic Publishers, Boston, 1996.

[2] Ravi P. Agarwal, *Difference Equations and Inequalities,* Marcel Dekker, Inc. , New York, 1992.

[3] Richard I. Avery, *Multiple Positive Solutions to Boundary Value Problems,* Dissertation, University of Nebraska-Lincoln, 1997.

[4] Sui Sun Cheng, *An underdeveloped research area: qualitative theory of functional partial difference equations,* International Mathematics Conference '94, World Sci. Publishing, River Edge, NJ, 65-75.

[5] R. Courant, K. Friedrichs and H. Lewy, *On partial difference equations of mathematical physics,* IBM J. Res. Develop. **11**(1967). 215-234.

[6] Klaus Deimling, *Nonlinear Functional Analysis,* Springer-Verlag, New York, 1985.

[7] Dajun Guo and V. Lakshmikantham, *Nonlinear Problems in Abstract Cones,* Academic Press, San Diego, 1988.

[8] Walter G. Kelley and Allan C. Peterson, *Difference Equations: An Introduction with Applications,* Academic Press, San Diego, 1991.

[9] Richard W. Leggett and Lynn R. Williams, *Multiple positive fixed points of nonlinear operators on ordered Banach spaces,* Indiana University Mathematics Journal **28**(1979). 673-688.

Proceedings of Dynamic Systems and Applications 3 (2001) 61-68

On Nonoscillatory Solutions of Forth-Order Differential Equations

Miroslav Bartušek and Monika Sobalová
Masaryk University, Brno, Czech Republic 662 95

ABSTRACT: In this paper the nonlinear fourth order differential equation in the self-adjoint form $y^{(4)} - (gy')' + rf(y) = 0$ is studied where $r \geq 0$ and $f(x)x > 0$ for $x \neq 0$. All nonoscillatory solutions are divided into the classes according to oscillatory/nonoscillatory characters of y', y'' and y''', and sufficient conditions are given for these classes to be empty.
AMS (MOS) subject classification. 34C10

1. INTRODUCTION

Consider the fourth-order differential equation in the self-adjoint form

$$y^{(4)} - (q(t)y')' + r(t)f(y) = 0 \tag{1}$$

where $q \in C^0(R_+)$, $R_+ = [0, \infty)$, $r \in C^0(R_+)$, $f \in C^0(R)$, $R = (-\infty, \infty), r(t) \geq 0$ on R_+ and $f(x)x > 0$ for $x \neq 0$.

Sometimes,

$$\int_0^\infty r(t)dt = \infty, \qquad \liminf_{|x|\to\infty} |f(x)| > 0. \tag{2}$$

will be supposed.

A continuous function $y : R_+ \to R$ is said to be an oscillatory function if there exists a sequence $\{t_n\}_1^\infty$ such that $\lim_{n\to\infty} t_n = \infty, y(t_n) = 0$ and $\sup_{T\leq t} |y(t)| > 0$ for every $T \in R_+$. A solution $y : R_+ \to R$ is called nonoscillatory if it is different from zero in a neighborhood of ∞. Denote the set of all nonoscillatory solutions by $\mathcal{N}$.

The qualitative behavior of solutions of (1) has been studied by Bartušek et al. [3] and Sobalová [9]. One possible approach to study Eq. (1) is used by Bartušek et al. [3] and Cecchi at al [5] assuming that the second order linear differential equation

$$h'' - q(t)h = 0 \tag{3}$$

is nonoscillatory. Indeed, in such a case, Eq. (1) can be written in the disconjugate form

$$y^{[4]} = \left(\frac{1}{h(t)}\left(h^2(t)\left(\frac{y'}{h(t)}\right)'\right)'\right)' + r(t)f(y) = 0 \tag{4}$$

on the intervals where a solution h of (3) is positive. The asymptotic behavior of solutions of (1) is based on the asymptotic behavior of the so-called quasiderivatives of a solution y of (4):

$$y^{[0]}(t) = y(t), \quad y^{[1]}(t) = \frac{y'(t)}{h(t)}, \quad y^{[2]}(t) = h^2(y^{[1]})' = y''(t)h(t) - y'(t)h'(t),$$

$$y^{[3]}(t) = \frac{1}{h}(y^{[2]}(t))' = y'''(t) - q(t)y'(t), \; y^{[4]}(t) = (y^{[3]}(t))' \tag{5}$$

and it is studied e.g. by Bartušek [1]. But using this method, it is necessary to know estimations of solutions of (3) and moreover, if (3) is oscillatory, the transformation into (4) is not possible.

Furthermore, we classify all nonoscillatory solutions according to the signs or oscillatory character of their derivatives. The set $\mathcal{N}$ can be divided, a-priori, into the following classes:

$$\begin{aligned}
W_1 &= \{y \in \mathcal{N} : y' \text{ is an oscillatory function}\},\\
W_2 &= \{y \in \mathcal{N} : y(t)y'(t) \neq 0 \text{ for large } t, y'' \text{ is an oscillatory function }\},\\
W_3 &= \{y \in \mathcal{N} : y(t)y'(t)y''(t) \neq 0 \text{ for large } t, y''' \text{ is an oscillatory function}\},\\
W_4 &= \{y \in \mathcal{N} : y^{(i)}(t)y(t) \geq 0, i = 0,1,2,\ y'''(t) \equiv 0, \text{ for large } t\},\\
N &= \{y \in \mathcal{N} : y(t)y'(t)y''(t)y'''(t) \neq 0 \text{ for large } t\}.
\end{aligned}$$

Nonoscillatory solutions from $W_i = 1,2,3$ are sometimes called weakly oscillatory solutions (see, e. g. Bartušek et al. [4]). It is useful to divide W_2 and W_3 into the subclasses:

$$\begin{aligned}
W_2^+ &= \{y \in W_2 : y(t)y'(t) > 0 \text{ for } t \geq T_y \in R_+\},\\
W_2^- &= \{y \in W_2 : y(t)y'(t) < 0 \text{ for } t \geq T_y \in R_+\},\\
W_3^{++} &= \{y \in W_3 : y(t)y^{(i)}(t) > 0, i = 1,2 \text{ for } t \geq T_y \in R_+\},\\
W_3^{+-} &= \{y \in W_3 : y(t)y'(t) > 0, y(t)y''(t) < 0 \text{ for } t \geq T_y \in R_+\},\\
W_3^{-+} &= \{y \in W_3 : y(t)y'(t) < 0, y(t)y''(t) > 0 \text{ for } t \geq T_y \in R_+\}.
\end{aligned}$$

We further divide all solutions of N into

$$\begin{aligned}
N_0 &= \{y \in N : y(t)y'(t) < 0,\ y(t)y''(t) > 0,\ y(t)y'''(t) < 0 \text{ for } t \geq T_y\},\\
N_1 &= \{y \in N : y(t)y'(t) > 0,\ y(t)y''(t) < 0,\ y(t)y'''(t) > 0 \text{ for } t \geq T_y\},\\
N_2 &= \{y \in N : y(t)y'(t) > 0,\ y(t)y''(t) > 0,\ y(t)y'''(t) < 0 \text{ for } t \geq T_y\},\\
N_3 &= \{y \in N : y(t)y^{(i)}(t) > 0,\ i = 1,2,3 \text{ for } t \geq T_y\}
\end{aligned}$$

where $T_y \in R_+$.

In this paper sufficient conditions are given under which the above given sets are empty.

2. CASE $q \leq 0$.

The following theorem reduces the possible types of nonoscillatory solutions.

Theorem 1. *Let (2) be valid and let $q \leq 0$. Then (1) has no solution y such that $y(t)y'(t) \geq 0$ holds for large t; i.e.,*

$$N_1 = N_2 = N_3 = W_2^+ = W_3^{++} = W_3^{+-} = W_4 = \emptyset.$$

Proof. Let y be a solution of (1) such that for simplicity

$$y(t) > 0,\ y'(t) \geq 0,\ t \geq T \geq 0. \tag{6}$$

By the integration of (1) and using (2) we have

$$y'''(t) - q(t)y'(t) = K - \int_T^t r(s)f(y(s))ds$$
$$\leq K - \min_{y(T)\leq s<\infty} f(s) \int_T^t r(s)ds \to -\infty \tag{7}$$

as $t \to \infty$ where $K = y'''(T) - g(T)y'(T)$. Since $y' \geq 0$ and $q \leq 0$, (7) yields $\lim_{t\to\infty} y'''(t) = -\infty$. Thus $\lim_{t\to\infty} y^{(j)}(t) = -\infty, j = 2, 1$. This contradiction to (6) proves the statement. □

Consequence 1. *Let (2) and $q \leq 0$ be valid. Then every nonoscillatory solution of (1) belongs to the set $N_0 \cup W_1 \cup W_2^- \cup W_3^{-+}$.*

The following examples show that equations (1) exist for which the sets N_0, W_2^- and W_3^{-+} are nonempty. Examples of solutions in W_1 are given by Kiguradze [7].

Example 1. The equation

$$y^{(4)} + (t^{1+\beta}y')' + \left[\frac{\beta^2}{4}t^{\beta+3} - \frac{\beta}{2}\left(\frac{\beta}{2}+1\right)\left(\frac{\beta}{2}+2\right)\left(\frac{\beta}{2}+3\right)\right]t^{-4}y = 0$$

has a solution $y \in N_0$ of the form $y = \frac{1}{t^{\frac{\beta}{2}}}$, where $\beta > 0, t \in [1, \infty)$.

Example 2. The equation $y^{(4)} + (t^3y')' + r(t)y = 0$, where

$$r(t) = \frac{t^4}{t^3 + 10\sin t}\left[-\left(\frac{1}{t} + \frac{10}{t^4}\sin t\right)^{(4)} - \left(t^3\left(\frac{1}{t} + \frac{10}{t^4}\sin t\right)'\right)'\right],$$

has a solution $y = W_3^{-+}$ of the form $y(t) = \frac{1}{t} + \frac{10}{t^4}\sin t, t \geq 10$.

Example 3. The equation $y^{(4)} + (t^4y')' + r(t)y = 0$, where

$$r(t) = \frac{t^3}{t^2 + 2\sin t}\left[-\left(\frac{1}{t} + \frac{2}{t^3}\sin t\right)^{(4)} - \left(t^4\left(\frac{1}{t} + \frac{2}{t^3}\sin t\right)'\right)'\right]$$

has a solution $y \in W_2^-$ of the form $y(t) = \frac{1}{t} + \frac{2}{t^3}\sin t, t$ large.

Theorem 2. *Let $q(t) \leq 0$ on R_+ and Eq. (3) is nonoscillatory. Then $W_1 = \emptyset$.*

Proof. Let h be a positive solution of (3) on $[T, \infty]$. Then (1) is equivalent to (4). Let $y \in W_1$. Then according to (5), $y \neq 0$ and $y^{(1)}$ is an oscillatory function for large t which is impossible (see Bartušek [1]). □

Consequence 2. *Let $q(t) \leq 0$ and $\int_0^\infty t|q(t)|dt < \infty$. Then $W_1 = \emptyset$.*

Proof. According to Hartman [6], Eq. (3) is nonoscillatory. □

Theorem 3. *Let $q(t) < 0$ on R_+ and let one of the following assumptions hold:*

(i) $|q(t)| \leq M < \infty$, *for* $t \geq T \geq 0$;
(ii) $|q(t)| \leq M_1 t^{1-\varepsilon}$, $t \geq T \geq 0, \varepsilon > 0$, $M_1 > 0$ *and* $|q|$ *is nondecreasing.*

Then $N_0 = W_2^- = W_3^{-+} = \emptyset$.

Proof. Let $y \in N_0 \cup W_2^- \cup W_3^{-+}$ and for simplicity let

$$y(t) > 0, \quad t \geq T_1 \geq T. \tag{8}$$

Let $\tau \geq T_1, \tau \geq T_y$ be an arbitrary zero of y''' (be an arbitrary number) if $y \in W_2^- \cup W_3^{-+}$ (if $y \in N_0$). Then it follows from the definitions of N_0, W_2^- and W_3^{-+} and from (8) that $K = y'''(\tau) - q(\tau)y'(\tau) < 0, y > 0$ is decreasing and $y' < 0$ on $[\tau, \infty)$.

Thus, there exists a sequence $\{t_k\}_1^\infty$, such that $\lim_{k\to\infty} t_k = \infty$ and

$$\lim_{k\to\infty} y'(t_k) = 0, \quad y''(t_k) \geq 0, \quad y'''(t_k) \leq 0, \quad k = 1, 2, \ldots . \tag{9}$$

If $y \in N_0$, it is an arbitrary sequence tending to ∞; if $y \in W_3^{-+}$, it is a subsequence of zeros of y'''. Let $y \in W_2^-$. If $y''(t) \geq 0$ for large t, then $\{t_k\}$ is the sequence of zeros of y'''. The case $y''(t) \leq 0$ for large t is impossible. If y'' changes the sign, then $\{t_k\}$ is the sequence of zeros of y'' that realize the local maxima of y'.

By the integration of (1), using (2) and (8), we have

$$y'''(t) - q(t)y'(t) = K - \int_\tau^t r(s)f(y(s))ds \leq K < 0, \; t \geq \tau. \tag{10}$$

Case (i). It follows from (10) and from the definition of $\{t_k\}_1^\infty$ that

$$0 > |q(t_k)|y'(t_n) \geq My'(t_k) \to 0$$

as $k \to \infty$. This contradiction proves the conclusion.

Case (ii). By the integration of (10) and by (2), (8) and (9),

$$\begin{aligned} -y''(\tau) + q(t_k)y(\tau) &\leq y''(t_k) - y''(\tau) - q(t_k)\,(y(t) - y(\tau)) \\ &\leq y''(t_k) - y''(\tau) - \int_\tau^{t_k} q(s)y'(s)ds \leq K(t_k - \tau). \end{aligned}$$

Since $|q(t_k)| \leq M_1 t_k^{1-\varepsilon}$, we get a contradiction for large k. □

Remark. According to Ex. 1, the statement of Theorem 3 (ii) cannot be extended to the assumption $|q(t)| \leq M_1 t^{1+\varepsilon}, \varepsilon > 0$.

Theorem 4. *Let* $q \leq 0$ *on* R_+ *and Eq. (3) be nonoscillatory. Then* $N_0 = W_3^{-+} = \emptyset$.

Proof. It is similar to the one of Theorem 3 (ii) as according to Cecchi at al [5], $\int_0^\infty q(s)ds < \infty$ if (3) is nonoscillatory. □

Remark. Theorems 1, 2 and 4 yield the following result.

If (H2) is valid, $q \leq 0$ on R_+ and (3) is nonoscillatory, then every nonoscillatory solution belongs to W_2^- (if it exists).

3. CASE $\mathbf{q \geq 0}$.

The following theorem can be proved similarly as Theorem 2.

Theorem 5. *Let $q \geq 0$. Then $W_1 = \emptyset$.*

The following examples show that the other sets may be nonempty with the possible exception of W_2^-; note that (2) is fulfilled.

Example 4. The equation $y^{(4)} - (t^3 y')' + r(t)y = 0$, where $r(t) = (-\sin t + 4t^3 + 3t^2 \sin t - t^3 \sin t)\left(\frac{t^2}{2} + \sin t\right)^{-1}, t \geq 2$, has the solution $y = \frac{t^2}{2} + \sin t$, that belongs to W_2^+.

Example 5. The equation $y^{(4)} - (t^4 y')' + r(t)y = 0$, where $r(t) = (-\sin t + 3t^5 - t^4 \sin t + 4t^3 \cos t)\left(\frac{t^3}{6} + \sin t\right)^{-1}, t \geq 2$, has the solution $y = \frac{t^3}{6} + \sin t$, that belongs to W_3^{++}.

Example 6. The equation $y^{(4)} - (ty')' + \frac{1}{t}y = 0, t \geq 1$ has the solution $y = t$ that belongs to W_4.

Example 7. The equation $y^{(4)} - (ty')' + \left(\frac{15}{16}t^{-4} + \frac{1}{4}t^{-1}\right)y = 0, t > 1$ has the solution $y = t^{\frac{1}{2}}$ that belongs to N_1.

Example 8. The equation $y^{(4)} - (t^2 y')' + \left(\frac{9}{16}t^{-4} + \frac{15}{4}\right)y = 0, t > 1$ has the solution $y = t^{\frac{3}{2}}$ that belongs to N_2.

Example 9. The equation $y^{(4)} - (ty')' + ty = 0, t \in R_+$ has the solution $y = e^t$ that belongs to N_3.

Example 10. The equation $y^{(4)} - (t^3 y')' + ry = 0$, where

$$r(t) = \frac{1}{t^{\frac{1}{2}} + t^{\frac{-5}{2}} \sin t} \left[\frac{15}{16}\, t^{-\frac{7}{2}} - \left(t^{-\frac{5}{2}} \sin t\right)^{(4)} + \left(t^3 \left(t^{\frac{1}{2}} + t^{-\frac{5}{2}} \sin t\right)'\right)' \right] > 0$$

for $t \geq 2$, has a solution $y = t^{\frac{1}{2}} + t^{\frac{-5}{2}} \sin t$ that belongs to W_3^+.

Example 11. The equation $y^{(4)} - y'' + \left[\frac{2}{t} - \frac{24}{t^3}\right] y^2 = 0, t \geq 4$ has the solution $y = \frac{1}{t}$ that belongs to N_0.

Example 12. The equation $y^{(4)} - y'' + r(t)y^3 = 0$, where

$$r(t) = \frac{t^9}{(t^2 - \sin t)^3} \left[-\left(\frac{2}{t} - \frac{\sin t}{t^3}\right)^{(4)} + \left(\frac{2}{t} - \frac{\sin t}{t^3}\right)'' \right] > 0$$

for large t, has the solution $y = \frac{2}{t} - \frac{\sin t}{t^3}$ that belongs to W_3^{-+}.

Theorem 6. *Let (2), $q \geq 0$ and $\int_0^\infty tq(t)dt < \infty$. Then $N_3 = W_3^{++} = W_4 = W_2^+ = \emptyset$.*

Proof. According to Marini & Zeza [8], Eq. (3) is nonoscillatory and there exists a solution h such that

$$h > 0,\ h' < 0 \quad \text{on} \quad R_+ \quad \text{and} \quad \lim_{t \to \infty} h(t) = h_0 > 0. \tag{11}$$

From this and from (2)

$$\int_0^\infty h(t)dt = \infty, \quad \int_0^\infty \frac{dt}{h^2(t)} = \infty, \quad \int_0^\infty r(t)dt = \infty. \tag{12}$$

Let $y \in N_3 \cup W_3^{++} \cup W_4$ with $y(t) > 0,\ y'(t) > 0,\ y''(t) \geq 0$ for $t \geq T_y$ (due to (1) and (2), $y(t)y'(t) \neq 0$ for large t if $y \in W_4$).

Eq. (1) can be transformed into (4), and (5) and (11) yield the existence of $T \geq T_y$ such that either

$$y^{[j]}(t) > 0, j = 0,1,2,3, \quad t \geq T, \tag{13}$$

or

$$y^{[i]}(t) > 0,\ j = 0,1,2,\ y^{[3]}(t) < 0,\ t \geq T, \tag{14}$$

since the case $y^{[j]}(t) > 0, j = 0,1,2,\ y^{[3]}(t)$ is an oscillatory function is impossible (see Bartušek [1]).

As (12) is valid, there exists no solution fulfilling (13) according to Bartušek [2], Theorem 4.

Furthermore, let y have the form (14). Then according to (4) and (5), $y^{[2]} > 0$ and $y^{[3]} < 0$ are decreasing for $t > T$, and thus, using (11),

$$\infty > y^{[2]}(T) - y^{[2]}(\infty) = -\int_T^\infty h(s)y^{[3]}(s)ds \geq |y^{[3]}(T)| \int_T^\infty h(s)ds = \infty.$$

This contradiction proves the conclusion in case $y \in N_3 \cup W_3^{++} \cup W_4$.

Let $y \in W_2^+$ with $y(t) > 0,\ y'(t) > 0$ for $t \geq T_y$. Thus (5) yields $y(t) > 0,\ y^{[1]}(t) > 0$. As $y^{[2]}$ and $y^{[3]}$ cannot oscillate (see Bartušek [1]) we obtain for a sequence $\{t_k\}_1^\infty$ of zeros of y'' that $y^{[2]}(t_k) > 0$. Thus either (13) or (14) must be valid and the proof is similar. □

Theorem 7. *Let $q(t) > 0$ and (2) be valid.*

(i) If $\lim_{t\to\infty} \frac{\int_0^t r(s)ds}{q(t)} = \infty$ *then* $N_1 = W_3^{+-} = \emptyset$.

(ii) If $\alpha > 0, M > 0, \alpha|x| \leq |f(x)|$ *for* $|x| \geq M$ *and* $\lim_{t\to\infty} \frac{\int_0^t sr(s)ds}{tq(t)} = \infty$, *then* $N_2 = \emptyset$.

Proof. (i) Let $y \in N_1$ with

$$y(t) > 0,\ y'(t) > 0,\ y''(t) < 0,\ y'''(t) > 0, t \geq T_y. \tag{15}$$

From this and an integration of (1), we have for large t

$$-q(t)y'(t) \leq K - \int_{T_y}^t r(s)f(y(s))ds \leq K - K_1 \int_{T_y}^t r(s)ds \to_{t\to\infty} \infty,$$

$$-q(t)y'(t) \leq -\frac{K_1}{2}\int_{T_y}^t r(s)ds, \quad K_1 = \min_{T_y \leq x < \infty} f(x) > 0, K = y''' - qy'|_{t=T_y}.$$

Thus, using assumption (i), $y'(t) \geq \frac{K_1}{2} \frac{\int_{T_y}^t r(s)ds}{q(t)} \to \infty$ as $t \to \infty$ which contradicts (15) since $y' > 0$ is decreasing.

Let $y \in W_3^{++}$. The proof is similar; we must only restrict the sequence $\{t_k\}_1^\infty$ to be zeros of y'''.

(ii) Let $y \in N_2$ with

$$y(t) > 0,\ y'(t) > 0,\ y''(t) > 0,\ y'''(t) < 0,\ t \geq T_y. \tag{16}$$

From this there exists $T \geq T_y$ such that

$$\begin{aligned} &y(t) \geq y'(T_y)(t - T_y) \geq \frac{y'(T_y)}{2} t, \\ &y'(t) \leq y'(T_y) + y''(T_y)(t - T_y) \leq 2y''(T_y)t, \quad t \geq T. \end{aligned}$$

From this, (2), and an integration of (1), we have

$$\begin{aligned} &- y'''(t) + 2q(t)y''(T_y)t \geq -y'''(t) + q(t)y'(t) = -K + \int_{T_y}^t r(s)f(y(s))ds \\ &\geq -K + \alpha \int_{T_y}^t r(s)y(s)ds \geq -K + \frac{\alpha}{2}\, y'(T_y) \int_{T_y}^t sr(s)ds \to \infty \end{aligned} \tag{17}$$

as $t \to \infty$. According to (16), there exists a sequence $\{t_k\}_1^\infty$ such that $\lim_{k\to\infty} t_k = \infty$, and $\lim_{k\to\infty} y'''(t_k) = 0$. From this and (17) we have for large k

$$2|y''(T_y)|q(t_k)t_k \geq \frac{\alpha}{4} y'(T_y) \int_{T_y}^{t_k} sr(s)ds,$$

and this contradicts the assumptions in (ii). □

Theorem 8. *Let (2),* $q \geq 0$ *and* $\int_0^\infty tq(t)dt < \infty$*. Then* $W_2^- = W_3^{-+} = \emptyset$.

Proof. According to Marini & Zeza [8], Eq. (3) is nonoscillatory and there exists a solution $h > 0$ such that $\lim_{t\to\infty} h(t) = h_0 \in (0, \infty)$; thus (1) can be transformed into (4). Let $y \in W_2^- \cup W_3^{-+}, y(t) > 0$ for $t \geq T_y$ and let $\{\tau_k\}_1^\infty$ be a sequence of zeros of y''', tending to ∞ and fulfilling $y''(\tau_k) \geq 0$. This is possible as $y''(t) \leq 0$ cannot be valid in case $y \in W_2$ (otherwise y becomes negative for large t). According to Bartušek [1], $y^{[i]}, i = 0, 1, 2, 3$ do not change the signs for large t, so using $y'''(\tau_k) = 0, y''(\tau_k) \geq 0$, (5) yields

$$y(t) > 0,\ y^{[1]}(t) < 0,\ y^{[2]}(t) > 0,\ y^{[3]}(t) > 0, \quad \text{for} \quad t \geq \bar{t} \geq T_y,$$

$|y^{[1]}|$ is nonincreasing and $y^{[2]}$ is nondecreasing. From this and the fact that $\lim_{t\to\infty} h(t) > 0$,

$$\infty > y^{[1]}(\infty) - y^{[1]}(\bar{t}) = \int_{\bar{t}}^\infty \left[y^{[1]}(s)\right]' ds = \int_{\bar{t}}^\infty \frac{y^{[2]}(s)}{h^2(s)} ds \geq y^{[2]}(\bar{t}) \int_{\bar{t}}^\infty \frac{ds}{h^2(s)} = \infty$$

and this contradiction proves the conclusion. □

ACKNOWLEDGMENT

This work was supported by the grant no. 201/99/0295 of the Grant Agency of the Czech Republic.

REFERENCES

1. Bartušek, M., On the structure of solutions of a system of four differential inequalities. Georg. Math. J., Vol 2 (1995). 225-236.
2. Bartušek, M., Oscillatory criteria for nonlinear n-th order differential equations with quasiderivatives. Georg. Math. J., Vol 3 (1996). 301-314.
3. Bartušek, M., Došlá, Z., & Graef, J.R., Limit-point type results for nonlinear fourth-order differential equations. Nonlinear Anal., Vol. 28 (1997). 779-792.
4. Bartušek, M., Cecchi, M., Došlá, Z., & Marini, M., On nonoscillatory solutions of third order nonlinear differential equation. Dynam. Systems Appl., to appear.
5. Cecchi, M., Marini, M., & Villari, G., Integral criteria for a classification of solutions of linear differential equations. J. Diff. Eq., Vol. 99 (1992). 381-397.
6. Hartman, Ph., Ordinary differential equations. Wiley, New York-London, 1964.
7. Kiguradze, I., An oscillation criterion for a class of ordinary differential equations. Diff. Urav., Vol. 28 (1992). 207-219.
8. Marini, M., & Zezza, P., On the asymptotic behaviour of the solutions of a class of second-order linear differential equations. J. Diff. Eqs., Vol 28, No 1 (1978). 1-17.
9. Sobalová, M., On existence of oscillatory solutions of the 4-th order differential equation. Proc. Int. Scient. Conf. of Math., Žilina 1998. 241-248.

Proceedings of Dynamic Systems and Applications 3 (2001) 69-76

Nonexistence of periodic solutions in a class of delay-differential equations

Edoardo Beretta[1] and Yasuhiro Takeuchi[2]
[1]Istituto di Biomatematica, Università di Urbino, I-61029 Urbino, Italy
[2]Department of Systems Engineering, Faculty of Engineering, Shizuoka University, Hamamatsu 432, Japan

ABSTRACT: Li and Muldowney[1] have obtained negative criteria for the existence of periodic solutions of systems of ordinary differential equations in R^n which generalize to the case $n > 2$ Bendixon and Dulac criteria. We propose to apply these criteria to the system of ordinary differential equations obtained from a system of differential equations with distributed delays when the delay kernels are suitably chosen as combinations of γ-functions. This provides negative criteria for existence of periodic solutions of a special class of delay equations. As an application we consider a chemostat model with delays in nutrient recycling and in the growth of biotic species.
AMS (MOS) subject classification. 34C25, 39B62

1. INTRODUCTION

Li and Muldowney[1] have obtained conditions for nonexistence of nonconstant periodic solutions for autonomous ordinary differential equations in R^n generalizing to the case $n > 2$ Bendixon and Dulac criteria. Furthermore Li and Wang[2] have studied the relations among the above criteria and stability matrices and further they obtained conditions in order that a matrix has no Hopf bifurcation points. Their results are in terms of logarithmic norms of the second additive compound matrix associated with the Jacobian. We propose to extend the results by [1] and [2] to a particular class of delay differential equations with distributed delays which, with a particular choice of delay kernel as γ-functions or their suitable combinations, can be transformed in an expanded system of ordinary differential equations. This method of transforming delay differential equations in o.d.e. is called "linear chain trick" and is frequently used to have a first insight on the effect that time delays can have on the stability (e.g. MacDonald[3]) and now, thanks to the above referred papers, even on existence of periodic solutions.

This paper is organized as follows: in Section 2 we recall the results by Li and Muldowney[1] and Li and Wang[2] recalling also the main features of linear chain trick. In Section 3 we illustrate how the Li and Muldowney criteria work by applying them to the chemostat model with delayed biotic growth and delayed recycling.

2. GENERAL THEORY

Assume to have a system of autonomous ordinary differential equations in R^n, say

$$\frac{dx}{dt} = f(x) \tag{1}$$

with $f : \mathrm{R}^n \to \mathrm{R}^n$ of class C^1 in some open subset D_0 of R^n. Li and Muldowney[1] investigated conditions for nonexistence of non-constant periodic solutions of (1), or more generally, they provided criteria for the nonexistence of simple closed curves which are invariant under the dynamics of (1). If A is an $n \times n$ matrix, we denote by $A^{[2]}$ its "second additive compound matrix" which is $\binom{n}{2} \times \binom{n}{2}$ and whose definition can be found in [1]. Here following we summarize their results. Let us denote by $\lambda_1 \geq \lambda_2 \geq \cdots \geq \lambda_{n-1} \geq \lambda_n$, the real eigenvalues of the symmetric real matrix

$$\frac{1}{2}\left(\left(\frac{\partial f}{\partial x}\right)^T + \left(\frac{\partial f}{\partial x}\right)\right)$$

where $\partial f/\partial x$ is the Jacobian matrix of f. Then, according to Li and Muldowney[1] we have:

Theorem 1 (Bendixon's Criterion in R^n) *A simple closed rectifiable curve which is invariant with respect to (1) cannot exist if any one of the following conditions is satisfied on* R^n*:*

(i) $\sup\left\{\frac{\partial f_r}{\partial x_r} + \frac{\partial f_s}{\partial x_s} + \sum_{q\neq r,s}\left(\left|\frac{\partial f_q}{\partial x_r}\right| + \left|\frac{\partial f_q}{\partial x_s}\right|\right) : 1 \leq r < s \leq n\right\} < 0;$

(ii) $\sup\left\{\frac{\partial f_r}{\partial x_r} + \frac{\partial f_s}{\partial x_s} + \sum_{q\neq r,s}\left(\left|\frac{\partial f_r}{\partial x_q}\right| + \left|\frac{\partial f_s}{\partial x_q}\right|\right) : 1 \leq r < s \leq n\right\} < 0;$

(iii) $\lambda_1 + \lambda_2 < 0;$

(iv) $\inf\left\{\frac{\partial f_r}{\partial x_r} + \frac{\partial f_s}{\partial x_s} - \sum_{q\neq r,s}\left(\left|\frac{\partial f_q}{\partial x_r}\right| + \left|\frac{\partial f_q}{\partial x_s}\right|\right) : 1 \leq r < s \leq n\right\} > 0;$

(v) $\inf\left\{\frac{\partial f_r}{\partial x_r} + \frac{\partial f_s}{\partial x_s} - \sum_{q\neq r,s}\left(\left|\frac{\partial f_r}{\partial x_q}\right| + \left|\frac{\partial f_s}{\partial x_q}\right|\right) : 1 \leq r < s \leq n\right\} > 0;$

(vi) $\lambda_{n-1} + \lambda_n > 0.$

We remark that if $y \in \mathrm{R}^n$ has the norm, $|y|_1 = \sum_{i=1}^n |y_i|$, $|y|_\infty = \sup_i |y_i|$ or $|y|_2 = (y^T y)^{1/2}$ then the Lozinskiĭ measure (or logarithmic norm) of the second additive compound matrix $(\partial f/\partial x)^{[2]}$ associated with the Jacobian matrix $\partial f/\partial x$ i.e. $\mu((\partial f/\partial x)^{[2]})$ is respectively the expression in (i), (ii), (iii), whereas $-\mu(-(\partial f/\partial x)^{[2]})$, according to the chosen norm, is respectively the expression in (iv), (v) and (vi). Hence, according to the chosen norm, (i), (ii), (iii) mean that

$$\mu\left(\left(\frac{\partial f}{\partial x}\right)^{[2]}\right) < 0$$

whereas (iv), (v), (vi) mean that

$$\mu\left(-\left(\frac{\partial f}{\partial x}\right)^{[2]}\right) < 0.$$

The same argument applies to any absolute norm where $|\cdot|$ is said to be "absolute" if $|y|$ is unchanged by replacing the components y_i of y by $|y_i|$.

Corollary 2 *Suppose that one of*

$$\mu\left(\left(\frac{\partial f}{\partial x}\right)^{[2]}\right) < 0, \quad \mu\left(-\left(\frac{\partial f}{\partial x}\right)^{[2]}\right) < 0$$

holds on R^n *where* μ *is a Lozinskiĭ measure (logarithmic norm) corresponding to an absolute norm* $|\cdot|$ *on* R^N, $N = \binom{n}{2}$. *Then, no simple closed rectifiable curve in* R^n *is invariant with respect to (1).*

Let D_0 be a simple connected subset of R^n and $x \mapsto A(x)$ be a C^1 nonsingular $\binom{n}{2} \times \binom{n}{2}$ matrix-real valued function on D_0. Furthermore denote by A_f the matrix obtained from A by replacing each entry a_{ij} of A by

$$(a_{ij})_f = \left(\frac{\partial a_{ij}(x)}{\partial x}\right)^T \cdot f.$$

Assume that the solutions of (1) exist for all $t \geq 0$. A subset D_1 of D_0 is said to be "absorbing" with respect to (1) if each bounded subset D of D_0 satisfies $x(t, D) \subset D_1$ for all sufficiently large t. Then, the following holds ([1]):

Theorem 3 *Suppose that*

(a) D_0 *is simply connected;*

(b) $\mu\left(A_f A^{-1} + A\left(\partial f/\partial x\right)^{[2]} A^{-1}\right) \leq b < 0$ *on a set* D_1 *which is absorbing with respect to (1).*

Then, there is no simple closed rectifiable curve in D_0 *which is invariant with respect to (1).*

Corollary 4 *If (1) has a compact global attractor* Ω *on which the following* $\mu(A_f A^{-1} + A(\partial f/\partial x)^{[2]} A^{-1}) < 0$ *is satisfied for some logarithmic norm, then there exists an absorbing neighborhood* D_1 *of* Ω *on which (b) holds. Further in* Ω *there is no simple closed rectifiable curve which is invariant with respect to (1).*

Remark that if in Theorem 3 and Corollary 4 we choose A as a real constant and nonsingular matrix, then $A_f = 0$ and (b) becomes $\mu(A(\partial f/\partial x)^{[2]} A^{-1})$. For example Corollary 4 could be rewritten as "if Ω is a compact global attractor for (1) on which $\mu(A(\partial f/\partial x)^{[2]} A^{-1}) < 0$ for some logarithmic norm and some arbitrary real constant nonsingular matrix A, then in Ω there is no simple rectifiable curve invariant with respect to (1)". Of course if we choose $A = I$, the identity $\binom{n}{2} \times \binom{n}{2}$ matrix, then $\mu((\partial f/\partial x)^{[2]}) < 0$ on Ω for some logarithmic norm. The arbitrarity of A can be usefully applied to improve the results.

In the following we denote by $M_n(\mathrm{R})$ the class of real n-dimensional matrices. Let $A \in M_n(\mathrm{R})$ and denote by $\sigma(A) = \{\lambda_i; i = 1, 2, \ldots, n\}$ its spectrum and by $s(A)$ its "stability modulus"

$$s(A) = \max\{\mathrm{Re}\ \lambda_i : \lambda_i \in \sigma(A)\}.$$

If $s(A) < 0$ we say that A is a stable matrix. The links between a stable matrix A and negativity of some Lozinskiĭ measure of its second additive compound matrix $A^{[2]}$ have been investigated by Li and Wang[2]. A first result(see [2]) is:

Theorem 5 *For $s(A) < 0$ it is necessary and sufficient that $s(A^{[2]}) < 0$ and $(-1)^n \det(A) > 0$.*

The following link between stability modulus and Lozinskiĭ measure holds:

Theorem 6 *$s(A) < 0$ if and only if $\mu(A) < 0$ for some Lozinskiĭ measure μ on $M_n(\mathrm{R})$.*

This result enables to replace Theor. 5 by the following:

Theorem 7 *Assume that $(-1)^n \det(A) > 0$. Then A is stable if and only if $\mu(A^{[2]}) < 0$ for some Lozinskiĭ measure μ on $M_N(\mathrm{R})$, $N = \binom{n}{2}$.*

All the conditions by Li and Muldowney[1] for non-existence of periodic solutions and their relation with stability (see [2]) can be applied to a special class of integro-differential equations with distributed infinite delays:

$$(2) \quad \dot{x}_i(t) = G_i(x_k(t), \int_{-\infty}^{t} \varphi_{ij}(t-\tau)x_j(\tau)d\tau), \quad i,k,j \in \mathrm{N}, \quad \mathrm{N} = \{1,2,...,n\},$$

where the delay kernels "φ_{ij}" can be assumed to be suitably normalized convex combinations:

$$(3) \qquad \varphi_{ij}(s) = \sum_{\rho=1}^{m_{ij}} c_{ij}^{(\rho)} \varphi_{ij}^{(\rho)}(s), \quad c_{ij}^{(\rho)} \geq 0, \quad \sum_{\rho=1}^{m_{ij}} c_{ij}^{(\rho)} = 1,$$

of γ-functions

$$(4) \qquad \varphi_{ij}^{(\rho)}(s) = \alpha_{ij}^{\rho} s^{\rho-1} \exp(-\alpha_{ij}s)/(\rho-1)!, \quad s \in [0,+\infty), \quad \alpha_{ij} \in R_+,$$

which satisfy

$$(5) \qquad \int_0^{+\infty} \varphi_{ij}^{(\rho)}(s)ds = 1, \quad \int_0^{+\infty} s\varphi_{ij}^{(\rho)}(s)ds = \frac{\rho}{\alpha_{ij}}.$$

Defining the functions

$$(6) \qquad y_{ij}^{(\rho)}(t) = \int_{-\infty}^{t} \varphi_{ij}^{(\rho)}(t-\tau)x_j(\tau)d\tau,$$

we obtain the linear o.d.e. system satisfying:

$$(7) \qquad \begin{cases} \dot{y}_{ij}^{(\rho)}(t) = \alpha_{ij}y_{ij}^{(\rho-1)}(t) - \alpha_{ij}y_{ij}^{(\rho)}(t), & \rho = 1,\dots,m_{ij} \\ y_{ij}^{(0)}(t) = x_j(t). \end{cases}$$

Hence, considering (3)- (7) we find that the integro-differential system (2) can be written as the expanded o.d.e. system :

$$(8) \qquad \begin{cases} \dot{x}_i(t) = G_i(x_k(t), \sum_{\rho=1}^{m_{ij}} c_{ij}^{(\rho)} y_{ij}^{(\rho)}(t)) \\ \dot{y}_{ij}^{(\rho)}(t) = \alpha_{ij}y_{ij}^{(\rho-1)}(t) - \alpha_{ij}y_{ij}^{(\rho)}(t) \\ y_{ij}^{(0)}(t) = x_j(t), \quad i,j,k \in \mathrm{N}, \quad \rho = 1,\dots,m_{ij}. \end{cases}$$

If we assume $G \in C^1$ both with respect to x_k and $y_{ij}^{(\rho)}$, we can apply the results of Li and Muldowney[1] and Li and Wang[2] to the o.d.e. system (8). Finally, remark that the initial conditions for (2), say at $t = 0$,

$$x_j(\tau) = \psi_j(\tau), \quad \tau \in (-\infty, 0], \quad j \in \mathrm{N}, \tag{9}$$

provide the initial conditions for (8):

$$\begin{cases} x_j(0) = \psi_j(0), & j \in \mathrm{N} \\ y_{ij}^{(\rho)}(0) = \int_{-\infty}^{0} \varphi_{ij}^{(\rho)}(-\tau)\psi_j(\tau)d\tau, & i, j \in \mathrm{N}, \quad \rho = 1, \ldots, m_{ij}. \end{cases} \tag{10}$$

3. APPLICATION

In this section, we apply Li and Muldowney criteria[1] to exclude the existence of periodic solutions in a chemostat model with delays. Beretta and Takeuchi[4] studied the properties of the chemostat model:

$$\begin{aligned} \dot{N}_1(t) &= D(N_1{}^0 - N_1) - aU(N_1)N_2 + b\gamma \textstyle\int_0^{+\infty} f(s)N_2(t-s)ds \\ \dot{N}_2(t) &= -(\gamma + D)N_2 + cN_2 \textstyle\int_0^{+\infty} g(s)U(N_1(t-s))ds \end{aligned} \tag{11}$$

with delayed nutrient recycling and delay in growth of the biotic species N_2 feeding upon the resource N_1. D is the chemostat wash-out rate constant, U is the resource uptake function that here we assume as the Michaelis-Menten function:

$$U(N_1) = N_1/(L + N_1), \quad N_1 \in R_{+0}, \tag{12}$$

and $N_1^0 > 0$ is the input constant concentration of resources. Furthermore, because of biological meaning, we assume that $a > c, b < 1$. The delay kernels $f, g : \mathrm{R}_{+0} \to \mathrm{R}_{+0}$ are assumed to be normalized, i.e. $\int_0^{+\infty} f(s)ds = \int_0^{+\infty} g(s)ds = 1$ and with finite up to the second moments, where

$$T_r \triangleq \int_0^{+\infty} sf(s)ds, \quad T_g \triangleq \int_0^{+\infty} sg(s)ds,$$

respectively are the average time delay in recycling (T_r) and in growth of biotic species (T_g).

At $t_0 \in R$, (11) is supplemented with the initial conditions:

$$N_{it_0} = \phi_i^*(\theta) \equiv \phi_i(t_0 + \theta), \quad \theta \in (-\infty, 0], \quad i = 1, 2, \tag{13}$$

where $N_i(t_0) = \phi_i^*(0)$, $i = 1, 2$, and $\phi^*(\theta) \in CB((-\infty, 0], R^2)$. If we use as a norm in R^2 : $|x|_1 = \sum_{i=1}^{2} |x_i|$ then the space of $CB((-\infty, 0])$ has the norm:$||\phi|| = \sup_{\theta \in (-\infty, 0]} |\phi(\theta)|_1$. Denote by $Q_H = \{\phi \in CB((-\infty, 0], R^2) : ||\phi|| < H\}$. Beretta & Takeuchi[4] proved the following boundedness property:

Theorem 8 *Assume the i.c. $\phi \in Q_H$ and T_g, T_r satisfying that $T_g \le (a - bc)/ac$, $T_r < 1/\gamma$. Then the solutions of (11) are uniformly bounded in R_{+0}^2, i.e. whenever be $t_0 \in R$ for any $t \ge t_0$*

$$N_1(t) + N_2(t) \le \bar{N} \equiv \frac{1}{b} \max\{2\eta H, \frac{(1 + aT_g/b)N_1^0}{1 - \gamma T_r}\}$$

where $\eta = \max\{1, b + b\gamma T_r + aT_g\}$.

We are interested in deriving sufficient conditions for nonexistence of periodic solutions of (11). Thus we remark that if $\gamma + D \geq c$, $\dot{N}_2(t) \leq -[(\gamma + D) - c]N_2 \leq 0$, i.e. $N_2(t)$ is a monotone decreasing function, and therefore (11) has no periodic solutions. Hence, herefollowing we assume

$$\gamma + D < c. \tag{14}$$

Therefore, the positive equilibrium of (11):

$$E_+ = \left(N_1{}^* = U^{-1}\left(\frac{\gamma + D}{c}\right), N_2{}^* = \frac{D(N_1{}^0 - N_1{}^*)}{aU(N_1{}^*) - b\gamma}\right)$$

exists provided that $N_1{}^* < N_1{}^0$, since $aU(N_1{}^*) - b\gamma > 0$. Here following we denote

$$\bar{u} = \sup_{N_1 \geq 0} U'(N_1)$$

that we assume to be finite. In case (12) is $\bar{u} = 1/L$.

To apply the Muldowney sufficient conditions for nonexistence of periodic solutions we transform (11) with linear chain trick assuming that the delay kernels are γ-functions of the first order:

$$f(s) = \alpha e^{-\alpha s}, \quad g(s) = \beta e^{-\beta s}, \quad s \in [0, +\infty) \tag{15}$$

where $\alpha, \beta \in R_+$. Therefore, herefollowing $T_r = 1/\alpha$ and $T_g = 1/\beta$. Furthermore, let us define:

$$N_2{}^{(1)}(t) \triangleq \int_0^{+\infty} \alpha e^{-\alpha s} N_2(t - s) ds, \tag{16}$$

$$U^{(1)}(t) \triangleq \int_0^{+\infty} \beta e^{-\beta s} U(N_1(t - s)) ds. \tag{17}$$

Remark that $U^{(1)}(t) < 1$ and furthermore, if Theorem 8 is applied, then $N_2{}^{(1)}(t) \leq \bar{N}, U^{(1)}(t) \leq U(\bar{N})$ for all $t \geq t_0$. From (11), (16) and (17) we get :

$$\begin{aligned} \dot{N}_1(t) &= DN_1{}^{(0)} - DN_1(t) - aU(N_1(t))N_2(t) + b\gamma N_2{}^{(1)}(t) \\ \dot{N}_2(t) &= -(\gamma + D)N_2(t) + cN_2(t)U^{(1)}(t) \\ \dot{U}^{(1)}(t) &= \beta U(N_1) - \beta U^{(1)}(t) \\ \dot{N}_2^{(1)}(t) &= \alpha N_2(t) - \alpha N_2{}^{(1)}(t). \end{aligned} \tag{18}$$

Remark that the i.c. (13) provide the i.c. for the expanded o.d.e. system (18) since $N_i(t_0) = \phi_i(t_0), i = 1, 2$, whereas

$$N_2{}^{(1)}(t_0) = \int_{-\infty}^0 \alpha e^{\alpha\theta} \phi_2(t_0 + \theta) d\theta; \quad U^{(1)}(t_0) = \int_{-\infty}^0 \beta e^{\beta\theta} U(\phi_1(t_0 + \theta)) d\theta.$$

Furthermore, if for $\phi \in Q_H$, $(N_1(t), N_2(t)) \in \Omega$ for all $t \geq t_0$, where $\Omega = \{(N_1, N_2) \in \mathrm{R}_{+0}{}^2 \mid N_1 + N_2 \leq \bar{N}\}$, then

$$0 \leq N_2(t) \leq \bar{N}, \quad 0 \leq U^{(1)}(N(t)) \leq U(\bar{N}) \tag{19}$$

for all $t \geq t_0$. In conclusion if $\phi \in Q_H$ and Theor.8 applies, the solutions of (18) in R^4_{+0} are uniformly bounded in the compact set K:

$$K = \{(N_1, N_2, U^{(1)}, N_2{}^{(1)}) \in R^4_{+0} \quad | \quad N_1 + N_2 \leq \bar{N}, U^{(1)} \leq U(\bar{N}), N_2{}^{(1)} \leq \bar{N}\}.$$

However, for the sake of simplicity, we will refer to the compact set Ω of R^2_{+0}, where it is implicitly assumed that (19) hold true.

Ordering the variables according to $(N_1, N_2, U^{(1)}, N_2{}^{(1)})$, the Jacobian matrix is:

$$(20) \qquad J = \begin{pmatrix} -(D + aU'(N_1)N_2) & -aU(N_1) & 0 & b\gamma \\ 0 & -(\gamma + D) + cU^{(1)} & cN_2 & 0 \\ \beta U'(N_1) & 0 & -\beta & 0 \\ 0 & \alpha & 0 & -\alpha \end{pmatrix}.$$

It is easy to check that at the positive equilibrium

$$\det J(E_+) = \alpha\beta c N_2{}^* U'(N_1{}^*)(aU(N_1{}^*) - b\gamma) > 0.$$

From the second additive compound matrix $J^{[2]}$ associated with (20) (see [1,2]), we can prove the following:

Theorem 9 *Assume (11) with the delay kernels (15). Assume that the uptake function satisfies that*

$$(21) \qquad \frac{c}{a} < U'(\bar{N}), \quad \bar{u} = \sup_{N_1 \geq 0} U'(N_1) < 1,$$

and

$$(22) \qquad aU(\bar{N}) < D.$$

Furthermore if T_g, T_r are sufficiently small to ensure

$$(23) \qquad \frac{1}{T_g} > \max\{b\gamma - (D - aU(\bar{N})), \frac{ac}{a - bc}\}$$

$$(24) \qquad \frac{1}{T_r} > \max\{c\bar{N} - \gamma - (D - aU(\bar{N})), \gamma\},$$

all solutions of (11) with i.c. $\phi \in Q_H$ are uniformly bounded in $\Omega = \{(N_1, N_2) \in R_{+0}{}^2 \ | \ N_1 + N_2 \leq \bar{N}\}$, and in Ω no periodic solutions can occur. Furthermore, whenever the positive equilibrium is feasible, then is (locally) asymptotically stable.

Poof. If (23), (24) hold true, Theor.8 implies that the solution with i.c. $\phi \in Q_H$ are uniformly bounded in Ω. Then it is enough to prove that in Ω one of the Lozinskiĭ norms of $J^{[2]}$ is negative. We choose $\mu_\infty(J^{[2]})$ and $\mu_\infty(J^{[2]}) < 0$ if we have diagonal dominance by row in $J^{[2]}$ (see also Theor.1,(ii)). Hence we obtain that in Ω :

$$(25) \qquad \begin{aligned} &\sup[-(\gamma + 2D + aU'(N_1)N_2) + cU^{(1)} + cN_2 + b\gamma] < 0 \\ &\sup[-(\beta + D + aU'(N_1)N_2) + aU(N_1) + b\gamma] < 0 \\ &\sup[-(\alpha + D + aU'(N_1)N_2) + aU(N_1) + \alpha] < 0 \\ &\sup[-(\beta + \gamma + D) + cU^{(1)} + \beta U'(N_1)] < 0 \\ &\sup[-(\alpha + \gamma + D) + cU^{(1)} + cN_2] < 0 \\ &\sup[-(\beta + \alpha) + \alpha + \beta U'(N_1)] < 0. \end{aligned}$$

Since U' is a monotone decreasing function of N_1 then in the first of (25)

$$-(aU'(N_1) - c)N_2 \leq -(aU'(\bar{N}) - c)N_2 < 0$$

because of (21). Furthermore, remark that in Ω, $U^{(1)} \leq U(\bar{N})$. Hence, (25) hold true if the following hold true:

$$(26) \quad \begin{aligned} &-(\gamma + D) - (D - cU(\bar{N})) + b\gamma < 0 \\ &-1/T_g - (D - aU(\bar{N})) + b\gamma < 0 \\ &-(D - aU(\bar{N})) < 0 \\ &-(1 - \bar{u})/T_g - \gamma - (D - cU(\bar{N})) < 0 \\ &-1/T_r - \gamma - (D - cU(\bar{N})) + c\bar{N} < 0 \\ &-(1 - \bar{u})/T_g < 0 \end{aligned}$$

where $D - cU(\bar{N}) > D - aU(\bar{N}) > 0$ because of (22) and $1 - \bar{u} > 0$ owing to the right side of (21). By these remarks it is easy to see that all the inequalities (26) hold true except for the second and fifth inequalities. These last two inequalities however hold true because of assumptions (23), (24). Then $\mu_\infty(J^{[2]}) < 0$ on Ω and no periodic solutions can occur in Ω. Finally, the Jacobian (20) satisfies that at positive equilibrium $(-1)^n \det J(E_+) > 0$, $E_+ \in \Omega$ and $\mu_\infty(J^{[2]}) < 0$ on Ω. Hence, this implies that $J(E_+)$ is a stable matrix by Theorem 7. This completes the proof.

The requirements (21), (22) of Theorem 9 are met if $L > 1$, $a/c > L$ and

$$\bar{N} < \frac{1}{b} \min\left\{ \left(\frac{La}{c}\right)^{1/2} - L, \ \frac{LD/a}{1 - D/a} \right\}.$$

Furthermore $\bar{N} = 2\eta H/b$ for some positive constant H if ${N_1}^0$ is chosen in such a way that:

$$(27) \qquad {N_1}^0 < \frac{2\eta H(1 - \gamma T_r)}{1 + aT_g/b}.$$

Therefore, let us choose a constant $H \in \mathrm{R}_+$ satisfying that:

$$H < (2\eta)^{-1} \min\left\{ \left(\frac{La}{c}\right)^{1/2} - L, \ \frac{LD}{a - D} \right\}$$

and, whenever T_g, T_r satisfy (23), (24), assume that ${N_1}^0$ satisfies (27). Then all the assumptions of Theorem 9 hold true.

REFERENCES

1. Li, Y., Muldowney, S., On Bendixon's criterion, J. Diff. Eqn. 106, (1993) 27-39.

2. Li, M. Y., Wang, L., A criterion for stability of matrices, J. Math. Anal. Appl. 255 (1998) 249-264.

3. MacDonald, N., Time Lags in Biological Models, Springer, Berlin, 1978.

4. Beretta, E., Takeuchi, Y., Qualitative properties of chemostat equations with time delays: boundedness, local and global asymptotic stability. Diff. Eqn. and Dynamical Systems, 2(1994)19-40.

Proceedings of Dynamic Systems and Applications 3 (2001) 77-84

Chaos in Vibrating Granular Flows

Denis Blackmore[1], Roman Samulyak[1] and Anthony Rosato[2]
[1] Department of Mathematical Sciences, New Jersey Institute of Technology, Newark, New Jersey 07102 - 1982
[2] Department of Mechanical Engineering and Particle Technology Center, New Jersey Institute of Technology, Newark, New Jersey 07102 - 1982

ABSTRACT: A continuum model derived recently for particle flow dynamics is used to study the behavior of granular flows in vibrating beds. Numerical solutions to this model are obtained which exhibit relatively stable patterns of vortex type. An asymptotic approximation of this model is derived and it is proved that it can have stable vortex dynamics and chaotic solutions. Melnikov theory is used to study bifurcations of the approximate model.
AMS(MOS) subject classification. 58F13, 58F05, 65N05, 34E05.

1. INTRODUCTION

Flows of granular materials have been studied intensively during the last two decades. Many efforts in the theory, such as [4, 6, 10], have been made to develop the main approaches based on continuum mechanics (transport theory) and kinetic theory (statistical mechanics). In these and other papers several models were derived for the description of granular dynamics. There have also been a number of simple, idealized models formulated by neglecting a variety of physical factors, but these tend to miss many of the features of granular flows of interest in applications. In this paper we employ numerical and analytic methods to study the dynamics of the model derived in [2] and show, in particular, that chos can occur.

2. PRELIMINARIES

We assume that the flow system is comprised of N identical inelastic spherical particles distributed throughout some region in $\mathbf{R}^3$. The common radius of all particles is a very small positive number. The motion of the particles may be described by a system of $3N$ second order ordinary differential equations expressing Newton's second law of motion:

$$m\ddot{x}^{(i)} = F_i := P_i + E_i + B_i \quad 1 \leq i \leq N, \tag{1}$$

where $\cdot \equiv d/dt$, m is the mass of each of the N identical particles, P_i is the force exerted by all the particles on the ith particle, E_i is the external or body force on the ith particle (which is usually just the gravitational force) and B_i is the transmitted boundary force exerted on the ith particle. Employing known particle-particle force models [4,6], the following model granular flows was derived in [2]:

$$\frac{\partial \mathbf{v}}{\partial t} + (\mathbf{v}\nabla)\mathbf{v} = \mathbf{G} - \nabla p + \nu\Delta\mathbf{v} + \lambda\nabla(\nabla\mathbf{v}), \tag{2}$$

where $\mathbf{G}$ and p are the external forces and the pressure, respectively, ν and λ depend only on the particle-particle force function and may be assumed to be constant in most applications, and ρ is the density of the granular flow.

3. SIMPLE VIBRATING BED MODEL

At $t = 0$ the particles are contained in the following region:

$$K_0 := \{(x_1, x_2) \in \mathbf{R}^2 : |x_1| < \sigma,\, x_2 > 0\}, \tag{3}$$

where $\sigma > 0$ is half of the width of the container. The bottom of the container, which is free to move up and down along the fixed side walls $x_1 = \pm\sigma$, is subject to a vertical oscillation of the form $a\,sin(\omega t)$, so that the particles are confined to the region

$$K_t := \{(x_1, x_2) \in \mathbf{R}^2 : |x_1| < \sigma,\, x_2 > a\,sin(\omega t)\}, \tag{4}$$

for each $t \geq 0$.

We assume that the particles initially fill the region

$$\Omega_0 := \{(x_1, x_2) \in \mathbf{R}^2 : |x_1| < \sigma,\, 0 \leq x_2 \leq h\}, \tag{5}$$

where $h > 0$ is the height of the particle continuum at time $t = 0$. The particle continuum at time $t \geq 0$ will be denoted by Ω_t.

It is assumed that prior to the initiation of motion, the particles in Ω_0 have settled under the force of gravity so that they initially move like parts of a rigid body. Hence we set $v_1 = 0$ and $v_2 = a\,\omega$ at $t = 0$.

Following the reasoning in [2], we take the boundary conditions on ∂K_t to be

$$v_n = w_n, \tag{6}$$

where v_n and w_n are, respectively, the velocity component of the particle and the wall in the direction normal to the wall at the particle position, and

$$\frac{\partial v_T}{\partial n} = -k v_T, \tag{7}$$

where v_T denotes the relative tangential component of velocity between the particle and ∂K_t, $\partial/\partial n$ is the partial derivative in the inner normal direction, and $k > 0$ is a constant that we call the *wall friction coefficient*. Note that $k = 0$ if ∂K_t is frictionless. A physically realistic boundary condition for a free-boundary point is that the shear stress is zero and so the normal derivative of the velocity vanishes:

$$\frac{\partial \mathbf{v}}{\partial n} = \nabla \mathbf{v} \cdot \hat{n} = (< \nabla v_1, \hat{n} >, < \nabla v_2, \hat{n} >) = 0, \tag{8}$$

where $\hat{n}$ is a unit normal to the free-boundary directed into the particle continuum. This free-boundary condition is consistent with simulation and experimental results [5,6,7].

We assume that at $t = 0$ the pressure is constant along the x_1-axis, the gravity force is balanced by the stationary pressure distribution along the x_2-axis and that the flow is driven by the boundary forces exerted by the vibrating walls at $t > 0$. Whence, the governing equations for the vibrating bed flow are:

$$\begin{aligned}\frac{\partial v_1}{\partial t}+v_1\frac{\partial v_1}{\partial x_1}+v_2\frac{\partial v_1}{\partial x_2} &= \nu\left(\frac{\partial^2 v_1}{\partial x_1^2}+\frac{\partial^2 v_1}{\partial x_2^2}\right)+\lambda\left(\frac{\partial^2 v_1}{\partial x_1^2}+\frac{\partial^2 v_2}{\partial x_1\partial x_2}\right)\\ \frac{\partial v_2}{\partial t}+v_1\frac{\partial v_2}{\partial x_1}+v_2\frac{\partial v_2}{\partial x_2} &= \nu\left(\frac{\partial^2 v_2}{\partial x_1^2}+\frac{\partial^2 v_2}{\partial x_2^2}\right)+\lambda\left(\frac{\partial^2 v_1}{\partial x_1\partial x_2}+\frac{\partial^2 v_2}{\partial x_2^2}\right)\end{aligned} \tag{9}$$

in $\Sigma := \{(x,t) : x \in \Omega_t,\ t > 0\}$;

$$v_1(x_1,x_2,0) = v_1^0(x_1,x_2), \qquad v_2(x_1,x_2,0) = v_2^0(x_1,x_2) \quad at \quad t = 0 \tag{10}$$

for all $x \in \Omega_0 = \{x \in \mathbf{R}^2 : |x_1| < \sigma,\ 0 < x_2 < h\}$, where the functions $v_1^0(x_1,x_2)$ and $v_2^0(x_1,x_2)$ determine the initial velocity distribution;

$$\frac{\partial v_1}{\partial x_2} = -kv_1, \quad v_2 = a\omega\, cos(\omega t) \tag{11}$$

for all particles on the bottom, $\{(x_1,x_2) : |x_1| < \sigma, x_2 = a\,sin(\omega t)\}$, of the bed when $t > 0$;

$$v_1 = 0, \quad , \frac{\partial v_2}{\partial x_1} = -kv_2 \tag{12}$$

for all particles on the left wall, $\{(x_1,x_2) : x_1 = -\sigma,\ x_2 > a\,sin(\omega t)\}$, of the container when $t > 0$;

$$v_1 = 0, \quad , \frac{\partial v_2}{\partial x_1} = -kv_2 \tag{13}$$

for all particles on the right wall, $\{(x_1,x_2) : x_1 = \sigma,\ x_2 > a\,sin(\omega t)\}$, of the container when $t > 0$; and

$$\frac{\partial \mathbf{v}}{\partial n} = 0 \tag{14}$$

for all particles on the free-boundary, consisting of all points in $\partial\Omega_t \backslash \partial K_t$, when $t > 0$.

4. NUMERICAL SOLUTIONS

In this section we consider some numerical solutions to the model (9) subject to the initial and the boundary conditions (10)-(14). To simplify our analysis, we shall ignore the effects of surface waves and free-boundary effects.

The governing equations in coordinates moving with the bed are

$$\begin{aligned} v_{1,t} &= \nu(v_{1,xx}+v_{1,yy}) + a\omega\, cos(\omega t)v_{1,y} - \alpha(v_1v_{1,x}+v_2v_{1,y}) + \lambda(v_{1,xx}+v_{2,xy}),\\ v_{2,t} &= -a\omega^2 sin(\omega t) + \nu(v_{2,xx}+v_{2,yy}) + a\omega\, cos(\omega t)v_{2,y} - \alpha(v_1v_{2,x}+v_2v_{2,y}) +\\ &+ \lambda(v_{1,xy}+v_{2,yy}), \end{aligned} \tag{15}$$

for $0 \le x \le 2$, $0 \le y \le 2$, where $x = x_1$ and $y = x_2 + a\sin\omega t$. The boundary conditions can be written as

$$\begin{aligned} v_1(0,y,t) &= 0, \qquad v_1(2,y,t) = 0,\\ v_2(x,0,t) &= 0, \qquad v_2(x,2,t) = 0, \end{aligned} \tag{16}$$

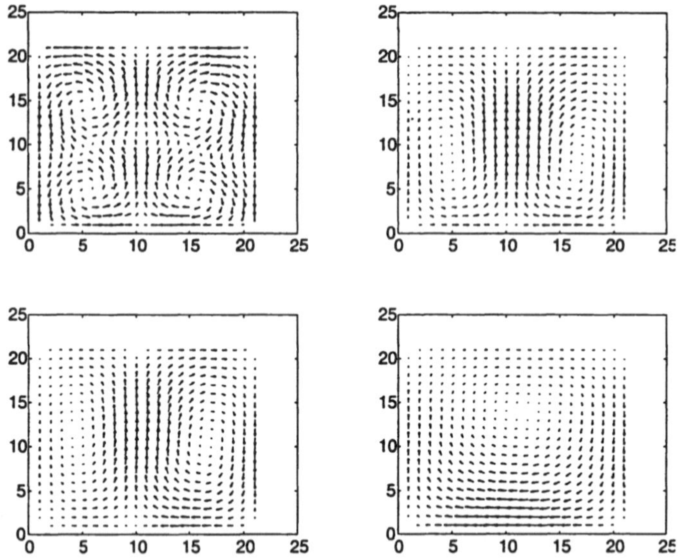

Figure 1: Motion of particles in bed (numerical solutions).

$$\frac{\partial v_1}{\partial y}(x,0,t) + kv_1(x,0,t) = 0, \quad \frac{\partial v_1}{\partial y}(x,2,t) = 0,$$

$$\frac{\partial v_2}{\partial x}(0,y,t) + kv_2(0,y,t) = 0, \quad \frac{\partial v_2}{\partial x}(2,y,t) + kv_2(2,y,t) = 0.$$

We investigated the system (15)-(16) using an explicit finite difference scheme with $\Delta x = \Delta y = 0.001$ and $\Delta t = 10^{-5}$ for both multi-vortex and random initial conditions. For certain ranges of the parameters the motion of the system starting with a multi-vortex configuration changes rapidly into a pair of vortices that persists for a relatively long time. The centers of this "stable" vortex pair oscillate with small amplitude synchronistically with the forced oscillations. An increase of ν results in an increase of the particle-particle friction and leads to damping of the vorticity (see Fig. 1). These types of particle dynamics are in agreement with experimental observations and computer simulation results [8,9]. Similar numerical results were obtained in [3] and [5].

5. ANALYSIS OF ASYMPTOTIC SOLUTIONS

In this section we shall find particular analytic approximate solutions to the system (15)-(16) using asymptotic methods. These solutions approximate the relatively stable patterns obtained numerically as well as the breakdown and diffusion of these patterns. We shall consider the system of equations (15) in the form

$$\mathbf{v}_t + (\mathbf{v}\nabla)\mathbf{v} = -a\omega^2 sin(\omega t)\hat{e}_2 + a\omega\, cos(\omega t)\mathbf{v}_y + \nu\nabla^2\mathbf{v} + \lambda\nabla(\nabla\mathbf{v}), \tag{17}$$

where $\hat{e}_2$ is a unit vector along the $y-$axis. It is natural to assume that the solution to the problem can be expanded as the sum of a time independent solution $\mathbf{v}^s$ and one oscillating with the frequency ω and amplitude function $\mathbf{v}^c$:

$$\mathbf{v}(x,y,t) = \mathbf{v}^s(x,y) + a\mathbf{v}^{(1)}\,cos(\omega t) + a\mathbf{v}^{(2)}\,sin(\omega t) + O(a^2), \tag{18}$$

where we assume a to be small. Substituting (18) into (17), retaining only the zeroth order terms with respect to a and assuming the divergence of $\mathbf{v}$ is small, it is not difficult to show that

$$\nabla \times (\mathbf{v}^s \times (\nabla \times \mathbf{v}^s)) = \nu \nabla \times (\nabla \times (\nabla \times \mathbf{v}^s)). \tag{19}$$

Equation (19) is used for fluid flows near obstacles [11]. We introduce a stream function $\Psi(x,y)$ such that

$$\mathbf{v}^s(x,y) = \nabla \times \Psi(x,y)\hat{e}_3, \tag{20}$$

where $\hat{e}_3$ is the unit vector along the z-direction. Hence, the velocity can be expressed as

$$\mathbf{v}^s(x,y) = \left(\frac{\partial \Psi}{\partial y}, -\frac{\partial \Psi}{\partial x}\right). \tag{21}$$

Using the stream function representation we can rewrite (19) as

$$\left[\frac{\partial}{\partial x}\left(\frac{\partial \Psi}{\partial y}\nabla^2\Psi\right) + \frac{\partial}{\partial y}\left(\frac{\partial \Psi}{\partial x}\nabla^2\Psi\right)\right] = \nu\Delta^2\Psi. \tag{22}$$

The simplest vortex type solution satisfying (22) is

$$\Psi = log\frac{(x-x_0)^2+(y-y_0)^2}{(x+x_0)^2+(y-y_0)^2} + kx, \tag{23}$$

where $(x_0,\, y_0)$ is the location of the vortex center (cf. [12]). Owing to (21), the stream function determines an integrable Hamiltonian system

$$v_1^s \equiv \frac{dx}{dt} = \frac{\partial \Psi}{\partial y}, \quad v_2^s \equiv \frac{dy}{dt} = -\frac{\partial \Psi}{\partial x}, \tag{24}$$

which in explicit form is

$$\begin{aligned} \frac{dx}{dt} &= (y-y_0)\left(\frac{1}{(x+x_0)^2+(y-y_0)^2} - \frac{1}{(x-x_0)^2+(y-y_0)^2}\right), \\ \frac{dy}{dt} &= \frac{x-x_0}{(x-x_0)^2+(y-y_0)^2} - \frac{x+x_0}{(x+x_0)^2+(y-y_0)^2} - k. \end{aligned} \tag{25}$$

The Hamiltonian function (23) is unbounded from the below in one vortex center point (x_0, y_0) and from the above in the other center $(-x_0, y_0)$. Therefore, these points must be excluded. There are two hyperbolic fixed points p_1 and p_2 at $x = 0$, $y = \pm\sqrt{2x_0/k - x_0^2}$ for $y_0 = 0$. The hyperbolic fixed points are connected by heteroclinic separatrices s_0, s_1, s_2. These limiting streamlines serve as the invariant stable and unstable manifolds for the fixed points; in particular, s_1 and s_2 comprise the stable manifolds and s_0 is a portion of the unstable manifold for p_1, and s_0 is a portion of the stable manifold and s_1 and s_2 comprise the unstable one for the point p_2.

Using (24) we can find a system of equations determining the functions $\mathbf{v}^{(1)}$, $\mathbf{v}^{(2)}$. This system can be solved numerically subject to the vibrating bed boundary conditions. But we are interested here in an analytical study of the behavior of a local solution. We can find the following simple particular solution for $\mathbf{v}^{(1)}$, $\mathbf{v}^{(2)}$ satisfying the system of equations:

$$v_1^{(1)} = 0, \quad v_2^{(1)} = \omega, \quad \mathbf{v}^{(2)} = 0. \tag{26}$$

Thus the system of equations for the perturbed trajectories of particles takes the following form:

$$\begin{aligned} \frac{dx}{dt} &= (y - y_0)\left(\frac{1}{(x + x_0)^2 + (y - y_0)^2} - \frac{1}{(x - x_0)^2 + (y - y_0)^2}\right), \\ \frac{dy}{dt} &= \frac{x - x_0}{(x - x_0)^2 + (y - y_0)^2} - \frac{x + x_0}{(x + x_0)^2 + (y - y_0)^2} - k + a\omega\, cos(\omega t). \end{aligned} \tag{27}$$

By using polar coordinates about the vortex centers and passing to the limit as $r \to 0$, it is straightforward to show that the center motion is

$$x_0 = 1, \quad y_0 = a\, sin(\omega t).$$

Substituting the solutions for the vortex centers motion into (26) we obtain

$$\begin{aligned} \frac{dx}{dt} &= \frac{y - a\, sin(\omega t)}{(x \pm 1)^2 + (y - a\, sin(\omega t))^2} - \frac{y - a\, sin(\omega t)}{(x - \mp 1)^2 + (y - a\, sin(\omega t))^2}, \\ \frac{dy}{dt} &= \frac{x \mp 1}{(x \mp 1)^2 + (y - a\, sin(\omega t))^2} - \frac{x \pm 1}{(x \pm 1)^2 + (y - a\, sin(\omega t))^2} - \\ &- k + a\omega cos(\omega t). \end{aligned} \tag{28}$$

To estimate the bifurcation values of the parameters we shall consider the case of small a. Expanding the right-hand side in a Taylor series and retaining into only the zero and first order terms we obtain

$$\begin{aligned} \frac{dx}{dt} &= y\left(\frac{1}{P_+} - \frac{1}{P_-}\right) - a\, sin(\omega t)\left[\frac{1}{P_+} - \frac{1}{P_-} + 2y^2\left(\frac{1}{P_-^2} - \frac{1}{P_+^2}\right)\right] \\ \frac{dy}{dt} &= \frac{x \mp 1}{P_-} - \frac{x \pm 1}{P_+} + a\left[\omega\, cos(\omega t) - 2y\, sin(\omega t)\left(\frac{x \pm 1}{P_+^2 -} - \frac{x \mp 1}{P_-^2}\right)\right], \end{aligned} \tag{29}$$

where $P_\pm = (x \pm 1)^2 + y^2$.

The dynamical system (28) has the form of a periodically perturbed integrable Hamiltonian system:

$$\frac{dx}{dt} = f_1(x, y) + a\, g_1(x, y, t, a, \omega), \; \frac{dy}{dt} = f_2(x, y) + a\, g_2(x, y, t, a, \omega), \tag{30}$$

where $a \ll 1$ and ω are parameters and the functions g_1 and g_2 are time-periodic. Hence the bifurcation parameters can be estimated using Melnikov theory [13,14].

The phase portrait of these equations at $a = 0$ consists of a pair of vortices (see Fig. 1). The periodic trajectories are associated with invariant tori of the integrable dynamical system (33). This picture changes in an essential way when $a \neq 0$. For small a the invariant tori still exist in a slightly deformed form for a wide range of parameters a and ω in accordance with the KAM theory [1,13,15]. The trajectories of particles are still periodic and lie between these tori. The picture is qualitively the same for some range of parameters a and ω when a is close to 1. But an increase in the amplitude a of the oscillations leads to new qualitative behavior of the particle trajectories - chaotic motion with extensive mixing.

In the case of small a Melnikov's method can be used to estimate bifurcation values. The Melnikov technique enables us to predict the behavior of the perturbed stable and unstable manifolds $W^s_{+,\varepsilon}$ and $W^u_{-,\varepsilon}$ by measuring the distance between $W^s_{+,\varepsilon}$ and $W^u_{-,\varepsilon}$. Let ε be the perturbation rate. The first order term in the Taylor series expansion of this distance is given by

$$d(t_0,\varepsilon) = \varepsilon \frac{M(t_0)}{\|\mathbf{f}(\mathbf{q}(-t_0))\|} + O(\varepsilon^2), \tag{31}$$

where $\mathbf{q}(t)$ is the particle trajectory of the unperturbed phase space lying on the heteroclinic separatrix s, t_0 defines a particular point of the heteroclinic orbit in a particular Poincaré section, and

$$\|\mathbf{f}(\mathbf{q}(-t_0))\| = [(f_1(\mathbf{q}(-t_0)))^2 + (f_1(\mathbf{q}(-t_0)))^2]^{1/2}. \tag{32}$$

The Melnikov function is defined to be

$$M(t_0) = \int_{-\infty}^{+\infty} [f_1(\mathbf{q}(t))g_2(\mathbf{q}(t), t+t_0) - f_2(\mathbf{q}(t))g_1(\mathbf{q}(t), t+t_0)]dt. \tag{33}$$

By the Melnikov theorem [13,14], simple zeros of $M(t_0)$ imply transverse intersection of the perturbed stable and unstable manifolds $W^s_{+,\varepsilon}$ and $W^u_{-,\varepsilon}$, signaling chaos.

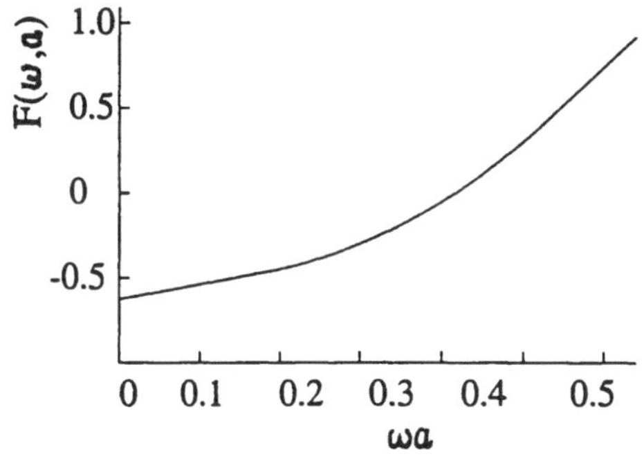

Figure 2: Amplitude of Melnikov function

It can be shown that

$$M(t_0,\omega,a) \sim F\left(\frac{1}{\omega a}\right) sin(\omega t_0). \tag{34}$$

M is periodic in t_0 for fixed ω. The amplitude, calculated numerically, for the different Poincaré sections is plotted in Fig. 2. This shows the existence of an infinite number of simple zeros of the function M and, therefore, the existence of a infinite number of transverse intersections of thc perturbed stable and unstable manifolds.

6. CONCLUDING REMARKS

Numerical solutions of a new dynamical systems model for vibrating granular flow were obtained. These solutions exhibited typical features observed in simulation and experimental studies of vibrating beds.In an effort to study the dynamics analytically, we used asymptotic expansions and a few simplifying assumptions. The approximate solutions share several important features with actual granular flows. Using this approach we showed the existence KAM-tori for the perturbed motion and bifurcation to chaotic motion by employing Melnikóv theory.

ACKNOWLEDGEMENTS

The research in this paper was supported in part by a grant from the NJCST.

REFERENCES

1. Arnold V.I., Mathematical Methods of Classical Mechanics, Springer Graduate Texts in Math. N. 60, Springer-Verlag, New York, 1989.
2. Blackmore, D., Rosato, A. and Samulyak, R., New mathematical models for particle flow dynamics, J. Nonlin. Math. Phys, Vol 6 (1999), 198-221.
3. Bourzutschky, M. and Miller, J., "Granular" convection in a vibrated fluid, Phys. Rev. Lett., Vol 74 (1995), 2216-2219.
4. Goldshtein, A., Shapiro, M., Mechanics of collisional motion of granular materials Part 1. General hydrodynamic equations, JFM, Vol 282 (1995), 75-114.
5. Hayakawa, H., Hong, D., Two hydrodynamical models of granular convection, Powders & Grains 97, Behringer and Jenkins (eds),Balkema, 1997, 417-420.
6. Lun, C., A kinetic theory for granular flow of dense, slightly inelastic, slightly rough spheres, JFM, Vol 233 (1991), 539-559.
7. Rosato, A., Lan, Y., Granular dynamics modeling of vibration-induced convection of rough inelastic spheres, Proc. 1st Int. Part. Tech. Forum, AIChE (1994), 446-454.
8. Lan, Y., Rosato, A., Convection related phenomena in vibrating granular beds, Phys. Fluids, Vol 9 (1996), 3615-3624.
9. Pöschel, T., Herrmann, H.J., Size segregation and convection, Europhys. Lett., Vol 29 (1995), 123-128.
10. Schaeffer, D., Instability in the evolution equations describing incompressible granular flow, J. Diff. Eq., Vol 66, 19-50.
11. Lamb, H., Hydrodynamics, Dover: New York, 1940.
12. Rom-Kedar, V., Leonard, A., Wiggins, S., An analytical study of transport, mixing and chaos in an unsteady vortical flow, JFM, Vol 214 (1990), 347-394.
13. Guckenheimer, J., Holmes, P., Nonlinear Oscillations, Dynamical Systems and Bifurcations of Vector Fields, Springer-Verlag: New York, 1997.
14. Melnikov, V.K., On the stability of the center for the periodic perturbations, Trans. Moscow Math., Vol 12 (1963), 1-57.
15. Wiggins, S., Global Bifurcations and Chaos – Analytical Methods, Springer-Verlag: New York, 1988.

Proceedings of Dynamic Systems and Applications 3 (2001) 85-90

A COUPLED PAIR OF EQUATIONS RELATED TO A HIERARCHICAL SIZE-STRUCTURED POPULATION MODEL

Kbenesh W. Blayneh
Department of Mathematics, Florida A&M University,
Tallahassee, FL 32307

Abstract: A coupled pair of partial and ordinary differential equations is conisdered. Existence and uniquness of solutions and some important properties of solutions are presented. Concluding remarks about how the model is used to analyze a hierarchical size - structured population model are briefly covered.

AMS subject classification. 35F25, 35M10, 92D25

1. INTRODUCTION

In some populations which are structured by (size, age, etc.), there is a hierarchy established by the structuring variable which determines an individual's activities such as, access to food, shelter, mate, etc. Consequently, the vital rates (growth, birth and death) of an individual in such a population can be affected by its rank in the hierarchy. This type of relation between the vital rates of an individual and its rank in the hierarchy has a dynamical consequence on the population, and is analyzed using hierarchical models.

The biological significance of different models based on a hierarchical organization of population is addressed by some authors (see Blayneh [1, 2, 3], Cushing[4, 5] and Henson et al [8]). To get analytical results, the complicated nature of these models requires transformation to a simpler (mathematically tractable) form. For instance, the hierarchical size-structured models in [1, 2, 4], which are similar to what is given by Eq.(13 - 15) in Section 3, can be reduced to a coupled pair of partial and ordinary differential equations in $S(t,s)$ and $B(t)$, respectively [2]. These equations are

$$\partial_t S + xX(t,S,B-S)\partial_x S = G(t,S,B-S); \quad (t,x)\notin \Gamma_1\cup\Gamma_2 \tag{1}$$

$$S(t,x) = S_0(x) = \int_{x_0}^{x} u\phi(u)du; \quad (t,x)\in\Gamma_1 \tag{2}$$

$$S(t,x) = 0; \quad (t,x)\in\Gamma_2, \tag{3}$$

Γ_1 and Γ_2 are boundary curves, we parametrize Γ_1 by $t=0$, $x=\sigma\geq x_0$ and Γ_2 by $t=z\geq 0, x=x_0$, and

$$B' = f(t,B) \tag{4}$$

$$B(0) = \int_{x_0}^{\infty} s\phi(s)ds, \tag{5}$$

Eq.(2) and Eq.(5) are initial conditions and Eq.(3) is a boundary condition. Without loss in generality, we assume x_o to be 1. For simplicity, throughout this paper we refer to the system given by Eq.(1 - 5) as " coupled pair ", and Eq.(13 - 15) as hierarchical model. The hierarchical model and the coupled pair are proved to be equivalent [2]. Consequently, an important result like the existence and uniqueness of solution to the hierarchical model is achieved. Also, this coupled pair facilitates a way to analyze the dynamics of some weighted integrals of the density of the hierarchical model.

Recently, some researchers are attracted to analyze and apply the hierarchical model to biological problems using the coupled pair of equations. Some of the reasons for this approach are; (i) the coupled pair of equations is relatively simple and mathematically tractable (ii) important questions like the existence of a globally defined solution to the hierarchical model as well as its asymptotic dynamics are answered using the coupled pair [2, 3]. Due to the nature of relation between the hierarchical and the coupled pair, we believe that understanding the dynamics of the coupled pair, is a significant step towards achieving remarkable analytical results about the hierarchical model.

As part of our goal, we prove that (i) the coupled pair has a unique pair of solutions using method of characteristic, (ii) the solution $S(t,x)$ of Eq.(1) is bounded from above by $B(t)$ and also that it approaches $B(t)$ asymptotically as $x \to \infty$. Next we make some technical assumptions. The assumptions are relatively simple so that several functions that arise in applications are not cut off.

Let $R_+ = [0, +\infty)$ and $R = (-\infty, +\infty)$.
(p1) : $G, X, f \in C^0(R^3_+, R)$, and $0 < X(t,x,y)$ for all $(t,x,y) \in R^3_+$.
(p2) : The partial derivatives of X, G, and f with respect to their arguments exist and are continuous on R^3_+, and $\partial_x X(.,x,y) - \partial_y X(.,x,y) \geq 0$.
(p3) : $\phi \in C^0(R_+, R_+)$, $s\phi \in L^1(R_+, R_+)$.

The first assumption in **p3** is made so that $S(t,s)$ has continuous initial curve and the second assumption implies that B(0), given by Eq.(5) is finite. In the following lemma, $Z = (x_1, x_2)$ and $|Z| = \sup_{(x_1,x_2)\in R^2}\{|x_1|, |x_2|\}$.

Lemma 1 *Assume that G satisfies the conditions in* **p1** *and* **p2**. *Let $B(t)$ be the solution of Eq.(4) on $[a,b)$, $0 < a < b$ with initial condition $B(a) \geq 0$. Suppose there is a scalar and nonnegative function $g_0(t,u)$, continuous on $[a,b) \times R$ with $|G(t,x_1,x_2)| \leq g_0(t,|Z|)$, $(t,x_1,x_2) \in [a,b) \times R^2$ and such that for any positive number k, the initial value problem $\dot{u} = g_0(t,u) + M_k$, $u(a) \geq 0$ has a unique solution $u(t)$ on $[a, b - \frac{1}{k})$, where M_k is an upper bound of $f(t,B)$ on $[a, b - \frac{1}{k}]$. Then the initial value problem*

$$\dot{y} = G(t,y,B-y), \quad |(y(a), B(a) - y(a))| \leq u(a) \tag{6}$$

has a unique solution on $[a,b)$.

Proof : Let $x_1 = y$, $x_2 = B - y$ then Eq.(6) is equivalent to

$$\begin{aligned} \dot{x_1} &= G(t, x_1,\ x_2) \\ \dot{x_2} &= -G(t, x_1,\ x_2) + f(t, x_1 + x_2). \end{aligned} \quad (7)$$

Let $h(t, x_1, x_2)$ be the right hand side of the system given by Eq.(7). Note that by the above relation between x_1 and x_2, $f(t, x_1 + x_2) = f(t, B)$ is continuous on $[a, b - \frac{1}{k}]$, $k > 0$, it is bounded by a positive constant M_k on the closed interval $[a, b - \frac{1}{k}]$. Hence, $|h(t, x_1, x_2)| \leq g_0(t, |Z|) + M_x$ for $(t, x_1, x_2) \in [a, b - \frac{1}{k}) \times R^2$. By hypothesis, $\dot{u} = g_0(t, u) + M_k$, $u(a) \geq 0$ has a unique solution on $[a, b - \frac{1}{k})$. Thus, by a theory on differential inequalities Hale [7], Eq.(6) has a unique solution on $[a, b - \frac{1}{k}]$. Since k is arbitrary, $x(t)$ exists on $[a, b)$ ⊡

2. ANALYTICAL RESULTS

We begin this section by defining two subregions R_1 and R_2 of $\{(t, x) : t \geq 0, x > 1\}$, the region bounded by Γ_1 and Γ_2. We restrict the proofs of Theorem 1 and Theorem 2 to one of the regions as similar results follow in the other region. If **p1** and **p2** hold and if G satisfies the hypothesis in Lemma 1, then the system of characteristic equations for Eq.(1) which is parametrized by w

$$\frac{dt}{dw} = 1 \quad (8)$$

$$\frac{dx}{dw} = xX(w, S, B - S) \quad (9)$$

$$\frac{dS}{dw} = G(w, S, B - S) \quad (10)$$

with initial conditions $t(0) = 0$, $x(0) = 1$ and $S(0) = 0$, has a unique solution $(t, x(t), S(t, x))$ on $[0, T)$. We will refer to the characteristic curve through $(0, 1)$ as the "main characteristic" and denote it by η, and the two subregions separated by η as R_1 and R_2, where

$$R_1 = \{(t, x) : 0 \leq t < T, x > \eta\} \quad \text{and} \quad R_2 = \{(t, x) : 0 \leq t < T, 1 < x < \eta\}$$

In the following theorem we assume that G satisfies the condition in Lemma 1 with $[a, b) = [0, T)$. Method of characteristic and the idea of differential inequlity are implemented in the proof of this theorem.

Theorem 1 *Assume that* **p1** *and* **p2** *hold. Let* $I = [0, T)$, $0 < T \leq \infty$ *be the maximal interval of existence for the solution* $B(t)$ *of the ordinary differential equation* (4) *with initial condition* (5) *then, Eq.(1 - 5) has a unique pair of solutions* $(S(t, x)$, $B(t))$ *on* $[0, T)$, $x > 1$.

Proof : We will prove that the coupled pair has a unique solution on the two regions separated by the main characteristic curve η. Let $t = \tilde{t}(w,\sigma)$, $s = \tilde{s}(w,\sigma)$ and $S = \tilde{S}(w,\sigma)$ denote the characteristic curve for Eq.(1) which is given by the system (8 - 10) parametrized by w with initial point $(0, \sigma, S_0(\sigma))$ on Γ_1 for $\sigma > 1$. Let $I = [0, T)$ be the maximal interval of existence for the solution B of Eq.(4 - 5). Since G satisfies the condition in Lemma 1, the solution $\tilde{S}(w,\sigma)$ is defined on $[0,T)$. Note that Γ_1 is parametrized by $t = 0, x = \sigma, \sigma \geq 1$ and since $\frac{\partial(t,x)}{\partial(w,\sigma)} \neq 0$, Γ_1 is noncharacteristic. It is clear that the solution of Eq.(9) is increasing in σ for each $w \geq 0$. This together with the assumption about X given in **p2** implies the invertibility of $x = \tilde{x}(w,\sigma)$, $t = \tilde{t}(w,\sigma)$ over R_1 giving C^1- functions $w = \tilde{w}(t,x)$ and $\sigma = \tilde{\sigma}(t,x)$. This allows us to define the solution $S_1(t,x) = \tilde{S}(\tilde{w}(t,x), \tilde{\sigma}(t,x))$. The uniqueness of this solution on R_1 follows from the fact that the system of characteristic given by Eq.(8 - 10) has a unique solution. A similar argument with initial condition on Γ_2 leads to a unique solution of Eq.(1 - 3) on R_2. There is no discontinuity at the point of intersection of Γ_1 and Γ_2. Hence, the solution $S(t,x)$ of Eq.(1 - 3) is continuous on the whole region $\overline{R}_1 \cup \overline{R}_2$ ⊡

From this theorem we conclude that if $B(t)$ is defined for all $t \geq 0$, then for each $x > 1$, $S(t,x)$ will also be defined for all $t \geq 0$. In order to get a globally defined solution of the coupled pair, we simply need to make some assumption on $f(t,B)$ so that the solution of Eq.(4) is defined for all $t \geq 0$. A globally defined solution is possible by making a biologically meaningful assumption about $f(t,B)$, see [2].

Before stating the next theorem, we define $f(t,B)$ and $G(t,S,B-S)$ as follows

$$f(t,B) \doteq \int_0^B \{Y(t,z,B-z) + X(t,z,B-z) - d(t,z,B-z)\}dz \qquad (11)$$

$$\begin{aligned} G(t,S,B-S) \;\doteq\; & \int_0^S \{X(t,z,B-z) - d(t,z,B-z)\}dz \\ & + \int_0^B Y(t,z,B-z)dz, \end{aligned} \qquad (12)$$

d, $Y \in C^{\circ}(R^3_+, R)$, and have continuous partial derivatives with respect to their arguments on R^3_+. The functions X, Y and d are used in defining the growth, birth and death rates, see Section 3.

Theorem 2 *Assume that* **p1** *and* **p2** *hold. Suppose that f and G are defined by Eq.(11) and Eq.(12), respectively. If $(S(t,x), B(t))$ is a solution of the coupled pair on an interval $[0,T)$, $T \geq 0$ for $1 < x < \eta$ and $x > \eta$, then*
(i) $0 \leq S(t,x) \leq B(t), \forall x \geq 1$, $t \in [0,T)$ and
(ii) $S(t,x) \to B(t)$ as $x \to \infty$ for all $t \in [0,T)$.

Proof : (i) $S(t,x) \geq 0$; to see this, observe that along a characteristic curve, $\frac{dS}{dw} \geq \int_0^S \{X(w+z,u,B-u) - d(w+z,u,B-u)\}du$, $z \geq 0$ in R_2, $\frac{dS}{dw} \geq \int_0^S \{X(w,u,B-$

$u) - d(w,u,B-u)\}du$ in R_1 and that the solution of $\frac{dS}{dv} = \int_0^S \{X(v,u,B-u) - d(v,u,B-u)\}du$ with initial condition $S(0) \geq 0$ is nonnegative. Define $U(w,\sigma) = B(w) - \tilde{S}(w,\sigma)$, $\sigma \geq 1, w \geq 0$, where $B(w)$ and $\tilde{S}(w,\sigma)$ are solutions of Eq.(4) and Eq.(10) respectively. It is easy to show that $U(w,\sigma)$ solves $\frac{dU}{dw} = \int_0^U \{X(w,B-v,v) - d(w,B-v,v)\}dv$, subject to the initial condition $U(0,\sigma) = \int_\sigma^\infty v\phi(v)dv$. Clearly the solution $U(w,\sigma) \geq 0$, for each $\sigma \geq 1$. Similar argument leads to the result $U(z,v) \geq 0$ in R_2 for each $z \geq 0$.

(ii) Since we are interested in the limit as $x \to \infty$ of $U(t,x)$, we take the region R_1. Take $t_1 \in (0,T)$ for each $w \in [0,t_1]$, $U(w,\sigma)$ is bounded from above by some positive number $k(w)$ (independent of sigma), this follows from the fact that $S(w,\sigma) \leq B(w)$ by part (i). Hence, the integral $\int_0^{U(w,\sigma)} \{X(w,B-v,v) - d(w,B-v,v)\}dv$ is bounded $\forall \sigma \geq 1$. Let $k(w)$ be its upper bound. From this and the result in part (i), we get $0 \leq U(w,\sigma) \leq U(0,\sigma)\exp\{k(w)\}$. Therefore, since $U(0,\sigma) \to 0$ as $\sigma \to \infty$, it follows that $U(w,\sigma) \to 0$ as $\sigma \to \infty$. But, $\sigma \to \infty$ if and only if $x \to \infty$, hence, $U(t,x) \to 0$ as $x \to \infty$. This implies that $S(t,x) \to B(t)$ as $x \to \infty$ for all $t \in [0,T)$ ⊡

3. CONCLUDING REMARKS

From Theorem 2 we learn that the solution of Eq.(1 - 3) is bounded from above by the solution of Eq.(4 - 5). Moreover, $S(t,x)$ approaches $B(t)$ asymptotically in x. These interesting properties of the solution are among a few reasons that make the coupled pair, a good mechanism to study the dynamics, and applications of the hierarchical size-structured population model. This model is given by the following equations.

$$\partial_t \rho(t,s) + \partial_s (sX(t,S,L)\rho(t,s)) = -d(t,S,L)\rho(t,s), \quad s > 1, t > 0 \tag{13}$$

$$sX(t,S,L)\rho(t,s)\,|_{s=1} = \int_1^\infty sY(t,S,L)\rho(t,s)ds, \quad t > 0 \tag{14}$$

$$\rho(0,s) = \phi(s), \quad s \geq 1, \tag{15}$$

where $\rho(t,s)$ is the density of individuals of size s at time t, $sX(t,S,L)$, $sY(t,S,L)$ and $d(t,S,L)$ are growth, birth and death rates of an individual of size s, respectively and $L = B - S$.

The mathematical details about the existence and uniqueness of solutions of this hierarchical model, and its reduction to the coupled pair are covered in [2]. The reduction of this model to the coupled pair is successful, because, the birth and growth rates of an individual are assumed to be proportional to its size. As a result of this reduction process, (i) the asymptotic dynamics of the density $\rho(t,s)$, is studied, by assuming that the population biomasses namely; $B(t)$ and $S(t,s)$, equilibrate when s is restricted to a compact set [3], (ii) using Eq.(4 - 5) the dynamics of the total population level quantity $B(t)$, of individuals practicing intraspecific predation and competition is studied Blayneh [1-3] and Cushing [4]. It is possible to see, by making additional assumption on the initial population density $\phi(t)$, that $B(t)$ and $S(t,x)$ defined by

$$B(t) = \int_1^\infty x\rho(t,x)dx \quad \text{and} \quad S(t,x) = \int_1^x u\rho(t,u)du \tag{16}$$

solve the coupled pair of equations where $\rho(t,s)$ is a solution of Eq.(13 - 15). Moreover, it is clear that conditions (i) and (ii) of Theorem 2 hold for $B(t)$ and $S(t,x)$.

Note that for many functions G, which arise from biological problems, it is not difficult to find g_0 satisfying the requirement in Lemma 1. One way of getting this result is by assuming the birth, growth and death rates of individuals to be bounded. To extend the solution of the coupled pair to the maximal interval where $B(t)$ is defined, we can also assume that G satisfies strong conditions like Lipschitz condition. However, we prefer not to make this assumption since it can force many functions that arise from application to be excluded.

ACKNOWLEDGMENT

The author acknowledges the support of the National Science Foundation Grant DMS-9306271. Thanks are to Dr. J. M. Cushing of the Applied Mathematics at the University of Arizona.

REFERENCES

1. Kbenesh W. Blayneh, Cannibalism in a seasonal environment, *Mathematical and Computer modelling,* (to appear)
2. Kbenesh W. Blayneh, A Hierarchical size-structured population model, (in preparation)
3. Kbenesh W. Blayneh, Asymptotic dynamics of the density of hierarchical size-structured population, *Nonlinear Studies,* (to appear)
4. J. M. Cushing, A size-structured model for cannibalism, *Theor. Popul. Biol.* **42(3)** *(1992), 347-361*
5. J. M. Cushing, The dynamics of a hierarchical age-structured populations, *J. Math. Biol.* **32** *(1994), 705-729*
6. M. E. Gurtin and R. C.MacCamy, Non-linear age-dependent population dynamics, *Arch. Rational. Mech. Anal.* **544** *(1974), 281-300*
7. J. K. Hale, Ordinary Differential Equations, New York, Wiley - Interscience, 1969.
8. S.M. Henson and J. M. Cushing, Hierarchical models of intra-specific competition: Scramble versus contest, J.Math. Biol. **34** (1996), 755-772

Proceedings of Dynamic Systems and Applications 3 (2001) 91-98

THE DISCRETE NONLINEAR NONSTATIONARY LOSSLESS SYSTEMS: STABILIZATION AND FEEDBACK EQUIVALENCE [1]

Andrey Yu. Bogdanov

Department of Mathematics an Mechanics,
Ulyanovsk State University, Leo Tolstoi st., 42,
Ulyanovsk, 432700, Russia

ABSTRACT: We consider the discrete feedback system with outputs

$$\begin{aligned} x(k+1) &= f(k, x(k)) + g(k, x(k))u(k) \\ y(k) &= h(k, x(k)) + J(k, x(k))u(k) \end{aligned} \tag{*}$$

The task of construction of the class of controls $\{u(k)\}$ which provide asymptotic, equi- and uniform asymptotic (local and global) stability of the zero solution of the system (*) is solved. The basis of investigation are the limiting equations and limiting Lyapunov functions methods which allow to extend and generalize many results obtained for the stationary systems of the similar classes.

AMS (MOS) subject classification: 93D15, 93C55, 93D20.

1. IMPLICATIONS OF LOSSLESS PROPERTY IN A CLASS OF DISCRETE NONLINEAR NONSTATIONARY SYSTEMS

Consider the problem of local and global stabilization of nonlinear nonstationary discrete system with feedback

$$\begin{aligned} x(k+1) &= f(k, x(k)) + g(k, x(k))u(k) \\ y(k) &= h(k, x(k)) + J(k, x(k))u(k), \end{aligned} \tag{1}$$

where $x \in R^n$, $u \in R^m$, $y \in R^m$; f, g, h, J are smooth mappings of appropriate dimensions satisfying the precompactness conditions of La Salle[1]. We shall suppose that $f(k,0) \equiv 0, h(k,0) \equiv 0$.

[1]This work was supported in part by RFFI (Grant 98-01-03284).

According to the works [2–5] we shall call the system (1) dissipative if there is a nonnegative Lyapunov function $V: Z_+ \times R^n \to R_+$, $V(k,0) \equiv 0$, and a function $W: Z_+ \times R^m \times R^m \to R^1$, such that for all $u \in R^m$ and for all $k \in Z_+$ the following estimate holds:

$$V(k+1, x(k+1)) - V(k, x(k)) \leq W(k, y(k), u(k)). \tag{2}$$

The function $V(k,x)$ is often named a storage function, and the function $W(k,y,u)$ is a supply rate. We assume that these functions satisfy the precompactness conditions too.

In the case when $W(k,y(k),u(k)) = y^T(k)u(k)$, the system (1) is called passive. It should be mentioned that for the case of strict inequality in (2) the task of system (1) stabilization can be solved by applying the Lyapunov-like theorem on asymptotic stability for discrete systems [6–7].

Consider the problem of system (1) stabilization in the case when there is an equality in (2) with $W = y^T u$. It is the case of so called lossless systems (thanks to C.I.Byrnes).

The necessary and sufficient conditions that the system (1) is lossless are determined in the following theorem.

Theorem 1.1 *The system (1) with C^2 Lyapunov function $V(k,x)$ is a lossless system if and only if*

A1)

$$V(k+1, f(k,x)) = V(k,x), \tag{3}$$

$$\left.\frac{\partial V(k+1,\alpha)}{\partial \alpha}\right|_{\alpha=f(k,x)} g(k,x) = h^T(k,x), \tag{4}$$

$$g^T(k,x)\left.\frac{\partial^2 V(k+1,\alpha)}{\partial \alpha^2}\right|_{\alpha=f(k,x)} g(k,x) = J^T(k,x) + J(k,x). \tag{5}$$

A2) *$V(k+1, f(k,x) + g(k,x)u)$ is quadratic in u.*

Proof. Necessity. Since system (1) is lossless then there exists nonnegative Lyapunov function $V(k,x)$ such that

$$V(k+1, f(k,x) + g(k,x)u) = V(k,x) + h^T(k,x)u + u^T\frac{J(k,x) + J^T(k,x)}{2}u,$$

$$\forall u \in R^m.$$

It is evident that letting $u \equiv 0$, we obtain (3). It is not difficult to see that

$$\frac{\partial V(k+1, f(k,x) + g(k,x)u)}{\partial u} = \left.\frac{\partial V(k+1,\alpha)}{\partial \alpha}\right|_{\alpha=f(k,x)+g(k,x)u} g(k,x) =$$

$$= h^T(k,x) + u^T[J(k,x) + J^T(k,x)].$$

$$\frac{\partial^2 V(k+1, f(k,x) + g(k,x)u)}{\partial u^2} =$$

$$= g^T(k,x)\left.\frac{\partial^2 V(k+1,\alpha)}{\partial\alpha^2}\right|_{\alpha=f(k,x)+g(k,x)u} g(k,x) = J^T(k,x) + J(k,x).$$

Then (4) and (5) are fulfilled immediately if $u \equiv 0$. The condition A2) is obvious.

Sufficiency. It follows from the assumption A2) that

$$V(k+1, f(k,x) + g(k,x)u) = A(k,x) + B(k,x)u + u^T C(k,x)u, \forall u \in R^m.$$

By applying Taylor formula we have

$$A(k,x) = V(k+1, f(k,x)) = V(k,x),$$

$$B(k,x) = \left.\frac{\partial V(k+1, f(k,x)+g(k,x)u)}{\partial u}\right|_{u=0} = \left.\frac{\partial V(k+1,\alpha)}{\partial\alpha}\right|_{\alpha=f(k,x)} g(k,x),$$

$$C(k,x) = \frac{1}{2}\left.\frac{\partial^2 V(k+1, f(k,x)+g(k,x)u)}{\partial u^2}\right|_{u=0} =$$

$$= \frac{1}{2}g^T(k,x)\left.\frac{\partial^2 V(k+1,\alpha)}{\partial\alpha^2}\right|_{\alpha=f(k,x)} g(k,x).$$

Due to assumption A1) and from (1) we can conclude that

$$V(k+1, f(k,x)+g(k,x)u) - V(k,x) = h^T(k,x)u + \frac{1}{2}u^T[J(k,x) + J^T(k,x)]u =$$

$$= y^T(k)u, \forall u \in R^m.$$

So, system (1) is lossless with Lyapunov function $V(k,x) \in C^2$.

Remark 1.1 In the case of linear stationary discrete systems theorem 1.1 is reduced to well known result of R.E. Kalman (see [8–9]). Linear system of the form

$$\begin{aligned} x(k+1) &= Ax(k) + Bu(k) \\ y(k) &= Cx(k) + Du(k) \end{aligned} \tag{6}$$

is a lossless system with Lyapunov function $V(x) = \frac{1}{2}x^T Px$, if and only if there is a matrix $P > 0$ such that

$$\begin{aligned} A^T PA = P, \quad & B^T PA = C \\ D + D^T &= B^T PB \end{aligned} \tag{7}$$

It implies from the condition (7) that if $D = 0$, then $C = 0$. It means that linear system (6) is lossless only if the output $y(k) = 0$. Moreover, if $rank B = m$, then the system (6) with $D = 0$ is lossless only if $P = 0$. In other words, such a linear system can never be a lossless system with positive definite Lyapunov function $V(x) = \frac{1}{2}x^T Px$.

Introduce the observability definition of the origin $x = 0$ for the system (1), which is a natural extension of corresponding concept from the theory of linear systems [10]. We define $f_k^0(x) = x, f_k^1(x) = f(k,x), ..., f_k^i(x) = f(k+i-1, f_k^{i-1}(x))$.

Then $\varphi(k,x,n_0) = f_{n_0}^{k-n_0}(x)$ is a solution of the system $x(k+1) = f(k,x(k)), x(n_0) = x, k \geq n_0$.

Definition 1.1 The system (1) is locally zero-obzervable if there is a neighbourhood U for $x = 0$ such that for all $x(n_0) = x \in U = U(n_0)$

$$y(k)|_{u(k)=0} = h(\varphi(k,x,n_0)) = 0, \forall k \geq n_0 \in Z_+ \text{ implies } x \equiv 0.$$

If $U \equiv R^n$, then system (1) is zero-obzervable.

Define the set $\tilde{S} = \{x \in R : h(n_0 + i, f_{n_0}^i(x)) = 0, \forall i, n_0 \in Z_+\}$. Then it is easy to see that system (1) is zero-observable if and only if $\tilde{S} = \{0\}$.

The necessary and sufficient conditions of zero-observability of the system (1) can be found in the following theorem.

Theorem 1.2 *Suppose that the system (1) is lossless with C^2 Lyapunov function $V(k,x)$. Define the set*

$$S = \{x \in R^n : \frac{\partial V(n_0+i+1,\alpha)}{\partial \alpha}\bigg|_{\alpha=f_{n_0}^{i+1}(x)} g(n_0+i, f_{n_0}^i(x)) = 0,$$
$$\forall i, n_0 \in Z_+\}$$

Then the system (1) is zero-observable if and only if $S = \{0\}$.

Proof. In view of definition 1.1 we need only to show the equivalence of the sets S and $\tilde{S}$. By applying the equality (4) we have:

$$h^T(n_0+i, x(n_0+i)) = h^T(n_0+i, f_{n_0}^i(x)) = (u \equiv 0) =$$
$$= \frac{\partial V(n_0+i+1,\alpha)}{\partial \alpha}\bigg|_{\alpha=f(n_0+i,f_{n_0}^i(x))=f_{n_0}^{i+1}(x)} g(n_0+i, f_{n_0}^i(x)) = 0, \forall x(n_0) = x,$$
$$\forall i \in Z_+, \forall n_0 \in Z_+.$$

So, $S = \{0\}$ if and only if $\tilde{S} = \{0\}$.

Remark 1.2 For stationary linear lossless systems this criteria of zero-observability can be reduced to well known in the literature on linear systems. For the linear system (6)

$$\frac{\partial V(n_0+i+1,\alpha)}{\partial \alpha}\bigg|_{\alpha=f_{n_0}^{i+1}(x)} g(n_0+i, f_{n_0}^i(x)) = x^T(A^{i+1})^T PB = 0$$

By applying Hamilton-Cayley theorem and considering $S = \{0\}$ we conclude that

$$rank[A^TPB,\ (A^2)^TPB,\ \ldots,(A^n)^TPB]^T = n$$

Since linear system (6) is lossless then from the conditions (7) we have

$$rank[C^T,\ A^TC,\ \ldots,(A^{n-1})^TC^T]^T = n$$

So the linear system (6) is classically observable.

Theorem 1.3 *Suppose that*

1) system (1) is lossless with Lyapunov function $V(k,x)$, satisfying the condition $w_1(\|x\|) \leq V(k,x) \leq w_2(\|x\|)$ in the neighbourhood of $x = 0$;

2) $\varphi : Z_+ \times R^m \to R^m$ is any smooth mapping satisfying the precompactness conditions and such that $\varphi(k,0) \equiv 0, y^T\varphi(k,y) \geq w(\|y\|), \forall y \neq 0, \forall k \in Z_+$;

3) system (1) is locally zero-observable and there is at least one limiting system

$$x(k+1) = f_0(k,x(k)) + g_0(k,x(k))u(k)$$

$$y(k) = h_0(k,x(k)) + J_0(k,x(k))u(k),$$

which is locally zero-observable at the moment $n_0 \geq 0$ (without loss of generality we may assume that $n_0 = 0$).

Then smooth output feedback as a control law

$$u = -\varphi(k,y) \tag{8}$$

locally stabilizes uniformly in x_0 the trivial solution $x = 0$ (this is equi-asymptotic stabilization).

If instead of the condition 3) we demand that the system (1) and one of the limiting systems are zero-observable, Lyapunov function $V(k,x)$ satisfies the condition $w_1(\|x\|) \leq V(k,x) \leq w_2(\|x\|)$ in R^n , then the control law (8) globally stabilizes the trivial solution $x = 0$ uniformly in x_0.

Proof. Any solution of closed system which is formed by (1) and (8) satisfies the nonstrict inequality

$$V(k+1,x(k+1)) - V(k,x(k)) = -y^T(k)\varphi(k,y(k)) \leq 0 \tag{9}$$

Consequently the closed system is uniformly stable in Lyapunov sense since $V(k,x)$ is not increasing along solutions of the closed system, $V(k,x)$ is positive definite and it "admits infinitely small higher limit" [7]. Prove that point $x = 0$ is also an attractor and the attractrion is uniform in x_0.

Suppose the contrary, i.e. absence of uniform in x_0 attraction. Let U_0 denotes the observation neighbourhood for the limiting system at the moment $n_0 = 0$. According to the uniform stability property we choose the neghbourhood U_δ of the point $x = 0$ such that all solutions of the closed system will satisfy the condition $\|x(k,n_0,x_0)\| < \varepsilon$, $\forall k \geq n_0$, if $x_0 \in U_\delta$, and besides ε is chosen so that $\bar{U}_\varepsilon \subset U_0$. Since we assume the absence of uniform in x_0 attraction, then there are: moment $\tilde{n}_0$, compact $K \subset U_\delta$, sequence $n_k \to +\infty$, sequence $\{x_0^{n_k}\} \in K$ and value $\varepsilon_0 > 0$ such that $\|x(\tilde{n}_0 + n_k, \tilde{n}_0, x_0^{n_k})\| \geq \varepsilon_0$. It is evident that $\varepsilon_0 \leq \varepsilon$.

From the property of uniform stability we choose $\delta_0 = \delta(\varepsilon_0/2) > 0$. Then $\delta_0 \le \|x(\tilde{n}_0+n, \tilde{n}_0, x_0^{\tilde{n}_k})\| \le \varepsilon$, $\forall k$, $\forall n \in [0, n_k]$. Let $\xi_k \to +\infty$ is the sequence for the limiting system determination. Choose the subsequence $\tilde{n}_k \to +\infty$ from the sequence $n_k \to +\infty$ that $\lim_{k\to+\infty}(\tilde{n}_k - \xi_k) = +\infty$. Let $\psi_k = x(\tilde{n}_0 + \xi_k, \tilde{n}_0, x_0^{\tilde{n}_k})$. Obviously $\delta_0 \le \|\psi_k\| < \varepsilon$. Consider the sequence of functions $\psi_k(n) = x(\tilde{n}_0 + \xi_k + n, \tilde{n}_0, x_0^{\tilde{n}_k}) = x(n, \tilde{n}_0 + \xi_k, \psi_k)$. We have $\|\psi_k(n)\| \ge \delta_0$ for $n \in [0, \tilde{n}_k - \xi_k]$.

Without loss of generality let $\psi_k \to \psi_0 \in U_0$. Then $\psi_k(n) \rightrightarrows \psi^*(n, 0, \psi_0)$ – a solution of the limiting system from the condition 3) of the theorem, and besides $\|\psi^*(n, 0, \psi_0)\| \ge \delta_0$, $\forall n \ge 0$.

Since $V_k[n] = V(n, x(n, \tilde{n}_0 + \xi_k, \psi_k))$ is not increasing and it is bounded from below then considering $k \to +\infty$ while (9) is true we have

$$V(n+1, \psi^*(n+1, 0, \psi_0)) - V(n, \psi^*(n, 0, \psi_0)) = c_0 - c_0 = 0 =$$

$$= -\tilde{y}^T(n)\tilde{\varphi}(n, \tilde{y}(n)) \le 0,$$

where the sign "$\sim$" means that we consider the corresponding limiting functions and $\tilde{y}(n) = h_0(n, \psi^*(n, 0, \psi_0)) + J_0(n, \psi^*(n, 0, \psi_0))\tilde{u}(n)$. Consequently, for all $n \in Z_+$ the output of the limiting system $\tilde{y}(n) = 0$. This means that $\tilde{u}(n) = -\tilde{\varphi}(n, \tilde{y}(n)) = 0, \forall n \in Z_+$. Since the limiting system is locally zero-observable at the moment $n_0 = 0$ and $\psi_0 \in U_0$, then we have to conclude that $\psi_0 = 0$. But this is a contradiction to the assumption $\|\psi_0\| \ge \delta_0$.

This contradiction completes the proof of theorem 1.3.

Remark 1.3 The uniform asymptotic stabilization of $x = 0$ for the lossless system (1) can be achieved if there is a unified neighbourhood of zero-observation for all limiting systems at all moments n_0.

2. THE FEEDBACK EQUIVALENCE OF THE DISCRETE SYSTEMS WITH OUTPUTS

In this section we shall solve the problem of feedback equivalence of an arbitrary discrete system (1) and a lossless system of the same structure.

Definition 2.1 We shall say that system (1) is regular at $x = 0$, if *det* $[J(k, 0)] \ge \delta_0 > 0$, $\forall\ k \in Z_+$. The system (1) is uniformly regular if $J(k, x)$ is uniformly inversable $\forall\ x \in R^n$, $\forall\ k \in Z_+$.

Suppose that system (1) is regular. Then by definition 2.1 there is a neighbourhood U of the origin $x = 0$ in R^n where $J^{-1}(k, x)$ is well defined. If we choose the control $u^*(k) = -J^{-1}(k, x(k))h(k, x(k))$, then $y(k) \equiv 0$, $\forall\ x(k) \in U \subset R^n$, where $x(k)$ is a trajectory of the dynamic system

$$x(k+1) = f^*(k, x(k)) = f(k, x(k)) + g(k, x(k))u^*(k)$$
$$\forall\, x(k) \in U \subset R^n$$

So, by contrast to the case of continuous stationary systems (Byrnes [2]) the zero dynamics of the discrete system (1) is governed by a system of dimension n, not $n-m$. Therefore this zero dynamics locally exists in U and it forms a smooth manifold of dimension n

$$Z^* = \{x \in U : y(k) = 0\} \equiv U \subset R^n.$$

If the system (1) is uniformly regular then

$$Z^* = \{x \in R^n : y(k) = 0\} \equiv R^n.$$

Definition 2.2 The regular system (1) has locally lossless zero dynamics if there is a C^2-smooth positive definite in certain neighbourhood U of the point $x = 0$ Lyapunov function $V(k, x)$ such that the following conditions hold:

1)$V(k+1, f^*(k, x)) = V(k, x)$;

2)$V(k+1, f^*(k, x) + g(k, x)u)$ is quadratic in u.

The uniformly regular system (1) has globally lossless zero dynamics if there is C^2-smooth positive definite Lyapunov function $V(k, x)$ satisfying the conditions 1) and 2).

Consider the following dynamic feedback

$$u(k) = \alpha(k, x(k)) + \beta(k, x(k))w(k), \quad \alpha(k, 0) \equiv 0$$

where functions $\alpha(k, x)$ и $\beta(k, x)$ are differentiable in x and defined either locally in the neighbourhood of $x = 0$ or globally. Also we suppose that $\beta(k, x)$ is uniformly inversable.

Theorem 2.1 (The main result) *Let $g(k, x)$ has locally uniform complete rank and Lyapunov function $V(k, x)$ is nondegenerative in $x = 0$. Then system (1) is locally equivalent by feedback to the lossless system with C^2-smooth positive definite Lyapunov function $V(k, x)$ if and only if system (1) is regular and it has locally lossless zero dynamics.*

Proof. The proof o this theorem is rather complicated but in many respects it is similar to anologous one of C.I. Byrnes for systems without time dependence. That is why and also in view of lack of space we omit it.

Remark 2.1 Consider the linear system (6) with $rank B = m$. The quadratic Lyapunov function $V(x) = \frac{1}{2}x^T P x$ with $P > 0$ is always nondegenerative and $V(Ax +$

Bu) is always quadratic in u. Consequently, by the theorem 2.1 the linear system (6) is equivalent by feedback to linear system with positive definite Lyapunov function $V(x) = \frac{1}{2}x^T Px$ if and only if there is a matrix $P > 0$ such that

$$(A - BD^{-1}C)^T P(A - BD^{-1}C) = P$$

The global version of theorem 2.1 can be proved if system (1) has globally lossless zero dynamics and it is uniformly regular, while $g(k, x)$ has globally uniform complete rank.

REFERENCES

1. LaSalle J.P. The Stability of Dynamical Systems / SIAM, Philadelphia, 1976. 76 p.
2. Byrnes C.I., Isidori A. Passivity, feedback equivalence, and the global stabilization of minimum phase nonlinear systems // IEEE Trans. Automat. Control. - 1991. Vol.36. P. 1228-1240.
3. Byrnes C.I., Lin W. Stabilization of discrete-time nonlinear systems by smooth state feedback // Syst. Contr. Letts. - 1993. Vol.21. P. 255-263.
4. Willems J.C. Dissipative dynamical systems part I: General theory // Arch. Ration, Mech. Anal. - 1972. Vol. 45. P. 325-351.
5. Hill D. Dissipativeness, stability theory and some remaining problems// in Analysis and Control of Nonlinear Systems. C.I. Byrnes, C.F. Martin and R.E. Sacks, Eds. North-Holland. - 1988. P. 443-452.
6. Furasov V.D. Stability and stabilization of discrete processes. - Moscow: Nauka, 1982.
7. Halanay A., Wexler D. Qualitative theory of sampling systems.- Moscow: Mir, 1971.
8. Kalman R., Falb P., Arbib M. Notes on mathematical theory of systems. Moscow: Mir, 1971.
9. Hitz L., Anderson B.D.O. Discrete positive real functions and their application to system stability // Proc. of IEEE. - 1969. Vol. 116. P. 153-155.
10. Nijmeijer H. Observability of autonomous discrete-time nolinear systems : A geometric approach // Int. J. Contr. - 1982. Vol. 36. P. 867-874.

Proceedings of Dynamic Systems and Applications 3 (2001) 99-104

Trigonometric Systems in Oscillation Theory of Difference Equations

Martin Bohner[1] and Ondřej Došlý[2]
[1]Department of Mathematics and Statistics, University of Missouri–Rolla
313 Rolla Building, Rolla, MO 65409-0020, U.S.A.
[2]Mathematical Institute, Czech Academy of Sciences,
Žižkova 22, CZ–61662 Brno, Czech Republic

ABSTRACT: We summarize the classical theory of the trigonometric transformation and of the Prüfer transformation for self-adjoint differential systems and present an overview on the discrete analogue of these concepts. We also offer some applications of these discrete transformations in oscillation theory of Sturm-Liouville second order difference equations.
AMS (MOS) subject classification. 39A10

1. PRÜFER AND TRIGONOMETRIC TRANSFORMATIONS

As a motivation, consider first the simple Sturm-Liouville second order differential equation

$$(r(t)x')' + c(t)x = 0 \tag{1.1}$$

with r and c continuous and $r(t) > 0$ in the interval under consideration. In oscillation theory of (1.1), trigonometric functions appear essentially in two different ways:

(i) *Prüfer transformation*: There exist differentiable functions ϱ and φ with $\varrho(t) > 0$ such that any nontrivial solution x and its quasiderivative rx' can be expressed in the form

$$x(t) = \varrho(t)\sin\varphi(t), \quad r(t)x'(t) = \varrho(t)\cos\varphi(t), \tag{1.2}$$

and ϱ, φ satisfy the first order system

$$\varrho' = \cos\varphi(t)\sin\varphi(t)\left(\frac{1}{r(t)} - p(t)\right)\varrho, \quad \varphi' = p(t)\sin^2\varphi + \frac{1}{r(t)}\cos^2\varphi, \tag{1.3}$$

(see [10] or [9, p. 332]).

(ii) *Trigonometric transformation*: If x and $\tilde{x}$ are two linearly independent solutions of (1.1), then there exist functions h and α, with $h(t) > 0$, such that x and $\tilde{x}$ can be expressed as

$$x(t) = h(t)\sin\alpha(t), \quad \tilde{x}(t) = h(t)\cos\alpha(t), \tag{1.4}$$

where h and α satisfy the differential equations

$$(r(t)h')' + c(t)h = \frac{\omega^2}{r(t)h^2}, \quad \alpha' = \frac{\omega}{r(t)h^2}, \tag{1.5}$$

$\omega := r(x'\tilde{x} - x\tilde{x}')$ being the (constant) Wronskian of x and $\tilde{x}$ (see [6]).

Both transformations are very useful when investigating oscillatory properties of (1.1). For example, the Sturm separation theorem for zeros of linearly independent solutions of (1.1) follows immediately from the trigonometric transformation (1.4), (1.5). The Prüfer transformation (1.2), (1.3) together with some elementary differential inequalities implies the Sturm comparison theorem.

In the extension of the Prüfer and the trigonometric transformation to linear Hamiltonian systems (LHS), the key role is played by so-called trigonometric systems. First recall that the LHS is a $2n$-dimensional first order differential system

$$x' = \mathcal{A}(t)x + \mathcal{B}(t)u, \quad u' = \mathcal{C}(t)x - \mathcal{A}^T(t)u, \tag{1.6}$$

where $x, u \in \mathbb{R}^n$ and $\mathcal{A}$, $\mathcal{B}$, and $\mathcal{C}$ are real $n \times n$-matrix-valued functions such that $\mathcal{B}(t)$ and $\mathcal{C}(t)$ are symmetric. A trigonometric system (TS) is a special LHS of the form

$$s' = \mathcal{Q}(t)c, \quad c' = -\mathcal{Q}(t)s, \tag{1.7}$$

where $s, c \in \mathbb{R}^n$ and $\mathcal{Q}$ is a real symmetric $n \times n$-matrix-valued function. The terminology "trigonometric system" is justified by the fact that in the scalar case $n = 1$, the pair $(s, c) = \left(\sin \int^t \mathcal{Q}(\tau)\mathrm{d}\tau, \cos \int^t \mathcal{Q}(\tau)\mathrm{d}\tau\right)$ solves (1.7). In the higher dimensional case, system (1.7) cannot be in general solved explicitly, but solutions of this system have many of the properties of the classical sine and cosine functions.

Simultaneously with vector LHS and TS, we will consider their matrix versions, where the vectors $x, u, s, c \in \mathbb{R}^n$ are replaced by $n \times n$-matrices denoted by the corresponding capital letters. A $2n \times n$-matrix-solution $\binom{X}{U}$ of (1.6) is called a conjoined basis if $X^TU = U^TX$ and $\operatorname{rank}\binom{X}{U} = n$. Differentiating, we can easily verify that any two solutions $\binom{X}{U}$ and $\binom{\tilde{X}}{\tilde{U}}$ satisfy $X^T\tilde{U} - U^T\tilde{X} = M$, where M is a constant $n \times n$-matrix. If $\binom{X}{U}$ and $\binom{\tilde{X}}{\tilde{U}}$ are conjoined bases with $M = I$, then we call them normalized conjoined bases.

Note also that we will use the concept of a trigonometric system for a slightly more general system than (1.7), namely for the system

$$S' = \mathcal{P}(t)S + \mathcal{Q}(t)C, \quad C' = -\mathcal{Q}(t)S + \mathcal{P}(t)C, \tag{1.8}$$

where $\mathcal{P}$ is antisymmetric, i.e., $\mathcal{P}^T + \mathcal{P} = 0$. However, the transformation $S = G(t)\tilde{S}$, $C = G^{T-1}\tilde{C}$, where $G' = \mathcal{P}(t)G$, i.e., $G^T = G^{-1}$ if $G(a)$ is orthogonal at some point $a \in \mathbb{R}$, transforms (without changing the system's oscillatory nature) (1.8) into (1.7), so solutions of (1.8) have essentially the same properties as those of (1.7). To consider a TS in the form (1.8) has the advantage that then we can define this system as the most general LHS with the property that the transformation $\binom{X}{U} = \begin{pmatrix} 0 & I \\ -I & 0 \end{pmatrix}\binom{\tilde{X}}{\tilde{U}}$ transforms (1.7) into itself (an LHS with this property is sometimes called self-reciprocal).

Barrett [2] and Reid [11, 12] showed that the Prüfer transformation (1.2), (1.3) extends to (1.6) as follows.

Proposition 1.1. *There exist a nonsingular differentiable $n \times n$-matrix-valued function H and a symmetric $n \times n$-matrix-valued function $\mathcal{Q}$ such that any conjoined basis $\binom{X}{U}$ of* (1.6) *can be expressed in the form*

$$X(t) = S^T(t)H(t), \quad U(t) = C^T(t)H(t),$$

where $\binom{S}{C}$ is a conjoined basis of (1.7) *satisfying $S^TS + C^TC = I$ and H solves the first order differential system*

$$H' = \left[S\mathcal{A}S^T + S\mathcal{B}C^T + C\mathcal{C}S^T - C\mathcal{A}^TC^T\right]H.$$

Concerning the trigonometric transformation (1.4), (1.5), its extension to the LHS (1.6) reads as follows (see [7]).

Proposition 1.2. *There exist $n \times n$-matrix-valued functions H and K with H being nonsingular and $K^T H = K^T H$, as well as a nonsingular $n \times n$-matrix-valued function $\mathcal{Q}$ such that the first components of a pair of normalized conjoined bases $\binom{X}{U}$ and $\binom{\tilde{X}}{\tilde{U}}$ of* (1.6) *can be expressed as*

$$X(t) = H(t)S(t), \quad \tilde{X}(t) = H(t)C(t),$$

where $\binom{S}{C}$ is a conjoined basis of (1.7) *satisfying $S^T S + C^T C = I$ and*

$$\mathcal{Q} = H^{-1}\mathcal{B}(H^T)^{-1} = H^T(\mathcal{C}H - \mathcal{A}^T K - K') - K^T(\mathcal{A}H + \mathcal{B}K - H').$$

The aim of this contribution is to present some recently established discrete versions of the Prüfer and trigonometric transformations of LHS and, as new results, to give some of their applications. The discrete counterpart of the LHS (1.6) is the so-called symplectic difference system (SΔS). This is the first order system

$$z_{k+1} = \mathcal{S}_k z_k, \quad z = \begin{pmatrix} x \\ u \end{pmatrix}, \quad \mathcal{S} = \begin{pmatrix} \mathcal{A} & \mathcal{B} \\ \mathcal{C} & \mathcal{D} \end{pmatrix} \tag{1.9}$$

where $x, u \in \mathbb{R}^n$, $\mathcal{A}, \mathcal{B}, \mathcal{C}, \mathcal{D} \in \mathbb{R}^n$ and the matrix $\mathcal{S}$ is symplectic, i.e., $\mathcal{S}^T \mathcal{J} \mathcal{S} = \mathcal{J}$ where $\mathcal{J} = \begin{pmatrix} 0 & I \\ -I & 0 \end{pmatrix}$. A discrete trigonometric system (ΔTS) is then the special SΔS

$$\begin{pmatrix} s_{k+1} \\ c_{k+1} \end{pmatrix} = \begin{pmatrix} \mathcal{P}_k & \mathcal{Q}_k \\ -\mathcal{Q}_k & \mathcal{P}_k \end{pmatrix} \begin{pmatrix} s_k \\ c_k \end{pmatrix}, \tag{1.10}$$

where the matrix $\mathcal{S}$ satisfies the additional identity $\mathcal{J}^T \mathcal{S} \mathcal{J} = \mathcal{S}$ (compare with the remark concerning self-reciprocal LHS (1.6)), which means that

$$\mathcal{P}^T \mathcal{P} + \mathcal{Q}^T \mathcal{Q} = I, \quad \mathcal{P}^T \mathcal{Q} - \mathcal{Q}^T \mathcal{P} = 0. \tag{1.11}$$

Some basic properties of systems (1.10) can be found in the paper of Anderson [1].

This paper is organized as follows. In the next section we present discrete Prüfer and trigonometric transformations of SΔS (1.9), in particular we show that the ΔTS (1.10) plays in discrete oscillation theory essentially the same role as its continuous counterpart (1.7). In the third section we present some applications of these transformations in oscillation theory of Sturm-Liouville equations.

2. DISCRETE TRANSFORMATIONS

In this section we present the recently established discrete trigonometric and Prüfer transformation of SΔS (1.9) (see [4, 5]).

Theorem 2.1 ([4, Theorem 3.1]). *Given a symplectic system* (1.9), *it is possible to find $n \times n$-matrix-valued sequences H and K such that H is nonsingular and $H^T K$ is symmetric, and the transformation*

$$\begin{pmatrix} s \\ c \end{pmatrix} = \begin{pmatrix} H^{-1} & 0 \\ -K^T & H^T \end{pmatrix} \begin{pmatrix} x \\ u \end{pmatrix} \tag{2.1}$$

transforms (1.9) *into a trigonometric system* (1.10) *with*

$$\begin{cases} \mathcal{P}_k = H_{k+1}^{-1}\mathcal{A}_k H_k + H_{k+1}^{-1}\mathcal{B}_k K_k \\ \mathcal{Q}_k = H_k^{-1}\mathcal{B}_k \left(H_k^T\right)^{-1}. \end{cases} \tag{2.2}$$

More precisely, if $Z = \binom{X}{U}$ and $\tilde{Z} = \binom{\tilde{X}}{\tilde{U}}$ are normalized conjoined bases of (1.9), *then the matrices H and K are given by*

$$HH^T = XX^T + \tilde{X}\tilde{X}^T \quad \text{and} \quad K = \left(UX^T + \tilde{U}\tilde{X}^T\right)\left(H^T\right)^{-1}$$

Furthermore, this transformation preserves oscillatory behavior and the matrix H can be chosen in such a way that the matrix $\mathcal{Q}$ is nonnegative definite.

In the continuous case, it is known that under some additional normality condition (see [11]), the continuous TS (1.7) is oscillatory (i.e., there exists a conjoined basis $\binom{X}{U}$ and a sequence $t_n \to \infty$ such that $\det X(t_n) = 0$) if and only if

$$\int^{\infty} \operatorname{Tr} Q(s) \mathrm{d}s = \infty, \tag{2.3}$$

where Tr stands for the trace of the matrix indicated. This statement coupled with trigonometric and Prüfer transformation has many applications in oscillation theory of LHS (1.6), see [7, 8].

Oscillation properties of SΔS (1.9) are defined using the concept of a focal point. An interval $(m, m+1]$ contains a focal point of the conjoined basis $\binom{X}{U}$ of (1.9) if

$$\operatorname{Ker} X_{m+1} \subset \operatorname{Ker} X_m \quad \text{and} \quad X_m X_{m+1}^{\dagger} \mathcal{B} \geq 0$$

fail to hold (see [3]). Here Ker and † denote the kernel and the Moore-Penrose inverse, respectively, ≥ 0 means positive semidefiniteness, and $\mathcal{B}$ is the $n \times n$ matrix in the upper right corner of the matrix $\mathcal{S}$ in (1.9). System (1.9) is said to be nonoscillatory if there exists $N \in \mathbb{N}$ such that the conjoined basis $\binom{X}{U}$ given by $X_N = 0$, $U_N = I$ has no focal point in (N, ∞); in the opposite case (1.9) is said to be oscillatory.

The following statement gives a necessary and sufficient condition for oscillation of a class of ΔTS (1.10) and can be viewed as a discrete version of (2.3).

Theorem 2.2 ([4, Theorem 4.3]). *The discrete TS (1.10) with $\mathcal{Q}_k$ positive definite for large k is oscillatory if and only if*

$$\sum^{\infty} \operatorname{arccot} \left[\lambda^{(1)}(\mathcal{Q}_k^{-1}\mathcal{P}_k)\right] = \infty,$$

where $\lambda^{(1)}(\cdot)$ stands for the least eigenvalue of the matrix indicated.

The discrete version of the Prüfer transformation reads as follows.

Theorem 2.3 ([5, Theorem 3.1]). *Let $Z = \binom{X}{U}$ be a $2n \times n$ matrix conjoined basis of (1.9). Then there exist a nonsingular $n \times n$ matrix H and $n \times n$ matrices S, C such that $\binom{X}{U}$ can be expressed in the form*

$$X_k = S_k^T H_k, \quad U_k = C_k^T H_k, \tag{2.4}$$

where $\binom{S}{C}$ is a conjoined basis of ΔTS (1.10) with

$$\begin{cases} \mathcal{P}_k &= (H_{k+1}^T)^{-1} \binom{X_k}{U_k}^T \mathcal{S}_k^T \binom{X_k}{U_k} H_k^{-1}, \\ \mathcal{Q}_k &= (H_{k+1}^T)^{-1} \binom{X_k}{U_k}^T \mathcal{S}_k^T \mathcal{J} \binom{X_k}{U_k} H_k^{-1} \end{cases} \tag{2.5}$$

and H solves the first order system $H_{k+1} = (\tilde{Z}_{k+1})^T (\mathcal{S}_k \tilde{Z}_k - \tilde{Z}_{k+1}) H_k$ with $\tilde{Z} = \binom{S^T}{C^T}$.

3. APPLICATIONS

In this section we offer two oscillation criteria for the Sturm-Liouville difference equation

$$\Delta(r_k \Delta x_k) + p_k x_{k+1} = 0 \tag{3.1}$$

with $r_k \neq 0$. Setting $u_k = r_k \Delta x_k$, this equation can be written in the form of the symplectic system (1.9) with $\mathcal{S} = \begin{pmatrix} 1 & \frac{1}{r} \\ -p & 1-\frac{p}{r} \end{pmatrix}$. Consequently, all statements concerning (1.9) apply also to (3.1).

The first criterion is based on the trigonometric transformation and Theorem 2.2.

Theorem 3.1. *Let $x, \tilde{x}$ be linearly independent solutions of* (3.1). *This equation is oscillatory if and only if*

$$\sum^{\infty} [\operatorname{arccot} r_k (x_k x_{k+1} + \tilde{x}_k \tilde{x}_{k+1})] = \infty.$$

In particular, if $r_k > 0$ eventually, $\sum^{\infty} \operatorname{arccot} r_k = \infty$ and all solutions of (3.1) *are bounded, then this equation is oscillatory.*

Proof. Trigonometric transformation (2.1) transforms (3.1) into scalar trigonometric system (1.10) with $\mathcal{P}_k = h_{k+1}^{-1} h_k^{-1} (x_k x_{k+1} + \tilde{x}_k \tilde{x}_{k+1})$, $\mathcal{Q}_k = h_{k+1}^{-1} h_k^{-1} r_k^{-1} > 0$ (compare (2.2)). Since (1.11) holds, there exists $\varphi \in [0, 2\pi)$ such that $\sin \varphi = \mathcal{Q}$, $\cos \varphi = \mathcal{P}$. Since $\mathcal{Q} > 0$, $\varphi \in (0, \pi)$ and hence $\varphi_k = \operatorname{arccot} \frac{\mathcal{P}_k}{\mathcal{Q}_k} = \operatorname{arccot} r_k (x_k x_{k+1} + \tilde{x}_k \tilde{x}_{k+1})$. The first statement now follows from Theorem 2.2 and the fact that trigonometric transformation preserves oscillatory properties. Finally, if all solutions of (3.1) are bounded and $r_k > 0$, there exists a constant $M > 0$ such that

$$\sum^{\infty} \operatorname{arccot} r_k (x_k x_{k+1} + \tilde{x}_k \tilde{x}_{k+1}) \geq \sum^{\infty} \operatorname{arccot} M r_k = \infty$$

since $\sum^{\infty} \operatorname{arccot} r_k = \infty$ if and only if $\sum \operatorname{arccot} M r_k = \infty$. The proof is complete. □

Theorem 3.2. *Suppose that $0 < p_k < 4 r_k$ holds for large k. Equation* (3.1) *is nonoscillatory provided*

$$\sum^{\infty} \operatorname{arccot} \left\{ \frac{1}{p_k} - \frac{r_k p_k}{\sqrt{p_k^2 + (1 - p_k r_k)^2} - 1 + r_k p_k - \frac{p_k^2}{2}} \right\} < \infty$$

and it is oscillatory if

$$\sum^{\infty} \operatorname{arccot} \left\{ \frac{1}{p_k} + \frac{r_k p_k}{\sqrt{p_k^2 + (1 - p_k r_k)^2} + 1 - r_k p_k + \frac{p_k^2}{2}} \right\} = \infty.$$

Proof. The assumption $0 < p_k < 4 r_k$ implies that $\mathcal{Q}_k$ given by (2.5) is positive for large k as can be verified by a direct computation. Since also $r_k > 0$, (3.1) is oscillatory if and only if the scalar trigonometric system (1.9) is oscillatory. Let φ_k be such that $\mathcal{P}_k = \cos \varphi_k$, $\mathcal{Q}_k = \sin \varphi_k$, i.e.,

$$\varphi_k = \operatorname{arccot} \frac{\mathcal{P}_k}{\mathcal{Q}_k} = \operatorname{arccot} \frac{\sin^2 \varphi_k + (\frac{1}{r} - p) \sin \varphi_k \cos \varphi_k + (1 - \frac{p}{r}) \cos^2 \varphi_k}{p \sin^2 \varphi_k + \frac{p}{r} \sin \varphi_k \cos \varphi_k + \frac{1}{r} \cos^2 \varphi_k}$$

(in the last equality we used (2.4), (2.5)). A solution of (1.9) is of the form $s_k = \sin \left(\sum^{k-1} \varphi_j \right)$, $c_k = \cos \left(\sum^{k-1} \varphi_j \right)$, so this system is nonoscillatory if and only if

$\sum^{\infty} \varphi_k < \infty$. The statement of the theorem now follows from the inequalities

$$\frac{1}{p_k} - \frac{r_k p_k}{\sqrt{p_k^2 + (1 - p_k r_k)^2} - 1 + r_k p_k - \frac{p_k^2}{2}}$$
$$\le \frac{\mu^2 + \left(\frac{1}{r_k} - p_k\right)\mu\nu + \left(1 - \frac{p_k}{r_k}\right)\nu^2}{p_k\mu^2 + \frac{p_k}{r_k}\mu\nu + \frac{1}{r_k}\nu^2} = \varphi_k$$
$$\le \frac{1}{p_k} + \frac{r_k p_k}{\sqrt{p_k^2 + (1 - p_k r_k)^2} + 1 - r_k p_k + \frac{p_k^2}{2}}$$

which hold for any μ, ν satisfying $\mu^2 + \nu^2 = 1$ as can be again verified by a direct computation. □

ACKNOWLEDGEMENT

The research of the second author is supported by Grant 201/98/0677 of the Czech Grant Agency.

REFERENCES

1. Anderson, D., Discrete trigonometric matrix functions. *Panamer. Math. J.*, 7 (1977), 39–54.
2. Barrett, J. H., A Prüfer transformation for matrix differential equations, *Proc. Amer. Math. Soc.*, 8 (1957), 510–518.
3. Bohner, M., Došlý, O., Disconjugacy and transformations for symplectic systems, *Rocky Mountain J. Math.*, 27 (1997), 707–743.
4. Bohner, M., Došlý, O., Trigonometric transformations of symplectic difference systems, *J. Differential Equations*, 1999, to appear.
5. Bohner, M., Došlý, O., The discrete Prüfer transformation, *Proc. Amer. Math. Soc.*, 2000, to appear.
6. Borůvka, O., Linear Differential Transformations of the Second Order. Oxford Univ. Press, London, 1971.
7. Došlý, O., On transformations of self-adjoint linear differential systems and their reciprocals, *Ann. Polon. Math.*, 50 (1990), 223–234.
8. Etgen, G. J., Oscillation properties of certain nonlinear matrix equations of second order, *Trans. Amer. Math. Soc.*, 122 (1966), 289–310.
9. Hartman, P., Ordinary Differential Equations. Wiley, New York, 1964.
10. Prüfer, H., Neue Herleitung der Sturm-Liouvilleschen Reihenentwicklung stetiger Funktionen, *Math. Ann.*, 95 (1926), 499–518.
11. Reid, W. T., A Prüfer transformation for differential systems, *Pacific J. Math.*, 8 (1958), 575–584.
12. Reid, W. T., Generalized polar coordinate transformations for differential systems, *Rocky Mountain J. Math.*, 1 (1971), 383–406.

Proceedings of Dynamic Systems and Applications 3 (2001) 105-112

On Generic oscillations for delay differential Equations

Baruch Cahlon and Darrell Schmidt
Department of Mathematics and Statistics
Oakland University
Rochester, MI 48309-4401

Abstract

In this paper we present a motivation for a generic oscillation theory for the solutions of linear delay differential equations, some known some new results, and open problems.

1980 Mathematics Subject Classification (1985 Revision): 34K15, 34K99

Keywords: generic, oscillation, delay, roots, characteristic equation, Laplace transform.

1. BACKGROUND AND MOTIVATION

The study of oscillatory solutions of delay differential equations (DDEs) has been the focus of numerous recent articles (see the texts [1], [2], [3] and their references as well as more recent articles [4], [5], [6]). The strongest results deal with linear delay differential equations with constant coefficients and constant delay. In this case, necessary and sufficient conditions are given for *all* solutions to be oscillatory (see [1]). For cases of nonconstant coefficients and/or delay and for nonlinear equations, typical results give conditions sufficient for *all* solutions to be oscillatory, or conditions sufficient for *some* solutions to be nonoscillatory. In this paper we examine oscillation theory in a different light. Rather than asking when all solutions are oscillatory or when there exist nonoscillatory solutions, we ask when a solution is *most likely* to be oscillatory or nonoscillatory.

Many physical and mechanical systems are modeled by a family of delay differential equations (for instances, when hysteresis is involved, see [7]) where it is known that there are oscillatory solutions and nonoscillatory solutions. In such cases it is important to understand whether the solutions are overwhelmingly of one or the other type. The thermostat model that led us to establish a generic oscillation theory is briefly given below and for more details we refer the reader to [7, 8].

Thermostats in cars are devices that control the operating temperature of the engine. The thermostat senses the coolant temperature and sends a larger or smaller flow of coolant through the radiator. In this way it is supposed to keep the coolant temperature nearly constant under normal operating conditions. They are very common devices, one per engine, and have been in operational use for many years; yet there is little literature on their dynamic behavior. Although they are conceptually simple, their dynamic behavior is not, since they exhibit hysteresis [7] (i.e., the way they open when the temperature rises differs from the way they close when the temperature falls) and since there is an inherent time delay due to flow from the thermostat to the radiator. A result is that in some auto engines and under certain operating conditions, the coolant temperature exhibits oscillatory behavior, and there is considerable interest in being able to predict whether engine will exhibit this phenomena before it goes into production.

In [7, 8], three mathematical models for the dynamic response of engine coolant temperature were derived and studied. Each involved a delay differential equation with hysteresis. In each model, the pertinent results were conditions under which all solutions of the models are oscillatory. Invariably, these conditions involved the size of the delay and the cooling efficiency. When these conditions are not satisfied, oscillatory and nonoscillatory solutions may exist, and of interest is whether the oscillatory solutions are so rare that they would be unobservable.

This leads us to consider the new concept of generically oscillatory solutions of delay differential equations (see [9], and [10]).

Consider the scalar delay differential equation

$$y'(t) = F(y(t-\tau_1), \dots, y(t-\tau_n)) \qquad (t > 0), \tag{1.1}$$

$$y(t) = \phi(t), \qquad -\tau \le t \le 0, \tag{1.2}$$

where $\tau_j \ge 0$ $(j = 1, \dots, n)$, $\tau = \max_{1 \le j \le n} \tau_j$, and $\phi \in C[-\tau, 0]$. We assume solutions of (1.1)-(1.2) exist, are unique, and extend to $[-\tau, \infty)$ for all $\phi \in C[-\tau, 0]$. We say that a solution y of (1.1)-(1.2) is oscillatory if y has arbitrarily large zeros.

Definition 1.1. *We say that the solutions of* (1.1)-(1.2) *are generically oscillatory if the set of all $\phi \in C[-\tau, 0]$ for which the corresponding solution y_ϕ of* (1.1)-(1.2) *is nonoscillatory is nowhere dense in $C[-\tau, 0]$.*

We say that the solutions of (1.1)-(1.2) *are generically nonoscillatory if the set of all $\phi \in C[-\tau, 0]$ for which the corresponding solution y_ϕ of* (1.1)-(1.2) *is oscillatory is nowhere dense in $C[-\tau, 0]$.*

A subset of $C[-\tau, 0]$ is nowhere dense in $C[-\tau, 0]$ if its closure has empty interior. As such if the solutions of (1.1)-(1.2) are generically oscillatory, then the set of initial functions that produce oscillatory solutions is overwhelming large compared to those that produce nonoscillatory solutions, and in this sense a solution is most likely oscillatory.

In the next Section, we consider the case with constant coefficients and constant delays and we present conditions based on the location of the roots of the characteristic equation that imply that the solutions are generically oscillatory or generically nonoscillatory. In Section 3, we give some new results and open problem

2. KNOWN RESULTS

We consider the linear problem with constant coefficients and delay

$$y'(t) + \sum_{j=1}^{n} a_j y(t - \tau_j) = 0 \qquad (t > 0), \tag{2.1}$$

$$y(t) = \phi(t) \qquad (-\tau \le t \le 0), \tag{2.2}$$

where $a_j \in R$, $\tau_j \ge 0$ $(j = 1, \ldots, n)$, $\phi \in C[-\tau, 0]$, and $\tau = \max_{1 \le j \le n} \tau_j$. Existence and uniqueness of solutions of (2.1)-(2.2) are known (see [11]). The purpose of this section is to give conditions sufficient for the solutions of (2.1)-(2.2) to be generically oscillatory or generically nonoscillatory in terms of the roots of the characteristic equation

$$z + \sum_{j=1}^{n} a_j e^{-z\tau_j} = 0. \tag{2.3}$$

Asymptotic stability of the trivial solution of (2.1) and oscillatory behavior of solutions of (2.1) are known (see [12], [13]). In particular, the zero solution of (2.1) is asymptotically stable if and only if all roots of (2.3) lie in the left half plane, and all solutions of (2.1) are oscillatory if and only if no roots of (2.3) are real. Criteria for generic oscillation or generic nonoscillation reduce to leading root(s) of (2.3) being real or nonreal. We call a root z_0 of (2.3) a leading root if $Re z \le Re z_0$ for all roots z of (2.3). It is known that that leading roots of (2.3) exist.

Theorem 2.1.

(a) *If the leading root of* (2.3) *is a single real number (of multiplicity possibly greater than one), then the solutions of* (2.1)-(2.2) *are generically nonoscillatory.*

(b) *If the leading roots of* (2.3) *form a single complex conjugate pair (of multiplicity possibly greater than one), then solutions of* (2.1)-(2.2) *are generically oscillatory.*

In the single delay case, Theorem 2.1 provides the following extension of the existing oscillation theory.

Theorem 2.2. *Let* $n = 1$, $a_1 = a \in R$, *and* $\tau = \tau_1 \ge 0$

(i) *If* $ae\tau > 1$, *then all solutions of* (2.1) *are oscillatory.*

(ii) *If* $ae\tau \le 1$, *then the solutions of* (2.1) *are generically nonoscillatory.*

In the case of several delays we have the following extension

Theorem 2.3. *Let $\tau_j \geq 0$ $(j = 1, \dots, n)$*

(i) *Suppose that $a_j > 0$ $(j = 1, \dots, n)$ and*

$$-\lambda + \sum_{j=1}^{n} a_j e^{\lambda \tau_j} > 0 \tag{2.4}$$

for all $\lambda > 0$. Then all solutions of (2.1) are oscillatory.

(ii) *Suppose that $a_j \in R\,(j = 1, \dots, n)$ and*

$$-\lambda_0 + \sum_{j=1}^{n} |a_j| e^{\lambda_0 \tau_j} < 0 \tag{2.5}$$

for some $\lambda_0 > 0$. Then the solutions of (2.1)-(2.2) are generically nonoscillatory.

(iii) *Suppose that $a_j > 0$ $(j = 1, \dots, n)$ and some $\tau_i > 0$. If*

$$-\lambda_0 + \sum_{i=1}^{n} a_i e^{\lambda_0 \tau_i} = 0 \tag{2.6}$$

for some $\lambda_0 > 0$ and

$$-\lambda + \sum_{j=1}^{n} a_j e^{\lambda \tau_j} \geq 0 \tag{2.7}$$

for all $\lambda > 0$, then the solutions of (2.1)-(2.2) are generically nonoscillatory.

Remark. In the case of one delay, Theorem 2.2 yields that if there is a nonoscillatory solution of (2.1)-(2.2), then the solutions are generically nonoscillatory. In the case of several delays and positive coefficients, Theorem 2.3 provides the same. Although this does not answer the concern for the engine coolant problem, it provides an encouraging development. One might ask whether it is at all possible for the solutions of (2.1)-(2.2) to be generically oscillatory while some solutions are nonoscillatory. In another words can we have nontrivial generic oscillations for (2.1)-(2.2). In [9], the authors obtained the results above as well as a general example which answers this question in the affirmative.

Select $a_j > 0$ and $\tau_j > 0\,(j = 1, \dots, n)$ so that

$$x + \sum_{j=1}^{n} a_j e^{-x\tau_j} > 0. \tag{2.8}$$

for all $x < 0$, and assume the leading root of (2.3) has multiplicity one. From inequality (2.8) all solutions of (2.1)-(2.2) are oscillatory and the function

$$F(z) = z + \sum_{j=1}^{n} a_j e^{-z\tau_j}$$

has two complex leading roots. Select $\tau = \max_{1\le j\le n} \tau_j$, and let

$$w(z) = z + \sum_{j=1}^{n} a_j e^{-z\tau_j} - \epsilon e^{-z\sigma} \tag{2.9}$$

where $\epsilon > 0$ and $\sigma > \tau$.

It can be shown that for $\epsilon > 0$ and sufficiently small (2.9) has a real root, but that the leading roots form a single complex pair. This implies that the solutions of

$$y'(t) + \sum_{j=1}^{n} a_j y(t-\tau_j) - \epsilon y(t-\sigma) = 0 \tag{2.10}$$

are generically oscillatory, while some solutions of (2.10) are nonoscillatory.

In [10], the authors considered the possibility of ties for leading roots. The first question deals with whether a tie between a real root and a complex pair is possible. In some cases, this is not possible.

Theorem 2.4. *If all the coefficients of (2.1) have the same sign, then there cannot be a tie for the leading roots of (2.3) between a real root and a complex pair.*

However, when the coefficients of (2.1) are not of the same sign such a tie can occur. In particular the leading roots of

$$w = \lambda + e^{\frac{-\pi\lambda}{6}} - e^{\frac{-11\pi\lambda}{6}} \tag{2.11}$$

are $0, i, -i$. In the case of real-complex ties we obtain

Theorem 2.5. *Assume that equation (1.3) has a tie of leading roots between x_0 and $x_0 \pm iy_0$, and x_0 has a multiplicity ℓ and $x_0 \pm iy_0$ have multiplicity k.*

1. *If $\ell > k$, then the solutions of* (1.1) *are generically nonoscillatory.*
2. *If $\ell < k$, the solutions of* (1.1) *are generically oscillatory.*
3. *If $\ell = k$, the solutions of* (1.1) *are neither generically oscillatory nor generically nonoscillatory.*

As such the solutions of

$$y'(t) + y(t - \frac{\pi}{6}) - y(t - \frac{11\pi}{6}) = 0$$

are neither generically oscillatory nor generically nonoscillatory.

3. NEW RESULTS AND OPEN PROBLEMS

In this section, we discuss extensions and open problems. It is unlikely that in the nonautonomous cases we can obtain a characterization of generic oscillation and generic nonoscillation quite like that in section 2. In addition, we may need to relax the requirements that the set of initial functions leading to oscillatory solutions or nonoscillatory solutions be nowhere dense. In nonautonomous or even nonlinear cases, we ask under what conditions does the existence of a nonoscillatory solution imply that this set of initial functions leading to oscillation solution be of first category in $C[-\tau, 0]$.

To this end, we consider the nonautonomous delay differential equation

$$y'(t) + \sum_{j=1}^{n} p_j(t)y(t - \tau_j(t)) = 0 \qquad (t > 0), \tag{3.1}$$

$$y(t) = \phi(t) \qquad (-\tau \leq t \leq 0), \tag{3.2}$$

where p_j, $\tau_j \in C([0, \infty), R^+)$ $(j = 1, \ldots, n)$, $\tau = \sup_{t \geq 0} \max_{1 \leq j \leq n} \tau_j(t)$ and $\phi \in C[-\tau, 0]$. It is known that if

$$\sum_{j=1}^{n} \int_{g(t)}^{t} p_j(s)ds < \frac{1}{e} \text{ for all } t \geq 0, \tag{3.3}$$

where $g(t) = \min_{1 \leq i \leq n}\{max\{0, t - \tau_i(t)\}\}$, then (3.1)-(3.2) has nonoscillatory solutions. (See [1, p.70]). Denote by y_ϕ the solution of (3.1)-(3.2), and for $k = 1, \ldots$ define $F_k : C[-\tau, 0] \to R$ by

$$F_k(\phi) = \begin{cases} \dfrac{1}{\min\limits_{2(k-1)\tau \leq t \leq 2k\tau} |y_\phi(t)|}, & \text{if } y_\phi(t) \neq 0 \text{ for all } t \in [2(k-1)\tau, 2k\tau] \\ 0, & \text{otherwise} \end{cases}$$

Under (3.3), it can be shown that each F_k is lower semicontinuous, $\lim_{k \to \infty} F_k(\phi) = \infty$ if y_ϕ is nonoscillatory, and $\left(F_k(\phi)\right)_{k=1}^{\infty}$ is bounded if y_ϕ is nonoscillatory. It follows from Osgood's theorem (see [14, p.148]) that the set of initial functions ϕ for which y_ϕ is oscillatory is of first catgeory in $C[-\tau, 0]$ or has nonempty inetrior. The authors believe that the second case can be eliminated.

We close with two open problems. The first deals with the autonomous case (2.1)-(2.2). In case there are ties for the leading roots of (2.3) involving more than one complex pair of roots, the problem is open. It can be shown (as in [10]), that if the tie involves one real leading root and several complex root and the real root has greater multiplicity than the complex roots, then the solutions of (2.1)-(2.3) are generically nonoscillatory. In case the leading roots of (2.3) are all complex with just two pairs or if the imaginary parts of the leading roots are form two or fewer commensurability classes, then the solutions are generically oscillatory. However, all other cases are still open.

In the nonautonomous case, (3.1)-(3.2) with all $p_j, \tau_j \in C([0, \infty), R^+)$ and τ_j bounded the authors conjecture that if one solution is nonoscillatory, then the solutions are generically nonoscillatory. This may have significant ramifications in the thermostat dynamics study.

REFERENCES

[1] Györi, I. and Ladas, G. Oscillation theory of delay differential equations with applications. Clarendon Press, Oxford, 1991.

[2] Gopalsamy, K. Stability and oscillation in delay differential equations of population dynamics. Kluwer Academic Publishers, Boston, 1992.

[3] Ladde, G.S., Lakshmikantham, V. and Zhang, B. G. Oscillation theory of differential equations with deviation arguments. Dekker, New-York, 1989.

[4] B. Thomas H., Oscillations in linear delay differential equations. J. Math. Anal. Appl., 195 (1995)., 261-277.

[5] Li, B. T., Oscillations of delay differential equations with variable coefficients. J. Math. Anal. Appl., 192 (1995). 312-321.

[6] Shustin, E. I., Super-high-frequency oscillations in a discontinuous dynamic system with time delay. Israel J. Math., 90 (1995). 199-219.

[7] Cahlon, B., Schmidt, D. , Shillor, M. and Zou, X., Analysis of Thermostat Models. European J. of Applied Mathematics, 8 (1997) 437-455.

[8] Cahlon, B., Schmidt, D., Shillor, M. and Zou, X., Oscillations in Thermostat Models. Advances in Mathematical Sciences and Applications 9 (1999). 25-49.

[9] Cahlon, B. and Schmidt, D., Generic Oscillations for Delay Differential Equations. Math. Anal. and Applic. 223 (1998). 288-301.

[10] Cahlon, B. and Schmidt, D., A Note on Generic Oscillations for Delay Differential Equations. Math. Anal. and Applic., to appear.

[11] Hale, J. K., Theory of functional differential equations. Springer Verlag, New York, 1977.

[12] Ladas, G. and Sficas, Y. G., Oscillations of delay differential equations with positive and negative coefficients. Proceedings of the International Conference on Qualitative Theory of Differential Equations, University of Alberta (1984). 232-240.

[13] Kuang, Yang, Delay differential equations with applications in population dynamics. Academic Press, San Diego, CA, 1993.

[14] Larsen, R., Functional Analysis. Marcel Dekker, Inc., New York, 1973.

Proceedings of Dynamic Systems and Applications 3 (2001) 113-120

POSITIVITY OF NONNEGATIVE SOLUTIONS FOR COOPERATIVE SEMIPOSITONE SYSTEMS

Alfonso Castro[1], C. Maya[2] and R. Shivaji[3]

[1]Department of Mathematics and Statistics
University of Texas at San Antonio, San Antonio, TX 78249

[2]Department of Mathematical Sciences
The University of North Carolina at Greensboro, Greensboro, NC 27402

[3]Department of Mathematics and Statistics
Mississippi State University, MS 39762

ABSTRACT: We establish that nonnegative solutions for cooperative semipositone systems in a ball are positive and hence radially symmetric.
AMS (MOS) subject classification. 35J55

1. INTRODUCTION

In this paper we consider a classical nonnegative solution $(u \geq 0, v \geq 0)$ for the system

$$-\Delta u = f(v) \quad \text{in } \Omega \tag{1.1}$$
$$-\Delta v = g(u) \quad \text{in } \Omega \tag{1.2}$$
$$u = 0 = v \quad \text{on } \partial\Omega, \tag{1.3}$$

where Ω is a ball in $\mathbb{R}^n$ for $n > 1$, $\partial\Omega$ its boundary, and $f, g : [0, \infty) \to \mathbb{R}$ are C^1 functions satisfying

$$f(0) < 0 \quad \text{and} \quad g(0) < 0 \quad \text{(semipositone)} \tag{1.4}$$

and

$$f' \geq 0 \quad \text{and} \quad g' \geq 0 \quad \text{(cooperative).} \tag{1.5}$$

We establish:

Theorem 1.1. *$u > 0$ and $v > 0$ for $x \in \Omega$.*

In [1], the single equation case

$$-\Delta w = h(w) \quad \text{in } \Omega$$
$$w = 0 \quad \text{on } \partial\Omega,$$

where $h : [0, \infty) \to \mathbb{R}$ is C^1 and $h(0) < 0$ was studied and it was proven that every nonnegative classical solution is strictly positive in Ω. This result proved that the conjecture in [4] that single equation semipositone problem may have nonnegative solution with interior zeros is false for the case when Ω is a ball in $\mathbb{R}^n$ for $n > 1$. For $n = 1$ there exist nonnegative solutions with interior zeros (see [2] and [3]).

Here we extend the result in [1] to cooperative systems. We note that our proofs also carry over to general cooperative systems of the form

$$\begin{aligned} -\Delta u_i &= f_i(u_1, u_2, \ldots, u_m) \quad \text{in } \Omega \text{ for } \quad i = 1, \ldots, m, \\ u_i &= 0 \quad \text{on } \partial\Omega, \end{aligned}$$

where Ω is a ball in $\mathbb{R}^n$ for $n > 1$ and $f_i : \underbrace{[0,\infty) \times [0,\infty) \ldots, [0,\infty)}_{m \text{ times}} \to \mathbb{R}$ are C^1 functions satisfying

$$f_i(0, 0, \ldots, 0) < 0, \qquad i = 1, 2, \ldots, m$$

and

$$\frac{\partial f_i}{\partial u_j} \geq 0, \qquad i \neq j.$$

We note that Theorem 1.1 holds also in the case when only one of $f(0)$ or $g(0)$ is negative. The positivity of nontrivial solutions in the case when both $f(0)$ and $g(0)$ are nonnegative follows by Maximum principle.

Theorem 1.1 is of great importance in the study of cooperative semipositone systems in a ball since positivity implies radial symmetry (see [3] & [5]). The proof of Theorem 1.1 follows by combining Lemma 4.2 in [5] (based on Maximum principle/Reflection arguments for cooperative systems) and analysis of solutions near the boundary.

In section 2, we discuss a version of Lemma 4.2 of [5] and the positivity of the components near the boundary. In section 3, we prove Theorem 1.1. In section 4, we state an open problem for bounded regions and a counterexample for the positivity of solutions in unbounded regions.

2. PRELIMINARY RESULTS

Without loss of generality we assume throughout this paper that Ω is a unit ball centered at the origin. First we recall some notations following [1]. Let

$$T_t := \{x \in \mathbb{R}^n : x.e = t\} \text{ and } \Sigma_t := \{x \in \Omega : x.e > t\},$$

where $e := (1, 0, \ldots, 0)$, and . denotes the usual inner product in $\mathbb{R}^n$. Let $\Sigma'(t)$ be the reflection of $\Sigma(t)$ across T_t. For $x \in \mathbb{R}^n$, define x^t to be the reflection of x across T_t. Here u_i and v_i will denote the partial derivative of u and v respectively with respect to x_i, similarly u_{ij} and v_{ij}. For $x \in \partial\Omega$, $\nu(x) := (\nu_1(x), \nu_2(x), \ldots, \nu_n(x))$ denote the outward unit normal and by the ϵ-neighborhood theorem there exists $\gamma > 0$ such that for every $0 < \epsilon < \gamma$, we can define (see [2])

$$\begin{aligned} D_\epsilon : &= \{x \in \Omega : dist(x, \partial\Omega) < \epsilon\} \\ &= \{x - t\nu(x) : x \in \partial\Omega \text{ and } t \in (0, \epsilon)\}. \end{aligned}$$

Now we state and prove lemmas that will be used to prove Theorem 1.1.

Lemma 2.1 (see also Lemma 4.2 in [5]). *Let (u, v) satisfy* (1.1) – (1.3) *and let $t \in [0, 1)$. Assume $z_1 = u$ and $z_2 = v$. Suppose for each $i = 1, 2$*

$$\frac{\partial z_i}{\partial x_1} \leq 0 \quad \forall x \in \Sigma_t \tag{2.1}$$

$$z_i(x) \leq z_i(x^t) \quad \forall x \in \Sigma_t \tag{2.2}$$

$$\text{and } z_i(x) \not\equiv z_i(x^t) \quad \text{in } \Sigma_t. \tag{2.3}$$

Then

$$z_i(x) < z_i(x^t) \quad \forall x \in \Sigma_t \tag{2.4}$$

and

$$\frac{\partial z_i}{\partial x_1} < 0 \quad \forall x \in T_t \cap \Omega, \quad \forall i = 1, 2. \tag{2.5}$$

Proof: Let $t \in [0, 1)$ and define $h_i(x) := z_i(x^t)$ for $i = 1, 2$ and $x \in \Sigma_t$. Then

$$\begin{aligned} -\Delta h_1(x) &= f(h_2(x)); \quad \Sigma_t \text{ and} \\ -\Delta h_2(x) &= g(h_1(x)); \quad \Sigma_t. \end{aligned}$$

Define $w_i(x) := h_i(x) - z_i(x)$ in Σ_t. Then

$$-\Delta w_1(x) = f(h_2(x)) - f(z_2(x)); \quad \Sigma_t \text{ and}$$

and

$$-\Delta w_2(x) = g(h_1(x)) - g(z_1(x)); \quad \Sigma_t.$$

From the Mean value theorem we get

$$-\Delta w_1(x) = f'(\theta_2(x))(h_2(x) - z_2(x)); \quad \Sigma_t \text{ and}$$

and

$$-\Delta w_2(x) = g'(\theta_1(x))(h_1(x) - z_1(x)); \quad \Sigma_t$$

where $\theta_i(x) \in [z_i(x), h_i(x)]$; $i = 1, 2$ for $x \in \Sigma_t$. Since $f' \geq 0, g' \geq 0$ and $w_i \geq 0$ in Σ_t we get $-\Delta w_i \geq 0$ in Σ_t. Note that $w_i = 0$ on $T_t \cap \Omega$. So by the Maximum principle $w_i > 0$ in Σ_t and $\frac{\partial w_i}{\partial x_1} > 0$ on $T_t \cap \Omega$. But on $T_t \cap \Omega$

$$\frac{\partial w_i}{\partial x_1} = \frac{\partial h_i}{\partial x_1} - \frac{\partial z_i}{\partial x_1} = -2\frac{\partial z_i}{\partial x_1}.$$

Therefore $\frac{\partial z_i}{\partial x_1} < 0$ on $T_t \cap \Omega$ and the proof is complete.

Lemma 2.2. *Let (u, v) be a solution to (1.1) – (1.3) and $z \in \partial\Omega$. If $\nu_1(z) > 0$ and there exists $\epsilon > 0$ such that $u(x) \geq 0$, $v(x) \geq 0$ for $x \in \overline{\Omega}$ with $\|x - z\| < \epsilon$, then there exists $\epsilon_0 \in (0, \epsilon)$ such that $u_1(x) < 0$ and $v_1(x) < 0 \quad \forall x \in \Omega$ with $\|x - z\| < \epsilon_0$.*

Proof: First we consider the case of u. Since $\nu_1(z) > 0$, without loss of generality we can assume that $\nu_1(x) > 0 \quad \forall x \in \partial\Omega$ with $\|x - z\| < \epsilon$. This and the assumption that $u \geq 0$ on $\|x - z\| < \epsilon$ imply that

$$u_1(x) \leq 0 \quad \forall x \in \partial\Omega \quad \text{with} \quad \|x - z\| < \epsilon.$$

If $u_1(z) < 0$ then the existence of $\epsilon_0 = \epsilon_0(u)$ follows from the continuity of u_1.
On the other hand if $u_1(z) = 0$, then $\nabla u(z) = 0$ since $\nabla u(z) = c\nu(z)$ for some c and $c = 0$ as $u_1(z) = 0$ and $\nu_1(z) > 0$. For $k \in \{2, 3, \ldots n\}$ we let $y := (-\nu_k(z), 0, \ldots, \underbrace{\nu_1(z)}_{kth}, 0, \ldots, 0)$. Then y is tangent to $\partial\Omega$ at z. So we can find $\phi : (-1, 1) \to \partial\Omega$ such that $\phi(0) = z$ and $\phi'(0) = y$. Thus $u_1(\phi(0)) = u_1(z) = 0$ and

$u_1(\phi(t)) \leq 0$ in a neighborhood of $t=0$. Therefore $u_1(\phi(t))$ has a local maximum at $t=0$. In particular

$$\begin{aligned} 0 &= \frac{d}{dt}(u_1(\phi(t))|_{t=0} \\ &= \nabla u_1(\phi(0)).\phi'(0) \\ &= \nabla u_1(z).y \\ &= -u_{11}(z)\nu_k(z) + u_{1k}(z)\nu_1(z). \end{aligned} \tag{2.6}$$

On the other hand we have $u(\phi(t)) = 0$. Calculating the second derivative of u with respect to t at $t=0$ we get

$$\begin{aligned} 0 &= \frac{d^2}{dt^2}u(\phi(t))|_{t=0} \\ &= \frac{d}{dt}[\nabla u(\phi(t)).\phi'(t)]|_{t=0} \\ &= u_{11}(z)\nu_k(z)^2 - 2u_{1k}(z)\nu_k(z)\nu_1(z) + u_{kk}(z)\nu_1(z)^2. \end{aligned}$$

By using (2.6) we get

$$u_{kk}(z)\nu_1(z)^2 = u_{11}\nu_k(z)^2. \tag{2.7}$$

Now by (2.7), we have

$$\begin{aligned} 0 < -f(0) = -f(v(z)) &= \Delta u(z) \\ &= \sum_{k=1}^{n} u_{kk}(z) \\ &= \sum_{k=2}^{n} u_{11}(z)\left(\frac{\nu_k(z)}{\nu_1(z)}\right)^2 + u_{11}(z) \\ &= u_{11}(z)\left[1 + \sum_{k=2}^{n}\left(\frac{\nu_k(z)}{\nu_1(z)}\right)^2\right]. \end{aligned}$$

Therefore $u_{11}(z) > 0$. By continuity of u_{11}, there exists $\epsilon_0 = \epsilon_0(u) \in (0,\epsilon)$ such that $u_{11}(x) > 0$ for $\|x - z\| < \epsilon_0(u)$ with $x \in \bar{\Omega}$.

Now since $\nu_1(x) > 0 \; \forall x \in \partial\Omega$ with $\|x - z\| < \epsilon_0(u)$, we can further assume that if $\|x - z\| < \epsilon_0(u)$ with $x \in \Omega$, then there exists $b(x) := x + \delta e$ $(\delta > 0)$ in $\partial\Omega$ with $x + se \in \Omega \; \forall s \in (0,\delta)$. Hence

$$0 > u_1(b(x)) - \int_0^{\delta} u_{11}(x+se)ds = u_1(x).$$

This completes the proof for u.

Again by considering v and following the same steps as for u one can show that there exists $\epsilon_0 = \epsilon_0(v) \in (0,\epsilon)$ such that $v_1(x) < 0 \; \forall x \in \Omega$ with $\|x - z\| < \epsilon_0(v)$. Choose $\epsilon_0 = \min\{\epsilon_0(u), \epsilon_0(v)\}$. This completes the proof of Lemma 2.2.

Lemma 2.3. *If (u,v) is a solution of (1.1)–(1.3) and $u \geq 0$, $v \geq 0$ on a neighborhood of $\partial\Omega$, then there exists $\mu > 0$ such that $u > 0$ and $v > 0$ on D_μ.*

Proof: Let $z \in \partial\Omega$. Since Δ is invariant under rotations , without loss of generality we can assume that

$$z = \nu(z) = e = (1, 0, \ldots, 0).$$

By Lemma 2.2 there exists $\epsilon_0 > 0$ such that $u(x) > 0$ and $v(x) > 0$ if $x \in \Omega$ and $\|x - z\| < \epsilon_0(z)$. Since $\partial\Omega$ is compact, there exists finite number of points $x_1, \ldots, x_m \in \partial\Omega$ such that

$$W := B(x_1, \epsilon_0(x_1)) \cup B(x_2, \epsilon_0(x_2)) \cup \ldots B(x_m, \epsilon_0(x_m)) \subset \partial\Omega.$$

As W is open in $\mathbb{R}^n$ and $\partial\Omega$ is closed, there exists $\mu > 0$ such that

$$\{x \in \mathbb{R}^n : \text{dist}(x, \partial\Omega) < \mu\} \subset W.$$

In particular, if $x \in \Omega$ and $\text{dist}(x, \partial\Omega) < \mu$ then $u(x) > 0$ and $v(x) > 0$.

3. PROOF OF THEOREM 1.1

We now establish the positivity of any nonnegative solution component wise. Suppose there exists $y \in \Omega$ such that at least one of the component of (u, v) becomes zero at y. Then $\|y\| < 1 - \mu$ due to Lemma 2.3. Without loss of generality we can assume y to have maximum norm. Also since Δ is invariant under rotations, we can assume that $y := (\delta, 0, \ldots, 0)$ with $\delta \in [0, 1 - \mu]$. Let α be defined by

$$\alpha := \inf\{s \in [0,1) : (2.1) - (2.3) \text{ hold on } \quad \Sigma_t \, \forall t \in (s, 1]\}. \tag{3.1}$$

By Lemmas 2.2-2.3 there exists $k \geq 2$ such that

$$\alpha \leq 1 - \frac{\epsilon_0}{k} \tag{3.2}$$

and since $\delta \geq 0$, we see that (by using Lemma 2.3)

$$\alpha \geq \frac{1}{4}. \tag{3.3}$$

Clearly (2.1) and (2.2) hold on Σ_α by continuity of u, u_1 and v, v_1. If (2.3) does not hold on Σ_α, then by Lemma 2.1 either $u(x) = u(x^\alpha)$ or $v(x) = v(x^\alpha) \, \forall x \in \Sigma_\alpha$. Without loss of generality assume $u(x) = u(x^\alpha) \, \forall x \in \Sigma_\alpha$. In particular, by the continuity of u we have

$$u(x^\alpha) = 0 \quad \text{if} \quad \|x\| = 1. \tag{3.4}$$

Thus setting $\sigma := \frac{\mu}{m}$ where μ is as in Lemma 2.3 and $m > 4\alpha$ is large enough so that $\alpha + \sigma < 1$, we obtain

$$\begin{aligned} 0 &= u(\alpha + \sigma, \sqrt{1 - (\alpha + \sigma)^2}, 0, \ldots, 0) \\ &= u(\alpha - \sigma, \sqrt{1 - (\alpha + \sigma)^2}, 0, \ldots, 0). \end{aligned} \tag{3.5}$$

But

$$\begin{aligned} \|\alpha - \sigma, \sqrt{1 - (\alpha + \sigma)^2}, 0, \ldots, 0\|^2 &= (\alpha - \sigma)^2 + 1 - (\alpha + \sigma)^2 \\ &= 1 - 4\alpha\sigma \\ &> 1 - m\sigma \\ &= 1 - \mu, \end{aligned}$$

hence (3.5) contradicts Lemma 2.3. Thus (2.3) hold on Σ_α and hence by Lemma 2.1, (2.4) and (2.5) also hold for $t = \alpha$.

However, using this we will now show that there exists $0 < \beta < \alpha$ such that (2.1) – (2.3) holds on $\Sigma_t \, \forall t \in (\beta, 1]$ contradicting the fact that α is the infimum. To do so we establish the following two claims:

Claim 1: There exists $\eta > 0$ such that if $\|x\| < 1$ and $x.e > \alpha - \eta$ then $u_1(x) < 0$

and $v_1(x) < 0$.
Proof of Claim 1: If not, there exists a sequence $\{x_n\}$ with $\|x_n\| < 1$, $\{x_n.e\}$ converging to α, and $u_1(x_n) \geq 0$ or $v_1(x_n) \geq 0$. Thus, without loss of generality we can assume that $u_1(x_n) \geq 0$ and that $\{x_n\}$ converges to $z \in T_\alpha \cap \bar{\Omega}$. If $z \in T_\alpha \cap \Omega$ then by (2.5) we have

$$0 > u_1(z) = \lim_{n\to\infty} u_1(x_n) \geq 0, \tag{3.6}$$

which is a contradiction. On the other hand if $z \in T_\alpha \cap \partial\Omega$, by Lemma 2.2 for n large $u_1(x_n) < 0$ which contradicts the definition of $\{x_n\}$.
Claim 2: There exists $\gamma \in (0, \eta)$ such that if $t \in (\alpha - \gamma, \alpha)$, $x \in \Sigma_t$ then $u(x^t) \geq u(x)$ and $v(x^t) \geq v(x)$.
Proof of Claim 2: We prove it again by contradiction. Suppose Claim 2 is not true. Without loss of generality we can assume that there is a sequence (t_n, x_n) with $t_n \to \alpha$, $x_n \in \Sigma_{t_n}$ and $x_n \to z$(say) such that $u(x_n^{t_n}) < u(x_n)$. Since $t_n \to \alpha$ we see that $z \in \bar{\Sigma}_\alpha$. If $z \in \Sigma_\alpha$, we have

$$u(z^\alpha) = \lim_{n\to\infty} u(x_n^{t_n}) \leq \lim_{n\to\infty} u(x_n) = u(z) \tag{3.7}$$

contradicting that (2.4) holds on Σ_α.

Thus either $\|z\| = 1$ or $z \in T_\alpha \cap \Omega$. If $z \in T_\alpha \cap \Omega$ then $x_n^{t_n} \to z$. By the Mean value theorem there exists a sequence $\{y_n\}$ with y_n in the segment joining $x_n^{t_n}$ with x_n such that $u_1(y_n) \geq 0$. Since for n large $y_n.e > \alpha - \eta$, by Claim 1 we have a contradiction. Thus $\|z\| = 1$ and $z.e > \alpha$. Now, for $x \in \bar{\Sigma}_\alpha$ we define

$$w_u(x) = u(x^\alpha) - u(x).$$

Then $w_u > 0$ on Σ_α since (2.4) holds on Σ_α. But by continuity of u we have

$$u(z^t) = \lim_{n\to\infty} u(x_n^{t_n}) \leq \lim_{n\to\infty} u(x_n) = u(z) = 0, \tag{3.8}$$

and hence $w_u(z) = 0$. Now

$$\begin{aligned} -\Delta w_u(x) &= f(v(x^\alpha)) - f(v(x));\ x \in \Sigma_\alpha \\ &= f'(\xi(x))(v(x^\alpha) - v(x)) \\ &\geq 0 \quad \text{since (2.4) holds on } \Sigma_\alpha \text{ and } f' \geq 0. \end{aligned}$$

Thus by Hopf's maximum principle we have $\frac{\partial w_u}{\partial x_1}(z) < 0$. Since $u \geq 0$, (3.8) implies $u_1(z^\alpha) = 0$. Further $u_1(z) \leq 0$ since $u \geq 0$. Hence

$$0 > w_1(z) = -u_1(z^\alpha) - u_1(z) \geq 0. \tag{3.9}$$

This contradiction proves Claim 2.
We have shown by Claim 2 that for some $\beta < \alpha$, (2.2) holds for all $t \in [\beta, \alpha)$. By taking $\beta \in (\alpha - \eta, \alpha)$, from Claim 1 we see that if $t \in [\beta, \alpha)$ then (2.1) as well as (2.3) hold. Hence this contradicts the definition of α. Thus neither u or v becomes zero at any $y \in \Omega$ and hence Theorem 1.1 is proven.

4. OPEN PROBLEMS

Positivity of solutions holds even if we consider Ω as the region between two balls or the union of two balls. However, the question of positivity of nonnegative solutions in general bounded region remains open including in the single equation case.

On the other hand, in unbounded regions there are nonnegative solutions with interior zeros. Indeed, consider the two point boundary value problem

$$-u'' = \lambda f(u) \quad \text{in } 0 < x < 1 \tag{4.1}$$

$$u(0) = 0 = u(1), \tag{4.2}$$

where $f(0) < 0$, $f'(u) > 0$ and $\lim_{u \to \infty} f(u) = \infty$. Then from [2] it follows that there exists an increasing sequence of positive numbers λ_n such that (4.1) – (4.2) has a nonnegative solution $u_n(x)$ with n interior zeros x_n in $(0,1)$. Now consider

$$-\Delta w = \lambda f(w) \text{ in } \quad \Omega := \{(x,y) : 0 < x < 1,\, y \in \mathbb{R}\} \tag{4.3}$$

$$w = 0 \quad \text{on } \partial\Omega := \{(z,y) : z \in \{0,1\}, y \in \mathbb{R}\}. \tag{4.4}$$

Clearly for $\lambda = \lambda_n$, $w_n(x,y) = u_n(x)$ is a nonnegative solution of (4.3)-(4.4) which vanishes on

$$\tilde{\Omega} := \{(x_n, y) : y \in \mathbb{R}, n = 1, 2, \ldots, \} \subset \Omega.$$

REFERENCES

1. Castro, Alfonso & Shivaji, R., *Nonnegative solutions to a Semilinear Dirichlet problem in a ball are positive and radially symmetric*, Comm. Partial Diff. Equa., **14**, (1989), 1091-1100.
2. Castro, Alfonso & Shivaji, R., *Nonnegative solutions for a class of nonpositone problems*, Proc. Roy. Soc. Edin., **108(A)**, (1988), 291-302.
3. Djairo, G. de Figueiredo, *Monotonicity and symmetry of solutions of elliptic systems in general domains*, NoDEA, **1**, (1994), 119-123.
4. Gidas, B., Ni, Wei-Ming & Nirenberg, L., *Symmetry and related properties via the Maximum Principle*, Comm. Math. Phys., **68**, (1979), 209-243.
5. Troy, William C., *Symmetry properties in systems of semilinear elliptic equations*, Journal of Differential Equations, **42**, (1981), 400-413.

Proceedings of Dynamic Systems and Applications 3 (2001) 121-128

QUENCHING AND COMPLETE BLOW-UP OF SOLUTIONS FOR DEGENERATE SEMILINEAR PARABOLIC EQUATIONS

Dedicated to the seventy-fifth birthday of Professor V. Lakshmikantham

C. Y. Chan and Jian Yang
Department of Mathematics, University of Louisiana at Lafayette
Lafayette, Louisiana 70504-1010, U.S.A.

ABSTRACT: Let q and a be constants such that $q \geq 0$ and $a > 0$, $T \leq \infty$, $D = (0, a)$, $\Omega = D \times (0, T)$, and $\bar{D}$ and $\bar{\Omega}$ denote the closures of D and Ω respectively. This article studies the degenerate semilinear parabolic initial-boundary value problem,

$$x^q u_t - u_{xx} = f(u) \text{ in } \Omega,$$

$$u(x, 0) = u_0(x) \text{ on } \bar{D},$$

$$u(0, t) = u(a, t) = 0, \ 0 < t < T.$$

If $\lim_{u \to c^-} f(u) = \infty$ for some positive constant c, then it is shown that it is impossible for u to quench in infinite time; furthermore for beyond quenching, it is shown that the problem has a weak solution

$$u \in C^{2,1}(\{u < c\} \cap \Omega) \cap C^{1,0}(\Omega) \cap C(\bar{\Omega}),\ u \leq c \text{ in } \Omega,$$

and that every element of the ω-limit set obtained from the minimal weak solution (among solutions obtained via regularization) is a weak steady-state solution. If $\lim_{u \to \infty} f(u) = \infty$, then criteria for u to blow up completely are given.
AMS (MOS) subject classification. 35K20, 35K60, 35K65, 35K57, 35K55

1. INTRODUCTION

Let a be a positive constant, $T \leq \infty$, $D = (0, a)$, $\Omega_T = D \times (0, T)$, and $\bar{D}$ and $\bar{\Omega}_T$ be their respective closures. As motivations of the problem we are going to study, let us consider some specific examples:

I. the semilinear parabolic problem,

$$u_t - u_{xx} = f(u) \text{ in } \Omega_T,$$

$$u(x, 0) = 0 \text{ on } \bar{D},$$

$$u(0, t) = u(a, t) = 0 \text{ for } 0 < t < T.$$

II. the degenerate semilinear parabolic problem,

$$xu_t - u_{xx} = f(u) \text{ in } \Omega_T,$$
$$u(x,0) = 0 \text{ on } \bar{D},$$
$$u(0,t) = u(a,t) = 0 \text{ for } 0 < t < T.$$

This problem represents the model studied by Ockendon [16] for the flow in a channel of a fluid whose viscosity depends on the temperature u with x and t denoting, respectively, the coordinates perpendicular and parallel to the channel walls.

Let L denote the degenerate parabolic operator,

$$x^q\frac{\partial}{\partial t} - \frac{\partial^2}{\partial x^2},$$

where q is any real number. As a generalization of the previous two problems, we consider the following degenerate semilinear parabolic initial-boundary value problem,

$$\left.\begin{aligned} Lu &= f(u) \text{ in } \Omega_T, \\ u(x,0) &= 0 \text{ on } \bar{D}, \\ u(0,t) = u(a,t) &= 0 \text{ for } 0 < t < T. \end{aligned}\right\} \tag{1}$$

2. INFINITE-TIME QUENCHING

Let $f \in C^1([0,c))$ for some positive constant c such that $\lim_{u\to c^-} f(u) = \infty$, $f(0) > 0$, and $f' > 0$.

A solution u is said to quench at time T if $\max_{x\in\bar{D}} u(x,t) \to c$ as $t \to T$. If $T < \infty$, then u quenches in a finite time; if $T = \infty$, then u quenches in infinite time. Chan and Kong [3] showed that the problem (1) has a unique classical solution before quenching occurs, and there exists a unique critical length a^* such that for $a < a^*$, u exists for all $t > 0$, and for $a > a^*$, u quenches in a finite time.

Does quenching occur in infinite time when $a = a^*$? So far in the literature, only the special case $q = 0$ has been studied. For the special case $q = 0$ and $f = (1-u)^{-\beta}$, where β is a positive constant, Levine [15] proved that quenching in infinite time does not occur by finding the explicit form of the minimum steady-state solution. Chan and Liu [6] showed that it is impossible for quenching to occur in infinite time in the case $q = 0$ and $\int_0^c f(s)ds < \infty$.

In this section, we consider L with q being a nonnegative real constant, and a function f without the restriction $\int_0^c f(s)ds < \infty$. We show that it is impossible for quenching to occur in infinite time. In fact, we [9] proved the following result.

Theorem 1. If $u(x,t)$ exists for all $t > 0$, then there exists a positive constant m ($< c$) such that $\sup_{(x,t)\in\bar{D}\times[0,\infty)} u(x,t) \le m$.

Physically, the above theorem for $q = 1$ says that Ockendon's model for the channel flow is valid even if the length of the channel is infinity long.

From this theorem, we can conclude as our special cases the one-dimensional results of Levine [15] and Chan and Liu [6] respectively.

3. BEYOND QUENCHING

Let q be a nonzero real constant greater than -1. We exclude the case $q = 0$ since it corresponds to the case where no degeneracy occurs. Let us consider the following problem,

$$Lu = f(u) \text{ in } \Omega_T,$$

$$u(x,0) = 0 \text{ on } \bar{D},$$

$$u(0,t) = u(a,t) = 0 \text{ for } 0 < t < T,$$

where $f \in C^2([0,c))$ for some positive constant c such that $f(0) > 0$, $f' > 0$, $f'' \geq 0$, and $lim_{u \to c^-} f(u) = \infty$. For $a > a^*$, u quenches in a finite time. What happens afterwards? To answer our question, let

$$\chi_S = \begin{cases} 1 \text{ if } u \in S, \\ 0 \text{ if } u \notin S, \end{cases}$$

and $\Omega = \Omega_\infty$. We study the following singular degenerate semilinear problem,

$$\left.\begin{aligned} Lu &= f(u)\chi_{\{\,u<c\,\}} \text{ in } \Omega, \\ u(x,0) &= u_0(x) \text{ on } \bar{D}, \\ u(0,t) = u(a,t) &= 0 \text{ for } 0 < t < \infty, \end{aligned}\right\} \tag{2}$$

where for some positive constants K_1 (> 1), K_2, K_3, and some constant $\gamma \in (0,2)$ such that

$$f'(u)\left(\frac{c-u}{f(u)}\right)^2 \leq K_1, \tag{3}$$

$$\int_u^c f(s)ds \leq \min\{K_2(c-u)f(u),\ K_3(c-u)^\gamma\}, \tag{4}$$

$$u_0 \in C^{2+\alpha}\left(\bar{D}\right) \text{ for some } \alpha \in (0,1),$$

$$0 \leq u_0 < c,\ u_0(0) = u_0'(0) = 0 = u_0(a).$$

We note that $u_0'' + f(u_0) \geq 0$ is not assumed. The proofs of the results in this section were given by Chan and Yang [10].

A function u is a weak solution of the problem (2) if and only if
(a) $u \in W_2^{1,0}(D \times [0, t_0]) \cap C(\bar{D} \times [0, t_0])$ for each $t_0 > 0$.
(b) for any $g(x,t) \in C^{2,1}(\bar{\Omega})$ such that g has a compact support with respect to t and $g(0,t) = g(a,t) = 0$,

$$\int_0^\infty \int_0^a uL^* g dx dt + \int_0^\infty \int_0^a f(u)\chi_{\{u<c\}} g dx dt = 0,$$

where $L^* = x^q \partial/\partial t + \partial^2/\partial x^2$ is the adjoint operator of L.

For any constant $\varepsilon \in (0,1)$, let

$$f_\varepsilon(u) = \frac{c-u}{\varepsilon f(u) + (c-u)} f(u).$$

We note that

$$\lim_{\varepsilon \to 0} f_\varepsilon(u) = f(u)\chi_{\{u< c\}}, \; f_\varepsilon(u) \le f(u).$$

Also, we can construct a nondecreasing function $\rho(x) \in C^3$ such that $\rho(x) = 0$ if $x \le 0$, and $\rho(x) = 1$ if $x \ge 1$. Let δ ($< a/2$) be a positive number,

$$\rho_\delta(x) = \begin{cases} 0, & x \le \delta, \\ \rho(\frac{x}{\delta} - 1), & \delta < x < 2\delta, \\ 1, & x \ge 2\delta, \end{cases}$$

$$u_{0\delta}(x) = \rho_\delta(x) u_0(x).$$

We show that the regularized problem,

$$L u_\delta^\varepsilon = f_\varepsilon(u_\delta^\varepsilon) \text{ in } (\delta, a) \times (0, \infty),$$
$$u_\delta^\varepsilon(x, 0) = u_{0\delta}(x) \text{ on } [\delta, a],$$
$$u_\delta^\varepsilon(\delta, t) = u_\delta^\varepsilon(a, t) = 0 \text{ for } 0 < t < \infty,$$

has a unique solution

$$u_\delta^\varepsilon \in C^{2+\alpha, 1+\frac{\alpha}{2}}([\delta, a] \times [0, \infty)).$$

Then, we show that the degenerate regularized problem,

$$L u^\varepsilon = f_\varepsilon(u^\varepsilon) \text{ in } \Omega,$$
$$u^\varepsilon(x, 0) = u_0(x) \text{ on } \bar{D},$$
$$u^\varepsilon(0, t) = u^\varepsilon(a, t) = 0 \text{ for } 0 < t < \infty,$$

has a unique nonnegative solution,

$$u^\varepsilon \in C(\bar{\Omega}) \cap C^{2,1}\left((0,a] \times [0,\infty)\right).$$

Theorem 2. *The problem (2) has a weak solution $u = \lim_{\varepsilon\to 0} u^\varepsilon$ having the property*

$$u \in C^{2,1}(\{u < c\} \cap \Omega) \cap C^{1,0}(\Omega) \cap C(\bar{\Omega}),$$

$$u \le c \text{ in } \Omega.$$

A function v is a weak steady-state solution of the problem (2) if and only if
(i) $v(x) \in W_2^1(D) \cap C(\bar{D})$.
(ii) for any $\psi(x) \in C^2\left(\bar{D}\right)$ vanishing at $x = 0$ and $x = a$,

$$\int_0^a v\psi'' dx + \int_0^a f(v)\chi_{\{v<c\}}\psi dx = 0.$$

The ω-limit set

$$\omega(u_0) \equiv \{v \in W_2^1(D) \cap C\left(\bar{D}\right) : \text{there is a sequence } \{t_n\} \text{ with } t_n \to \infty$$
$$\text{such that } u(x, t_n; u_0) \to v \text{ uniformly as } n \to \infty\}.$$

Theorem 3. *The ω-limit set is nonempty. Every element of the ω-limit set is a weak steady-state solution.*

For the multidimensional version of the problem (2) with $q = 0$, and

$$f(u) = (1-u)^{-\sigma}, \tag{5}$$

where σ is a constant in $(0,1)$, existence of a weak solution μ with continuous μ_x was studied by Phillips [17] while Fila, Levine and Vazquez [12] showed that every element of the ω-limit set obtained from the minimal weak solution (among solutions via regularization) is a weak steady-state solution. We note that (5) satisfies (3) and (4). Thus, Theorem 3 extends the one-dimensional result of Fila, Levine and Vazquez.

By using a completely different approach, Chan and Kong [4] showed that the problem (2) with $u_0 \equiv 0$ has a weak solution such that

$$u \in C^{2,1}(\{u < c\} \cap \Omega) \cap C^{1,0}(\Omega) \cap C(\bar{\Omega}),$$

$$u \le c \text{ in } \Omega,$$

$$u \text{ is nondecreasing with respect to } t \text{ in } \{u < c\} \cap \Omega,$$

and that as t tends to infinity, all weak solutions (not necessarily obtained via regularization) having the above properties tend to a unique weak steady-state solution, which is independent of q.

Corollary 4. *For $a > a^*$, the ω-limit set (obtained from the minimal weak solution among solutions via regularization) consists of the unique weak steady-state solution U of the problem,*

$$-U'' = f(U) \text{ for } 0 < x < b,\ U(0) = 0,\ U(b) = c,$$

$$U(x) = c \text{ for } b < x < B,$$

$$-U'' = f(U) \text{ for } B < x < a,\ U(B) = c,\ U(a) = 0,$$

where

$$b = \frac{1}{\sqrt{2}} \int_0^c \left(\int_y^c f(s)ds \right)^{-\frac{1}{2}} dy,\ B = a - b.$$

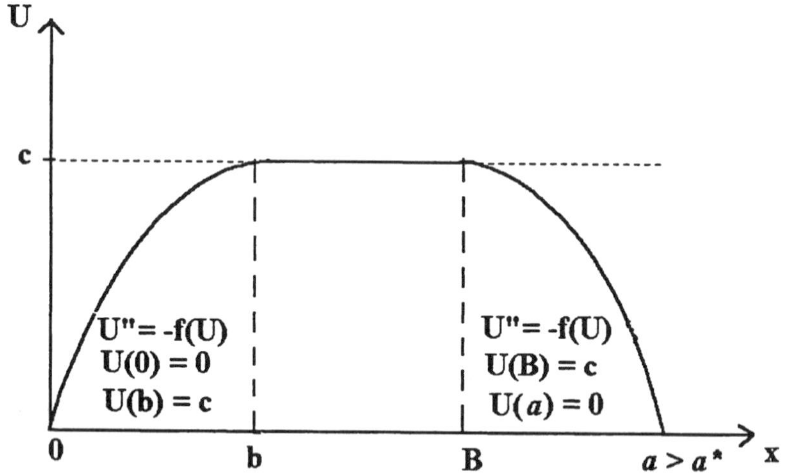

The values b and B are determined by explicit formulae.

Figure 1

4. COMPLETE BLOW-UP

Let q and x_0 be constants with $q \geq 0$ and $0 < x_0 < a$. We consider the following degenerate semilinear parabolic initial-boundary value problem,

$$\left. \begin{aligned} Lu &= f\left(u\left(x_0, t\right)\right) \text{ in } \Omega_T, \\ u\left(x, 0\right) &= u_0\left(x\right) \geq 0 \text{ on } \bar{D}, \\ u\left(0, t\right) &= u\left(a, t\right) = 0 \text{ for } 0 < t < T, \end{aligned} \right\} \quad (6)$$

where $f \in C^2\left([0, \infty)\right)$, $f\left(0\right) \geq 0$, $f'\left(s\right) > 0$ and $f''\left(s\right) \geq 0$ for $s > 0$, $\int_{z_0}^{\infty} f^{-1}\left(s\right) ds < \infty$ for some $z_0 > 0$, and $u_0\left(x\right) \in C^{2+\alpha}\left(\bar{D}\right)$ for some constant $\alpha \in (0, 1)$. This describes the situation where the nonlinear reaction takes place only at a single point x_0.

A solution u is said to blow up at $x = \widetilde{x}$ if there exists a sequence $\{(x_n, t_n)\}$ such that $(x_n, t_n) \to (\widetilde{x}, T)$ and $\lim_{n\to\infty} u(x_n, t_n) \to \infty$. Furthermore, if u blows up at every point $x \in \bar{D}$, then the complete blow-up occurs.

Chan and Liu [7] and Floater [13] investigated the blow-up of u in the case $f(u(x_0, t))$ replaced by u^p. In the case $q = 0$, and $f(u(x_0, t))$ replaced by $f(u(x, t))$, the complete blow-up was studied by Baras and Cohen [1], and Lacey and Tzanetis [14]. Chadam, Peirce and Yin [2] studied the complete blow-up of the solution of the problem (6) with $q = 0$.

The proofs of the following results were given by Chan and Yang [8]. Let $\delta < \min\{x_0, a/2\}$. The problem,

$$\begin{gathered} Lu_\delta = f(u_\delta(x_0, t)) \text{ in } (\delta, a) \times (0, t_0), \\ u_\delta(x, 0) = u_{0\delta}(x) \geq 0 \text{ on } [\delta, a], \\ u_\delta(0, t) = u_\delta(a, t) = 0 \text{ for } 0 < t < t_0, \end{gathered}$$

is shown to have a unique nonnegative solution,

$$u_\delta \in C^{2+\alpha, 1+\alpha/2}([\delta, a] \times [0, t_0]).$$

Theorem 5. *There exists some t_0 $(< T)$ such that the problem (6) has a unique nonnegative solution,*

$$u \in C(\bar{\Omega}_{t_0}) \cap C^{2,1}((0, a] \times [0, t_0]).$$

Let T be the supremum over t_0 for which there is a unique solution,

$$u \in C(\bar{\Omega}_{t_0}) \cap C^{2,1}((0, a] \times [0, t_0]).$$

Then, the problem (6) has a unique nonnegative solution,

$$u \in C(\bar{D} \times [0, T)) \cap C^{2,1}((0, a] \times [0, T)).$$

If $T < \infty$, then $u(x_0, t)$ is unbounded in $(0, T)$.

Theorem 6. *If $u_0(x)$ is sufficiently large in a neighborhood of x_0, then the solution of the problem (6) blows up in a finite time.*

Theorem 7. *If the solution of the problem (6) blows up in a finite time T, then the blow-up set is $\bar{D}$.*

We remark that the impulsive effects on quenching for problems involving L were studied by Chan and Kong [5], and those on the blow-up of solutions were investigated by Chan and Yuen [11]. Quenching rates were studied by Yuen [18].

REFERENCES

1. Baras, P., & Cohen, L., Complete blow-up after $T_{\max}$ for the solution of a semilinear heat equation. J. Funct. Anal. 71 (1987). 142-174.

2. Chadam, J. M., Peirce, A., & Yin, H. M., The blowup property of solutions to some diffusion equations with localized nonlinear reactions. J. Math. Anal. Appl. 169 (1992). 313-328.

3. Chan, C. Y., & Kong, P. C., Quenching for degenerate semilinear parabolic equations. Appl. Anal. 54 (1994). 17-25.

4. Chan, C. Y., & Kong, P. C., Solution profiles beyond quenching for degenerate reaction-diffusion problems. Nonlinear Anal. 24 (1995). 1755-1763.

5. Chan, C. Y., & Kong, P. C., Impulsive quenching for degenerate parabolic equations. J. Math. Anal. Appl. 202 (1996). 450-464.

6. Chan, C. Y., & Liu, H. T., Quenching in infinite time on the N-dimensional ball. Dynam. Contin. Discrete Impuls. Systems 2 (1996). 303-315.

7. Chan, C. Y., & Liu, H. T., Global existence of solutions for degenerate semilinear parabolic problems. Nonlinear Anal. 34 (1998). 617-628.

8. Chan, C. Y., & Yang, J., Complete blow-up for degenerate semilinear parabolic equations. J. Comput. Appl. Math., in press.

9. Chan, C. Y., & Yang, J., No quenching in infinite time for degenerate singular semilinear parabolic equations. Appl. Math. Comput., to appear.

10. Chan, C. Y., & Yang, J., Beyond quenching for degenerate singular semilinear parabolic equations. Appl. Math. Comput., to appear.

11. Chan, C. Y., & Yuen, S. I., Impulsive effects on global existence of solutions for degenerate semilinear parabolic equations. Appl. Math. Comput. 90 (1998). 97-116.

12. Fila, M., Levine, H. A., & Vazquez, J. L., Stabilization of solutions of weakly singular quenching problems. Proc. Amer. Math. Soc. 119 (1993). 555-559.

13. Floater, M. S., Blow-up at the boundary for degenerate semilinear parabolic equations. Arch. Rat. Mech. Anal. 114 (1991). 57-77.

14. Lacey, A. A., & Tzanetis, D., Complete blow-up for a semilinear diffusion equation with a sufficiently large initial condition. IMA J. Appl. Math. 41 (1988). 207-215.

15. Levine, H. A., Quenching, nonquenching and beyond quenching for solutions of some parabolic equations. Ann. Mat. Pura Appl. (4) 155 (1989). 243-260.

16. Ockendon, H., Channel flow with temperature-dependent viscosity and internal viscous dissipation. J. Fluid Mech. 93 (1979). 737-746.

17. Phillips, D., Existence of solutions of quenching problems. Appl. Anal. 24 (1987). 253-264.

18. Yuen, S. I., Quenching rates for degenerate semilinear parabolic equations. Dynam. Systems Appl. 6 (1997). 139-151.

Proceedings of Dynamic Systems and Applications 3 (2001) 129-134

COMPETITION BETWEEN HEAT INPUT AND OUTPUT ON QUENCHING

Dedicated to the seventy-fifth birthday of Professor V. Lakshmikantham

C. Y. Chan[1] and S. I. Yuen[2]
[1]Department of Mathematics, University of Louisiana at Lafayette
Lafayette, Louisiana 70504-1010, U.S.A.

[2]Department of Mathematics, Lakeland Community College
Kirtland, Ohio 44094-5198, U.S.A.

ABSTRACT: Let p and q be positive constants, and $T \leq \infty$. This paper studies the following problem:

$$u_t - u_{xx} = 0,\ 0 < x < a,\ 0 < t < T,$$

$$u(x,0) = u_0(x),\ 0 \leq x \leq a,$$

$$u_x(0,t) = (1 - u(0,t))^{-p},\ u_x(a,t) = (1 - u(a,t))^{-q},\ 0 < t < T,$$

where $u_0(x)$ is a given function such that $0 \leq u_0(x) < 1$, and satisfies the compatibility conditions at $x = 0$ and $x = a$ respectively. It is shown that u attains its value 1 in a finite time only at a. When this occurs, u_t blows up. Criteria for whether u attains the value 1 in a finite time are also given.

AMS (MOS) subject classification. 35K60, 35K20

1. INTRODUCTION

Let a be a positive constant, $D = (0, a)$, $\bar{D}$ be the closure of D, $T \leq \infty$, $\Omega = D \times (0, T)$,

$$\partial\Omega = \left(\bar{D} \times \{0\}\right) \cup (\{0, a\} \times (0, T)),$$

and $Hu = u_t - u_{xx}$. The concept of quenching was introduced in 1975 by Kawarada [7] through the study of a polarization phenomenon in ionic conductors. Let us consider the singular first initial-boundary value problem:

$$Hu = f(u) \text{ in } \Omega,\ u = 0 \text{ on } \partial\Omega, \tag{1}$$

where $f \in C^2[0, c)$ for some positive constant c such that $f(0) > 0$, $f' > 0$, $f'' \geq 0$, and $\lim_{u \to c^-} f(u) = \infty$. The solution u is said to quench if there exists a finite time T such that

$$\sup\{|u_t(x,t)| : 0 \leq x \leq a\} \to \infty \text{ as } t \to T^-. \tag{2}$$

Here, T is called the quenching time. Since u_t can be shown to be positive, a necessary condition for (2) to hold is

$$\max\{u(x,t) : 0 \leq x \leq a\} \to c^- \text{ as } t \to T^-. \tag{3}$$

The spatial point where u reaches c is called a quenching point.

As their special cases, Chan and Kong [2, 3] showed that (3) implies (2) under the conditions $\int_0^c f(u)du < \infty$ and $\int_0^c f(u)du = \infty$ respectively. This can be seen intuitively as follows: The profile of u at each time t_i is symmetric about the line $x = a/2$. A sketch of the profile of u at each time t_i is given in the following figure:

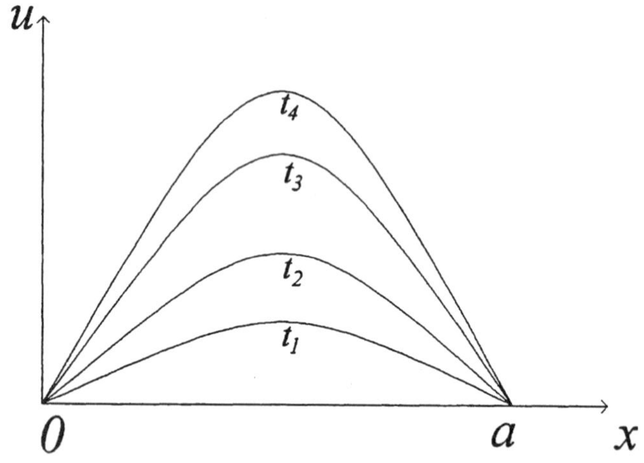

Figure 1

When (3) holds, the right-hand side of (1) tends to ∞. Thus, either u_{xx} tends to $-\infty$ or u_t tends to ∞. Since $u < c$ (which is finite), u_{xx} cannot tend to $-\infty$, and hence u_t tends to ∞.

Since (3) implies (2) under certain conditions, the necessary condition (3) is also used in studying quenching.

Another direction in quenching is the study of the critical length a^*, which is the length such that the solution exists globally for $a < a^*$, and quenching (according to the necessary condition) occurs for $a > a^*$. Existence of a unique critical length and its computational method were studied by Chan and Chen [1], and Chan and Kwong [4]. The critical length is determined by the corresponding steady-state problem. Intuitively, we can see this as follows: when the length a is small, the boundary conditions $u = 0$ prevents u from reaching the value c. As a increases, the boundary conditions have lesser effect so that u reaches the value c at $x = a/2$. Thus, a^* exists. Since for $a < a^*$, u exists for all $t > 0$, it follows that a^* is determined by the steady-state problem.

Quenching problems described by the homogeneous heat equations subject to nonlinear boundary conditions were studied by Levine [8] and Cobb [6] for the one-dimensional case, and by Levine and Lieberman [9] for the multidimensional case. Here, we would like to study a quenching problem due to a nonlinear input and a nonlinear output of heat through the boundaries. Let p and q be positive constants. We consider a one-dimensional heat conduction rod of length a subject to a nonnegative initial temperature $u_0(x)$; at $x = 0$, heat is taken away at a rate $(1 - u(0,t))^{-p}$; at $x = a$, heat goes in at a rate $(1 - u(a,t))^{-q}$. Mathematically, the temperature u of the rod is given by the following problem:

$$\left.\begin{aligned} Hu &= 0 \text{ in } \Omega, \\ u(x,0) &= u_0(x) \text{ on } \bar{D}, \\ u_x(0,t) &= (1 - u(0,t))^{-p},\ 0 < t < T, \\ u_x(a,t) &= (1 - u(a,t))^{-q},\ 0 < t < T, \end{aligned}\right\} \tag{4}$$

where $u_0(x)$ is a given function such that $0 \le u_0 < 1$, and satisfies the compatibility conditions at $x = 0$ and $x = a$ respectively. The problem is said to quench if there exists a finite time T such that

$$\max\{u(x,t) : 0 \le x \le a\} \to 1^- \text{ as } t \to T^-.$$

We study the interaction among the diffusion, the initial temperature, and the rates of heat input and output.

2. SINGLE QUENCHING POINT AND BLOW-UP OF u_t

The proofs of the results in this article were given by Chan and Yuen [5]. Let $u(x,t;u_0)$ and $v(x,t;v_0)$ be solutions of the problem (4) with initial data given respectively by $u_0(x)$ and $v_0(x)$. Our main results depend on the following comparison result, whose proof is nontrivial.

THEOREM 1. *If* $u_0 \le v_0 < 1$, *then*

$$u(x,t;u_0) \le v(x,t;v_0) \text{ on } \bar{\Omega}.$$

A function z is a lower solution of the problem (1) if and only if

$$Hz \le 0 \text{ in } \Omega,\ z(x,0) \le u_0(x) \text{ on } \bar{D},$$

$$z_x(0,t) \ge (1 - z(0,t))^{-p} \text{ for } t > 0,$$

$$z_x(a,t) \le (1 - z(a,t))^{-q} \text{ for } t > 0.$$

It is an upper solution when the inequalities are reversed.

For any nonnegative initial function u_0 to be a lower solution, $u_0'(0) \geq 1$. Since $u_0''(x) \geq 0$, we have $u_0'(x) \geq 1$. Therefore, $u_0(x) \geq x$. From this and the fact that $u \leq 1$, we have $a \leq 1$.

LEMMA 2. *If u_0 is a lower solution, then $u_t > 0$ in Ω.*

THEOREM 3. *If u_0 is a lower solution, then the problem (4) has a unique quenching point a.*

THEOREM 4. *If $\lim_{t\to T} u(a,t) = 1$ for some finite time T, then u_t blows up.*

3. NONQUENCHING AND QUENCHING

Let us consider the positive steady states of the problem (4):

$$\left.\begin{aligned} U'' &= 0, \\ U'(0) &= (1 - U(0))^{-p}, \\ U'(a) &= (1 - U(a))^{-q}. \end{aligned}\right\} \tag{5}$$

We have $U = l + mx$, where

$$m = (1 - l)^{-p}, \; m = (1 - l - ma)^{-q}.$$

From these,

$$U = l + (1 - l)^{-p} x, \tag{6}$$

where

$$(1 - l)^{-p} = \left[1 - l - (1 - l)^{-p} a\right]^{-q},$$

which gives

$$a(l) = (1 - l)^{p+1} - (1 - l)^{p+p/q}. \tag{7}$$

Now, $a'(l) = 0$ implies

$$l = 1 - \left[\frac{q(1+p)}{p(1+q)}\right]^{q/(p-q)}. \tag{8}$$

We note that $a(l) > 0$ for $0 < l < 1$ implies $p > q$. Since $a(0) = 0 = a(1)$ and $a(l) > 0$, it follows that (8) gives $\max_{0 \leq l \leq 1} a(l)$. We denote this value by A. From (7),

$$A = \left[\frac{q(1+p)}{p(1+q)}\right]^{q(1+p)/(p-q)} - \left[\frac{q(1+p)}{p(1+q)}\right]^{p(1+q)/(p-q)}.$$

Since $a(0) = 0 = a(1)$ and $a(l) > 0$ for $0 < l < 1$, the graph of $a(l)$ given by (7) is concave downwards with its maximum attained at A.

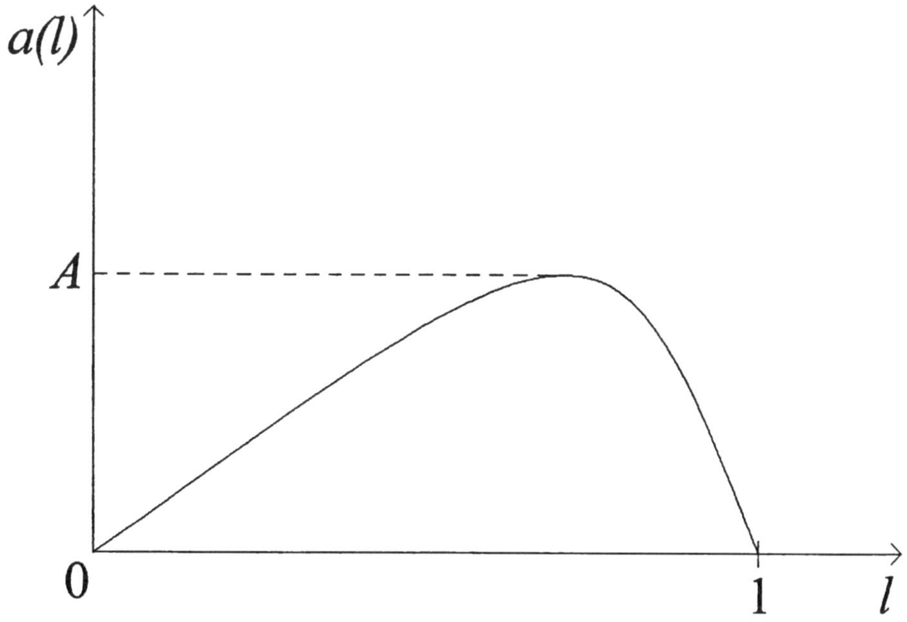

Figure 2

Let l_- and l_+ with $l_- < l_+$ denote, respectively, the values of l corresponding to each value of $a \in (0, A)$. Also, let U_- and U_+ denote, respectively, the solutions of the problem (5) corresponding to l_- and l_+. From (6), $U_- < U_+$.

THEOREM 5 (i) *If $q \geq p$, and u_0 is a lower solution, then u quenches in a finite time.*

(ii) *If $p > q$ and $a \in (0, A]$, then*

(a) u exists globally, provided $u_0 \leq U_+$;

(b) for any positive number ε, there exists u_0 with $|u_0 - U_+| < \varepsilon$ such that u quenches in a finite time, provided $qa \geq 1$;

(c) u quenches in a finite time, provided $a > A$.

REFERENCES

1. Chan, C. Y., & Chen, C. S., A numerical method for semilinear singular parabolic quenching problems. Quart. Appl. Math. 47 (1989). 45-57.

2. Chan, C. Y., & Kong, P. C., Quenching for degenerate semilinear parabolic equations. Applicable Anal. 54 (1994). 17-25.

3. Chan, C. Y., & Kong, P. C., Channel flow of a viscous fluid in the boundary layer. Quart. Appl. Math. 55 (1997). 51-56.

4. Chan, C. Y., & Kwong, M. K., Existence results of steady-states of semilinear reaction-diffusion equations and their applications. J. Differential Equations 77 (1989). 304-321.

5. Chan, C. Y., & Yuen, S. I., Parabolic problems with nonlinear absorptions and releases at the boundaries. Appl. Math. Comput., to appear.

6. Cobb, S., Quenching of a parabolic system with a nonlinear boundary condition. Dynam. Systems Appl. 1 (1992). 251-256.

7. Kawarada, H., On solutions of initial-boundary problem for $u_t = u_{xx} + 1/(1-u)$. Publ. Res. Inst. Math. Sci. 10 (1975). 729-736.

8. Levine, H. A., The quenching of solutions of linear parabolic and hyperbolic equations with nonlinear boundary conditions. SIAM J. Math. Anal. 14 (1983). 1139-1153.

9. Levine, H. A., & Lieberman, G. M., Quenching of solutions of parabolic equations with nonlinear boundary conditions in several dimensions. J. Reine Angew. Math. 345 (1983). 23-38.

Proceedings of Dynamic Systems and Applications 3 (2001) 135-138

HYPERBOLIC QUENCHING WITH NONLINEAR DAMPING AND SOURCE TERMS

Dedicated to the seventy-fifth birthday of Professor V. Lakshmikantham

C. Y. Chan[1] and J. K. Zhu[2]

[1]Department of Mathematics
University of Louisiana at Lafayette
Lafayette, Louisiana 70504-1010

[2]Department of Mathematics and Physics
Fort Valley State University
Fort Valley, Georgia 31030-3298

ABSTRACT: Let D be a bounded connected domain in $\mathbb{R}^n$ with a piecewise smooth boundary ∂D, $\Omega = D \times (0,T)$, $T \leq \infty$, and $\partial\Omega = \partial D \times [0,T)$. This article studies the hyperbolic initial-boundary value problem,

$$u_{tt} - \Delta u + au_t\,|u_t|^{q-1} = b\left(\frac{1}{1-u}\right)^p \text{ in } \Omega,$$

$$u\,(x,0) = u_0\,(x)\,,\ u_t\,(x,0) = u_1\,(x) \text{ for } x \in D;\ u\,(x,t) = 0 \text{ on } \partial\Omega,$$

where p, q, a and b are given positive constants such that $p > 1$ and $q > 1$. A criterion for a solution to reach the value 1 somewhere in D in a finite time is established. Similar results for second and third boundary conditions are also obtained respectively.

AMS (MOS) subject classification. 35Q05, 35L70, 35L20

1. INTRODUCTION

The concept of quenching was introduced in 1975 by Kawarada [4] through a first initial-boundary value problem for a semilinear heat equation. Chang and Levine [2] extended the concept to a first initial-boundary value problem for a semilinear wave equation in 1981. Because of the lack of a maximum principle for hyperbolic equations as useful as that for parabolic equations, quenching phenomena for hyperbolic equations have not been studied as extensively as for parabolic equations.

Two types of quenching phenomena for hyperbolic initial-boundary value problems have been studied: one with singular nonlinearities in the differential equations, and the other with singular nonlinearities in the boundary conditions. The former type was studied by Chang and Levine [2] and Axtell [1] for the one-dimensional case, and Smith [8] and Levine and Smiley [6] for the multi-dimensional case. For the latter type, Levine [5] considered the one-dimensional case while Rammaha [7] for the multi-dimensional case.

Let p, q, a and b be given positive constants such that $p > 1$ and $q > 1$,

$$Lu \equiv u_{tt} - \Delta u + a u_t \left|u_t\right|^{q-1},$$

and D be a bounded connected domain in $\mathbb{R}^n$ with a piecewise smooth boundary ∂D, $\Omega = D \times (0,T)$, $T \leqslant \infty$, and $\partial\Omega = \partial D \times [0,T)$. Recently, Georgiev and Todorova [3] studied the blow-up of solutions for the initial-boundary value problem with nonlinear damping and source terms:

$$\begin{gathered} Lu = bu\left|u\right|^{p-1} \text{ in } \Omega, \\ u(x,0) = \varphi(x),\; u_t(x,0) = \psi(x) \text{ for } x \in D, \\ u(x,t) = 0 \text{ on } \partial\Omega. \end{gathered}$$

Here, we would like to study the corresponding quenching phenomena. We consider the problem with the first boundary condition:

$$\left.\begin{gathered} Lu = b\left(\frac{1}{1-u}\right)^p \text{ in } \Omega, \\ u(x,t) = u_0(x),\; u_t(x,0) = u_1(x) \text{ for } x \in D, \\ u(x,t) = 0 \text{ on } \partial\Omega, \end{gathered}\right\} \tag{1.1}$$

where $u_0(x) = 0$ on ∂D. Then we discuss the problem with the second boundary condition:

$$\left.\begin{gathered} Lu = b\left(\frac{1}{1-u}\right)^p \text{ in } \Omega, \\ u(x,t) = u_0(x),\; u_t(x,0) = u_1(x) \text{ for } x \in D, \\ \frac{\partial u(x,t)}{\partial n} = 0 \text{ on } \partial\Omega, \end{gathered}\right\} \tag{1.2}$$

where n denotes the outward unit normal, and $\partial u_0(x)/\partial n = 0$ on ∂D. Finally, we study the problem with the third boundary condition:

$$\left.\begin{gathered} Lu = b\left(\frac{1}{1-u}\right)^p \text{ in } \Omega, \\ u(x,t) = u_0(x),\; u_t(x,0) = u_1(x) \text{ for } x \in D, \\ k_1\frac{\partial u}{\partial n}(x,t) + k_2 u(x,t) = 0 \text{ on } \partial\Omega, \end{gathered}\right\} \tag{1.3}$$

where k_1 and k_2 are given positive constants, and

$$k_1 \frac{\partial u_0}{\partial n}(x) + k_2 u_0(x) = 0 \text{ on } \partial D.$$

2. FIRST BOUNDARY CONDITIONS

Let $\int_D dx = |D|$, $\beta = p/(p-1)$, and α be any positive constant less than $[p/(p+1)]^{1/\beta}$. The proofs of the results in this article will appear elsewhere.

LEMMA 1. *The system*

$$\varepsilon = \frac{p\left(2b|D|\right)^{\beta-1} c^{\beta}}{(p-1)^{\beta}\left[p-(p+1)\alpha^{\beta}\right]},$$

$$c = \frac{aq|D|^{\frac{1}{q+1}}\varepsilon + \left(2|D|\right)^{\beta-1}(q+1)\varepsilon^{\beta}}{a(q+1)},$$

has a unique positive solution ε and c.

Let

$$E(t) = \frac{b}{p-1}\left\|\frac{1}{1-u}\right\|_{p-1}^{p-1} - \frac{1}{2}\|u_t\|_2^2 - \frac{1}{2}\|\nabla u\|_2^2,$$

$$h(t) = \int_D u u_t dx,$$

$$G(t) = cE(t) + \varepsilon h(t),$$

$$\gamma = \frac{a\varepsilon|D|^{\frac{1}{q+1}}}{(q+1)} + 2^{\beta-1}\varepsilon^{\beta}|D|^{\beta} + \frac{b\varepsilon(p+1)|D|}{p(p-1)\alpha^{p}}.$$

THEOREM 2. *If*

$$G^{\beta}(0) > 2^{\beta}\gamma + \frac{2\gamma}{\beta-1},$$

$$E(0) > \max\left\{0, -\frac{\varepsilon h(0)}{c}\right\},$$

$$p \geqslant \frac{1+q}{q}, \tag{2.1}$$

then a solution of the problem (1.1) quenches in a finite time.

3. OTHER BOUNDARY CONDITIONS

We now study other boundary conditions.

THEOREM 3. *If*

$$G^{\beta}(0) > \gamma 2^{\beta} + \frac{2a\varepsilon |D|^{\frac{1}{q+1}}}{(\beta-1)(q+1)},$$

$$E(0) > \max\left\{0, -\frac{\varepsilon h(0)}{c}\right\},$$

and (2.1) holds, then a solution of the problem (1.2) quenches in a finite time.

THEOREM 4. *Under the hypotheses of Theorem 2, a solution of the problem (1.3) quenches in a finite time.*

REFERENCES

1. Axtell, J., A numerical study of the derivatives of solutions of the wave equation with a singular forcing term at quenching. Numer. Methods Partial Differential Equations 1 (1989). 53-76.

2. Chang, P. H., & Levine, H. A., The quenching of solutions of semilinear hyperbolic equations. SIAM J. Math. Anal. 12 (1982). 893-903.

3. Georgiev, V., & Todorova, G., Existence of a solution of the wave equation with nonlinear damping and source terms. J. Differential Equations 109 (1994). 295-308.

4. Kawarada, H., On solutions of initial-boundary problem for $u_t = u_{xx} + 1/(1-u)$. Publ. Res. Inst. Math. Sci. 10 (1975). 729–736.

5. Levine, H. A., The quenching of solutions of linear parabolic and hyperbolic equations with nonlinear boundary conditions. SIAM J. Math. Anal. 14 (1983). 1139-1153.

6. Levine, H. A., & Smiley, M. W., Abstract wave equations with a singular nonlinear forcing term. J. Math. Anal. Appl. 103 (1984). 409-427.

7. Rammaha, M. A., On the quenching of solutions of the wave equation with a nonlinear boundary condition. J. Reine Angew. Math. 407 (1990). 1-18.

8. Smith, R. A., On a hyperbolic quenching problem in several dimensions. SIAM J. Math. Anal. 20 (1989). 1081-1094.

Proceedings of Dynamic Systems and Applications 3 (2001) 139-146

WAVELETS AND HYPOTHESIS TESTING

R. Chandramouli and K.M. Ramachandran

Dept. of Electrical and Computer Engg., Iowa State University, IA 50011
Department of Mathematics, University of South Florida, FL 33620

ABSTRACT: In this paper, wavelets based hypothesis testing techniques are discussed. First, a level-dependent hypothesis test is derived. This is followed by the fusion of the multiple decisions at various levels of the wavelet decomposition to obtain a global decision. Theoretical analysis of two fusion rules are given.

AMS (MOS) subject classification: 35K60, 35K57

1. INTRODUCTION

Applications such as radar target detection motivate the study of signal detection using hypothesis tests. The observed data are noisy and also incomplete in many cases. The aim is to detect the presence/absence of a signal of interest with a given probability of error from the noisy observations. In general, it is easier to design hypothesis tests for independent and identically distributed (i.i.d.) observations. Since the i.i.d. requirement need not be satisfied by the noisy observations, studying the equivalent problem in the frequency domain is often helpful. Of course, we know that the Karhunen-Loeve transform (KLT) [1] decorrelates the data at an enormous computational cost. And, if the observations are Gaussian, this implies that the transform co-efficients are i.i.d. However, a good approximation to the KLT can be achieved via wavelet transforms. The wavelet transform co-efficients of the observed Gaussian distributed samples behave as i.i.d. samples. This fact has been widely used by many researches studying the detection of transient signals. If a hypothesis test is performed on a level-by-level basis in the wavelet domain, it is not clear how to combine the outcomes of the hypothesis tests at each level of the wavelet transform into one global decision. This motivates the present study. Even if the observations are dependent, their wavelet coefficients within and across the scales can be assumed to be nearly independent. This facilitates the likelihood ratio based level-dependent hypothesis test. The idea is to first develop a level dependent hypothesis test followed by a fusion rule that combines these decisions within an acceptable decision error probability.

Suppose a sequence of observations $\{r_i\}_{i=1}^{N}$ is given, where $r_i \in \Re, \forall i$ which obeys the regression equation

$$r_i = \sqrt{E_0} s_i + \sigma z_i \tag{1}$$

$\{s_i\}$ is an unknown regression sequence (signal) and $\{z_i\}$ is the noise sequence. $E_0 \geq 0$ controls the energy of the signal and σ^2 is the variance of the noise. Then the aim in detection is to find if $E_0 = 0$ or not. That is, we detect the presence ($E_0 > 0$) or absence

($E_0 = 0$) of the signal from the noisy observations using statistical hypothesis testing. The problem is posed as a binary hypothesis test [1]. Mostly, only the detection of transient signals using wavelets has been studied by many researchers. In [2] a nonparametric hypothesis test in the wavelet domain is studied. The null hypothesis corresponds to the signal that is identically zero and the alternative is composite and nonparametric: functions from the alternative set are bounded away from zero in the L^2 norm and also possess some smoothness properties. The adaptive wavelet hypothesis test is shown to be nearly minimax. Testing the linearity of regression function using its noisy observations is studied in [3]. The hypothesis test corresponds to the function being linear versus non-linear. The paper develops an adaptive testing method using Haar wavelets that is computationally simple. The universal threshold is used as the decision threshold to test the hypothesis.

The paper is organized as follows. The problem is introduced in Section 2. Section 3 and 4 deal with the proposed decision fusion rules. Asymptotic analysis of the proposed detector is given in Section 5 followed by a summary of the paper in Section 6.

2. PROBLEM DEFINITION

In this paper the problem of detecting a known, deterministic signal buried in noise is studied. The problem is posed as a hypothesis test. Assume that there are two possible hypotheses, H_0 and H_1, corresponding to two possible probability distributions P_0 and P_1, respectively, on the measurable space $(\Re^N, \mathcal{B}^N)$; where $\mathcal{B}^N$ is the Borel σ-algebra over $\Re^N$. The hypothesis test is defined as the test between

$$\begin{aligned} H_1 : r_i &= \sqrt{E_0} s_i + \sigma z_i \\ H_0 : r_i &= \sigma z_i, \quad i = 1, \ldots, N \end{aligned} \tag{2}$$

where the vectors $\mathbf{r} = (r_1, r_2, \ldots, r_N)$, $\mathbf{s} = (s_1, s_2, \ldots, s_N)$, and $\mathbf{z} = (z_1, z_2, \ldots, z_N)$ denote the observations, the deterministic signal, and stationary, zero-mean, unit variance white Gaussian noise. $\sqrt{E_0}$ is the signal to noise ratio parameter. Further, we assume the *normalization condition*, namely, $\sum_{i=1}^{N} s_i^2 = 1$. Clearly, the additive noise distribution is determined by the marginal density $p \sim \mathcal{N}(0, \sigma^2)$ on $\Re^N$. We use the Neyman-Pearson optimality criterion [1] to find a solution to this problem in the wavelet domain. N is assumed to be a power of 2, say, $N = 2^J$ for an integer J. Taking the wavelet transform of the observation vector, we get

$$\begin{aligned} H_1 : \bar{r}_{j,i} &= \sqrt{E_0} \bar{s}_{j,i} + \sigma \eta_{j,i} \\ H_0 : \bar{r}_{j,i} &= \sigma \eta_{j,i}, \quad i = 1, 2, \ldots, J; \; i = 1, 2, \ldots, N_j = N/2^j \end{aligned} \tag{3}$$

where $\bar{r}_{j,i}$, $\bar{s}_{j,i}$ and $\eta_{j,i}$ denote the wavelet co-efficients corresponding to the observed noisy samples, original signal samples and the noise respectively. At level j of the wavelet decomposition the log-likelihood ratio is given by

$$\begin{aligned} l_j(\bar{r}_{j,1}, \bar{r}_{j,2}, \ldots, \bar{r}_{j,N_j}) &= log \prod_{i=1}^{N_j} \frac{p(\bar{r}_{j,i} - \sqrt{E_0}\bar{s}_{j,i})}{p(\bar{r}_{j,i})} \\ &= log \prod_{i=1}^{N_j} \frac{\frac{1}{2\sigma^2} e^{-\left(\bar{r}_{j,i} - \sqrt{E_0}\bar{s}_{j,i}\right)^2 / 2\sigma^2}}{\frac{1}{2\sigma^2} e^{-\bar{r}_{j,i}^2 / 2\sigma^2}} \\ &= \frac{\sqrt{E_0}}{\sigma^2} \sum_{i=1}^{N_j} \bar{r}_{j,i} \bar{s}_{j,i} - \frac{E_0}{2\sigma^2} \sum_{i=1}^{N_j} \bar{s}_{j,i}^2 \end{aligned} \tag{4}$$

The *decision function* $\phi_j : \Re \rightarrow \{0,1\}$ measurable with respect to the σ-algebra generated by $\{\bar{r}_{j,1}, \bar{r}_{j,2}, \ldots, \bar{r}_{j,N_j}\}$, gives the final decision of the hypothesis test at level j. Here, H_0 and H_1 are denoted by $\{0,1\}$ respectively. We define the false alarm probability, $\alpha = P_0(\text{decide } H_1)$, and the miss probability, $\beta = P_1(\text{decide } H_0)$. Noting that the second term in Eq. (4) is a constant. The Neyman-Pearson decision rule is given by

$$L_j = \frac{1}{\sigma^2}\sum_{i=1}^{N_j} \bar{r}_{j,i}\bar{s}_{j,i} \begin{cases} \geq & T_j \text{ decide } H_1 \\ < & T_j \text{ decide } H_0 \end{cases} \tag{5}$$

where T_j is the decision threshold. T_j is computed based on the constant false alarm constraint, *i.e.*, α is fixed. The above test statistic can be re-written as $\phi_j = 1(L_j \geq T_j)$ where $1(.)$ is the indicator function. The conditional means of the test statistic, L_j are given by

$$\begin{aligned} \tilde{\mu}_{j,1} &= E[L_j|H_1] \\ &= \frac{1}{\sigma^2} E\left[\sum_{i=1}^{N_j} \bar{r}_{j,i}\bar{s}_{j,i}|H_1\right] \\ &= \frac{1}{\sigma^2}\sum_{i=1}^{N_j} E\left[\sqrt{E_0}\bar{s}_{j,i}^2 + \bar{s}_{j,i}\eta_{j,i}\right] \\ &= \frac{\sqrt{E_0}}{\sigma^2}\sum_{i=1}^{N_j} \bar{s}_{j,i}^2 \end{aligned} \tag{6}$$

$$\begin{aligned} \tilde{\mu}_{j,0} &= E[L_j|H_0] \\ &= \frac{1}{\sigma^2}\sum_{i=1}^{N_j} E\left[\eta_{j,i}\bar{s}_{j,i}\right] \\ &= 0 \end{aligned} \tag{7}$$

and the conditional variances are,

$$\begin{aligned} \tilde{\sigma}_{j,1}^2 &= Var[L_j|H_1] \\ &= \frac{1}{\sigma^4}\sum_{i=1}^{N_j} Var\left[\sqrt{E_0}\bar{s}_{j,i}^2 + \bar{s}_{j,i}\right] \\ &= \frac{1}{\sigma^2}\sum_{i=1}^{N_j} \bar{s}_{j,i}^2 \end{aligned} \tag{8}$$

$$\begin{aligned} \tilde{\sigma}_{j,0}^2 &= Var[L_j|H_0] \\ &= \frac{1}{\sigma^2}\sum_{i=1}^{N_j} \bar{s}_{j,i}^2 \end{aligned} \tag{9}$$

Let $\tilde{\sigma}_{j,1}^2 = \tilde{\sigma}_{j,0}^2 = \tilde{\sigma}_j$. We also note that $L_j \sim N(\tilde{\mu}_j, \tilde{\sigma}_j^2)$ under both H_1 and H_0. Further, we assume that $\{L_j\}_{j=1}^J$ are independent due to the decorrelating nature of the wavelet decomposition. To simplify the computation of the decision rule, let $T_j = T, \forall j$. We study two decision fusion rules that combine the decisions, $\{\phi_j\}_{j=1}^J$, into a global decision. The first is the OR rule and the second one is the AND rule.

3. OR DECISION FUSION

In the OR rule, the global decision is H_1 if it is accepted by at least one $\phi_j, j = 1, 2, \ldots, J$, *i.e.*, the global decision, $\phi_{OR}^* = 1(\cup_j[\phi_j = 1])$. The false alarm probability is therefore given by

$$\begin{aligned}\alpha_{OR} &= 1 - P_0(\phi_1 = 0, \phi_2 = 0, \ldots, \phi_J = 0) \\ &= 1 - \prod_{j=1}^{J} P_0(\phi_j = 0) \text{ (due to independence)}\end{aligned} \tag{10}$$

Let $f_{j,k}$ denote the probability density function (pdf) of L_j conditioned on hypothesis H_k, $k = 0, 1$. Then,

$$\begin{aligned}P_0(\phi_j = 0) &= \int_{-\infty}^{T} f_{j,0}(l)dl \\ &= \int_{-\infty}^{T} \frac{1}{\sqrt{2\pi\bar{\sigma}_j^2}} e^{-l^2/2\bar{\sigma}_j^2} dl \\ &= \frac{1}{2} + \frac{1}{2} erf\left[\frac{T}{\sqrt{2}\bar{\sigma}_j}\right]\end{aligned} \tag{11}$$

where $erf(x) = \frac{2}{\sqrt{\pi}} \int_0^x e^{-t^2} dt$. Therefore, from Eq. (10) we get

$$\alpha_{OR} = 1 - \prod_{j=1}^{J} \left[\frac{1}{2} + \frac{1}{2} erf\left[\frac{T}{\sqrt{2}\bar{\sigma}_j}\right]\right] \tag{12}$$

3.1. Choice of T

Solving Eq. (12) to get a closed-form solution for the optimal value of T is difficult. Therefore we use approximations to solve for T. The trade-off as a result of the approximations is also addressed. Due to the normalization condition we see that $\sum_{j=1}^{J} \sum_{i=1}^{N_j} \bar{s}_{j,i}^2 = 1$. Hence, from Eq. (9) we observe that

$$\begin{aligned}\bar{\sigma}_j^2 &= \frac{1}{\sigma^2} \sum_{i=1}^{N_j} \bar{s}_{j,i}^2 \\ &\leq \frac{1}{\sigma^2}\end{aligned} \tag{13}$$

We approximate $\frac{1}{\bar{\sigma}_j}$ by σ in Eq. (12). Using the fact that the error function, $erf(.)$ is an increasing function we see that the new false alarm probability, say, $\tilde{\alpha}_{OR} \geq \alpha_{OR}$. So, it may seem that this approximation resulting in a higher false alarm probability may not be useful. However, we will show below that there is an increase in the power of the test corresponding to the loss in the false alarm. Since the power of the test is more important in many applications, it is worthwhile to explore the proposed method of computing the decision threshold. Now,

$$\begin{aligned}\alpha_{OR} &\leq 1 - \prod_{j=1}^{J} \left[\frac{1}{2} + \frac{1}{2} erf\left[\frac{\sigma T}{\sqrt{2}}\right]\right] \\ \Rightarrow \left[\frac{1}{2} + \frac{1}{2} erf\left[\frac{\sigma T}{\sqrt{2}}\right]\right]^J &\leq 1 - \alpha_{OR} \\ \Rightarrow T &\leq \frac{\sqrt{2}}{\sigma} erf^{-1}[2(1 - \alpha_{OR})^{1/J} - 1]\end{aligned} \tag{14}$$

The above inequality can be taken as an equality to choose the decision threshold, T. It is natural to study the effect of this approximation on the power of the hypothesis test. The power of the test is equal to $1 - \beta_{OR}$ and

$$\begin{aligned} 1-\beta_{OR} &= 1 - P_1(\bigcap_j[\phi_j = 0) \\ &= 1 - \prod_{j=1}^{J} P_1(\phi_j = 0) \;\; \text{(due to independence)} \\ &= 1 - \prod_{j=1}^{J}\left[\frac{1}{2} + \frac{1}{2}erf\left[\frac{T-\tilde{\mu}_{j,1}}{\sqrt{2}\tilde{\sigma}_j}\right]\right] \end{aligned} \tag{15}$$

Using the approximation $\sigma^2 \leq \frac{1}{\tilde{\sigma}_j^2}$ in the above equation and calling the resulting power $1-\tilde{\beta}_{OR}$, we see that $1-\beta_{OR} \leq 1-\tilde{\beta}_{OR}$ due to the increasing nature of the error function. The power of the test increases by the choice of T given in Eq. (14). Next, we study the AND decision fusion.

4. AND DECISION FUSION

The AND fusion is the second fusion rule that we study to make global decision regarding the true hypothesis. Here, we accept H_1 only if it is accepted at all the levels, $j = 1, 2, \ldots, J$. This can be written as $\phi^*_{AND} = 1(\bigcap_j[\phi_j = 1])$. The false alarm probability can be computed as

$$\begin{aligned} \alpha_{AND} &= P_0(\phi_1 = 1, \phi_2 = 1, \ldots, \phi_J = 1) \\ &= \prod_{j=1}^{J} P_0(\phi_j = 1) \\ &= \prod_{j=1}^{J}\left[\frac{1}{2} - \frac{1}{2}erf\left[\frac{T}{\sqrt{2}\tilde{\sigma}_j}\right]\right] \end{aligned} \tag{16}$$

Again, substituting $\sigma^2 \leq \frac{1}{\tilde{\sigma}_j}$ in the above equation we see that the resulting false alarm $\tilde{\alpha}_{AND} \leq \alpha_{AND}$. Therefore,

$$\begin{aligned} \alpha_{AND} &\leq \left[\frac{1}{2} - \frac{1}{2}erf\left[\frac{\sigma T}{\sqrt{2}}\right]\right]^J \\ \Rightarrow T &\leq \frac{\sqrt{2}}{\sigma}erf^{-1}\left[1 - 2\alpha_{AND}^{1/J}\right] \end{aligned} \tag{17}$$

Thus, T can be computed. Now, the power of the test is given by

$$\begin{aligned} 1-\beta_{AND} &= P_1(\phi_1 = 1, \phi_2 = 1, \ldots, \phi_J = 1) \\ &= \prod_{j=1}^{J} P_1(\phi_j = 1) \\ &= \prod_{j=1}^{J}[1 - P_1(\phi_j = 0)] \end{aligned} \tag{18}$$

where

$$\begin{aligned} P_1(\phi_j = 0) &= \int_{-\infty}^{T} f_{j,1}(l)dl \\ &= \frac{1}{2} + \frac{1}{2} erf\left[\frac{T - \tilde{\mu}_{j,1}}{\sqrt{2}\tilde{\sigma}_j}\right] \end{aligned} \tag{19}$$

By using Eq. (18), Eq. (19), and $\sigma \leq 1/\tilde{\sigma}_j$ we obtain

$$\begin{aligned} 1 - \beta_{AND} &= \prod_{j=1}^{J} \left[\frac{1}{2} - \frac{1}{2} erf\left[\frac{T - \tilde{\mu}_{j,1}}{\sqrt{2}\tilde{\sigma}_j}\right]\right] \\ &\leq \prod_{j=1}^{J} \left[\frac{1}{2} - \frac{1}{2} erf\left[\frac{\sigma(T - \tilde{\mu}_{j,1})}{\sqrt{2}}\right]\right] = 1 - \tilde{\beta}_{AND} \end{aligned} \tag{20}$$

We observe again that using the approximation to compute the value of the decision threshold results in a loss in the false alarm but a gain in the power of the test.

5. ASYMPTOTIC ANALYSIS

In this section we study the asymptotic performance of the two decision fusion rules as the number of observations becomes infinitely large.

Theorem 1 $\tilde{\alpha}_{OR} \to 0$ *and* $\tilde{\beta}_{OR} \to 1$ *as* $N \to \infty$.

Proof. When Eq. (14) is used to compute the value of T, we get

$$\begin{aligned} T &= \frac{\sqrt{2}}{\sigma} erf^{-1}\left[2(1 - \alpha_{OR})^{1/J} - 1\right] \\ &= \frac{\sqrt{2}}{\sigma} erf^{-1}\left[2(1 - \alpha_{OR})^{1/logN} - 1\right] \end{aligned} \tag{21}$$

As $N \to \infty$, $2(1 - \alpha_{OR})^{1/logN} - 1 \to 1$ and therefore $T \to \infty$. Therefore,

$$\begin{aligned} \tilde{\alpha}_{OR} &= 1 - \prod_{j=1}^{J} \left[\frac{1}{2} + \frac{1}{2} erf\left[\frac{\sigma T}{\sqrt{2}}\right]\right] \\ &= 0 \text{ as } N \to \infty \end{aligned} \tag{22}$$

Similarly,

$$\begin{aligned} \tilde{\beta}_{OR} &= \prod_{j=1}^{J} \left[\frac{1}{2} + \frac{1}{2} erf\left[\frac{(T - \tilde{\mu}_{j,1})\sigma}{\sqrt{2}}\right]\right] \\ &= 1 \text{ as } N \to \infty \end{aligned} \tag{23}$$

We have an analogous result for the AND decision fusion.

Theorem 2 $\tilde{\alpha}_{AND} \to 1$ *and* $\tilde{\beta}_{AND} \to 0$ *as* $N \to \infty$.

Proof. From Eq. (17),

$$\begin{aligned} T &= \frac{\sqrt{2}}{\sigma} erf^{-1}\left[1 - 2\alpha_{AND}^{1/logN}\right] \\ &= \frac{\sqrt{2}}{\sigma} erf^{-1}[-1] \text{ as } N \to \infty \\ &= -\infty \end{aligned} \tag{24}$$

Therefore,

$$\begin{aligned} \tilde{\alpha}_{AND} &= \left[\frac{1}{2} - \frac{1}{2} erf\left[\frac{\sigma T}{\sqrt{2}}\right]\right]^{logN} \\ &= 1 \text{ as } N \to \infty \end{aligned} \tag{25}$$

and

$$\begin{aligned} \tilde{\beta}_{AND} &= 1 - \prod_{j=1}^{J}\left[\frac{1}{2} - \frac{1}{2} erf\left[\frac{(T - \tilde{\mu}_{j,1})\sigma}{\sqrt{2}}\right]\right] \\ &= 0 \end{aligned} \tag{26}$$

6. SUMMARY

A Neyman-Pearson optimal level dependent hypothesis test is presented. For a given false alarm probability, a Neyman-Pearson test is performed at each level of the wavelet decomposition. The decisions of these tests are then combined using two decision fusion rules to obtain a global decision. The OR and AND decision fusion rules are discussed. A method to compute the decision thresholds and the asymptotic properties of the hypothesis test are derived.

7. REFERENCES

[1] H.L. Van Trees, *Detection, estimation, and modulation theory, Part I*, New York, John Wiley and Sons, 1968.

[2] V.G. Spokoiny, "Adaptive hypothesis testing using wavelets," *Anns. of Statistics*, vol. 24, no. 6, pp. 2477-2498, 1996.

[3] V.G. Spokoiny, "Testing a linear hypothesis using Haar transform," *Preprint*, Weierstrass Institute for Applied Analysis and Stochastics, 1997.

Proceedings of Dynamic Systems and Applications 3 (2001) 147-154

Asymptotic Behavior of Inhomogeneous Iterates for Nonlinear Operators in Ordered Banach Spaces

Yong-Zhuo Chen
Department of Mathematics, Computer Science and Engineering
University of Pittsburgh at Bradford, Bradford, PA 16701
yong@imap.pitt.edu

Abstract. This paper considers a discrete dynamical system defined by a sequence of operators $\{f_n\}$, which are defined on the positive cone of an ordered Banach space. Suppose that

$$y \geq tx \quad \text{implies} \quad f_n(y) \geq \phi(t) f_n(x)$$

for all $n \geq 1$ and $t \in (0,1]$, where $\phi : (0,1] \to (0,1]$ satisfies $\phi(t) > t$. We discuss the asymptotic behavior of the inhomogeneous iterates of f_n through the projective metrics. By using the convexity of projective metrics, our results improve the relevant results in the literature.

AMS(MOS) subject classification. 47H07, 47H09

1. Introduction

Let a discrete dynamical system be defined by an operator f on a subset of the positive closed convex cone of an ordered Banach space. Suppose f satisfies

$$f(tx) \geq \phi(t) f(x), \tag{1}$$

where $t \in (0,\ 1)$ and ϕ is a positive function on $(0,\ 1)$. Several authors discussed the asymptotic behavior of the iterates f^n under various assumptions on ϕ. For example, M. A. Krasnosel'skiĭ [10] investigated u_0-concave operator, where $\phi(t) = [1+\eta(x,t)]t$ with $\eta(x,t) > 0$, A. J. B. Potter [15] initiated the study of α-concave mapping, where $\phi(t) = t^\alpha$ with $\alpha \in (0,\ 1)$, and U. Krause's [11] [12] studied the ascending operators, where $\phi : [0,\ 1] \to [0,\ 1]$ is continuous and $\lambda < \phi(\lambda)$ for $\lambda \in (0,\ 1)$. In a series of papers, [2], [3], [4], [5] and [6], we considered the monotone operator f, where $\phi(t) = t^{\alpha(a,b)}$ and $\alpha(a,\ b) \in (0,\ 1)$ for all $t \in [a,\ b] \subset (0,\ 1)$.

All of the previous works noticed that projective metrics, Hilbert metrics and its variant, Thompson's metric [16], are powerful tools in studying the operators satisfying (1). However, they did not fully use the geometric properties of projective metrics. Recently, R. D. Nussbaum [14] proved an interesting property, the "metrically convexity", for the Hilbert metric and Thompson's metric. Using his result, we are able to relax the various assumptions imposed on ϕ in (1) and improve the relevant results in the literatures.

In this paper we will consider problems in a more general setting. Suppose the discrete dynamical system is nonautonomous, i.e., it changes in time. So that it is described by a sequence of operators $\{f_n\}$. We are to study the asymptotic behaviour of the inhomogeneous iterates $f_n \circ f_{n-1} \circ \cdots \circ f_1$, as $n \to \infty$, by means of Thompson's metric.

2. Results in Metrically Convex Metric Space

A meric space $(X,\ d)$ is said to be metrically convex if for each $x,\ y \in X$ with $x \neq y$, there is a z which is different from x and y such that $d(x,y) = d(x,z) + d(z,y)$.

Lemma 2.1[6, Lemma 2.1]. Let X be a complete metrically convex space and $T_n :\ X \to X$ satisfies

$$d(T_n x, T_n y) \le \phi(d(x,y)),$$

where $\phi : [0,\infty) \to [0,\infty)$, $\phi(t) < t$ for all $t > 0$ and $\phi(0) = 0$. Then there exists an $\psi : [0,\infty) \to [0,\infty)$ such that

(i) $\psi(t) < t$, $t > 0$.

(ii) ψ is nondecreasing.

(iii) ψ is subadditive.

(iv) ψ is upper semicontinuous from the right.

(v) $d(T_n x, T_n y) \le \psi(d(x,y))$.

Sketch of Proof. Let

$$\psi(t) = \sup\ \{d(T_n x, T_n y) \mid x,\ y \in X,\ \ d(x,y) \le t,\ \ n = 1,\ 2,\ 3,\ \cdots\}.$$

Using (2) and the fact that each two distinct points of a complete convex metric space can be joined by a metric segment of the space (see [1]), it can be verified that ψ satisfies (i)-(iv). □

We also need the following result on a recursive inequality, which is proved by Jachymski [9, Lemma 3].

Lemma 2.2. Let sequences $\{a_n\}$ and $\{b_n\}$ satisfy

$$a_{n+1} \leq \phi(a_n) + b_n \,, \tag{2}$$

where a_n , $b_n \geq 0$ and $b_n \to 0$ as $n \to \infty$. Suppose that $\phi : [0,\infty) \to [0,\infty)$ is nondecreasing, subadditive and $\phi(t) < t$ for $t > 0$. Then $\lim_{n\to\infty} a_n = 0$.

An operator T on X is called a Lipschitz operator if $d(Tx,Ty) = r\, d(x,y)$ for all x, $y \in X$, where $r > 0$ is the Lipshitz constant.

Theorem 2.3. Let (X,d) be a complete metrically convex space and $T_n : X \to X$, $n \geq 1$, be a sequence of Lipschitz operators with the same Lipshitz constant r, which converges uniformly to T_0 on X. Suppose that for some $l > 1$, we have

$$d(T_0^l x, T_0^l y) \leq \phi(d(x,y)), \tag{3}$$

where $\phi : [0,\infty) \to [0,\infty)$, $\phi(t) < t$ for all $t > 0$, and $\phi(0) = 0$. Then for any $x_1 \in X$, we have $\lim_{n\to\infty} T_n \circ T_{n-1} \circ \cdots \circ T_1(x_1) = x_0$, where x_0 is the unique fixed point of T_0.

Proof. As in [7], we define the sequence of lumped operators: $S_m = T_{m+l-1} \circ T_{m+l-2} \circ \cdots \circ T_{m+1} \circ T_m$ and $S_0 = T_0^l$. By Lemma 1.1, there exists an $\psi : [0,\infty) \to [0,\infty)$ satisfying condition (i) - (iv) in that lemma, and

$$d(S_0 x,, S_0 y) \leq \psi(d(x,y)). \tag{4}$$

Theorem 2 in [1] tells us that S_0 has a unique fixed point $x_0 \in X$ and $\lim_{k\to\infty} S_0^k(x) = x_0$ for any $x \in X$. Note that the assumptions imply T_0 is also a lipshitz function with constant r, hence is continuous. We have $T_0(x_0) = T_0(\lim_{k\to\infty} S_0^k(x)) = \lim_{k\to\infty} T_0^{kl+1}(x) = x_0$, i.e., x_0 is also a unique fixed point of T_0.

For any point $x_1 \in X$, let $x_2 = T_1 x_1$, and $x_{n+1} = T_n x_n$, $n = 1, 2, \cdots$. If we let $y_1 = x_1$ and $y_m = x_{(m-1)l}$ for $m \geq 2$, then

$$\begin{aligned} d(x_0, y_{m+1}) &= d(x_0, S_m y_m) \\ &\leq d(x_0, S_0 y_m) + d(S_0 y_m, S_m y_m) \\ &\leq \psi(d(x_0, y_m)) + b_m \,, \end{aligned}$$

where $b_m = d(S_0 y_m, S_m y_m)$. That T_n converges uniformly to T_0 on X implies that $\{S_m\}$ converges uniformly to S_0, hence $\lim_{m\to\infty} b_m = 0$. Using Lemma 2.2, we have

$$\lim_{n\to\infty} d(x_0, x_m) = 0. \tag{5}$$

For any $n > 0$, there exist nonnegative integers $m(n)$ and $i(n)$ such that $n = m(n)l + i(n)$ with $0 \leq i(n) \leq l$. We use induction on $i(n)$.

$$\begin{aligned} & d(x_0, x_{m(n)l+1}) \\ \leq\ & d(T_0^{m(n)l+1} x_0, T_0 S_{m(n)} y_{m(n)}) + d(T_0 S_{m(n)} y_{m(n)}, T_{m(n)l+1} S_{m(n)} y_{m(n)}) \\ \leq\ & r\, d(x_0, S_{m(n)} y_{m(n)}) + d(T_0 S_{m(n)} y_{m(n)}, T_{m(n)l+1} S_{m(n)} y_{m(n)}. \end{aligned}$$

Hence $\lim_{n\to\infty} d(x_0, x_{m(n)l+1}) = 0$ due to (6) and the uniform convergence of $\{T_n\}$. Now assume that $\lim_{n\to\infty} d(x_0, x_{m(n)l+(i-1)}) = 0$. Then

$$\begin{aligned} & d(x_0, x_{m(n)l+i}) \\ \le\ & d(T_0^{m(n)l+i} x_0, T_0 x_{m(n)+(i-1)}) + d(T_0 x_{m(n)+(i-1)}, T_{m(n)l+i} x_{m(n)+(i-1)}) \\ \le\ & r\, d(x_0, x_{m(n)+(i-1)}) + d(T_0 x_{m(n)+(i-1)}, T_{m(n)l+i} x_{m(n)+(i-1)}) \end{aligned}$$

implies that $\lim_{n\to\infty} d(x_0, x_{m(n)l+i}) = 0$ by the induction hypothesis and the uniform convergence of $\{T_n\}$. Therefore we can conclude $\lim_{n\to\infty} x_n = x_0$. □

3. Inhomogeneous Iterates in Ordered Banach Spaces

Let $(B, \|\cdot\|)$ be a real Banach space which is partially ordered by a closed convex cone P. $\overset{\circ}{P}$ stands for the interior of P. We assume that $\overset{\circ}{P}$ is nonempty and the norm is monotone, i.e., $x \le y$ implies that $\|x\| \le \|y\|$, which implies that the cone is normal. For $r > 0$, we denote $B_r(u) = \{x \in B : \|x - u\| < r\}$.

$x, y \in P - \{0\}$ are called comparable if there exist positive numbers λ and μ such that $\lambda x \le y \le \mu x$. This splits $P - \{0\}$ into disjoint components of P. $\overset{\circ}{P}$ is a component of P if P is solid. Let x and y be comparable, and put $M(x/y) = \inf\{\lambda \in R : x \le \lambda y\}$. The Thompson's metric is defined by

$$\bar{d}(x, y) = \ln\{\max[M(x/y),\ M(y/x)]\}.$$

Lemma 3.1 ([13, Lemma 2.3]). (i) Let $x, y \in \overset{\circ}{P}$ and $r > 0$ be a number such that the closed ball of radius r and center x and y, respectively, is contained in P. Then

$$\bar{d}(x, y) \le \ln\left(1 + \frac{\|x - y\|}{r}\right). \tag{6}$$

(ii) If P is a normal cone and the norm is monotone on P, then for $x, y \in P - \{0\}$,

$$\|x - y\| \le (2e^{\bar{d}(x,y)} - e^{-\bar{d}(x,y)} - 1) \min\{\|x\|, \|y\|\}. \tag{7}$$

Theorem 3.2. Let $f_n : \overset{\circ}{P} \to \overset{\circ}{P}$, $n \ge 1$, be a sequence of mappings, which converges uniformly to f_n on $\overset{\circ}{P}$. Suppose f_0 satisfies the following:

(i) For some positive integer l, $y \ge tx$ implies that $f_0^l(y) \ge \phi(t) f_0^l(x)$, where $\phi : (0,1) \to (0,1)$, $\phi(t) > t$ for all $t \in (0,1)$, $\phi(1) = 1$, and $x, y \in \overset{\circ}{P}$.

(ii) There exists $\delta > 0$ such that $B_\delta(f_0(x)) \subset \overset{\circ}{P}$ for all $x \in \overset{\circ}{P}$.

Then f_0 has a unique fixed point $x_0 \in \overset{\circ}{P}$, and $\lim_{n\to\infty} f_n \circ f_{n-1} \circ \cdots \circ f_1(x_1) = x_0$ for any initial point $x_1 \in \overset{\circ}{P}$.

Proof. By the uniform convergence, there exists $N > 0$ such that $f_n \in B_{\frac{\delta}{2}}(f_0(x))$ for all $n > N$ and $x \in \overset{\circ}{P}$. It follows that $B_{\frac{\delta}{2}}(f_n(x)) \subset \overset{\circ}{P}$ for all $n > N$ and $x \in \overset{\circ}{P}$. Using (7), the uniform convergence of $\{f_n\}$ to f_0 in norm implies the uniform convergence of $\{f_n\}$ to f_0 in $\bar{d}$.

Let $x,\ y \in \overset{\circ}{P}$ and $x \neq y$. Without loss of generality, assume $M(x/y) \geq M(y/x)$, i.e., $\bar{d}(x,y) = \ln(M(x/y))$ and $M(x/y) > 1$. Since $x \geq M(y/x)^{-1}y \geq M(x/y)^{-1}y$,

$$f_0^l(x) \geq \phi(M(x/y)^{-1})f_0^l(y)).$$

Hence $M(f_0^l(y)/f_0^l(x)) \leq \phi(M(x/y)^{-1})^{-1}$. On the other hand, since $y \geq M(x/y)^{-1}x$,

$$f_0^l(y) \geq \phi(M(x/y)^{-1})f_0^l(x)),$$

which implies $M(f_0^l(x)/f_0^l(y)) \leq \phi(M(x/y)^{-1})^{-1}$. Therefore we have

$$\bar{d}(f_0^l(x), f_0^l(y)) \leq \ln \phi(M(x/y)^{-1})^{-1} = -\ln \phi(e^{-\bar{d}(x,y)}).$$

Define $\bar{\phi}(t) = -\ln \phi(e^{-t})$, $t \geq 0$. For $t > 0$, $\phi(e^{-t}) > e^{-t}$, so that $\bar{\phi}(t) < t$. Now we have $\bar{d}(f_0^l(x), f_0^l(y)) \leq \bar{\phi}(\bar{d}(x,y))$.

By a similar argument as above and using (9), we can prove that $\{f_n\}$ is a sequence of Lipschitz operators with respect to $\bar{d}$ and the same Lipschitz constant r. Since $(\overset{\circ}{P}, d)$ is metrically convex metric space, by Theorem 2.3, there exists a unique fixed point $x_0 \in \overset{\circ}{P}$ of f_0 such that $\lim_{n\to\infty} f_n \circ f_{n-1} \circ \cdots \circ f_1(x_1) = x_0$ for any initial point $x_1 \in \overset{\circ}{P}$. An application of (8) concludes the proof. □

Corollary 3.3. Let $f : \overset{\circ}{P} \to \overset{\circ}{P}$ satisfy:

$$y \geq tx \quad \text{implies that} \quad f(y) \geq \phi(t)f(x),$$

where $\phi : (0,1) \to (0,1)$, $\phi(t) > t$ for all $t \in (0,1)$, $\phi(1) = 1$, and $x,\ y \in \overset{\circ}{P}$. Then f has a unique fixed point $x_0 \in \overset{\circ}{P}$, and $\lim_{n\to\infty} f^n(x_1) = x_0$ for any initial point $x_1 \in \overset{\circ}{P}$.

Corollary 3.3 is an improvement of [8, Theorem 2.2.6].

4. Applications

Let $\Omega \subset R^n$ be a compact set. $C(\Omega)$ stands for the Banach space consists of all real valued continuous function on Ω with the sup norm and is partially ordered by the cone

$$C_+(\Omega) = \{x \in C(\Omega) : x(s) \geq 0,\ s \in \Omega\}.$$

Then

$$\overset{\circ}{C}_+ (\Omega) = \{x \in C(\Omega) : x(s) > 0,\ s \in \Omega\}.$$

Let us consider the sequence of Hammerstein integral operators

$$A_n(x)(t) = \int_\Omega K_n(t,s)\, f_n(s, x(s))\, ds, \quad n = 1, 2, \cdots,$$

where $K_n \in C(\Omega \times \Omega)$ are strictly positive, $x \in C(\Omega)$, and K_n converge uniformly to a strictly positive function K on $\Omega \times \Omega$. Suppose f_n are strictly positive continuous functions on $\Omega\times \overset{\circ}{C}_+ (\Omega)$ and converge uniformly to f on $\Omega\times \overset{\circ}{C}_+ (\Omega)$, where f possesses the following properties:

(i) $f_s(x) = f(s, x)$ is nondecreasing in x, where $x \in\overset{\circ}{C}_+ (\Omega)$. $f_s^{[l]}$ stands for the lth iterates of f_s.

(ii) $\frac{f_s^{[l]}(x)}{g(s,x)}$ is nonincreasing in x, where g is a strictly positive continuous function on $\Omega\times \overset{\circ}{C}_+ (\Omega)$ and

$$\min_{s\in\Omega} \frac{g(s,\ \lambda x(s))}{g(s,\ x(s))} \geq \phi(\lambda)$$

for all $x \in\overset{\circ}{C}_+ (\Omega)$ and $\lambda \in (0,\ 1)$, where $\phi : (0,1) \to (0,1)$, and $\phi(\lambda) > \lambda$.

(iii) There exist $a > 0$ such that $\int_\Omega K(t,s)f(s,\ x(s)) \geq a$ for all $x \in\overset{\circ}{C}_+ (\Omega)$.

Let

$$A(x)(t) = \int_\Omega K(t,s)\, f(s, x(s))\, ds.$$

Then A_n converges uniformly to A on $\overset{\circ}{C}_+ (\Omega)$. Assumption (i) implies A is nondecreasing. For $\lambda \in (0,\ 1)$,

$$A^l(\lambda x)(t) = \int_\Omega K(t,s)\, f_s^{[l]}(\lambda x(s))\, ds$$

$$\begin{aligned}
&= \int_\Omega K(t,s)\,\frac{f_s^{[l]}(\lambda x(s))}{g(s,\lambda x(s))}\cdot g(s,\lambda x(s))\,ds \\
&\geq \int_\Omega K(t,s)\,\frac{f_s^{[l]}(x(s))}{g(s,x(s))}\cdot g(s,\lambda x(s))\,ds \\
&\geq \phi(\lambda)\int_\Omega K(t,s)\,f(s,x(s))\,ds \\
&= \phi(\lambda)\,A(x)(t).
\end{aligned}$$

Assumption (iii) implies that $B_\alpha(Ax) \subset \overset{\circ}{C}_+(\Omega)$ for all $x \in \overset{\circ}{C}_+(\Omega)$. By Theorem 3.2, there exists $x_0 \in \overset{\circ}{C}_+(\Omega)$ such that $A(x_0) = x_0$, and $\lim_{n\to\infty} A_n \circ A_{n-1} \circ \cdots \circ A_1(x_1) = x_0$ for any $x_1 \in \overset{\circ}{C}_+(\Omega)$.

References

1. D. W. Boyd and J. S. W. Wong, *On nonlinear contractions*, Proc. Amer. Math. Soc. 20 (1969), 458-464.
2. Y.-Z. Chen, *Thompson's metric and mixed monotone operators*, J. Math. Anal. Appl. 117 (1993), 31-37.
3. Y.-Z. Chen, *Inhomogeneous iterates of contraction mappings and nonlinear ergodic theorems*, Nonlinear Anal., TMA, 39 (2000), 1-10.
4. Y.-Z. Chen, *Continuation method for α-sublinear mappings*, to appear in Proc. Amer. Math. Soc. .
5. Y.-Z. Chen, *Path stability and nonlinear weak ergodic theorems*, to appear in Trans. Amer. Math. Soc..
6. Y.-Z. Chen, *Metric convexity and inhomogeneous iterates of mappings in ordered Banach spaces*, submitted.
7. T. Fujimoto and U. Krause, *Asymptotic properties for inhomogeneous iterates of nonlinear operators*, SIAM J. Math. Anal. 19 (1988), 841-853.
8. D. Guo and V. Lakshimikantham, *Nonlinear Problems in Abstract Cones*, Academic Press, New York, 1988.
9. J. R. Jachymski, An extension of A. Ostrowski's theorem on the round-off stability of iterations, Aequ. Math., 53 (1997), 242-253.
10. M. A. Krasnosel'skiĭ and P. P. Zabreĭko, *Geometrical Methods of Nonlinear Analysis*, Springer-Verlag, Berlin, 1984.
11. U. Krause, *A nonlinear extension of the Birkhoff-Jentzsch theorem*, J. Math. Anal. Appl. 114 (1986), pp. 552-568.
12. U. Krause, *Positive nonlinear systems: some results and applications*, Proceedings of the First World Congress of Nonlinear Analysts 1992, W. de Gruyter, Berlin 1994.

13. U. Krause and R. D. Nussbaum, *A limit set trichotomy for self-mappings of normal cones in Banach spaces*, Nonlinear Anal., TMA, 20 (1993), 855-870.

14. R. D. Nussbaum, *Iterated nonlinear maps and Hilbert's projective metric*, Mem. Amer. Math. Soc. Vol.75, No.391 (Sept. 1988).

15. A. J. B. Potter, *Application of Hilbert's projective metric to certain classes of non-homogeneous operators*, Quar. J. Math. Oxford (2), 28 (1977), 93-99.

16. A. C. Thompson, *On certain contraction mappings in a partially ordered vector space*, Proc. Amer. Math. Soc., 14 (1963), 438-443.

Proceedings of Dynamic Systems and Applications 3 (2001) 155-162

Coherent Structures ("Dissipative Solitons") in Nonlinear Systems with Dissipation

C. I. Christov[1] and M. G. Velarde[2]

[1]*Dept. of Mathematics, University of Southwestern Louisiana
Lafayette, LA 70504-1010, USA*

[2]*Instituto Plurisdisciplinar, Universidad Complutense
Paseo Juan XXIII, No1, Madrid 28040, SPAIN*

Generalized Wave Equations containing dispersion, dissipation and energy-production (GDWE) are introduced *en lieu* of dissipative NEE as more suitable models for two-way interaction of localized waves. The particle-like behavior and the long-time evolution of coherent structures upon take-over and head-on collisions are investigated numerically by means of an adequate difference scheme which represents faithfully the balance/conservation laws.

1. MODELS OF THIN-LAYER FLOWS WITH FREE SURFACE

Boussinesq [2, 3] showed that a balance between nonlinearity and dispersion was possible for the surface waves on a shallow inviscid layer and derived the first generalized wave equation (BE). He found analytically the localized solution [3] of his equation and demonstrated that the solitary wave propagates without change of shape similarly to the propagation of the linear waves. Korteweg and de Vries, [13] derived in 1895 a Nonlinear Evolution Equation (KDV) in the moving frame. The two models read:

$$\text{BE}: \quad u_{tt} = \left[\gamma^2 u + \alpha u^2 - \beta u_{xx}\right]_{xx}, \qquad \text{KDV}: \quad u_t + u u_x + u_{xxx} = 0\,.$$

The particle-like behavior of the solitary waves of KDV was revealed numerically by Zabusky & Kruskal in 1965 [14] who coined the name *soliton* to designate the nonlinear waves with *corpuscular* comportment.

Kapitza [12] initiated the investigation of the role of viscosity and surface tension in thin-layer flows. Benney [1] carried out the boundary-layer simplifications for the viscous case and Homsy [11] derived the correct weakly-nonlinear approximation

$$u_t + 2\alpha_1 u u_x + \alpha_2 u_{xx} + \alpha_4 u_{xxxx} = 0\,,$$

known now as Kuramoto–Sivashinsky KS equation.

KS represents the extreme case when the dispersion is negligible in comparison with dissipation while KDV is the other extreme. In between these two extreme cases is the dissipation modified Korteweg-de Vries equation (KDV-KS):

$$u_t + 2\alpha_1 u u_x + \alpha_2 u_{xx} + \alpha_3 u_{xxx} + \alpha_4 u_{xxxx} = 0\,.$$

derived first in [10] and treated extensively in the literature since (see, e.g., [4, 9]). The numerical simulations [8] show that the localized solutions of KDV-KS behave in many instances as quasi-solitons which warrants the coinage "dissipative solitons".

2. TWO-WAVE GENERALIZATION OF KDV-KS

Dissipation was added to BE in somewhat heuristic fashion in [6] arriving at the following generalized wave equation

$$u_{tt} = \left[\gamma^2 u + \alpha_1 u^2 + \alpha_3 u_{xx} - \alpha_4 u_{xxt} - \alpha_2 u u_t\right]_{xx} . \tag{1}$$

For $\alpha_2 = \alpha_4 = 0$ (no energy dissipation and/or energy input) the last equation reduces to BE. On the other hand for slowly evolving motions in the frame moving to the right with phase speed γ one has $t_1 = \frac{1}{2}t$, $x_1 = x - \gamma t$, $u(t,x) = u_1(t_1, x_1)$. Then

$$\frac{\partial u}{\partial x} = \frac{\partial u_1}{\partial x_1}, \quad \frac{\partial u}{\partial t} = \frac{1}{2}\frac{\partial u_1}{\partial t_1} - \gamma\frac{\partial u_1}{\partial x_1}, \quad \frac{\partial^2 u}{\partial t^2} = \frac{1}{4}\frac{\partial^2 u_1}{\partial t_1^2} - \gamma\frac{\partial^2 u_1}{\partial t_1 \partial x_1} + \gamma^2\frac{\partial^2 u_1}{\partial x_1^2}, \tag{2}$$

$$\left|\frac{\partial^m u_1}{\partial t_1 \partial x_1^{m-1}}\right| \ll \gamma \left|\frac{\partial^m u_1}{\partial x_1^m}\right|, \quad \left|\frac{\partial^2 u_1}{\partial t_1^2}\right| \ll \gamma \left|\frac{\partial^2 u_1}{\partial t_1 \partial x_1}\right|, \tag{3}$$

and upon neglecting the higher-order terms one gets

$$-\gamma\frac{\partial^2 u_1}{\partial x_1 \partial t_1} = \frac{\partial^2}{\partial x_1^2}\left[\alpha_1 u_1^2 + \alpha_2\gamma\frac{\partial u_1}{\partial x_1} + \alpha_3\frac{\partial^2 u_1}{\partial x_1^2} + \alpha_4\gamma\frac{\partial^3 u_1}{\partial x_1^3}\right] .$$

After one integration with respect to the spatial variable x_1 the last equation recasts to the KDV-KS equation, namely

$$u_t + 2\bar{\alpha}_1 u u_x + \bar{\alpha}_2 u_{xx} + \bar{\alpha}_3 u_{xxx} + \bar{\alpha}_4 u_{xxxx} = 0 . \tag{4}$$

Thus (1) encompasses both BE and KDV-KSV.

3. MASS CONSERVATION AND ENERGY BALANCE LAW

Upon introducing an auxiliary function q the eq.(1) is transformed to the following b.v.p.

$$\begin{aligned} u_t &= q_{xx} , \\ q_t &= \gamma^2 u + \alpha_1 u^2 - \beta u_{xx} - \alpha_4 u_{xxt} - \alpha_2 u_t , \\ & u = q_x = 0 , \quad \text{for} \quad x = -L_1, L_2 , \end{aligned} \tag{5}$$

where $-L_1, L_2$ are the values of spatial coordinate at which we truncate he infinite interval (so-called "actual infinities"). For the "mass" and energy of the wave system

$$M \stackrel{\text{def}}{=} \int_{-L_1}^{L_2} u dx , \qquad E \stackrel{\text{def}}{=} \int_{-L_1}^{L_2} \frac{1}{2}\left[\gamma^2 u^2 + q_x^2 + \frac{1}{3}\alpha_1 u^3 + \beta u_x^2\right] dx ,$$

the following conservation and balance laws are derived in [6]

$$\frac{dM}{dt} = 0 , \qquad \frac{dE}{dt} = \alpha_2 \int_{-L_1}^{L_2} u_t^2 dx - \alpha_4 \int_{-L_1}^{L_2} u_{xt}^2 dx . \tag{6}$$

We consider a uniform grid in $[-L_1, L_2]$ with spacing h and total number of points N: $x_i = -L_1 + (i-1)h$, $h = (L_1 + L_2)/(N-1)$, and make use of the implicit difference scheme from [6] which represents faithfully the balance (conservation) laws (6).

4. RESULTS AND DISCUSSION FOR PURELY DISSIPATIVE CASE ($\alpha_3=0$)

It is well known now that the dispersion stabilizes the structures in KDV-KS [4, 8, 9]. So that the dispersionless case $\alpha_3 = 0$ is the hardest to treat and we succeed to demonstrate quasi-soliton behavior it will be much more so for the cases with dispersion. For definiteness we choose the rest of the parameters as follows: $\alpha_1 = 3, \alpha_2 = \alpha_4 = 1$.

The shape of a stationary coherent structure $u(x,t) = v(\xi)$, $\xi = x - ct$ is governed by the the following b.v.p.

$$(\gamma^2 - c^2)v + \alpha_1 v^2 - \alpha_4 cv''' + \alpha_2 cv' = 0\,, \qquad v \to 0 \quad \text{for} \quad \xi \to \pm\infty\,. \tag{7}$$

Denoting $v = cF(\xi)$ one gets

$$-sF + \alpha_1 F^2 + \alpha_4 F''' + \alpha_2 F' = 0\,, \quad s = -(\gamma^2 - c^2)c^{-1}\,.$$

In [5] it is shown that only two solutions of homoclinics type exist, say F_+ and F_-. The respective eigen-values for Cc are $s = \pm 2\hat{s}$, $2\hat{s} \approx 1.216$ and the two solutions are mirror images to each other, namely $F_-(\xi) = -F_+(-\xi)$. These are shown in Figure 1.

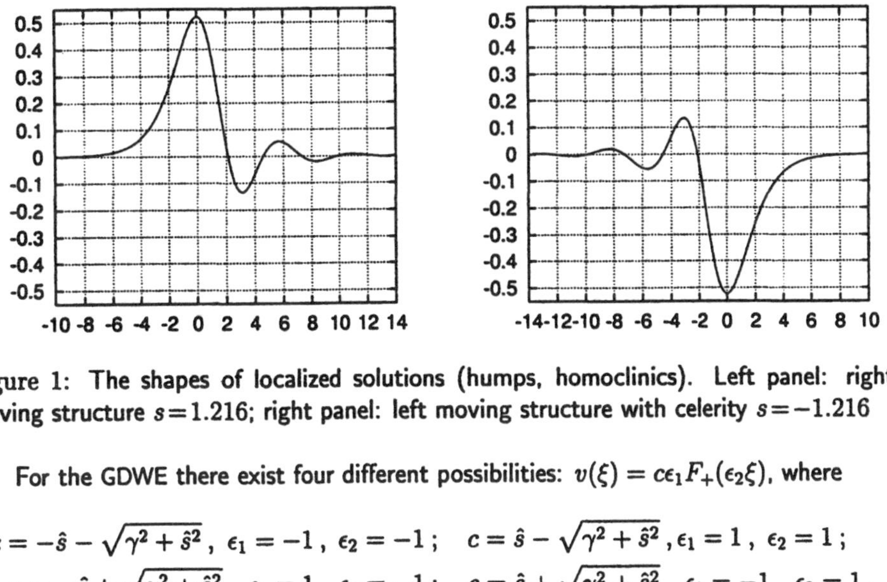

Figure 1: The shapes of localized solutions (humps, homoclinics). Left panel: right moving structure $s=1.216$; right panel: left moving structure with celerity $s=-1.216$

For the GDWE there exist four different possibilities: $v(\xi) = c\epsilon_1 F_+(\epsilon_2 \xi)$, where

$$c = -\hat{s} - \sqrt{\gamma^2 + \hat{s}^2}\,,\ \epsilon_1 = -1\,,\ \epsilon_2 = -1\,; \quad c = \hat{s} - \sqrt{\gamma^2 + \hat{s}^2}\,, \epsilon_1 = 1\,,\ \epsilon_2 = 1\,;$$
$$c = -\hat{s} + \sqrt{\gamma^2 + \hat{s}^2}\,,\ \epsilon_1 = 1\,,\ \epsilon_2 = -1\,; \quad c = \hat{s} + \sqrt{\gamma^2 + \hat{s}^2}\,,\ \epsilon_1 = -1\,,\ \epsilon_2 = 1\,.$$

From here sixteen different initial wave configurations can be constructed.

The first case we consider is the take-over collision of two coherent structures depicted in Figure 2. The evolution is indeed slow in the moving frame and the physical situation is the one for which the NEE of type of KS is derived. Yet there is a principal difference between the GDWE considered here and the NEE: in KS case the two structures travel at the same phase speed and can form bound states and no take-over collision can be investigated. On can see in the Figure that the slower structure does not survive the interaction while the faster reappear from it pretty much in tact. Yet the interaction triggers the energy-input mechanism and after long enough time the long-wave instability transform the motion to an apparently chaotic one.

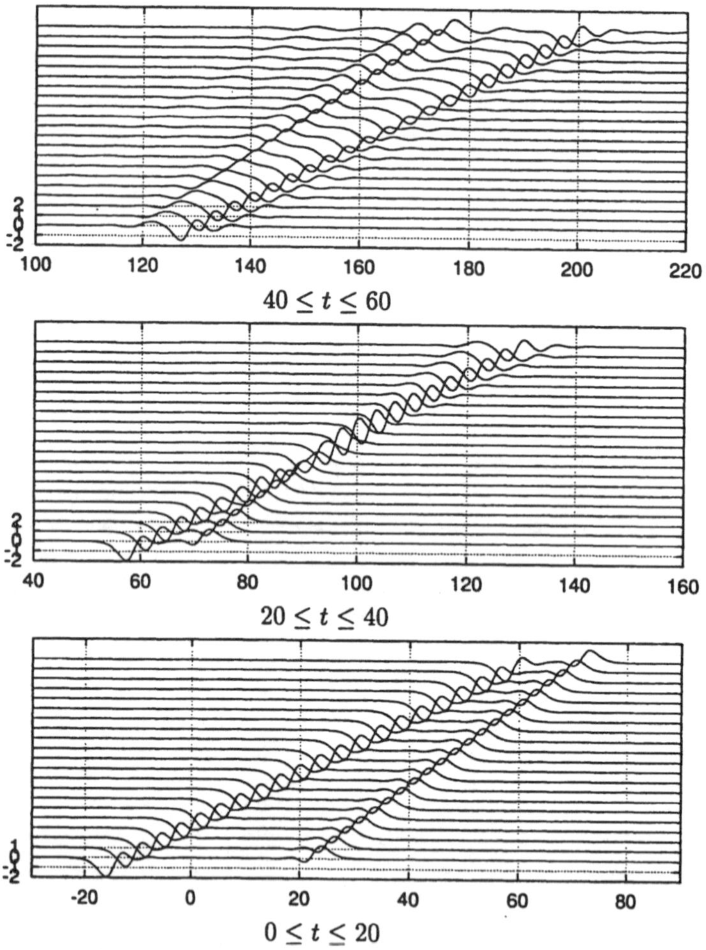

Figure 2: Take-over interaction of a supercritical ($c = 3.667$) and a subcritical ($c = 2.453$) coherent structures for $\gamma = 3$.

In Figure 3 an important case of head-on collision of two structures of the same shape and energy (both supercritical in the sense that $c > \gamma$) is shown. This case corresponds roughly to the head-on collision in KDV-KS treated in [8] where the coherent structures are shown to break down after the collision. The behavior observed in the present work is drastically different showing that the coherent structures may retain their identity after the collision. This means that retaining the Newtonian inertia (the second time derivative) is essential for the adequate modelling of the head-on collisions. Note that the phase shift is rather small in this purely inelastic model. It means that the notion of "inelastic" collision defined solely on the base of the appearance of a phase shift needs refinement.

The same conclusion can be drawn from Figure 4 where the head-on collision of two different coherent structures (supercritical and subcritical one) is shown. The structures retain their shapes and the phase shift exhibit the same properties as for the conservative system (see [7]), namely the smaller structure is shifted more.

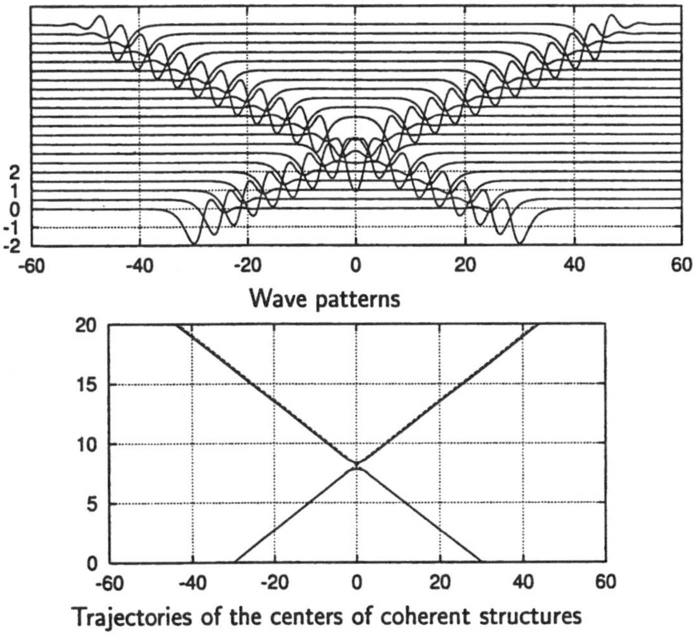

Figure 3: Head-on collision of supercritical coherent structures ($c = \pm 3.667$) for $\gamma = 3$.

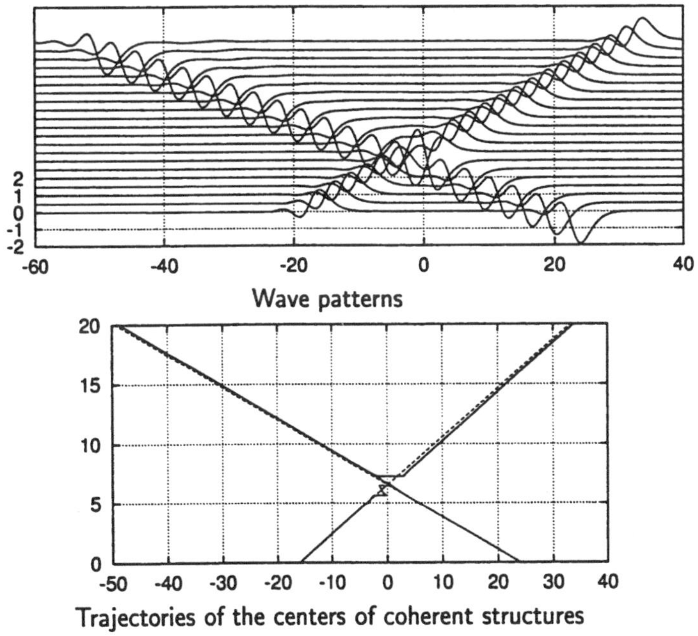

Figure 4: Head-on collision of a supercritical ($c = -3.667$) and a subcritical ($c = 2.453$) structures for $\gamma = 3$.

Figures 3 and 4 suggest that the soliton-like behaviour and the almost-elastic collision can be observed in purely inelastic system of dissipation/excitation type. Note that this is only true for supercritical structures.

The above described behavior is even better observed for smaller characteristic speed (say $\gamma = 3$) in Fig. 5

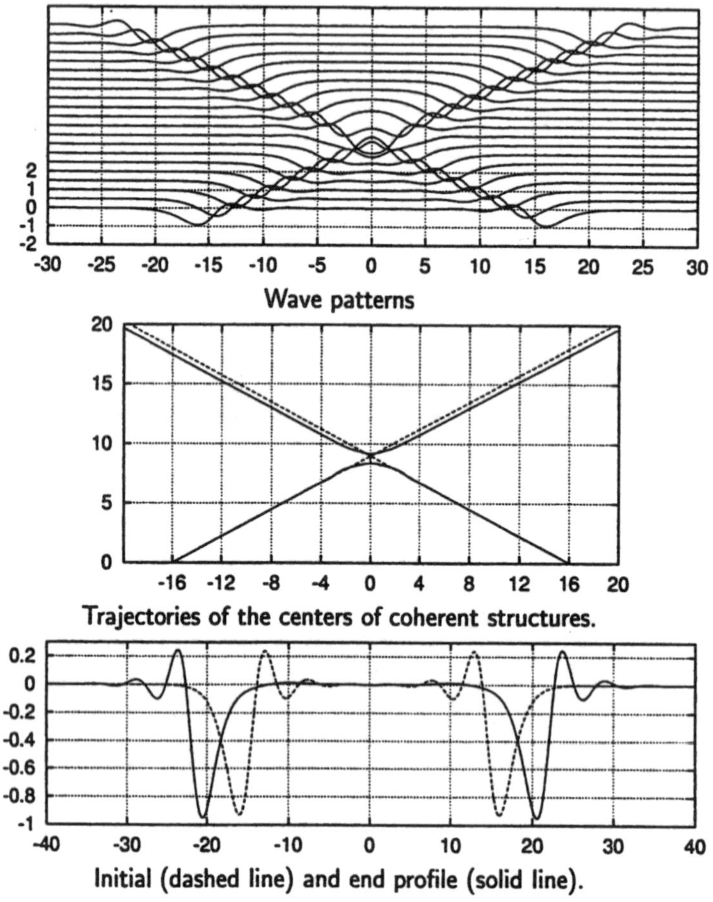

Figure 5: Head-on collision of two supercritical humps: $c = \pm 1.778, \gamma = 1$.

Completely different is the situation with the head-on collision of subcritical structures. Upon collision they stick to each other (feature observed in [8] for hydraulic jumps) and form a single bulge. Then this wave shape grows and blows-up in finite time as shown in Figure 6

5. CONCLUSIONS

The second-time derivative (the linear Newton inertia) in GDWE stabilizes the shapes of the coherent structures. The deformation of the wave shapes during collisions does not immediately trigger the long-wave instability and a clear cut quasi-particle behavior is

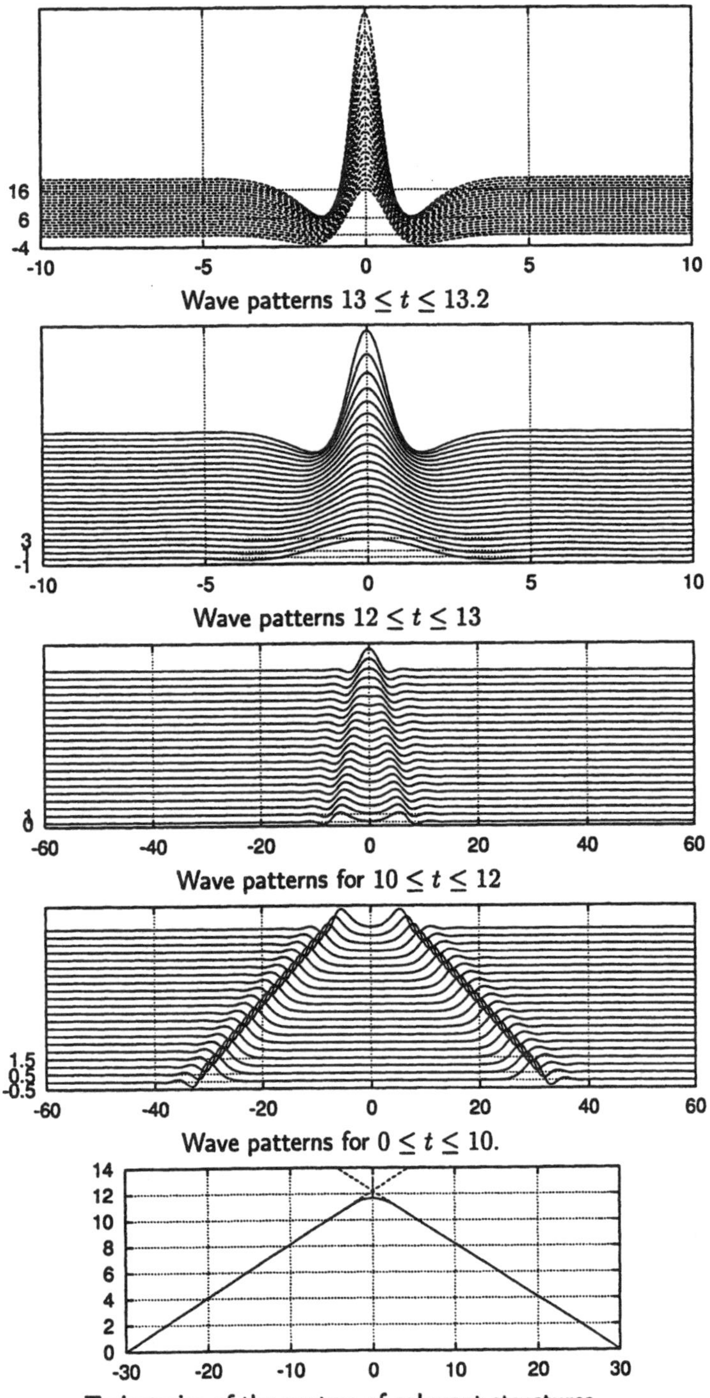

Figure 6: Blow-up in the head-on collision of two subcritical humps: $c = \pm 2.453, \gamma = 3$.

observed at long times. Similarly to the NEE the energy-input mechanism always brings about the long-wave instability given enough time. Unlike NEE, the quasi-particle behavior is unequivocally demonstrated for moderate times. In take-over collisions the supercritical structures survive while the subcritical evolve into wave shapes which increase in time.

In head-on collisions the supercritical coherent structures (or a super- and sub-critical) behave like quasi-particles and retain identity after collision. In the latter case the development of the long-wave instability is faster. After a head-on collision the subcritical coherent structures form a single "bulge" whose shape triggers a nonlinear blow-up in finite time.

ACKNOWLEDGEMENT

This work is supported by Grant LEQSF(1999-2002)-RD-A-49 from the Louisiana Board of Regents.

References

[1] D. J. Benney, Long waves in liquid film, J. Maths & Phys., Vol 45 (1966), 150–155.

[2] J. V. Boussinesq, Théorie de l'intumescence liquide appelée onde solitaire ou de translation, se propageant dans un canal rectangulaire, Comp. Rend. Hebd. des Seances de l'Acad. des Sci., Vol 72 (1871), 755–759.

[3] J. V. Boussinesq, Théorie des ondes et des remous qui se propagent le long d'un canal rectangulaire horizontal, en communiquant au liquide contenu dans ce canal des vitesses sensiblement pareilles de la surface au fond, Journal de Mathématiques Pures et Appliquées, Vol 17 (1872), 55–108.

[4] H.-C. Chang, Evolution of nonlinear waves on vertically falling films – a normal form analysis, Chem. Engn. Sci., Vol 42 (1987), 515–533.

[5] C. I. Christov & M. G. Velarde, On localized solutions of an equation governing Bénard-Marangoni convection, Appl. Math. Modelling, Vol 17 (1993), 311–320.

[6] C. I. Christov & M. G. Velarde, Evolution and interaction of solitary waves (solitons) in nonlinear dissipative systems, Physica Scripta, Vol T55 (1994), 101–106.

[7] C. I. Christov & M. G. Velarde, Inelastic interaction of Boussinesq solitons, J. Bifurcation & Chaos, Vol 4 (1994), 1095–1112.

[8] C. I. Christov & M. G. Velarde, Dissipative solitons, Physica, Vol D 86 (1995), 323–347.

[9] C. Elphick, G. R. Ierley, O. Regev & E. A. Spiegel, Interacting localized structures with Galilean invariance, Phys. Rev., Vol A44 (1991), 1110–1122.

[10] P. L. García-Ybarra, J. L. Castillo & M. G. Velarde, Bénard–Marangoni convection with a deformable interface and poorly conducting boundaries, Phys. Fluids, Vol 30 (1987), 2655–2661.

[11] G. M. Homsy, Model equations for wavy viscous film flow, Lect. in Appl. Math., Vol 15 (1974), 191–194.

[12] P. L. Kapitza, Wave flow in thin layer of viscous liquid, Zh. Exp. Theor. Fiz., Vol 18 (1948), 3–18. in Russian.

[13] D. J. Korteweg & G. de Vries, On the change of form of long waves advancing in a rectangular canal, and on a new type of long stationary waves, Phil. Mag., ser.5, Vol 39 (1895), 422–443.

[14] N. J. Zabusky & M. D. Kruskal, Interaction of 'solitons' in collisionless plasma and the recurrence of initial states, Phys. Rev. Lett., Vol 15 (1965), 57–62.

Proceedings of Dynamic Systems and Applications 3 (2001) 163-168

EXISTENCE OF TRIPLE POSITIVE SOLUTIONS FOR A NONLINEAR IMPULSIVE BOUNDARY VALUE PROBLEM

John M. Davis[1], P.W. Eloe[2] and M.N. Islam[2]

[1]Department of Mathematics
Baylor University, Waco, TX 76798 USA
[2]Department of Mathematics
University of Dayton, Dayton, OH 45469-2316 USA

ABSTRACT: We prove the existence of at least three positive solutions of the nonlinear impulsive boundary value problem $x''(t) = f(t, x(t))$, $t \in [0,1] \setminus \{t_1, \ldots, t_m\}$, $x(0) = x(1) = 0$, $\Delta x(t_k) = p_k(x(t_k))$, $\Delta x'(t_k) = q_k(x(t_k))$, $k = 1, \ldots, m$. Cone theoretic arguments and a nonlinear fixed point theorem due to Leggett and Williams are used.

AMS (MOS) Subject Classification. 34A37, 34B15

1. INTRODUCTION

Let $0 = t_0 < t_1 < \cdots < t_m < t_{m+1} = 1$ be given. Define the impulse

$$\Delta x(t_k) = x(t_k^+) - x(t_k^-)$$

where $x(t_k)$ indicates $x(t_k^-)$, $k = 1, 2, \ldots, m$. We establish the existence of at least three positive solutions to the following nonlinear impulsive boundary value problem (IBVP)

$$x''(t) = f(t, x(t)), \; t \in [0,1] \setminus \{t_1, \ldots, t_m\}, \tag{1.1}$$

$$x(0) = x(1) = 0, \tag{1.2}$$

$$\begin{cases} \Delta x(t_k) = p_k(x(t_k)) \\ \Delta x'(t_k) = q_k(x(t_k)) \qquad k = 1, \ldots m, \end{cases} \tag{1.3}$$

where $f : [0,1] \times [0, \infty) \to (-\infty, 0]$, $q_k : [0, \infty) \to (-\infty, 0]$, and

$$t_k q_k(y) \le p_k(y) \le (t_k - 1) q_k(y), \qquad y \ge 0, \; k = 1, \ldots, m. \tag{1.4}$$

We assume throughout that the functions f, p_k, and q_k are continuous.

Recent attention has been directed toward obtaining the existence of multiple positive solutions (in particular, triple positive solutions) for boundary value problems for ordinary differential equations. We refer the reader to Chyan & Davis [4], Chyan et al [5], Davis et al [6], Davis & Henderson [7], Henderson & Thompson [13], Guo & Lakshmikantham [11], and the references therein for details.

Impulsive differential equations have also been of interest lately. See the books by Bainov et al [1] as well as Bainov & Simeonov [2,3] for a comprehensive treatment of this subject. Eloe & Islam [10] work with monotone methods for IBVPs. The papers by Eloe & Henderson [9] and Guo & Liu [12] deal with the existence of multiple positive solutions for IBVPs but each uses techniques very different from those presented here.

Each of the papers [4, 5, 7, 13] on the existence of triple positive solutions appeals to a theorem by Leggett and Williams developed using the fixed point index in ordered Banach spaces. We remark that none of the previously cited papers has made use of this Leggett-Williams Fixed Point Theorem in the context of IBVPs. Leggett & Williams [14] applied their nonlinear fixed point theorem to prove the existence of at least three positive solutions for Hammerstein integral equations of the form

$$y(x) = \int_\Omega G(x,s)g(s,y(s))\,ds, \qquad \Omega \subset R^N$$

by making use of suitable inequalities imposed on G and g.

The layout of the paper is as follows. In Section 2, we provide some definitions and background results and we state the Leggett-Williams Fixed Point Theorem. Then in Section 3, we impose growth conditions on the nonlinearity which allow us to apply the Leggett-Williams Fixed Point Theorem to obtain the existence of at least three positive solutions of (1.1)-(1.3).

2. LEGGETT-WILLIAMS THEOREM

For the convenience of the reader, we include here the necessary definitions and the Leggett-Williams Theorem from cone theory in Banach spaces.

Definition 2.1 Let $\mathcal{B}$ be a Banach space over R. A nonempty, closed set $\mathcal{P} \subset \mathcal{B}$ is said to be a cone provided

(a) $\alpha\mathbf{u} + \beta\mathbf{v} \in \mathcal{P}$ for all $\mathbf{u}, \mathbf{v} \in \mathcal{P}$ and all $\alpha, \beta \geq 0$, and

(b) $\mathbf{u}, -\mathbf{u} \in \mathcal{P}$ implies $\mathbf{u} = \mathbf{0}$.

Definition 2.2 A Banach space $\mathcal{B}$ is called a partially ordered Banach space if there exists a partial ordering $\preceq$ on $\mathcal{B}$ satisfying

(a) $\mathbf{u} \preceq \mathbf{v}$, for $\mathbf{u}, \mathbf{v} \in \mathcal{B}$ implies $t\mathbf{u} \preceq t\mathbf{v}$, for all $t \geq 0$, and

(b) $\mathbf{u}_1 \preceq \mathbf{v}_1$ and $\mathbf{u}_2 \preceq \mathbf{v}_2$, for $\mathbf{u}_1, \mathbf{u}_2, \mathbf{v}_1, \mathbf{v}_2 \in \mathcal{B}$ imply $\mathbf{u}_1 + \mathbf{u}_2 \preceq \mathbf{v}_1 + \mathbf{v}_2$.

Let $\mathcal{P} \subset \mathcal{B}$ be a cone and define $\mathbf{u} \preceq \mathbf{v}$ if and only if $\mathbf{v} - \mathbf{u} \in \mathcal{P}$. Then $\preceq$ is a partial ordering on $\mathcal{B}$ and we will say that $\preceq$ is the partial ordering induced by $\mathcal{P}$. Moreover, $\mathcal{B}$ is a partially ordered Banach space with respect to $\preceq$.

We also state the following definitions for future reference.

Definition 2.3 The map α is a nonnegative continuous concave functional on $\mathcal{P}$ provided $\alpha : \mathcal{P} \to [0, \infty)$ is continuous and

$$\alpha(tx + (1-t)y) \geq t\alpha(x) + (1-t)\alpha(y)$$

for all $x, y \in \mathcal{P}$ and $0 \leq t \leq 1$.

Definition 2.4 Let $0 < a < b$ be given and α be a nonnegative continuous concave functional on the cone $\mathcal{P}$. Define the convex sets $\mathcal{P}_r$ and $\mathcal{P}(\alpha, a, b)$ by

$$\mathcal{P}_r = \{y \in \mathcal{P} : ||y|| < r\} \quad \text{and} \quad \mathcal{P}(\alpha, a, b) = \{y \in \mathcal{P} : a \leq \alpha(y),\ ||y|| \leq b\}.$$

Next we state the Leggett-Williams Fixed Point Theorem. The proof can be found in Deimling's text [8] and utilizes the fixed point index in ordered Banach spaces.

Theorem 2.1 (Leggett-Williams Fixed Point Theorem) Let $c > 0$ be given and let $\mathcal{A} : \bar{\mathcal{P}}_c \to \bar{\mathcal{P}}_c$ be a completely continuous operator and let α be a nonnegative continuous concave functional on $\mathcal{P}$ such that $\alpha(y) \le ||y||$ for all $y \in \bar{\mathcal{P}}_c$. Suppose there exist $0 < a < b < d \le c$ such that

(C1) $\{y \in \mathcal{P}(\alpha, b, d) : \alpha(y) > b\} \neq \emptyset$ and $\alpha(\mathcal{A}y) > b$ for $y \in \mathcal{P}(\alpha, b, d)$,

(C2) $||\mathcal{A}y|| < a$ for $||y|| \le a$, and

(C3) $\alpha(\mathcal{A}y) > b$ for $y \in \mathcal{P}(\alpha, b, c)$ with $||\mathcal{A}y|| > d$.

Then $\mathcal{A}$ has at least three fixed points y_1, y_2, and y_3 such that $||y_1|| < a$, $b < \alpha(y_2)$, and $||y_3|| > a$ with $\alpha(y_3) < b$.

3. TRIPLE POSITIVE SOLUTIONS

We will apply the following lemma to solutions of (1.1) on each subinterval $[t_k, t_{k+1}]$, $k = 0, \ldots, m$. See Eloe & Henderson [9] for details.

Lemma 3.1 Suppose $x(t) \in C^2[a, b]$, $x''(t) \le 0$ on $a \le t \le b$ with $x(a) \ge 0$, $x(b) \ge 0$, $a < b$. Then

$$x(t) \ge ||x||/4, \qquad (3a + b)/4 \le t \le (a + 3b)/4.$$

Our goal is to apply Theorem 2.1 to a completely continuous operator whose kernel is the Green's function, $G(t, s)$, for

$$x''(t) = 0, \qquad 0 \le t \le 1,$$
$$x(0) = x(1) = 0$$

(It is well known that $G(t, s) \le 0$ for each $(t, s) \in [0, 1] \times [0, 1]$.) In preparation of doing so, define the Banach space

$$\mathcal{B} = \{x \in PC[0, 1] : x(t) \in C[t_k, t_{k+1}],\ k = 0, \ldots, m\}$$

(where $PC[0, 1]$ denotes the piecewise continuous functions on $[0, 1]$) with the norm

$$||x|| = \max_{0 \le k \le m} ||x||_k \qquad \text{where} \qquad ||x||_k = \max_{t_k \le t \le t_{k+1}} |x(t)|.$$

For each $k = 0, \ldots, m$, define the subintervals

$$S_k = [(3t_k + t_{k+1})/4, (t_k + 3t_{k+1})/4]$$

and define the cone $\mathcal{P} \subset \mathcal{B}$ by

$$\mathcal{P} = \left\{x \in \mathcal{B} : x(t) \ge 0,\ \min_{t \in S_k} x(t) \ge ||x||_k/4,\ k = 0, \ldots, m\right\}.$$

For each function $u \in \mathcal{B}$, define the operator $\mathcal{T}$ by

$$\mathcal{T}u(t) = I(t,u) + \int_0^1 G(t,s)f(s,u(s))\,ds,$$

where the impulse, $I(t,u)$, is given by

$$I(t,u) = \sum_{k=1}^{m} I_k(t,u)$$

and

$$I_k(t,u) = \begin{cases} t(-p_k(u(t_k)) - (1-t_k)q_k(u(t_k))), & 0 \le t \le t_k, \\ (1-t)(p_k(u(t_k)) - t_k q_k(u(t_k))), & t_k \le t \le 1. \end{cases}$$

Because of (1.4), $u(t) \ge 0$ implies $I_k(t,u(t)) \ge 0$, $0 \le t \le 1$. Hence $\mathcal{T}u \ge 0$ for each $u \in \mathcal{P}$. Moreover, if $u \in \mathcal{P}$, then $\mathcal{T}u$ satisfies the conditions of Lemma 3.1 on each subinterval $[t_k, t_{k+1}]$, $k = 0, \ldots, m$, and therefore $\mathcal{T} : \mathcal{P} \to \mathcal{P}$. Standard applications of the Arzela-Ascoli Theorem on each subinterval $[t_k, t_{k+1}]$ show that $\mathcal{T} : \mathcal{B} \to \mathcal{B}$ is completely continuous. Finally, if $x \in \mathcal{P}$ is a fixed point of $\mathcal{T}$, then x is a nonnegative solution of (1.1)-(1.3).

We will also need the following inequalities. Eloe & Henderson [9] showed

$$|I(t,x)| \le m\left(\max_{1 \le k \le m} |q_k(x(t_k))|\right). \tag{3.1}$$

Also, Davis & Henderson [7] proved

$$\max_{t\in[0,1]} \int_0^1 |G(t,s)|\,ds \le 1/6. \tag{3.2}$$

We now present the main result of the paper.

Theorem 3.1 Let $0 < a < b < c$ be such that

(i) $|f(t,w)| < 3a$ and $\max_{1\le k\le m} |q_k(w)| \le a/2m$ for $0 \le w \le a$, $t \in [0,1]$,

(ii) $|f(t,w)| \ge Lb$, for $b \le w \le c$, $t \in [0,1]$, where

$$L = \left(\min_{0\le k\le m} \min_{t\in S_k} \int_{S_k} -G(t,s)\,ds\right)^{-1},$$

(iii) $|f(t,w)| \le 3c$ and $\max_{1\le k\le m} |q_k(w)| \le c/2m$ for $0 \le w \le c$, $t \in [0,1]$.

Then the IBVP (1.1)-(1.3) has at least three positive solutions x_1, x_2, and x_3 satisfying

$$||x_1|| < a, \qquad b < \min_{0\le k\le m}\left\{\min_{t\in S_k} |x_2(t)|\right\},$$

$$||x_3|| > a, \qquad b > \min_{0\le k\le m}\left\{\min_{t\in S_k} |x_3(t)|\right\}.$$

Proof: We begin by defining the nonnegative continuous concave functional α by

$$\alpha(x) = \min_{0 \le k \le m} \left\{ \min_{t \in S_k} x(t) \right\}, \qquad \text{for } x \in \mathcal{P}.$$

We will prove that the conditions of Theorem 2.1 are satisfied.

First, we show that $\mathcal{T} : \bar{\mathcal{P}}_c \to \bar{\mathcal{P}}_c$. To do so, choose $x \in \bar{\mathcal{P}}_c$. Then $||x|| \le c$ and by assumption (iii), $|f(t, x(t))| \le 3c$, $t \in [0,1]$. Employing (3.1) and (3.2), we see that

$$\begin{aligned}
||\mathcal{T}x|| &= \max_{t \in [0,1]} \left| I(t,x) + \int_0^1 G(t,s) f(s, x(s))\, ds \right| \\
&\le \max_{t \in [0,1]} \left\{ |I(t,x)| + \int_0^1 |G(t,s)||f(s,x(s))|\, ds \right\} \\
&\le \max_{t \in [0,1]} \left\{ m \left(\max_{1 \le k \le m} |q_k(x(t_k))| \right) + 3c \cdot 1/6 \right\} \\
&\le c/2 + c/2 \\
&= c.
\end{aligned}$$

Hence $\mathcal{T} : \bar{\mathcal{P}}_c \to \bar{\mathcal{P}}_c$.

In a similar way, using assumption (i), one can show that $||x|| \le a$ implies $||\mathcal{T}x|| < a$. Consequently, condition (C2) of Theorem 2.1 is fulfilled.

To verify property (C1) of Theorem 2.1, we note that if we let $y(t) \equiv c$ then $y \in \mathcal{P}(\alpha, b, c)$ and $\alpha(y) = c > b$. Therefore

$$\{x \in \mathcal{P}(\alpha, b, c) : \alpha(x) > b\} \neq \emptyset.$$

For $x \in \mathcal{P}(a, b, c)$, we know $\alpha(x) \ge b$ and $||x|| \le c$ by definition. Thus $b \le |x(s)| \le c$ for $s \in S_k$, $k = 0, \ldots, m$. As a result, (ii) implies $|f(x, x(s)| \ge Lb$ for each $s \in S_k$. Thus

$$\begin{aligned}
\alpha(\mathcal{T}x) &= \min_{0 \le k \le m} \min_{t \in S_k} \left(I(t,x) + \int_0^1 G(t,s) f(s,x(s))\, ds \right) \\
&\ge \min_{0 \le k \le m} \min_{t \in S_k} \left(\int_0^1 G(t,s) f(s,x(s))\, ds \right) \\
&> \min_{0 \le k \le m} \min_{t \in S_k} \left(\int_{S_k} G(t,s) f(s,x(s))\, ds \right) \\
&\ge Lb \min_{0 \le k \le m} \min_{t \in S_k} \left(\int_{S_k} -G(t,s)\, ds \right) \\
&= b.
\end{aligned}$$

Hence, condition (C1) of Theorem 2.1 is satisfied.

Notice that in the statement of our theorem, we are using the special case of Theorem 2.1 where $d = c$. Because of this, the argument for showing (C3) holds is completely analogous to the argument for showing (C1) holds.

REFERENCES

1. D.D. Bainov, V. Lakshmikantham, and P.S. Simeonov "Theory of Impulsive Differential Equations," World Scientific, Singapore, 1989.

2. D.D. Bainov and P.S. Simeonov "Impulsive Differential Equations: Periodic Solutions and Applications," Longman Scientific and Technical, Essex, 1993.

3. D.D. Bainov and P.S. Simeonov "Systems with Impulse Effects," Ellis Horwood Limited, Chichister, 1989.

4. C.J. Chyan and J.M. Davis, Existence of triple positive solutions for (n,p) and (p,n) boundary value problems, *Commun. Appl. Anal.*, in press.

5. C.J. Chyan, J.M. Davis and W.K.C. Yin, Existence of triple positive solutions for $(k, n-k)$ right focal boundary value problems, *Nonlinear Stud.*, in press.

6. J.M. Davis, P.W. Eloe, and J. Henderson, Triple positive solutions for multipoint conjugate boundary value problems, *Georgian Math. J.*, in press.

7. J.M. Davis and J. Henderson, Triple positive symmetric solutions for a Lidstone boundary value problem, *Differential Equations Dynam. Systems*, in press.

8. K. Deimling, "Nonlinear Functional Analysis," Springer-Verlag, New York, 1985.

9. P.W. Eloe and J. Henderson, Positive solutions of boundary value problems for ordinary differential equations with impulse, *Dynam. Contin. Discrete Impuls. Systems* **4** (1998), 285–294.

10. P.W. Eloe and M.N. Islam, Monotone methods and fourth order Lidstone boundary value problems with impulse effects, *Commun. Appl. Anal.*, in press.

11. D. Guo and V. Lakshmikantham, "Nonlinear Problems in Abstract Cones," Academic Press, San Diego, 1988.

12. D. Guo and X. Liu, Multiple positive solutions of boundary value problems for impulsive differential equations, *Nonlinear Anal.* **25** (1995), 327–337.

13. J. Henderson and H.B. Thompson, Existence of multiple solutions for some nth order boundary value problems, *Proc. Amer. Math. Soc.*, in press.

14. R. Leggett and L. Williams, Multiple positive fixed points of nonlinear operators on ordered Banach spaces, *Indiana Univ. Math. J.* **28** (1979), 673–688.

Proceedings of Dynamic Systems and Applications 3 (2001) 169-176

SOLVING STRONGLY SINGULAR INTEGRAL EQUATIONS

JOHAN H DE KLERK

School for Computer, Statistical and Mathematical Sciences
Potchefstroom University for CHE, Potchefstroom
South Africa

ABSTRACT: It has previously been shown that integral equations can be solved effectively by means of an L_1-approximation technique. This method has also successfully been applied to singular integral equations. In the present discussion attention is paid to strongly singular integral equations, and in particular to the case of strongly singular integral equations containg Hadamard integrals.

AMS(MOS) 45Exx

1. THE PROBLEM

Integral equations are encountered in many mathematical models and a variety of numerical methods for obtaining solutions to integral equations has been proposed during the last few decades. Various kinds of orthogonal polynomials have been used in some of these methods to solve the given integral equation by means of collocation methods (among others, that of Nied [1], who uses Chebyshev polynomials of the second kind). The basis of these methods is the reduction of the integral equation to a set of algebraic equations by means of orthogonal polynomials. When collocation is used, the number of terms, N, say, in the polynomial expansion is equal to the number of collocation points. However, as Nied mentions: "It is not difficult to show that for large values of N, the system of equations ... becomes badly ill-conditioned." Other methods are based on successive approximations (for example, the pseudo-analytical method of Hanna and Brown [2]) or on iterative methods (for example, the method of Kleinman and Van den Berg [3]).

Still another strategy for solving integral equations is a technique based on L_p approximations. By reformulating the problem of solving an integral equation as a continuous L_p approximation problem, it could be solved numerically by a minimization technique, obtaining an approximate solution to the original problem. The L_p approximation technique described in the present paper is related to well-known methods in the sense that minimization of the L_2 norm of the residue, for example, is equivalent to a weighted Galerkin technique. Minimization of the residue in other norms will only in special cases reduce to the Galerkin method. In general, these norms will of course lead to a different result for the numerical solution on a given basis. This fact makes it more difficult to compare different techniques with each other.

This approximation method is discussed in detail in De Klerk [4] and De Klerk, Eyre and Venter [5]. In the last paper attention is paid to the case of singular integral equations consisting of integrals of Cauchy type. The purpose of the present paper, however, is to show that the L_p approximation method (and in particular the special case of an L_1 approximation method) could also be extended to the case of strongly singular integral equations.

Typeset by $\mathcal{A}_{\mathcal{M}}\mathcal{S}$-TEX

The integrals encountered in this study are in general of the form

$$\oint_a^b \frac{k(x,t)f(t)}{(t-x)^2}dt, \quad a < x < b,$$

which is to be understood in the Hadamard or finite part sense (see for example Kaya and Erdogan[6]).

In section 2 the mathematical fundamentals are discussed and a brief description of the general integral equation problem is provided and how to reformulate it as an approximation problem. In section 3 attention is paid to the numerical procedure, and in section 4 an application will be presented with appropriate remarks.

2. MATHEMATICAL FUNDAMENTALS

2.1 The general integral equation problem

In general our concern is to solve a Fredholm integral equation (linear or non-linear) with typical form

$$\phi(x) - \lambda \int_a^b K(x,t,\phi(t))dt = \psi(x),$$

with $x \in [a,b]$.

In this formulation we will assume that the so-called driving function ψ and the kernel K are known functions, while the function ϕ is the unknown function to be found.

2.2 The general continuous L_p approximation problem

For our purposes we will formulate the general continuous L_p approximation problem as follows: Assume that a function f, defined on a real interval X, is given, together with an approximation function $F(A)$, which depends on the n parameters $A = (a_1, a_2, \ldots, a_n)$ in R_n. For the continuous L_p approximation problem, one then has to find a vector $\hat{A} \in R_n$ such that

$$\int_X \left|f(x) - F(\hat{A},x)\right|^p dx \leq \int_X |f(x) - F(A,x)|^p dx,$$

for all $A \in R_n$. In this discussion attention will in particular be paid to the case $p = 1$.

For the case where $F(A)$ is a linear function in A, that is, $F(A)$ is of the form

$$F(A,x) = \sum_{i=1}^{n} a_i \phi_i(x),$$

with the a_i, $i = 1, \ldots, n$ unknown constants and the ϕ_i, $i = 1, \ldots, n$, given basis functions, it can be shown under certain assumptions (see for example Rice [7]) that for the continuous L_1 approximation problem, there exists a solution and that it is in fact a unique solution.

In the case of $F(A)$ being non-linear in A, it is generally not possible to provide an answer to the question of either the existence or uniqueness of a solution. For special cases with constraints on $F(A)$, answers can be provided. However, Rice [8] remarks: "One can analyze this question based on a variety of assumptions concerning F, but the conclusions will have little import on the other problems which arise naturally in approximation theory." However, making a convexity and differentiability assumption, Rice proceeds to prove a characterization theorem for non-linear L_1 approximations.

2.3 The reformulation of the integral equation problem as an approximation problem

The problem of solving an integral equation can now be reformulated as follows as an L_p approximation problem. Assume that the solution function ϕ of the integral equation of section 2.1 can be approximated by the function $F(A)$, which can be either linear or non-linear. Substitution of this function in the given integral equation leads to the approximation

$$F(A,x) - \lambda \int_a^b K(x,t,F(A,t))dt \approx \psi(x), \qquad x \in [a,b].$$

One can then express the approximation problem in the L_p norm as

$$\min_A \int_a^b \left| F(A,x) - \lambda \int_a^b K(x,t,F(A,t))dt - \psi(x) \right|^p dx$$

where $F(A,x)$ is defined on R_n with respect to A and on $[a,b]$ with respect to x.

Note that the linearity or non-linearity with respect to A of the above function depends on the approximation function $F(A)$ as well as on the kernel K. It is of little concern whether the function to be minimized is linear or non-linear. This results in a bigger and freer choice of approximation functions than would be the case in for example collocation methods.

3. THE NUMERICAL PROCEDURE

For a discussion of the specific technique, the interested reader is again referred to De Klerk, Eyre and Venter [5]. However, to summarize, in principle the technique consists of two numerical procedures, namely:

- A minimization procedure; and
- an integration procedure.

The minimization is performed by using a standard optimization algorithm, for example UMPOL in the IMSL Library, which is based on the downhill simplex method of Nelder and Mead (see, for example Press *et al.*, [9]).

The numerical integration is performed by means of a technique developed by Venter and Laurie [10]. For this procedure the standard adaptive integration routine QAGE (from the Quadpack subroutine package of Piessens *et al.* [11]), was modified to allow for the efficient evaluation of integrals of the general form $\int_a^b |f(x)|dx$.

It is shown in De Klerk [4] that satisfactory results could be obtained by means of this strategy. In De Klerk *et al.*, [5] special attention is paid to the case of singular integrals with a Cauchy type singularity, that is, with an integral of the general form

$$\int_a^b \frac{k(x,t)f(t)}{t-x}dt.$$

In this case it was also shown that singular integral equations of this type can be solved effectively by using the approximation method described above. However, the more difficult problem of integal equations consisting of *strongly* singular integrals, was not addressed.

In the present paper the specific purpose is to pay attention to this case: namely the case of strongly singular integrals, where the integral part of the equation is of the general form

$$\int_a^b \frac{k(x,t)f(t)}{(t-x)^2}dt, \qquad a < x < b.$$

By using a real life problem, we would like to show that the L_1 approximation technique yields good results.

4. AN APPLICATION

In contrast with some other methods, the approximation technique discussed here has amongst others the following advantages: The approximation function $F(A)$ is not prescribed beforehand. For a given problem a variety of choices regarding norm, form, linearity/non-linearity and accuracy level can thus be made and applied.

We would like to discuss the following example of a strongly singular integral equation which is solved by the above-mentioned L_1 approximation technique. The example has some major *numerical* pitfalls.

The analysis of cracks in brittle solids, is a topic of particular interest (see for example Hori and Nemat-Nasser [12] and Ladopoulos [13]). Such problems can generally be formulated mathematically by an integral equation of the general form

$$f(x) - \int_a^b k(x,t)\frac{f(t)}{(t-x)^2}dt = g(x), \qquad a < x < b,$$

where g and k are known functions, while f is the unknown function to be found.

In the present example the elasticity problem for an infinite array of periodic cracks in a half-plane is considered. This problem was examined by Nied [1] in order to determine among others the stress intensity factors.

For this example an infinite array of stacked internal cracks, uniformly spaced apart, a distance h is considered. The usual approach to this problem (again with reference to Nied) is to formulate it in terms of a singular integral equation of Cauchy type. Nied, however, follows an alternative approach where the unknown function is taken to be the crack surface displacement. A hypersingular integral equation then results, which contains a singularity of order $1/x^2$ – that is, a strongly singular integral equation containing a Hadamard integral.

Consider an infinite stacked array of periodically spaced straight cracks in the half-plane $x \geq 0$. Assume that the cracks all have the same form, length and orientation. Assume that the x-axis is aligned lengthwise with one of the cracks and that the y-axis is perpendicular to all the cracks. Assume that the left and right end points of the cracks are given by $x = a$ and $x = b$ respectively, with h the distance between any two cracks in the direction of the y-axis. Assume that $V(x)$ is the crack surface displacement; also assume that all the crack surfaces are subjected to an equal and symmetric loading, specified by $p(x)$.

The strongly singular integral equation for the crack surface displacement function $V(x)$ is then expressed by

$$\oint_a^b \frac{V(t)dt}{(t-x)^2} + \int_a^b V(t)\left[K_1(t,x) + K_2(t,x,h)\right]dt = \frac{-\pi(\kappa+1)}{2\mu}p(x),$$

where κ and μ are constants, $p(x)$ is the loading function and where K_1 and K_2 are respectively given by

$$K_1(t,x) = \frac{-1}{(t+x)^2} + \frac{12x}{(t+x)^3} + \frac{-12x^2}{(t+x)^4}$$

and

$$\begin{aligned} K_2(t,x,h) = \frac{1}{h^2}\Bigg\{ & -\frac{1}{\alpha^2} + \frac{1}{\beta^2} + 3\pi^2 \operatorname{cosech}^2\pi\alpha + 5\pi^2 \operatorname{cosech}^2\pi\beta \\ & - 2\pi^3\alpha \operatorname{cosech}^2\pi\alpha \coth\pi\alpha - 6\pi^3\beta \operatorname{cosech}^2\pi\beta \coth\pi\beta \\ & + (\beta^2 - \alpha^2)\left[2\pi^4 \operatorname{cosech}^2\pi\beta \coth^2\pi\beta + \pi^4 \operatorname{cosech}^4\pi\beta - \frac{3}{\beta^4}\right]\Bigg\}, \end{aligned}$$

with $\alpha = \frac{t-x}{h}$ and $\beta = \frac{t+x}{h}$.

Consider in the rest of the discussion the case of internal cracks, that is, $a > 0$. Using the expressions

$$t = \frac{b-a}{2}r + \frac{b+a}{2}, \qquad x = \frac{b-a}{2}s + \frac{b+a}{2} \qquad \text{and} \qquad V(t) = \frac{b-a}{2}\tilde{V}(r),$$

the integral equation becomes,

$$\oint_{-1}^1 \frac{\tilde{V}(r)dr}{(r-s)^2} + \int_{-1}^1 \tilde{V}(r)\left(\frac{b-a}{2}\right)^2 \left[K_1(t,x) + K_2(t,x,h)\right]dr = \frac{-\pi(\kappa+1)}{2\mu}p(x),$$

with the function $\tilde{V}$ to be found.

Choosing an appropriate form for the approximation function, the subsequent problem can be reduced greatly. By choosing $\tilde{V}(r) = F(r)\sqrt{1-r^2}$ and setting as linear approximation function

$$F(r) = \sum_{i=0}^{\infty} a_i U_i(r) \approx \sum_{i=0}^{N} a_i U_i(r),$$

where $U_i(t)$ is the Chebyshev polynomial of the second kind, the problem can be written as an L_p approximation problem. The following remarks concerning the solution could be made:

- A variety of norm and form choices could be made.
- Different methods could be used to find the minimum value of the optimization problem.
- Different methods could also be used for evaluating the function values needed in the method of Nelder and Mead.
- A variety of strategies could of course be used to evaluate the argument of the absolute value function. In the form that it is mentioned above, there is in principle no problem to calculate the absolute value, but the following is worth mentioning:
  * To calculate the inner integral, a Gauss quadrature method was used. However, major numerical difficulties were encountered in the calculation of the functions $K_1(t,x)$ and $K_2(t,x,h)$. For example, notice that since $\alpha = \frac{t-x}{h}$, it may happen on the one hand that α tends to 0 (and therefore that $\frac{1}{\alpha^2}$ tends to infinity), and on the other hand that α tends to a large value (and therefore that the hyperbolic functions tend to either 0 or infinity, which in practice means that it could not be calculated numerically). In the end, these computational problems were overcome by dividing the interval under discussion into four subintervals, namely $[0.0, 0.00001]$, $[0.00001, 0.001]$, $[0.001, 7.0]$ and $[7.0, 28.0]$ and using a different computing strategy on each.

The most interesting aspect of this problem is the so-called *normalized stress intensity factor* $k(a)$ and $k(b)$. These factors are given by

$$k(a) \approx \left(\frac{2\mu}{1+\kappa}\right)\sqrt{\frac{b-a}{2}}\left[A_0 - 2A_1 + 3A_2 - \ldots (-1)^N (N+1) A_N\right],$$

and

$$k(b) \approx \left(\frac{2\mu}{1+\kappa}\right)\sqrt{\frac{b-a}{2}}\left[A_0 + 2A_1 + 3A_2 + \ldots + (N+1) A_N\right].$$

To compare the results obtained for this problem by means of the discussed L_1 approximation technique with those obtained by Nied, it is necessary to compare the different values of $k(a)$ and $k(b)$. These are given in Table 1 (with $l = b - a$ and $N = 4$) and the comparisons are done for $\frac{l}{a} = 1, 10, 100$ and $\frac{l}{l+h} = 0.01, 0.1, 0.3, 0.6, 0.7, 0.9, 0.96$. It should be mentioned that Nied's results are given in his paper in graph form only. For the purpose of comparison the appropriate values were read off from his graph to the best of our ability.

As can be seen, there is a good agreement between most of the values, especially in the cases when $\frac{l}{a}$ and $\frac{l}{l+h}$ tend to the smaller values. Some of the bigger differences might be ascribed to the fact that Nied's values were graphically found.

It is clear that, on average, a good solution could be obtained by using an approximation technique for an integral equation with strongly singular behaviour. There is, of course, always the danger that one could jump to an incorrect conclusion by considering only a finite number of examples. Still, this example gives one some certainty and satisfaction concerning the method.

TABLE 1

Comparison 1 $(\frac{l}{a} = 1)$

$\frac{l}{l+h}$	0.010	0.100	0.300	0.500	0.700	0.900	0.960
$k(a)$ (Nied)	1.031	1.002	0.828	0.572	0.378	0.200	0.100
$k(a)$ (L_1-method)	1.034	1.002	0.835	0.570	0.372	0.151	0.061
$k(b)$ (Nied)	1.021	1.000	0.833	0.573	0.361	0.181	0.090
$k(b)$ (L_1-method)	1.024	0.994	0.833	0.570	0.373	0.151	0.061

Comparison 2 $(\frac{l}{a} = 10)$

$\frac{l}{l+h}$	0.010	0.100	0.300	0.500	0.700	0.900	0.960
$k(a)$ (Nied)	1.459	1.400	0.963	0.591	0.378	0.200	0.100
$k(a)$ (L_1-method)	1.478	1.409	0.969	0.591	0.381	0.152	0.061
$k(b)$ (Nied)	1.167	1.115	0.839	0.573	0.361	0.181	0.090
$k(b)$ (L_1-method)	1.162	1.109	0.840	0.571	0.373	0.152	0.061

Comparison 3 $(\frac{l}{a} = 100)$

$\frac{l}{l+h}$	0.010	0.100	0.300	0.500	0.700	0.900	0.960
$k(a)$ (Nied)	2.854	2.698	1.650	0.814	0.433	0.200	0.100
$k(a)$ (L_1-method)	2.617	2.464	1.513	0.754	0.414	0.151	0.061
$k(b)$ (Nied)	1.302	1.224	0.843	0.573	0.361	0.181	0.090
$k(b)$ (L_1-method)	1.305	1.230	0.853	0.573	0.378	0.151	0.061

REFERENCES

1. H.F. Nied, Periodic array of cracks in a half-plane subjected to arbitrary loading, *Transactions of the ASME*, 54:642-648(1987).
2. O.T. Hanna and L.F. Brown, A new method for the numerical solution of Fredholm integral equations of the first kind, *Chemical engineering science*, 46:2749-2753(1991).
3. R.E. Kleinman and P.M. van den Berg, Iterative methods for solving integral equations, *Radio science*, 26:175-181(1991).
4. J.H. de Klerk, Solutions of integral equations via L_1 approximations, *International journal of control*, 48:2121-2128(1988).
5. J.H. de Klerk, D. Eyre and L.M. Venter, L_p approximation method for the numerical solution of singular integral equations, *Applied mathematics and computation*, 72:285-300(1995).
6. A.C. Kaya and F. Erdogan, On the solution of integral equations with strongly singular kernels, *Quarterly of applied mathematics*, 45:105-122(1987).
7. J.R. Rice, *The approximation of functions*, vol 1, Addison-Wesley, London, 1964.
8. J.R. Rice, On nonlinear L_1 approximation, *Archive for rational mechanics and analysis*, 17:61-66(1964).
9. W.H. Press, B.P. Flannery, S.A. Teukolsky and W.T. Vetterling, *Numerical recipes in Pascal: The art of scientific programming*, Cambridge University Press, Cambridge, 1989.
10. A. Venter and D.P. Laurie, Automatic quadrature of functions of the form $g(|f(x)|)$, *BIT*, 36:387-394(1996).
11. R. Piessens, E. de Doncker-Kapenga, C.W. Überhuber and D.K. Kahaner, QUAD PACK – *A subroutine package for automatic integration* (Springer-Verlag, Berlin, 1983).
12. M. Hori and S Nemat-Nasser, Asymptotic solution of a class of strongly singular integral equations, *SIAM Journal of Applied Mathematics*, 50:716-725(1990).
13. E.G. Ladopoulos, Systems of finite-part singular integral equations in L_p applied to crack problems, Engineering Fracture Mechanics, 48:257-266(1994).

Proceedings of Dynamic Systems and Applications 3 (2001) 177-184

Evolving Cellular Automata for Natural Phenomena Modelling

A. Dobnikar[1], A. Likar[2], S. Vavpotič[1]
[1] Faculty of Computer and Information Science, University of Ljubljana, Tržaška 25, 1001 Ljubljana, Slovenia
[2] Faculty of Mathematics and Physics, University of Ljubljana, Jadranska 19, 1001 Ljubljana, Slovenia
Email: {Andrej.Dobnikar,Simon.Vavpotic}@fri.uni-lj.si, Andrej.Likar@fmf.uni-lj.si

ABSTRACT: Natural phenomena modelling (as a special case of dynamic systems modelling) with evolving cellular automata are described in this paper. The basic idea is to use stochastic cellular automata together with the local evolving algorithms in order to model dynamic systems representing certain natural phenomena and on the other hand to search for the optimal local parameters that enable the close fit of the model processing with the actual sequence of the phenomena. With the help of a case study of the forest fire spread problem, we show the value of the approach.

1. INTRODUCTION

The modelling of natural phenomena dynamics is one of the most exciting challenges of science today. The reason is that natural phenomena are time- and space-dependent, where time dependencies are normally given with corresponding differential equations while the influences of space are captured within unknown space-dependent parameters. What we expect to be known is the sequence of the states of the phenomena under observation (also called the measuring data). Typical examples of the fields that fit into this kind of reasoning are meteorology, water or air pollution, spread of fires, etc.

In this paper we want to demonstrate a new approach to the problem of process modelling. We suppose that the physical interpretation of the natural phenomena under investigation is known. We also presume that the sequence of the states describing the dynamics of the phenomenon is at our disposal. The idea is to use cellular automata (CA), together with a Local Evolving Algorithm (LEA) based on the local genetic operators. With CA we want to realise the modelling of a dynamic system representing certain natural phenomena and with LEA the search for the proper values of the space-dependent parameters. As a case study we use the forest-fire prediction problem, where a diffusion equation as a physical interpretation of the spread of fire is used and where we expect to have the sequence of the states describing the actual process.

2. THEORETICAL BACKGROUND

Stochastic cellular automata (SCA) is defined with the two-dimensional array of finite automata cells, $SCA = (SCA(r); 0<=r<MaxX, MaxY)$. Each cell is described with $SCA(r)=(S,p,L,G,f,F)$ where S is the set of all possible states of the cell indicated by 2D vector r, p is the probability vector of the state transitions for each neighbouring (5-cell neighbourhood) combination, L is the set of local and G the set of global parameters that belong to cell r, f is a current cell's fitting value, and F is the set of flags of the cell,

which are problem-dependent. Each component of p corresponds to a neighbourhood configuration and acts as a gene. The set of all probability components that belongs to one particular cell represents the rule genome. There are as many genomes as cells in the CA structure. The complete number of genomes corresponds to the so-called blueprint or chromosome of the SCA.

As mentioned above, the LEA originates in the CPA, which means that it follows the same basic "big loop", [2]:

BIG LOOP:

```
BEGIN  {Genetic Algorithm}
   g:=0;  {generation counter}
   Initialize_Population_P(g);
   Evaluate_Population_P(g);
   WHILE not done DO
      g:=g+1;
      Select_P(g) from P(g-1);
      Crossover_P(g);
      Mutate_P(g);
      Evaluate_P(g);
   END;
END.
```

Its local features are essentially the same as within the CPA, [2]. There are however some important differences. The mutation operator in the LEA replaces the probability value (gene) that belongs to the current neighbourhood combination with a randomly selected probability (value within a closed interval [0..1]) instead of inverting a binary value attached to a randomly chosen combination within a CA rule (genome). The mutation rate within LEA is also an important parameter that should be picked up so that the convergence is appropriate and that the results are accurate enough. The crossover operator changes genomes in the same way as the CPA does. The only difference is that it moves probabilities instead of binary values.

The searching of the optimal local parameters with an evolving procedure (LEA) is controlled with the help of a so-called fitting function. The optimal local parameters, when used within the physical model and/or its cellular implementation, should fit best to the sequence of the measured data of the observed phenomenon. The fitting function belongs to every cell within *SCA* and is a cumulative one, which means that for each comparison of the cell's state with the corresponding pixel taken from the observed picture within the measured sequence, a weighed contribution is calculated. The order of the pictures in the sequence, as well as the state of the cell, influences its contribution.

3. SCA MODEL OF DIFFUSION (FOREST FIRE SPREAD PROBLEM)

3.1 Original *SCA* Diffusion Model

To test the approach described above, we tackled as a case study the important yet still unresolved problem of wild land fire spread. The starting-point is the basic equation of energy law:

$$\frac{\partial h}{\partial t} = -div\ \bar{j} + \sigma \tag{1}$$

where h is specific enthalpy of the medium, j energy flux density and σ the specific power of heat generation. Assuming that the flux density is proportionate to the sum of the gradient of the temperature and the advective term from the bulk motion of hot air due to wind with velocity ω [4],

$$\bar{j} = -\alpha(\bar{r})\ grad\ T + \rho\, c_p T \bar{\varpi} \tag{2}$$

where ρ is air density, $\alpha(r)$ conductivity field, c_p specific heat of air at constant pressure and that the specific enthalpy h and heat generation are functions of temperature as well as co-ordinates, one arrives at the simplest diffusion model for the temperature T as a function of time and co-ordinates r, as follows:

$$\frac{\partial T}{\partial t} = \frac{1}{\rho\, c_p}\frac{\partial h}{\partial t} = \frac{1}{\rho\, c_p} div\,(\alpha(\bar{r})\ grad\, T) - div\,(\bar{\omega} T) + \frac{\sigma}{\rho\, c_p} \tag{3}$$

Several essential features can be learned from the last differential equation. First, we observe the locality principle, which means that the temperature profile is determined from the vicinity of the observed cell. In the simplest case we take into account the first neighbours of the cell at r, which we denote as $N(r)$. We later increase this vicinity to the next ones simply in order to enlarge the propagation velocity span of the CA. Next, we observe that equation (3) possesses so-called "diffusive wave solutions" in which local disturbances spread in time and space with a finite speed.

$$\frac{\partial T}{\partial t} = \beta\ div\ (\alpha(\bar{r})\ grad\ T) - div\ (\bar{\omega}\, T) + Q \tag{4}$$

The equation (4) was derived from equation (3) and serves as a guide for the construction of a CA structure and the rules appropriate for modelling the forest fire. It is important to realise that the parameter fields α, ß and Q are not known and, furthermore, that the equation is only a rough representation of the actual circumstances, which may require non-linear dynamic equations. The precise temperature field is never required in fire spread forecasting. What is important is the propagation velocity of the fire front in different areas, which depends on fuel distribution, wind direction and the geographical details of the terrain [3]. The CA should therefore mimic front propagation with different speeds at different areas, and take into account the influence of the wind. In our approach we also tried to avoid the well-known co-ordinate ghosting [3], where the form of the fire front closely follows the form of the mesh representing the cells of the fuel bed.

The following CA incorporates most of the basic facts built into the equation (3), as well as arguments based on common sense. The fuel bed is discretized and could be envisaged as trees in the forest. Instead of observing temperature, we observe a tree as our cellular automaton (cell), which has three states represented by variable s values 0, 1 and 2. The value 0 represents a tree that has not yet been reached by the forest fire (low temperature, fuel available). The value 1 represents a burning tree (temperature is rising, fuel is being burned, possible consequential ignition of neighbouring trees). The last state that is represented by the value 2 marks a burned tree (no fuel left, temperature is falling, consequential ignition of neighbouring trees impossible).

The ignition is regulated by probabilistic criteria based on the state of the vicinity *N(r)*. The fire from a burning tree ignites a tree in its vicinity with a probability *v(r)*. The probabilities *v(r)* form a field and incorporate the fuel resources, as well as the diffusivity and influence of the wind. The ignition is supposed to be certain if more than three neighbouring trees are on fire.

The velocity of a fire front is regulated by the proper choice of the probability field *v(r)*. In real-time applications the field is regulated by several global parameters in order to synchronise with the observed patterns of the fire fronts. With low probabilities *v(r)* the fire front can extinguish spontaneously.

The effects of the global wind are introduced through a local re-calibration of the field *v(r)*. The probability for the ignition of trees in *N(r)* is increased in the direction along the wind and decreased in the opposite direction. Smooth interpolation is assumed for other directions. The velocity of the wind is another global parameter to be adjusted during the synchronisation period. Due to the probabilistic formulation of the CA, the approach avoids the problems of co-ordinate ghosting for probabilities, which are not close to unity.

3.2. Inverted *SCA* Diffusion Model

The above diffusion model that spreads fire from a burning tree to neighbouring trees is far from suitable for cellular implementation. We can obtain a cellular diffusion algorithm by inverting the original diffusion model. Instead of spreading fire (original diffusion model), the cell's probability of being ignited from its vicinity is calculated. After ignition, no further investigation of its vicinity is needed. This is due to the fact that we are mostly interested in the dynamics of the fire front.

Although the effect of both algorithms is virtually the same, there are two important differences. The probability of ignition *v(r)* for each cell in the inverted model must be higher, because the spreading of the fire from one (distant tree) cell gives the tree less chance to ignite than to be ignited from the whole vicinity. Second, we have already mentioned that the *SCA* cell has 3 states. Observing the states in the original model, state 1 (burning) is an active state in the sense that it influences the ignition of the neighbouring cells with some probability, while in the inverted model, state 0 (unburned) is active due to the possibility of its being burned. State 2 in both models means a burned cell.

Vicinity is defined with 8 cells that surround the observed cell with radius 1, together with the cell itself, plus the limited number of cells that surround the observed cell with radius 2.

4. LOCAL EVOLVING ALGORITHM

The main feature of the Local Evolving Algorithm (LEA) for the Forest Fire Spread (FFS) problem is in combining the inverted diffusion model with a stochastic version of a Cellular Programming Algorithm (SCPA). This approach enables one to adapt the theoretical diffusion model to the actual sequence of events.

This is achieved by replacing the default *SCA* ignition probabilities *v(r)* with adapted ignition probabilities *w(r)*, which are defined as follows:

$$q(\bar{r}) = v(\bar{r}) + 2\cdot(p(\bar{r})_z - 1/2)$$
$$w(\bar{r}) = \begin{cases} 0 & q(\bar{r}) < 0 \\ q(\bar{r}) & 0 \le q(\bar{r}) \le 1 \\ 1 & q(\bar{r}) > 1 \end{cases} \tag{5}$$

w(r) is obtained from *q(r)* by constraining *q(r)* to the interval [0..1]. *q(r)* depends on a constant probability of ignition *v(r)* and the dynamic, neighbourhood-dependent probability of ignition $p(r)_z$, where index z denotes neighbourhood dependence.

SCPA is an extension of the binary standard Cellular Programming Algorithm (CPA), [2] commonly used for obtaining optimal cellular automata rules. SCPA deals with probabilities from the interval [0..1] instead of using binary values 0 and 1. The two genetic operators (mutation and crossover) are adapted in the same way, although it is worth noting that the mutation has to undergo additional changes, which is due to the scaling of the changes of probability values instead of just mutating 0 with 1 or 1 with 0. One possible solution is given with the equation:

$$p(\bar{r})_z = p(\bar{r})_z\cdot(1-scale) + (random(0..1) - (1-scale)/2)\cdot scale \tag{6}$$

Since the parameter *scale* is constrained to the interval [0..1], we can observe two extreme cases. When *scale* is 0 there is no change to the rule. On the other hand, when *scale* is 1 the new rule combination value is randomly chosen regardless of the previous value.

Before the experimental results are given, one important detail has to be explained – that is, the results of the learning mode have to be spread into the area where one expects fire propagation. This is done so for different regions that have similar geographical and/or meteorological data (or, in short, local parameters); the average of the learned parameters are correspondingly extrapolated.

In order to make prediction, LEA loads the last picture from the measured sequence into the SCA. The number of prediction steps depends on how far we want to forecast. The time period between the two states corresponds to the time between the two pictures from the measured sequence. In this way we can establish future fire spread or its possible extinguishing.

5. EXPERIMENTAL RESULTS

We test the above idea on the forest fire problem, where we presume that the sequence of pictures showing the time propagation is at our disposal. We also suppose that the individual regions of the area of the fire are equipped with typical local data such as humidity, type of vegetation and therefore the degree of flammability or probability of ignition, average wind, etc. The actual sequence is obtained in an artificial way, with the help of original diffusion and a number of realistic local parameters. Our goal is to use *SCA* with typical (geographical) parameters and LEA in order to find the parameters that make a close fit between the measured sequence and the sequence obtained with *SCA* and an inverse diffusion algorithm. We show also that the convergence of the local

Fig. 1. The actual terrain map

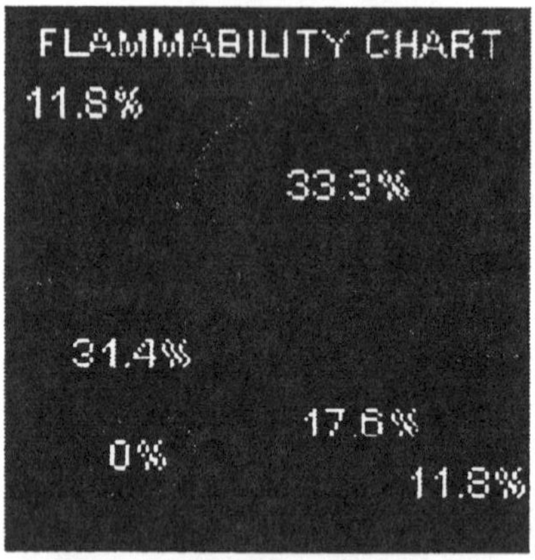

Fig. 2. Typical flammability chart

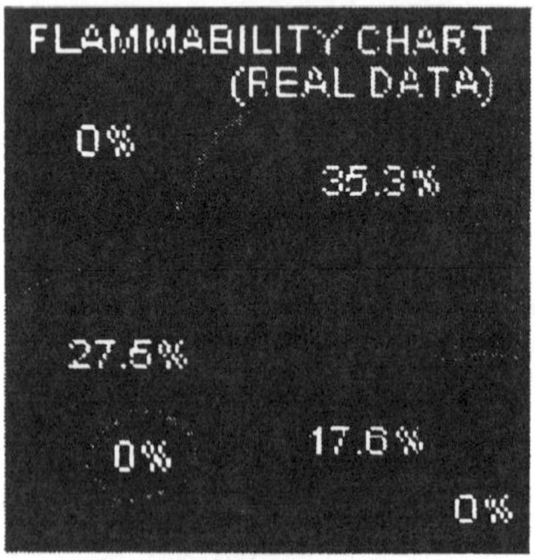

Fig. 3. Realistic flammability chart

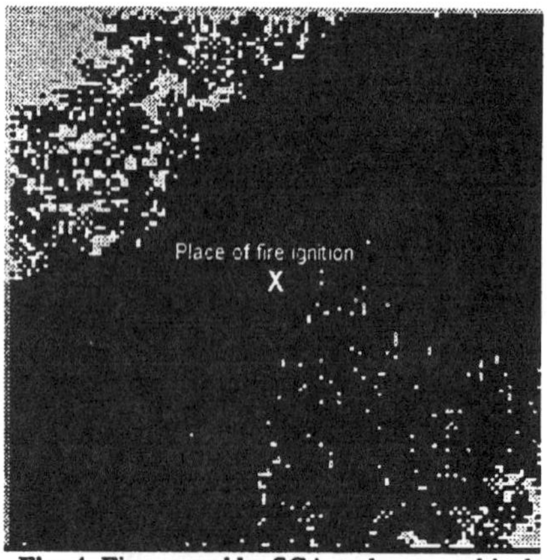

Fig. 4. Fire spread by SCA and geographical flammability chart

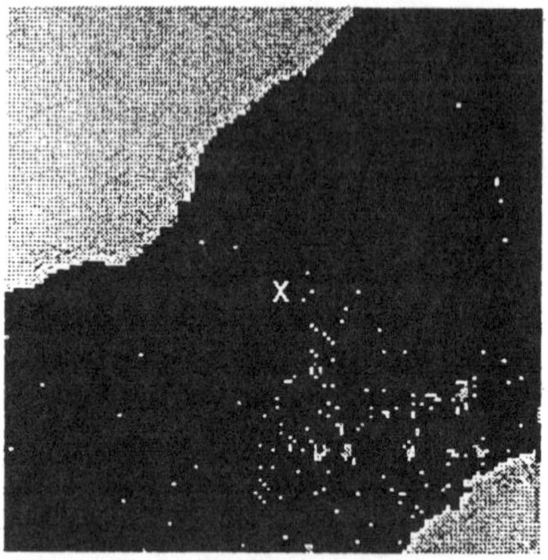

Fig. 5. Result of SCA with help of LEA

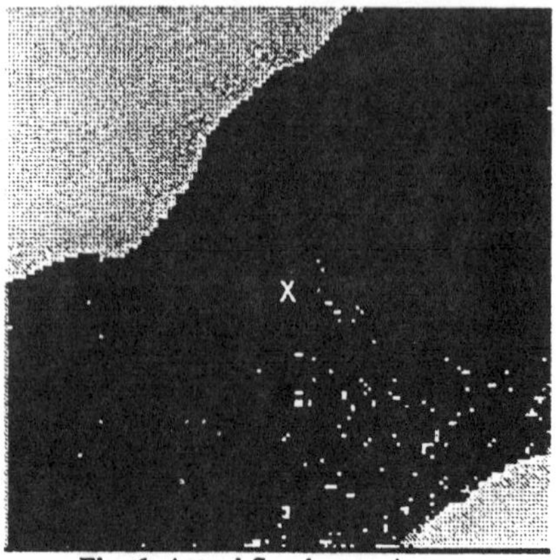

Fig. 6. Actual fire devastation

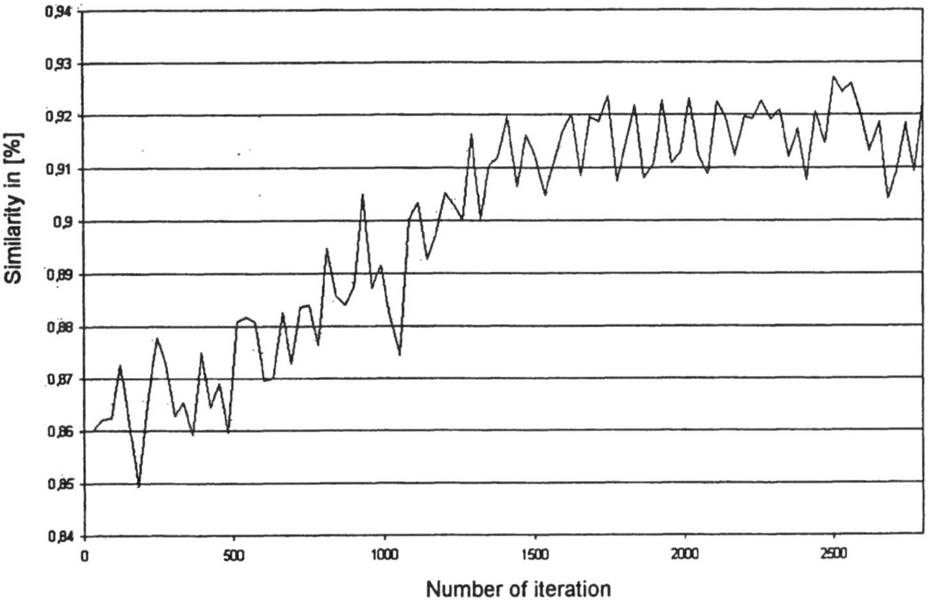

Fig. 7. Increase of similarity between the last picture and state

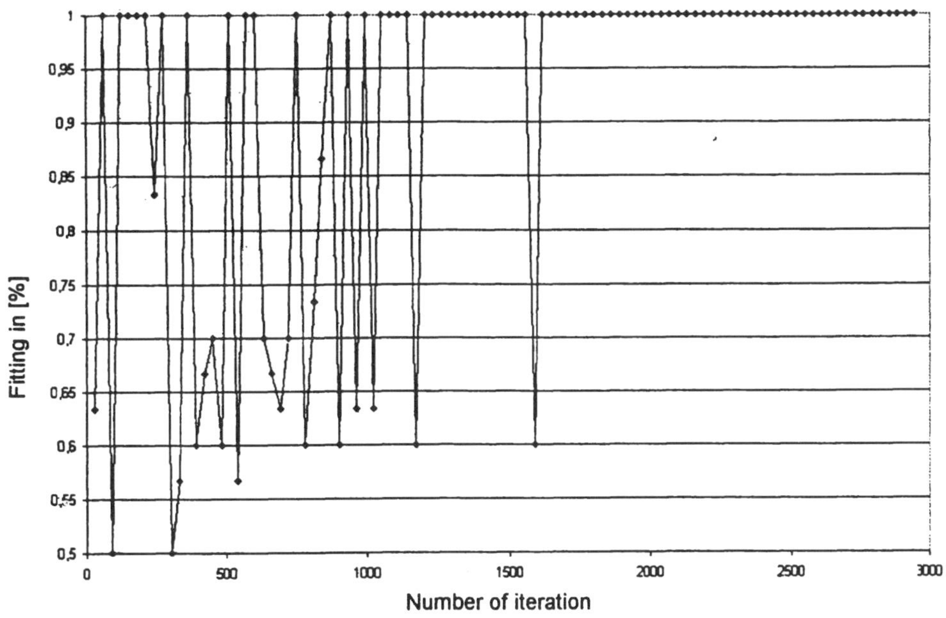

Fig. 8. The fitting value of a typical SCA cell

parameters strongly indicates the feature of learning, which is due to the LEA, based on local genetic operators.

Figure 1 shows the actual terrain map, where there are five areas with different flammability coloured with natural colours. Place of the initial fire ignition is marked with a black cross.

Figure 2 shows a typical (geographical) flammability chart where all the influential data is combined into the percentage (probability) of ignition.

Figure 3 gives the realistic (actual) data that is not known to us.

Figure 4 shows the result of the SCA using the inverted diffusion algorithm with typical local parameters, Figure 5 gives the result of the SCA with the learned data obtained with the help of the LEA and Figure 6 indicates the fire spread obtained with the inverted diffusion based on the actual data. The site of the initial ignition of the fire is marked with a white cross. It is evident that the last two figures are almost identical and that the processing using inverted diffusion on the typical parameters does not guarantee good prediction.

Figure 7 shows an increase in the similarity between the last picture of the learning sequence and the last state of the evolving *SCA* in percentages during the learning sequence.

Figure 8 finally outlines the convergence of the fitting value of a typical cell within *SCA*. It confirms the feature of learning during evolutionary processing. Although the curve fluctuates, it can clearly be seen that it quickly converges to the final state of 100% fitness. Fluctuations are also becoming less common and smaller before they completely disappear.

6. CONCLUSION

Natural Phenomena as complex dynamic systems are modelled with the help of SCA and LEA. This enables us to bring together theoretical and experimental work related to natural phenomena and also to make predictions of its future progression. In our future work we intend to automate the transformation between differential equation describing time dependency of the phenomena and its cellular implementation.

7. REFERENCES

1. T. Toffoli, N. Margolus, Cellular Automata Machines, MIT Press, 1988
2. M. Sipper, Evolution of Parallel Cellular Machines, Springer, 1997
3. D. D. Richards, Int. J. Wildland Fire 5(2): 63-72, 1995
4. R. O. Weber, Int. J. Wildland Fire 1(4): 245-248, 1991
5. E. Sanchez, M. Tomassini, Towards Evolvable Hardware, Springer, 1996

Proceedings of Dynamic Systems and Applications 3 (2001) 185-192

CHAOS IN A CONTINUOUS-TIME BOOLEAN NETWORK

R. Edwards

Department of Mathematics and Statistics
University of Victoria
P.O.Box.3045, STN CSC, Victoria, BC, Canada V8W 3P4

Abstract: Continuous-time systems with switch-like behaviour occur in chemical kinetics, gene regulatory networks and neural networks. Networks with hard switching, as a limiting case of smooth sigmoidal switching, retain the richest possible range of behaviors but are mathematically more tractable. The form of an underlying discrete (fractional-linear) map encodes information on existence, stability and exact periods of periodic orbits. In richly connected structures with four or more variables, aperiodic behaviour can occur. We investigate a simple 4-dimensional example with Boolean interaction terms in which a Smale horseshoe-like object reveals chaotic dynamics.

AMS(MOS) subject classification: 34C35, 92B20, 94C10

1. INTRODUCTION

Aperiodic behavior in systems of three or more ordinary differential equations (ODEs) often requires careful tuning of parameters and there are few general principles that enable us to say when a given system can have such behaviour. Here we investigate a class of systems of ODEs in which aperiodic behavior results from a particular structure of coupling between the variables rather than finely tuned parameter values.

For systems of interacting quantities that vary continuously but are dominated by switch-like behavior, Glass [4,5,6] proposed an approach, using structural equivalence classes based on state transition diagrams on an n-cube, to aid in relating dynamical behaviour to the structure of interactions in the system. Glass & Pasternack [8,9] then showed that periodic oscillations of various types occur with particular structural classes. This approach is exact in the case of 'hard' switching (interaction terms depend only on whether the other variables are above or below threshold), but appears to be a good model for the case of steep sigmoidal switching. The methods of analysis work only for the case of identical decay rates for all variables.

These networks were first investigated (Glass [4,5,6]) in the context of chemical and biological oscillations: kinetics of interacting chemical species, interacting biological species, gene and enzyme regulatory networks and neural networks. We have recently argued that networks of this type may underlie transitions between irregular and regular tremor generated by the brain's motor circuitry in Parkinson's disease (Edwards *et al* [3]). As a consistent body of theory has emerged in the case of identical decay rates and hard switching, and as the existing nomenclature is confusing, we propose to call these systems 'Glass networks' after their originator. The hard switching makes them remarkably tractable, yet they remain very rich in dynamical possibilities.

The structural equivalence classes have a particularly simple representative in which the values of the interaction terms are just Boolean functions ('Boolean Glass networks'). It is not clear to what extent these are representative of the dynamics of their respective classes, but they do permit complex behavior.

Aperiodic behavior appears to be common in Glass networks with many (6 or more) variables (Lewis & Glass [10]; Mestl *et al* [11]; Glass & Hill [7]; Edwards *et al* [3]) and chaos was found in one Glass network of 4 variables with a particular set of parameters (Mestl *et al* [13]). However, numerical simulations suggest that aperiodic behavior is not so rare even among 4-dimensional Boolean Glass networks, and here we show how available techniques can be used to analyze the behavior of one such network, via an object resembling the Smale horseshoe.

2. GLASS NETWORKS

A Glass network is a system of the form

$$\dot{y}_i = -y_i + F_i(\tilde{y}_1, \tilde{y}_2, \ldots, \tilde{y}_n)\,, \quad i = 1, \ldots, n\,, \tag{1}$$

where

$$\tilde{y}_i = \begin{cases} 0 & \text{if } y_i < 0 \\ 1 & \text{if } y_i > 0 \end{cases}\,. \tag{2}$$

Note that all thresholds are 0 (Equation 2), and F_i depends only on the signs of the variables y_i. Systems with non-zero thresholds and values of $\tilde{y}_i$ other than 0 and 1, possibly different for each variable, can be reduced to the above form by appropriate transformations. Similarly, a decay rate parameter may be put in the equations, and this may even depend on the $\tilde{y}_i$, but the analysis below will only apply if they are uniform among the variables of the system. This network can be considered a limiting case of smooth networks in which the step function (Equation 2) is replaced by a sigmoid, $\tilde{y}_i = g(y_i)$, whose range is the open interval $(0, 1)$.

Since for a Glass network there are a finite number of values F_i ($n2^n$ of them), it is clear that solutions are globally bounded. In the sigmoidal case, we would need the additional assumption that F_i is bounded on $(0,1)^n, \forall i$. Boolean Glass networks are defined as Glass networks for which $F_i(\tilde{\mathbf{y}}) = \pm 1$ for all i and all $\tilde{\mathbf{y}}$, so that the interactions are Boolean functions (note that we could equivalently have made the input values, $F_i \in \{0, 1\}$ if the thresholds had been set at $\frac{1}{2}$). All the theory we will use applies to the general Glass networks, but our chaotic example will Boolean.

While weaker conditions suffice for some of the following theory, we will assume:

Condition 1: $F_i \neq 0\,, \quad \forall i, \forall \tilde{\mathbf{y}}\,,$ and

Condition 2: $F_i(\tilde{y}_1, \ldots \tilde{y}_i = 0, \ldots, \tilde{y}_n) = F_i(\tilde{y}_1, \ldots \tilde{y}_i = 1, \ldots, \tilde{y}_n)$.

where we use $\tilde{\mathbf{y}}$ for the vector $(\tilde{y}_1, \tilde{y}_2, \ldots, \tilde{y}_n)'$, and similarly use bold face for other vectors, the $'$ denoting matrix transposition. Condition 2 states that F_i does not depend on $\tilde{y}_i$, *i.e.*, that there is no self-input in the network.

3. CYCLES AND PERIODIC ORBITS

The analysis of Glass networks was begun by Glass & Pasternack [9] and was further developed mainly by Mestl *et al* [12] and Mestl *et al* [13]. What follows is a brief summary of this work though some results are new.

The main property of Glass networks that makes them tractable is that trajectories are piecewise-linear. For $\mathbf{y} = (y_1, y_2, \ldots, y_n)$ in one orthant of phase space (and therefore with one fixed sign structure) the solution to Equation 1 (in vector form) is

$$\mathbf{y}(t) = \mathbf{f} + (\mathbf{y}(0) - \mathbf{f})e^{-t}\,, \tag{3}$$

which describes exponential approach to $\mathbf{f} = (f_1, f_2, \ldots, f_n) = \mathbf{F}(\tilde{\mathbf{y}})$ in a straight line. Thus, each orthant of phase space (with sign structure $\tilde{\mathbf{y}}$) has an associated focal point, $\mathbf{f}$, somewhere in $\mathbf{R}^n$. If trajectories in an orthant are directed to a focal point $\mathbf{f}$ within that orthant then once the orthant is entered no further switchings take place and $\mathbf{f}$ is a stable fixed point of the network dynamics. Otherwise, trajectories are formed of piecewise-linear segments between orthant boundaries, with sharp corners at the boundaries. Under Condition 2, there is no ambiguity in the direction of flow across an orthant boundary so trajectories are well defined there.

We now denote by $\mathbf{y}^{(k)}$ the kth such orthant boundary crossing on a trajectory and assume that $\mathbf{f}^{(k)}$, the focal point associated with the orthant being entered, does not lie in that orthant. The map from one boundary to the next can be represented as an operator $(M^{(k)} : \mathbf{R}^n \to \mathbf{R}^n)$:

$$\mathbf{y}^{(k+1)} = M^{(k)}\mathbf{y}^{(k)} = \frac{B^{(k)}\mathbf{y}^{(k)}}{1 + \langle \psi^{(k)}, \mathbf{y}^{(k)} \rangle}, \quad B^{(k)} = I - \frac{\mathbf{f}^{(k)}\mathbf{e}_j'}{f_j^{(k)}}, \quad \psi^{(k)} = \frac{-\mathbf{e}_j}{f_j^{(k)}}, \tag{4}$$

where j is the variable that switches at the kth step, $\mathbf{e}_j$ denotes the standard basis vector in $\mathbf{R}^n$ and the angle brackets denote the Euclidean inner product ($\langle \psi, \mathbf{y} \rangle = \psi'\mathbf{y}$). Thus, $M^{(k)}$ is a fractional-linear map with a vector numerator and scalar denominator. The composition of such maps is again a fractional-linear map of the same form. Also, since these maps are between orthant boundaries where one of the y_i's is always 0, they can be reduced by one dimension, by removing the appropriate row and column in each $B^{(k)}$, $\mathbf{y}^{(k)}$ and $\psi^{(k)}$. For a cycle, (a trajectory that returns to its initial orthant boundary), we arrive at (dropping the superscripts)

$$M\mathbf{y} = \frac{A\mathbf{y}}{1 + \langle \phi, \mathbf{y} \rangle}, \tag{5}$$

where A is $(n-1) \times (n-1)$, $\phi \in \mathbf{R}^{n-1}$ and $\mathbf{y} \in \mathbf{R}^{n-1}$. This discrete map, along with the crossing times, contains all information in the full continuous-time dynamics.

The structure of the dynamics of an n-dimensional network may be represented by a state transition diagram, a directed graph on an n-cube, where nodes (vertices, labelled by the Boolean vectors, $\tilde{\mathbf{y}}$) represent orthants of phase space and edges represent transitions across orthant boundaries. Under Conditions 1 and 2, the flow across orthant boundaries is unambiguous and determines a direction on the edge between the corresponding nodes on the n-cube. If a periodic orbit exists for the network, then it must follow a directed cycle of edges on the n-cube. The converse is not necessarily true, as we will see below.

We now list without proof key properties of the cycle map M (Equation 5) and corresponding periodic orbits.

Proposition 1 *Trajectories starting at different points on a given ray through the origin remain on a common ray under iteration of M (though the time taken on the two continuous trajectories will differ). Under Condition 2, such trajectories converge as $t \to \infty$ (though this convergence may be to the origin).*

Proposition 2 *Linear subspaces are mapped to linear subspaces by M. In particular, straight lines are mapped to straight lines, and planes are mapped to planes.*

Along a cycle on the n-cube, there may be branching nodes, *i.e.*, nodes with more than one outgoing edge. These correspond to orthants from which trajectories can exit by more than one boundary hyperplane, depending on which variable reaches zero first. Alternate exit variables impose constraints on the region of an orthant boundary that maps forwards through a specified sequence of boundaries. These constraints take the form of linear inequalities, and the restricted regions are the interiors of 'proper cones' (Berman & Plemmons [1], p.6).

Proposition 3 *Given an n-cube cycle and initial orthant boundary, $\mathcal{O}$, the cone from which trajectories follow the cycle and return to $\mathcal{O}$ is given by $C = \{\mathbf{y} \in \mathcal{O} | R\mathbf{y} \geq 0\}$, where R is a matrix with one row for each alternate exit variable, $y_i^{(k)}$, around the cycle, each row being*

$$R_{i,\cdot} = -\frac{\mathbf{e}_i'}{f_i^{(k)}} B^{(k)} B^{(k-1)} \ldots B^{(0)} . \tag{6}$$

We allow equality, $R\mathbf{y} = 0$, (trajectories for which two variables cross simultaneously) as limiting cases. Many of the inequalities generated by Equation 6 will be redundant and can be weeded out in computation.

The domain of definition of M is only $C \subset \mathcal{O}$. Trajectories starting outside of C, but in $\mathcal{O}$, eventually branch away from the given cycle. Note also that M maps C into $\mathcal{O}$, not necessarily into C. However, a fixed point of the map lying inside C continues to return and corresponds to a periodic orbit for the differential equations. If C is empty, no periodic orbit corresponding to this n-cube cycle exists.

Proposition 4 *Any non-zero (real) fixed point of M (Equation 5) in C is a (real) eigenvector of A with eigenvalue > 1. Conversely, if $\mathbf{v}$ is a real eigenvector of A with eigenvalue $\lambda > 1$, and $\mathbf{v} \in C$, then*

$$\mathbf{y}^* = \frac{(\lambda - 1)\mathbf{v}}{\langle \phi, \mathbf{v} \rangle} \tag{7}$$

is a fixed point of M, unique in the span of $\mathbf{v}$. If $\lambda = 1$, then the only fixed point in the span of $\mathbf{v}$ is $\mathbf{0}$.

Proposition 5 *A fixed point, $\mathbf{y}_i^*$, of M corresponding to the eigenvalue λ_i of A, is asymptotically stable if $\lambda_i > |\lambda_j|, \forall j \neq i$, neutrally stable if $\lambda_i \geq |\lambda_j|, \forall j \neq i$, but equality holds for some j, and unstable otherwise.*

Proposition 6 *A periodic orbit with cycle map M has period $P = \log(\lambda)$, where λ is the eigenvalue of the matrix A associated with the fixed point on the orbit.*

4. APERIODIC BEHAVIOR

Aperiodic behavior in Glass networks appears common for large n. Numerical evidence for ergodicity via an invariant measure for a 6-dimensional network was given by Lewis & Glass [10]. For this and a similar 6-dimensional network discussed by Edwards *et al* [3], aperiodic behavior appeared for certain values of a parameter. Mestl *et al* [12] showed that chaos cannot exist in 3-dimensional Glass networks, but a 4-dimensional network with a special set of parameters was shown by Mestl *et al*

orthant ($\tilde{\mathbf{y}}$)	focal pt. ($\mathbf{F}$)			
0 0 0 0	−1	1	1	1
0 0 0 1	−1	1	−1	1
0 0 1 0	1	−1	1	−1
0 0 1 1	1	−1	−1	−1
0 1 0 0	−1	1	1	1
0 1 0 1	−1	1	1	1
0 1 1 0	1	−1	1	−1
0 1 1 1	1	−1	1	−1
1 0 0 0	−1	1	−1	−1
1 0 0 1	−1	1	−1	−1
1 0 1 0	1	1	−1	−1
1 0 1 1	1	−1	−1	−1
1 1 0 0	−1	1	−1	1
1 1 0 1	−1	1	1	1
1 1 1 0	1	1	−1	1
1 1 1 1	1	−1	1	1

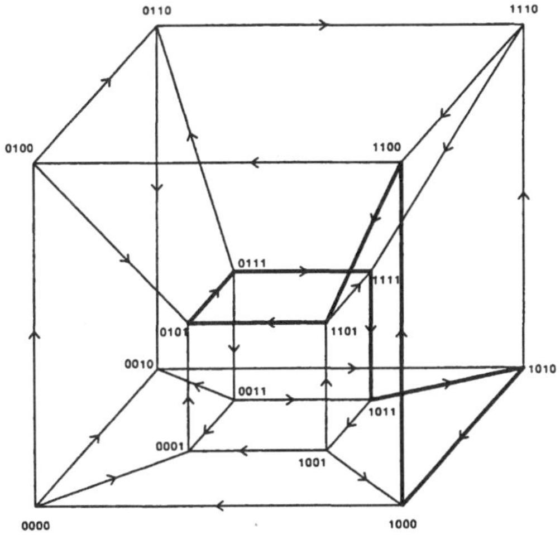

Figure 1: 4-cube structure of a Boolean Glass network with chaotic behavior. The cycle marked with bold lines corresponds to an unstable periodic orbit.

[13] to have aperiodic dynamics, though it was not proven that there was a chaotic attractor. None of these examples were Boolean Glass networks and it was not known whether 4-dimensional networks of this class could exhibit aperiodicity.

Numerical experiments that we performed on randomly generated 4-dimensional Boolean Glass networks produced examples for which no fixed point or periodic orbit was detected. We investigate one of these, defined by the interaction function, $\mathbf{F}$, in Fig. 1. One way to express this network as a system of ODEs is as follows:

$$\begin{aligned}
\dot{y}_1 &= -y_1 + 2[\tilde{y}_3] - 1 \\
\dot{y}_2 &= -y_2 + 2[1 - \tilde{y}_3 + \tilde{y}_1\tilde{y}_3 - \tilde{y}_1\tilde{y}_3\tilde{y}_4] - 1 \\
\dot{y}_3 &= -y_3 + 2[(1 - \tilde{y}_1)(1 - \tilde{y}_4) + \tilde{y}_2\tilde{y}_4] - 1 \\
\dot{y}_4 &= -y_4 + 2[(1 - \tilde{y}_1)(1 - \tilde{y}_3) + \tilde{y}_1\tilde{y}_2] - 1 .
\end{aligned}$$

A projection of an example 4-dimensional trajectory is shown in Fig. 2a.

Consider the two cycles,

$$\begin{aligned}
&0101 \to 0111 \to 1111 \to 1011 \to \mathbf{1001} \to 1000 \to 1100 \to 1101 \quad \text{and} \\
&0101 \to 0111 \to 1111 \to 1011 \to \mathbf{1010} \to 1000 \to 1100 \to 1101 .
\end{aligned}$$

Both are feasible and starting from the $(0{+}{-}{-}{+})$ boundary (*i.e.*, the boundary between 1101 and 0101) they have return maps M_0 and M_1 defined respectively by

$$A_0 = \begin{pmatrix} 1 & 0 & 0 \\ -2 & 5 & 2 \\ 0 & 2 & 1 \end{pmatrix}, \quad \phi_0 = (4, -4, 0)^T ,$$

$$A_1 = \begin{pmatrix} 1 & -2 & -2 \\ -2 & -3 & -6 \\ 0 & -2 & -3 \end{pmatrix}, \quad \phi_1 = (4, -4, 0)^T .$$

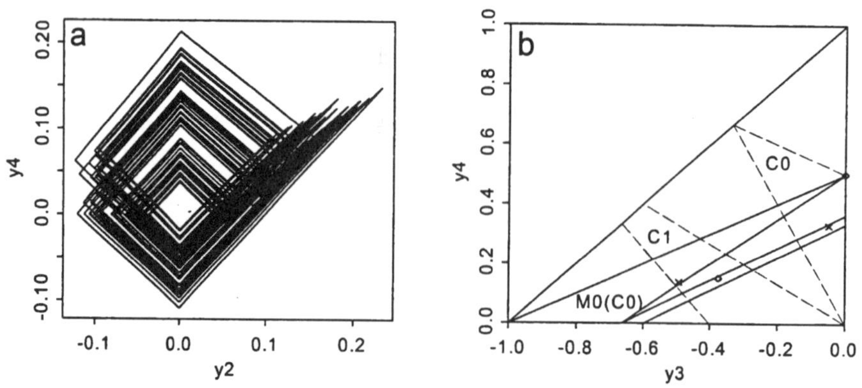

Figure 2: (a) Projection onto the y_2–y_4 plane of a trajectory for the 4-dimensional network of Fig. 1. The last 500 of 1000 orthant boundary transitions are shown. (b) Projections of returning cones for two cycles and their images. The triangle $(-1,0),(0,1),(0,0)$ is a projection of the orthant boundary $(0+-+)$. The regions indicated by dotted lines, labelled C_0 and C_1, are the returning cones. The regions indicated by solid lines cross-cutting C_0 and C_1 are the images of these cones under one iteration of their respective maps, M_0 and M_1. $M_1(C_1)$ (unlabelled) is the narrower region, containing two marked points. The diamonds represent eigenvectors of M_0 and M_1 and the crosses represent eigenvectors of the composite maps, M_0M_1 and M_1M_0.

Their eigenvalues (λ_i) and corresponding eigenvectors ($\mathbf{v}_i$) are for A_0,

$$\lambda_1 \approx 5.8284\,, \quad \lambda_2 = 1.0000\,, \quad \lambda_3 \approx 0.1716$$

$$\mathbf{v}_1 \approx \begin{pmatrix} 0.0000 \\ 0.7071 \\ 0.2929 \end{pmatrix}, \quad \mathbf{v}_2 = \begin{pmatrix} 0.5000 \\ 0.0000 \\ 0.5000 \end{pmatrix}, \quad \mathbf{v}_3 \approx \begin{pmatrix} 0.0000 \\ -0.2929 \\ 0.7071 \end{pmatrix},$$

and for A_1,

$$\lambda_1 \approx -6.8709\,, \quad \lambda_2 \approx 1.9457\,, \quad \lambda_3 \approx -0.0748$$

$$\mathbf{v}_1 \approx \begin{pmatrix} 0.2026 \\ 0.5257 \\ 0.2716 \end{pmatrix}, \quad \mathbf{v}_2 \approx \begin{pmatrix} 0.4728 \\ -0.3754 \\ 0.1518 \end{pmatrix}, \quad \mathbf{v}_3 \approx \begin{pmatrix} -0.2590 \\ -0.4401 \\ 0.3009 \end{pmatrix}.$$

Neither A_0 nor A_1 has its dominant eigenvector in $(+-+)$, and therefore, neither cycle has a stable periodic orbit. Fig. 2b shows the returning cones for these two cycles in the $(0+-+)$ boundary. In order to depict these 3-dimensional objects in a plane figure, we have projected the cones onto the plane $y_2 - y_3 + y_4 = 1$, which is the part of the unit l^1 ball in $\mathbf{R}^3$ that lies in the $(+-+)$ octant, and then plotted y_3 vs. y_4. Trajectories starting in the region labelled C_0 follow the first cycle above, M_0, and return to the $(0+-+)$ boundary. Similarly, trajectories starting in C_1 follow M_1.

The images of these two regions under their respective maps are also shown in Fig. 2b. The marked point at $\left(0, \frac{1}{2}\right) \in C_0$ represents the eigenvector $\mathbf{v}_2$ of A_0 ($\lambda_2 > 0$) so the ray through $\mathbf{v}_2$ in $\mathbf{R}^3$ is invariant under M_0 (and thus so is its projection in the figure), but since the corresponding eigenvalue is 1, no non-zero point on the ray is actually fixed and the origin attracts. The stretching and contraction in the directions of the other two eigenvectors (which are perpendicular and lie in the y_3–y_4 plane) are clearly visible. Since both eigenvalues are positive, there is no inversion of the image in either direction. Note that the eigenvector $\mathbf{v}_3$ with projected coordinates $(\frac{\sqrt{2}}{2} - 1, \frac{\sqrt{2}}{2}) \approx (-0.2929, 0.7071)$ lies outside C_0.

The returning region C_1 for map M_1 is also subject to stretching and contracting in similar directions. The eigenvector $\mathbf{v}_2$ associated with eigenvalue $\lambda_2 \approx 1.9457$ lies inside C_1, so the ray through this point is invariant, and there is an unstable fixed point of M_1 on this ray at $(0.1318, -0.1046, 0.0423)$, which appears on the projection at $(-0.3754, 0.1518)$. The period of the corresponding unstable orbit is $\log(\lambda_2) \approx 0.6656$. The other two eigenvalues are negative, so in the other two directions (in which the C_1 region is stretched and contracted) we also have inversion.

The combined map, defined by M_0 and M_1, restricted to the projected plane, therefore contains something similar to a Smale horseshoe [2]. It is not exactly topologically equivalent to the horseshoe, but retains many of its properties. Points getting mapped out of $C_0 \cup C_1$ will go elsewhere but there is nevertheless a Cantor set, Λ, in the projected plane, consisting of points whose trajectories remain in $C_0 \cup C_1$ both forwards and backwards in time, and an infinite set of unstable periodic points. The main difference from the Smale horseshoe is that no points in the region $M_1(C_1) \cap C_0$ are mapped (by M_0) into C_0 (this is easy to check by finding the images of the vertices of $M_1(C_1) \cap C_0$ and joining them up by straight lines to find the images of the boundary edges, since straight lines are mapped to straight lines, even when projected). Thus, no points of Λ, aside from $\left(\frac{1}{2}, 0, \frac{1}{2}\right)$ itself, lie in $M_0(C_0) \cap C_0$ and trajectories of periodic points never follow the M_0 cycle twice in a row, *i.e.*, their symbolic trajectories do not contain the string '00'. Compositions of M_0 and M_1 for which M_0 is not repeated twice produce unstable periodic orbits. For example, the fixed points corresponding to M_0M_1 and M_1M_0 are marked in Fig. 2b.

In order to deal with the radial direction that is suppressed in the above projection, we need only confirm that trajectories from points in the Cantor set, Λ, (or on the rays through points of Λ) do not converge to the origin. In this case convergence in the radial direction ensures that asymptotically we can ignore the radial component and we approach the 2-dimensional chaotic dynamics of Λ.

Boundedness away from the origin can be shown by finding a neighbourhood of the origin in which points on rays through Λ always move away from the origin under the map. Simple but tedious calculations show that for each of the corners, Q_i, $i = 1, \ldots, 12$, of the three regions $M_0(C_0) \cap C_1$, $M_1(C_1) \cap C_1$ and $M_1(C_1) \cap C_0$, the l^1 norm of the appropriate map, M_j, satisfies $\|M_j(kQ_i)\|_1 > \|kQ_i\|_1 = k$ if $0 < k < \frac{3}{22}$. Since planes are mapped to planes by each of M_0 and M_1, the same is true for rays through any point in the interior of one of these three regions (which contain Λ). Thus, points closer to the origin than $\|\mathbf{y}\|_1 = \frac{3}{22}$ move away and there is no possibility of convergence to the origin on any ray through Λ.

5. DISCUSSION

We have shown that chaotic dynamics exist for the network of Fig. 1. We have not shown that orbits in Λ are attracting and, in fact, numerical evidence suggests that they are not. When these equations are integrated, trajectories *do* return repeatedly to the orthant boundary $(0+-+)$, but have itineraries including several other cycles besides M_0 and M_1. These other cycles have returning cones in the gaps left by C_0 and C_1. Thus, the dynamics are actually more complicated than suggested by the analysis above.

It is surprising that complex dynamics are possible in such a simple network of only 4 variables with Boolean interactions. It certainly is not possible in 4-element discrete-time switching networks. Nevertheless, the techniques discussed here allow considerable progress in analysis of these Glass networks. An important unsolved question is to find necessary or sufficient conditions on the connection structure for chaotic dynamics to occur.

ACKNOWLEDGEMENTS

This work was partially supported by grants from the University of Victoria and the Natural Sciences and Engineering Research Council of Canada.

REFERENCES

1. Berman, A., & Plemmons, R. J., Nonnegative Matrices in the Mathematical Sciences. Academic Press, New York, 1994.
2. Devaney. R. L., An Introduction to Chaotic Dynamical Systems. Addison-Wesley, New York, 1989.
3. Edwards, R., Beuter, A. & Glass, L. Parkinsonian tremor and simplification in network dynamics. Bull. Math. Biol., Vol. 61 (1999) pp. 157–177.
4. Glass, L., Classification of biological networks by their qualitative dynamics. J. Theor. Biol., Vol. 54 (1975a) pp. 85–107.
5. Glass, L., Combinatorial and topological methods in nonlinear chemical kinetics. J. Chem. Phys., Vol. 63 (1975b) pp. 1325–1335.
6. Glass, L., Combinatorial aspects of dynamics in biological systems, in: Landman, U. (ed.), Statistical Mechanics and Statistical Methods in Theory and Application. Plenum, New York, 1977, pp.585–611.
7. Glass, L., & Hill, C., Ordered and disordered dynamics in random networks. Europhys. Lett., Vol. 41 (1998) pp. 599–604.
8. Glass, L., & Pasternack, J. S., Prediction of limit cycles in mathematical models of biological oscillations. Bull. Math. Biol., Vol. 40 (1978a) pp. 27–44.
9. Glass, L., & Pasternack, J. S., Stable oscillations in mathematical models of biological control systems. J. Math. Biol., Vol. 6 (1978b) pp. 207–223.
10. Lewis, J. E., & Glass, L., Nonlinear dynamics and symbolic dynamics of neural networks. Neural Computation, Vol. 4 (1992) pp. 621–642.
11. Mestl, T., Bagley, R. J., & Glass, L., Common chaos in arbitrarily complex feedback networks. Phys. Rev. Lett., Vol. 79 (1997) pp. 653–656.
12. Mestl, T., Plahte, E., & Omholt, S.W., Periodic solutions in systems of piecewise-linear differential equations. Dynamics and Stability of Systems, Vol. 10 (1995) pp. 179–193.
13. Mestl, T., Lemay, C., & Glass, L., Chaos in high-dimensional neural and gene networks. Physica D, Vol. 98 (1996) pp. 33–52.

Proceedings of Dynamic Systems and Applications 3 (2001) 193-200

Riccati Equations on a Measure Chain

Lynn Erbe[1] and Allan Peterson[2]
[1,2]Department of Mathematics and Statistics
University of Nebraska-Lincoln
Lincoln, NE 68588-0323

ABSTRACT: We obtain some necessary and sufficient conditions for disconjugacy of a second order linear equation on a measure chain in terms of an associated quadratic functional. We also provide a simple oscillation criterion.
AMS Subject Classification. 34B10,39A10.

Let $\mathbb{T}$ be a measure chain. We are concerned with the second order self-adjoint equation

$$Lx(t) := [p(t)x^{\Delta}(t)]^{\Delta} + q(t)x^{\sigma}(t) \quad = \quad 0, \tag{1}$$

where $p(t) > 0$ and continuous on $\mathbb{T}^{\kappa}$ and $q(t)$ is right-dense continuous on $(\mathbb{T}^{\kappa})^2$. To understand this so called differential equation (1) on a measure chain (time scale) $\mathbb{T}$ we refer the reader to the references Agarwal & Bohner [1], Erbe & Hilger [2], Erbe & Peterson [3], & Hilger [4] and for a detailed discussion of the delta derivative, the forward and backward jump operators, as well as some other notation and references.

Theorem 1 *If $x(t)$ is a solution of the self-adjoint equation $Lx(t) = 0$ with $x(t)x^{\sigma}(t) > 0$ on $\mathbb{T}$, and if we make the substitution*

$$z(t) := \frac{p(t)x^{\Delta}(t)}{x(t)} \tag{2}$$

for $t \in \mathbb{T}^{\kappa}$, then $z(t)$ is a solution of the Riccati equation

$$Rz(t) := z^{\Delta}(t) + q(t) + \frac{z^2(t)}{p(t) + \mu(t)z(t)} = 0,$$

on $\mathbb{T}^{\kappa}$ and

$$p(t) + \mu(t)z(t) > 0 \tag{3}$$

is satisfied on $\mathbb{T}^{\kappa}$.

Proof

Assume $x(t)$ is a solution of the self-adjoint equation $Lx(t) = 0$ with $x(t)x^{\sigma}(t) > 0$ on $\mathbb{T}$. If we make the so-called *Riccati substitution* (2)

$$z(t) := \frac{p(t)x^{\Delta}(t)}{x(t)}$$

for $t \in \mathbb{T}^\kappa$. Then

$$\begin{aligned} z^\Delta(t) &= \frac{x(t)\left[p(t)x^\Delta(t)\right]^\Delta - p(t)\left(x^\Delta(t)\right)^2}{x(t)x^\sigma(t)} \\ &= \frac{-q(t)x(t)x^\sigma(t) - p(t)\left(x^\Delta(t)\right)^2}{x(t)x^\sigma(t)} \\ &= -q(t) - \frac{x(t)}{p(t)x^\sigma(t)}\left[\frac{p(t)x^\Delta(t)}{x(t)}\right]^2 \\ &= -q(t) - \frac{x(t)}{p(t)x^\sigma(t)} z^2(t). \end{aligned}$$

But

$$\frac{p(t)x^\sigma(t)}{x(t)} = \frac{p(t)\left[x(t) + \mu(t)x^\Delta(t)\right]}{x(t)}.$$

Hence

$$\frac{p(t)x^\sigma(t)}{x(t)} = p(t) + \mu(t)z(t) > 0$$

for $t \in \mathbb{T}^\kappa$. Substituting this we have

$$z^\Delta(t) = -q(t) - \frac{z^2(t)}{p(t) + \mu(t)z(t)}$$

for $t \in \mathbb{T}^\kappa$. Hence $z(t)$ is a solution of the Riccati equation

$$Rz(t) := z^\Delta(t) + q(t) + \frac{z^2(t)}{p(t) + \mu(t)z(t)} = 0, \tag{4}$$

on $\mathbb{T}^\kappa$ satisfying (3) for $t \in \mathbb{T}^\kappa$. □

Note that if $\mathbb{T} = \mathbb{R}$, then $\mu(t) = 0$ and the Riccati equation (4) is the well studied Riccati equation

$$Rz(t) = z'(t) + q(t) + \frac{1}{p(t)} z^2(t) = 0.$$

On the other hand if $\mathbb{T} = \mathbb{Z}$, then $\mu(t) = 1$ and the Riccati equation (4) is the well-known equation (see, for example, Chapter 6 in Kelley & Peterson [5])

$$Rz(t) := z^\Delta(t) + q(t) + \frac{z^2(t)}{p(t) + z(t)} = 0$$

Theorem 2 *The self-adjoint equation $Lx(t) = 0$ has a positive solution on $\mathbb{T}$ iff the Riccati equation $Rz = 0$ has a solution $z(t)$ on $\mathbb{T}^\kappa$ satisfying (3) on $\mathbb{T}^\kappa$.*

Proof First assume that $Lx(t) = 0$ has a positive solution on $\mathbb{T}$. Then by Theorem 1 the Riccati equation $Rz = 0$ has a solution $z(t)$ on $\mathbb{T}$ satisfying (3) on $\mathbb{T}^\kappa$. Conversely assume $z(t)$ is a solution of the Riccati equation $Rz(t) = 0$ on $\mathbb{T}^\kappa$ such that (3) holds on $\mathbb{T}^\kappa$. Let $a \in \mathbb{T}^\kappa$ and consider the initial value problem (IVP)

$$x^\Delta(t) = \frac{z(t)}{p(t)} x(t), \quad x(a) = 1. \tag{5}$$

Since

$$1 + \frac{z(t)}{p(t)}\mu(t) = \frac{p(t) + \mu(t)z(t)}{p(t)} > 0$$

this IVP has a unique solution $x(t)$. In fact $x(t)$ is given by

$$x(t) = e^{\int_a^t \xi_{\mu(r)}\left(\frac{z(r)}{p(r)}\right)\Delta r}$$

where $\xi_h(z)$ is the cylinder transformation which is defined by $\xi_h(z) = \frac{1}{h}Log(1 + hz)$, $if\ h > 0$ and $\xi_0(z) = z$. Note that $x(t) > 0$ on $\mathbb{T}$. By (5) it follows that

$$p(t)x^{\Delta}(t) = z(t)x(t).$$

Taking the delta derivative of both sides we get

$$\begin{aligned}
\left[p(t)x^{\Delta}(t)\right]^{\Delta} &= x^{\sigma}(t)z^{\Delta}(t) + z(t)x^{\Delta}(t) \\
&= x^{\sigma}(t)\left[-q(t) - \frac{z^2(t)}{p(t) + \mu(t)z(t)}\right] + \frac{z^2(t)x(t)}{p(t)} \\
&= -q(t)x^{\sigma}(t).
\end{aligned}$$

Hence $x(t)$ is a positive solution of $Lx(t) = 0$. □

Theorem 3 *Assume* $\sup\mathbb{T} = \infty$, $t_0 \in \mathbb{T}$, $h_0 := \inf\{\mu(t) : t \in [t_0, \infty)\} > 0$, $p(t) > 0$ *is bounded above on* $[t_0, \infty)$ *and*

$$\int_{t_0}^{\infty} q(t)\Delta t = \infty.$$

Then the equation $Lx = 0$ *is oscillatory on* $[t_0, \infty)$.

Proof Assume $Lx = 0$ is nonoscillatory on $[t_0, \infty)$. Then there is a $t_1 \geq t_0$ such that $Lx = 0$ has a positive solution $x(t)$ on $[t_1, \infty)$. We then make the Riccati substitution

$$z(t) = \frac{p(t)x^{\Delta}(t)}{x(t)}.$$

By Theorem 1 $z(t)$ is a solution of the Riccati equation $Rz(t) = 0$ and (3) is satisfied on $[t_1, \infty)$. Integrating both sides of the Riccati equation from t_1 to t we get

$$z(t) = z(t_1) - \int_{t_1}^{t} q(t)\Delta t - \int_{t_1}^{t} \frac{z^2(s)}{p(s) + \mu(s)z(s)}\Delta s.$$

Letting $t \to \infty$ we get

$$\lim_{t\to\infty} z(t) = -\infty. \tag{6}$$

But from (3) we have that

$$z(t) > -\frac{p(t)}{\mu(t)} \geq -\frac{M}{h_0}$$

for $t \in [t_1, \infty)$ where M is an upper bound on $p(t)$. But this contradicts (6) and the proof is complete. □

We define $\mathbb{A}$ to be the set of functions $\mathbb{A} = \{u : [a, \sigma(b)] \to \mathbb{R}$ *is continuous,* u^{Δ} *is right-dense continuous on* $[a, \sigma(b)]$, $u(a) = 0,\ u(\sigma^2(b)) = 0\}$. Then we define the quadratic functional $\mathbb{Q}$ on $\mathbb{A}$ by

$$\mathbb{Q}u = \int_a^{\sigma^2(b)} \left\{p(t)\left[u^{\Delta}(t)\right]^2 - q(t)u^2(\sigma(t))\right\} \Delta t.$$

Definition We say that $\mathbb{Q}$ is positive definite on $\mathbb{A}$ provided $\mathbb{Q}u \geq 0$ for all $u \in \mathbb{A}$ and $\mathbb{Q}u = 0$ iff $u = 0$.

We next state and prove the important completing the square lemma.

Lemma 4 *Assume $z(t)$ is a solution of the Riccati equation $Rz(t) = 0$ on $[a, \sigma^2(b)]$ with $p(t) + \mu(t)z(t) > 0$ on $[a, \sigma(b)]$ and assume that $u \in \mathbb{A}$. Then we get the "completing the square" formula:*

$$\left[z(t)u^2(t)\right]^{\Delta} = \left\{p(t)\left[u^{\Delta}(t)\right]^2 - q(t)u^2(\sigma(t))\right\} - \left\{\frac{z(t)u^{\sigma}(t)}{\sqrt{p(t)+\mu(t)z(t)}} - \sqrt{p(t)+\mu(t)z(t)}\,u^{\Delta}(t)\right\}^2,$$

$t \in [a, \sigma(b)]$.

Proof Consider, for $t \in [a, \sigma(b)]$,

$$\begin{aligned}
&\left[z(t)u^2(t)\right]^{\Delta}\\
=\;& u^2(\sigma(t))z^{\Delta}(t) + u^{\sigma}(t)u^{\Delta}(t)z(t) + u^{\Delta}(t)u(t)z(t)\\
=\;& u^2(\sigma(t))\left[-q(t) - \frac{z^2(t)}{p(t)+\mu(t)z(t)}\right] + u^{\sigma}(t)u^{\Delta}(t)z(t) + u^{\Delta}(t)\left[u(\sigma(t)) - \mu(t)u^{\Delta}(t)\right]z(t)\\
=\;& \left\{p(t)\left[u^{\Delta}(t)\right]^2 - q(t)u^2(\sigma(t))\right\} - \frac{z^2(t)u^2(\sigma(t))}{p(t)+\mu(t)z(t)} + 2u^{\sigma}(t)u^{\Delta}(t)z(t) - \left[p(t)+\mu(t)z(t)\right]\left[u^{\Delta}(t\right.\\
=\;& \left\{p(t)\left[u^{\Delta}(t)\right]^2 - q(t)u^2(\sigma(t))\right\} - \left\{\frac{z(t)u^{\sigma}(t)}{\sqrt{p(t)+\mu(t)z(t)}} - \sqrt{p(t)+\mu(t)z(t)}\,u^{\Delta}(t)\right\}^2
\end{aligned}$$

□

Theorem 5 *Assume $x(t)$ is a solution of the self-adjoint equation $Lx(t) = 0$ on $[a, \sigma^2(b)]$ and let*

$$u(t) := \begin{cases} 0 & \text{if } a \leq t \leq c\\ x(t) & \text{if } c < t \leq d\\ 0 & \text{if } d < t \leq \sigma^2(b) \end{cases}$$

where $a \leq c \leq \sigma(c) < d \leq \sigma(b)$. Then

$$\mathbb{Q}u = C + D,$$

where

$$C := \begin{cases} -p(c)x(c)x^{\Delta}(c) & \text{if } \mu(c) = 0 \\ \frac{p(c)x(c)x^{\sigma}(c)}{\mu(c)} & \text{if } \mu(c) > 0 \end{cases}$$

and

$$D := \begin{cases} p(d)x(d)x^{\Delta}(d) & \text{if } \mu(d) = 0 \\ \frac{p(d)x(d)x^{\sigma}(d)}{\mu(d)} & \text{if } \mu(d) > 0. \end{cases}$$

Proof Let $u(t)$ be as in the statement of this theorem and consider

$$\begin{aligned} \mathfrak{Q}u &= \int_a^{\sigma^2(b)} \left\{p(t)\left[u^{\Delta}(t)\right]^2 - q(t)u^2(\sigma(t))\right\}\Delta t \\ &= \int_c^{\sigma(d)} p(t)\left[u^{\Delta}(t)\right]^2 \Delta t - \int_c^d q(t)u^2(\sigma(t))\Delta t \\ &= \int_c^{\sigma(c)} p(t)\left[u^{\Delta}(t)\right]^2 \Delta t + \int_d^{\sigma(d)} p(t)\left[u^{\Delta}(t)\right]^2 \Delta t - \int_c^{\sigma(c)} q(t)u^2(\sigma(t))\Delta t \\ &+ \int_{\sigma(c)}^d \left\{p(t)\left[x^{\Delta}(t)\right]^2 - q(t)x^2(\sigma(t))\right\}\Delta t. \end{aligned}$$

Using integration by parts on the first term in the last integral we get

$$\begin{aligned} \mathfrak{Q}u &= \int_c^{\sigma(c)} p(t)\left[u^{\Delta}(t)\right]^2 \Delta t + \int_d^{\sigma(d)} p(t)\left[u^{\Delta}(t)\right]^2 \Delta t - \int_c^{\sigma(c)} q(t)u^2(\sigma(t))\Delta t \\ &+ \left[p(t)x^{\Delta}(t)x(t)\right]_{\sigma(c)}^d - \int_{\sigma(c)}^d Lx(t)x^{\sigma}(t)\Delta t. \\ &= C + D, \end{aligned}$$

where

$$C = \int_c^{\sigma(c)} p(t)\left[u^{\Delta}(t)\right]^2 \Delta t - \int_c^{\sigma(c)} q(t)u^2(\sigma(t))\Delta t - p(\sigma(c))x^{\Delta}(\sigma(c))x(\sigma^2(c))$$

and

$$D = \int_d^{\sigma(d)} p(t)\left[u^{\Delta}(t)\right]^2 \Delta t + p(d)x^{\Delta}(d)x(\sigma(d)).$$

If $\mu(c) = 0$, then $C = p(c)x(c)x^{\Delta}(c)$ as desired. Assume that $\mu(c) > 0$, then

$$\begin{aligned} C &= p(c)\frac{x^2(\sigma(c))}{\mu^2(c)}\mu(c) - q(c)x^2(\sigma(c))\mu(c) - p(\sigma(c))x^{\Delta}(\sigma(c))x(\sigma^2(c)) \\ &= \frac{p(c)x^2(\sigma(c))}{\mu(c)} - \left\{\frac{p(\sigma(c))x^{\Delta}(\sigma(c)) - p(c)x^{\Delta}(c)}{\mu(c)} + q(c)x^{\sigma}(c)\right\}x^{\sigma}(c)\mu(c) - p(c)x^{\Delta}(c)x^{\sigma}(c) \\ &= \frac{p(c)x^2(\sigma(c))}{\mu(c)} - Lx(c)x^{\sigma}(c)\mu(c) - p(c)\frac{x^{\sigma}(c) - x(c)}{\mu(c)}x^{\sigma}(c) \\ &- \frac{p(c)x(c)x^{\sigma}(c)}{\mu(c)}. \end{aligned}$$

If $\mu(d) = 0$, then $D = p(d)x(d)x^{\Delta}(d)$. Finally, if $\mu(d) > 0$, then

$$\begin{aligned} D &= p(d)\left[u^{\Delta}(d)\right]^2 \mu(d) + p(d)x^{\Delta}(d)x(d) \\ &= p(d)\frac{x^2(d)}{\mu^2(d)}\mu(d) + p(d)\frac{x^{\sigma}(d) - x(d)}{\mu(d)}x(d) \\ &= \frac{p(d)x(d)x^{\sigma}(d)}{\mu(d)}. \end{aligned}$$

□

Theorem 6 *The self-adjoint equation $Lx(t) = 0$ is disconjugate on $[a, \sigma^2(b)]$ iff $\mathbb{Q}$ is positive definite on $\mathbb{A}$.*

Proof Assume that $Lx(t) = 0$ is disconjugate on $[a, \sigma^2(b)]$. Then there is a solution $x(t)$ without generalized zeros in $[a, \sigma^2(b)]$. If we make the Riccati substitution

$$z(t) = \frac{p(t)x^{\Delta}(t)}{x(t)}$$

then by Theorem 1 $z(t)$ is a solution of the Riccati equation $Rz(t) = 0$ on $[a, \sigma(b)]$ satisfying (??). By making appropriate definitions we can assume $z(t)$ is a solution of the Riccati equation $Rz(t) = 0$ on $[a, \sigma^2(b)]$. Hence if $u \in \mathbb{A}$ we have by Lemma 4 that

$$\left[z(t)u^2(t)\right]^{\Delta} = \left\{p(t)\left[u^{\Delta}(t)\right]^2 - q(t)u^2(\sigma(t))\right\} - \left\{\frac{z(t)u^{\sigma}(t)}{\sqrt{p(t) + \mu(t)z(t)}} - \sqrt{p(t) + \mu(t)z(t)}u^{\Delta}(t)\right\}^2.$$

Integrating from a to $\sigma^2(b)$ and using $u(a) = 0 = u(\sigma^2(b))$ we get

$$\mathbb{Q}u = \int_a^{\sigma^2(b)} \left\{\frac{z(t)u^{\sigma}(t)}{\sqrt{p(t) + \mu(t)z(t)}} - \sqrt{p(t) + \mu(t)z(t)}u^{\Delta}(t)\right\}^2 \Delta t.$$

It follows that $\mathbb{Q}u \geq 0$ for all $u \in \mathbb{A}$. Now assume that $\mathbb{Q}u = 0$. Then

$$\frac{z(t)u^{\sigma}(t)}{\sqrt{p(t) + \mu(t)z(t)}} = \sqrt{p(t) + \mu(t)z(t)}u^{\Delta}(t)$$

for $t \in [a, \sigma(b)]$. It follows that $u(t)$ solves the initial value problem

$$u^{\Delta}(t) = \frac{z(t)}{p(t) + \mu(t)z(t)}u^{\sigma}(t), \quad u(a) = 0.$$

This implies $u(t) = 0$ for all $t \in [a, \sigma^2(b)]$ and hence $\mathbb{Q}$ is positive definite on $\mathbb{A}$.

Conversely assume $\mathbb{Q}$ is positive definite on $\mathbb{A}$. Assume $Lx = 0$ is not disconjugate on $[a, \sigma^2(b)]$. Then there are points $a \leq c \leq \sigma(c) < d \leq \sigma(b)$ and a solution $x(t)$ of $Lx = 0$ such that $x(c) = 0$ if $\mu(c) = 0$ and $p(c)x(c)x^{\sigma}(c) \leq 0$ if $\mu(c) > 0$, $x(d) = 0$ if $\mu(d) = 0$ and $p(d)x(d)x^{\sigma}(d) \leq 0$ if $\mu(d) > 0$, and $x(t) \neq 0$ in (c, d). Let $u(t)$ be defined as in Theorem 6. Then since $\mathbb{Q}$ is positive definite on $\mathbb{A}$, $\mathbb{Q}u > 0$. But by Theorem 6,

$$\mathbb{Q}u = C + D \leq 0$$

which is a contradiction. □

Theorem 7 *Assume $x \in \mathbb{D}$ has no generalized zeros in $\mathbb{T}$ and z is defined by the Riccati substitution (2) for $t \in \mathbb{T}^\kappa$. Then (3) holds for $t \in \mathbb{T}^\kappa$ and*

$$Lx(t) = x^\sigma(t)Rz(t)$$

for $t \in (\mathbb{T}^\kappa)^2$.

Proof Assume $x \in \mathbb{D}$ has no generalized zeros in $\mathbb{T}$ and z is defined by the Riccati substitution (2) for $t \in \mathbb{T}^\kappa$. For $t \in \mathbb{T}^\kappa$ consider

$$\begin{aligned} x^\sigma(t)z^\Delta(t) &= x^\sigma(t)\left[\frac{p(t)x^\Delta(t)}{x(t)}\right]^\Delta \\ &= \frac{x(t)\left[p(t)x^\Delta(t)\right]^\Delta - p(t)\left[x^\Delta(t)\right]^2}{x(t)} \\ &= \left[p(t)x^\Delta(t)\right]^\Delta - \frac{p(t)}{x(t)}\left[x^\Delta(t)\right]^2 \\ &= Lx(t) - q(t)x^\sigma(t) - x^\sigma(t)\frac{x(t)}{p(t)x^\sigma(t)}z^2(t). \end{aligned}$$

Hence by (3)

$$x^\sigma(t)z^\Delta(t) = Lx(t) - q(t)x^\sigma(t) - x^\sigma(t)\frac{z^2(t)}{p(t) + \mu(t)z(t)}.$$

Solving for $Lx(t)$ we get the desired result

$$Lx(t) = x^\sigma(t)Rz(t)$$

for $t \in (\mathbb{T}^\kappa)^2$. □

REFERENCES

1. R. Agarwal & M. Bohner, Basic calculus on time scales and some of its applications, preprint.
2. L. Erbe & S. Hilger, Sturmian Theory on Measure Chains, Differential Equations and Dynamical Systems, 1 (1993), 223-246.
3. L.H. Erbe & A. Peterson, Green's functions and comparison theorems for differential equations on measure chains, Dynamics of Continuous, Discrete and Impulsive Systems, 6 (1999),121-137.
4. S. Hilger, Analysis on measure chains-a unified approach to continuous and discrete calculus, Results in Mathematics, 18 (1990), 18-56.
5. W. Kelley & A. Peterson, Difference Equations: An Introduction with Applications, Academic Press, 1991.

Proceedings of Dynamic Systems and Applications 3 (2001) 201-208

Quantum Gravity

E. E. Escultura
School of Statistics, University of the Philippines, Q. C. 1101, Philippines

ABSTRACT. Fundamental physical principles of grand unified theory of nature called flux theory of gravitation are stated. Matter has two fundamental states – dark and visible. Dark matter's basic constituent is superstring; that of visible matter is primum. Superstring is thin, extended, helical and loop and has toroidal flux. Suitable agitation converts its segment to primum; suitable de-agitation induces evolution towards infinitesimal torus. Theory's most major achievement is unification of macro gravity, dynamics of fluxes of superstrings, and quantum gravity, dynamics of superstring's induced fluxes, and, therefore, all of nature's interactions. Quantum gravity includes clustering and interaction from primal through biological scales including proton, neutron, neutrino, atom and electromagnetism, matter-anti-matter interaction, surface tension, osmosis and capillary action in cells of living organism.
AMS(MOS) subject classification. 35K60, 35K57

1. INTRODUCTION

The key discovery prompted by the resolution of Fermat's last theorem Escultura[10] is the undecidability of any physical principle in mathematics, an implicit criticism of physic's reliance on mathematics for fundamental principles, Peat [17]. Remedy: search for fundamental physical principles undertaken in Escultura [6, 8, 9, 13].

2. FUNDAMENTAL PHYSICAL PRINCIPLES

The reader is referred to Escultura [6, 8, 9, 13] for fundamental physical principles of the theory which we summarize. Basic constituent of matter is superstring: thin, extended, helical, loop, has oriented flux, called flux torus, repository of latent energy. Dark matter consists of non-agitated superstrings whose cycles have lengths less than 10^{-14} meters (frequency greater than 10^{22} cycles/sec). Suitable agitation raises cycle lengths of segment beyond 10^{-14} meters to form primum, basic constituent of visible matter. Photon is primum that breaks off from superstring with its flux torus and rides on cosmic wave at speed equal to forward flux rate. Non-agitated superstring is 100 billion billion times smaller than proton, Kaku & Thompson [15]. In any interaction matter minimizes energy loss, maximizes energy gain and preserves total energy (latent plus kinetic energy); De-agitation (at resonance with simple Cosmic waves) shrinks its cycles, enhances latent energy and induces evolution towards infinitesimal torus. Superstring is path of flux torus in dark matter; flux torus has flux torus, the latter has flux torus, ... Thus, superstring is nested fractal sequence of superstrings.

Energy conservation has equivalent forms: order, sysmmetry, economy, least action, optimality, efficiency, stability, self-similarity (fractal), coherence, resonance, quantization, smoothness, uniformity, motion-symmetry balance, evolution to infinitesimal configuration and, in biology, stability and diversity and complexity of functions and configuration. Natural configuration, process and interaction have maximal oscillatory properties.

Cosmic waves are caused by diverse activity of Cosmos (e.g., supernovae, core vortex spin, flux turbulence). Shock wave is great concentration of cosmic waves (energy) in small space, Escultura [13(Fig. 4(a))]; nature stores huge energy in fractal, e.g. superstring, shock wave.

Dark matter is fundamental chaos or mixture of diverse order; suitable boundary condition induces flux coherence called turbulence. Fundamental chaos includes composites of cosmic waves that allows alteration of direction and characteristic of motion of primum or photon in flight (e.g., superconductivity [13]). Superstring has independent infinitesimal higher-dimensional motion due to pounding of flux torus by diverse cosmic waves that induces infinitesimal coherent flux of same direction along helical neighborhood. When superstring shrinks, due to de-agitation, its latent energy rises since flux torus travels shorter path.

Energy conservation requires coherence of superstring toroidal fluxes; flux coherence in primum is called spin. Spin settles puzzle in quantum mechanics regarding charge distribution: charge is distributed coherently along cycles. Prima's induced fluxes govern electromagnetic and matter-anti-matter interactions in accordance with this dynamics: two prima of opposite spins and, therefore, same induced flux direction (compatible) attract and those with same spin and, therefore, opposite induced flux directions (incompatible), repel each other.

When cosmic shock wave of suitable energy hits superstring, one of these may occur: (1) outer torus breaks, remaining superstring dissipates into dark matter with its flux torus (latent energy), (2) segment of outer superstring bulges (acquires kinetic energy) and forms primum, Escultura [13(Fig. 1(a))], (3) primum is plucked from its loop with flux torus, rides on cosmic wave of suitable configuration, becomes planar and propelled as photon, (4) primum or coupled prima rides on cosmic wave of suitable configuration and propelled as cosmic ray and (5) when hit by suitable shock wave, non-agitated superstring is projected through helical path and forms superstring loop; helical path configuration is due to internal structure of dark matter which is principal over external factor.

Stable superstring has at most one primum; two prima may pull it apart and break the loop since they acquire autonomy from dark matter. Suitable agitation may loosen up cycles but still in dark region. Their de-agitation creates low pressure that induces formation of flux vortices, essential dynamics of macro gravity. Flux torus in flight (photon) that fails to reach forward flux speed or comes to rest disintegrates and dissipates into dark matter.

Freed from dark viscosity, visible matter acquires visible properties: kinetic energy, charge (visible induced flux), called charge, mass, momentum and autonomous motion. Visible matter has components in dark matter: its superstring loops. Among known simple prima are electron (unit negative charge) positron (unit positive charge) and up and down quarks (+2/3 and - 1/3 charges, respectively, Escultura & Gudkov). Proton, neutron, neutrino, anti-neutrino are coupled prima. Up quark is magnet, polarity subject to right-hand rule; electron has reverse polarity. Up quark has same mass as down quark, 600 times mass of electron. Mathematical models of up and down quarks, electron, positron, proton, neutron, neutrino and anti-neutrino and their sizes, masses and latent energy are given in Escultura & Gudkov [14].

When electron and positron come close at their rims, attractive momenta of approach force cycles to overlap; being congruent and with opposite fluxes, cycle for cycle, forming destructive coupling and resulting in propagation of two photons in opposite directions. While two prima with same spin are repulsive their fluxes add up to coherent flux when coupled, Escultura [13(Fig. 2(a))]; prima with equal but opposite spins are attractive but have 0 coherent induced fluxes, Escultura [13(Fig. 2(b))]. Proton is cluster of three prima, two up quarks and down quark whose cross-section forms three petals joined at origin of moving coordinate; mathematical model is in Escultura & Gudkov [14]. Proton is also magnet, polarity determined by net flux. Two protons are attractive when coupled north-south or south-north, this coupling is non-destructive. Two protons may be coupled also by attaching down quark of one to up quark of another at their rims. This makes gluon unnecessary; like graviton, gluon does not exist. Moreover, any number of protons can be coupled this way limited only by instability of large nuclei. Neutron consists of proton, electron and anti-neutrino, Escultura & Gudkove [14]. Although proton and electron have opposite but equal charge, they are not anti-matter of each other, the quark's cycles being much larger than the electron's that their coupling is not destructive. Having no coherent induced flux, neutron neither attracts nor repels primum; however, like any superstring or cluster of superstrings, it is pulled by any coherent flux of sufficient strength. Thus, neutrons and protons are drawn together by proton fluxes to coexist in nucleus of atom.

3. QUANTUM GRAVITY

At visible region flux and pressure are interrelated; for example, initial chaotic rush of air molecules from high to low pressure stabilizes as coherent flux or turbulence. Conversely, flux of air molecules creates low pressure and sucks air molecules around it. A simple experiment will verify this: place piece of paper above air passage, say, between two parallel rectangular blocks on flat surface and shoot air jet stream through that passage. One would think that the paper would fly off. No; it will be sucked by jet stream. This follows from the Flux-Pressure Complementarity in Escultura [13].

origin of projected superstring torus is point of low pressure which induces it to retrace its path. Retracing occurs at any point in path. Flux coherence is motive force for clustering at astronomical, biological, atomic and primal scales. At astronomical scale (macro), random clusters of low pressure caused by big bang are motive force for fractal formation of vortices which is essence of gravitation at this scale and accounts for formation of galaxies and stellar and planetary systems Escultura [8, 9, 13]. At primal, atomic and biological scales flux coherence also induces formation of primal, atomic, molecular and biological clusters. This is the essence of quantum gravity that includes surface tension essential for biological processes; it is the basis of osmosis and capillary action in the cells of living organism.

Light, energetic and uncharged prima like neutrinos do not resonate with and, therefore, are unaffected by flux; they go through it at negligible resistance and dissipation of energy, the reason they ride on cosmic waves across galaxies.

Charge is kinetic energy and subject to energy conservation. When positive and negative charges are equal in atomic fluxes, it has no coherent flux beyond electron orbits. Electrons neutralize net charge by their spins but flux pressure and sucking effect remains. When there are less orbital electrons than protons in nucleus, atom is positive ion. From oscillation universality and energy conservation principles both electron orbits and flux streamlines follow oscillatory patterns subject to quantization.

With this flux dynamics as boundary conditions, we have the genesis of atom with initial phase of our universe consisted of superstrings as boundary conditions. While clustering is stochastic, Esucltura [9, 13,] at both fundamental states, structure and fluxes of superstrings, as boundary conditions, broaden forms and motion of visible matter. Agitation by big bang yields prima as basic ingredients of diverse clusters. As soon as early universe cools sufficiently, light clusters like neutrino, proton and neutron form. Further cooling induces clusters of simple and coupled prima to form light nuclei like hydrogen, star's main ingredient.

Consider cluster of protons or neutrons as nucleus and disregard coupled prima with 0 coherent flux, e.g., neutrons, to obtain, in effect, cluster of protons alone. Since protons have same net spin no coherent flux exists in cluster's interior, Escultura [13(Fig. 4(c))]. However, coherent flux exists around nucleus, its power depending on number of protons contributing to it. This flux which forms vortex with nucleus at eye does these: (a) attracts clusters of superstrings with or without prima, (b) throws electrons into orbit along this flux and keeps them there and (c) shields the nucleus from charged and less energetic prima. Suitably energetic prima penetrate flux and get sucked by low pressure in eye. Nucleus, being calm sanctuary at vortex eye, induces superstrings towards infinitesimal tori and nurtures mini-black hole, quiet storage of de-agitated dark matter. At astronomical scale, every body vortex has black hole at core. Thus, quantum gravity is self-similar to macro gravity which accounts for some literature on our universe as fractal.

Inward flux builds up superstrings in eye, induces more spin to nucleus by its momentum, agitates and converts prima for clustering in cooler inner layer of eye, this is basis of complex nuclear formation, aside from fusion and stochastic clustering of prima and superstrings. A long cosmological process is verified by existence of isotopes of same element.

Quantum gravity in solids is verified experimentally and reported by *Scientific American* in its amateur scientist column some years ago, Escultura [12]. We have not only explained quantum gravity but also brought natural clustering and interactions under same flux theory.

4. THE MATHEMATICS AND PHYSICS OF BIOLOGY AND INTELLIGENCE

We summarize fundamental biological principles in Escultura [9]. Physical processes proceed stochastically towards higher order and complexity subject to already attained configurations and other boundary conditions and energy conservation. When one physical system with suitable internal structure interacts with another, a process is unraveled nature of which is determined jointly by all factors with internal structure principal over external factor. Structure and characteristics of stable biological organism is encoded in genes as sequence of DNA base molecules; this genetic code is modified in offspring's embryo by combination of parents' genes; modified genes produce RNA which is copy of modified DNA, except for replacement of DNA base uracil by RNA base thymine; modified genes dictate development, structure and characteristics of offspring. Diverse motion of dark matter induces infinitesimal changes in configuration of organism which, together with combination of genes from parents, induces infinitesimal evolutionary changes in species; shock waves and chemical agents may also induce mutation.

Mathematics cannot be fully understood without understanding nature of thought or intelligence. The brain, center of thought in organism, is result of 500 million years of biological evolution Escultura [9]. While humans are known for most advanced intelligence, other species have this capability also at varying levels of complexity from primitive form like reflex and instincts through the most advanced forms like intuition, creativity and analytical capability. Even in primitive form, thought warns organism of danger and guides search for food. This is evident even in primitive organism with no differentiated brain.

Thought is advanced level of consciousness and consciousness is induced by cosmic waves impinging on dark componenents of atoms and neural cells, Escultura [9, 13]. In lower organisms consciousness manifests itself as reflexes and instincts; in advanced organisms it is enhanced by training. Aside from humans, dolphins, like ants and bees, are known to have fairly developed intelligence.

Being higher capability, thought has prior stages of development: (1) differentiation of brain and command over organism, (2) development of sense organs, (3) transmission of signals along nerve cells to appropriate region of brain, (4) encoding of signals, (4) retrieval, and (5) creativity. To illustrate mechanics of sensation, consider sweetness. Vibration of sugar molecule resonates with taste bud for sweetness causing the latter to vibrate with same characteristics. Vibration encodes characteristics on electro-chemical waves along nerve cells to appropriate region of brain's cortex where they are encoded on neural clusters. Sensation of sweetness is experienced during encoding. Wave signals with these characteristics resonate with and induce corresponding configuration on neural cluster. Through resonance this cluster vibrates until signals stop coming. Then encoding is complete and cluster is de-agitated but retains configuration. This mechanics of sensation is theoretical basis of discovery of artificial sugar like saccharin whose molecular configuration is similar to that of sugar and induces same sensation and cluster characteristics. Brain can distinguish saccharin from sugar by comparing signals and information emanating from various sense organs.

Physically, encoding involves clustering and activation of neural interconnections with configuration determined by resonance with incoming signals. Signals resonate only with appropriate region of cortex and goes through other regions with neither resonance nor encoding. Encoded cluster may have components at different regions of cortex depending on complexity of event being recorded. For instance, it may have sight, smell, touch and sound components. Brain makes model of the physical event with appropriate cluster configuration that encodes signals. When brain waves of suitable intensity agitates this cluster it vibrates and sends out brain waves of exactly the same characteristics as the waves that configured it. At this point the organism recalls the event and forms concept. Sufficient evidence (e.g., athletic skills) suggests this modeling by brain is quite precise.

A concept is not simply mechanical translation of encoded events; it is synthesized and refined through process of inferences, comparisons and optimization, a process called brain's integrative function. For instance, when one observes hammer at a distance hitting something observed sound and motion do not correspond due to difference in speeds between sound and light. Yet, this discrepancy is not integrated in the concept *hammer*.

Mental activity ranges from concept formation through inferences and creative activity and manipulation of concepts according to certain logic through the integrative function of brain subject to energy conservation in the form of optimization Escultura [9]. Logic, however, is not inherent in one's brain; it is learned in the same way values form. Learning consolidates one's way of thinking including creativity and analytical capability. Therefore, intelligence as individual capability is synthesized accumulated experience; society integrates and synthesizes the collective experiences as the foundation of culture. At the same time, culture profoundly affects one's way of thinking without obliterating stamp of individuality Escultura [9].

Brain's integrative function continues to develop through training during life time of individual. Unless brain suffers from disorder (e.g., autism) it has control of application of cosmic energy (brain waves) by way of concentration, screening irrelevant signals and disbanding unwanted clusters. A person, for instance, may listen only to something he wants to hear without even being conscious of it. Understanding this capability improves mental skills such as insights, memory, retrieval, creativity and analytical capability. However, same capability is also basis of orthodoxy. At this point new ideas appear strange.

Not all concepts have their origin in experience. Creativity is acquired by training. The human brain, at least, has learned to create new concepts with or without signals from external world. Its integrative capability allows it to form arbitrary cluster configuration. It can even create contradictory concepts such as a set that is its element and a set that is not its element both of which are contradictory being self-referent.

These are the biological foundations of thought. Biology and intelligence, however, is further anchored on physics [9, 13]. At the same time, this dynamics of biological process has its trigger in dark matter, in the tremendous latent energy of superstrings and the powerful dynamo for latent to kinetic energy conversion, namely, the great superstring agitators: cosmic waves which resonate with atoms, cells and neurons through their dark components..

Vibration of neural clusters propagates brain waves that travel along nerve cells. Brain commands other parts of human body through these waves and chemical messengers. This dynamics between cosmic energy and motion of atoms and molecules and, therefore, also organs and their cells in an organism explains apparent spontaneous and autonomous behavior of biological processes such as the so-called free will. Continual vibration of atoms and molecules in organism is observable but that of cosmic waves which induce it as well as atomic, molecular and cellular dark components are not, Escultura [8, 9, 13].

The mathematics of gravitation developed in Escultura [6, 8, 9, 11, 12, 13] include: (a) new nonstandard analysis (b) fractals and (c) the generalized integral. Still to be written down but whose foundations are already in the works is (d) the mathematics of chaos and turbulence. However, standard mathematics is also crucial in this theory and includes: (1) mathematics of brain waves which partially but not exclusively belongs to differential equations; (2) upgraded Pontrjagin maximum principle, Pontrjagin et al, Young [18, 36] crucial in the solution of the gravitational n-body problem, Escultura [6] and useful in the study of turbulence and inverse problems; (3) topology and combinatorics of DNA, Escultura[9]; and simulation and modeling.

In conclusion we single out three major findings of the flux theory of gravitation.

(1) Fractal nature of our universe and unification of macro and quantum gravity and, therefore, all of nature's interactions.

(2) Only bodies with cosmological history may exert gravitation force on other bodies; gravitation is either *pull* (opposite spins) or *push* (same spin); the latter governs ocean tide dynamics, Escultura [9].

(3) Thermonuclear explosion (release of photons and prima) is due not to fusion of deuteron but to agitation of superstrings by trigger atom bomb converting latent to kinetic energy as photons and prima; the agitated elements lose mass and lighter elements tend to fuse; therefore, fusion follows the explosion contrary to conventional belief.

REFERENCES

1. Astronomy, 88 - 93, August 1995.
2. Davies, P. C. & Brown, J., Superstring: A Theory of Everything? Cambridge University Press, New York, 1988.
3. Discover, February 1996, 38 -45.
4. Escultura, E. E, Diophantus: Introduction to Mathematical Philosophy (With Solution of Fermat and Other Applications), Kalikasan, Manila, 1993.
5. Escultura, E. E, (1996) Probabilistic Mathematics and Applications to Dynamic Systems Including Fermat's Last Theorem, *Proc.* Second International Conference on Dynamic Systems and Applications, Dynamic Publishers, Inc., Atlanta, 147 - 152.
6. Escultura, E. E, The Solution of the Gravitational n-Body Problem, Nonlinear Analysis, Vol. 30(1997), 5021 - 5032.
7. Escultura, E. E, The Mathematics of Singularities. *Trans.* National Academy of Science and Technology, Voll. 14(1992), Manila, 141 - 153.
8. Escultura, E. E, Superstring Loop Dynamics and Applications to Astronomy, Nonlinear Analysis, Vol. 35(1999), 959 - 985.
9. Escultura, E. E, The Physics of Biology and Intelligence, Nonlinear Analysis, accepted, 1997.
10. Escultura, E. E, Exact Solutions of Fermat's Equation (Definitive Resolution of Fermat's Last Theorem, Nonlinear Studies, Vol. 5 1998, 227 - 254.
11. Escultura, E. E, The Reconstruction of the Reals, Nonlinear Analysis, accepted, 1998.
12. Escultura, E. E, Foundations of Materialist Mathematics. Nonlinear Analysis, submitted.
13. Escultura, E. E, From Quantum to Macro-Gravity, Nonlinear Analysis, submitted, 1998.
14. Escultura, E. E. & Gudkov, V. V, Mathematical Models on the way from Superstring to Photon, Nonlinear Analysis, submiitted, 2000.
15. Kaku, M. & Thompson, J., Beyond Einstein, Anchor Books, 1995.
16. New Scientist, 12 July 1997, 28 - 32.
17. Peat, F. D. (1970) Superstring and the Search for Theories of Everything, Abacus, London. 1970.
18. Pontrjagin, L. S., Boltyanski, V. V., Gamkrelidze, R R. V., & Mischenko, E. F. The Mathematical Theory of Optimal Processes, K. N. Trirogoff (Tran.), L. W. Neustadt (Ed.), Interscience Publishers, New York, 1962.
19. Robinson, A. Nonstandard Analysis, North-Holland, Amsterdam, 1966.
20. Royden, H. L. Real Analysis, MacMillan, 3rd ed., New York, 1987.
21. Scientific American, April 1983, 708 - 745.
22. Scientific American, April 1995, 11 - 14.
23. Smoot, G. and Davidson, K. Wrinkles in Time, Avon Books, New York, 1988.
24. Time, November 1995, 40 - 49.
25. Young, L. C. Generalized curves and the existence of an attained absolute minimum in the calculus of variations, Compt. Rend. Sci. Lettr., Varsovie, Cl III Vol. 30(1937), 211 - 234.
26. Young, L. C. (1969) Lectures on the Calculus of Variations and Optimal Control Theory, W. B. Saunders, Philadelphia.

Proceedings of Dynamic Systems and Applications 3 (2001) 209-216

The Schrödinger Equation on Helical Manifolds: Bound States

Jagannathan Gomatam and Anthony J Mulholland
Department of Mathematics
Glasgow Caledonian University, Cowcaddens Road
Glasgow G4 OBA
Scotland

Abstract

The Schrödinger equation on orthogonal helical manifolds was derived in the early seventies by one of the authors with a view to understanding the quantum dynamics of macromolecular systems with helical symmetry. In recent years, advances in nanotechnology have facilitated fabrication of quantum systems of prescribed geometry. Therefore it is of interest to analyse the quantum dynamics of such systems. We present classes of global orthogonal co-ordinate systems with helical symmetry, internal, annular and external to singular cylinders whose radii are determined by the helical pitches. All these co-ordinate systems are invertible. We show that a quantum particle confined to move in a tubular domain of small cross-section, in the neighbourhood of a curve, experiences an attractive potential proportional to the square of the curvature, *without requiring the curvature to be small or a smooth function.*

AMS(MOS) subject classification: 81S99, 35K20

1. INTRODUCTION

The importance of the study of quantum mechanics in a variety of curved manifolds has been well emphasized in a number of publications in recent years (Garavaglia and Gomatam [1]-Clarke [6]). In addition to elucidating the quantum dynamics implied by nonseparable solutions of the Shrödinger equation, these investigations also alert us to the possible onset of quantum effects in products of miniaturization technology (see Exner and Seba [4]). Until recently these problems have been analysed within the framework of a local co-ordinate system, culminating in the demonstration of the relationship between attractive geometric potential generated by the mean and the Gaussian curvatures of the surface, (da Costa [3]), or the torsion and the curvature of the path on which the particle is constrained to move (Clarke [6]). One of the main assumptions employed in these investigations (da Costa [3]-Exner [4]) is that the curvature is small and varies slowly with the arc length. As was demonstrated by Goldstone and Jaffe [5], these assumptions could be relaxed and the existence of bound states established. All these investigations (da Costa [3]-Clarke[6]) were carried out in the framework of a local co-ordinate system, such as the Frenet-Serret co-ordinate system. We present classes of global, orthogonal, invertible coordinate systems with helical symmetry, internal, annular and external to singular cylinders whose radii are determined by the helical pitches. This is in itself a non-trivial problem in differential geometry of surfaces. Even though the Schrödinger equation in these coordinate systems do not posses separable solutions in general, it is possible to obtain separation by introducing the standard, narrow confininement in terms of delta-function potential walls in appropriate degrees of freedom. The effective potential manifests itself in the remaining degree of freedom, and, is not in the main, functionally dependent on the curvature alone. However there appears an attractive component which is equal to -(1/4)(curvature)^2 as has been demonstrated under more restrictive assumptions by da Costa [3]. In section 2, we discuss briefly the annular helical system and its invertibility. Section 3 is devoted to a discussion of the existence of bound states in the internal helical system. An analytical expression for the energy eigenvalues due to the geometric potential in the external system is the subject matter of section 4. Conclusions are outlined in section 5.

2. ANNULAR HELICAL COORDINATE SYSTEMS

We have discussed in detail the derivation (Garavaglia and Gomatam [1]) of the Schrödinger equation in external and internal orthogonal helical coordinate systems

$$\Sigma_i(\rho,\phi,z) = c_i z + d_i\phi + M_i(\rho), i = 1,2,3, \tag{1}$$

with $c_3 = 0$ and $d_3 = 0$, respectively, where (ρ, ϕ, z) are the cylindrical polar co-ordinates. When none of the c_i and d_i's is allowed to vanish, but chosen carefully with definite signature, we obtain an orthogonal system defined inside an annular cylindrical domain.

The annular orthogonal helical system, hereinafter denoted by $\{\Sigma, \Lambda, \Omega\}$ is defined in terms $\{\rho, \phi, z\}$, and the parameters $a_i, q_i > 0$:

$$\Sigma = -a_1 z + q_1\phi + a_1 \int d\rho \frac{\sqrt{(\rho^2 - a^2)(\rho^2 + c^2)}}{\rho\sqrt{b^2 - \rho^2}}, \tag{2}$$

$$\Lambda = a_2 z + q_2\phi + a_2 \int d\rho \frac{\sqrt{(\rho^2 - a^2)(b^2 - \rho^2)}}{\rho\sqrt{\rho^2 + c^2}}, \tag{3}$$

$$\Omega = a_3 z - q_3\phi + a_3 \int d\rho \frac{\sqrt{(b^2 - \rho^2)(\rho^2 + c^2)}}{\rho\sqrt{\rho^2 - a^2}}, \tag{4}$$

where

$$a^2 = q_1q_2/a_1a_2, b^2 = q_2q_3/a_2a_3, c^2 = q_3q_1/a_3a_1, \tag{5}$$

and

$$\rho_{\min} \in Max\ (a^2, c^2).$$

For definiteness we assume that $a^2 > c^2$. A simple substitution $\sigma = \rho^2$ transforms all the integrals into linear combinations of elliptic integrals of the first and the third kinds, (Gradshteyn and Ryzhik [7, p220]).

The invertibility of equations (2)-(4) is ascertained by noting that

$$\begin{aligned} L &= \left[\frac{b^2\Sigma}{a_1(b^2+c^2)} + \frac{c^2(b^2-a^2)\Lambda}{a_2(b^2+c^2)(c^2+a^2)} + \frac{a^2\Omega}{a_3(c^2+a^2)}\right]\frac{2}{(b^2-a^2)}, \qquad (6)\\ &= \int \frac{d\rho}{\sqrt{(\sigma-c^2)(b^2-\sigma)(\sigma-a^2)}}, \\ &= \frac{2}{\sqrt{(b^2+c^2)}}F(\chi, k), \qquad (7) \end{aligned}$$

where,

$$k = \sqrt{\frac{(b^2 - a^2)}{(b^2 + c^2)}}.$$

and

$$\chi = \arcsin\sqrt{\frac{(b^2 - \rho^2)}{(b^2 + a^2)}}, b^2 > \rho^2 \geq a^2 \tag{8}$$

$$= \arcsin\sqrt{\frac{(b^2 + c^2)(\rho^2 - a^2)}{(b^2 - a^2)(\rho^2 + c^2)}}, b^2 \geq \rho^2 > a^2. \tag{9}$$

In this paper we discuss the case (9) where the limits of integration are σ and b. Equation (2)-(4) imply $\rho = \sqrt{b^2 - (b^2 - a^2)Sn^2(u,k)}$, with

$$u = \frac{L\sqrt{(b^2 + c^2)}}{2}$$

and $Sn(u,k)$ is the Jacobi Elliptic function, (Gradshteyn and Ryzhik [7]) of modulus k. Further details on this general system can be found in Gomatam and Mulholland [8].

3. THE GEOMETRIC POTENTIAL IN INTERNAL HELICAL CO-ORDINATE SYSTEM

We will illustrate the effect of geometric potential on particle behaviour by first considering internal helical co-ordinate system defined by

$$\xi = B_1 z + g_1\phi + B_1\left[\sqrt{b^2 - \rho^2} + bLn\frac{\rho}{b+\sqrt{b^2 - \rho^2}}\right], \tag{10}$$

$$\eta = B_2 z - g_2\phi + B_2\left[\sqrt{b^2 - \rho^2} + bLn\frac{\rho}{b+\sqrt{b^2 - \rho^2}}\right], \tag{11}$$

$$\lambda = B_3\left[z + \sqrt{b^2 - \rho^2}\right]. \tag{12}$$

This system is invertible, with

$$\rho = b\,\mathrm{sech}\frac{v}{b},\ \phi = \frac{B_2\xi - B_1\eta}{D},\ D = g_2 B_1 + g_1 B_2,\ z = \frac{\lambda}{B_3} - b\tanh\frac{v}{b},\ b^2 = g_1 g_2/B_1 B_2,$$

where $v = \lambda/B_3 - (g_2\xi + g_1\eta)/D$. The metric tensor is $(B_3 = 1)$:

$$g_{11} = \frac{g_2^2}{B_1 D}\mathrm{sech}^2\frac{v}{b},\quad g_{22} = \frac{g_1}{B_2 D}\mathrm{sech}^2\frac{v}{b},\quad g_{33} = \tanh^2\frac{v}{b}.$$

The Schrödinger equation is:

$$D\cosh^2\frac{v}{b}\left[\frac{B_1}{g_2}\frac{\partial^2\Psi}{\partial\xi^2} + \frac{B_2}{g_1}\frac{\partial^2\Psi}{\partial\eta^2}\right] + \cosh^2\frac{v}{b}\frac{\partial^2\Psi}{\partial\lambda^2}$$

$$-\frac{1}{b}\coth\frac{v}{b}\left[B_1\frac{\partial\Psi}{\partial\xi}+B_2\frac{\partial\Psi}{\partial\eta}\right]-\frac{1}{b}\coth\frac{v}{b}(1+\coth^2\frac{v}{b})\frac{\partial\Psi}{\partial\lambda}+\frac{2M}{h^2}(E-V)=0. \quad (13)$$

The substitution

$$\Psi=\frac{1}{\sqrt{tanh\frac{v}{b}}}\chi(\xi,\eta,\lambda), \quad (14)$$

leads to

$$D\,[\frac{B_1}{g_2}\frac{\partial^2\chi}{\partial\xi^2}+\frac{B_2}{g_1}\frac{\partial^2\chi}{\partial\eta^2}]+\mathrm{sech}^2\frac{v}{b}[\coth^2\frac{v}{b}\frac{\partial^2\chi}{\partial\lambda^2}-\frac{2}{b}\coth^3\frac{v}{b}\frac{\partial\chi}{\partial\lambda}+$$
$$(\frac{1}{b^2}+\frac{5}{4b^2}\mathrm{csch}^2\frac{u}{b}+\frac{1}{4b^2}\mathrm{csch}^4\frac{u}{b}(4\cosh^2\frac{u}{b}\ +1))\chi+\frac{2M}{h^2}(E-V)\chi]\ =\ 0. \quad (15)$$

Note that there is no restriction on the domain of $\{\xi,\eta,\lambda\}$ so far. We assume that in the domain Δ_{internal}, bounded by symmetrically located, impenetrable helical surfaces

$$\xi=-\xi_0/2,\quad \xi=\xi_0/2,\quad \eta=-\eta_0/2,\quad \eta=\eta_0/2,$$

where ξ_0 and η_0 are small such that

$$|\ g_2\xi_0+\ g_1\eta_0\ |\quad <<\quad 2\lambda D, \quad (16)$$

we have

$$V(\xi,\eta,\lambda)\left\{\begin{array}{lll} = & 0 & \forall\,\xi,\eta\in\Delta_{\text{internal}} \\ \rightarrow & \infty & \text{otherwise,}\end{array}\right.$$

in order to confine the particle to the neighbourhood of the curve defined by the intersection of $\xi=0$ and $\eta=0$ and to facilitate separability. It is easily verified that this logarithmic curve has the curvature $K_{\text{in}}=\frac{-\rho}{b\sqrt{b^2-\rho^2}}$ which increases in magnitude as $\rho\rightarrow b$. In the vicinity of $\xi=0$ and $\eta=0$ for small ξ_0 and η_0, v tends to λ. As λ increases from a sufficienly small value to ∞, $|K_{\text{in}}|$ decreases from a large value to 0. The inequality represented by (16) does not in any way restrict the magnitude of the curvature $|K_{\text{in}}|$ to be small. Within the framework of this approximation it is possible to obtain a separable solution of the form

$$\chi=\chi_1(\xi)\chi_2(\eta)\chi_3(\lambda). \quad (17)$$

Using(17) in (15) and imposing the boundary conditions,

$$\chi_1(-\xi_0/2)=0,\quad \chi_1(\xi_0/2)=0,\quad \chi_2(-\eta_0/2)=0,\quad \chi_2(\eta_0/2)=0$$

we obtain:

$$\chi_1(\xi)=A_M\sin(\frac{2M\pi\xi}{\xi_0}),\quad \chi_2(\eta)=B_N\sin(\frac{2N\pi\eta}{\eta_0}),$$

where

$$E_3=4D\,[\frac{B_1}{g_2}(\frac{M\pi}{\xi_0})^2+\frac{B_2}{g_1}(\frac{N\pi}{\eta_0})^2].$$

With

$$\chi_3(\lambda)=G(\rho)\sqrt{b^2-\rho^2}/\rho$$

we obtain

$$\frac{\partial^2G(\rho)}{\partial\rho^2}+\left[\frac{2ME}{\hbar^2}\frac{b^2}{\rho^2}-(\ E_3\frac{b^4}{\rho^4}\quad-\quad\frac{b^2}{4\rho^2(b^2-\rho^2)}\ )\right]G(\rho)=0 \quad (18)$$

To obtain the correct interpretation of the geometric potential in terms of the curvature $K_{in} = \frac{-\rho}{b\sqrt{b^2-\rho^2}}$ of the curve defined by $\xi = 0$ and $\eta = 0$, we let

$$\rho = b\exp[-\sigma/b] \quad \text{and} \quad G(\rho) = \exp[-2\sigma/b]H(\sigma),$$

and arrive at

$$\frac{\partial^2 H(\sigma)}{\partial \sigma^2} + (\frac{2ME}{\hbar^2} - E_3 \exp[2\sigma/b] + \frac{1}{4b^2(\exp[2\sigma/b]-1)})H(\sigma) = 0 \tag{19}$$

and $K_{in} = \frac{-1}{b\sqrt{\exp[2\sigma/b]-1}}$. The geometric potential turns out to be $-\frac{K_{in}^2}{4}$ as expected. The main result here is that no assumption was made on the smallness of K_{in}. In fact as $\sigma \to 0$, $\rho \to b$ and K_{in} becomes very large. In the absence of E_3 term, the equation for $H(\sigma)$ has the Hulthén potential, (Flügge [10]), admitting in this instance, no bound states. With E_3 present, as σ varies from 0 to ∞, the potential varies from $-\infty$ to ∞, leading to an effective confinement problem. With the transformation $\tau = \rho/b$ and $G(\rho) = P(\tau)$, we obtain

$$\frac{\partial^2 P(\tau)}{\partial \tau^2} + \left[\frac{2ME}{\hbar^2}\frac{b^2}{\tau^2} - E_3\frac{b^2}{\tau^4} + \frac{1}{4\tau^2(1-\tau^2)}\right]P(\tau) = 0 \tag{20}$$

As there are no known exact solutions available at present, we solve the differential equation for $P(\tau)$ numerically with $P(0) = G(.99999) = 0$. The reason for this is not only the singularity of the geometric potential at $\tau = 1$, but the need to maintain the validity of the approximation given by equation (16). We display a selection of eigenvalues for the case $B_1 g_1 = B_2 g_2$, $\xi_0 = \eta_0 = \frac{2g_1}{B_2}\sqrt{(B_1^2 + B_2^2)}$ in Table1. Note that $b^2 E_3 = M^2 + N^2$ and the inequality represented equation (16) now becomes

$$\frac{g_1}{B_2}(B_1 + B_2) << .004\sqrt{(B_1^2 + B_2^2)}$$

These calculation employed the solution $P(\tau) \approx \tau\exp[-\sqrt{M^2+N^2}/\tau]$, near $\tau \approx 0.001$,integrating forward to satisfy the boundary condition $P(.99999) = 0$. In all these computations the absolute relative error of $P(.99999)/[P(\rho)]_{MAX}$ is of the order of 10^{-2}. A more rigorous and extensive analysis will be necessary to assess the character of the complete energy spectrum.

Table 1

Eigenmode	1	2	3	1	2	3	1	1	2	3
(M, N)	(1,1)	(1,1)	(1,1)	(1,2)	(1,2)	(1,2)	(1,3)	(4,4)	(4,4)	(4,4)
$\frac{2ME}{\hbar^2}b^2$	12.8	28.2	46.0	22.3	44.9	70.0	35.3	81.0	134.7	189.9

4. THE GEOMETRIC POTENTIAL IN EXTERNAL HELICAL CO-ORDINATE SYSTEM

The external system (Garavaglia and Gomatam [1]) is given by the following equations:

$$\alpha = -A_1 z + l_1\phi - l_1\left[\frac{\sqrt{\rho^2 - a^2}}{a} - \arccos\frac{a}{\rho}\right], \tag{21}$$

$$\beta = A_2 z + l_2\phi - l_2\left[\frac{\sqrt{\rho^2 - a^2}}{a} - \arccos\frac{a}{\rho}\right], \quad (22)$$

$$\gamma = \phi + \arccos\frac{a}{\rho}, \quad (23)$$

where $a^2 = \frac{l_1 l_2}{A_1 A_2}$; and the inverse relations are

$$\rho = a\sqrt{w^2+1},\ \ z = \frac{(A_1\beta - A_2\alpha)}{F},\ \ \phi = \gamma - \arccos\frac{w}{\rho},$$

where

$$w = \gamma - \frac{(A_1\beta + A_2\alpha)}{F} \text{ and } F = (l_2 A_1 + l_1 A_2).$$

The metric tensor is $g_{11} = \frac{l_2}{A_1 F}, g_{22} = \frac{l_1}{A_2 F}, g_{33} = a^2 w^2$. The transformation $\Psi = \frac{1}{\sqrt{w}}\,\chi(\alpha,\beta,\gamma)$ leads to the Schrödinger equation

$$F\frac{A_1}{l_2}\frac{\partial^2\chi}{\partial\alpha^2} + F\frac{A_2}{l_1}\frac{\partial^2\chi}{\partial\beta^2} + \frac{1}{a^2w^2}\frac{\partial^2\chi}{\partial\gamma^2} - \frac{2}{a^2w^3}\frac{\partial\chi}{\partial\gamma} + \left[\frac{2M}{\hbar^2}(E-V) + \frac{(5+w^2)}{4a^2w^4}\right]\chi = 0. \quad (24)$$

In equation (24), we first assume that

$$|\,\alpha_0 A_{1+}\beta_0 A_2\,| \ll 2\gamma F \quad (25)$$

so that $w \to \gamma$. In the domain $\Delta_{external}$ symmetrically bounded by impenetrable helical surfaces,

$$\alpha = -\alpha_0/2, \quad \alpha = \alpha_0/2, \quad \beta = -\beta_0/2, \quad \beta = \beta_0/2$$

introduce the delta-function potential

$$V(\alpha,\beta,\gamma)\begin{cases} = & 0 \quad \text{for} \alpha,\beta \in \Delta_{external} \\ \to & \infty \quad \text{for} \alpha,\beta \notin \Delta_{external} \end{cases}$$

which confines the particle to the neighbourhood of the curve defined by the intersection of $\alpha = 0$ and $\beta = 0$. Note that this curve is spiralling outwards, lies entirely on the plane $z = 0$; hence the required curve is also defined by $z = 0$ and $\alpha = 0$ with the curvature $K_{ext} = \frac{1}{\sqrt{\rho^2 - a^2}}$. In the vicinity of $\alpha = 0$ and $\beta = 0$ for small α_0 and β_0, we replace w by γ. We can now separate equation (24). With $\chi = \chi_1(\alpha)\chi_2(\beta)\chi_3(\gamma)$, the separation of variables follows immediately

$$F\frac{A_1}{l_2}\frac{1}{\chi_1(\alpha)}\frac{\partial^2\chi_1(\alpha)}{\partial\alpha^2} + F\frac{A_2}{l_1}\frac{1}{\chi_2(\alpha)}\frac{\partial^2\chi_2(\beta)}{\partial\beta^2} + \frac{2M}{\hbar^2}E + \ s = 0$$

leading to

$$\chi_1(\alpha) = A_M\sin(\frac{2M\pi\alpha}{\alpha_0}), \quad \chi_2(\beta) = BN\sin(\frac{2N\pi\beta}{\beta_0}),$$

and

$$\frac{2M}{\hbar^2}E \quad = 4F\left[\frac{A_1}{l_2}\frac{M^2}{\alpha_0^2} + \frac{A_2}{l_1}\frac{N^2}{\beta_0^2}\right]\pi^2 \quad - s, \quad s > 0,$$

where s is determined by the boundary value problem

$$\frac{\partial^2\chi_3(\gamma)}{\partial\gamma^2} - \frac{2}{\gamma}\frac{\partial\chi_3(\gamma)}{\partial\gamma} + \left[\frac{1}{4} + \frac{5}{4\gamma^2} - s\ a^2\gamma^2\right]\chi_3(\gamma) = 0.$$

The boundary condition on $\chi_3(\gamma)$ is discussed below. As before, in order to relate the geometric potential to K_{ext} we let $\gamma = \sqrt{2\tau/a}$ and $\chi_3(\gamma) = \tau^{1/4}P(\tau)$ and arrive at

$$\frac{\partial^2 P(\tau)}{\partial\tau^2} + [\ -s + 1/(8a\tau)]P(\tau) = 0$$

We note that $-1/(8a\tau) = -1/(4(\rho^2 - a^2) = -\frac{1}{4}K^2_{ext}$. It can be shown that the equation for $P(\tau)$ can be solved exactly by the substitution

$$P(\tau) = \tau\exp[-\tau\sqrt{s}]f(2\sqrt{s}\tau)$$

leading to the Kummer equation (Abramowitz and Stegun [9]):

$$\frac{\partial^2 f(u)}{\partial u^2} + (\frac{2}{u} - 1)\frac{\partial f(u)}{\partial u} + \left[\frac{\frac{1}{(16a\ \sqrt{s})} - 1}{u}\right] f(u) = 0.$$

The regular solution is given by

$$M\left[1 - 1/(16a\ \sqrt{s})\ \mid 2 \mid\ 2\sqrt{s}\tau\ \right],$$

the Kummer function or the Confluent Hypergeometric function. The approximation (25) implies that $\gamma_0 \leq \gamma < \infty$, and hence both the regular and the irregular solutions are needed. Because of the fact that the middle argument is 2, we have to construct the second solution (Abramowitz and Stegun [9]) with great care. This is given by

$$U[1 - 1/(16a\sqrt{s})\ \mid 2 \mid\ 2\sqrt{s}\tau].$$

which has logarithmic singularitiy at $\gamma = 0$. The complete solution is given by:

$$f[a\ \sqrt{s}\gamma^2] = A\ M[1 - 1/(16\ a\ \sqrt{s}) \mid 2 \mid\ a\sqrt{s}\gamma^2] + B\ U[1 - 1/(16\ a\ \sqrt{s})\ \mid 2 \mid a\ \sqrt{s}\gamma^2].$$

Boundedness of both the solutions as $\gamma \to \infty$ is guaranteed by the stipulation that

$$1 - 1/(16\ a\ \sqrt{s}) = -L, \quad L = 0, 1, 2, \ldots$$

Locating an infinite potential barrier at $\gamma = \gamma_0$ demands that

$$\frac{A}{B} = \frac{-1U\left[-1/(16\ a\ \sqrt{s})\ \mid\ 2\ \mid a\sqrt{s}\gamma_0^2\ \right]}{\gamma_0^2 M\left[1 - 1/(16\ a\ \sqrt{s})\ \mid\ 2\ \mid\ a\ \sqrt{s}^2\gamma_0^2\ \right]}.$$

The energy eigenvalues are given by

$$\frac{2M}{\hbar^2}E \quad = 4F\left[\frac{A_1}{l_2}\frac{M^2}{\alpha_0^2} + \frac{A_2}{l_1}\frac{N^2}{\beta_0^2}\right]\pi^2 - \frac{1}{(16(L+1)a)^2}.$$

Notice that the term responsible for geometric quantisation is overshadowed by the "box quantisation" term as $L \to \infty$. Further for sufficiently small cross sections determined by $\{\alpha_0, \beta_0\}$, the"box quantisation " term dominates. In our model, α_0 and β_0 are constrained by (25). For negative values of s, the separation constant, we have a discrete spectrum (generated by confining potentials) embedded in a continuum.. The corresponding Kummer functions with, s replaced by is, are expressible in terms of Coulomb wave functions (Abramowitz and Stegun, [9]).

5. CONCLUSIONS

We have developed a global orthogonal co-ordinate system which will provide a convenient frame work for perturbative and computational investigation of boundary value problems with helical geometry. In particular this can be employed in the analysis of the bound states of the confined Schrödinger equation in nanostructures with helical geometry. We have demonstrated the existence of bound states for the internal and external tubular domains of nano cross section, bounded by helical walls harbouring delta-function potentials. In the external case we were able to obtain the eigenvalue spectrum explicitly.

References

[1] T Garavaglia T and J Gomatam, The Schrodinger equation in helical co-ordinates, Ann. Phys., 89, pp 1-10 (1975)

[2] T Garavaglia T and J Gomatam, Exact solutions of the Schrodinger equation in biconical co-ordinates, Acta Phys. Aus. 53, pp 237-241(1981)

[3] R C T da Costa, Quantum mechanics of a constrained particle, Phys Rev A, 23, 1982-1987 (1981)

[4] P Exner and P Seba, Bound states in curved quantum waveguides, J Math Phys. 30, pp 2574-2580 (1989)

[5] J Goldstone and R L Jaffe, Bound states in twisting tubes, Phys Rev B, 45, pp 100-107 (1992)

[6] Iain J Clarke and Anthony Bracken, Bound States with tubular waveguides with Torsion, J Phys A;: Math. Gen.29, pp 4727-4535 (1996)

[7] I S Gradshteyn and I M Ryzhik, Tables of integrals, series and products, Academic Press (1980)

[8] J Gomatam and A J Mullholland, The Schrödinger equations on helical manifolds: bounded states, Technical Report 98-103 Glasgow Caledonian University (1998)

[9] M Abramowitz and I A Stegun, Handbook of Mathematical Functions, Dover Publications, New York (1965)

[10] S Flügge, Practical Quantum Mechanics I, p 175, Springer-Verlag, N.Y. (1971)

Proceedings of Dynamic Systems and Applications 3 (2001) 217-224

Positive Solutions of a Boundary Value Problem for Fourth Order Nonlinear Differential Equations

John R. Graef[1] and Bo Yang[2]
[1]Department of Mathematics, University of Tennessee at Chattanooga
Chattanooga, TN 37403 USA
[2]Department of Mathematics and Statistics, Mississippi State University
Mississippi State, MS 39762

ABSTRACT: In this paper, the authors consider the fourth order ordinary differential equation

$$u'''' = \lambda a(t) f(u), \quad 0 < t < 1,$$

together with the boundary condition

$$u(0) = u'(1) = u''(0) = u''(1) = 0.$$

Some existence results for this problem are obtained, and an example to illustrate the results is included.

AMS (MOS) subject classification. 34B15

1. INTRODUCTION

We consider the fourth order ordinary differential equation

$$u'''' = \lambda a(t) f(u), \quad 0 < t < 1, \tag{1}$$

together with the boundary conditions

$$u(0) = u'(1) = u''(0) = u''(1) = 0, \tag{2}$$

where $a : [0,1] \to [0,\infty)$ and $f : [0,\infty) \to [0,\infty)$ are continuous, and $\lambda > 0$ is a parameter. Boundary value problems for differential equations are important both in applications as well from a theoretical perspective. Discussions of the variety of physical problems that lead to boundary value problems for second and higher order equations can be found, for example, in the works of Agarwal and Wong [1], Eloe and Henderson [2], Henderson and Wang [4], and Wong [9]. The problem (1)–(2) can be used to describe, for example, the deflection of an elastic beam that is simply supported at the left hand endpoint and fastened with a sliding clamp at the right hand end point. If $f(0) = 0$, then the problem (1)–(2) always has the trivial solution, but in this paper we are only interested in positive solutions, i.e., a solution $u(t)$ of (1)–(2) such that $u(t) > 0$ on $(0,1)$. In the remainder of this paper, we assume that

(i) $0 < \int_0^1 a(t)dt$

and

(ii) $f_0 = \lim_{x\to 0^+}(f(x)/x)$ and $f_\infty = \lim_{x\to\infty}(f(x)/x)$ exist.

Many other authors require that the function f be either superlinear at 0 and ∞, i.e., $f_0 = 0$ and $f_\infty = \infty$, or be sublinear at 0 and ∞, i.e., $f_0 = \infty$ and $f_\infty = 0$.

We make no assumptions here about the values of f_0 and f_∞ except that if $f_0 = 0$, then Theorem 2 should be used, and if $f_\infty = 0$, Theorem 3 should be applied. However, if f is superlinear (sublinear), then Theorem 2 (Theorem 3) guarantees that the problem (1)–(2) has at least one positive solution for all $0 < \lambda < \infty$. We also wish to point out that we do not ask that $a(t)$ not vanish identically on any subinterval of $[0, 1]$ as is usually the case.

We will use the following theorem in the proof of our results.

Theorem 1. (Krasnosel'skii's Fixed Point Theorem [5].) *Let $\mathcal{B}$ be a Banach space and let $\mathcal{P} \subset \mathcal{B}$ be a cone in $\mathcal{B}$. Assume that Ω_1 and Ω_2 are open subsets of $\mathcal{B}$ with $0 \in \Omega_1 \subset \overline{\Omega_1} \subset \Omega_2$, and let*

$$L : \mathcal{P} \cap (\overline{\Omega_2} - \Omega_1) \to \mathcal{P}$$

be a completely continuous operator such that either

(i) $||Lu|| \leq ||u||$ *if* $u \in \mathcal{P} \cap \partial\Omega_1$, *and* $||Lu|| \geq ||u||$ *if* $u \in \mathcal{P} \cap \partial\Omega_2$, *or*
(ii) $||Lu|| \geq ||u||$ *if* $u \in \mathcal{P} \cap \partial\Omega_1$, *and* $||Lu|| \leq ||u||$ *if* $u \in \mathcal{P} \cap \partial\Omega_2$.

Then L has a fixed point in $\mathcal{P} \cap (\overline{\Omega_2} - \Omega_1)$.

2. PRELIMINARY RESULTS

The Green's function $G : [0, 1] \times [0, 1] \to [0, \infty)$ for the problem

$$y'' = 0, \quad y(0) = y(1) = 0 \tag{3}$$

is given by

$$G(t, s) = \begin{cases} t(1-s), & t \leq s, \\ s(1-t), & s \leq t, \end{cases}$$

and the Green's function $J : [0, 1] \times [0, 1] \to [0, \infty)$ for

$$y'' = 0, \quad y(0) = y'(1) = 0 \tag{4}$$

is given by

$$J(t, s) = \begin{cases} t, & t \leq s, \\ s, & s \leq t. \end{cases}$$

It follows that the Green's function $Z : [0, 1] \times [0, 1] \to [0, \infty)$ for the problem

$$y'''' = 0, \quad y(0) = y'(1) = y''(0) = y''(1) = 0 \tag{5}$$

has the form

$$Z(t, s) = \int_0^1 J(t, v) G(v, s) dv.$$

Solving the boundary value problem (1)–(2) is then equivalent to solving the integral equation

$$u(t) = \lambda \int_0^1 Z(t, s) a(s) f(u(s)) ds, \ 0 \leq t \leq 1, \tag{6}$$

and it is also equivalent to solving the problem

$$u''(t) = -\lambda \int_0^1 G(t,s)a(s)f(u(s))ds, \quad u(0) = u'(1) = 0.$$

In order to apply Krasnosel'skii fixed point theorem, we let

$$\mathcal{B} = \{x \mid x \in C^2[0,1],\ x(0) = x'(1) = 0\}$$

be our Banach space with the norm

$$||x|| = \max_{t\in[0,1]} |x''(t)|,\ x \in \mathcal{B}.$$

We define

$$\mathcal{P} = \{x \in \mathcal{B} \mid x''(t) \le 0 \text{ and concave up on } [0,1]\}$$

and observe that $\mathcal{P}$ is a cone in $\mathcal{B}$. Next, we define the operator $\mathcal{T} : \mathcal{P} \to \mathcal{B}$ by

$$(\mathcal{T}u)(t) = \lambda \int_0^1 Z(t,s)a(s)f(u(s))ds,\ 0 \le t \le 1.$$

Clearly, $\mathcal{T}(\mathcal{P}) \subset \mathcal{P}$ and $\mathcal{T} : \mathcal{P} \to \mathcal{P}$ is a completely continuous operator.

Finding a solution of the integral equation (6) is equivalent to finding a solution of

$$\mathcal{T}u = u,\ u \in \mathcal{P}.$$

That is, in order to solve the boundary value problem (1)–(2), we only need to find a fixed point of $\mathcal{T}$ in $\mathcal{P}$.

The proof of the following lemma is straightforward and we omit the details here.

Lemma 1. *If $x \in \mathcal{P}$, then $x(t) \le \frac{1}{2}t(2-t)||x|| \le \frac{1}{2}||x||$ on $[0,1]$.*

Let

$$A_1 = \max_{t\in[0,1]} \int_0^1 G(t,s)a(s)(s(2-s)/2)ds.$$

Next, we define a function $g : [0,1] \times [0,1] \to [0,\infty)$ by

$$g(t,0) = 1-t,$$
$$g(t,1) = t,$$

and for $h \in (0,1)$,

$$g(t,h) = \begin{cases} t/h, & 0 \le t \le h, \\ (1-t)/(1-h), & h \le t \le 1. \end{cases}$$

It is easy to see that for each $(t,h) \in [0,1] \times [0,1]$, we have

$$g(t,h) \ge \min\{t, 1-t\} \ge t(1-t).$$

Lemma 2. *If $x \in C^2[0,1]$ with $x(t) \geq 0$ and $x''(t) \leq 0$ on $[0,1]$, and $t_0 \in [0,1]$ is such that $x(t_0) = \max_{0 \leq t \leq 1} x(t)$, then*

$$x(t) \geq g(t,t_0)x(t_0), \quad \text{for } t \in [0,1].$$

Lemma 2 is a direct consequence of the properties of the functions x and g. Now define the function $L : [0,1] \times [0,1] \to [0,\infty)$ by

$$L(t,h) = \int_0^1 J(t,s)g(s,h)ds$$

and the function $b : [0,1] \to [0,\infty)$ by

$$b(t) = \int_0^1 J(t,s)\min\{s, 1-s\}ds.$$

Then we see that

$$b(t) = \begin{cases} -t^3/6 + t/4, & 0 \leq t \leq 1/2, \\ t^3/6 - t^2/2 + t/2 - 1/24, & 1/2 \leq t \leq 1, \end{cases}$$

and, moreover, $L(t,h) \geq b(t) > 0$ for $t \in (0,1)$.

The proof of the following lemma is also straightforward and will be omitted.

Lemma 3. *If $x \in \mathcal{P}$ and $|x''(t_0)| = ||x||$, then*

$$-x''(t) \geq g(t,t_0)||x||,$$

and

$$x(t) \geq L(t,t_0)||x||.$$

Finally, we define the function $K : [0,1] \times [0,1] \to [0,\infty)$ by

$$K(t,h) = \int_0^1 G(t,s)a(s)L(s,h)ds.$$

For every fixed $h \in [0,1]$, we can choose a $\tau = \tau(h) \in (0,1)$ such that

$$K(\tau,h) = \max_{0 \leq t \leq 1} K(t,h).$$

Then $\tau = \tau(h)$ is a function defined on $[0,1]$. We note that $\tau = \tau(h)$ is not necessarily continuous. Clearly,

$$K(\tau(h),h) \geq \int_0^1 G(1/2,s)a(s)b(s)ds, \; h \in [0,1].$$

We let

$$A_2 = \min_{0 \leq h \leq 1} K(\tau(h),h)$$

and see that

$$A_2 \geq \int_0^1 G(1/2,s)a(s)b(s)ds > 0. \tag{7}$$

While our results in the next section require the use of A_2, the estimate (7) can provide a useful approximation, and, in certain cases, be much easier to calculate.

3. MAIN RESULTS

Theorem 2. *If*

$$(A_2 f_\infty)^{-1} < \lambda < (A_1 f_0)^{-1},$$

Then the problem (1)–(2) has at least one positive solution.

Proof. Choose $\epsilon > 0$ such that

$$(f_0 + \epsilon)\lambda A_1 \leq 1.$$

There exists an $H_1 > 0$ such that

$$f(x) \leq (f_0 + \epsilon)x \text{ for } 0 < x \leq H_1.$$

Then, if $u \in \mathcal{P}$ with $||u|| = H_1$, we have

$$\begin{aligned}
|(\mathcal{T}u)''(t)| &= \lambda \int_0^1 G(t,s)a(s)f(u(s))ds \\
&\leq \lambda \int_0^1 G(t,s)a(s)(f_0 + \epsilon)u(s)ds \\
&\leq \lambda(f_0 + \epsilon)||u|| \int_0^1 G(t,s)a(s)(s(2-s)/2)ds \\
&\leq \lambda(f_0 + \epsilon)||u||A_1 \\
&\leq ||u||.
\end{aligned}$$

This means $||\mathcal{T}u|| \leq ||u||$, so if we let

$$\Omega_1 = \{u \in B| \ ||u|| < H_1\},$$

then

$$||\mathcal{T}u|| \leq ||u||, \text{ for } u \in \mathcal{P} \cap \partial\Omega_1.$$

Now choose $c \in (0, 1/4)$ and $\epsilon_1 > 0$ such that

$$\lambda > \left(\min_{0 \leq h \leq 1} \int_c^1 G(\tau(h), s)a(s)L(s,h)ds(f_\infty - \epsilon_1)\right)^{-1}.$$

There exists H_2' such that

$$f(x) \geq (f_\infty - \epsilon_1)x \text{ for } x \geq H_2'.$$

Let

$$H_2 = \max\{\frac{H_2'}{b(c)}, 2H_1\}$$

and let $u \in \mathcal{P}$ with $||u|| = H_2$. Then there exists $t_0 \in [0,1]$ such that $u''(t_0) = -||u||$, and we have

$$\begin{aligned} |(Tu)''(\tau(t_0))| &\geq \lambda \int_c^1 G(\tau(t_0), s)a(s)f(u(s))ds \\ &\geq \lambda \int_c^1 G(\tau(t_0), s)a(s)(f_\infty - \epsilon_1)u(s)ds \\ &\geq \lambda(f_\infty - \epsilon_1)||u|| \int_c^1 G(\tau(t_0), s)a(s)L(s, t_0)ds \\ &\geq ||u||, \end{aligned}$$

i.e., $||\mathcal{T}u|| \geq ||u||$. Hence, if we let

$$\Omega_2 = \{u \in \mathcal{B} |\ ||u|| < H_2\},$$

then

$$||\mathcal{T}u|| \geq ||u||, \text{ for } u \in \mathcal{P} \cap \partial\Omega_2.$$

The hypotheses of Theorem 1(i) are now satisfied, so there is a fixed point of $\mathcal{T}$ in $\mathcal{P}$. This completes the proof of the theorem.

Theorem 3. *If*

$$(A_2 f_0)^{-1} < \lambda < (A_1 f_\infty)^{-1},$$

then problem (1)–(2) has at least one positive solution.

Proof. Choose $\epsilon > 0$ such that

$$(f_0 - \epsilon)\lambda A_2 \geq 1.$$

There exists H_1 such that

$$f(x) \geq (f_0 - \epsilon)x \text{ for } 0 < x \leq H_1.$$

Now for $u \in \mathcal{P}$ with $-u''(h) = ||u|| = H_1$, we have

$$\begin{aligned} |(\mathcal{T}u)''(\tau(h))| &= \lambda \int_0^1 G(\tau(h), s)a(s)f(u(s))ds \\ &\geq \lambda(f_0 - \epsilon) \int_0^1 G(\tau(h), s)a(s)u(s)ds \\ &\geq \lambda(f_0 - \epsilon)||u|| \int_0^1 G(\tau(h), s)a(s)L(s, h)ds \\ &\geq \lambda A_2(f_0 - \epsilon)||u|| \\ &\geq ||u||, \end{aligned}$$

and so $||\mathcal{T}u|| \geq ||u||$. With

$$\Omega_1 = \{u \in \mathcal{B} |\ ||u|| < H_1\},$$

we see that

$$||\mathcal{T}u|| \geq ||u||, \text{ for } u \in \mathcal{P} \cap \partial\Omega_2.$$

Next, we choose $\epsilon_1 \in (0,1)$ such that

$$((f_\infty + \epsilon_1)A_1 + \epsilon_1)\lambda \leq 1.$$

There exists H_2 such that

$$f(x) \leq (f_\infty + \epsilon_1)x \text{ for } x \geq H_2.$$

Let

$$M = \max_{0 \leq x \leq H_2} f(x)$$

and choose $p \in (0, 1/2)$ such that

$$\max_{0 \leq t \leq 1} \int_0^p G(t,s)a(s)ds \leq \frac{\epsilon_1}{2M}.$$

Let

$$H_3 = \max\{2H_1, \frac{H_2}{b(p)}, 1\}.$$

If $u \in \mathcal{P}$ with $||u|| = H_3$, then $u(t) \geq b(t)H_3$, and so there exists $\beta \in (0,p)$ such that

$$0 \leq u(t) \leq H_2 \text{ for } t \in (0, \beta),$$

and

$$u(t) \geq H_2 \text{ for } t \in (\beta, 1).$$

We then have

$$\begin{aligned}
|(\mathcal{T}u)''(t)| &\leq \lambda \int_0^1 G(t,s)a(s)f(u(s))ds \\
&\leq \lambda(\epsilon_1 + \int_\beta^1 G(t,s)a(s)f(u(s))ds) \\
&\leq \lambda(\epsilon_1 + (f_\infty + \epsilon_1) \int_\beta^1 G(t,s)a(s)u(s)ds) \\
&\leq \lambda(\epsilon_1 + H_3(f_\infty + \epsilon_1) \int_0^1 G(t,s)a(s)(s(2-s)/2)ds) \\
&\leq \lambda(\epsilon_1 + A_1 H_3(f_\infty + \epsilon_1)) \\
&\leq H_3 \\
&= ||u||,
\end{aligned}$$

i.e., $||\mathcal{T}u|| \leq ||u||$. Letting

$$\Omega_2 = \{u \in B | \ ||u|| < H_3\},$$

then

$$||\mathcal{T}u|| \leq ||u||, \text{ for } u \in \mathcal{P} \cap \partial\Omega_2.$$

Now from Theorem 1(ii), we see that the problem (1)–(2) has at least one positive solution.

4. EXAMPLE

As an illustration of our results, we give the following example.

Example. Consider the boundary value problem

$$u''''(t) = \lambda a(t) f(u(t)), \tag{8}$$

$$u(0) = u'(1) = u''(0) = u''(1) = 0, \tag{9}$$

where

$$a(t) = 1 + 5t^6 \quad \text{and} \quad f(u) = u(1 + 20u)/(1 + u).$$

Here we see that $f_0 = 1$ and $f_\infty = 20$. Using *Mathematica* or another appropriate software, it is easy to check that $A_1 < 0.07$ and $A_2 > 0.017$. By Theorem 2, if $2.94118 < \lambda < 14.2857$, then the problem (8)–(9) has at least one positive solution. None of the results in [1–4, 6–9] apply to this problem. For example, Ruyun and Wang consider only the superlinear and sublinear cases (see [6; Theorem 2]).

ACKNOWLEDGMENT

The research of the first author was supported by the Mississippi State University Biological and Physical Sciences Research Institute.

REFERENCES

1. Agarwal, R., & Wong, F. H., Existence of positive solutions for higher order boundary value problems. Nonlinear Studies, Vol 5 (1998) pp. 15–24.

2. Eloe, P. W., & Henderson, J., Positive solutions and nonlinear multipoint conjugate eigenvalue problems. Electron. J. Differential Equations, Vol 1997 (1997) pp. 1–11.

3. Eloe, P. W., Henderson, J., & Wong, P. J. Y., Positive solutions for two–point boundary value problems. Proc. Dynamic Systems Appl. Vol 2 (1996) pp. 135–144.

4. Henderson, J., & Wang, H., Positive solutions for nonlinear eigenvalue problems. J. Math. Anal. Appl. Vol 208 (1997) pp. 252–259.

5. Krasnosel'skii, M. A., Positive Solutions of Operator Equations. Noordhoff, Groningen, 1964.

6. Ruyun, M., & Wang, H., On the existence of positive solutions of fourth-order ordinary differential equations. Applicable Anal. Vol 59 (1995) pp. 225-231.

7. Ruyun, M., Zhang, J., & Fu, S., The method of lower and upper solutions for fourth-order two-point boundary value problems. J. Math. Anal. Appl. Vol 215 (1997) pp. 415-422.

8. Wang, H., On the existence of positive solutions for semilinear elliptic equation in the annulus. J. Differential Equations Vol 109 (1994) pp. 1–7.

9. Wong, P. J. Y., Triple positive solutions of conjugate boundary value problems. Comput. Math. Appl. Vol 36 (1998) pp. 19–35.

Proceedings of Dynamic Systems and Applications 3 (2001) 225-232

Nonlinear Systems with Impulse Perturbations

John R. Graef[1] and János Karsai[2]
[1]Department of Mathematics, University of Tennessee at Chattanooga
Chattanooga, TN 37403, USA
[2]Department of Medical Informatics, A. Szent-Györgyi Medical University
Szeged, Korányi fasor 9, 6720, HUNGARY

ABSTRACT: We consider the nonlinear system with impulsive damping

$$(\phi_\beta(x'))' + f(x) = 0 \quad (t \neq t_n), \quad x'(t_n + 0) = b_n x'(t_n)$$

where $(n = 1, 2 \dots)$, $\phi_\beta(u) = |u|^\beta \operatorname{sgn} u$ for $\beta > 0$, and $0 \leq b_n \leq 1$. We prove criteria to guarantee the nonoscillation of solutions and to exclude overdamping, that is, every nonoscillatory solution must tend to zero as $t \to \infty$.

AMS (MOS) subject classification. 34D05, 34D20, 34C15

1. INTRODUCTION

Consider the impulsively damped system

$$\begin{aligned} &(\phi_\beta(x'))' + f(x) = 0, \quad t \neq t_n, \\ &x(t_n + 0) = x(t_n), \quad x'(t_n + 0) = b_n x'(t_n), \end{aligned} \tag{1}$$

where $0 \leq t_1 < t_2, \dots, t_n < t_{n+1}$, $t_n \to \infty$ as $n \to \infty$, $0 \leq b_n \leq 1$ for $n = 1, 2 \dots$, $\phi_\beta(u) = |u|^\beta \operatorname{sgn} u$ for $\beta > 0$, $f : \mathbf{R} \to \mathbf{R}$ is continuous, and $uf(u) > 0$ for $u \neq 0$. For the sake of simplicity, we assume that f is an odd function.

When $\phi_\beta(u) = u$, system (1) is an impulsive analogue of the nonlinear oscillator equation

$$x'' + a(t)x' + f(x) = 0 \tag{2}$$

with a nonnegative damping coefficient $a(t)$. For a detailed description of this analogy, see Graef and Karsai [4, 5]. A negative b_n results in a beating effect, which has no continuous analogue (see the discussions of the case $\phi_\beta(u) = u$ in [4, 6]). Due to space limitations, we do not deal with this case in detail and only indicate the possible generalizations.

It is known that if the function $a(t)$ is small, but large enough in some sense, then the solutions of (2) oscillate and tend to zero. This situation is called the small damping case. For large enough $a(t)$, the solutions are monotone decreasing in magnitude, and this is sometimes called the large damping case. If the system is overdamped, that is, $a(t)$ grows too fast to infinity as t tends to infinity, then the solutions decrease in magnitude but may not tend to zero (Ballieu and Peiffer [2]).

The problem of attractivity for system (1) and its special cases was investigated by the present authors in [4, 5, 6]. In this paper, we formulate conditions for the nonoscillation of the solutions, and these turn out to be necessary and sufficient conditions for linear systems with constant impulses. We also give sufficient conditions for the nonoscillatory solutions to tend to zero as $t \to \infty$.

2. DEFINITIONS AND LEMMAS

Throughout the paper, we assume that every solution can be continued to ∞. It is clear that the solutions are differentiable and $\phi_\beta(x'(t))$ is piecewise continuous and continuous from the left hand side at every $t > 0$.

We investigate the solutions with the help of the energy function

$$V(x,y) = y\phi_\beta(y) - \int_0^y \phi_\beta(s)\,ds + \int_0^x f(s)\,ds =: \Phi(y) + F(x), \tag{3}$$

where $\Phi(y) = \frac{\beta}{\beta+1}|y|^{\beta+1}$ and the function F are both even and positive definite. To simplify the formulation of our results, we also assume that

$$\lim_{u\to\pm\infty} F(u) = \infty. \tag{4}$$

Without this assumption, the boundedness of all solutions does not follow from the boundedness of V. It is easy to verify that $V(t) = V(x(t), x'(t))$ is constant along the solutions of the equation without impulses

$$(\phi_\beta(x'))' + f(x) = 0. \tag{5}$$

The change in the energy along the solutions of (1) is

$$\begin{aligned} V(t_{n+1}) - V(t_n) &= V(t_n + 0) - V(t_n) \qquad (6)\\ &= \Phi(x'(t_n+0)) + F(x(t_n+0)) - \Phi(x'(t_n)) - F(x(t_n))\\ &= |b_n|^{\beta+1}\Phi(x'(t_n)) - \Phi(x'(t_n)) = -\Phi(x'(t_n))(1 - |b_n|^{\beta+1}). \end{aligned}$$

Note that $V(t)$ is nonincreasing if $|b_n| \le 1$, independent of the sign of b_n and the magnitude of β, and is constant between any t_n and t_{n+1}. The case $b_n = 0$ $(a_n = 1)$ is critical since we lose the uniqueness of the solutions at t_n from the left. Moreover, in this case, there exist solutions that are identically zero for $t > t_n$.

The following lemma is fundamental for estimating $\phi_\beta(x'(t))$; the proof is a modification of the proof Corollary 2.4 of Bainov and Simeonov [1]. We introduce the notation

$$\Psi(s,t) := \prod_{s\le t_i<t} b_i. \tag{7}$$

Lemma 1. *Let $x(t)$ be a nonzero solution of (1) and $T \ge 0$. Then*

$$\phi_\beta(x'(t)) = \phi_\beta(x'(T))\Psi^\beta(T,t) - \int_T^t \Psi^\beta(s,t) f(x(s))ds. \tag{8}$$

The next lemma classifies the solutions of the system (1) as being either monotonic or oscillatory on some interval $[T,\infty)$.

Lemma 2. *Let $x(t)$ be a solution of (1) that is not identically zero on any interval $[T,\infty)$, and let s_1 and s_2 be consecutive zeros of $x'(t)$. Then there exists $\tilde{t} \in (s_1, s_2)$ such that $x(\tilde{t}) = 0$. Hence, solutions of (1) are either oscillatory or monotonic.*

The proof of this lemma is similar to the proof of Lemma 13 in [4] and so we omit the details here.

Note that if $b_n = 0$, then the solution can become zero from t_n forward. We also note that if $b_n < 0$ for some n, the solutions can be nonoscillatory and not monotonic; these are sometimes called "saw-tooth solutions" (see Graef and Karsai [6] in the special case $\phi_\beta(u) = u$).

To investigate either the oscillatory or attractivity properties of the solutions, an understanding of the basic behavior of the solutions of equation (5) is essential. It is easy to see that every nonzero solution of (5) is periodic. Since f and ϕ_β are odd, the length of the time intervals on which $x(t)x'(t) \geq 0$ or $x(t)x'(t) \leq 0$ are equal. For a solution $x(t)$ with $V(t) \equiv r$, denote the the distance between two consecutive zeros of $x'(t)$, i.e., the *half-period*, by $\Delta_{f,\phi_\beta}(r)$ or simply by $\Delta(r)$. In the case $\phi_\beta(u) = u$, the behavior of the solutions was investigated, for example, in Reissig, Sansone and Conti [8]. To obtain $\Delta_{f,\phi_\beta}(r)$ formally, let $x(t)$ be a solution of (5) with $F(x(0)) = 0$ and $\Phi(x'(0)) = r$. Since $V(t) = F(x(t)) + \Phi(x'(t)) \equiv r$, we have $F(x(\frac{\Delta_{f,\phi_\beta}(r)}{2})) = r$, $\Phi(x'(\frac{\Delta_{f,\phi_\beta}(r)}{2})) = 0$, and $x'(t) = \Phi^{-1}(r - F(x(t)))$. Dividing by the right-hand side of the last equality and integrating, we obtain

$$\frac{1}{2}\Delta_{f,\phi_\beta}(r) = \int_0^{\frac{\Delta_{f,\phi_\beta}(r)}{2}} \frac{x'(t)\,dt}{\Phi^{-1}(r - F(x(t)))}. \tag{9}$$

Hence, using the substitution $x = x(t)$, the quantity $\Delta_{f,\phi_\beta}(r)$ can be expressed in the form

$$\Delta_{f,\phi_\beta}(r) = 2\int_0^{F^{-1}(r)} \frac{dx}{\Phi^{-1}(r - F(x))} = 2r\int_0^1 \frac{dv}{f(F^{-1}(rv)\Phi^{-1}(r(1-v))} \tag{10}$$

where $u = F(x)$, $v = u/r$, and F^{-1} and Φ^{-1} are the inverses of F and Φ on $[0, \infty)$, respectively. In the special case where $f(x) = \phi_\alpha(x)$ with $\alpha > 0$, we obtain that

$$\Delta(r) = r^{\frac{1}{\alpha+1} - \frac{1}{\beta+1}} \frac{2\,\beta^{\frac{1}{1+\beta}}\,\Gamma(\frac{1}{1+\alpha})\,\Gamma(\frac{\beta}{1+\beta})}{(1+\alpha)^{\frac{\alpha}{1+\alpha}}\,(1+\beta)^{\frac{1}{1+\beta}}\,\Gamma(\frac{1+2\beta+\alpha\beta}{1+\alpha+\beta+\alpha\beta})}. \tag{11}$$

If $\alpha = \beta$, then Δ is the constant

$$\Delta_\beta = \frac{2\,\beta^{\frac{1}{1+\beta}}\,\Gamma(\frac{1}{1+\beta})\,\Gamma(\frac{\beta}{1+\beta})}{1+\beta} = \frac{2\beta^{\frac{1}{1+\beta}}}{(1+\beta)\sin\frac{\pi}{1+\beta}}\pi, \tag{12}$$

and in particular, $\Delta_\beta = \pi$ for $\beta = 1$.

If $\alpha > \beta$, then $\lim_{r\to 0}\Delta(r) = \infty$, and if $\alpha < \beta$, then $\lim_{r\to 0}\Delta(r) = 0$. The case $\alpha = \beta$ is called *halflinear* (see Elbert [3]), so we may refer to the other cases as *super-* and *sub-halflinear*, respectively.

3. NONOSCILLATION CRITERIA

Let $\varphi(t)$ be the solution of the equation

$$(\phi_\beta(x'))' + \phi_\beta(x) = 0 \tag{13}$$

without impulses for which

$$\varphi(0) = 1 \text{ and } \varphi'(0) = 0. \tag{14}$$

Clearly, $V(\varphi(t)) \equiv 1/(\beta+1)$. For this equation, $F_\beta(x) := \int_0^x \phi_\beta = \Phi(x)/\beta$ and the half-period of $\varphi(t)$ is Δ_β defined by (12). Note that solutions with such special initial conditions are commonly used in the theory of half-linear systems (see [3]).

The following theorem yields a nonoscillation result by comparing (5) with (13).

Theorem 1. *Let* $|f(F^{-1}(u))| \leq |\phi_\beta(F_\beta^{-1}(u))| = (\beta+1)^{\frac{\beta}{\beta+1}}|u|^{\frac{\beta}{1+\beta}}$ *for* $0 \leq r \leq r_0$. *Assume that there exist numbers* $N > 0$ *and* $0 < \delta < \frac{\Delta_\beta}{4}$ *such that* $d_n = t_n - t_{n-1} \leq \frac{\Delta_\beta}{2} - 2\delta$ *and the inequality*

$$\frac{\varphi^{\beta+1}(\frac{\Delta_\beta}{2} - \delta)}{b_n^{\beta+1} + (1 - b_n^{\beta+1})\varphi^{\beta+1}(\frac{\Delta_\beta}{2} - \delta)} \geq \varphi^{\beta+1}(\delta) \tag{15}$$

holds for every $n > N$. *Then every solution of the system (1) with* $F(x(0)) + \Phi(x'(0)) \leq r_0$ *is eventually nonoscillatory.*

Proof. Assume that (15) holds for $n > N$ and let $x(t)$ be a nontrivial solution of (1) such that $F(x(0)) + \Phi(x'(0)) \leq r_0$. The energy $V(t) = F(x(t)) + \Phi(x'(t))$ is constant on each interval (t_{n-1}, t_n), so denote this value by ε_n. Suppose that $x(t)$ has no zero on some (t_{n-1}, t_n) for $n > N$ and

$$F(x(t_n - 0)) \geq \varepsilon_n \varphi^{\beta+1}(\frac{\Delta_\beta}{2} - \delta) =: \varepsilon_n \sigma_1, \tag{16}$$

where $\varphi(t)$ is the solution of (13) with the initial conditions (14). We will show that (16) implies

$$F(x(t_n - 0)) = F(x(t_n + 0)) \geq \varepsilon_{n+1}\varphi^{\beta+1}(\delta) =: \varepsilon_{n+1}\sigma_2. \tag{17}$$

Moreover, we will prove that (17) and $d_{n+1} \leq \frac{\Delta_\beta}{2} - 2\delta$ imply that $x(t)$ has no zero on the interval (t_n, t_{n+1}) and

$$F(x(t_{n+1} - 0)) \geq \varepsilon_{n+1}\sigma_1. \tag{18}$$

By induction, we would then have that $x(t)$ is nonoscillatory on $[t_{n-1}, \infty)$.

From (16) and $V(t)$, we have

$$\begin{aligned}
\varepsilon_{n+1} &= F(x(t_n - 0)) + b_n^{\beta+1}\Phi(x'(t_n - 0)) = F(x(t_n - 0)) + b_n^{\beta+1}(\varepsilon_n - F(x(t_n - 0))) \\
&= b_n^{\beta+1}\varepsilon_n + F(x(t_n - 0))(1 - b_n^{\beta+1}) \geq b_n^{\beta+1}\varepsilon_n + (1 - b_n^{\beta+1})\varepsilon_n\sigma_1 \\
&= \varepsilon_n\left[b_n^{\beta+1} + (1 - b_n^{\beta+1})\sigma_1\right].
\end{aligned}$$

We can estimate $F(x(t_n + 0))$ by

$$\begin{aligned}
F(x(t_n + 0)) &= \varepsilon_{n+1} - b_n^{\beta+1}\Phi(x'(t_n - 0)) = \varepsilon_{n+1} - b_n^{\beta+1}(\varepsilon_n - F(x(t_n + 0))) \\
&\geq b_n^{\beta+1}F(x(t_n + 0)) + \varepsilon_{n+1}\left(1 - \frac{b_n^{\beta+1}}{b_n^{\beta+1} + (1 - b_n^{\beta+1})\sigma_1}\right) \\
&\geq b_n^{\beta+1}F(x(t_n + 0)) + \varepsilon_{n+1}(1 - b_n^{\beta+1})\sigma_2,
\end{aligned}$$

where in the last step we used condition (15). Simplifying, we obtain the desired condition $F(x(t_n + 0)) \geq \varepsilon_{n+1}\sigma_2$.

To show that (17) and $d_{n+1} \leq \frac{\Delta_\beta}{2} - 2\delta$ imply (18), we have to prove that d_{n+1} is less than the time elapsed by changing F from $\varepsilon_n\sigma_1$ to $\varepsilon_{n+1}\sigma_2$ at the energy value ε_{n+1}. Similar to deriving (9) and (10), this time can be estimated by making several changes of variable and using the homogeneity of ϕ_β as follows:

$$\int_{F^{-1}(\varepsilon_{n+1}\sigma_1)}^{F^{-1}(\varepsilon_{n+1}\sigma_2)} \frac{dx}{\Phi^{-1}(\varepsilon_{n+1} - F(x))} = \int_{\sigma_1}^{\sigma_2} \frac{\varepsilon_{n+1} dv}{f(F^{-1}(\varepsilon_{n+1} v))\Phi^{-1}(\varepsilon_{n+1}(1-v))}$$

$$\geq \int_{\sigma_1}^{\sigma_2} \frac{\varepsilon_{n+1} dv}{\phi_\beta(F_\beta^{-1}(\varepsilon_{n+1} v))\Phi^{-1}(\varepsilon_{n+1}(1-v))} = \int_{\sigma_1}^{\sigma_2} \frac{dv}{\phi_\beta(F_\beta^{-1}(v))\Phi^{-1}(1-v)}$$

$$= \int_{(\beta+1)F_\beta(\varphi(\frac{\Delta_\beta}{2}-\delta))}^{(\beta+1)F_\beta(\varphi(\delta))} \frac{\frac{1}{\beta+1} dv}{\phi_\beta(F_\beta^{-1}(\frac{v}{\beta+1}))\Phi^{-1}(\frac{1-v}{\beta+1})} = \int_{\varphi(\frac{\Delta_\beta}{2}-\delta)}^{\varphi(\delta)} \frac{dq}{\Phi^{-1}(\frac{1}{\beta+1} - F_\beta(q))} = \frac{\Delta_\beta}{2} - 2\delta.$$

Finally, we have only to prove that there exists a t_n with $n > N$, satisfying (16). Assume that the solution $x(t)$ has a zero $s_0 \geq t_N$. Let s_0' be the first zero of $x'(t)$ to the right of s_0. (Since $x(t)$ is a nontrivial solution, it cannot become identically zero on (s_0', ∞).) Then, the condition on d_n implies that the first t_n after s_0' will satisfy inequality (16). □

The condition $|f(F^{-1}(u))| \leq (\beta+1)^{\frac{\beta}{\beta+1}} |u|^{\frac{\beta}{1+\beta}}$ holds for $f(x) = \phi_\alpha(x)$, $\alpha > \beta$, for

$$|u| \leq \frac{(\beta+1)^{\frac{\beta(\alpha+1)}{\alpha-\beta}}}{(\alpha+1)^{\frac{\alpha(\beta+1)}{\alpha-\beta}}}.$$

In this case, we cannot expect every solution to be nonoscillatory. Indeed, if $d_n = d > 0$, then the solutions with the properties $\Delta(F(x_k(t_1))) = d/k$ for $k = 1, 2, \ldots$, and $x'(t_1) = 0$ are periodic. By (10), we know that $\lim_{k\to\infty} F(x_k(t_1)) = \infty$. On the other hand, if $0 < \alpha < \beta$, and $d_n \geq d > 0$, then solutions satisfying the inequality $\Delta(V(0)) < d$ are oscillatory on $[T, \infty)$.

Remark 1. *If $f(x) = \phi_\beta(x)$, then Theorem 1 guarantees that every solution of (1) is nonoscillatory.*

Applying the above theorem to the case $b_n = b$, $d_n = d$, we see condition (15) is very sharp. In the linear case, $\varphi(t) = \cos t$, and (15) becomes

$$\frac{\sin^2 \delta}{b^2 + (1 - b^2)\sin^2 \delta} \geq \cos^2 \delta. \tag{19}$$

It can be shown that this condition is equivalent to the necessary and sufficient condition given in Theorem 4.1 in [6].

4. ASYMPTOTIC DECAY OF NONOSCILLATORY SOLUTIONS

The results in this section describe the asymptotic behavior of the nonoscillatory solutions of the impulsive system (1).

Theorem 2. *Every nonoscillatory solution of (1) tends to zero as $t \to \infty$ if and only if*

$$\int_0^\infty \left(\int_0^s \Psi^\beta(u, s)du\right)^{\frac{1}{\beta}} ds = \infty, \tag{20}$$

where $\Psi(s,t) := \prod_{s \leq t_i < t} b_i$ as in (7).

Proof. Suppose that (20) holds and there exists a positive nonoscillatory solution $x(t)$ that does not tend to zero. This solution is bounded, and by Lemma 2, it is monotonic, so $f(x(t)) \geq \delta > 0$ for some $\delta > 0$ and all sufficiently large t. First, we prove that $x(t)$ is nonincreasing on $[T, \infty)$ for some $T \geq 0$. Suppose the contrary. Then, by Lemma 1 and the fact that $\Psi(T,t) \leq \Psi(s,t)$, we have

$$\begin{aligned} 0 \leq \phi_\beta(x'(t)) &\leq \phi_\beta(x'(T))\Psi^\beta(T,t) - \delta \int_T^t \Psi^\beta(s,t)ds \\ &\leq \phi_\beta(x'(T))\Psi^\beta(T,t) - \delta\Psi^\beta(T,t)(t-T). \end{aligned}$$

The right hand side tends to $-\infty$ as $t \to \infty$, and this contradicts to the nonnegativity of $\phi_\beta(x'(t))$, so $x'(t) \leq 0$ for $t \geq T$.

Since $x'(t) \leq 0$ on $[T, \infty)$, we have

$$\phi_\beta(x'(t)) \leq -\delta \int_T^t \Psi^\beta(s,t)\, ds$$

and so

$$x'(t) \leq -\phi_\beta^{-1}(\delta) \left(\int_T^t \Psi^\beta(s,t)\, ds \right)^{\frac{1}{\beta}}.$$

Hence,

$$x(T) - x(t) \geq \phi_\beta^{-1}(\delta) \int_T^t \left(\int_T^u \Psi^\beta(s,u)ds \right)^{\frac{1}{\beta}} du.$$

Condition (20) yields a contradiction to $x(t) > 0$.

Now suppose that (20) is not satisfied; we will construct a bounded nonoscillatory solution that does not tend to zero. Take $x(T) = x_0 > 0$ and $\phi_\beta(x'(T)) = 0$. By (6), $V(t)$ is nonincreasing and $|x(t)| \leq x_0$, so choose $K = \max\{|f(x)| : -x_0 \leq x \leq x_0\}$. Next, choose T such that

$$\int_T^\infty \left(\int_T^u \Psi^\beta(s,u)ds \right)^{\frac{1}{\beta}} du < \frac{x_0}{2K^{1/\beta}}.$$

Then, taking ϕ_β^{-1} of both sides of (8) and integrating, we obtain

$$\begin{aligned} x(t) &= x(T) - \int_T^t \left(\int_T^u f(x(s))\Psi^\beta(s,u)ds \right)^{\frac{1}{\beta}} du \\ &\geq x_0 - K^{1/\beta} \int_T^\infty \left(\int_T^u \Psi^\beta(s,u)ds \right)^{\frac{1}{\beta}} du > \frac{x_0}{2}. \end{aligned}$$

This completes the proof of the theorem. □

The above theorem is analogous to some results of Hatvani et al. [7] and Smith [9] who studied the damped harmonic oscillator $x'' + a(t)x' + x = 0$. It also generalizes the results in [5].

It is not easy to check condition (20). Modifying an approach used in [5], we can give a much simpler necessary condition, as well as a sufficient condition, for nonoscillatory solutions to converge to zero.

Theorem 3. *If*

$$\sum_{n=1}^{\infty} (t_{n+1} - t_n)^{1+\frac{1}{\beta}} = \infty, \tag{21}$$

then every nonoscillatory solution tends to zero as $t \to \infty$. On the other hand, if

$$\limsup_{n\to\infty} b_n < 1 \text{ and } \sum_{n=1}^{\infty}(t_{n+1}-t_n)^{1+\frac{1}{\beta}} < \infty, \tag{22}$$

then there exists a bounded nonoscillatory solution that tends to a nonzero limit as $t \to \infty$.

Proof. Let $t_0 = 0$ and $d_n = t_n - t_{n-1}$ for $n \geq 1$. Observe that if $t_{n-1} < s < t \leq t_n$, then $\Psi(s,t) = 1$ for $n = 1, 2, \ldots$, while if $t_{n-1} < s \leq t_n < t \leq t_{n+1}$, then $\Psi(s,t) = b_n$. In general, if $t_{n-1} < s \leq t_n < \cdots < t_{n+k-1} < t \leq t_{n+k}$, then $\Psi(s,t) = \prod_{j=n}^{n+k-1} b_j$. Let

$$I_0 = \sum_{n=1}^{\infty}\int_{t_n}^{t_{n+1}}\left(\int_{t_n}^{t}\Psi^{\beta}(s,t)\,ds\right)^{\frac{1}{\beta}}dt = \frac{1}{1+\frac{1}{\beta}}\sum_{n=1}^{\infty} d_n^{1+\frac{1}{\beta}}, \tag{23}$$

and for $k \geq 1$,

$$I_k = \sum_{n=1}^{\infty}\int_{t_{n+k-1}}^{t_{n+k}}\left(\int_{t_{n-1}}^{t_n}\Psi^{\beta}(s,t)\,ds\right)^{\frac{1}{\beta}}dt = \sum_{n=1}^{\infty}\prod_{j=n}^{n+k-1} b_j\, d_n^{\frac{1}{\beta}} d_{n+k}. \tag{24}$$

Clearly, $\sum_{k=0}^{\infty} I_k$ is equal to the integral in (20). Now, by (21), $I_0 = +\infty$, so (20) hold and the first part of the theorem is proved.

To prove the second part, we need an upper estimate for the integrals I_k. Let $0 \leq b_n \leq \sigma < 1$ for every $n = 1, 2, \ldots$. Then,

$$I_0 < \sum_{n=1}^{\infty} d_n^{1+\frac{1}{\beta}},\ I_1 \leq \sigma\sum_{n=1}^{\infty} d_n^{\frac{1}{\beta}} d_{n+1},\ I_2 \leq \sigma^2\sum_{n=1}^{\infty} d_n^{\frac{1}{\beta}} d_{n+2},\ \ldots,\ I_k \leq \sigma^k\sum_{n=1}^{\infty} d_n^{\frac{1}{\beta}} d_{n+k},\ \ldots$$

Then, using Hölder's inequality, we obtain

$$\begin{aligned}\sum_{k=1}^{\infty} I_k &< \sum_{k=0}^{\infty}\sigma^k\left(\sum_{n=1}^{\infty} d_n^{\frac{1}{\beta}} d_{n+k}\right) \leq \sum_{k=0}^{\infty}\sigma^k\left(\sum_{n=1}^{\infty} d_n^{1+\frac{1}{\beta}}\right)^{\frac{1}{\beta+1}}\left(\sum_{n=1}^{\infty} d_{n+k}^{1+\frac{1}{\beta}}\right)^{\frac{\beta}{\beta+1}} \\ &\leq \left(\sum_{k=0}^{\infty}\sigma^k\right)\left(\sum_{n=1}^{\infty} d_n^{1+\frac{1}{\beta}}\right) = \frac{1}{1-\sigma}\sum_{n=1}^{\infty} d_n^{1+\frac{1}{\beta}},\end{aligned}$$

which completes the proof of the theorem. □

By a refinement of the proof of Theorem 2, we can estimate the asymptotic decay of the solutions.

Theorem 4. *Let $\omega(t)$ be a positive, locally integrable function. If*

$$\lim_{t\to\infty}\left(\int_0^t\int_0^s \omega(u)\Psi^{\beta}(u,s)du\right)^{\frac{1}{\beta}} ds = \infty,$$

then for every bounded nonoscillatory solution $x(t)$ of (1),

$$\liminf_{t\to\infty}\frac{|f(x(t))|}{\omega(t)} = 0.$$

Proof. Let $x(t)$ a positive bounded nonoscillatory solution for which

$$\liminf_{t\to\infty}\frac{f(x(t))}{\omega(t)} = 2\delta > 0.$$

If t is large enough, $f(x(t))/\omega(t) > \delta$. From this point on, the proof is the same as the first part of the proof of Theorem 2. □

Similarly, the first part of Theorem 3 can also be given in a more general form.

Theorem 5. *Let $\omega(t)$ be a positive, nonincreasing function. If*

$$\sum_{n=1}^{\infty} \omega^{\frac{1}{\beta}}(t_{n+1})(t_{n+1}-t_n)^{1+\frac{1}{\beta}} = \infty, \tag{25}$$

then for every bounded nonoscillatory solution $x(t)$ of (1),

$$\liminf_{t\to\infty} \frac{|f(x(t))|}{\omega(t)} = 0.$$

ACKNOWLEDGEMENTS

The research of the first author was supported by the Mississippi State University Biological and Physical Sciences Research Institute. This research was completed during the visit of the second author to Mississippi State University with the support of a Hungarian Eötvös Fellowship.

REFERENCES

1. Bainov, D. D., Simeonov, P. S., Systems with Impulse Effect, Stability, Theory and Applications, Ellis Horwood Ltd., 1989.
2. Ballieu, R. J., Peiffer, K., Attractivity of the origin for the equation $\ddot{x}+f(t,x,\dot{x})|\dot{x}|^{\alpha}\dot{x}+g(x)=0$, *J. Math. Anal. Appl.*, 65 (1978), 321–332.
3. Elbert, Á., A half-linear second order differential equation, in Differential Equations: Qualitative Theory, Colloq. Math. Soc. J. Bolyai, Vol 30, North Holland, Amsterdam, 1979, 153–181.
4. Graef, J. R., Karsai, J., On the asymptotic behavior of solutions of impulsively damped nonlinear oscillator equations, *J. Comput. Appl. Math.*, 71 (1996), 147–162.
5. Graef, J. R., Karsai, J., Intermittent and impulsive effects in second order systems, *Nonlinear Anal.*, 30 (1997), 1561–1571.
6. Graef, J. R., Karsai, J., Asymptotic properties of nonoscillatory solutions of impulsively damped nonlinear oscillator equations, *Dynam. Contin. Discrete Impuls. Systems*, 3 (1997), 151–165.
7. Hatvani, L., Krisztin, T., Totik, V., A necessary and sufficient condition for the asymptotic stability of the damped oscillator, *J. Differential Equations*, 119 (1995), 209–223.
8. Reissig, R., Sansone, G., Conti, R., Qualitative Theory of Nonlinear Differential Equations, Izd. Nauka, Moskow, 1974 (in Russian).
9. Smith, R. A., Asymptotic stability of $x''+a(t)x'+x=0$, *Quart. J. Math. Oxford (2)*, 12 (1961), 123–126.

Proceedings of Dynamic Systems and Applications 3 (2001) 233-240

Some New Results and Open Problems on Oscillation of Nonlinear Difference Equations

John R. Graef[1], L. Ramuppillai[2], and E. Thandapani[2]
[1]Department of Mathematics, University of Tennessee at Chattanooga
Chattanooga, TN 37403 USA
[2]Department of Mathematics, Periyar University
Salem 636011, Tamil Nadu, India

ABSTRACT: Consider the damped nonlinear difference equation

$$\Delta(a_n \Delta y_n) + \phi(n, y_n, \Delta y_n) + q_n f(y_{n+1}) = 0,$$

where $a_n > 0$, $q_n > 0$, $uf(u) > 0$ for $u \neq 0$, and there are nonnegative sequences $\{P_n\}$ and $\{p_n\}$ such that $-P_n v^2 \le v\phi(n,u,v) \le p_n v^2$. The authors give sufficient conditions for all solutions to oscillate and illustrate their results with some examples. Open problems and suggestions for future research are also included.

AMS (MOS) subject classification. 39A10

1. INTRODUCTION

In this paper, we are concerned with second order damped difference equations of the form

$$\Delta(a_n \Delta y_n) + \phi(n, y_n, \Delta y_n) + q_n f(y_{n+1}) = 0, \quad n \in \mathbb{N}, \tag{1}$$

where $\mathbb{N} = \{0, 1, 2, \dots\}$, Δ is the difference operator defined by $\Delta y_n = y_{n+1} - y_n$, and the real sequences $\{a_n\}$ and $\{q_n\}$ and the functions f and ϕ satisfy the following conditions:

- (C_1) $a_n > 0$ and $q_n > 0$ for all $n \ge n_0 \in \mathbb{N}$;
- (C_2) $\phi : \mathbb{N} \times \mathbb{R} \times \mathbb{R} \to \mathbb{R}$ is continuous and there are nonnegative real sequences $\{P_n\}$ and $\{p_n\}$ such that $-P_n v^2 \le v\phi(n,u,v) \le p_n v^2$ for all $(n,u,v) \in \mathbb{N} \times \mathbb{R} \times \mathbb{R}$ and $\phi(n,u,0) = 0$;
- (C_3) $f : \mathbb{R} \to \mathbb{R}$ is continuous, satisfies $uf(u) > 0$ for all $u \neq 0$, and there is a constant $M_1 \ge 1$ such that $\frac{f(u)}{u} \ge M_1$ for all $u \neq 0$.

By a solution of equation (1), we always mean a real sequence $\{y_n\}$ satisfying (1) for all $n \in \mathbb{N}$ and for which $\sup\{|y_n| : n \ge s\} > 0$ for any $s \in \mathbb{N}$. A solution is called *nonoscillatory* if it is eventually of constant sign, and it is called *oscillatory* otherwise.

In the past several years, qualitative properties of solutions of difference equations have received a great deal of attention in the literature, and we cite the works [1-7, 9, 10] as recent contributions. Equations of the form

$$\Delta(a_n \Delta y_n) + p_n \Delta y_n + q_n f(y_{n+1}) = 0 \tag{e_1}$$

and

$$\Delta^2 y_n + p_n \phi(y_n, \Delta y_n) + q_n f(y_{n+1}) g(\Delta y_n) = 0 \tag{e_2}$$

are considered in the papers [6] and [7], respectively, and the authors obtain criteria for the oscillation of all solutions under the condition

$$\sum_{n=n_0}^{\infty} \frac{1}{a_n} \left(\prod_{s=n_0}^{n-1} \left(1 - M\frac{p_s}{a_s} \right) \right) = \infty, \tag{2}$$

where $M = 1$ for equation (e_1) and $0 < v\phi(u,v) \leq Mv^2$ for equation (e_2). In this paper, we obtain some criteria for the oscillation of all solutions of equation (1) when condition (2) is not satisfied. What also distinguishes our results from those of other authors is that we allow for "negative" damping, that is, we do not require a condition of the form

$$v\phi(n,u,v) \geq 0$$

as is usually the case. For example, the function $\phi(n,u,v) = v \sin u$ satisfies condition (C_2) but not the above inequality. Also observe that if ϕ does satisfy the above inequality, then we can take the sequence $\{P_n\}$ in (C_2) to be identically zero.

2. MAIN RESULTS

We begin with the following theorem.

Theorem 1. *Assume that*

$$a_n - p_n > 0 \text{ for all } n \geq n_0 \in \mathbb{N}, \tag{3}$$

$$\rho_{n_0} = \sum_{n=n_0}^{\infty} \frac{1}{a_n} \left(\prod_{s=n_0}^{n-1} \left(1 - \frac{p_s}{a_s} \right) \right) < \infty, \quad n_0 \in \mathbb{N}_0, \tag{4}$$

and

$$\liminf_{n\to\infty} q_n \rho_{n+1} > M_1. \tag{5}$$

If there exists a positive nonincreasing sequence $\{h_n\}$ such that

$$\sum_{n=n_0}^{\infty} (M_1 q_n - P_n) h_n = \infty, \tag{6}$$

then all solutions of equation (1) are oscillatory.

Proof. Let $\{y_n\}$ be a nonoscillatory solution of equation (1), and without loss of generality, we assume that $y_n > 0$ for all $n \geq n_0 \in \mathbb{N}$. The proof for the case $y_n < 0$ eventually is similar and will be omitted. We consider the following cases for the behavior of $\{\Delta y_n\}$.

Case 1: $\{\Delta y_n\}$ is oscillatory. *(a)* First, suppose there exists an integer $n_1 \geq n_0$ such that $\Delta y_{n_1} < 0$. Now from (1),

$$\Delta(a_{n_1} \Delta y_{n_1}) \Delta y_{n_1} = -q_{n_1} f(y_{n_1+1}) \Delta y_{n_1} - \phi(n_1, y_{n_1}, \Delta y_{n_1}) \Delta y_{n_1},$$

or

$$\Delta y_{n_1} (a_{n_1+1} \Delta y_{n_1+1} - a_{n_1} \Delta y_{n_1}) > -p_{n_1} (\Delta y_{n_1})^2,$$

and hence,

$$a_{n_1+1}\Delta y_{n_1+1}\Delta y_{n_1} > (a_{n_1} - p_{n_1})(\Delta y_{n_1})^2 > 0.$$

Thus, we have $\Delta y_{n_1+1} < 0$, and so by induction, we obtain $\Delta y_n < 0$ for all $n \geq n_1$.

(b) Now suppose $\Delta y_{n_1} = 0$. Then, equation (1) implies $\Delta y_{n_1+1} < 0$ since $\phi(n_1, y_{n_1}, 0) = 0$, and we are back to case *(a)*. Thus, in either situation, we obtain $\Delta y_n < 0$ for all $n \geq n_1 + 1$, and this contradicts the assumption that $\{\Delta y_n\}$ oscillates. Thus, $\{\Delta y_n\}$ has fixed sign.

Case 2: $\Delta y_n < 0$ for all $n \geq n_1$ for some $n_1 \geq n_0$. From (1) and (C_2), we have

$$\Delta u_n + \frac{p_n}{a_n} u_n \geq 0 \tag{7}$$

where $u_n = -a_n \Delta y_n$. From (7), we obtain

$$u_n \geq u_{n_1} \prod_{s=n_1}^{n-1} \left(1 - \frac{p_s}{a_s}\right),$$

or

$$\Delta y_n \leq -\frac{u_{n_1}}{a_n} \prod_{s=n_1}^{n-1} \left(1 - \frac{p_s}{a_s}\right).$$

Summing the last inequality from n_1 to $n-1$, we obtain

$$y_n - y_{n_1} \leq a_{n_1}\Delta y_{n_1} \sum_{s=n_1}^{n-1} \frac{1}{a_s} \left[\prod_{t=n_1}^{s-1} \left(1 - \frac{p_t}{a_t}\right)\right] \tag{8}$$

for $n \geq n_1$. From (8), we have

$$y_{n_1} \geq -a_{n_1}\Delta y_{n_1} \sum_{s=n_1}^{n-1} \frac{1}{a_s} \left[\prod_{t=n_1}^{s-1} \left(1 - \frac{p_t}{a_t}\right)\right]$$

for $n \geq n_1$. Letting $n \to \infty$, we obtain

$$y_{n_1} \geq -(a_{n_1}\Delta y_{n_1})\rho_{n_1}. \tag{9}$$

In view of (C_2), equation (1) implies

$$\Delta(a_n \Delta y_n)\Delta y_n + p_n(\Delta y_n)^2 + q_n f(y_{n+1})\Delta y_n \geq 0,$$

or

$$a_{n+1}\Delta y_{n+1} - (a_n - p_n)(\Delta y_n) + q_n f(y_{n+1}) \leq 0. \tag{10}$$

Since $(a_n - p_n)\Delta y_n < 0$ for all $n \geq n_1$, (10) yields

$$a_{n+1}\Delta y_{n+1} + q_n f(y_{n+1}) < 0 \tag{11}$$

for all $n \geq n_1$. In view of (C_3) and (9), (11) implies

$$a_{n+1}\Delta y_{n+1} - q_n M_1 \rho_{n+1} a_{n+1} \Delta y_{n+1} < 0,$$

so

$$q_n\rho_{n+1} < \frac{1}{M_1}$$

for $n \geq n_1$, which contradicts condition (5).

Case 3: $\Delta y_n > 0$ for all $n \geq n_1 \geq n_0$. Now define $z_n = h_{n-1}a_n\frac{\Delta y_n}{y_n}$. The monotonicity of $\{h_n\}$ and conditions (C_2) and (C_3) imply

$$\Delta z_n \leq -M_1h_nq_n + P_nh_n.$$

Summing from n_1 to n, we obtain

$$z_{n+1} \leq z_{n_1} - \sum_{s=n_1}^{n}(M_1q_s - P_s)h_s.$$

Condition (6) implies that z is eventually negative, and this contradiction completes the proof of the theorem.

Example 1. Consider the difference equation

$$\Delta(n\Delta y_n) + \frac{(\Delta y_n)^3}{|\Delta y_n|^2+4} + \frac{4n+1}{2}(y_{n+1} + y_{n+1}^3) = 0, \quad n \geq 2. \qquad (E_1)$$

Here, $a_n = n$, $q_n = (4n+1)/2$, $\phi(n,u,v) = \frac{v^3}{|v|^2+4}$, $f(u) = u + u^3$, and we can take $P_n \equiv 0$ and $p_n \equiv 1$. We have

$$\rho_n = \sum_{j=n}^{\infty}\frac{1}{j}\left[\prod_{i=n}^{j-1}\left(1-\frac{1}{i}\right)\right] = \sum_{j=n}^{\infty}\frac{n-1}{j(j-1)} = (n-1)\sum_{j=n}^{\infty}\Delta\left(-\frac{1}{j-1}\right) = 1.$$

Now it is easy to see that all conditions of Theorem 1 with $h_n \equiv 1$ are satisfied, so every solution of (E_1) is oscillatory. One such solution is $\{y_n\} = \{(-1)^n\}$.

In the following theorem, we obtain oscillation criteria for equation (1) by using a generalized Riccati transformation due to Yu [8].

Theorem 2. *Suppose conditions (3)–(5) hold. If there exists a real positive sequence $\{h_n\}$ such that*

$$\sum_{n=n_0}^{\infty}\left[\psi_n - \frac{a_{n+1}h_{n+1}^2r_n^2}{4h_n}\right] = \infty, \quad n_0 \in \mathbb{N}, \qquad (12)$$

where $r_n = \dfrac{\Delta h_n}{h_{n+1}} + 2h_n$ *and* $\psi_n = h_n(M_1q_n - P_n + a_{n+1}h_{n+1}^2 - \Delta(a_nh_n))$, *then all solutions of equation (1) are oscillatory.*

Proof. Proceeding exactly as in the proof of Theorem 1, we have that $\Delta y_n > 0$ for all $n \geq n_1$ for some $n_1 \geq n_0$. Define $z_n = h_n\left[\dfrac{a_n\Delta y_n}{y_{n+1}} + a_nh_n\right]$ for $n \geq n_1$; then for $n \geq n_1$, conditions (C_2) and (C_3) imply

$$\Delta z_n \leq \frac{z_{n+1}\Delta h_n}{h_{n+1}} + h_n\left[\frac{P_n\Delta y_n}{y_{n+1}} - M_1q_n - \frac{a_{n+1}(\Delta y_{n+1})^2}{y_{n+1}y_{n+2}} + \Delta(a_nh_n)\right].$$

Due to the monotonicity of $\{y_n\}$, this becomes

$$\Delta z_n \le \frac{z_{n+1}\Delta h_n}{h_{n+1}} + h_n\left[P_n - M_1 q_n - \frac{1}{a_{n+1}}\left(\frac{z_{n+1}}{h_{n+1}} - a_{n+1}h_{n+1}\right)^2 + \Delta(a_n h_n)\right]$$

$$= -\psi_n + r_n z_{n+1} - \frac{h_n}{a_{n+1}h_{n+1}^2} z_{n+1}^2$$

$$= -\psi_n - \frac{h_n}{a_{n+1}h_{n+1}^2}\left[z_{n+1} - \frac{a_{n+1}h_{n+1}^2 r_n}{2h_n}\right]^2 + \frac{a_{n+1}h_{n+1}^2 r_n^2}{4h_n}$$

$$\le -\psi_n + \frac{a_{n+1}h_{n+1}^2 r_n^2}{4h_n}$$

for $n \ge n_1$. Summing the last inequality from n_1 to n, we have

$$\sum_{s=n_1}^{n}\left(\psi_s - \frac{a_{s+1}h_{s+1}^2 r_s^2}{4h_s}\right) \le z_{n_1} - z_{n+1} \le z_{n_1},$$

which contradicts (12). This completes the proof of the theorem.

We conclude this section with some examples.

Example 2. Consider the equation

$$\Delta(n(n+1)(n+2)\Delta y_n) - (n+1)(n+2)\Delta y_n + 2(n+1)(n+2)^2(y_{n+1} + y_{n+1}^{1/3}) = 0, \quad n \ge 2. \quad (E_2)$$

This equation provides us with an example where we have $p_n \equiv 0$. Here, we have $a_n = n(n+1)(n+2)$, $P_n = (n+1)(n+2)$, $q_n = 2(n+1)(n+2)^2$, $f(u) = u + u^3$, and

$$\rho_n = \sum_{j=n}^{\infty}\frac{1}{j(j+1)(j+2)} = \frac{1}{2}\sum_{j=n}^{\infty}\Delta\left(-\frac{1}{j(j+1)}\right) = \frac{1}{2n(n+1)}.$$

The hypotheses of Theorem 1 are satisfied with $h_n \equiv 1$, and so every solution of (E_2) is oscillatory. One such solution is $\{y_n\} = \{(-1)^n\}$. Equation (E_2) also satisfies the conditions of Theorem 2 with $h_n \equiv 1$, $r_n = 2$, and $\psi_n = 3(n+1)^2(n+2)$.

Example 3. Consider the equation

$$\Delta(n(n+1)\Delta y_n) + (-1)^n(n+1)\Delta y_n + (n+1)[2(n+1) + (-1)^{n+1}](y_{n+1} + y_{n+1}^5) = 0, \quad n \ge 2. \quad (E_3)$$

With $h_n \equiv 1$ and

$$\rho_n = \sum_{j=n}^{\infty}\frac{1}{j(j+1)}\left[\prod_{i=n}^{j-1}\left(1 - \frac{1}{i}\right)\right] = \sum_{j=n}^{\infty}\frac{1}{j(j+1)(j-1)}$$

$$= \frac{1}{2}\sum_{j=n}^{\infty}\Delta\left(-\frac{1}{j(j-1)}\right) = \frac{1}{2n(n-1)},$$

we see that all the conditions of Theorem 1 are satisfied. Hence, every solution of equation (E_3) is oscillatory, and one such solution is $\{y_n\} = \{(-1)^n\}$.

3. EXTENSIONS AND OPEN PROBLEMS

An important type of nonlinear difference equation that has received a lot of attention in the last few years is the class of half-linear equations

$$\Delta(a_n|\Delta y_n|^{\alpha-1}\Delta y_n) + q_n|y_{n+1}|^{\alpha-1}y_{n+1} = 0, \quad \alpha > 0, \tag{H-L}$$

which has the linear equation ($\alpha = 1$)

$$\Delta(a_n\Delta y_n) + q_n y_{n+1} = 0$$

as a special case. There is a growing literature on this interesting class of equations. While some authors have considered damped versions of equation (H-L) such as

$$\Delta(a_n|\Delta y_n|^{\alpha-1}\Delta y_n) + b_n|\Delta y_n|^{\alpha-1}\Delta y_n + q_n|y_{n+1}|^{\alpha-1}y_{n+1} = 0,$$

with $b_n \geq 0$, there are no known oscillation results for equations of this type with negative damping.

Another interesting problem would be to obtain a sharp lower bound on the size of the negative damping that still ensures the oscillation of all solutions. Also, there are no known comparison theorems for equations with negative damping. For example, such a result might take the form:

If all solutions of the equation

$$\Delta(a_n\Delta y_n) + \phi(n, y_n, \Delta y_n) + q_n f(y_{n+1}) = 0$$

with

$$v\phi(n, u, v) \leq 0$$

are oscillatory and

$$v\phi(n, u, v) \leq v\Phi(n, u, v) \leq 0,$$

then all solutions of the equation

$$\Delta(a_n\Delta y_n) + \Phi(n, y_n, \Delta y_n) + q_n f(y_{n+1}) = 0,$$

are oscillatory.

Finally, it would be interesting to see results for forced or perturbed equations with negative damping. In such a case, it is reasonable to ask whether equation (1) could be viewed as a perturbation of the equation

$$\Delta(a_n\Delta y_n) + q_n f(y_{n+1}) = 0.$$

Perturbation terms, however, are usually controlled by y_n or by the nonlinear term $f(y_n)$, not by Δy_n as was done above.

ACKNOWLEDGMENT

The research of the first author was supported by the Mississippi State University Biological and Physical Sciences Research Institute.

REFERENCES

1. Agarwal, R. P., Difference Equations and Inequalities. Marcel Dekker, New York, 1992.
2. Agarwal, R. P., Pandian, S., & Thandapani, E., Oscillatory property for second order nonlinear difference equations via Lyapunov's second method. In: Advances in Nonlinear Dynamics, pp. 11–21, Gordon and Breach, New York, 1997.
3. Hooker, J. W., & Patula, W. T., A second order nonlinear difference equation: oscillation and asymptotic behavior. J. Math. Anal. Appl. Vol 91 (1983) pp. 9–24.
4. Kulenovic, M. R. S., & Budencevic, M., Asymptotic analysis of nonlinear second order difference equation. An. Stiint. Univ. "Al. I. Cuza" Iasi Sect. I a Mat. (N.S.) Vol 30 (1984) pp. 39–52.
5. Thandapani, E., Asymptotic and oscillatory behavior of solutions of nonlinear second order difference equations. Indian J. Pure Appl. Math. Vol 24 (1993) pp. 365–372.
6. Thandapani, E., Oscillation theorems for second order damped nonlinear difference equations. Czech. Math. J. Vol 45 (1995) pp. 327–335.
7. Thandapani, E., & Pandian, S., Oscillation theorems for nonlinear second order difference equations with a nonlinear damping term. Tamkang J. Math. Vol 26 (1995) pp. 49–58.
8. Yu, Y. H., Higher type oscillation criterion and Sturm type comparison theorem. Math. Nachr. Vol 153 (1991) pp. 485–496.
9. Zhang, B. G., Oscillation and asymptotic behavior of second order difference equations J. Math. Anal. Appl. Vol 173 (1993) pp. 58–68.
10. Zhang, B. G., & Chen, G. D., Oscillation of second order nonlinear difference equations. J. Math. Anal. Appl. Vol 199 (1996) pp. 827–841.

Proceedings of Dynamic Systems and Applications 3 (2001) 241-248

Parallel preconditioned algorithms for solving special tridiagonal systems

George A. Gravvanis
University of the Aegean, Department of Mathematics
GR 832 00 Karlovasi, Samos, Greece
email : gag@aegean.gr

ABSTRACT : A new class of parallel approximate inverse matrix algorithmic techniques based on the concept of sparse approximate LU-type factorization procedures is introduced for computing classes of explicitly approximate inverses without inverting the decomposition factors. Explicit preconditioned conjugate gradient-type schemes in conjunction with approximate inverse matrix techniques are presented for the efficient solution of special tridiagonal systems. Applications of the proposed method on special tridiagonal linear systems is discussed and numerical results are given.

Keywords : Special tridiagonal systems, approximate factorization procedures, approximate inverse matrix techniques, preconditioning, parallel iterative methods.

AMS(MOS) : 65F10, 65F50, 65W05; **C.R. categories** : F.2.1, G.1.0, G.1.3, G.1.8.

1. INTRODUCTION

In recent years research efforts have been directed on the production of numerical software for solving sparse linear systems of algebraic equations, which occur in many engineering and scientific problems, on uniprocessor or multiprocessor systems, Axelsson [2], Dongarra et al [5], Golub & Loan [11], Ortega [24], Saad [25]. Hence sparse matrix computations, which have inherent parallelism, are therefore of central importance in scientific and engineering computing and furthermore the need for high performance computing, which is about 70% of supercomputer time, has had some effect on the design of modern computer systems, Dongarra et al [5], Evans & Sutti [7], Golub & Loan [11], Ortega [24].

Current research efforts are focused on the development, testing and analysis of new numerical algorithms, that are needed to exploit multiprocessor systems and can be classified as fundamental "parallel numerical algorithms". Such research work has been concentrated on parallel machines, i.e. vector or array processors and systolic arrays, Akl [1], Ortega [24].

An important achievement over the last decades is the appearance and use of Preconditioning methods for the numerical solution of sparse systems. The well known Preconditioning methods based on Incomplete Factorization or Successive Over-relaxation or Approximate Inverses by minimizing the Frobenious norm of the error for fixed sparsity pattern, Axelsson [2], Golub & Loan [11], Saad [25], are very difficult to implement them on parallel systems, Dongarra et al [5], Evans & Sutti [7], Golub & Loan [11], Ortega [24], Saad [25].

Recently, explicit preconditioned iterative methods, Axelsson [2], Axelsson & Kolotilina [3], Golub & Loan [11], Ortega [24], Saad [25], have been extensively used for solving efficiently sparse linear systems, resulting from the finite element or finite difference discretization of partial differential equations in two and three space variables, on multiprocessor systems, Gravvanis [13], [14], [15], [16], [17], [18], Lipitakis & Gravvanis [21], [22]. The effectiveness of the explicit preconditioned iterative methods based on adaptive approximate inverse matrix (AIM) techniques is related to the fact that the approximate inverse exhibit a similar "fuzzy" structure as the coefficient matrix.

Our attention is now focused on the numerical solution of a special class of linear systems (considered as special tridiagonal linear systems), i.e.,

$$A\,u = s \qquad (1)$$

where A is a sparse special tridiagonal (n×n) matrix, cf. (2), where $b \neq 0$ and $|a/b| > 2$. The coefficient matrix A has a special form : it is triadiagonal ($\rho=0$) and a Toeplitz matrix ($\mu=\nu=1$, i.e. constant diagonals), Boisvert [4]. Computational techniques on the solution of special tridiagonal linear systems have been presented in Boisvert [4], Evans [9], Fisher et al [10], Golub & Loan [11], Malcolm & Palmer [23], Widlund [27].

$$A = \begin{bmatrix} a/\mu & b & & & & \rho b \\ b & a & b & & & \\ & \ddots & \ddots & \ddots & 0 & \\ & & \ddots & \ddots & \ddots & 0 \\ & 0 & & b & a & b \\ \rho b & & 0 & & b & a/\nu \end{bmatrix} \qquad (2)$$

Such systems occur when solving certain constant-coefficient elliptic partial differential equations by the Fourier method, Boisvert [4], or using finite difference methods to solve linear constant-coefficient boundary value problems, where the constants μ, ν, ρ are determined from the boundary conditions along oposite sides of the domain, as shown in the following Table, Boisvert [4], Golub & Loan [11], viz :

Boundary Conditions	μ	ν	ρ
Dirichlet-Dirichlet	1	1	0
Dirichlet-Neumann	1	2	0
Neumann-Dirichlet	2	1	0
Neumann- Neumann	2	2	0
Periodic	1	1	1

The derivation of suitable parallel methods was the main objective for which several forms of an approximate inverse of a given matrix, based on adaptive approximate LU-type factorization procedures, have been proposed, Gravvanis [13], [14], Lipitakis & Gravvanis [21], [22]. The main motive for the derivation of the AIM techiques lies in the fact that they can be used in conjunction with explicit preconditioned iterative schemes and are suitable for solving linear systems on parallel and vector processors. An important feature of these techniques is the provision of both explicit direct and preconditioned iterative methods for solving linear systems, such that the best method for the given problem to be selected, Gravvanis [18].

Finally, the performance and applicability of the new parallel preconditioned conjugate gradient schemes is discussed by solving special tridiagonal systems and numerical results are given.

2. APPROXIMATE INVERSE MATRIX TECHNIQUES

In this section we present algorithmic procedures for computing the elements of the approximate inverse, based on sparse approximate LU-type factorization procedures, Gravvanis [13], [14], Gravvanis & Lipitakis [19], [20]. According to the structure of the coefficient matrix A, cf. (2), "fill-in" terms are required during the decomposition process.

Let us now assume the approximate factorization of the coefficient matrix A, i.e.

$$A \approx L_r U_r \tag{3}$$

retaining r non-zero entries, i.e. r "fill-in" terms, by applying the so-called "position-principle" in the factorization process, where L_r and U_r, cf. (4), are sparse strictly lower and upper (with main diagonal unity elements) triangular matrices of the same profile as the coefficient matrix A, cf. (2).

Then, the elements of the decomposition L_r and U_r factors can be obtained by a Special Tridiagonal Approximate **LU**-type **FA**ctorization algorithmic procedure (henceforth called the **STALUFA** algorithm).

$$L_r \equiv \begin{bmatrix} w_1 & & & & 0 \\ d_1 & w_2 & & & \\ & \ddots & \ddots & & \\ & 0 & & \ddots & \\ e_1 & \text{---} & e_r & d_{n-1} & w_n \end{bmatrix} \quad U_r \equiv \begin{bmatrix} 1 & g_1 & & 0 & h_1 \\ & 1 & \ddots & & \vert \\ & & \ddots & \ddots & h_r \\ & 0 & & \ddots & g_{n-1} \\ & & & & 1 \end{bmatrix} \tag{4}$$

The memory requirements of the **STALUFA** algorithm is $O(5n)$ words and the computational work required by the factorization process is $\approx O(2n+2r)$ multiplicative operations for r>1. Specifically, when r=1 the memory requirements is $O(3n)$ words, while the computational work required is $\approx O(2n)$ multiplicative operations. The

STALUFA algorithm can be implemented on multiprocessor systems by following certain parallel decomposition techniques, Akl [1], Ortega [24].

Let $M^{\delta l} \equiv \left(\mu_{i,j}\right)$, $i \in [1,n]$, $j \in [\max(1, i-\delta l+1), \min(n, i+\delta l-1)]$, an $[n\times(2\delta l\text{-}1)]$ matrix, be the approximate inverse of the coefficient matrix A. The elements of the approximate inverse can be determined by retaining a certain number of elements of the inverse, i.e. only δl elements in the lower part and in the upper part of the inverse (by applying the so-called "position-principle"), next to the main diagonal, the remaining elements not being computed at all, and its elements can be computed by solving recursively the following systems :

$$M_r^{\delta l} L_r = U_r^{-1} \quad \text{and} \quad U_r M_r^{\delta l} = L_r^{-1} \quad \delta l \in [1, ..., n) \tag{5}$$

without inverting the decomposition factors L_r and U_r, Gravvanis [13], [14], [15].

It should be noted that the computation of the elements $\mu_{i,j}$ of the inverse, using a "fish-bone" computational procedure, can be successively determined as follows : From the equations of (5) for i=n, ..., 1 and j= max(1, i-δl+1), ..., min(n, i+δl-1) respectively, we can obtain the elements of the approximate inverse, Gravvanis [15], Lipitakis & Gravvanis [21].

In order to solve efficiently large order linear systems a moving window shifted from bottom to top is used, such that only [n×(2δl-1)]-vectors are retained in storage, Gravvanis [17], [18]. This **O**ptimized form of the **S**pecial **T**ridiagonal **A**pproximate **I**nverse **M**atrix algorithm procedure (henceforth called the **OSTAIM** algorithm) is particularly effective for solving "narrow-banded" sparse systems of very large order , i.e. δl << n/2.

The memory requirements of the **OSTAIM** algorithm are ≈[n×(2δl-1)] words and the computational work involved is $\approx O(3\delta l n)$ multiplicative operations. It should be noted that when δl=1, then $\approx O(n)$ multiplicative operations are required.

It should be also noted that according to the proposed computational strategy this class of approximate inverses can be considered that includes various families of approximate inverses having in mind the desired requirements of accuracy, storage and computational work as can be seen by the following diagrammatic relationship, i.e.,

$$\begin{array}{ccccccc} & & & \underline{\text{class I}} & & \underline{\text{class II}} & & \underline{\text{class III}} \\ A^{-1} \equiv M & \leftarrow & & M^{\delta l} & \leftarrow & M_r^{\delta l} & \leftarrow & M_{r,n}^{\delta l} \end{array} \tag{6}$$

where the entries of $M^{\delta l}$ have been computed based on the diagonal, sub-diagonal and super-diagonal elements of the L_r and U_r decomposition factors, while the entries of $M_r^{\delta l}$ have been computed and retained during the computational procedure of the (approximate) inversion. The entries of $M_{r,n}^{\delta l}$ have been retained after the computation of the approximate inverse by retaining additionally the n-th row and column of the approximate inverse.

3. EXPLICIT PRECONDITIONED ITERATIVE METHODS

In this section we present a class of explicit preconditioned iterative schemes based on the **OSTAIM** techniques of section 2 for solving special tridiagonal linear system, cf. (1).

The **Explicit Preconditioned Generalized Conjugate Gradient Square (EPGCGS)** algorithm for solving linear systems, can be expressed by the following compat scheme :

Let u_0 be an arbitrary initial approximation to the solution vector u. Then,

set $u_0 = 0$ and $e_0 = 0$, solve $r_0 = M_r^{\delta l}(s - Au_0)$, (7)

set $\sigma_0 = r_0$ and $p_0 = (\sigma_0, r_0)$. (8)

Then, for i=0, 1, ..., (until convergence) compute the vectors $u_{i+1}, r_{i+1}, \sigma_{i+1}$ and the scalar quantities α_i, β_{i+1} as follows :

form $q_i = A\sigma_i$, calculate $\alpha_i = p_i \Big/ \left(\sigma_0, M_r^{\delta l} q_i\right)$, (9)

compute $e_{i+1} = r_i + \beta_i e_i - \alpha_i M_r^{\delta l} q_i$, (10)

$d_i = r_i + \beta_i e_i + e_{i+1}$, and $u_{i+1} = u_i + \alpha_i d_i$, (11)

form $q_i = Ad_i$, compute $r_{i+1} = r_i - \alpha_i M_r^{\delta l} q_i$, (12)

set $p_{i+1} = (\sigma_0, r_{i+1})$, evaluate $\beta_{i+1} = p_{i+1}/p_i$, (13)

compute $\sigma_{i+1} = r_{i+1} + 2\beta_{i+1} e_{i+1} + \beta_{i+1}^2 \sigma_i$. (14)

The computational complexity of the **EPGCGS** method, assuming that $M_r^{\delta l}$ can be compactly stored in n×(2δl-1) diagonal vectors, is ≈O[(4δl+15)n mults + 8n adds]v operations, where v denotes the number of iterations required for convergence to a predetermined tolerance level.

In the following we present a modified form of the van der Vorst **BI-CGSTAB** method, Vorst [26], using the explicit approximate inverse $M_r^{\delta l}$. This modified method, henceforth called the **Explicit Preconditioned Biconjugate Conjugate Gradient-STAB (EPBI-CGSTAB)** method, Gravvanis [13], can be expressed by the following compat scheme :

Let u_0 be an arbitrary initial approximation to the solution vector u. Then,

set $u_0 = 0$ compute $r_0 = s - Au_0$, (15)

set $r_0' = r_0$, $\rho_0 = \alpha = \omega_0 = 1$ and $v_0 = p_0 = 0$. (16)

Then, for i=0, 1, ..., (until convergence) compute the vectors u_{i+1}, r_{i+1} and the scalar quantities α, β, ω_i as follows :

calculate $\rho_i = \left(r_0', r_{i-1}\right)$, and $\beta = \left(\rho_i / \rho_{i-1}\right) / \left(\alpha / \omega_{i-1}\right)$, (17)

compute $p_i = r_{i-1} + \beta\left(p_{i-1} - \omega_{i-1} v_{i-1}\right)$ (18)

form $y_i = M_r^{\delta l} p_i$, $v_i = A y_i$, and $\alpha = \rho_i / \left(r_0', v_i\right)$, (19)

compute $x_i = r_{i-1} - \alpha v_i$, form $z_i = M_r^{\delta l} x_i$ and $t_i = A z_i$, (20)

set $\omega_i = \left(M_r^{\delta l} t_i, M_r^{\delta l} x_i\right) / \left(M_r^{\delta l} t_i, M_r^{\delta l} t_i\right)$, (21)

compute $u_i = u_{i-1} + \alpha y_i + \omega_i z_i$. and $r_i = x_i - \omega_i t_i$. (22)

The computational complexity of the **EPBI-CGSTAB** method is ≈O[(6δl+16)n mults + 6n adds]ν operations, where ν denotes the number of iterations required for convergence to a predetermined tolerance level.

The convergence analysis of similar explicit approximate inverse preconditioning has been presented in Gravvanis [15], [17].

4. NUMERICAL RESULTS

In this final section we present a class of problems which indicate the performance and applicability and effectiveness of the derived explicit preconditioned conjugate gradient-type schemes based on AIM techniques. Let us consider the following model problem :

<u>**Model problem**</u> : The non-zero elements of A, cf. (2), are set to $a \rangle 2.0, b=-1.0$ with μ=ν=ρ=1, i.e. the periodic case. The special tridiagonal linear systems is scaled by 1/b.

The rhs vector of the linear system, cf. (1), was chosen as the product of the coefficient matrix A by the solution vector, with its components equal to unity. The **EPGCGS** and the **EPBI-CGSTAB** method, was terminated when $\|r_i\|_\infty \langle 10^{-6}$.

The computational approach adopted for this class of problem is to produce LU-factors of exactly the same profile as the coefficient matrix A, i.e. r=1, thus requiring O(2n) multiplicative operations. Further we compute the approximate inverse with δl=1, i.e. $M^{\delta l=1}$, involving only the diagonal vector $w_1, w_2, \ldots, w_n$ of the L_r decomposition factor, and the requiried computational work is O(n). Thus the total computational work is O(3n) multiplicative operations.

Numerical results are presented in Table 1 for the **EPGCGS** and the **EPBI-CGSTAB** methods for several values of order n with the retention parameter δl=1, i.e. the $M^{\delta l=1}$ class of approximate inverse.

It should be mentioned that the convergence behavior of the **EPBI-CGSTAB** method is in general better than the **EPGCGS** method in conjunction with the approximate inverse $M^{\delta l=1}$, especially when the value of the diagonal elements $a \rightarrow 2.0$.

Method	EPGCGS using $M^{\delta l=1}$						EPBI-CGSTAB using $M^{\delta l=1}$					
n \ a	2.001	2.01	2.10	2.5	10.0	50.0	2.001	2.01	2.10	2.5	10.0	50.0
64	61	41	22	11	4	3	48	34	22	12	4	2
128	81	65	23	11	4	3	77	52	24	12	4	2
256	145	97	23	11	5	3	127	51	23	12	4	2
512	204	68	24	11	5	3	217	57	21	12	4	2
1024	>500	67	23	11	5	3	262	65	24	12	4	2

Table 1 : The convergence behaviour of the **EPGCGS** and **EPBI-CGSTAB** scheme.

Assuming a PRAM linear array model with n processors is used, the computation of the elements of the class I approximate inverse $M^{\delta l=1}$ can be performed in O(1), i.e. constant time, since the inversion of the diagonal elements, i.e. $w_1, w_2, \ldots, w_n$, of L_r is required. Additionally, in the implementation of the explicit preconditoned conjugate gradient-type schemes the inner product can be performed in O(logn), using the prefix computation model. Thus the proposed iterative method results in a fast explicit preconditioned scheme.

Finally, it should be stated that the proposed explicit preconditoned conjugate gradient type schemes, based on the derived classes of the approximate inverse, are efficient for the solution of special tridiagonal linear systems.

REFERENCES

1. **Akl S.G.** (1997). Parallel Computation models and methods, Prentice-Hall.
2. **Axelsson O.** (1994). Iterative solution methods, CambridgeUniversity Press.
3. **Axelsson O., Kolotilina L.** (1990). Preconditioned Conjugate Gradient methods, Lecture Notes in Mathematics 1457, Springer.
4. **Boisvert R.F. (1991).** Algorithms for special tridiagonal systems, SIAM J. Sci. Stat. Comput. 12, 423-442.
5. **Dongarra J.J., Duff I.S., Sorensen D.C., van der Vorst H.A. (1991).** Solving linear systems on vector and shared memory computers, SIAM, Philadelphia.
6. **Duff I.S., Erisman A.M., Reid J.K. (1986).** Direct methods for sparse matrices, Oxford University Press, London.
7. **Evans D.J., Sutti C. (1988).** Parallel Computing : Methods, Algorithms and Applications, Proceedings of the International Meeting on Parallel Computing, Adam Hilger.

8. **Evans D.J. (1985)** : Iterative methods for sparse systems, In Sparsity and its Applications, Evans D.J., ed., Cambridge University Press.
9. **Evans D.J. (1972).** An algorithm for the solution of certain tridiagonal systems of linear equations, Comput. J. 15, 356-359.
10. **Fisher D., Golub G., Hald O., Leiva C., Widlund O. (1974).** On Fourier-Toeplitz methods for seperable elliptic problems, Math Comp 28, 349-368.
11. **Golub G.H., van Loan C. (1989).** Matrix Computations, The Johns Hopkins University Press.
12. **Gragg B., Harrod W. (1984).** The numerically stable reconstruction of Jacobi matrix from spectral data, Numer. Math. 44, 317-355.
13. **Gravvanis G.A. (1999).** Approximate inverse banded matrix techniques, Engineering Computations 16, No 3, 337-346.
14. **Gravvanis G.A. (1998).** An approximate inverse matrix technique for arrowhead matrices, I. J. Comp. Math. 70, 35-45.
15. **Gravvanis G.A. (1998).** Parallel matrix techniques, Computational Fluid Dynamics 98, eds, K.D. Papailiou, D. Tsahalis, J. Periaux, C. Hirsch, M. Pandolfi, vol. 1, 472-477, Wiley.
16. **Gravvanis G.A. (1997).** On the numerical modelling and solution of non-linear boundary value problems, Proc. of the 10th Inter. Confer. on Numer. Methods in Thermal Problems, eds, R. W. Lewis and J.T. Cross, vol X, 898-909, Pineridge Press.
17. **Gravvanis G.A. (1996).** The rate of convergence of explicit approximate inverse preconditioning, I. J. Comp. Math. 60, 77-89.
18. **Gravvanis G.A. (1995).** Explicit preconditioned methods for solving 3D boundary value problems by approximate inverse finite element matrix techniques, I. J. Comp. Math. 56, 77-93.
19. **Gravvanis G.A., Lipitakis E.A. (1996).** A three dimensional explicit preconditioned solver, Comp. Math. with Appl. 32, 111-131.
20. **Gravvanis G.A., Lipitakis E.A. (1995).** On the numerical modelling and solution of initial/boundary value problems, Proc. of 9th Inter. Confer. on Numer. Meth. in Thermal Problems, eds, R.W. Lewis and P. Durbetaki, Vol. IX, Part 2, 782-793, Pineridge Press.
21. **Lipitakis E.A., Gravvanis G.A. (1995).** Explicit preconditioned iterative methods for solving large unsymmetric finite element systems, Computing 54, 167-183.
22. **Lipitakis E.A., Gravvanis G.A. (1992).** A class of explicit preconditioned conjugate gradient methods for solving large finite element system, I. J. Comp. Math. 44, 189-206.
23. **Malcolm M.A., Palmer J. (1974).** A fast method for solving a class of tridiagonal linear systems, Comm. A.C.M. 17, 14-17.
24. **Ortega J.M. (1988).** Introduction to parallel and vector solution of linear systems, In series : Frontiers of Computer Science, ed. A. L. Rosenberg, Plenum Press, N.Y.
25. **Saad Y. (1996).** Iterative methods for sparse linear systems, PWS Publishing, New York.
26. **van der Vorst H.A. (1992).** BI-CGSTAB : A fast and smoothly converging variant BI-CG for the solution of nonsymmetric linear systems, SIAM J. Sci. Stat. Comp. 13, 631-644.
27. **Widlund O. (1972).** On the use of fast methods for seperable finite difference equations for the solution of general elliptic problems, In Sparse Matrices and Applications, eds, D.J. Rose and R.A. Willoughby, 121-131, Plenum Press.

Proceedings of Dynamic Systems and Applications 3 (2001) 249-256

A Priori Estimates and Solvability of a Generalized Multi-Point Boundary Value Problem.

Chaitan P. Gupta
Department of Mathematics
University of Nevada, Reno
Reno, NV 89557 USA

Abstract

Let $f : [0,1] \times R^2 \longmapsto R$ be a function satisfying Caratheodory's conditions and $e(t) \in L^1[0,1]$. Let $\xi_i, \tau_j \in [0,1]$, a_i, $b_j \in R$, $i = 1, 2, \cdots, m$, $j = 1, 2, \cdots, n$, $0 \leq \xi_1 < \xi_2 < \cdots < \xi_m \leq 1$, $0 \leq \tau_1 < \tau_2 < \cdots < \tau_n \leq 1$ be given. This paper is concerned with the problem of existence of a solution for the generalized multi-point boundary value problems

$$x''(t) = f(t, x(t), x'(t)) + e(t), 0 < t < 1,$$
$$\sum_{i=1}^{m} a_i x(\xi_i) = 0, \ \sum_{j=1}^{n} b_j x'(\tau_j) = 0.$$

The author had studied such problems in [4], [5]. One needs a priori estimates of Poincaré type of the form

$$\| x \|_\infty \leq C \| x' \|_\infty , \ \| x' \|_\infty \leq C \| x'' \|_1 ,$$

for $x(t) \in W^{2,1}(0,1)$ satisfying the boundary conditions $\sum_{i=1}^{m} a_i x(\xi_i) = 0$, $\sum_{j=1}^{n} b_j x'(\tau_j) = 0$, in order to apply Leray Schauder continuation theorem to the problem of existence of a solution of the above boundary value problem. The a priori estimates obtained in this paper are considerably better and general than those in [4], [5]. This, in turn, enables us to obtain better existence theorems for such problems. It is also indicated how the a priori esimates for the above boundary value problem can be used to study the boundary value problem

$$x''(t) = f(t, x(t), x'(t)) + e(t), 0 < t < 1,$$
$$\sum_{i=1}^{m} a_i x(\xi_i) = 0, \ \sum_{j=1}^{n} b_j x(\tau_j) = 0.$$

Keywords and Phrases: generalized multi-point boundary value problem, a priori estimates, Leray Schauder Continuation theorem, Caratheodory's conditions, Arzela-Ascoli Theorem.

AMS(MOS) Subject Classification: 34B10, 34B15, 34G20.

1 Introduction.

Let $f : [0,1] \times R^2 \longmapsto R$ be a function satisfying Caratheodory's conditions and $e : [0,1] \longmapsto R$ be a function in $L^1[0,1]$, $\xi_i, \tau_j \in [0,1]$, a_i, $b_j \in R$, $i = 1, 2, \cdots, m$, $j = 1, 2, \cdots, n$, $0 \leq \xi_1 < \xi_2 < \cdots < \xi_m \leq 1$, $0 \leq \tau_1 < \tau_2 < \cdots < \tau_n \leq 1$ be given. We study the problem of existence of solutions for the generalized multi-point boundary value problem

$$x''(t) = f(t, x(t), x'(t)) + e(t),\ 0 < t < 1,$$
$$\sum_{i=1}^{m} a_i x(\xi_i) = 0,\ \sum_{j=1}^{n} b_j x'(\tau_j) = 0. \tag{1}$$

We also show that the generalized multi-point boundary value problem

$$x''(t) = f(t, x(t), x'(t)) + e(t),\ 0 < t < 1,$$
$$\sum_{i=1}^{m} a_i x(\xi_i) = 0,\ \sum_{j=1}^{n} b_j x(\tau_j) = 0. \tag{2}$$

can be reduced to a generalized multi-point boundary value problem of the type (1). Accordingly, an existence theorem for (2) can be obtained from an existence theorem for (1). We obtain new and natural a priori estimates for the boundary value problem (1) in section 2.

The author had earlier studied several generalized multi-point boundary value problems in [4] and [5]. It can be easily seen that each one of the generalized multi-point boundary value problems studied in [4] and [5] can be easily reduced to either (1) or (2). One can then use the improved a priori estimates obtained here for the generalized multi-point boundary value problems in [4] and [5].

The study of multi-point boundary value problems for linear second order ordinary differential equations was initiated by V. A. Il'in and E. A. Moiseev in [7], [8] motivated by the work of Bitsadze and Samarski on non-local linear elliptic boundary problems, [1], [2], [3].

2 A PRIORI ESTIMATES.

Proposition 1 :- *Let $f : [0,1] \times R^2 \longmapsto R$ be a function satisfying Caratheodory's conditions and $e : [0,1] \longmapsto R$ be a function in $L^1[0,1]$, $\xi_i, \tau_j \in [0,1]$, a_i, $b_j \in R$, $i = 1, 2, \cdots, m$, $j = 1, 2, \cdots, n$, $0 \le \xi_1 < \xi_2 < \cdots < \xi_m \le 1$, $0 \le \tau_1 < \tau_2 < \cdots < \tau_n \le 1$ be given. Then the generalized multi-point boundary value problem (1) is a non-resonant problem if*

$$(\sum_{i=1}^{m} a_i)(\sum_{j=1}^{n} b_j) \neq 0.$$

The proof of this proposition is straightforward and is left to the reader to save space.

Theorem 2 :- *Let $\xi_i, \tau_j \in [0,1]$, a_i, $b_j \in R$, $i = 1, 2, \cdots, m$, $j = 1, 2, \cdots, n$, $0 \le \xi_1 < \xi_2 < \cdots < \xi_m \le 1$, $0 \le \tau_1 < \tau_2 < \cdots < \tau_n \le 1$ be given such that $(\sum_{i=1}^m a_i)(\sum_{j=1}^n b_j) \neq 0$. Let $x(t) \in W^{2,1}(0,1)$ be such that $\sum_{i=1}^m a_i x(\xi_i) = 0$, $\sum_{j=1}^n b_j x'(\tau_j) = 0$. Then there exist μ_1, $\mu_2 \in [0,1)$ such that*

$$\| x \|_\infty \le \frac{1}{1-\mu_1} \| x' \|_\infty \,, \quad \| x' \|_\infty \le \frac{1}{1-\mu_2} \| x'' \|_1 \,. \tag{3}$$

Indeed, $\mu_1 = \min\{\frac{\sum_{i=1}^m (a_i)_+}{\sum_{i=1}^m (a_i)_-}, \frac{\sum_{i=1}^m (a_i)_-}{\sum_{i=1}^m (a_i)_+}\}$ *and* $\mu_2 = \min\{\frac{\sum_{j=1}^n (b_j)_+}{\sum_{j=1}^n (b_j)_-}, \frac{\sum_{j=1}^n (b_j)_-}{\sum_{j=1}^n (b_j)_+}\}$.

Proof:- Let for $a \in R$, $a_+ = \max\{a, 0\}$, $a_- = \max\{-a, 0\}$ so that $a = a_+ - a_-$ and $| a | = a_+ + a_-$. Let, now, $\sigma_+ = \sum_{i=1}^m (a_i)_+$ and $\sigma_- = \sum_{i=1}^m (a_i)_-$. Now, we see from our assumptions that $\sigma_+ \neq \sigma_-$ and $\sum_{i=1}^m (a_i)_+ x(\xi_i) = \sum_{i=1}^m (a_i)_- x(\xi_i)$ so that there exist λ_1, $\lambda_2 \in [0,1]$ such that $\sigma_+ x(\lambda_1) = \sigma_- x(\lambda_2)$. If, now, either σ_+ or σ_- is zero than one of $x(\lambda_2)$ or $x(\lambda_1)$ has to be zero and it follows immediately that $\| x \|_\infty \le \| x' \|_\infty$. On the other hand, let both σ_+ and σ_- be non-zero. Then either $x(\lambda_1) = x(\lambda_2) = 0$, in which case again we immediately get $\| x \|_\infty \le \| x' \|_\infty$, or $x(\lambda_1) \neq 0$, $x(\lambda_2) \neq 0$ with $x(\lambda_1) \neq x(\lambda_2)$, since $\sigma_+ x(\lambda_1) = \sigma_- x(\lambda_2)$ and $\sigma_+ \neq \sigma_-$. In this case, it follows immediately from the equations $x(t) = x(\lambda_1) + \int_{\lambda_1}^t x'(s)ds = \frac{\sigma_-}{\sigma_+} x(\lambda_2) + \int_{\lambda_1}^t x'(s)ds$, $x(t) = x(\lambda_2) + \int_{\lambda_2}^t x'(s)ds = \frac{\sigma_+}{\sigma_-} x(\lambda_1) + \int_{\lambda_2}^t x'(s)ds$ that $\| x \|_\infty \le \frac{1}{1-\mu_1} \| x' \|_\infty$ with $\mu_1 = \min\{\frac{\sigma_+}{\sigma_-}, \frac{\sigma_-}{\sigma_+}\} < 1$. Thus we see that there exists a $\mu_1 \in [0,1)$ such that $\| x \|_\infty \le \frac{1}{1-\mu_1} \| x' \|_\infty$ with $\mu_1 = \min\{\frac{\sum_{i=1}^m (a_i)_+}{\sum_{i=1}^m (a_i)_-}, \frac{\sum_{i=1}^m (a_i)_-}{\sum_{i=1}^m (a_i)_+}\}$. Similarly, we see that there exists a $\mu_2 \in [0,1)$ such that $\| x' \|_\infty \le \frac{1}{1-\mu_2} \| x'' \|_1$ with $\mu_2 = \min\{\frac{\sum_{j=1}^n (b_j)_+}{\sum_{j=1}^n (b_j)_-}, \frac{\sum_{j=1}^n (b_j)_-}{\sum_{j=1}^n (b_j)_+}\}$. This completes the proof of the theorem.//

Remark 3 :- *For an $x(t) \in W^{2,1}(0,1)$ as in Theorem 2 we have the estimate*

$$\| x \|_\infty \le \frac{1}{(1-\mu_1)(1-\mu_2)} \| x'' \|_1 \,.$$

The following theorem improves this estimate if we use Taylor's threorem with remainder after the second term to estimate $\| x \|_\infty$.

Theorem 4 :- *Let $\xi_i, \tau_j \in [0,1]$, $a_i, b_j \in R$, $i = 1, 2, \cdots, m$, $j = 1, 2, \cdots, n$, $0 \le \xi_1 < \xi_2 < \cdots < \xi_m \le 1$, $0 \le \tau_1 < \tau_2 < \cdots < \tau_n \le 1$ be given such that $(\sum_{i=1}^m a_i)(\sum_{j=1}^n b_j) \neq 0$. Let $x(t) \in W^{2,1}(0,1)$ be such that $\sum_{i=1}^m a_i x(\xi_i) = 0$, $\sum_{j=1}^n b_j x'(\tau_j) = 0$. Then*

$$\| x \|_\infty \le K \| x'' \|_1 , \text{ where } \tag{4}$$

$$K = \min\{M, \frac{1}{(1-\mu_1)(1-\mu_2)}\} \text{ with}$$

$$M = \max\{M_1, M_2\} \text{ where}$$

$$\begin{aligned} M_1 &= \max\{\sum_{i=1}^m (\frac{a_i}{\sum_{i=1}^m a_i})_+ \xi_i, \sum_{i=1, i\neq j}^m (\frac{a_i}{\sum_{i=1}^m a_i})_+ \mid \xi_i - \xi_j \mid, \\ &\sum_{i=1}^m (\frac{a_i}{\sum_{i=1}^m a_i})_+ \mid \xi_i - 1 \mid, j = 1, \cdots, m\} \end{aligned}$$

and

$$\begin{aligned} M_2 &= \max\{\sum_{i=1}^m (\frac{1}{1-\mu_2}(\frac{a_i}{\sum_{i=1}^m a_i})_+ \xi_i + \frac{\mid a_i \mid}{\mid \sum_{i=1}^m a_i \mid} \xi_i), \\ &\sum_{i=1}^m (\frac{1}{1-\mu_2}(\frac{a_i}{\sum_{i=1}^m a_i})_+ (1-\xi_i) + \frac{\mid a_i \mid}{\mid \sum_{i=1}^m a_i \mid}(1-\xi_i))\}, \end{aligned}$$

where μ_1, μ_2 are as given in theorem 2.

Proof:- Let $c \in [0,1]$ be such that $\| x \|_\infty = \mid x(c) \mid$. We may assume, without any loss in generality, that $x(c) > 0$. Let us first suppose that $c \in (0,1)$ so that $x'(c) = 0$. Now, for $i = 1, \cdots, m$ we have

$$x(\xi_i) = x(c) + r_i, \text{ where } r_i = \int_c^{\xi_i} (\xi_i - s) x''(s) ds \le 0.$$

Now, by assumption, we have $\sum_{i=1}^m a_i \neq 0$. It, next, follows from the above equation and the assumption that $\sum_{i=1}^m a_i x(\xi_i) = 0$ that

$$(\sum_{i=1}^m a_i) x(c) = -\sum_{i=1}^m a_i r_i.$$

Accordingly, we get that

$$\begin{aligned} x(c) &= -\sum_{i=1}^m \frac{a_i}{\sum_{i=1}^m a_i} r_i = -\sum_{i=1}^m (\frac{a_i}{\sum_{i=1}^m a_i})_+ r_i + \sum_{i=1}^m (\frac{a_i}{\sum_{i=1}^m a_i})_- r_i \\ &\le -\sum_{i=1}^m (\frac{a_i}{\sum_{i=1}^m a_i})_+ r_i, \text{ since } r_i \le 0 \text{ for every } i = 1, \cdots, m. \end{aligned}$$

It thus follows that

$$\begin{aligned}
\| x \|_\infty = x(c) &\le \sum_{i=1}^{m}\Big(\frac{a_i}{\sum_{i=1}^m a_i}\Big)_+ \Big| \int_c^{\xi_i} (\xi_i - s)x''(s)ds \Big| \\
&\le \sum_{i=1}^{m}\Big(\frac{a_i}{\sum_{i=1}^m a_i}\Big)_+ \Big| \int_c^{\xi_i} | \xi_i - s || x''(s) | ds \Big| \\
&\le \sum_{i=1}^{m}\Big(\frac{a_i}{\sum_{i=1}^m a_i}\Big)_+ | \xi_i - c ||| x'' \|_1 \\
&\le \max\{\sum_{i=1}^{m}\Big(\frac{a_i}{\sum_{i=1}^m a_i}\Big)_+ | \xi_i - t |: \ t \in [0,1]\}\cdot \| x'' \|_1 .
\end{aligned}$$

Clearly,

$$\begin{aligned}
&\max\{\sum_{i=1}^{m}\Big(\frac{a_i}{\sum_{i=1}^m a_i}\Big)_+ | \xi_i - t |: t \in [0,1]\} = M_1 \\
&= \max\{\sum_{i=1}^{m}\Big(\frac{a_i}{\sum_{i=1}^m a_i}\Big)_+ \xi_i, \sum_{i=1, i\neq j}^{m}\Big(\frac{a_i}{\sum_{i=1}^m a_i}\Big)_+ | \xi_i - \xi_j |, \\
&\qquad \sum_{i=1}^{m}\Big(\frac{a_i}{\sum_{i=1}^m a_i}\Big)_+ | \xi_i - 1 |, j = 1,\cdots,m\}.
\end{aligned}$$

Thus, if $c \in (0,1)$ we have

$$\| x \|_\infty \le M_1 \| x'' \|_1 \tag{5}$$

We, next, consider the case when $c = 0$, that is $\| x \|_\infty = x(0)$. In this case we have $x'(0) \le 0$ and 0

$$x(\xi_i) = x(0) + x'(0)\xi_i + \int_0^{\xi_i} (\xi_i - s)x''(s)ds, \ i = 1,\cdots,m.$$

It, then, follows that

$$\begin{aligned}
x(0) &= -\sum_{i=1}^{m}\Big(\frac{a_i\xi_i}{\sum_{i=1}^m a_i}\Big)x'(0) - \sum_{i=1}^{m}\Big(\frac{a_i}{\sum_{i=1}^m a_i}\Big)\int_0^{\xi_i} (\xi_i - s)x''(s)ds \\
&\le \sum_{i=1}^{m}\Big(\frac{a_i\xi_i}{\sum_{i=1}^m a_i}\Big)_+ | x'(0) | + \sum_{i=1}^{m} \Big| \frac{a_i}{\sum_{i=1}^m a_i} \Big| \int_0^{\xi_i} \xi_i | x''(s) | ds \\
&\le \sum_{i=1}^{m}\Big(\frac{a_i\xi_i}{\sum_{i=1}^m a_i}\Big)_+ \| x' \|_\infty + \sum_{i=1}^{m} \Big| \frac{a_i\xi_i}{\sum_{i=1}^m a_i} \Big| \| x'' \|_1 .
\end{aligned}$$

Using, now, theorem 2 we get from above that

$$\| x \|_\infty \le \Big(\sum_{i=1}^{m} \frac{1}{1-\mu_2}\Big(\frac{a_i\xi_i}{\sum_{i=1}^m a_i}\Big)_+ + \Big| \frac{a_i\xi_i}{\sum_{i=1}^m a_i} \Big|\Big) \| x'' \|_1 .$$

Similarly, we get when $c = 1$ that

$$\| x \|_\infty \le \Big(\sum_{i=1}^{m} \frac{1}{1-\mu_2}\Big(\frac{a_i(1-\xi_i)}{\sum_{i=1}^m a_i}\Big)_+ + \Big| \frac{a_i(1-\xi_i)}{\sum_{i=1}^m a_i} \Big|\Big) \| x'' \|_1 .$$

We thus see that

$$\| x \|_\infty \leq M_2 \| x'' \|_1 , \tag{6}$$

where

$$\begin{aligned} M_2 &= \max\{\frac{1}{1-\mu_2}\sum_{i=1}^{m}((\frac{a_i}{\sum_{i=1}^m a_i})_+\xi_i + \frac{| a_i |}{| \sum_{i=1}^m a_i |}\xi_i), \\ &\frac{1}{1-\mu_2}\sum_{i=1}^{m}((\frac{a_i}{\sum_{i=1}^m a_i})_+(1-\xi_i) + \frac{| a_i |}{| \sum_{i=1}^m a_i |}(1-\xi_i))\}. \end{aligned}$$

It, now, follows from estimates (5) and (6) that

$$\| x \|_\infty \leq M \| x'' \|_1 , \text{ where } M = \max\{M_1, M_2\}.$$

Also we see from theorem 2 that

$$\| x \|_\infty \leq \frac{1}{(1-\mu_1)(1-\mu_2)} \| x'' \|_1 .$$

Estimate (4) is now immediate. This completes the proof of the theorem.//

The following proposition shows how one can reduce a generalized boundary value problem of type (2) to a generalized boundary value problem of type (1).

Proposition 5 *:- Let $\xi_i, \tau_j \in [0,1]$, a_i, $b_j \in R$, $i = 1, 2, \cdots, m$, $j = 1, 2, \cdots, n$, $0 \leq \xi_1 < \xi_2 < \cdots < \xi_m \leq 1$, $0 \leq \tau_1 < \tau_2 < \cdots < \tau_n \leq 1$ be given. Let $x(t) \in W^{2,1}(0,1)$ be such that $\sum_{i=1}^m a_i x(\xi_i) = 0$, $\sum_{j=1}^n b_j x(\tau_j) = 0$. Then there exist $\sigma_{ij} \in [0,1]$, $i = 1, 2, \cdots, m$, $j = 1, 2, \cdots, n$ such that $\sum_{i=1}^m \sum_{j=1}^n a_i b_j(\xi_i - \tau_j)x'(\sigma_{ij}) = 0$.*

Proof:- Applying mean-value theorem we see that for every i, j there exist a $\sigma_{ij} \in [0,1]$ such that

$$x(\xi_i) - x(\tau_j) = (\xi_i - \tau_j)x'(\sigma_{ij}).$$

It, now, follows easily from our assumptions that

$$\sum_{i=1}^{m}\sum_{j=1}^{n} a_i b_j(\xi_i - \tau_j)x'(\sigma_{ij}) = 0.$$

This completes the proof of the proposition.//

Remark 6 *:- It is easy to see that the generalized boundary value problem (2) is a non-resonant problem if $(\sum_{i=1}^m a_i\xi_i)(\sum_{j=1}^n b_j) - (\sum_{i=1}^m a_i)(\sum_{j=1}^n b_j\tau_j) \neq 0$. Note that this condition implies that at least one of $\sum_{i=1}^m a_i$ or $\sum_{j=1}^n b_j$ is different from zero. In view of this, it follows from the above proposition that the generalized boundary value problem of type (1) to which the problem (2) reduces to is also a non-resonant problem.*

3 MAIN RESULTS.

Theorem 7 *:- Let $f : [0,1] \times R^2 \longmapsto R$ be a function satisfying Caratheodory's conditions. Assume that there exist functions $p(t)$, $q(t)$, $r(t)$ in $L^1(0,1)$ such that*

$$|f(t,x_1,x_2)| \leq p(t)\,|x_1| + q(t)\,|x_2| + r(t) \tag{7}$$

for a.e. $t \in [0,1]$ and all $(x_1,x_2) \in R^2$. Let $\xi_i, \tau_j \in [0,1]$, a_i, $b_j \in R$, $i = 1, 2, \cdots, m$, $j = 1, 2, \cdots, n$, $0 \leq \xi_1 < \xi_2 < \cdots < \xi_m \leq 1$, $0 \leq \tau_1 < \tau_2 < \cdots < \tau_n \leq 1$, with $(\sum_{i=1}^{m} a_i)(\sum_{j=1}^{n} b_j) \neq 0$ be given. Then the multi-point boundary value problem (1) has at least one solution in $C^1[0,1]$ provided

$$(K\,\|p\|_1 + \frac{1}{1-\mu_2}\,\|q\|_1) < 1, \tag{8}$$

where K and μ_2 are as in Theorem 4.

Proof:- Let X denote the Banach space $C^1[0,1]$ and Y denote the Banach space $L^1(0,1)$ with their usual norms. It is easy to define a linear mapping $L : D(L) \subset X \longmapsto Y$ and a nonlinear mapping $N : X \longmapsto Y$ such that that $x \in C^1[0,1]$ is a solution of the boundary value problem (1) if and only if x is a solution to the operator equation

$$Lx = Nx + e.$$

Next, it is easy to define a linear mapping $M : Y \to X$ is such that for $y \in Y$, $My \in D(L)$ and $LMy = y$; and for $u \in D(L)$, $MLu = u$. Furthermore, it is easy to show using the Arzela-Ascoli Theorem that MN maps a bounded subset of X into a relatively compact subset of X so that $MN : X \longmapsto X$ is a compact mapping and that the operator equation $Lx = Nx + e$ is equivalent to the equation

$$x = MNx + Ke.$$

We apply the Leray-Schauder Continuation theorem (see, e.g. [6], Corollary IV.7) to obtain the existence of a solution for $x = MNx + Me$ or equivalently to the boundary value problem (1).

To do this, it suffices to verify that the set of all possible solutions of the family of equations

$$x''(t) = \lambda f(t, x(t), x'(t)) + \lambda e(t),\ 0 < t < 1,$$
$$\sum_{i=1}^{m} a_i x(\xi_i) = 0,\ \sum_{j=1}^{n} b_j x'(\tau_j) = 0 \tag{9}$$

is, a priori, bounded in $C^1[0,1]$ by a constant independent of $\lambda \in [0,1]$. This is easy to show using the fact that if $x \in W^{2,1}(0,1)$, with $\sum_{i=1}^m a_i x(\xi_i) = 0$, $\sum_{j=1}^n b_j x'(\tau_j) = 0$ we have the inequalities $\| x \|_\infty \leq K \| x'' \|_1$ and $\| x' \|_\infty \leq \frac{1}{1-\mu_2} \| x'' \|_1$, where K and μ_2 are as in Theorem 4, and our assumption (8).

This completes the proof of the theorem.//

References

[1] A. V. Bitsadze, *On the theory of nonlocal boundary value problems*, Soviet Math. Dokl. 30(1984), No.1, p. 8-10.

[2] A. V. Bitsadze, *On a class of conditionally solvable nonlocal boundary value problems for harmonic functions*, Soviet Math. Dokl. 31(1985), No. 1, p. 91-94.

[3] A. V. Bicadze and A. A. Samarskiĭ, *On some simple generalizations of linear elliptic boundary problems*, Soviet Math. Dokl. 10(1969), No. 2, p. 398-400.

[4] C. P. Gupta, *A Generalized Multi-Point Boundary Value Problem for Second Order Ordinary Differential Equations*, Applied Mathematics and Computation, Vol. 89, (1998), pp 133-146.

[5] C. P. Gupta, *Solvability of a Generalized Multi-Point Boundary Value Problem of Mixed type for Second Order Ordinary Differential Equations*, Proceedings of Dynamic Systems and Applications, Vol. 2, (1996), pp. 215-222.

[6] J. Mawhin, *Topological degree methods in nonlinear boundary value problems*, in "NSF-CBMS Regional Conference Series in Math." No. 40, Amer. Math. Soc., Providence, RI, 1979.

[7] V.A. Il'in, and E. I. Moiseev, *Nonlocal boundary value problem of the first kind for a Sturm-Liouville operator in its differential and finite difference aspects*, Differential Equations, 23(1987), No. 7, p. 803-810.

[8] V.A. Il'in, and E. I. Moiseev, *Nonlocal boundary value problem of the second kind for a Sturm-Liouville operator*, Differential Equations, 23(1987), No. 8, p. 979-987.

Proceedings of Dynamic Systems and Applications 3 (2001) 257-264

ON THE HEATING MECHANISM OF THE SOLAR ATMOSPHERE BY THE RESONANCE ABSORPTION OF THE ALFVEN WAVES IN AN ISOTHERMAL ATMOSPHERE(I)

BY

Alkahby Hadi[1], Jalbout Fouad[2] and Mohammad Mahrous[3]

[1]Department of Mathematics, Dillard University, New Orleans, LA 70122 USA
[2] Department of Physics, Dillard University, New Orleans, LA 70122 USA
[3] Department of Mathematics, University of New Orleans, New Orleans, LA 70148

ABSTRACT: In this paper, we will consider the heating process of the solar atmosphere by the resonance absorption of the Alfven waves in an isothermal atmosphere. We will consider three cases. In the first case we will investigate an ideal atmosphere permeated by a uniform horizontal magnetic field, in the second case the atmosphere is ideal and permeated by a non-uniform vertical magnetic field with Larmor effect, and in the third case the atmosphere is viscous and permeated by a uniform horizontal magnetic field. It is shown that if the viscosity dominates the oscillatory process in ***a high (low) - β plasma***, it creates an absorbing and a reflecting barrier and the heating process is ***acoustic (magnetoacoustic)***. When the magnetic field dominates the oscillatory motion it produces a reflecting but not an absorbing layer because of the dissipationless nature of the magnetic field. As a result, the heating process is ***magnetoacoustic***. An estimate for the values of the kinetic and magnetic energies of the wave are obtained. Equations for the resonance and formula for the magnitude of the reflection coefficients are derived. The conclusions are presented in connection with heating process of the solar atmosphere.

Keywords : Magnetoacoustic waves, Newtonian Cooling Coefficient, Wave Propagation, Reflection Coefficient, Resonance Absorption

1991 AMS SUBJECT CLASSIFICATION CODES : 76N, 76Q

1. INTRODUCTION

The theory of the formation of the solar chromosphere and corona has as its main challenging goal the specification of the nature of the solar heating mechanism. For the specification of the heating process of the solar atmosphere, many models have been formulated and analyzed Priest [11]. In addition, the propagation of atmospheric waves, both in isothermal and in non-isothermal atmospheres, has been investigated extensively in recent years in connection with heating process of the solar atmosphere. Moreover, the discovery of hydromagnetic waves was followed by an extensive study of magnetoacoustic waves in an isothermal atmosphere. Much of the motivation of these studies comes from their relevance to phenomena in compressible ionized fluids, such as solar, stellar, earth's atmospheres and to certain phenomena in ocean dynamics (see Alkahby[2-6], Yanowitch [12] for references]). Early theories of coronal heating were essentially based on the dissipation mechanisms of acoustic waves or shock waves. Recent theories involve magnetic energy dissipation as the source of thermal energy. One of these mechanisms is the resonance absorption of Alfven waves, which was suggested as a process for the heating of fusion plasma nearly 35 years ago Alkahby [3]. Ionson [8] and

Hollweg [10] both concluded that the resonance absorption could explain the observed heating in the solar corona. Davila [9] calculated the heating rate at the resonance layer to determine the energy dissipation in the plasma.

The aim of this work is to investigate the heating of the solar atmosphere by the resonance absorption of the Alfven waves and to calculate the values of the kinetic and magnetic energies of an upward and a downward propagating magnetoacoustic wave in an isothermal atmosphere. We will consider three cases. In the first case we investigate an ideal atmosphere permeated by a uniform horizontal magnetic field, in the second case the atmosphere is ideal but permeated by a non-uniform magnetic field with Larmor effect, and in the third case the atmosphere is viscous permeated by a uniform horizontal magnetic field. It is shown (in all the three cases) that when the magnetic dominates the motion it creates a non absorbing reflecting layer because of dissipationless nature of the magnetic field. As a result, the heating mechanism is ***magnetoacoustic*** and resonance will occur for infinitely many values of the magnetic field and of the frequency of the wave if the frequency of the wave is matched with the strength of the magnetic field. At the resonance frequencies, the kinetic and magnetic energies increase to very large values and this may account for the heating process of the solar atmosphere. On the contrary, the dominance of the viscosity, in the third case, produces an absorbing and a reflecting layer. Consequently, the heating process is ***acoustic***. Finally, equations for resonance and formula for the magnitude of the reflection coefficient are derived and analyzed.

2. MATHEMATICAL FORMULATION OF THE PROBLEMS

The hydromagnetic equations of motion for pulsating stars consist of the momentum equation, the continuity equation, induction equation, pressure equation, energy equations, Maxwell equation, which can be written as follows:

$$\rho[\tfrac{\partial \boldsymbol{V}}{\partial t}+(\boldsymbol{V}\cdot\nabla\boldsymbol{V})]+\nabla P=\rho\ \mathbf{g}+\tfrac{4}{3}\mu\nabla^2\boldsymbol{V}+\tfrac{\eta}{4\pi}[\nabla x\boldsymbol{H}x\boldsymbol{H}], \qquad (1)$$

$$\rho_t+\nabla(\rho\cdot\boldsymbol{V})=0, \qquad (2)$$

$$\frac{\partial \boldsymbol{H}}{\partial t}+\nabla\times(\boldsymbol{H}\times\boldsymbol{V})+\frac{c}{4\pi eN}\nabla\times[(\nabla\times\boldsymbol{H})\times\boldsymbol{H}]=0, \qquad (3)$$

$$P=R\rho T, \qquad (4)$$

$$\nabla\times\boldsymbol{H}=\tfrac{4\pi}{c}\boldsymbol{J}, \qquad (5)$$

$$\boldsymbol{E}=\tfrac{\mu}{ceN}\boldsymbol{J}\times\boldsymbol{H}=\frac{\mu}{4\pi eN}[(\nabla\times H)\times H], \qquad (6)$$

$$\frac{\partial \boldsymbol{H}}{\partial t}+\nabla\times(H\times\boldsymbol{V})+\tfrac{c}{\mu}\nabla\times\boldsymbol{E}=\mathbf{0}. \qquad (7)$$

We have to indicate that equation (3) is derived from equation (7) and equation (6). In the above equations (1-7), ρ is the mass density, $\boldsymbol{V}$ is the fluid vertical velocity, P is the pressure, $\boldsymbol{g}$ is the gravitational acceleration, c is the speed of light in vacuum, e is the electron charge, N is the number of ions per unit volume, $\boldsymbol{E}$ is the electric field, $\boldsymbol{J}$ is the current, μ is the permeability and η is the dynamic viscosity coefficient, $\boldsymbol{H}$ is the magnetic field strength, R is the gas constant. The equations of motion form a system of nonlinear partial differential equations which, in most cases, cannot be solved. For small-amplitude oscillations the dependent variables can be written as the sum of a mean value and a small perturbation. The equations are then simplified to a linear system by neglecting all products of perturbation terms. Let P, ρ, V, and $\boldsymbol{b}$ be the perturbations in the pressure, density, vertical velocity, and the magnetic field strength and P_0, ρ_0, T_0, $\boldsymbol{B}_0$ be the equilibrium quantities where $\boldsymbol{H} = \boldsymbol{B}_0(z) + \boldsymbol{b}(t, z)$. Also we restrict our investigation to an isothermal atmosphere which has an infinite electrical conductivity. In addition, we investigate small oscillations about equilibrium which depend only on the time t and on the vertical coordinate $z \geq 0$. As a result of the above restriction, equilibrium pressure, temperature and density satisfy the gas law $P_0 = R\rho_0 T_0$ and the hydrostatic equation $P_0' + g\rho_0 = 0$. Consequently, the pressure and density are given by

$$P_0(z)/P_0(0) = \rho_0(z)/\rho_0(0) = \exp(-z/H_{ds}), \tag{8}$$

where H_{ds} is the density scale height and defined by $H_{ds} = RT_o/g$. The subscripts z and t denote the differentiation of the independent variables with respect to z and t respectively. We consider solutions which are harmonic in time i.e. $V(z,t) = W(z)\exp(-i\sigma t)$,where σ denotes the frequency of the wave. It is more convenient to rewrite the equation of motion in dimensionless form: $z^* = z/H$, $\sigma_a = c_s/2H_{ds}$ is the adiabatic cut off frequency, where $c_s^2 = \gamma RT_0 = \gamma gH_{ds}$ is the adiabatic sound speed, $W^* = W/c_s$, $t^* = t\sigma_a$, $\eta^* = 2\eta/3\rho_0 cH_{ds}$, $\sigma^* = \sigma/\sigma_a$, ,$a_1 = a_A^2/c^2$, where $a_A = \mathrm{B}_0/\sqrt{4\pi\rho_0(0)}$ is the Alfven speed, $a = a_1 + i\sigma^*\eta^*$. The star can be omitted, since all variables are written in dimensionless form from now on.

Boundary Condition: To obtain a unique solution, physically relevant conditions must be imposed. When $\boldsymbol{H} = \boldsymbol{0}$ and $\eta = 0$, there is no justification for any physical mechanism to be used to determine a unique solution. In this case, the only boundary condition is the radiation condition which will ensure a unique solution. When $\eta = 0$ and $\boldsymbol{H} \neq \boldsymbol{0}$, the acoustic waves are only influenced by the effect of the magnetic field. As a result, there is no dissipative mechanism and the only condition which will be used to ensure a unique solution is the magnetic energy condition. This condition can be expressed mathematically in the following form :

$$\int_0^\infty |\nabla \boldsymbol{H}|^2 d\tau < \infty. \tag{9}$$

When $\eta \neq 0$, the dissipative mechanism exist because of the effect of the viscosity. As a result, a unique solution is obtained from the requirement that the average (per period) rate of energy dissipation in a column of fluid should be finite. Since the dissipation function depends on the square of the velocity gradients, this implies

$$\mu \int_0^\infty |W_z|^2 dz < \infty. \tag{10}$$

The boundary condition at the ground can always be made by $W(0) = 1$. Moreover, for vertically spinning waves (section 4) we assume that $G(0,\sigma) = 1$ by suitable normalizing $G(z,\sigma)$, where $G(z,\sigma)$, is the magnetic field perturbation spectrum for a wave of frequency σ and at altitude z. It will be seen that the upper boundary conditions in connection with boundary conditions at z = 0, will determine a unique solution for all the three cases which will be consider in this paper.

3. THE RESONANCE PHENOMENA IN AN IDEAL ATMOSPHERE PERMEATED BY A UNIFORM HORIZONTAL MAGNETIC FIELD

In this section, we will investigate the resonance occurrence in an isothermal atmosphere, derive an equation for the resonance and estimate the values of the kinetic and magnetic energies. The differential equation for this case can be derived from equation (1- 4), by setting $\eta = 0$, and can be written in the following form

$$\rho_0(z)V_{zz}(t,z) - 4(\rho_0(z)V_z(t,z))_z - a_1 V_{zz}(t,z) = 0. \tag{11}$$

We will consider solutions which are harmonic in time t, i.e. $V(z,t) = W(z)\exp(-i\sigma t)$, then $W(z)$ satisfies the following differential equation

$$4(1 + a_1 \exp(z))D^2 W(z) - 4DW(z) + \sigma^2 W(z) = 0, \tag{12}$$

where $D = d/dz$. Introducing a new dimensionless variable ξ defined by $\xi = -\exp(-z)/a_1$ which transforms the differential equation (12) to the following differential equation

$$[\xi(1-\xi)D^2 + (1-2\xi)D - \sigma^2/4]W(\xi) = 0, \tag{13}$$

where $D = d/d\xi$ and $\arg(\xi) = 0$. Comparing equation (13) with the standard form of the hypergeometric differential equation

$$[\xi(1-\xi)D^2 + (c-(a+b+1)\xi)D - ab]W(\xi) = 0. \tag{14}$$

we find that $c = 1,\ a + b = 1\ and\ ab = \sigma^2/4$. Solving for the parameters a and b we obtain $a = (1 + \sqrt{\sigma^2 - 1})/2$ and $b = (1 - \sqrt{\sigma^2 - 1})/2$. For $|\xi| < 1$, equation (13) has two linearly independent solution of the form [11, 15.5.16, 15.5.17]

$$W_1(\xi) = F(a,b,c,\xi), \tag{15}$$

$$W_2(\xi) = W_1(\xi)\ln\xi + \sum_{n=1}^{\infty} \frac{(a)_n(b)_n}{(n!)^2}\xi^n[\psi(a+n) - \psi(a) + \psi(b+n) - \psi(a) - 2\psi(n+1) + \psi(n)], \tag{16}$$

The solution $W_2(\xi)$ will be eliminated by the magnetic energy condition because it increases like a linear function of z. The final solution of the differential equation (13) is a multiple of $W_1(\xi)$,

$$W(\xi) = CW_1(\xi) = CF(a,b,c,\xi), \tag{17}$$

where C is constant and will be determined later. For $|\xi| > 1$, and $|\arg(-\xi)| < \pi$, the analytic continuation of the solution of equation (13) defined by equation (17) can be written like

$$W(\chi) = C[\frac{\Gamma(c)\Gamma(b-a)}{\Gamma(b)\Gamma(c-a)}(-\xi)^{-a}F(a, 1-c+a, 1-b+a, \xi^{-1}) + \frac{\Gamma(c)\Gamma(a-b)}{\Gamma(a)\Gamma(c-b)}(-\xi)^{-b}F(b, 1-c+b, 1-a+b, \xi^{-1})], \tag{18}$$

Retaining the most significant terms of equation (18) and reintroducing the variable z via ξ, the solution of the differential equation (13) can be written in the following form

$$W(z) = C.\{\exp[1/2 + ik]z + R\exp[1/2 - ik]z\} \tag{19}$$

where $k = (\sqrt{\sigma^2 - 1})/2$ denotes the wave number and R denotes the reflection coefficient which is defined by,

$$R = [\frac{\Gamma(a-b)\Gamma^2(b)}{\Gamma^2(a)\Gamma(b-a)}]\exp(-2i\,k\,\delta) = \exp(i\theta - 2i\,k\,\delta) = \exp(i\,\phi)\ , \tag{20}$$

and $\delta = \ln a_1, \theta = 2\arg\Gamma(a-b) + 4\arg\Gamma(b)$ and $\phi = \arg(R)$ It is clear that the magnitude of the reflection coefficient is one. Also the magnitude of the reflection coefficient can be determined from the kinetic energy, which can be written in the following form

$$(K.E.) = \rho|\,W(z)|^2 = \frac{1 + R^2 + 2|R|\cos[2k(z+\delta) - \theta]}{2|1+R|^2}\ . \tag{21}$$

Moreover the maximum and minimum values of the kinetic energy can be written like

$$M = \max(\rho|W|^2 = [\frac{1+|R|}{|1+R|}]^2, \qquad m = \min(\rho|W|^2) = [\frac{1-|R|}{|1+R|}]^2, \tag{22}$$

where the location of these quantities are at

$$z_M = \frac{\theta}{2k} \pm \frac{n\pi}{k} - \delta, \quad z_m = \frac{\theta}{2k} \pm \frac{(n+1/2)\pi}{k} - \delta\ , \ n = \pm 1, \pm 2, \pm 3.......... . \tag{23}$$

Using the induction equation, the same formulae for the magnetic energy of the wave and the locations of the maximum and minimum quantities can be derived in an isothermal atmosphere. Discussion of the above results will presented in section 6.

4. THE RESONANCE PHENOMENA IN AN IDEAL ATMOSPHERE PERMEATED BY A NON-UNIFORM VERTICAL MAGNETIC FIELD WITH LARMOR EFFECT

The differential equation for this case can be obtained from equations (1-4) using equations (5-7) and setting $\eta = 0$. The magnetic field $\boldsymbol{B}(z)$ is anti parallel to $\boldsymbol{g}$ because it is vertical. As a result, we obtain $\boldsymbol{B}\times\boldsymbol{g} = \boldsymbol{0}$. Since the velocity and the magnetic field perturbations depend only on the transverse coordinates the wave are incompressible and satisfy $\nabla.\boldsymbol{H} = \boldsymbol{0}$. The velocity $\boldsymbol{V}$ can be eliminated from equation (1), and the resulting equation for the magnetic field perturbation can be written in the following form

$$\frac{\partial^2 \boldsymbol{b}}{\partial t^2} + \nabla\times[(\frac{c_A}{B})^2(\boldsymbol{B}\times(\nabla\times\boldsymbol{b})\times\boldsymbol{B})] + \nabla\times[(\frac{c_A^2}{g_f B})[\nabla\times\frac{\partial\boldsymbol{b}}{\partial t}\times\boldsymbol{B}]] = 0\ , \tag{24}$$

where $c_A^2 = \frac{\mu B}{4\pi\rho_0(0)}$, $g_f = \frac{\mu MeB(z)}{cm_i}$ and $H_d = \frac{c_A^2}{g_f}$ denote the Alfven speed, gyrofrequency of the Larmor spiraling of ions of mass m_i and charge number M, and Hall diffusivity respectively. Introducing the complex magnetic field perturbation $\boldsymbol{b}(t,z) = \boldsymbol{b}_x(t,z) \pm \boldsymbol{b}_y(t,z)$, where $\pm$ correspond to counterclockwise and clockwise polarization. In addition, we can use Fourier decomposition

$$b(t,z) = \int_{-\infty}^{\infty} G(t,z)\exp(-i\sigma t)d\sigma,$$

where G(t,z) is the magnetic field perturbation spectrum for wave of frequency σ at height z. As a result, the differential equation (23) can written like

$$c_A^2\,[D\,G(t,z)]^2 \pm \sigma\, D\,[H_d\, D\, G(t,z)] + \sigma\, G(t,z) = 0. \tag{25}$$

where D = d/dz. Using the properties of the isothermal atmosphere [equation (1)], then the magnetic field and gyrofrequency decay exponentially with altitude twice the density scale height while Hall diffusivity increases with the altitude half the density scale height , i.e $B(z) = c_1 \exp(\frac{-z}{2})$, $g_f = g_f(0)\exp(\frac{-z}{2})$, and $H_d = H_d(0)\exp(\frac{z}{2})$. In addition, letteing $\in = \sigma/g_f(0)$, $r = \sigma K/c_A$, and introducing the dimensionless variable x which is defined by $\psi = \pm\exp(-z/2)/\in$,and $G = \psi\Phi(\psi)$, then the differential equation (24) can be written in the following form

$$[\psi(1-\psi)D^2 + (2-3\psi)D - (1+4r^2)]\Phi(\psi) = 0. \tag{26}$$

This differential equation is a special case of the hypergeometric differential equation (14), which has two linearly independent solutions. Following the same procedure as in section 3, we have one solution will be eliminated by the magnetic energy condition and the second one can be written, after introducing the variable z via the variable ψ, in the following form

$$\psi(z) = \{Cons./(1+R_1)\}\,[\exp(1+2ir)z + R_1 \exp(1-2ir)z \tag{27}$$

where $r$denotes the wave number and R_1 denotes the reflection coefficient, which can be written like

$$R_1 = \exp(i\theta_2),\;\; \theta_1 = 2\arg\Gamma(4ir) + 4\arg\Gamma(1-2ir) - 2r\ln\in \tag{28}$$

It is clear that the magnitude of the reflection coefficient is one and the same equations for the kinetic and magnetic energies (as in section 3) can be derived. The final conclusions and observation about the results of this section will be introduced in section 6.

5. THE RESONANCE PHENOMENA IN A VISCOUS ATMOSPHERE PERMEATED BY A UNIFORM HORIZONTAL MAGNETIC FIELD

For this case the differential equation can be obtained from the differential equations (1-4) and the resulting differential equation can be written in the following form

$$\rho_0(z)V_{zz}(t,z) - 4(\rho_0(z)V_z(t,z))_z + aV_{zz}(t,z) = 0 \tag{29}$$

We will consider solutions which are harmonic in time t, i.e. $V(z,t) = W(z)\exp(-i\sigma t)$, then $W(z)$ satisfies the following differential equation

$$4(1-a\exp(z))D^2W(z) - 4DW(z) + \sigma^2 W(z) = 0, \tag{30}$$

where $D = d/dz$. Introducing a new dimensionless variable χdefined by $\chi = \exp(-z)/a$ which transforms the differential equation (29) into

$$[\chi(1-\chi)D^2 + (1-2\chi)D - \sigma^2/4]W(\chi) = 0. \tag{31}$$

where $D = d/d\chi$ and $0 < \arg(\chi) < -\pi/2$. Once again this differential equation is a special case of the hypergeometric differential equation. It has two linearly independent solutions, one of them will be eliminated by the dissipation condition and the second one can be written in the following form

$$W(z) = C.\{\exp[1/2 + ik]z + R_2 \exp[1/2 - ik]z\}. \quad (32)$$

where $k = (\sqrt{\sigma^2 - 1})/2$ denotes the wave number of the wave and R denotes the reflection coefficient and defined by

$$R_2 = \exp(-2k \arg a + i\theta_2), \quad \theta_2 = 2\arg\Gamma(2ik) + 4\arg\Gamma(1 - ik) - 2k\ln|a|. \quad (33)$$

It is clear that the magnitude of the reflection coefficient $|R_2| = \exp(-\pi k)$ when and $|R_2| = 1$, when $\arg a = 0$.The conclusions of this section will be presented in section 6.

6. CONCLUSIONS AND GENERAL REMARKS

It clear from equations (19), (27) and (32) that (in each one of the above three cases) the solution, below the reflecting layer) can be written as a linear combination of an upward and a downward propagating wave with equal wave number (with equal wavelength). Equation (21) represents a sinusoidal oscillations with wavelength of $\lambda/2$, where $\lambda = 2\pi/k$, and equation (23) represents the locations of the maximum and minimum values of the wave energies. Similar equations for the magnetic energy (M.E) can be derived from the induction equation. As a result, all conclusions obtained below concerning the kinetic energy (K.E) could be applied just as well to values of the magnetic energy. To put these conclusions in a more physical setting, it is convenient to defined $\beta = \frac{2}{\gamma}(c_s/c_A)^2 = P_0/P_m$, where P_m is the magnetic pressure. From equation equations (21) and (22) we have the following observations:

(A) For the first two cases we have $|R| = |R_1| = 1$.Thus the magnetic field creates a non absorbing and reflecting layer near $z_1 = -\log a_1$, *and* $z_2 = -\log\ \epsilon$respectively. Moreover, mini(K.E) = 0, and Max(K.E) = $1/(1 + \cos\theta)$.

(B) As a result of **(A)**, Max(K.E)→Max(M.E)→ ∞if $\theta = \pm(2n + 1)\pi$, i.e, when

$$2\delta = [\frac{\theta}{2\pi} \mp (n + \tfrac{1}{2})]\lambda. \quad (34)$$

We call this equation **"resonance equation"** which indicates that the resonance may occur for infinite many values of the magnetic field and of the frequency of the wave if the wavelength of the wave is matched with the strength of the magnetic field.

(C) It follows from **(B)** that the resonance will occur when $\beta > 2/\gamma$. As a result, the mechanism for the coronal heating is **magnetoacoustic**. On the contrary the heating mechanism for the corona is **acoustic** when $\beta \leq 2/\gamma$.

(D) The location of the maximum and minimum and the values of the kinetic energy depends on $\ln a_1$, which indicates the influence of the magnetic field on the heating process of the solar atmosphere. At the same time the ratio M/m depends on the magnitude of the reflection coefficient and uniquely determines the magnitude of the reflection coefficient |R| in the above three cases.

(E) In the third case the above conclusions are conditional. When the viscosity dominates the oscillatory process we have $\arg a = \pi/2$ and magnitude of the reflection coefficient is $\exp(-\pi k)$.On the other hand, when the magnetic field dominates the motion, we have $\arg a = 0$,and the magnitude of the reflection coefficient |R| = 1. Consequently, in a viscous atmosphere, the heating process in regions of strong magnetic field (like the sun spots) is **magnetoacoustic**, while in regions of a viscous atmosphere the heating process is **acoustic**.

REFERENCES

1. Abramowitz, M. and Stegun, I., (1964) *Handbook of Mathematical Functions,* National Bureau of Standards, Washington, D.C.

2. Alkahby, H., (1988) "Reflection and dissipation of vertically propagating acoustic gravity waves in an isothermal atmosphere," Ph.D. *Thesis*, Adelphi University.

3. Alkahby, H., (1993a) "On the coronal heating mechanism by the resonant absorption of Alfven waves," *Internat.J. Math & Math. Sci.* **16**, No. 4, 811-816.

4. Alkahby, H., (1993b) "On the heating of the solar corona by the resonant absorption of Alfven waves,"*Applied Math.Letters* **6**, No. 6, 59-64.

5. Alkahby, H., (1993c) "Reflection and dissipation of hydromagnetic waves in a viscous and thermally conducting isothermal atmosphere,"*Geophys. Astrphys. Fluid Dynam.* **72**, 197-207.

6. Alkahby, H.Y., (1998a) Effect of Newtonian cooling on upward propagating magnetoacoustic waves in the solar atmosphere. *Proc. Of the Fourteenth Annual Conf. On Appl. Math. Univ. of Central Oklahoma.* 57-65.

7. Alkahby, H.Y. and M. A Mahrous, (1999a) On the reflection and dissipation of oblique Alfven waves in an isothermal atmosphere. *Internat. J. Math & Math Sci.* Vol. 22, No. 1 161-169.

8. Ioson, J.A., (1982) Resonance electromagnetic heating of stellar loops, AP. J Vol. 254, 318 - 334.

9. Davila, J. M., (1987) Heating of the solar corona by the resonance absorption of Alfven waves, AP. J. Vol. 317, 514 - 521.

10. Hollweg, J. V., (1984) Resonance of the coronal loops, AP. J., Vol 277, 392 - 403.

11. Priest, E. R., (1984) *Solar Magnetohydrodynamics*, D. Reidl Pub. Co.

12. Yanowitch, M., (1967) The effect of viscosity on vertically oscillations of an isothermal atmosphere, Can. J. Phys. Vol. 45, 2003 - 2008.

Proceedings of Dynamic Systems and Applications 3 (2001) 265-272

NUMERICAL STUDIES OF VERTICALLY PROPAGATING ACOUSTIC AND MAGNETOACOUSTIC WAVES IN AN ISOTHERMAL ATMOSPHERE (II)

Alkahby Hadi[1], Jalbout Fouad[2] and Mohammad Mahrous[3]

[1]Department of Mathematics , Dillard University, New Orleans, LA 70122
[2] Department of Physics, Dillard University, New Orleans, LA 70122
[3] Department of Mathematics, University of New Orleans, New Orleans, LA 70148

ABSTRACT: In this numerical study, we investigate the effects of Newtonian cooling, viscosity and thermal conduction on upward propagating acoustic waves in an isothermal atmosphere. We will consider only the case of small Prandtl number. The aim of this study is to determine, numerically, the effect of Newtonian cooling on the magnitude of the reflection coefficients which were created by the effects of the viscosity or thermal conduction and to compare the results of the numerical computations with those of the asymptotic evaluations. This problem leads to a system of differential equations for the temperature and velocity. Using central differences, the system of differential equations was replaced by a set of difference equations. The results of the numerical computations were sketched in figures 1 -6 and they are in agreement with those of the asymptotic evaluations for five places.

Keywords : Acoustic Waves, Newtonian Cooling Coefficient, Reflection Coefficient

1991 AMS SUBJECT CLASSIFICATION CODES : 76N, 76Q

1. INTRODUCTION

The propagation of atmospheric waves, both in isothermal and in non-isothermal atmospheres, has been investigated extensively in recent years. Much of the motivation of these studies comes from their relevance to phenomena in the solar atmosphere and to certain phenomena in ocean dynamics Alkahby[1-3], Yanowitch [6-8]. It is well known that the solar corona is extremely hot, typical temperatures are 10^6 K compared with $5x10^3$ K in the photosphere. Consequently, thermal energy must be continually supplied to maintain this temperature against radiative cooling. As a result, these two questions must be answered: how is the energy supplied to the corona, and how is it dissipated? To answer these questions, many mathematical models and dissipative mechanisms are suggested Alkahby[1-3], Alkahby & Yanowitch [4], Yanowitch [6-8].

In this paper, we investigate numerically, the combined effect of viscosity, thermal conduction and Newtonian cooling on an upward and a downward propagating acoustic wave in an isothermal atmosphere. In addition, we compare the results of the numerical computations with those of the asymptotic evaluations for practical purposes. It is shown that the presence of a small viscosity creates a layer which acts like an absorbing and a reflecting barrier for waves generated below it. The oscillatory process is an adiabatic one below and above the reflecting layer. On the contrary, the presence of thermal conduction produces a partially reflecting layer and the motion is adiabatic, below the reflecting layer, and an isothermal one above it. In these two cases the atmosphere can be divided into two distinct regions. When the oscillatory process is dominated by the effect of thermal diffusivity, in the presence of the viscosity, i.e. for small Prandtl number, the atmosphere can be divided into three distinct regions. In the lower region, the motion is adiabatic, it is isothermal in the middle region and in the upper region the motion decays exponentially with altitude. The two lower regions are connected by a partially reflecting layer, which is produced by the effect of thermal conduction, while the upper two regions are connected by an absorbing and a reflecting barrier, which is created by the effect of the viscosity. The presence of two reflecting layers will influence the motion in the lower region and the final conclusions depend on the relative location of the two reflecting layers. The addition of Newtonian cooling affects mainly the lower region and transforms the motion from an adiabatic form to an isothermal one. Moreover, it produces waves attenuation and alters the wavelength. If the Newtonian cooling coefficient is large compared with the adiabatic cutoff frequency of the wave, the temperature in the lower regions evens out and the wave motion approaches an isothermal one. Consequently, the effect of thermal conduction and the attenuation in the wave amplitude are eliminated since the isothermal region is dissipationless. The results of the numerical computation and asymptotic evaluations are compared and they are in agreement for five places. The results of the computations of the reflection coefficient were sketched in figures 1 - 6.

2. MATHEMATICAL FORMULATION OF THE PROBLEMS

Suppose an isothermal atmosphere, which is viscous and thermally conducting, occupies the upper half-space $z > 0$.We will investigate small oscillations about equilibrium which depend only on the time t and on the vertical coordinate z. Let P, ρ, W, and T be the perturbations in the pressure, density, vertical velocity and temperature, and P_0, ρ_0, T_0 be the equilibrium quantities. Equilibrium pressure, temperature (T_0 = constant)and density satisfy the gas law $P_0 = R\rho_0 T_0$ and the hydrostatic equation $P_0' + g\rho_0 = 0$ (g is the gravitational acceleration and R = gas constant). Consequently, the pressure and density are given by

$$P_0(z) = P_0(0)\exp(-z/H), \qquad \rho_0(z) = \rho_0(0)\exp(-z/H)$$

where H is the density scale height and defined by $H = RT_o/g$. Moreover, the linearized equations of motion (conservation of momentum and mass, the heat equation and the gas law) can be written like

$$\rho_0\, W_t + P_z + \rho g = \tfrac{4}{3}\mu W_{zz}, \tag{1}$$

$$\rho_t + (\rho_0 W)_z = 0, \tag{2}$$

$$\rho_0[c_V T_t + qT + gHW_z] = \kappa T_{zz}, \tag{3}$$

$$P = R(\rho_o T + T_0\rho). \tag{4}$$

Here c_V is the specific heat at constant volume, μ is the dynamic viscosity coefficient, which is assumed to be constant, q is the Newtonian cooling coefficient which refers to the heat exchange between hot and cold regions in the solar atmosphere and κ is the thermal conductivity coefficient. The subscripts z and t denote the differentiation of the independent variables with respect to z and t respectively. Moreover, to simplify the problem formulation we will consider only solutions which are harmonic in time t, i. e.

$W(z,t) = W^*(z)\exp(-i\sigma t)$ *and* $T(z,t) = T^*(z)\exp(-i\sigma t)$ where σ denotes the frequency of the wave. It is more convenient to rewrite the differential equations of motion in dimensionless form: $z^* = z/H$, $\sigma_a = c/2H$ is the adiabatic cut off frequency of the wave, where $c^2 = \gamma RT_0 = \gamma gH$ is the adiabatic sound speed and γ denotes the ratio of the specific heats, $W^* = W/c$, $\mu^* = 2\mu/3\rho_0 cH$, $\sigma^* = \sigma/\sigma_a$, $t^* = t\sigma_a$, $T^* = T/2\gamma T_0$, $\kappa^* = 2\kappa/c_V cH\rho_o(0)$, $q^* = q/\sigma_a$,. The star can be omitted, since all variables are written in dimensionless form from now on. In addition, we introduce the dimensionless parameter $P_r = c_p\mu/\kappa$ which denotes the Prandtl number and c_p is the specific heat at constant pressure. It is clear that Prandtl number measure the relative strength of the viscosity with respect to the thermal conduction. Substituting all these quantities in the differential equations (1 - 4) and eliminating the pressure and density we obtain the following system of differential equations for the velocity and the temperature:

$$(D^2 - D + \gamma\omega^2/4)W(z) - i\sigma\kappa P_r e^z D^2 W(z) + i\gamma(D-1)T(z) = 0, \tag{5}$$

$$(\gamma - 1)DW(z) = \gamma(i\omega - q)T(z) + \gamma\kappa e^z D^2 T(z) \tag{6}$$

where D denotes the derivative of the indicated variable with respect to z. If furthermore, the velocity W is eliminated from equation (5), one obtains a single fourth-order differential equation for the temperature T:

$$[\gamma\sigma(D^2 - D + \sigma^2/4) + iq(D^2 - D + \gamma\sigma^2/4) - i\kappa e^z D^2(D^2 + D + \gamma\sigma^2/4)$$

$$-iP_r\gamma(\sigma + iq)\kappa e^z D(D+1) - \gamma P_r(\kappa e^z)^2 D^2(D+1)(D+2)]T(z) = 0 \tag{7}$$

Moreover, the first two terms of the differential equation (7) can be combined to give:

$$[(D^2 - D + \tau\sigma^2/4) - i(\kappa/m)e^z D^2(D^2 + D + \gamma\sigma^2/4) - iP_r m\tau(\kappa/m)e^z D(D+1)$$

$$-(\gamma P_r m)(\kappa e^z/m)^2 D^2(D+1)(D+2)]T(z) = 0, \tag{8}$$

where $\tau = \gamma(\sigma + iq)/(\gamma\sigma + iq) = \gamma(\sigma + iq)/m$ and $m = \gamma\sigma + iq$ (9)

Boundary Condition: To obtain a unique solution for the differential equation (8), physically relevant conditions must be imposed. The first is the dissipation condition (DC), which requires that the rate of change of the energy dissipation in an infinite column of fluid of unit cross-section to be finite. Since the dissipation function depends on the squares of the velocity gradients, the dissipation condition is equivalent to

$$\mu \int_0^\infty |W_z|^2\, dz < \infty. \tag{10}$$

The second one is the entropy condition (EC), which is determined by the equation for the rate of the entropy, from which it follows that

$$\kappa \int_0^\infty |T_z|^2\, dz < \infty. \tag{11}$$

If these boundary conditions are not sufficient or unablicable conditions to determine a unique solution then we will impose the radiation condition. Boundary conditions are required at z = 0, and we shall adopt the lower boundary condition (LBC): In a fixed interval $0 < z < z_0$, the solution of the differential equation should approach some solution of the limiting differential equation as $\kappa \to 0$ and $\mu \to C$, i.e $[D^2 - D + \tau\sigma^2/4]T(z) = 0$. For numerical computations, we assume that

$$T(0) = 1, \qquad \text{and} \qquad W(0) = 1. \tag{12}$$

The boundary conditions (10) and (11) are necessary and sufficient, as upper boundary conditions, to ensure a unique solution.

3. COMBINED EFFECT OF NEWTONIAN COOLING AND THERMAL CONDUCTION

The differential equation for this case can be obtained from the differential equation (8) by setting $P_r = 0$. Thus we have

$$[(D^2 - D + \tau\sigma^2/4) - i(\kappa/m)e^z D^2(D^2 + D + \gamma\sigma^2/4)]T(z) = 0 \tag{13}$$

To solve this differential equation, when $|\ \kappa/m\ |\ e^z \ll 1$ and for an arbitrary value of q, it is convenient to introduce a new dimensionless variable ξ defined by

$$\xi = \exp(-z)/(i\kappa/m) = \exp[-z - \log\ |\ \kappa/m\ |\ + i\theta_m + 3\pi i/2\ , \tag{14}$$

where $\theta_m = \arg(m)$. As a result, the differential equation (13) is transformed into

$$[\xi(\theta^2 + \theta + \tau\sigma^2/4) - \theta^2(\theta^2 - \theta + \gamma\sigma^2/4)]T(\xi) = 0, \tag{15}$$

where $\theta = \xi\frac{d}{d\xi}$. The solution of this differential equation, which satisfies the prescribed boundary condition, can be written like

$$T(z) \sim const\{[\exp(1/2 - r(q) + i\beta)z] + RC_1[\exp(1/2 + r(q) - i\beta)z], \tag{16}$$

where RC_1 is the reflection coefficient, which is defined by

$$RC_1 = \exp(G_* - iH_*)R_*[r(q), \beta, \beta_i]S_*[r(q), \beta, \beta_i], \tag{17}$$

where

$G_* = 2r(q)\log\ |\ \kappa/m\ |\ -2\kappa\theta_m,\ H_* = 2\beta\log\ |\ \kappa/m\ |\ +\pi r(q) + 2r(q) + 2r(q)\theta_m,$
$R_* = A_{11}/B_{11},\ A_{11} = \Gamma^2[\frac{1}{2} + r(q) - i\beta]\Gamma[r(q) - i(\beta_i + \beta)\Gamma[-2r(q) + 2i\beta]\Gamma[1 + r(q) + i(\beta_i - \beta)]$
$B_{11} = \Gamma^2[\frac{1}{2} - r(q) + i\beta]\Gamma[-r(q) + i(\beta_i + \beta)\Gamma[2r(q) - 2i\beta]\Gamma[1 - r(q) - i(\beta_i - \beta),$
$S_* = e^{\pi\beta}[\sinh\pi(\beta_i - \beta) - i\tan[\pi r(q)]\cosh\pi(\beta_i - \beta)/\Theta]$
$\Theta = \sinh\pi(\beta_i + \beta) + i\tan[\pi r(q)]\cosh\pi(\beta_i + \beta)$.
Here $2\beta = \sqrt{\tau\sigma^2 - 1}$, $2\beta_i = \sqrt{\gamma\sigma^2 - 1}$. It clear from equation (16) that the first term of the solution, on the right, represents an upward traveling wave, its amplitude decaying with the altitude like $\exp(-r(q)z)$, while the second term is a downward traveling wave decaying in the same rate. When q = 0, we have r(q) = 0, $m = \gamma\sigma$ and the wave number $\beta = \beta_a = \sqrt{\sigma^2 - 1}/2$. Consequently, the magnitude of the reflection coefficient can be written like:

$$|RC_1| \to \ | RC_2 \ | = e^{\pi\beta_a}[\sinh \pi(\beta_i - \beta_a) \ / \sinh \pi(\beta_i + \beta_a)]. \qquad (18)$$

It is clear that $|RC_2| < |RC_3| = \exp(-\pi\beta_a)$. On the other hand when $q \to \infty$, we have r(q) = 0, $\tau \to \gamma$, and $\beta \to \beta_i$. Consequently, $| RC_2 \ | \to \ | RC_4 \ | = 0$. As a result, the reflection layer, which was created by the effect of thermal conduction is eliminated completely, the wavelength is changed from the adiabatic value to the isothermal one and the oscillatory process is transformed from the adiabatic form to the isothermal one. This result also indicates that the process of the heat transfer from cold to hot regions in the solar atmosphere are more dominant than those of thermal conduction. This conclusion is supported by many observations Alkahby [1 - 3], Yanowitch [7 - 8]. The results of the numerical computations were sketched in figure 3. In addition, figure 3 shows the magnitude of the reflection coefficient for q = 0.2 and q = 0.4 and these two values are denoted by $|RC_{11}|$and $|RC_{22}|$ respectively

4. COMBINED EFFECTS OF NEWTONIAN COOLING, THERMAL CONDUCTION AND VISCOSITY

In the lower region and for $| \kappa/m | e^{-} << 1, \sigma > \sigma_a$ and small value of the Newtonian cooling coefficient q, the solution of the differential equation (8) can be written in the following form:

$$T(z) \sim \text{Const. } \{\exp[(1/2 - r(q) + i\beta \ )] \ + RC_5 \exp[1/2 + r(q) - i\beta)z]\} \qquad (19)$$

where RC_5 denotes the reflection coefficient, which is defined by

$$RC_5 = \exp[-\pi\beta + A_* - iB_*]R_* \frac{M_1 - M_2 \exp[-2\beta_i \log P_r + i\theta^* + 2\beta_i\theta_m]}{(M_3 - M_4 \exp[-2\beta_i \log P_r + i\theta^* + 2\beta_i\theta_m]}, \qquad (20)$$

where

$$A_* = 2r(q)\log \ | \kappa/m \ | -2\beta\theta_m, \ B_* = 2\beta \log \ | \kappa/m \ | + \pi r(q) + 2r(q)\theta_m,$$
$$M_1 = \exp(2\pi\beta_i) - \exp(2\pi\beta)[\cos(2\pi r(q)) \ + i\sin(2\pi r(q))]$$
$$M_2 = \exp(2\pi\beta)[\cos(2\pi r(q)) \ + i\sin(2\pi r(q))] \ - \exp(-2\pi\beta_i)$$
$$M_3 = \exp(2\pi\beta_i) - \exp(2\pi\beta)[\cos(2\pi r(q)) \ - i\sin(2\pi r(q))]$$
$$M_4 = \exp(-2\pi\beta)[\cos(2\pi r(q)) \ - i\sin(2\pi r(q))] \ - \exp(-2\pi\beta_i)$$
$$D_* = D_{11}/D_{22},$$
$$D_{11} = -(1/2 - i\beta_i)\Gamma^2(2i\beta_i)\Gamma(1 + r(q) - i(\beta_i - \beta)\Gamma(1 - r(q) - i(\beta_i - \beta)),$$
$$D_{22} = (1/2 + i\beta_i)\Gamma^2(-2i\beta_i)\Gamma(1 - r(q) + i(\beta_i + \beta)\Gamma(1 + r(q) + i(\beta_i - \beta)),$$
$$\theta^* = \arg D_* - 2\beta_i \log \gamma$$

In addition to the conclusions of section 3, we have the following observations. **(A)** When q = 0, we have r(q) = $A_* = \theta_m = 0$ and $B_* = 2\beta_a \log(\kappa)$, and we recover the result obtained in Lyons and Yanowitch [9]. In this case the reflection process is more complicated because of the existence of two reflecting layers and the final conclusion depends on their relative locations. The magnitude of the reflection coefficient is $| RC_5 \ | \le |RC_3|$ (as a result of oscillatory nature of the reflection coefficient, $|RC_5|$ may be less than $|RC_2|$). The results of the numerical computation are sketched in figure 4. **(B)** For a fixed value of the thermal conductivity coefficient κ, small values of the Newtonian cooling coefficient q and $\sigma > \sigma_a$ the solution of the differential equation (8) is given by equation (19) and its behavior is described in section (3). In the present case the atmosphere can be divided into three distinct regions. The lower region is approximately adiabatic and the middle one is isothermal. In the upper region, the solution is influenced by the combined effects of the thermal diffusivity and the kinematics viscosity and it will decay exponentially with the altitude. The upper two regions are connected by an absorbing and reflecting barrier created by the combined effect of the viscosity and thermal conduction. while the lower two regions are connected by a partially reflecting layer produced by the effect of thermal conduction. The magnitude of the reflection coefficient, $|RC_2| < |RC_5| < |RC_3|$ ($|RC_5|$may be less than $|RC_2|$because of the oscillatory nature of the reflection coefficient). The results of the numerical computation, for the magnitude of the reflection coefficient, are sketched in figure 5. (C) On the other hand when $q \to \infty$ and $\sigma > \sigma_a$, one obtains $M_1 = M_4 = r(q) = \theta_m = A_* = 0,\ \beta_a \to \beta_i, C_* D_* \to \theta_i,$ *and* $M_2 \to M_3$. Consequently, the magnitude of the reflection coefficient becomes $|RC_5| \to |RC_6| = \exp(-\pi\beta_i)$, and the results of the numerical computation are sketched in figure 6. The lower reflecting layer which was created by thermal conduction, will be eliminated because the wavelengths below and above this layer become equal. From the above conclusions, it is clear that, in the solar photosphere the temperature fluctuations could be smoothed out as a result of heat transfer by radiation between any two regions, with different temperatures. In addition the heat transfer by radiation is more dominant than that of the conduction. In addition, the reflecting layer created by the viscosity will not be eliminated but the magnitude of the reflection coefficient is reduced to the isothermal value. Consequently, the nature of the reflecting layer, which was created by the viscosity, will be influenced greatly by the heat radiation.

5. COMPUTING SCHEME AND RESULTS OF COMPUTATIONS

The results of the previous section are asymptotically valid as Pr $\to 0$. In order to determine the range of Pr in which they are reasonably accurate, the boundary-value problem (defined by the system of equations 5 and 6) was solved numerically for several values of Pr and β. For the numerical integration it is convenient to deal with the system (5 and 6) with boundary conditions (12) at z = 0. The upper boundary conditions are replaced by these two conditions: $DW \to\ DT \to 0$, as $z \to \alpha$. Using centered differences, we can replace the system of differential equations (5 and 6) by a set of difference equations, $A_n\Phi_{n+1} + B_n\Phi_n + C_n\Phi_{n-1} = 0$, where Φ_n is the column vector with components $iT(z_n)$ and $W(z_n)$, and A_n, B_n and $C_n$are 2 x 2 matrices. The system was solved in an interval $0 \le z \le L$ sufficiently large to allow T and W to reach their limiting values, and the boundary condition $DW \to\ DT \to 0$, as $z \to \alpha$, was replaced by $\Phi_N = \Phi_{N-1}$, where N is the index of the point $z = L$. The problem was solved by the standard method in which the $\Phi_n$are computed from a linear equation $\Phi_n = \alpha_n\Phi_{n+1} + \beta_n$ by backward integration, while the matrices $\alpha_n = -[B_n + C_n\alpha_{n-1}]^{-1}A_n$, and the vectors $\beta_n = -[B_n + C_n\alpha_{n-1}]^{-1}C_n\beta_{n-1}$ are computed by forward integration. Moreover, let M and m denote the maximum and minimum values of the oscillation amplitude and $d = \sqrt{M/m}$, then the magnitude of the reflection coefficient can be computed from the formula $|RC_i| = (d-1)/(d+1), i = 1,2,3,4,5,6$ The results of the computations are shown in figures 1, 2, 3, 4, 5, and 6, where figures 1 and 2 indicate the nature of the decaying factor in the amplitude of the wave and the change in the wave number respectively.

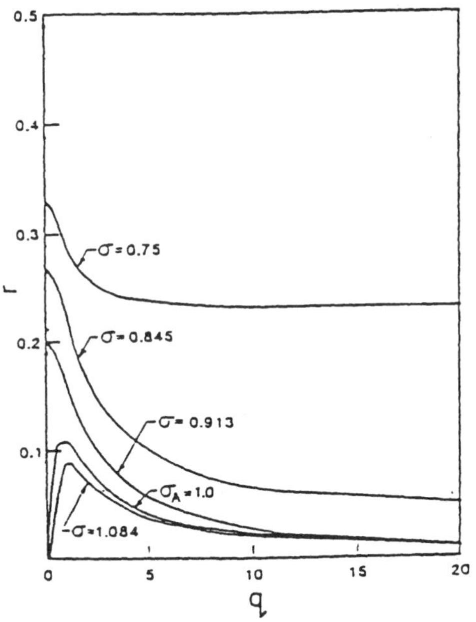

Fig. 1. Decay constant d as a function of q for several values of σ, $\gamma = 1.4$.

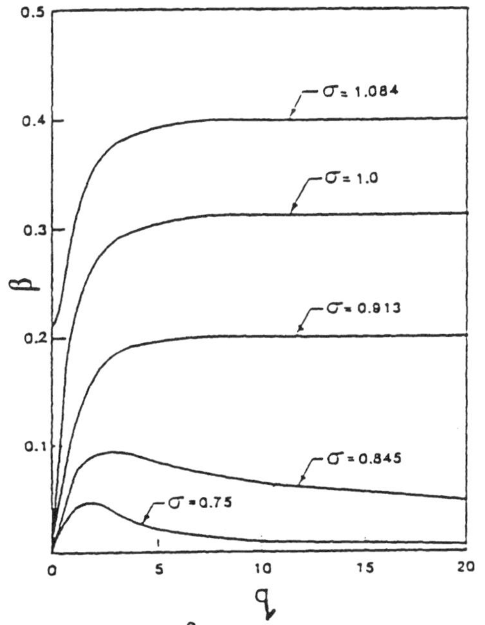

Fig. 2. Wave number β as a function of q for several values of σ, $\gamma = 1.4$.

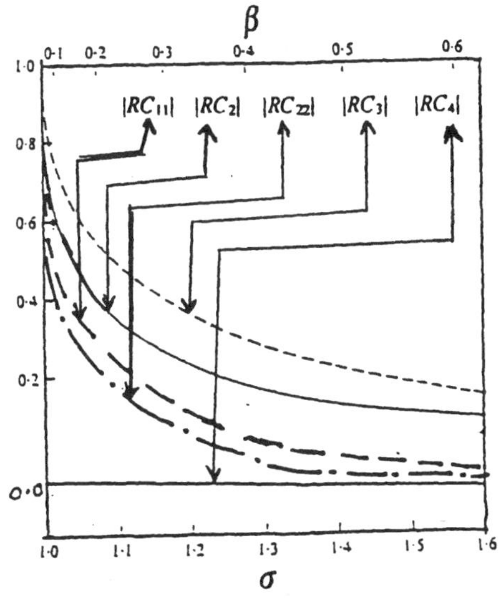

Fig. 3. Magnitude of the reflection coefficient for fixed value of κ and $q \geq 0$.

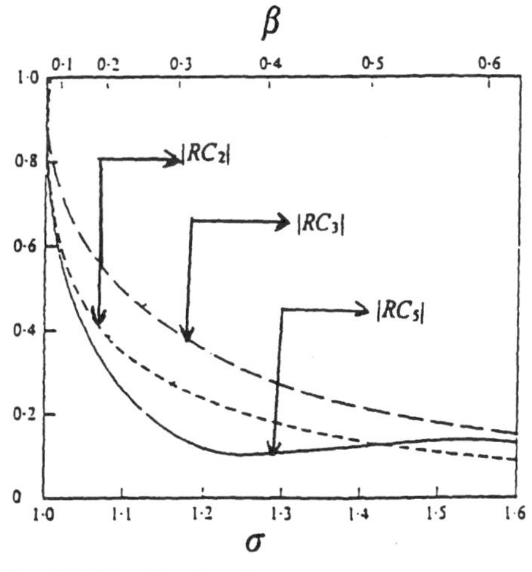

Fig. 4. Magnitude of the reflection coefficient for $P_r = 10^{-4}$, $q = 0.0$.

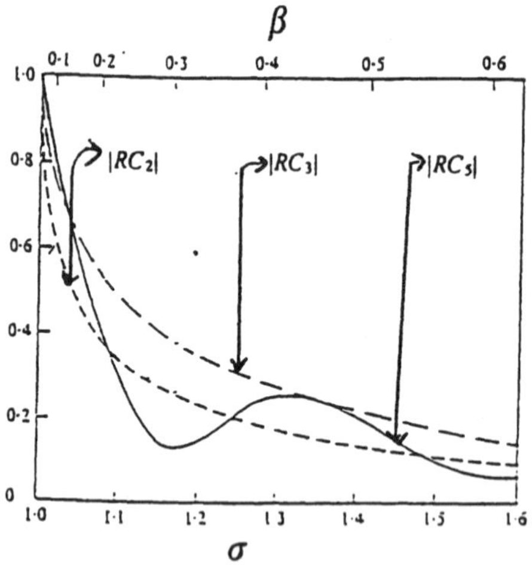

Fig. 5. Magnitude of the reflection coefficient for $P_r = 10^{-6}$, $q = 0.1$.

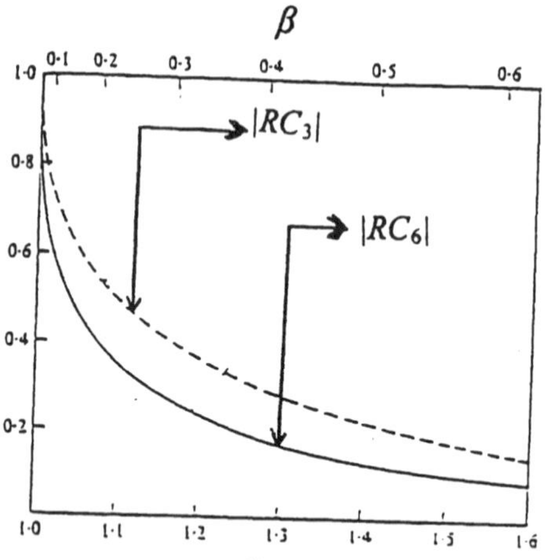

Fig. 6. Magnitude of the reflection coefficient for $P_r = 10^{-6}$, *and* $q \to \infty$

REFERENCES

1. Alkahby, H., On the coronal heating mechanism by the resonant absorption of Alfven waves, Internat. J. Math & Math. Sci., Vol. 16, No. 4(1993a), 811-816.

2. Alkahby, H., On the heating of the solar corona by the resonant absorption of Alfven waves, Applied Math. Letters.. Vol. 6, No. 6(1993b), 59-64.

3. Alkahby, H., Reflection and dissipation of hydromagnetic waves in a viscous and thermally conducting isothermal atmosphere, Geophys. Astrphys. Fluid Dynam., Vol.72(1993c), 197-207.

4. Alkahby, H., & Yanowitch, M., The effect of Newtonian cooling on the reflection of vertically propagating acoustic waves in an isothermal atmosphere, Wave Motion 11, (1989), 419-426.

5. Alkahby, H.Y. & Mahrous, M. A., On the reflection and dissipation of oblique Alfven waves in an isothermal atmosphere. Internat. J. Math & Math Sci.Vol. 22(1999a), No.1, 161-169.

6. Yanowitch, M., Effect of viscosity on vertical oscillations of an isothermal atmosphere, Can J. Phys., Vol. 45 (1967a), 2003-2008.

7. Yanowitch, M., Effect of viscosity on gravity waves and the upper boundary conditions, J. Fluid Mech., Vol. 29 (1967b), 209-231.

8. Yanowitch, M., Vertically propagating hydromagnetic waves in an isothermal atmosphere with a horizontal magnetic field, Wave Motion, Vol. 1(1979), 123-125.

9. Lyons, P. & Yanowitch, M., Vertical oscillations in a viscous and thermally conducting isothermal atmosphere, J. Fluid Mech., Vol.66 (1974), 273-288.

Proceedings of Dynamic Systems and Applications 3 (2001) 273-278

ON THE ASYMPTOTIC BEHAVIOR OF A DELAY FIBONACCI EQUATION

Dedicated to Professor Lakshmikantham on his 75th birthday

Yoshihiro HAMAYA

Department of Mathematical Information Science
Okayama University of Science
Japan

ABSTRACT.
We consider the property of convergence of ratios of successive values to Fibonacci sequences with delay, and are studied by applying the prime theorem. For the classical Fibonacci sequence, this property is the theorems of Poincare and Perron which deal with the asymptotic behavior of linear difference equations with constant coefficients.

1. Introduction

For the Fibonacci sequence, it has several well-known that the ratio of successive terms approaches a fixed limit and others (cf.[8]). In this paper, we consider the property of convergence of ratios of successive values to the Fibonacci sequence with delay:

$$x_n = x_{n-x_{n-1}} + x_{n-x_{n-2}}, \quad n \geq 3,$$

and are studied by applying the prime number theorem. To the standard Fibonacci sequence, this property is the theorems of H. Poincare and O. Perron which known as the asymptotic behavior of the ratios of successive terms to solutions of linear difference equations with (constant) coefficients.
Now, before the main contents of this paper, we prepare a review of the standard Fibonacci sequence.

2. The review of Fibonacci sequence

The numbers

$$1, 1, 2, 3, 5, 8, 13, 21, 34, 55, 89, 144, 233, 377, 610, \cdots,$$

well-known as Fibonacci numbers, have been called by the nineteenth-century French mathematician E. Lucas after Leonard Fibonacci of Pisa, one of the best mathematicians of the Middle Ages, who referred to them in his book Liber Abaci (1202) in the connection with his rabbit problem (cf.[2]). This problem may be stated as follows: How many pairs of offspring does one pair of mature rabbits produce in a year, if each pair of rabbits gives birth to a new pair each month starting when it reaches its maturity age of two months? (See [3, p70-71 and 7, p78]). For convenience, we shift from $n+2$ to n in start point. Table shows the number of pairs of rabbits by the end of each month. The first pair has offspring at the end of the first month, and thus we have two pairs. At the end of the second month only the first pair has offspring, and thus we have three pairs. At the end of the third month, the first and second pairs will have offspring and hence we have five pairs. Continuing this procedure, we arrive at Table. If $f(n)$ is the number of pairs of rabbits at the end of n months, then the recurrence relation that represents this model is given by the second order linear difference equation. The Fibonacci sequence given by

$$f(n) = f(n-1) + f(n-2) \quad for \quad n \geq 3 \tag{1}$$

with initial value problems $f(1) = f(2) = 1$, where $f(n)$ is a function with both range and domain the positive integers.
We define the characteristic equation of Eq. (1) by

$$\lambda^2 - \lambda - 1 = 0.$$

Then, the characteristic roots are

$$\lambda_1 = \frac{1+\sqrt{5}}{2} \quad and \quad \lambda_2 = \frac{1-\sqrt{5}}{2}.$$

We have the general solution of Eq. (1)

$$f(n) = \alpha(\frac{1+\sqrt{5}}{2})^n + \beta(\frac{1-\sqrt{5}}{2})^n, \qquad n \geq 1.$$

Using the initial value problem $f(1) = 1$ and $f(2) = 1$, we obtain

$$\alpha = \frac{1}{\sqrt{5}} \quad and \quad \beta = \frac{-1}{\sqrt{5}}.$$

Therefore, we have

$$\begin{aligned} f(n) &= \frac{1}{\sqrt{5}}[(\frac{1+\sqrt{5}}{2})^n - (\frac{1-\sqrt{5}}{2})^n] \\ &= \frac{1}{\sqrt{5}}(\lambda_1^n - \lambda_2^n), \qquad n \geq 1. \end{aligned}$$

For the standard Fibonacci sequence $f(n)$, it is very interesting to note that

$$\lim_{n\to\infty} \frac{f(n+1)}{f(n)} = \lambda_1.$$

This number is called the 'golden section,' which was considered by the ancient Greeks to be the most aesthetically pleasing ratio for the length of a rectangle to its width. Also, the Fibonacci sequence is very interesting to mathematicians; in fact, an entire publication, "The Fibonacci Quarterly [8]," dwells on the intricacies of this fascinating sequence. In the next problem, we consider to extend to the more general theory of

Fibonacci equation, that is a linear homogeneous difference equation with constant coefficients. We consider the kth order difference equation

$$f(n+k) + p_1 f(n+k-1) + \cdots + p_k f(n) = 0, \tag{2}$$

where the p_i's are constants. Then we have the following Poincare and Perron types theorems for constant coeffcients.

Theorem A. (Poincare's Theorem)
If the roots $\lambda_1, \cdots, \lambda_k$ of characteristic equation for (2)

$$\lambda^k + p_1\lambda^{k-1} + \cdots + p_k = 0$$

satisfy

$$|\lambda_1| > |\lambda_2| > \cdots > |\lambda_k|,$$

then every nontrivial solution $f(n)$ of Eq. (2) satisfies

$$\lim_{n\to\infty} \frac{f(n+1)}{f(n)} = \lambda_i$$

for some i.

Theorem B. (Perron's Theorem)
In addition to the assumptions of Theorem A, suppose that $p_k \neq 0$ for each k. Then there are fundamental solutions $f_1(n), \cdots, f_k(n)$ of Eq. (2) that satisfy

$$\lim_{n\to\infty} \frac{f_i(n+1)}{f_i(n)} = \lambda_i, \qquad 1 \le i \le k.$$

See [1,3,7] for proofs of theorem A and B.

3. The delay Fibonacci equation

In this talk, we shall consider the asymptotic behavior of the ratio of successive terms of solutions of the New delay Fibonacci equation of

$$f(n) = f(n - f(n-1)) + f(n - f(n-2)) \quad for \quad n \ge 3 \tag{3}$$

with initial valu problems $f(1) = f(2) = 1$, where $f(n)$ is a function with both range and domain the positive integers. This equation proposed by Hofstadter (cf.[4,5]) as a purely unsolved problems in Number Theory [5]. First of all, we make the following Table and this corresponds to the table of standard Fibonacci sequence.

$$1, 1, 2, 3, 3, 4, 5, 5, 6, 6, 6, 8, 8, 8, 10, 9, 10, 11, 11,$$
$$12, 12, 12, 12, 16, 14, 14, 16, 16, 16, 16, \cdots.$$

We can consider three problems for Eq.(3) as follows:

1. Does $f(n)$ miss infinitely many integers such as $7, 13, 15, \cdots$?
What is it's behavior in general ?

2. Is there a limit $\alpha < \infty$ such that

$$\frac{f(n+1)}{f(n)} \to \alpha \qquad as \qquad n \to \infty \quad ?$$

What is this α ?

3. Study to make the theorem of Poincare and Perron type of recurrence with three terms and over for Eq.(3), for example,

$$\begin{aligned} f(n) &= f(n - f(n - f(n-1))) + f(n - f(n - f(n-2))) \\ &+ f(n - f(n - f(n-3))), \qquad n \geq 4 \end{aligned}$$

and we have to set the adequate initial condition

$$f(1), f(2), f(3) = ?$$

In this paper, we, particularly, only consider the theorem as a partial-answer to problem 2.
The main result is the following:

Theorem . For Eq.(3), under the above assumptions, we have

$$\lim_{n\to\infty} \frac{f(n+1) - [(n+1)/2]}{f(n) - [n/2]} = 1, \tag{4}$$

where $[n]$ means the greatest integer $\leq n$.

See Hamaya [6] for the proof of this theorem, however in a supplementary of this theorem, we only set the following the sequence of $f(n+1)/f(n)$:

$$1, 1, 2, \frac{3}{2}, 1, \frac{4}{3}, \frac{5}{4}, 1, \frac{6}{5}, 1, 1, \frac{4}{3}, 1, 1, \frac{5}{4}, \frac{9}{10}, \frac{10}{9}, \frac{11}{10}, \\ 1, \frac{12}{11}, 1, 1, 1, \frac{4}{3}, \frac{7}{8}, 1, \frac{8}{7}, 1, 1, 1, \cdots.$$

Moreover, we have a conjecture for problem 1 as follows:

Conjecture. $f(n)$ can not miss infinitely many integers.

Remark. If the following equality yields, we can easily obtain the result of theorem and another problems. It is certain that there is a constant c for which

$$f(x) = \pi(x) + (c + ord(1)) \frac{x^{3/4}}{(\log x)^{3/2}},$$

where $ord(1)$ is the small order of Landau, $\pi(x)$ denotes the number of primes $\leq x$, and moreover $x \in \mathbf{R}^+$ and x is sufficiently large. However, this has never been proved so far as we know that.
For the problem 3, unfortunately, our method [6] can not apply directly it. we mayby need the new idea and tools.

References

[1] R. P. Agarwal, *Difference Equations and Inequalities,* Marcel Dekker, Inc., New York, Basel and Hong Kong, 1992.

[2] G. E. Bergum, A. N. Philippou and A. F. Horadam, *Applications of Fibonacci Numbers,* Vol 1,2,3,4,5,6,7., Kluwer Academic Publishers.

[3] S. N. Elaydi, *An Introduction to Difference Equations,* Springer-Verlag New York Inc., 1996.

[4] P. Erdos and R. L. Graham, *Old and New Problems and Results in Combinatorial Number Theory,* Monographie de L'Enseignement Mathematique No. 28, 1980, 83-84.

[5] R. K. Guy, *Unsolved Problems in Number Theory,* Springer-Verlag New York Inc., 1981.

[6] Y. Hamaya, *On the asymptotic behavior of delya Fibonacci equations,* to appear.

[7] W. G. Kelley and A. C. Peterson, *Difference Equations-An Introduction with Applications-,* Academic Press Inc., 1991.

[8] The Fibonacci Quarterly, The Official Journal of The Fibonacci Association, South Dakota State Univ. and Santa Clara Univ., 1963-.

Proceedings of Dynamic Systems and Applications 3 (2001) 279-286

On Parameter Size for Existence of Periodic Solutions by the Method of Averaging

Maoan Han
Department of Mathematics, Shanghai Jiao Tong University
Shanghai 200030, PR China

ABSTRACT: The method of averaging plays an important role in nonlinear studies. Using the averaged equation we can find periodic solutions of a given periodic perturbed system with a sufficiently small parameter. However, the standard averaging theorem does not offer a definite parameter size for the existence of periodic solutions. We give a general averaging theorem which contains certain inequalities. These inequalities provide an estimation of parameter size for the existence of periodic solutions. Some applicatons are presented.

AMS(MOS) subject classification. 34C29, 34C25

1. EXISTENCE OF PERIODIC SOLUTIONS

Consider first a linear periodic system of the form

$$\dot{x} = \epsilon[Ax + f(t)], \tag{1.1}$$

where $x \in R^n$, $n \geq 1$, A is an $n \times n$ constant matrix, $f : R \to R^n$ is a continuous periodic function with period $T > 0$, and ϵ is a small parameter. Let $\mathcal{P}_T$ denote the set of all continuous periodic functions with period $T > 0$. Then, as we know, $\mathcal{P}_T$ is a linear normed space with a norm

$$|f| = \sup_{t\in R} \max_{1\leq j\leq n} |f_j(t)|, \tag{1.2}$$

where $f = (f_1, \cdots, f_n)$. For an $n \times n$ matrix $Q = (q_{ij})$ we define its norm as

$$|Q| = \max_i \sum_{j=1}^{n} |q_{ij}|. \tag{1.3}$$

One can prove that $|Qf| \leq |Q| \cdot |f|$. The following lemma is fundamental.

Lemma 1.1 *Suppose* $\det A \neq 0$. *Then (1.1) has a unique periodic solution* $\mathcal{K}_\epsilon f \in \mathcal{P}_T$ *satisfying the following.*

(i)

$$(\mathcal{K}_\epsilon f)(t) = \epsilon(e^{-\epsilon AT} - I)^{-1} \int_0^T e^{-\epsilon As} f(t+s)ds.$$

(ii) $|\mathcal{K}_\epsilon f| \leq M(A,\epsilon)|f|$, *where*

$$M(A,\epsilon) = |\epsilon| \int_0^T |(e^{-\epsilon AT} - I)^{-1} e^{-\epsilon As}|ds.$$

(iii) If $|\epsilon|T|A| < 1$, *then* $M(A,\epsilon) < \frac{|A^{-1}|(1+|\epsilon|T|A|)}{1-|\epsilon|T|A|}$.

Proof The expression of $\mathcal{K}_\epsilon f$ in (i) is direct by the formula of constant variation. By (i), it is obvious that

$$|\mathcal{K}_\epsilon f| \leq |\epsilon| \int_0^T |(e^{-\epsilon AT} - I)^{-1} e^{-\epsilon As}| \cdot |f| ds = M(A, \epsilon)|f|.$$

Now let $|\epsilon| T |A| < 1$. We have

$$e^{-\epsilon AT} - I = \sum_{k \geq 1} \frac{(-\epsilon AT)^k}{k!} = -\epsilon AT(I + L),$$

where

$$L = \sum_{k \geq 1} \frac{(-\epsilon AT)^k}{(k+1)!}. \tag{1.4}$$

Hence,

$$\epsilon(e^{-\epsilon AT} - I)^{-1} = -\frac{1}{T} A^{-1}(I + L)^{-1}. \tag{1.5}$$

By (1.4), we have

$$|L| \leq \sum_{k \geq 1} \frac{(|\epsilon||A|T)^k}{(k+1)!} = \frac{1}{|\epsilon|T|A|} \left[e^{|\epsilon|T|A|} - 1 - |\epsilon|T|A| \right] < |\epsilon|T|A|.$$

It follows that

$$|(I + L)^{-1}| = |\sum_{j \geq 0} (-1)^j L^j| \leq \sum_{j \geq 0} |L|^j = \frac{1}{1 - |L|} < \frac{1}{1 - |\epsilon|T|A|}.$$

Hence, by (1.5) we obtain

$$|\epsilon(e^{-\epsilon AT} - I)^{-1}| < \frac{|A^{-1}|}{T(1 - |\epsilon|T|A|)}. \tag{1.6}$$

Note that

$$\int_0^T |e^{-\epsilon As}| ds \leq \int_0^T e^{|-\epsilon As|} ds = T \frac{e^{|\epsilon|T|A|} - 1}{|\epsilon|T|A|} < T(1 + |\epsilon|T|A|). \tag{1.7}$$

Then the conclusion (iii) follows from (1.6) and (1.7).

Now consider a nonlinear system of the form

$$\dot{x} = \epsilon f(t, x, \epsilon), \tag{1.8}$$

where $x \in R^n$, $\epsilon > 0$, $f : R \times R^n \times R \to R^n$ is continuous and periodic in t with period $T > 0$. We also suppose f is C^1 with respect to x and satisfies

$$f(t, x, \epsilon) = f_1(t, x) + |\epsilon|^p f_2(t, x, \epsilon), \; p > 0. \tag{1.9}$$

By [1, 2, 3], as we know, a near-identity transformation of the form $x = y + \epsilon W(t,y)$ carries (1.8) into the system

$$\dot{y} = \epsilon[f_1(t,y) - W_t + F(t,y,\epsilon)], \tag{1.10}$$

where

$$F(t,y,\epsilon) = (I + \epsilon W_y)^{-1}[f(t, y+\epsilon W, \epsilon) - (I + \epsilon W_y) f_1(t,y)]. \tag{1.11}$$

Note that $(I + \epsilon W_y)^{-1}$ exists if $|\epsilon W_y| < 1$. We now determine the function W using the equations

$$f_1(t,y) - \bar{f}_1(y) = W_t, \quad \int_0^T W(t,y)dt = 0, \tag{1.12}$$

where

$$\bar{f}_1(y) = \frac{1}{T}\int_0^T f_1(t,y)dt. \tag{1.13}$$

Then (1.10) can be rewritten as

$$\dot{y} = \epsilon[\bar{f}_1(y) + F(t,y,\epsilon)]. \tag{1.14}$$

We can prove now

Theorem 1.1 *Let (1.9), (1.11) and (1.12) hold. Suppose there exist $y_0 \in R^n$, $b > 0$, $L_1(\epsilon, b) > 0$, $L_2(b) > 0$ and $N(\epsilon) > 0$ such that the following are satisfied:*
(i) $\bar{f}_1(y_0) = 0$, $\det A \neq 0$, where $A = \bar{f}_{1y}(y_0)$;
(ii) $|\epsilon||W_y| < 1$, $|\bar{f}_1(y) - A| \leq L_2(b)$ for $|y - y_0| \leq b$, $t \in R$;
(iii) $|F(t,y_0,\epsilon)| \leq N(\epsilon)$, $|F_y(t,y,\epsilon)| \leq L_1(\epsilon,b)$ for $|y - y_0| \leq b$, $t \in R$;
(iv) $M(A,\epsilon)[N(\epsilon) + b(L_1(\epsilon,b) + L_2(b))] \leq b$, where

$$M(A,\epsilon) = |\epsilon| \int_0^T |(e^{-\epsilon AT} - I)^{-1} e^{-\epsilon As}| ds.$$

Then (1.14) has a unique periodoc solution $y(t)$ which satisfies $|y(t) - y_0| \leq b$.

Proof Set

$$G(t,y,\epsilon) = F(t,y,\epsilon) + \bar{f}_1(y) - A(y - y_0). \tag{1.15}$$

Then (1.14) becomes

$$\dot{y} = \epsilon[A(y - y_0) + G(t,y,\epsilon)]. \tag{1.16}$$

Let

$$\mathcal{P}_{T,b} = \{y \in \mathcal{P}_T : |y - y_0| \leq b\}.$$

By lemma 1.1, for any $g \in \mathcal{P}_{T,b}$, the equation

$$\dot{y} = \epsilon[A(y - y_0) + g(t)]$$

has a unique solution $\mathcal{K}_\epsilon g$ in $\mathcal{P}_T$ with

$$|\mathcal{K}_\epsilon g| \leq M(A,\epsilon)|g|. \tag{1.17}$$

For any $y \in \mathcal{P}_{T,b}$ let

$$(\mathcal{K}^* y)(t) = (\mathcal{K}_\epsilon G(\cdot, y(\cdot), \epsilon))(t).$$

Then by (i)-(iv), (1.15), (1.17) and the mean value theorem, we have

$$|\mathcal{K}^* y| \leq M[N + b(L_1 + L_2)] \leq b, \text{ for } y \in \mathcal{P}_{T,b},$$

$$|\mathcal{K}^* y_1 - \mathcal{K}^* y_2| \leq M(L_1 + L_2)|y_1 - y_2|, \text{ for } y_i \in \mathcal{P}_{T,b}, \ i = 1, 2.$$

Note that $M(L_1 + L_2) < 1$ by (iv). Hence, $\mathcal{K}^* : \mathcal{P}_{T,b} \to \mathcal{P}_{T,b}$ is a contracting map. Therefore, it has a unique fixed point y^* in $\mathcal{P}_{T,b}$. That is, $y^* = \mathcal{K}_\epsilon G(\cdot, y^*, \epsilon)$, and y^* is the only periodic solution of (1.16) in $\mathcal{P}_{T,b}$. This ends the proof.

Remark 1.1 *Since $F(t, y, 0) = 0$ we can choose $N(0) = 0$, $L_1(0, b) = 0$, $L_2 = O(b)$, and $b = O(N(\epsilon))$ by lemma 1.1(iii) such that all conditions of theorem 1.1 are satisfied for $|\epsilon| > 0$ small if $\det A \neq 0$. This gives the usual averaging theorem.*

Remark 1.2 *As we have seen, the idea of the proof of theorem 1.1 is not new. But the statement here is new and contains more information than the usual one. We can use it to find a value $\epsilon_0 > 0$ such that (1.18) has a periodic solution for $0 < |\epsilon| < \epsilon_0$. This kind of examples will be given in the next section.*

For the linear periodic system

$$\dot{x} = \epsilon[B(t)x + g(t)], \tag{1.18}$$

where B and g are continuous and periodic with period $T > 0$, using theorem 1.1 we can prove

Theorem 1.2 *Suppose*

(i) $\det A \neq 0$, *where* $A = \frac{1}{T}\int_0^T B(t)dt$;

(ii)

$$|B_0(t) - \bar{B}_0| \leq a_1, \ |B(t)(B_0(t) - \bar{B}_0) - (B_0(t) - \bar{B}_0)B(t)| \leq a_2,$$

where

$$B_0(t) = \int_0^t (B(s) - A)ds, \ \ \bar{B}_0 = \frac{1}{T}\int_0^T B_0(t)dt;$$

(iii)

$$0 < |\epsilon|a_1 < 1, \ \ \frac{|\epsilon|a_2}{1 - |\epsilon|a_1} M(A, \epsilon) < 1,$$

where a_1, a_2 are positive constants. Then (1.18) has a unique periodic solution.

Proof By (1.12), it is easy to see that $W_y = B_0(t) - \bar{B}_0$. Hence, by (1.11)

$$F_y = (I + \epsilon W_y)^{-1}\epsilon[B(t)W_y - W_y B(t)].$$

By (ii), we have $|W_y| \leq a_1$, $|BW_y - W_y| \leq a_2$. Note that

$$|(I + \epsilon W_y)^{-1}| \leq \frac{1}{1 - |\epsilon|a_1}.$$

It follows that

$$|F_y| \le \frac{|\epsilon|a_2}{1-|\epsilon|a_1} = L_1, \quad L_2 = 0.$$

Then the conclusion follows from theorem 1.1 by choosing $b > 0$ sufficiently large. The proof is completed.

2. APPLICATIONS

In this section we apply theorems 1.1 and 1.2 to some periodic systems to find definite intervals of the parameter ϵ for the existence of periodic solutions.

First, consider

$$\dot{x} = \epsilon\left[\begin{pmatrix} 1+\cos t & \sin t \\ \cos t & 1+\sin t \end{pmatrix} x + \begin{pmatrix} \cos t \\ \sin t \end{pmatrix}\right]. \tag{2.1}$$

By (1.3), it is direct that $T = 2\pi$, $A + I$, $a_1 = a_2 = 2$, and $M(A,\epsilon) = 1$ since

$$B_0 = \begin{pmatrix} \sin t & 1-\cos t \\ \sin t & 1-\cos t \end{pmatrix},$$

$$B(t)(B_0(t)-\bar{B}_0) - (B_0(t)-\bar{B}_0)B(t) = \begin{pmatrix} 1 & -1 \\ 1 & -1 \end{pmatrix}.$$

Hence we have immediately from theorem 1.2

Proposition 1.1 *For $0 < |\epsilon| < \frac{1}{4}$, (2.1) has a unique 2π-periodic solution.*

Second, consider a scalar equation of the form

$$\dot{x} = \epsilon[a_0(t) + a_1(t)x + a_2(t)x^2], \tag{2.2}$$

where a_i, $i = 0, 1, 2$, are continuous 2π-periodic functions. Let

$$\bar{a}_i = \frac{1}{2\pi}\int_0^{2\pi} a_i(t)dt,$$

$$a_i^*(t) = \int_0^{*}(a_i(s)-\bar{a}_i)ds - \frac{1}{2\pi}\int_0^{2\pi}\int_0^{*}(a_i(s)-\bar{a}_i)dsdt.$$

By (1.12), we can choose $W(t,y) = a_0^* + a_1^* y + a_2^* y^2$. Using (1.11) we have

$$F(t,y,\epsilon) = \frac{\epsilon}{1+\epsilon W_y}[A_{10}(t) + 2A_{20}(t)y + A_{21}(t)y^2 + \epsilon a_2 W^2],$$

where $A_{ij} = a_i a_j^* - a_j a_i^*$, $0 \le j < i \le 2$. Then it is easy to know that

$$F_y = \frac{\epsilon}{(1+\epsilon W_y)^2}[b_0 + \epsilon b_1 + \epsilon^2 b_2], \tag{2.3}$$

where

$$b_0 = 2(A_{20} + A_{21}y),\ b_1 = 2A_{21}W,$$
$$b_2 = 2a_2W[(a_1^*)^2 - a_2^*a_0^* + a_1^*a_2^*y + 3(a_2^*y)^2]. \tag{2.4}$$

If $\bar{a}_0 = 0$, then $\bar{f}_1(0) = 0$ and

$$F(t, 0, \epsilon) = \frac{\epsilon}{1 + \epsilon a_1^*}[A_{10}(t) + \epsilon a_2(a_0^*)^2]. \tag{2.5}$$

By (2.3-(2.5) we can find a number $\epsilon_0 > 0$ such that (2.2) has a periodic solution for $0 < |\epsilon| < \epsilon_0$. For example, we have

Proposition 1.2 *The equation*

$$\dot{x} = \epsilon[\sin t + (1 + \cos t)x + x^2 \cos t]$$

has a 2π*-periodic solution for* $0 < |\epsilon| < \frac{1}{11}$.

Proof We have

$$\bar{a}_0 = 0,\ \bar{a}_1 = 1,\ \bar{a}_2 = 0,\ a_0^* = -\cos t,\ a_1^* = a_2^* = \sin t.$$

Further, by (2.4),

$$W = -\cos t + y\sin t + y^2 \sin t,\quad b_0 = -2(1 + y\sin t),$$

$$b_1 = -2W\sin t,\quad b_2 = W\sin 2t[\sin t + \cos t + y\sin t + 3y^2 \sin t].$$

Hence,

$$|W| \le 1 + |y| + y^2,\quad |W_y| \le 1 + 2|y|,\quad |b_0| \le 2(1 + |y|),$$
$$|b_1| \le 1 + 2(|y| + y^2),\quad |b_2| \le (1 + |y| + y^2)(\sqrt{2} + |y| + 3y^2).$$

Let $|y| \le 1$. Then $|b_0| \le 4$, $|b_1| \le 5$, $|b_2| < 27$. Thus, from (2.3) and (2.5)

$$|F_y| \le \frac{|\epsilon|}{(1 - 2|\epsilon|)^2}[4 + 5|\epsilon| + 27|\epsilon|^2] < \frac{|\epsilon|}{1 - 4|\epsilon|}[4 + 5|\epsilon| + 27|\epsilon|^2] = L_1,$$

$$|F(t, 0, \epsilon)| \le \frac{|\epsilon|}{1 - |\epsilon|}(2 + |\epsilon|) < \frac{2|\epsilon|}{1 - 4|\epsilon|} = N(\epsilon).$$

Also, $L_2 = 0$, $M = 1$. Note that if $b = 1$ and $|\epsilon| \le \frac{1}{11}$, then

$$10|\epsilon| + 5|\epsilon|^2 + 27|\epsilon|^3 \le 1,$$

or

$$\frac{2|\epsilon|}{1 - 4|\epsilon|} + \frac{|\epsilon|}{1 - 4|\epsilon|}[4 + 5|\epsilon| + 27|\epsilon|^2] \le 1.$$

The conclusion follows by theorem 1.1 and the proof is completed.

Proposition 1.3 *The scalar equation*

$$\dot{x} = \epsilon[\frac{x^2}{2(1+x^2)} - \sin x + \cos t]$$

has infinitely many periodic solutions for $0 < |\epsilon| \leq \frac{1}{57}$.

Proof Let $a(x) = \frac{x^2}{2(1+x^2)}$. Then $\bar{f}_1(x) = a(x) - \sin x$, $W = \sin t$. Since $0 \leq a(x) < 1/2$, $\bar{f}_1$ has infinitely many simple roots x_k, $k = 0, \pm 1, \cdots$, with $x_0 = 0$, $\lim_{k\to\pm\infty} x_k = \pm\infty$. By (1.11), we have

$$F = \frac{1}{1+\epsilon \sin t}[a(y+\epsilon \sin t) - a(y) - \sin(y + \epsilon \sin t) + \sin y].$$

Hence, for $0 < |\epsilon| < 1$, by (1.2) and (1.3)

$$|F| \leq \frac{|\epsilon|}{1-|\epsilon|}[\sup|a'(y)| + 1], \ |F_y| \leq \frac{|\epsilon|}{1-|\epsilon|}[\sup|a''(y)| + 1],$$

$$|\bar{f}_{1y}(y) - \bar{f}_{1y}(x_k)| \leq (\sup|a''(y)| + 1)|y - x_k|.$$

It is easy to see that $|a'| \leq 1/3$, $|a''| \leq 1$. We can choose

$$N(\epsilon) = \frac{4|\epsilon|}{3(1-|\epsilon|)}, \ L_1(\epsilon, b) = \frac{2|\epsilon|}{1-|\epsilon|}, \ L_2(b) = 2b.$$

Note that

$$\sin x_k = a(x_k) < \frac{1}{2}, \ |\cos x_k| = \sqrt{1 - \sin^2 x_k} > \frac{\sqrt{3}}{2}.$$

This yields that $|A| = |a'(x_k) - \cos x_k| \geq \frac{\sqrt{3}}{2} - \frac{1}{3} > \frac{1}{2}$, and $M = |A|_{-1} < 2$. By theorem 1.1 it suffices to show

$$2\left[\frac{4|\epsilon|}{3(1-|\epsilon|)} + b\left(\frac{2|\epsilon|}{1-|\epsilon|} + 2b\right)\right] \leq b,$$

or

$$8|\epsilon| + 3(5|\epsilon| - 1)b + 12(1 - |\epsilon|)b^2 \leq 0.$$

The above inequality has a solution in $b > 0$ if $\Delta = 9 - 522|\epsilon| + 657|\epsilon|^2 \geq 0$, which is satisfied when $0 < |\epsilon| \leq \frac{1}{57}$. The proof is completed.

By taking

$$L_1 = N = \frac{|\epsilon|}{1-|\epsilon|}, \ L_2 = b, \ M = \frac{2}{\sqrt{3}},$$

in the same way we can prove

Proposition 1.4 *For* $0 < |\epsilon| \leq \frac{1}{3}$, *the toral equation*

$$\dot{x} = \epsilon[-\sin x + \cos t]$$

has exactly two 2π*-periodoc solutions on the torus.*

Finally, we consider a space quadratic system of the form

$$\dot{x} = y - \epsilon xz,\ \dot{y} = -x - \epsilon yz,\ \dot{z} = \epsilon(2x^2 - 1 + z). \tag{2.6}$$

We can prove

Proposition 1.5 *For* $0 < |\epsilon| \le \frac{1}{21\pi}$, *(2.6) has a limit cycle.*

Proof Let $(x, y, z) = (\sqrt{h_1}\sin\theta, \sqrt{h_1}\cos\theta, h_2)$. We can obtain from (2.6)

$$\frac{dh}{d\theta} = \epsilon(-2h_1h_2, 2h_1\sin^2\theta - 1 + h_2)^{\mathrm{T}} \equiv \epsilon f(\theta, h),$$

where $h = (h_1, h_2)$. The average of f is $\bar{f}(h) = (-2h_1h_2, h_1 + h_2 - 1)^{\mathrm{T}}$. By (1.11) and (1.12), we have

$$W(\theta, h) = (0, \frac{1}{2}h_1\sin 2\theta)^{\mathrm{T}},\ (I + \epsilon W_h)^{-1} = \begin{pmatrix} 1 & 0 \\ \frac{1}{2}\epsilon\sin 2\theta & 1 \end{pmatrix},$$

$$F(\theta, h, \epsilon) = (\epsilon h_1^2\sin 2\theta, \epsilon h_1\sin 2\theta(-h_2 - \frac{1}{2} + \frac{1}{2}\epsilon h_1\sin 2\theta))^{\mathrm{T}}.$$

Let $h_0 = (1, 0)$, $b = \max\{|h_1 - 1|, |h_2|\}$. Then noting that

$$D\bar{f}(h) = \begin{pmatrix} -2h_2 & -2h_1 \\ 1 & 1 \end{pmatrix},\quad (D\bar{f}(h_0))^{-1} = \begin{pmatrix} \frac{1}{2} & 1 \\ -\frac{1}{2} & 0 \end{pmatrix},$$

$$F_h = \begin{pmatrix} 2\epsilon h_1\sin 2\theta & 0 \\ \epsilon\sin 2\theta(-h_2 - \frac{1}{2} + \epsilon h_1\sin 2\theta) & -\epsilon h_1\sin 2\theta \end{pmatrix},$$

by (1.2) and (1.3) we have

$$|A| = |D\bar{f}(h_0)| = 3,\ |A^{-1}| = 1,\ |D\bar{f}(h) - D\bar{f}(h_0)| \le 4b = L_2,$$

$$|F(\theta, h_0, \epsilon)| \le |\epsilon| = N,\ |F_h| \le |\epsilon|[3b + 5/2 + |\epsilon|(1 + b)] = L_1.$$

By lemma 1.1(iii) we have

$$M(A, \epsilon) \le \frac{1 + 6\pi|\epsilon|}{1 - 6\pi|\epsilon|} \le \frac{9}{5}$$

if $|\epsilon| \le \frac{1}{21\pi}$. Also, it is easy to see that

$$9[|\epsilon| + b(4b + \frac{5}{2}|\epsilon| + 3b|\epsilon| + |\epsilon|^2(1 + b))] \le 5b$$

if $b = 4|\epsilon|$ and $|\epsilon| \le \frac{1}{21\pi}$. Hence, the conclusion follows from theorem 1.1. The proof is completed.

REFERENCES

1. Guckenheimer,J., Holmes,P. Nonlinear Oscillations, Dynamical Systems, and Bifurcations of Vector Fields, Springer-Verlag, New York, 1983.
2. Farkas,M., Periodic Motions. Springer-Verlag, New York, 1994.
3. Hale,J.K., Ordinary Differential Equations. Robert E. Krieger Publishing Co., Malabar, Florida, 1980.

Proceedings of Dynamic Systems and Applications 3 (2001) 287- 296

Discretised Models of Impacting Beams

M. C. Harrison, R. P. Menday and W. A. Green
Department of Mathematical Sciences,
Loughborough University, Loughborough, LE11 3TU, UK

ABSTRACT: The study of the dynamics of a partial delamination in a vibrating beam [2], leads to the problem of modelling the impact of the delaminated region with the main body of the beam. In an attempt to tackle this problem without solving the full elastodynamic equations, the delaminated portion is discretised as a set of elastic slugs joined together by elastic shear springs, whilst the impact region of the main beam is discretised as a set of elastic rods. The system is assumed to be moving in its fundamental mode when each slug impacts its corresponding rod, at which stage compressive stress waves are set up in both. The stress waves in the slugs propagate with non-constant profiles and the behaviour following the arrivals of the reflected pulses at the contact faces depends on the stiffnesses of the shear springs. The effective coefficient of restitution can be determined in terms of the spring stiffnesses. Results are obtained for the cases of slugs connected (a) to their nearest neighbours only and (b) to nearest neighbours and to next nearest neighbour.
AMS(MOS) subject classification 34C15, 70K40, 73D35

1 INTRODUCTION

The initial motivation for this work was to model the growth of delaminations in a damaged laminate due to dynamic stresses arising from forced vibrations. The first approach [2] was to consider a cantilever beam undergoing periodic forced vibrations and having a partially detached flap at its free end. The flap is free to deform in the direction away from the main beam and when it impacts the beam it will either rebound or it will travel in contact with the beam for some period of its motion. This simple approach treats the impact as an elastic collision with the rebound being determined in terms of an effective coefficient of restitution e between the impacting bodies. Here we develop more sophisticated models of the impact to determine the extent to which the coefficient of restitution model is accurate and reasonable and to suggest ways of predicting an effective coefficient of restitution e.

We can obtain an improved model of the impacting behaviour by considering the propagation of stress waves introduced by the impact. A similar situation occurs when a slug of elastic material impacts a rod of elastic material. The impact sets up two compressive stress waves one travelling down the rod and the other travelling back along the slug, each stress wave having a constant profile. When the receding wave meets the free surface at the back of the slug it is reflected as a tensile stress wave which travels with constant profile up the slug to the contact face. Then either, the magnitude of the tensile stress wave is greater than the compressive stress at the contact face, in which case the slug and the rod move apart with the slug moving in the opposite direction to its original motion or, the tensile wave is not sufficient to reduce the compressive stress to zero and the slug and rod remain in contact with two new stress waves being initiated, one in the rod and the other travelling backwards in the slug. In the latter case the slug and rod remain in contact, with additional waves being set up following each passage backwards and forwards in the slug, until the first wave set up in the rod has been reflected back as a tensile wave from the free surface and arrives at the contact surface. At this time the systems part and the slug keeps moving along its original direction but with a reduced speed.

In the case of the impacting beams the stress waves set up would travel through the depth of each beam in the direction away from the contact surfaces. In an attempt to determine the subsequent motion without solving the full elastodynamic equations in both the main beam and the delaminated portion, the delaminated portion is discretised as a set of elastic slugs joined together by elastic shear springs, whilst the impact region of the main beam is discretised as a set of elastic rods as shown in Figure 1.

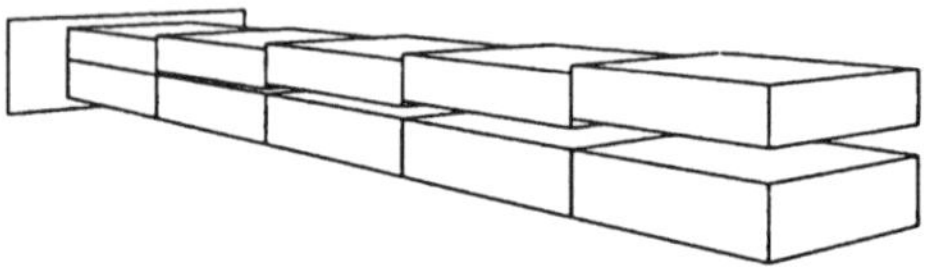

Figure 1: A discretised model of the delamination and the main beam.

First, we consider three single degree-of-freedom models: (a) a single 'slug' impacting a rod of semi-infinite length, (b) a single 'slug' impacting a rod of finite length, and (c) a single 'slug', coupled to stationary neighbouring slugs on either side, impacting a rod of semi-infinite length. Second, we consider some multi-degree of freedom systems, in particular discretised models of beams where a set of slugs are joined together with shear springs.

2 SINGLE DEGREE OF FREEDOM MODELS

2.1 A slug impacts a semi-infinite rod

First consider a slug of depth h_s travelling at speed V impacting a stationary semi-infinite rod at time $t = 0$ as shown in Figure 2. Using non-dimensional variables so that the slug has depth 1, let $x(X_0, t)$ and $\bar{x}(X_0, t)$ be the position at time t of the cross-section which is at location X_0 at $t=0$ of the slug and rod respectively. Let the functions $u(X_0, t)$ and $\bar{u}(X_0, t)$ denote the additional non-dimensional displacement following the impact, then the position of the two bodies is,

$$x = X_0 + \frac{V}{c}\Big(t + u(X_0, t)\Big)\ , \quad \bar{x} = X_0 + \frac{V}{c}\alpha\bar{u}(X_0, t) \tag{1}$$

where $\alpha = c/\bar{c}$, and c and $\bar{c}$ are the wave speeds in the slug and rod respectively.

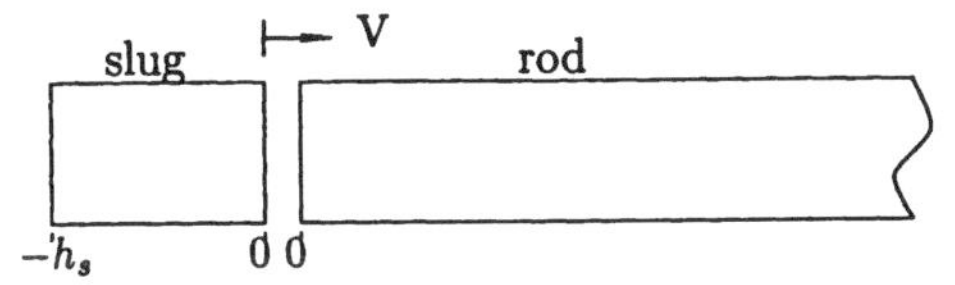

Figure 2: Impact between the slug and the rod

Variation in stress throughout the slug or rod is caused by wave propagation at impact. In this case there are no waves in the rod returning to the interface. It is the reflected waves in the slug which can cause the slug and rod to move apart, if at all. In the slug and rod, $u(X_0,t)$ and $\bar{u}(X_0,t)$ satisfy the one-dimensional wave equations with the general solutions,

$$u(X_0,t) = f(X_0 - t) + g(X_0 + t)\ , \quad \bar{u}(X_0,t) = \bar{f}(X_0 - \tfrac{1}{\alpha}t) + \bar{g}(X_0 + \tfrac{1}{\alpha}t) \tag{2}$$

where $f(...)$, $g(...)$, $\bar{f}(...)$ and $\bar{g}(...)$ are arbitrary functions.

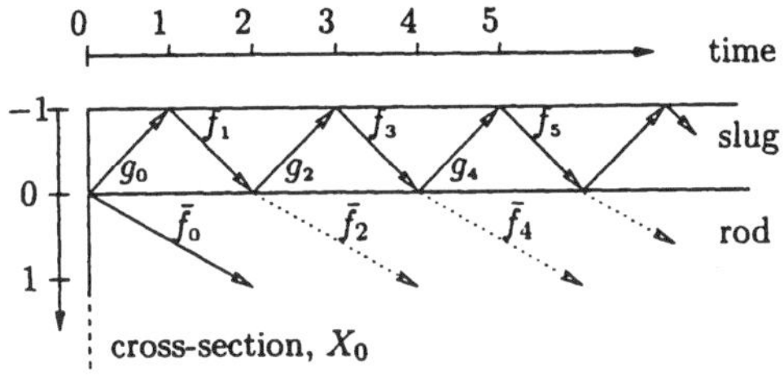

Figure 3: Wave propagation in the two bodies

Figure 3 represents the two bodies in contact after the impact has occured. The vertical (downward) axis indicates the cross-section in either body, and the horizontal axis the non-dimensional time. At $t = 0$ a stress wave in each body propagates from the interface. After non-dimensionalising, the wave speed in the slug is 1, and in the rod it is $\frac{1}{\alpha}$ or $\frac{\bar{c}}{c}$. In the diagram, the wave speed in the slug is greater than the wave speed in the rod.

At impact the interface conditions state that the particle velocities and the stress in the slug and the rod must be equal. If we substitute the general solutions (2) into the interface conditions, integrate and solve, we obtains the solutions for the new waves $u(X_0,t)$ and $\bar{u}(X_0,t)$. Hence the overall motion of the two bodies is given by,

$$x = X_0 + \frac{V}{c}\Big(t - \frac{1}{1+z}(t + X_0)\mathsf{H}[t + X_0]\Big) \qquad \text{for } t < 1 \tag{3}$$

$$\bar{x} = X_0 + \frac{V}{c}\frac{z}{1+z}(t - \alpha X_0)\mathsf{H}[t - \alpha X_0] \qquad \text{for } t < 2 \tag{4}$$

where $z = \frac{c\rho}{\bar{c}\bar{\rho}}$, ρ and $\bar{\rho}$ are the density of the slug and rod respectively and $\mathsf{H}[...]$ is the Heaviside function.

The solution in the slug is valid up to time $t = 1$ when the wave $g_0(X_0 + t)$ reaches the boundary $X_0 = -1$ of the slug. At this time a wave travelling in the opposite (positive) direction must be set up in order to maintain zero stress at $X_0 = -1$; this wave is shown in Figure 3 as $f_1(X_0 - t)$. The stress at the boundary $X_0 = -1$ is zero. Application of this boundary condition enables us to find the new wave and hence the solution in the slug,

$$x = X_0 + \frac{V}{c}\left(t - \frac{1}{1+z}\left\{(t+X_0)\mathsf{H}[t+X_0] + (t - X_0 - 2)\mathsf{H}[t - X_0 - 2]\right\}\right) \tag{5}$$

which is valid for $1 < t < 2$. The solution in the rod remains unchanged.

When the wave f_1 reaches the interface, a negative travelling wave g_2 is set up in the slug and a positive travelling wave $\bar{f}_2$ in the rod. These waves maintain the boundary conditions at the interface, i.e. the particle velocities and stresses must be equal. We find,

$$\begin{aligned} x = X_0 + \frac{V}{c}\Big(t - \frac{1}{1+z}\Big\{(t+X_0)\mathsf{H}[t+X_0] + (t - X_0 - 2)\mathsf{H}[t - X_0 - 2]\Big\} \\ + \frac{(1-z)}{(1+z)^2}(t + X_0 - 2)\mathsf{H}[t + X_0 - 2]\Big) \end{aligned} \tag{6}$$

$$\begin{aligned} \bar{x} = X_0 + \frac{V}{c}\Big(\frac{z}{1+z}(t - \alpha X_0)\mathsf{H}[t - \alpha X_0] \\ - \frac{2z}{(1+z)^2}(t - \alpha X_0 - 2)\mathsf{H}[t - \alpha X_0 - 2]\Big) \end{aligned} \tag{7}$$

We can compute the stress at the interface at time $t = 2$ to be $V c\rho\frac{1-z}{(1+z)^2}$. If $z \leq 1$, this is positive and the slug and rod part company at $t = 2$, i.e. when the reflected wave hits the interface. On the point of separation, we can find the velocity of the interface of each body from equations (5) and (4) to be,

$$\left.\frac{\partial x}{\partial t}\right|_{X_0=0} = \frac{V}{c}\left(\frac{z-1}{z+1}\right), \quad \left.\frac{\partial \bar{x}}{\partial t}\right|_{X_0=0} = \frac{V}{c}\left(\frac{z}{z+1}\right) \tag{8}$$

In the slug at $t = 2$ the velocity is constant throughout the body, and therefore the mean slug velocity is equal to the velocity of the interface. In the semi-infinite rod the mean velocity is always zero, even when the velocity of the interface is non-zero. The coefficient of restitution is the ratio of the relative mean speed of separation to the relative mean speed of approach. Hence, we deduce that $e = \frac{1-z}{1+z}$; this is independent of the slug speed.

If $z > 1$ it can be shown that the condition for separation is never satisfied, and the slug and rod remain in contact. In this case, the stress at the interface is always negative, however as $t \to \infty$, the stress approaches zero and the slug comes to rest. We could argue that the effective coefficient of restitution is then zero.

2.2 A slug impacts a finite rod

If we now consider a slug impacting a rod of finite length ah_s there is a new boundary condition at $X_0 = a$; in practice, $a > 1$. This implies that there are positive and negative travelling waves in both bodies. A wave returning to the interface creates new waves in both bodies. Now waves returning to the interface in either body can cause separation. It takes 1 time unit for a wave to propagate through the slug and αa time units to propagate through the rod. We expect the materials of the slug and the rod to have comparable wave propagation speeds, and therefore α to be approximately equal to one. As $a > 1$, we can assume that $\alpha a > 1$. This implies that of the two waves set up at $t = 0$, the one in the slug is the first one to return back to the interface. If $z \leq 1$, the slug and the rod will again part at $t = 2$. If $z > 1$, then assuming that the wave in the slug has returned to the interface N times before the return of the first wave in the rod, the stress in the rod at the time of arrival of this wave is given by

$$\sigma = \frac{Vc\rho}{(z+1)^2}\left(2z - (z+1)\left(\frac{z-1}{z+1}\right)^N\right) \tag{9}$$

Since $z > 1$, then $0 < \left(\frac{z-1}{z+1}\right)^N < 1$ for all $N > 0$ and $2z - (z+1) > 0$, hence σ is positive and the slug and the rod part company at $t = 2\alpha a$, i.e. when the first reflected wave in the rod reaches the interface, irrespective of the number of multiple reflections in the slug.

The mean velocities of both bodies, and hence the coefficient of restitution, can be found by computing the definite integral of the velocity with respect to X_0 over the entire length of the body. This depends on the number of internal reflections and the length of the rod. Note that if $z = 1$, the mean velocity of the slug is 0 and the mean velocity of the rod is $\frac{1}{\alpha a}$. This implies that $e = \frac{1}{\alpha a}$ if $z = 1$. Figure 4 shows the value of e as a function of z for different values of αa. If the two bodies are constructed from the same material,

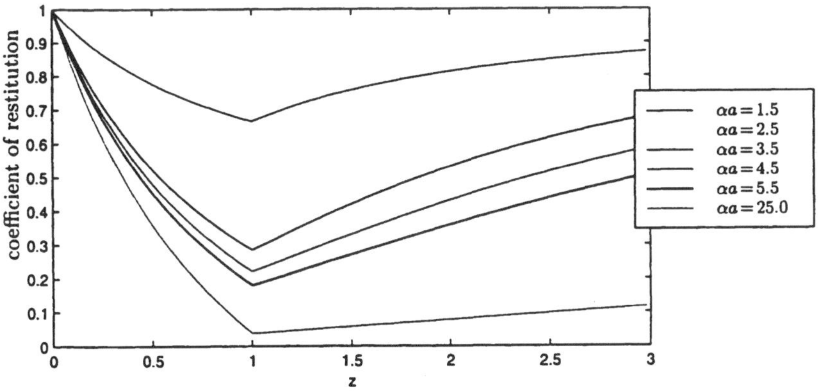

Figure 4: Graph of the coefficient of restitution, e against z

then $z = 1$. Consider therefore the area surrounding the parameter choice $z = 1$. Although this is not a single degree-of-freedom system, in the case of a delamination and main beam constructed from the same material, the quantity αa will be large since the thickness of the beam will be large compared with the thickness of the delamination. For large αa ($\alpha a > 5$) the coefficient of restitution is within the range $0 < e < 0.3$, for $0.8 < z < 1.3$. The effective coefficient of restitution for the experimental apparatus [1] was in the region $0 < e < 0.3$, particularly when the delamination was constructed from thin steel foil (where $z \approx 1$).

2.3 A slug with side coupling impacts a semi-infinite rod

Next consider a slug impacting a stationary semi-infinite rod as in section (2.1). However now the slug is attached through continuously distributed shear springs to two stationary slugs, one on either side. If the coupling force is proportional to the relative displacement of the cross-section X_0 of both bodies, the non-dimensional equation of motion for the slug becomes,

$$\frac{\partial^2 u}{\partial t^2} = \frac{\partial^2 u}{\partial X_0^2} - q^2(t+u)\ , \quad q^2 = \frac{h_s^2}{c^2}\frac{2kb}{A\rho} \tag{10}$$

where k is the spring stiffness per unit area, bh_s is the area of adjacent sides of the slug and A is the area of the impacting surfaces. In the rod $\bar{u}$ satisfies the wave equation, as before.

To solve the equation of motion for the slug (10), we first take its Laplace transform with respect to t. We then impose the boundary conditions in the transform domain, viz. stress free ends, continuity of particle velocities at the interface and continuity of stress at the interface to obtain the transform solution for the slug displacement,

$$\hat{u}(X_0, s) = \frac{s\cosh(\lambda(1+X_0))}{(s^2+q^2)(-z\lambda\sinh(\lambda) - s\cosh(\lambda))} - \frac{q^2}{s^2(s^2+q^2)} \tag{11}$$

where $\lambda = \sqrt{(s^2+q^2)}$ and s is the transform parameter. We can then find [3], in the frequency domain, the slug stress and slug mean velocity. These expressions do not have straight-forward analytical inverse Laplace transforms and must be inverted numerically.

One of the implications of the side coupling is that the waves in the slug no longer propagate with constant profiles. Without coupling, the stress at the interface is a series of constant step functions. Waves in the slug arriving at the interface, cause these step functions and they occur at time $t = 2$ and subsequent multiples of 2. With coupling, the time when the bodies part company is dependent on the strength of the coupling. When there is zero coupling and $z > 1$, the slug and the rod do not part company. However, the presence of the coupling causes the bodies to always part company and not only when the reflected waves meet the interface. The time at which parting occurs depends on the strength of the coupling.

Figure 5 shows the parting times and the mean parting velocity of the slug, for varying values of z and some sample values of q. When calculating the mean velocity we take $V/c=1$. Note, the mean velocity of the rod is always zero, however we can obtain positive mean parting velocities of the slug. However, on parting, the pointwise velocity of the rod at the interface must be greater than that of the slug. Examining the results for $z<1$, we see that for all q, parting occurs at $t=2$, and there is an increase in the mean slug speed as the coupling strength increases. This speed increase however is relatively small. This corresponds to an increase in e. For $z>1$, the strength of the coupling has greater influence on the mean velocity. As the coupling increases, the time of parting decreases. This now gives a non-zero e.

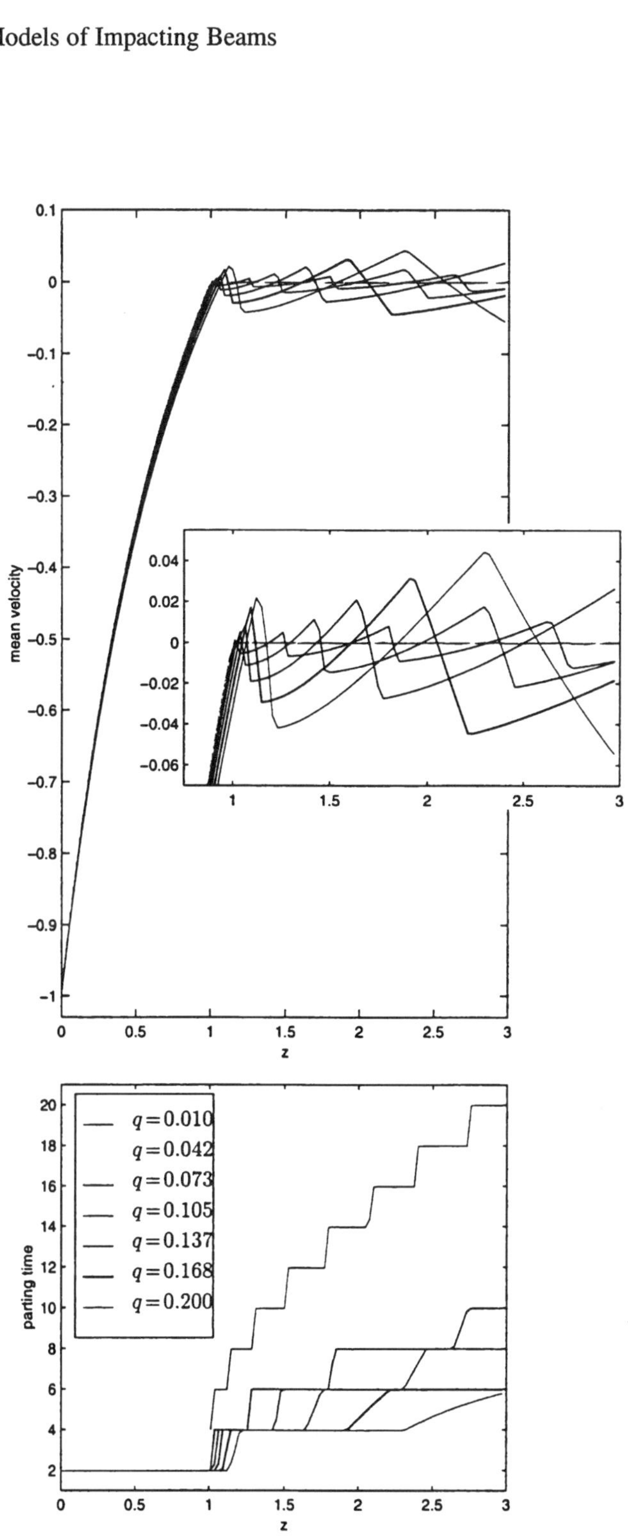

Figure 5: Slug departing times and mean velocities

3 MULTI DEGREE OF FREEDOM MODELS

We have considered models where the delamination is constructed from a series of slugs, which are coupled by shear springs to their nearest neighbour, as in Figure 6. The impacting region of the main beam is modelled as a series of rods, where impacts occur between corresponding pairs of slugs and rods in the models. As the number of slugs, n, approaches infinity, we obtain a closer approximation to the continuous model. The effect of the coupling is to introduce a finite number of natural modes, equal to the number of slugs in the discretisation. When the slugs move as rigid bodies, for nearest neighbour coupling the continuous structure behaves like a vibrating string.

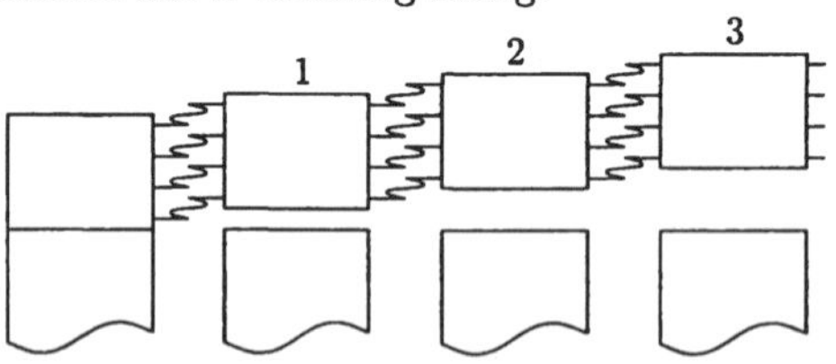

Figure 6: A system of elastic slugs connected in a 'train' by a series of shear springs

We assume that the delamination is moving in the fundamental mode of vibration when an impact occurs between the delamination and the main beam. Each of the slugs impacts its corresponding rod, at which stage the compressive stress waves are set up in each slug and in each rod. Because of the effect of the shear springs, the stress waves in the slugs propagate with non-constant profiles and the nature of the behaviour following the arrivals of the reflected pulses at the contact faces depends on the stiffnesses of the shear springs.

When the impact takes place in the fundamental mode, we find that the fundamental beam mode gives the only non-zero contribution to the rebound motion. In addition, the motion is determined by the solution for the first normal mode of the single slug/rod model and we can determine the results from Figure 5, if we replace the parameter q by $\sqrt{-\mu_1 q_n}$, where μ_j are the eigenvalues of a tridiagonal $n \times n$ matrix (see[3]). For a beam of rectangular cross-section with shear modulus K, $q_n = \sqrt{\frac{K}{\rho c^2}}n\left(\frac{h_s}{l}\right)$. Note that the characteristics of the impact can be found by choosing q according to the strength of the coupling and q_n behaves like $\frac{h_s}{l}$. In effect, this is the same as for connection to 2 static slugs and the effective coefficient of restitution can be determined in terms of the spring stiffnesses; e now depends on the number of slugs.

We have also considered models where we couple each slug both to its nearest neighbour and to its next-nearest neighbour. As the number of slugs in the discretisation approaches infinity, we obtain the beam bending equation for the rigid body movement of the slugs. The continuous structure behaves like a beam in bending when the slugs move as rigid bodies. Again the motion is determined by the frequency of the first normal mode of the single slug/rod model and we can read results from Figure 5, if we now replace q by $\sqrt{-\xi_1 q_n}$ where ξ_j are again the eigenvalues of an $n \times n$ matrix given in [3]. For a beam of rectangular cross-section, $q_n = \frac{n^2}{\sqrt{12}}\left(\frac{h_s}{l}\right)^2$ and now q_n behaves like $\left(\frac{h_s}{l}\right)^2$.

4 CONCLUSIONS

For the *uncoupled* single degree of freedom models, if $z \leq 1$, parting occurs at $t=2$ when the first wave in the slug returns to the interface. For $z > 1$, separation is caused by the first wave in the rod returning to the interface, and we only observe parting if we make the rod of finite length. For $z \approx 1$, and for $a\alpha > 5$, we predict an effective coefficient of restitution in the range $0 < e < 0.3$. For the *coupled* system, parting always occurs. If $z < 1$, parting occurs at $t=2$. For $z > 1$, the coupling causes separation to occur. The stronger the coupling, the shorter the time before parting, and the slug always has a small mean parting velocity.

For the multi degree of freedom models, after an impact in the fundamental mode, the rebound is in the fundamental mode. The rebound motion is determined by the frequency of the 1st normal mode of the single slug/rod model. We can therefore determine the results from Figure 5 where we, in effect, choose q according to the strength of the coupling.

In the impacting beam model, it is assumed that when the impact takes place the main beam is sufficiently massive for its motion not to be disturbed. Thus the ratio of the rebound speed of the delamination to its impact speed is equal to the coefficient of restitution, e. In terms of the discrete model of coupled slugs impacting elastic rods, this is equivalent to the assumption made in the model that the rods are of semi-infinite extent. For a delaminated beam, the material of the flap is identical with the material of the remainder of the beam, which corresponds to taking $z = 1$. In the simple model of a slug impacting a semi-infinite rod, this would result in the coefficient of restitution, e, being zero.

In the case of coupled slugs, our results show that the effect of the coupling for the situation when $z = 1$ and the system is impacting in its fundamental mode, is to induce the system to rebound with a small speed in the same mode. This corresponds to a small, but non-zero, effective coefficient of restitution, e. This effective coefficient of restitution is independent of the impact velocity but is strongly dependent on the ratio of the depth of the delamination to its length. For the system with nearest neighbour coupling the effective parameter q which determines the rebound, is directly proportional to this ratio. For the system with both nearest and next-nearest coupling the parameter is proportional to the square of this ratio.

REFERENCES

1. Harrison, M.C., Kane, S.J., Green, E.R. and Morrison, C.J., Idealised Impact Behaviour of a Delamination Beam under Periodic Forcing, AD-Vol.46, Dynamic Response and Behaviour of Composites, ASME, 1995.
2. Harrison, M.C., Kane, S.J. and Green, W.A., Forced Impact Vibration of a Delaminated Beam, Proceedings of Dynamic Systems and Applications, Vol. 2, Dynamic Publishers, USA, 1996.
3. Menday, R.P., The Forced Vibration of a Partially Delaminated Beam, Doctoral Thesis, Loughborough University, Loughborough, UK, 1999.

Proceedings of Dynamic Systems and Applications 3 (2001) 297-304

On Lyapunov's Direct Method for Nonautonomous Functional Differential Equations

László Hatvani
Bolyai Institute, Aradi vértanúk tere 1, Szeged, Hungary, H-6720

Dedicated to Professor V. Lakshmikantham on his 75th birthday

ABSTRACT: Sufficient conditions for the asymptotic stability (A.S.) and uniform asymptotic stability (U.A.S.) of the zero solutions of FDE's with finite delays are formulated by the method of Lyapunov functionals. The derivatives of the functionals with respect to the equations are negative semidefinite in terms of either $|x(t)|$ or L_2-norm of segment x_t and may depend explicitly on time t. The theorems do not require boundedness of the right-hand side of the equation. The results are motivated by and applied to nonautonomous FDE with distributed delays.
AMS (MOS) subject classification. 34K20

1. INTRODUCTION

Consider the nonautonomous system of functional differential equations

$$x'(t) = F(t, x_t) \qquad (F(t,0) \equiv 0), \tag{1.1}$$

where we use the standard notations [6]: $C = C([-h,0];\mathbf{R}^n)$ denotes the space of continuous functions from $[-h,0]$ into $\mathbf{R}^n$; $0 < h = \text{const.}$; $F : \mathbf{R}_+ \times C \to \mathbf{R}^n$ is continuous and maps bounded sets into bounded sets. For any solution $x : \mathbf{R}_+ \to \mathbf{R}^n$, segment $x_t \in C$ is defined by $x_t(s) := x(t+s)$, $-h \le s \le 0$. We will use two norms in space C:

$$\|\varphi\|_\infty := \max_{-h\le s\le 0} |\varphi(s)|, \qquad \|\varphi\|_2 := \left(\int_{-h}^{0} \varphi^2(s)\, ds\right)^{1/2}.$$

For $0 < H = \text{const.}$ the ball C_H in C is defined by $C_H := \{\varphi \in C : \|\varphi\|_\infty < H\}$. For $t_0 \in \mathbf{R}_+$, $\varphi \in C$ we denote by $x(t) = x(t; t_0, \varphi)$ any solution satisfying the initial condition $x_{t_0} = \varphi$. The segment of this solution at t is denoted by $x_t = x_t(t_0, \varphi)$. A continuous, strictly increasing function $W : \mathbf{R}_+ \to \mathbf{R}_+$ with $W(0) = 0$ is called a *wedge*. Wedges will be denoted consistently by $W, W_1, W_2, \ldots$.

In this paper we give some sufficient conditions for the asymptotic stability and uniform asymptotic stability of the zero solution of (1.1).

Definition 1 [6, 12]. The zero solution of (1.1) is called:

a) *stable*, if for every $\varepsilon > 0$ and $t_0 \in \mathbf{R}_+$ there is a $\delta(\varepsilon, t_0) > 0$ such that $\|\varphi\|_\infty < \delta$ implies $|x(t; t_0, \varphi)| < \varepsilon$ for all $t \geq t_0$.

b) *uniformly stable*, if it is stable with $\delta = \delta(\varepsilon)$ independent of $t_0 \in \mathbf{R}_+$;

c) *attractive*, if for every $t_0 \in \mathbf{R}_+$ there is a $\sigma(t_0) > 0$ such that $\|\varphi\|_\infty < \sigma$ implies

$$\lim_{t \to \infty} x(t; t_0, \varphi) = 0; \tag{1.2}$$

d) *uniformly attractive*, if the convergence in (1.2) is uniform with respect to t_0, φ; i.e., there exists a constant $\sigma > 0$ and, for every $\eta > 0$, there is a $T(\eta)$ such that $t \geq t_0 + T(\eta)$ implies $|x(t; t_0, \varphi)| < \eta$ for all $t_0 \in \mathbf{R}_+$, $\varphi \in C_\sigma$;

e) *asymptotically stable* (A.S.), if it is stable and attractive;

f) *uniformly asymptotically stable* (U.A.S.), if it is uniformly stable and uniformly attractive.

These definitions are straightforward generalizations of those for ordinary differential equations introduced by A.M. Lyapunov [15]. However, not only the definitions but the direct method of Lyapunov has been generalized to FDE's, serving as the basic tool for stability investigations for these equations, too [1–14, 15–18]. The natural generalization of the Lyapunov function from ODE's to FDE's is the Lyapunov functional. To illuminate this, let us consider the nonautonomous functional differential equation with distributed delays

$$x'(t) = -a(t)x(t) + b(t) \int_{t-h}^{t} \lambda(u)x(u)\, du, \tag{1.3}$$

where the given functions $a, b : \mathbf{R}_+ \to \mathbf{R}$, $\lambda[-h, \infty) \to \mathbf{R}$ are continuous and $a(t) \geq 0$ for all $t \geq 0$. Define a Lyapunov functional by

$$V(t, x_t) := |x(t)| + \int_{-h}^{0} \int_{t+s}^{t} |b(u-s)||\lambda(u)||x(u)|\, du\, ds, \tag{1.4}$$

whose derivative with respect to (1.3) admits the estimate

$$V'(t, x_t) \leq -\left[a(t) - |\lambda(t)| \int_{t}^{t+h} |b(r)|\, dr\right] |x(t)|.$$

It is negative definite provided that

$$a(t) \geq |\lambda(t)| \int_{t}^{t+h} |b(r)|\, dr \quad \text{for all} \quad t \geq 0, \tag{1.5}$$

which can be interpreted in the way that the negative instantaneous feedback dominates the delayed one.

It is worth observing that the Lyapunov functional V has a special structure. It is the sum of two members. One of them depends only on the value $x_t(0) = x(t)$, the other one uses the whole segment x_t. V. Lakshmikantham and his collaborators [13–14] have developed a theory based upon differential inequalities and the comparison technique involving Lyapunov functionals of this structure, i.e., functionals V : $\mathbf{R}_+ \times \mathbf{R}^n \times C$. Here we deal with the cases of applications of such functionals when there is no explicit inequality between the Lyapunov functional V and its derivative V'.

The first theorem of this type was

Theorem A (N. Krasovskii [12], see also [6]). *Suppose that there exists a continuous functional $V : R_+ \times C \to \mathbf{R}_+$ satisfying the conditions*

(i) $W_1(|x(t)|) \leq V(t, x_t)$;

(ii) $V(t, x_t) \leq W_2(\|x_t\|_\infty)$;

(iii) $V'(t, x_t) \leq -W(|x(t)|)$;

(iv) *$F([0, \infty) \times K)$ is bounded for every bounded set $K \subset C$*

for every solution $x : [-h, \infty) \to \mathbf{R}^n$ of (1.1) and for all $t \geq 0$.

Then the zero solution of (1.1) is U.A.S.

By this theorem, if functions a, b and λ are bounded, and

$$a(t) - |\lambda(t)| \int_t^{t+h} |b(r)|\, dr \geq c_1 > 0 \qquad (t \geq 0) \tag{1.6}$$

with an appropriate constant c_1, then the zero solution of (1.3) is U.A.S. It is obvious, that the boundedness of a is not necessary for U.A.S.: the larger $a(t)$ is the better from the point of U.A.S. This can be deduced from

Theorem B (N. Krasovskii [12]). *Suppose that there exists a continuous functional $V : \mathbf{R}_+ \times C \to \mathbf{R}_+$ satisfying conditions*

(i) $W_1(|x(t)|) \leq V(t, x_t)$;

(ii) $V(t, x_t) \leq W_2(|x(t)|) + W_3(\|x_t\|_2)$;

(iii) $V'(t, x_t) \leq -W(|x(t)|)$

for every solution $x : [-h, \infty) \to \mathbf{R}^n$ of (1.1) and for all $t \geq 0$.

Then the zero solution of (1.1) is U.A.S.

Changing the order of integration in Lyapunov functional (1.4) one can get the estimate

$$V(t,x_t) \le |x(t)| + \left(\int_t^{t+h} |b(r)|\,dr\right)\left(\int_{t-h}^{t} \lambda^2(u)\,du\right)\|x_t\|_2^2, \tag{1.7}$$

whence it follows that condition (1.6) and the boundedness of $\int_t^{t+h}|b|$, $\int_{t-h}^{t}\lambda^2$ imply U.A.S. for (1.3).

Though Theorem B has the advantage of not requiring a direct boundedness conditions of the right-hand side F of (1.1), the application of the theorem to (1.3) needs the integral boundedness of $|b|$ and λ^2. These conditions can be weakened by introducing a pseudo norm instead of $\|x_t\|_2$ in Theorem B (see [8]). Moreover, applying Theorem B one has to face another problem: condition (iii) is not a real "nonautonomous" one in the sense that the right-hand side cannot depend on time t explicitly. This is the reason of requiring (1.6). The following question arises. Instead of (1.5) and (1.6) assume

$$a(t) - |\lambda(t)| \int_t^{t+h} |b(r)|\,dr \ge \eta(t) \ge 0 \quad (t \ge 0) \tag{1.8}$$

and ask: under what condition on η can we state U.A.S., perhaps only A.S. for (1.3)?

In this paper we formulate improvements of Theorem B allowing the right-hand side of inequality (iii) in the theorem to depend on t explicitly. In other words, we give sufficient conditions for A.S. and U.A.S. using Lyapunov functionals with negative semidefinite derivatives instead of those with negative definite derivatives.

2. THE RESULTS

Comparing Theorems A and B one can see that Krasovskii could drop condition (iv) from Theorem A by means of a finer upper estimate for the Lyapunov functional V. Some applications inspired us to introduce finer upper estimate also for the derivative of the Lyapunov functional [3]. E.g., consider equation (1.3) again, but now with the new Lyapunov functional

$$V(t,x_t) := |x(t)| + \beta \int_{-h}^{0}\int_{t+s}^{t} |b(u-s)||\lambda(u)||x(u)|\,du\,ds, \quad 1 \le \beta = \text{const.} \tag{2.1}$$

Simple computations similar to the case of (1.4) give the estimates

$$V(t,x_t) \le |x(t)| + \beta\left(\int_t^{t+h} |b|\right)\left(\int_{t-h}^{t} \lambda^2\right)\|x_t\|_2^2; \tag{2.2}$$

$$V'(t,x_t) \le \left[a(t) - \beta|\lambda(t)|\int_t^{t+h}|b|\right]|x(t)| - (\beta-1)|b(t)|\int_{t-h}^{t}|\lambda(u)||x(u)|\,du. \quad (2.3)$$

By putting $\beta = 1$ and using (1.6), the last inequality is of type (iii) in Theorem B. However, if $\beta > 1$, then at the moment t the whole segment x_t takes part in driving the function $V(t,x_t)$ to zero as $t \to \infty$. The following two theorems handle these different situations, which will be formulated generally by the inequalities

$$V'(t,x_t) \le -\eta(t)W(|x(t)|), \qquad V'(t,x_t) \le -\eta(t)W(\|x_t\|_2), \quad \eta(t) \ge 0. \qquad (2.4)$$

The derivative V' is only negative semidefinite, so the measurable coefficient $\eta : \mathbf{R}_+ \to \mathbf{R}_+$ may vanish on a subset of $\mathbf{R}_+$. Of course, in order to guarantee A.S. or U.A.S. we have to require of η to have a "large total effect", i.e., a certain kind of divergence of the integral $\int_H \eta(t)\,dt$ over "sufficiently dense" sets $H \subset \mathbf{R}_+$.

Definition 2. A measurable set $H \subset \mathbf{R}_+$ is said to be *h-dense* if there exists a $\kappa > 0$ such that $\mu(H \cap [t, t+h]) \ge \kappa$ for all $t \ge 0$. (Here, and in the sequel, $\mu(\cdot)$ denotes the Lebesgue measure).

Theorem 1. *Assume that there exists a continuous functional $V : \mathbf{R}_+ \times C \to \mathbf{R}_+$ satisfying the conditions*

(i) $W_1(|x(t)|) \le V(t,x_t)$;

(ii) $V(t,x_t) \le W_2(|x(t)|) + W_3(\|x_t\|_2)$;

(iii) $V'(t,x) \le -\eta(t)W(|x(t)|)$ *with a measurable function* $\eta : \mathbf{R}_+ \to \mathbf{R}_+$ *for every solution* $x : [-h,\infty) \to \mathbf{R}^n$ *of (1.1) and for all* $t \ge 0$.

A [4]. *If*

$$\int_H \eta(t)\,dt = \infty \qquad (2.5)$$

for every h-dense set $H \subset \mathbf{R}_+$, then the zero solution of (1.1) is A.S.

B. *Suppose, in addition, that the divergence in (2.5) is uniform in the sense that for every h-dense set $H \subset \mathbf{R}_+$*

$$\lim_{(T-t_0)\to\infty} \int_{[t_0,T)\cap H} \eta(t)\,dt = \infty \qquad (2.6)$$

uniformly with respect to $t_0 \in \mathbf{R}_+$.

Then the zero solution of (1.1) is U.A.S.

The following theorem shows that the second inequality in (2.4) can guarantee A.S. and U.A.S. under weaker conditions on η.

Theorem 2. *Assume that there exists a continuous functional* $V : \mathbf{R}_+ \times C \to \mathbf{R}_+$ *satisfying conditions (i)-(ii) in Theorem 1 and*

(iii) $V'(t,x) \leq -\eta(t) W(\|x_t\|_2)$ *with a measurable function* $\eta : \mathbf{R}_+ \to \mathbf{R}_+$ *for every solution* $x : [-h, \infty) \to \mathbf{R}^n$ *of (1.1) and for all* $t \geq 0$.

A [4]. *If*

$$\int_0^\infty \eta(t)\, dt = \infty, \tag{2.7}$$

then the zero solution of (1.1) is A.S.

B. *If*

$$\lim_{(T-t_0)\to\infty} \int_{t_0}^T \eta(t)\, dt = \infty \tag{2.8}$$

uniformly with respect to $t_0 \in \mathbf{R}_+$, *then the zero solution of (1.1) is U.A.S.*

3. EXAMPLES, APPLICATIONS

At first we illuminate conditions (2.5)-(2.8) by examples. Define the step functions

$$\eta_1(t) := \begin{cases} 1 & \text{if } \; 3kh \leq t \leq (3k+1)h \\ 0 & \text{otherwise} \end{cases} \qquad \eta_2(t) := \begin{cases} 1 & \text{if } \; k^2 h \leq t \leq (k^2+1)h \\ 0 & \text{otherwise} \end{cases}$$

where $k = 0, 1, 2, \ldots$. It is easy to see that

a) η_1 satisfies (2.6);

b) η_2 satisfies (2.5), but it does not satisfy (2.6).

Let

$$\eta_3(t) := \begin{cases} 1 & \text{if } \; kh \leq t \leq (k+\frac{1}{2})h, \\ & \qquad k = 0, 1, 2, \ldots \\ 0 & \text{otherwise} \end{cases} \qquad H := \bigcup_{k=0}^{\infty} \left((k + \frac{1}{2})h, (k+1)h \right).$$

Then η_3 satisfies (2.8), but it does not satisfy (2.5). In fact, H is obviously h-dense ($\kappa = h/2$); nevertheless, $\int_H \eta_3(t)\, dt = 0 \neq \infty$. At the same time, we have

$$\lim_{(T-t_0)\to\infty} \int_{t_0}^T \eta_3(t)\, dt \geq \lim_{(T-t_0)\to\infty} \left(\frac{T - t_0}{h} - 1 \right) \frac{h}{2} = \infty$$

uniformly with respect to $t_0 \in \mathbf{R}_+$.

Applying Theorem 1 to equation (1.3) and Lyapunov functional (1.4) we get

Corollary 1. *Suppose that functions* $t \mapsto \int_t^{t+h} |b(s)|\, ds$, $t \mapsto \int_t^{t+h} |\lambda(x)|\, ds$ *are bounded on* $\mathbf{R}_+$.

A. *If*

$$\int_H \left[a(t) - |\lambda(t)| \int_t^{t+h} |b| \right] dt = \infty$$

for every h-dense set $H \subset \mathbb{R}_+$, then the zero solution of (1.3) is A.S.

B. *If*

$$\lim_{(T-t_0)\to 0} \int_{[t_0,T]\cap H} \left[a(t) - |\lambda(t)| \int_t^{t+h} |b| \right] dt = \infty$$

for every h-dense set $H \subset \mathbb{R}_+$ uniformly with respect to $t_0 \in \mathbb{R}_+$, then the zero solution is U.A.S.

If we want to apply Theorem 2 to equation (1.3) and functional (1.4) we face a new problem. To meet condition (iii) in Theorem 2 we would have to estimate the integral $\int_{t-h}^{t} |\lambda(u)||x(u)|du$ in (2.3) from *below* by $\|x_t\|_2$. The Schwarz inequality would give an estimate from *above*. (Unfortunately, in [4, Ex. 3.1] we made a mistake: the estimate for V' on p. 290 is correct only with $\lambda(t) \equiv \lambda_0$ =const.) This difficulty can be overcome by introducing a new pseudo norm $\int_{t-h}^{t} |\lambda(s)||x(s)|ds$ instead of $\|x_t\|_2$ [8].

Finally, let us compare our results with recent ones of Bo Zhang [18]. He generalized Theorem B to nonautonomous systems (1.1) requiring inequalities (2.4) with $\eta(t) \equiv 1$ on the union $I = \bigcup_{i=1}^{\infty}[t_i - h, t_i]$, respectively $I = \bigcup_{i=1}^{\infty}[t_i - \sigma, t_i]$, $(\sigma > 0)$ of the sequence of discrete intervals. In our language this means, that he restricted the investigation to the step function coefficients

$$\eta(t) := \begin{cases} 1 & \text{if } t \in I \\ 0 & \text{otherwise} \end{cases}.$$

As we saw it in examples, conditions (2.5)-(2.8) can be easily checked in this class of functions. E.g., if $I = \bigcup_{i=1}^{\infty}[t_i - h, t_i]$, then (2.5) is automatically satisfied, and (2.6) is implied by the existence of a constant J with $t_{i+1} - t_i \le J$ $(i = 1, 2, \ldots)$, which is Zhang's condition for U.A.S. Therefore, Zhang's theorems are special cases of our results. On the other hand, if

$$V'(t, x_t) \le -\left(\sin^2 \frac{\pi}{h} t\right) W(|x(t)|) \quad (t \ge 0),$$

then Zhang's condition cannot be satisfied. At the same time, condition (2.6) is obviously met by the h-periodic function $\eta(t) := \sin^2(\pi t/h)$ $(t \ge 0)$; therefore, our Theorem 2 guarantees U.A.S.

ACKNOWLEDGEMENT

Research was supported by the Hungarian Foundation for Scientific Research, Grant No. T029188.

REFERENCES

1. Burton, T. A., Volterra integral and differential equations. Academic Press, New York, 1983

2. Burton, T. A., Stability and periodic solutions of ordinary and functional differential equations. Academic Press, New York, 1985.
3. Burton, T. A. & Hatvani, L., Stability theorems for nonautonomous functional differential equations by Liapunov functionals. Tôhoku Math. J., Vol. 41 (1989) pp. 65–104.
4. Burton, T.& Hatvani, L., On nonuniform asymptotic stability for nonautonomous functional differential equations. Differential and Integral Equations, Vol. 3 (1989) pp. 285–293.
5. Burton, T. A. & Makay, G., Asymptotic stability for functional differential equations. Acta Math. Hungar., Vol. 65(3) (1994) pp. 243–251.
6. Hale, J. K. & Verduyn Lunel, S. M., Introduction to functional differential equations. Springer-Verlag, New York, 1993.
7. Hatvani, L., On the asymptotic stability of the solutions of functional differential equations. In: Qualitative Theory of Differential Equations (Ed. by B. Sz.-Nagy and L. Hatvani) (Colloquia Math. Soc. J. Bolyai Vol 53) North Holland, Amsterdam, 1990; pp. 227–238.
8. Hatvani, L., On Lyapunov's direct method for nonautonomous FDE's. Functional Differential Equations, Vol. 5 (1998) pp. 315–323.
9. Kato, J., Liapunov's second method in functional differential equations. Tôhoku Math. J., Vol. 32 (1980) pp. 487–497.
10. Kobayashi, Katsumasa, Asymptotic stability in functional differential equations. Nonlinear Anal., Vol. 20 (1993) pp. 359–364.
11. Kolmanovskii, V. B., Torelli, L. & Vermiglio, R., Stability of some test equations with delay. SIAM J. Math. Anal., Vol. 25 (1994) pp. 948–961.
12. Krasovskii, N. N., Stability of motion. Stanford University Press, Stanford, 1963.
13. Lakshmikantham, V., Leela, S. & Sivasundaram, S., Lyapunov functions on product space and stability theory of delay differential equations. J. Math. Anal. Appl., Vol. 154 (1991) pp. 391-402.
14. Lakshmikantham, V., Matrosov, V. M. & Sivasundaram, S., Vector Lyapunov functions and stability analysis of nonlinear systems. Kluwer Academic Publ., Dordrecht, 1991.
15. Liapounoff, A. M., Problème général de la stabilité du mouvement. Annals of Math. Studies 17, Princeton Univ. Press, Princeton, 1949.
16. Wang, Tingxiu, Asymptotic stability and the derivatives of solutions of functional differential equations. Rocky Mountain J. Math., Vol. 24 (1994) pp. 403–427.
17. Wang, Tingxiu, Weakening the condition $W_1(|\varphi(0)|) \leq V(t,\varphi) \leq W_2(\|\varphi\|)$ for uniform asymptotic stability. Nonlinear Anal., Vol. 23 (1994) pp. 251–264.
18. Zhang, Bo, Asymptotic stability on functional differential equations. Trans. Amer. Math. Soc., Vol. 347 (1995) pp.1375–1382.

Proceedings of Dynamic Systems and Applications 3 (2001) 305-312

The Properties of Zeros for a Class of Exponential Polynomials and Applications in Economical Systems

Ming He*, Xiaoyun Ma** and Weijiang Zhang*

*Department of Applied Mathematics, Shanghai Jiao Tong University, Shanghai, 200030 P. R. China
**Department of Mathematics and Computer Science, University of San Diego, 5998 Alcala Park, San Diego, CA 92110 U. S. A.

ABSTRACT: In this paper, we consider a class of exponential polynomials with the form $H(z) = f(z)e^z + q$, where $f(z)$ is a degree n polynomial with real coefficients, and q is a constant. The theorem of necessary and sufficient conditions under which all zeros of $H(z)$ lie on the left of the imaginary axis is established. It is useful to obtain a death solution of systems of coupled neural oscillators with time lag by using the above theorem.

We will apply the result to some economic models, like Kalecki's macro – dynamical models of business cycles, whose characteristic equations have the same form as $H(z)$. Analysis shows that the period of the business cycle can be determined by some economic parameters.

AMS (MOS) subject classification 34K20

1. Introduction

In many systems, the mathematical models deal with differential difference equations. Some characteristic equations of differential difference equations contain exponential terms, which are more complex. If $f(z)$ is a degree n polynomial with real coefficients, then it has only n zeros. But for the exponential polynomial $H(z) = f(z)e^z + q$, where q is a constant, it may have infinitely many zeros. When considering the stability of a system, we have to investigate if the zeros of the characteristic equations are on the left of the imaginary axis. So it is useful to study the properties of zeros of exponential polynomials.

In this paper, we consider a class of exponential polynomials with the form $H(z) = f(z)e^z + q$, and investigate the properties of its zeros. This research is motivated by the study of coupled oscillators with time-delay coupling and some economical models with time-delay relationships.

In section 2, we will give some background for models of neural oscillators and business cycles. In section 3, we establish a necessary and sufficient condition for stability of zeros for a class of exponential polynomials. This result is the generalization of the results in [9]. In section 4, we obtain the death solutions for a chain of oscillators with one direction time-delay coupling on a cycle. This result is the generalization of the result in [5]. In section 5, we consider a more complicated case -- a coupled chain of cyclic chains of oscillators with time-delay coupling. In section 6, we give a theoretic period of a business cycle in economics, by analyzing a low degree exponential polynomial.

2. Systems of coupled oscillators and models of business cycles

Coupled oscillators arise in many areas of biology. The research on the coupling oscillators is derived from the theory of central pattern generators (CPG)[1]. The phenomena of phase-locking and death of coupled oscillators are more developed from the middle of 80's. Chains of oscillators with nearest-neighbor coupling were first investigated in [2]-[5]. The structures of nearest multiple couplings were then studied in [6]-[7]. A pattern of long-range coupling between oscillators was investigated in [8]. The death solution of two oscillators coupled with a time lag was obtained in [5]. In this paper, we are going to investigate the death phenomena of coupled oscillators.

The mathematical model of systems of coupled oscillators has the following form:

$$\dot{\theta}_i(t) = \omega_i + \sum_{j\in A_i} h_{ij}(\theta_i(t), \theta_j(t)), \ i = 1,2,\dots,n. \tag{2.1}$$

Where $\theta_i(t)$ is the phase of the ith oscillator, ω_i is the inherent frequency of the ith oscillator, h_{ij} is the coupling function between the ith and jth oscillators.

If $A_i = \{i-1, i+1\}$, the coupling in (2.1) is the nearest-neighbor coupling. If $A_i = \{i-m, \dots, i-1, i+1, \dots, i+m\}$, the coupling in (2.1) is the nearest multiple couplings. An asymptotically stable constant-value solution of (2.1) is called a death solution of (2.1). When the solution orbits of (2.1) approach gradually to the time-independent solution, the rhythmic oscillations of oscillators vanish gradually, i.e., the phases $\{\theta_i(t)\}$ will tend to the constant values $\{\theta_i\}$.

Suppose that the oscillators nearly couple with the others in the same direction with a time lag and assume the coupling is weak, then the mathematical model is:

$$\begin{cases} \dot{\theta}_1(t) = \omega_1 + h_1(\theta_1(t), \theta_2(t-\tau)) \\ \dot{\theta}_2(t) = \omega_2 + h_2(\theta_2(t), \theta_3(t-\tau)) \\ \cdots\cdots\cdots \\ \dot{\theta}_n(t) = \omega_n + h_n(\theta_n(t), \theta_1(t-\tau)) \end{cases} \tag{2.2}$$

where $\tau \geq 0$ is time lag.

The characteristic equation of the linearization of (2.2) is an exponential polynomial with the form $H(z) = f(z)e^z + q$. We need to investigate the zeros of the characteristic equation to obtain the death solution of (2.2).

Another motivation for us to consider is that the exponential polynomial comes from the study of business cycles in economics.

Suppose that $I(t)$ is the investment rate, U is the rate of depreciation of capital goods, then the Kalecki's macro-dynamical model of business cycle has the form:

$$I'(t) = m[I(t) - I(t-1)] - n[I(t-1) - U] \tag{2.3}$$

This is a differential difference equation, and its characteristic equation is an exponential polynomial. The main eigenvalue of (2.3) will determine the behavior of the system. We will discuss it in section 6.

3. Stability of zeros for a class of exponential polynomials

In paper [9], the stability conditions of zeros for an exponential polynomial $h(z, e^z)$, i.e. the conditions for all zeros to have negative real parts, were discussed. We are going to give the necessary and sufficient conditions for $H(z) = f(z)e^z + q$ to have all zeros lying on the left of the imaginary axis. The cases when $f(z)$ is 1st or 2nd order polynomials were discussed in [9]. We will extend the result in this paper.

Theorem 3.1 Let $H(z) = f(z)e^z + q$, where q is a real number. Assume that

$f(z)=z^n+p_1z^{n-1}+\cdots+p_{n-1}z+p_n$ is a degree *n* polynomial with real coefficients and its zeros have negative real parts. For any real parameter λ, let $f(i\lambda)=u(\lambda)+iv(\lambda)$, $H(i\lambda)=F(\lambda)+iG(\lambda)$, where $u(\lambda)$ and $v(\lambda)$ are the real part and imaginary part of $f(i\lambda)$ respectively, and $F(\lambda)$ and $G(\lambda)$ are the real part and imaginary part of $H(i\lambda)$ respectively. Then necessary and sufficient conditions for all zeros of *H(z)* to lie on the left of imaginary axis are:

(1) All the zeros of function $G(\lambda)$ are real;

(2) For any real zero $\bar{\lambda}$ of $G(\lambda)$, one of the following two conditions is satisfied:

a) $q\sin\bar{\lambda}/v(\bar{\lambda})<1,\ v(\bar{\lambda})\neq 0$; b) $-q\cos\bar{\lambda}/u(\bar{\lambda})<1,\ v(\bar{\lambda})=0$.

Under the same hypotheses, we will prove theorem 3.1 by following lemmas:

Lemma 3.1 The zeros of $u(\lambda)$ and $v(\lambda)$ (when they exist) are real, simple and alternating. For any $\lambda\in\mathbf{R}$, $u(\lambda)v'(\lambda)-u'(\lambda)v(\lambda)>0$.

Proof : $f(i\lambda)=u(\lambda)+iv(\lambda)$, $f(-i\lambda)=u(\lambda)-iv(\lambda)$, $if'(i\lambda)=u'(\lambda)+iv'(\lambda)$. Let $\alpha_k+i\beta_k$ be a zero of $f(\lambda)$, *k=1,2,...,n*, where $\alpha_k<0$, then

$$f(z)=\prod_{k=1}^{n}(z-\alpha_k-i\beta_k)=\prod_{k=1}^{n}(z-\alpha_k+i\beta_k).$$

Let $x+iy$ be a zero of $u(\lambda)=[f(i\lambda)+f(-i\lambda)]/2$, then

$$\prod_{k=1}^{n}[i(x+yi)-\alpha_k-i\beta_k]=-\prod_{k=1}^{n}[-i(x+yi)-\alpha_k-i\beta_k].$$

Hence

$$\prod_{k=1}^{n}[(\alpha_k+y)+i(\beta_k-x)]=-\prod_{k=1}^{n}[(\alpha_k-y)+i(\beta_k+x)]=-\prod_{k=1}^{n}[(\alpha_k-y)+i(x-\beta_k)].$$

Thus

$$\prod_{k=1}^{n}[(\alpha_k+y)^2+(\beta_k-x)^2]=\prod_{k=1}^{n}[(\alpha_k-y)^2+(\beta_k-x)^2].$$

Because $\alpha_k<0,\ k=1,\cdots,n$, the above formula holds iff *y=0*. It implies that the zeros of $u(\lambda)$ all are real. Similarly, the zeros of $v(\lambda)$ are also real.

$$u(\lambda)u'(\lambda)+v(\lambda)v'(\lambda)+i[u(\lambda)v'(\lambda)-v(\lambda)u'(\lambda)]=if(-i\lambda)f'(i\lambda)$$

$$=i\prod_{k=1}^{n}(-i\lambda-\alpha_k-i\beta_k)\cdot\prod_{k=1}^{n}(i\lambda-\alpha_k-i\beta_k)\cdot\sum_{k=1}^{n}\frac{1}{i\lambda-\alpha_k-i\beta_k}$$

$$=\prod_{k=1}^{n}[\alpha_k^2+(\lambda+\beta_k)^2]\cdot\sum_{k=1}^{n}\frac{\lambda+\beta_k-i\alpha_k}{\alpha_k^2+(\lambda+\beta_k)^2}.$$

Hence, for $\lambda\in\mathbf{R}$, $u(\lambda)v'(\lambda)-u'(\lambda)v(\lambda)>0$, so $u(\lambda)$ and $u'(\lambda)$ can not be zero simultaneously. Therefore, the zeros of $u(\lambda)$ are simple, and so are zeros of $v(\lambda)$. Furthermore, $u(\lambda)$ and $v(\lambda)$ can not vanish at the same point.

$$\frac{d}{d\lambda}\left[\frac{v(\lambda)}{u(\lambda)}\right]=\frac{u(\lambda)v'(\lambda)-u'(\lambda)v(\lambda)}{u^2(\lambda)}>0,\ \forall\lambda,\ u(\lambda)\neq 0.$$

It implies that $v(\lambda)/u(\lambda)$ strictly increases from $-\infty$ to $+\infty$ between two successive zeros of $u(\lambda)$. Thus the zeros of $u(\lambda)$ and $v(\lambda)$ alternate along the λ axis and are interspersed.

Lemma 3.2 The zeros of $G(\lambda)$ are all real.

Proof: $H(i\lambda) = f(i\lambda)e^{i\lambda} + q = [u(\lambda) + iv(\lambda)]\cdot(\cos\lambda + i\sin\lambda) + q$
$= u(\lambda)\cos\lambda - v(\lambda)\sin\lambda + q + i[u(\lambda)\sin\lambda + v(\lambda)\cos\lambda]$.

So $F(\lambda) = u(\lambda)\cos\lambda - v(\lambda)\sin\lambda + q$, $G(\lambda) = u(\lambda)\sin\lambda + v(\lambda)\cos\lambda$.

If n is even, then $G(\lambda)$ has the principal term $\lambda^n \sin\lambda$. Let $J_k = (-2k\pi + \pi/2, 2k\pi + \pi/2)$ for sufficiently large *k*. If *n* is odd, then $G(\lambda)$ has the principal term $\lambda^n \cos\lambda$. Let $J_k = (-2k\pi, 2k\pi)$ for sufficiently large *k*. We will show that there are *4k+n* real zeros of $G(\lambda)$ in the interval J_k for sufficiently large *k*.

There are two cases of the real zeros of $G(\lambda)$: (I) $v(\lambda) \neq 0, \sin\lambda \neq 0, G(\lambda) = 0$; (II) $v(\lambda) = 0, \sin\lambda = 0, G(\lambda) = 0$. In case (I), the zero of $G(\lambda)$ satisfies the equation $\cot\lambda = -u(\lambda)/v(\lambda)$.

Lemma 3.1 tells that the zeros of $v(\lambda)$ are all real. Then the zeros of $v(\lambda)$ can be arranged as $\lambda_{-m}, \lambda_{-m+1}, \cdots, \lambda_{-1}, \lambda_0, \lambda_1, \cdots, \lambda_{m-1}, \lambda_m$, $n=2m+1$, or $2m+2$, $m \geq 0$. Suppose *k* is sufficiently large so that $\{\lambda_j\}_{j=-m}^{m}$ is contained in J_k. Let λ_{-m-1} and λ_{m+1} be the left and right end-points of J_k respectively. Since

$$\frac{d}{d\lambda}\left[-\frac{u(\lambda)}{v(\lambda)}\right] = \frac{u(\lambda)v'(\lambda) - u'(\lambda)v(\lambda)}{v^2(\lambda)} > 0,\ \forall\lambda \in (\lambda_j, \lambda_{j+1}),\ j = -m-1, \ldots, m,$$

then $-u(\lambda)/v(\lambda)$ is strictly increasing from negative values to positive values in $(\lambda_j, \lambda_{j+1})$, $j = -m, \ldots, m-1$, and intersects with every branch of $\cot\lambda$ inside $(\lambda_j, \lambda_{j+1})$ at a unique point. The value of λ at the intersection point is a real zero of $G(\lambda)$ of case (I). A detailed analysis gives the same result in $(\lambda_{-m-1}, \lambda_{-m})$ and $(\lambda_m, \lambda_{m+1})$.

Suppose that there are *s* zeros of $v(\lambda)$ in the set $\{l\pi \mid l = 0, \pm1, \pm2, \ldots\}$, i.e., there are s zeros of $G(\lambda)$ in case (II). If $\lambda_j \notin \{l\pi \mid l = 0, \pm1, \pm2, \ldots\}$, then the same branch of $y = \cot\lambda$ will occur in $(\lambda_{j-1}, \lambda_j)$ and $(\lambda_j, \lambda_{j+1})$ simultaneously. So this branch will be repeatedly counted, when we count the branch number of $\cot\lambda$ in $\bigcup_{j=-m-1}^{m}(\lambda_j, \lambda_{j+1})$. If $n = 2m+1$, $G(\lambda)$ has $4k+(n-s)$ real zeros in $J_k = (-2k\pi, 2k\pi)$; If $n=2m+2$, $G(\lambda)$ has $(4k+1)+(n-1-s)$ real zeros in $J_k = (-2k\pi + \pi/2, 2k\pi + \pi/2)$, k is sufficiently large. Hence, $G(\lambda)$ always has *4k+n* real zeros whenever *n* is odd or even. Then Theorem 13.3 in [9] tells that the zeros of $G(\lambda)$ are all real.

Lemma 3.3[9] The zeros of $H(z)$ are on the left of imaginary axis if and only if all the zeros of $G(\lambda)$ are real and $G'(\bar{\lambda})F(\bar{\lambda}) > 0,\ \forall\bar{\lambda} : G(\bar{\lambda}) = 0$.

Lemma 3.4 For any real zero $\bar{\lambda}$ of $G(\lambda)$, the inequality $G'(\bar{\lambda})F(\bar{\lambda}) > 0$ holds if

and only if:
(1) $p_n + q > 0$;
(2) For any positive zero $\bar{\lambda}$ of $G(\lambda)$, one of the following two conditions is satisfied:
a) $q \cdot \sin\bar{\lambda} / v(\bar{\lambda}) < 1$, and $v(\bar{\lambda}) \neq 0$; b) $-q \cdot \cos\bar{\lambda} / u(\bar{\lambda}) < 1$, and $v(\bar{\lambda}) = 0$.
Proof is omitted due to the length limitation of this paper.

Proof of Theorem 3.1: Combine Lemmas 3.1-3.4 together.

4. Death of oscillators on a cycle with time-delay coupling in one direction

The death solution of two oscillators with time-delay coupling was obtained in [5]. They used a necessary and sufficient condition[9] about the stability of zeros for an exponential polynomial to get stability of the death solution. Based on section 3, we now consider a chain of oscillators on a cycle with time-delay coupling. Moreover, the results can be generalized to the case of a chain of cycles.

Suppose that the oscillators nearly couple with the others in the same direction with time lag, and assume the coupling is weak, then the mathematical model is:

$$\begin{cases} \dot{\theta}_1(t) = \omega_1 + h_1(\theta_1(t), \theta_2(t-\tau)) \\ \dot{\theta}_2(t) = \omega_2 + h_2(\theta_2(t), \theta_3(t-\tau)) \\ \cdots\cdots\cdots \\ \dot{\theta}_n(t) = \omega_n + h_n(\theta_n(t), \theta_1(t-\tau)) \end{cases} \tag{4.1}$$

where $\tau \geq 0$ is time lag.

Theorem 4.1 Suppose that (4.1) satisfies the following conditions:
(H1') There exist n constant numbers $\theta_1^0, \theta_2^0, \cdots, \theta_n^0$, such that

$0 = \omega_k + h_k(\theta_k^0, \theta_k^0)$, $k = 1, 2, \ldots, n$;

(H2') Suppose that there exists an interval J, such that $\{\theta_k^0\}$ is contained in J, and for any θ and $\bar{\theta}$ in J, $(\partial_1 h_k + \partial_2 h_k)(\theta, \bar{\theta}) < 0$ and $\partial_2 h_k(\theta, \bar{\theta}) > 0$.
Then (4.1) has a unique stable time-independent solution.

Proof : If $\tau = 0$, then system (4.1) has no time-delay. As in section 2, we can construct the upper and lower solutions to obtain the unique time-independent solution $\{\theta_k\}$, which is stable when $\tau = 0$. Now we will show that it is also stable as $\tau > 0$.

Suppose that the linearization of (4.1) at $\{\theta_k\}$ is

$$\begin{cases} \dot{\xi}_1(t) = a_{11}\xi_1(t) + a_{12}\xi_2(t-\tau) \\ \dot{\xi}_2(t) = a_{22}\xi_2(t) + a_{23}\xi_3(t-\tau) \\ \cdots\cdots\cdots \\ \dot{\xi}_n(t) = a_{nn}\xi_n(t) + a_{n1}\xi_1(t-\tau) \end{cases} \tag{4.2}$$

where $\xi_k(t) = \theta_k(t) - \theta_k$, $a_{kk} = \partial_1 h_k(\theta_k, \theta_{k+1})$, $a_{k,k+1} = \partial_2 h_k(\theta_k, \theta_{k+1})$, $k = 1, 2, \ldots, n$, (view $n+1$ as 1). The characteristic equation of (4.2) is

$$(\lambda - a_{11})(\lambda - a_{22}) \cdots (\lambda - a_{nn}) - a_{12}a_{23} \cdots a_{n1}e^{-n\tau\lambda} = 0. \tag{4.3}$$

Let $z = n\tau\lambda$, $b_{ij} = n\tau a_{ij}$, then the characteristic equation changes to

$$(z - b_{11})(z - b_{22}) \cdots (z - b_{nn})e^{z} - b_{12}b_{23} \cdots b_{n1} = 0. \tag{4.4}$$

Let $f(z)=(z-b_{11})(z-b_{22}).. (z-b_{nn})$, $q = -b_{12}b_{23} .. b_{n1}$, then the above characteristic equation becomes $H(z) = f(z)e^z - q = 0$. (H2') gives $b_{jj} < 0$, $0 < b_{j,j+1} < -b_{jj}$, $j=1, 2,...,n$. Hence, the zeros, $b_{jj}, j=1,...,n$, of $f(z)$ are all negative.

Let $\bar{\lambda}$ be a real zero of $G(\lambda)$, then

$$| \, q\sin\bar{\lambda} / v(\bar{\lambda}) \, | < -q / \sqrt{u^2(\bar{\lambda})+v^2(\bar{\lambda})}, \text{ when } v(\bar{\lambda}) \neq 0;$$

$$| \, -q\cos\bar{\lambda} / u(\bar{\lambda}) \cdot | < -q / \sqrt{u^2(\bar{\lambda})+v^2(\bar{\lambda})}, \text{ when } v(\bar{\lambda}) = 0.$$

Also,

$$u^2(\bar{\lambda})+v^2(\bar{\lambda}) = f(i\bar{\lambda})f(-i\bar{\lambda}) = \prod_{j=1}^{n}(\bar{\lambda}^2+b_{jj}^2) > \prod_{j=1}^{n} b_{jj}^2 > q^2.$$

Thus, the conditions in theorem 3.1 are all satisfied. So, all zeros of characteristic equation (4.4) have negative real parts. Therefore, when $\tau > 0$, $\{\theta_k\}$ is also a unique stable constant solution of (4.1).

5. The chains of cycles with time-delay coupling

In this section, we treat a chain of cycles of oscillators coupling with time lag. Suppose there are m cycles of oscillators ($0 \le m \le n$), which have coupled structure as in section 4. Then m cycles couple one by one with time lag. The corresponding mathematical model is:

$$\begin{cases} \dot{\theta}_{ik}(t) = \omega_{ik} + h_{ik}^{+}(\theta_{ik}(t), \theta_{i,k+1}(t-\tau)), \\ \quad k = 1,2,\cdots,n_i - 1; \; i = 1,2,\cdots,m; \; n_i \ge 1; \; n_1 + n_2 + \cdots + n_m = n, \\ \dot{\theta}_{in_i}(t) = \omega_{in_i} + h_{in_i}^{+}(\theta_{in_i}(t), \theta_{i1}(t-\tau)) + h_{in_i}^{-}(\theta_{in_i}(t), \theta_{i+1,1}(t-\tau)) \\ \quad i = 1,2,\cdots,m-1, \\ \dot{\theta}_{mn_m}(t) = \omega_{mn_m} + h_{mn_m}^{+}(\theta_{mn_m}(t), \theta_{m1}(t-\tau)). \end{cases} \tag{5.1}$$

Theorem 5.1 Suppose that

(H1") There are n constant numbers θ_{ik}^0 ($k = 1,\cdots,n_i$; $i = 1,\cdots,m$), such that

$$0 = \omega_{ik} + h_{ik}^{+}(\theta_{ik}^0, \theta_{ik}^0), \; k = 1,\cdots,n_i - 1; \; i = 1,\cdots,m,$$

$$0 = \omega_{in_i} + h_{in_i}^{+}(\theta_{in_i}^0, \theta_{in_i}^0) + h_{in_i}^{-}(\theta_{in_i}^0, \theta_{in_i}^0), \; i = 1,\cdots,m-1,$$

$$0 = \omega_{mn_m} + h_{mn_m}^{+}(\theta_{mn_m}^0, \theta_{mn_m}^0);$$

(H2") There exists an interval J, such that $\{\theta_{ik}^0\}$ is contained in J and for any $\theta, \bar{\theta}$ in J,

$$(\partial_1 h_{ik}^{\pm} + \partial_2 h_{ik}^{\pm})(\theta, \bar{\theta}) < 0, \text{ and } \partial_2 h_{ik}^{\pm}(\theta, \bar{\theta}) > 0, \; k = 1,\cdots,n_i; i = 1,\cdots,m.$$

Then (5.1) has a unique stable time-independent solution.

Proof: Consider $\tau = 0$ first. System (5.1) has no time lag. We can obtain a unique stable time-independent solution $\{\theta_{ik}\}$, which is also a unique time-independent solution of (5.1) when $\tau > 0$.

When $\tau > 0$, the characteristic equation of the linearization of (5.1) at $\{\theta_{ik}\}$ is

$$\prod_{j=1}^{m}[f_j(z)e^z + q_j] = 0, \tag{5.2}$$

where $\quad f_j(z) = (z - b_{11}^j)(z - b_{22}^j)\cdots(z - b_{n_j n_j}^j)$, $q_j = -b_{12}^j b_{23}^j \cdots b_{n_j 1}^j$, $j = 1,2,\cdots,m$,

$b_{kk}^{j} = n_j \tau \partial_1 h_{jk}^{+}(\theta_{jk}, \theta_{j,k+1})$, $b_{k,k+1}^{j} = n_j \tau \partial_2 h_{jk}^{+}(\theta_{jk}, \theta_{j,k+1})$. Using (H2") of Theorem 5.1 and Theorem 3.1, we conclude that all zeros of (5.2) lie on the left of the imaginary axis. Hence, the unique time-independent solution $\{\theta_{ik}\}$ of (5.1) is stable when $\tau > 0$.

6. Kalecki's model of business cycles

In this section, we shall treat a mathematical model of business cycles. The model is designed to demonstrate that the length of the business cycle could be estimated by a function of some economic parameters.

Suppose that *I(t)* is the investment rate, U is the constant rate of depreciation of capital goods, and *A(t)* is the production rate. Then

$$A(t) = \int_{t-1}^{t} I(x)\,dx. \tag{6.1}$$

Assume that *K(t)* is the capital at time t, and *L(t)* is the rate of delivery of finished goods. Then we have

$$K'(t) = L(t) - U. \tag{6.2}$$

From the preceding, we also see that *L(t)* = *I(t-1)*, i.e., the delivery rate at time *t* is equal to the investment rate at time *t- 1*. Thus

$$K'(t) = I(t-1) - U. \tag{6.3}$$

Suppose that the investment rate, the quantity of capital, and the production rate have the following linear relationship:

$$I(t) = m[c + A(t)] - nK(t), \tag{6.4}$$

where *m, c,* and *n* are positive constants, which can be roughly determined by economic materials.

Differentiate (6.4) and let *u(t)=I(t)-U,* we obtain

$$u'(t) = m[u(t) - u(t-1)] - nu(t-1). \tag{6.5}$$

The characteristic equation of (6.5) is:

$$(z-m)e^{z} + (m+n) = 0. \tag{6.6}$$

This exponential polynomial does not satisfy the conditions described in Theorem 3.1.

If $m+n \le e^{m-1}$, then it is easy to show that (6.6) has one or two real positive root. This situation can not occur in economic systems.

When $m+n > e^{m-1}$, we let $z=x+iy$ and look for complex roots of (6.6). In this case, (6.6) contains infinitely many complex roots. A detailed analysis shows that when $m+n > e^{m-1}$, cyclical variations occur in our economic system (6.4). Also, the solution of (6.5) will have the following form when time *t* is sufficiently far away from the initial interval $(0, \delta)$:

$$u(t) \approx F_1 e^{(m-x_1)t} \sin(y_1 t), \quad t >> 0, \tag{6.7}$$

where $x_1 + i y_1$ is the main eigenvalue of (6.6), i.e., $x_1 > 0$ is the smallest. Thus the period of business cycle model (6.4) is equal to $T = \dfrac{2\pi}{y_1}$, and the investment rate *I(t)*is

$$I(t) \approx F_1 e^{(m-x_1)t} \sin(y_1 t) + U. \tag{6.8}$$

The amplitude of fluctuations is decreasing, remains constant, or increasing, according to $x_1 > m$, $x_1 = m$, or $x_1 < m$.

Reference

1. N. Kopell, and O.B. Ermentrout, Coupled oscillators and the design of central pattern generators, Math Bios., 1988, 90: 87~109.
2. G.B. Ermentrout, and N. Kopell, Frequency plateaus in a chain of weakly coupled oscillators, SIAM J. Math. Anal. 1984, 15: 215~237.
3. N. Kopell, and G.B. Ermentrout, Symmetry and phase-locking in chains of weakly coupled oscillators, Comm. Pure and Appl. Math. 1986, 39: 623~600.
4. N. Kopell, and G.B. Ermentrout Phase transitions and other phenomena in chains of oscillators, SIAM J. Appl. Math., 1990, 50: 1014~1052.
5. G.B. Ermentrout, and N. Kopell, Oscillator death in systems of coupled neural oscillators, SIAM J Math, 1990, 50: 125~146.
6. N. Kopell, W. Zhang, and G.B. Ermentrout, Multiple coupling in chain of oscillators, SIAM J Math Anal, 1990, 4: 935~953.
7. W. Zhang and M. He, Oscillation death of chains of oscillators with multiple coupling, Mathematica Applicata, 1995, 8: 333-338.
8. G.B. Ermentrout, and N. Kopell, Inhibition-produced patterning in chains of coupled nonlinear oscillators, SIAM J. Appl. Math., 1994, 54(2): 478~507.
9. R. Bellman, and K. L. Cooke, Differential difference equations, New York: Academic Press, 1963, 440~449.
10. L. S. Pontryagin, "On the zeros of some elementary transcendental functions", Amer. Math. Soc. Transl., Ser. 2, Vol. 1, 1955, pp.95~110. (Translated from: Izv. Akad. Nauk SSSR, Ser. Mat., Vol.6, 1942, pp.115-134).
11. R. Frisch, and H. Holme, The characteristic solutions of a mixed difference and differential equation occurring in economic dynamics, Econometrica, 1935, 3: 225~239.
12. M. Kalecki, A macrodynamic theory of business cycles, Econometrica, 1935, 3: 327~344.
13. R. H. Strotz, J. C. McAnulty, and J. B. Naines, Jr., Goodwin's nonlinear theory of the business cycles: An electroanalog solution, 1953, 21: 390~415.
14. W. Brock, and C. L. Sayers, Is the business cycle characterized by deterministic chaos, Working Paper 8617, Social Systems Research Institute, University of Wisconsin-Madison, 1986.

Proceedings of Dynamic Systems and Applications 3 (2001) 313-318

ON FUNCTIONAL DIFFERENTIAL EQUATIONS IN ORDERED BANACH SPACES

S. HEIKKILÄ

Department of Mathematical Sciences, University of Oulu,
P.O. Box 3000, FIN-90014, University of Oulu, Finland

Dedicated to Professor V. Lakshmikantham for his 75'th birthday

ABSTRACT: In this paper we derive existence and comparison results to the implicit functional problem
$u'(t) - g(t,u) = f(t,u,u'(t) - g(t,u))$ a.e. in $[t_0, t_1]$, $u(t) = B(t,u)$, $t \in [t_0 - r, t_0]$
in an ordered Banach space whose bounded and increasing sequences have weak limits.

AMS (MOS) subject classification. 34K30

1. INTRODUCTION

We study the implicit functional problem

$$\begin{cases} u'(t) - g(t,u) = f(t,u,u'(t) - g(t,u)) \text{ a.e. in } J = [t_0, t_1], \\ u(t) = B(t,u), \ t \in J_0 = [t_0 - r, t_0]. \end{cases} \tag{1.1}$$

in an ordered Banach space $E = (E, \leq, \|\cdot\|)$ whose bounded and increasing sequences have weak limits. Denoting $\mathcal{F} = C([t_0 - r, t_1], E)$, and assuming that $f\colon J \times \mathcal{F} \times E \to E$, $g\colon J \times \mathcal{F} \to E$ and $B\colon J_0 \times \mathcal{F} \to E$, the dependence of f, g and B on their second arguments can be functional. This allows also interdependence between equations of (1.1) (cf. Ex. 3.1).

We prove existence and comparison results for extremal solutions of (1.1) in the set

$$V = \{u \in \mathcal{F} \mid u_{|J} \text{ is absolutely continuous and a.e. differentiable}\}.$$

As a special case we obtain results for the explicit functional initial function problem

$$u'(t) = g(t,u) + q(t) \text{ a.e. in } J, \quad u(t) = \alpha(t), \ t \in J_0. \tag{1.2}$$

No continuity hypotheses are imposed on the functions f, g and $B(t, \cdot)$.

2. HYPOTHESES AND MAIN RESULTS

Assume that $\mathcal{F}$ and $C(J_0, E)$ are ordered pointwise, and $L^1(J, E)$ a.e. pointwise. The following hypotheses are imposed on the functions $f\colon J \times \mathcal{F} \times E \to E$, $g\colon J \times \mathcal{F} \to E$ and $B\colon J_0 \times \mathcal{F} \to E$.

(g1) $g(\cdot, u) \in L^1(J, E)$ for each $u \in \mathcal{F}$, problem (1.2) has for all $q \in L^1(J, E)$ and $\alpha \in C(J_0, E)$ the least solution in V, and it is increasing in q and α.

(f1) $f(\cdot, u, v(\cdot))$ is strongly measurable whenever $u \in \mathcal{F}$ and $v \in L^1(J, E)$, and $f(t, u, x)$ is increasing in u and x for a.e. $t \in J$.

(f2) There exist $h_0 \in L^1_+(J)$ and $q_0 \in L^1(J, E)$ such that $\|f(t, u, x)\| \le h_0(t)$ and $q_0(t) \le f(t, u, x)$ for a.e. $t \in J$ and all $u \in \mathcal{F}$, $x \in E$,

(B1) $B(t, \cdot)$ is increasing for each $t \in J_0$ and $\{B(\cdot, u) \mid u \in \mathcal{F}\}$ is equicontinuous.

(B2) There exist $b_0 \in C(J_0, \mathbb{R}_+)$ and $\alpha_0 \in C(J_0, E)$ such that $\|B(t, u)\| \le b_0(t)$ and $\alpha_0(t) \le B(t, u)$ for all $u \in \mathcal{F}$ and $t \in J_0$.

Theorem 2.1. *Assume that the hypotheses (g1), (f1), (f2), (B1) and (B2) hold. Then problem (1.1) has the least solution u_* in V in the sense that if $u \in V$ is any solution of (1.1), then $u_* \le u$. Moreover, u_* increasing with respect to f and to B.*

Proof. We shall first reduce problems (1.1) and (1.2) to operator equations. Denote $X = L^1(J, E) \times C(J_0, E)$ with componentwise ordering, and define

$$Lu := (u' - g(\cdot, u), u_{|J_0}), \quad Nu := (f(\cdot, u, u' - g(\cdot, u)), B(\cdot, u)), \; u \in V. \tag{2.1}$$

$L, N\colon V \to X$, problem (1.1) is reduced to the operator equation $Lu = Nu$, and (1.2) is equivalent to $Lu = (q, \alpha)$. The hypothesis (g1) can be restated as follows.
(L) The least solution u of $Lu = x$ exists in V and is increasing in $x = (q, \alpha) \in X$.
Next we reduce equation $Lu = Nu$ to a fixed point problem. Define $G\colon X \to X$ by

$$Gx := Nu, \quad \text{where } u \text{ is the least solution of } Lu = x \text{ in } V. \tag{2.2}$$

G has the following properties when $a = (q_0, \alpha_0)$:
(G) If $x_1 \le x_2$, then $a \le Gx_1 \le Gx_2$, and well-ordered sets of $G[X]$ have supremums.

These properties imply by Theorem 1.2.1 of [2] that G has the least fixed point x_*, and that

$$x_* = \min\{x \in X \mid Gx \le x\}. \tag{2.3}$$

Denoting by u_* the least solution of $Lu = x_*$, we then have $Lu_* = x_* = Gx_* = Nu_*$. In particular, u_* is a solution of problem (1.1). To prove that u_* is the least solution of (1.1), let $u \in V$ be a solution of (1.1). Then $Lu = Nu$. Denoting by $\hat{u}$ the least of the solutions v of $Lv = x = Nu$, we have $\hat{u} \le u$ and $L\hat{u} = Lu$, whence

$Gx = N\hat{u} \leq Nu = x$. Thus $x_* \leq x$, i.e. $Lu_* \leq L\hat{u}$. This result and property (L) imply that $u_* \leq \hat{u}$. This and $\hat{u} \leq u$ imply that $u_* \leq u$.

The assertion that u_* is increasing with respect to f and B can be proved by applying property (2.3) and definitions (2.1) and (2.2).

Proof of (G): Let $x_i = (q_i, \alpha_i)$, $i = 1, 2$, satisfy $x_1 \leq x_2$, and let u_i be the least solutions of $Lu = x_i$, $i = 1, 2$. Then $q_1 \leq q_2$, and $u_1 \leq u_2$ by property (L), whence the hypotheses (f1), (f2), (B1) and (B2) imply that

$$(q_0, \alpha_0) \leq (f(\cdot, u_1, q_1(\cdot)), B(\cdot, u_1)) \leq (f(\cdot, u_2, q_2(\cdot)), B(\cdot, u_2)).$$

The above inequalities can be rewritten as $a \leq Nu_1 \leq Nu_2$, i.e. $a \leq Gx_1 \leq Gx_2$.

Let W be a well-ordered set in $G[X] \subseteq N[V] \subset X = L^1(J, e) \times C(J_0, E)$. Applying (2.1), (f2), (B1) and (B2) we see that the projection W_1 of W into $L^1(J, E)$ is a well-ordered subset of $L^1(J, E)$, a.e. pointwise bounded by $h_0 \in L^1_+(J)$, and that the projection W_2 of W into $C(J_0, E)$ is a well-ordered, pointwise bounded and equicontinuous subset of $C(J_0, E)$. These properties, and the assumption that increasing and bounded sequences of E have weak limits ensure (cf. [1]) that both W_1 and W_2, and hence also W, have supremums. □

3. SPECIAL CASES AND AN EXAMPLE

In the case when $g(t, u) \equiv 0$ problems (1.1) and (1.2) can be rewritten as

$$u'(t) = f(t, u, u'(t)) \text{ a.e. in } J, \quad u(t) = B(t, u), \ t \in J_0, \tag{3.1}$$

$$u'(t) = q(t) \text{ a.e. in } J, \quad u(t) = \alpha(t), \ t \in J_0. \tag{3.2}$$

(3.2) has for all fixed $q \in L^1(J, E)$ and $\alpha \in C(J_0, E)$ a unique solution

$$u(t) = \begin{cases} \alpha(t_0) + \int_{t_0}^t q(s)\,ds, & t \in J, \\ \alpha(t), & t \in J_0, \end{cases}$$

and it is increasing with respect to q and α. Thus the hypothesis (g1) holds, whence Theorem 2.1 implies the following result.

Corollary 3.1. *If the hypotheses (f1), (f2), (B1), and (B2) are valid, then problem (3.1) has the least solution, and it is increasing with respect to f and to B.*

As dual results of Theorem 2.1 and Corollary 3.1 we get existence and comparison results for the greatest solution of problems (1.1) and (3.1).

If $B(t, u) = \alpha(t)$ and $f(t, u, x) = g(t, u) + q(t)$, then (3.1) is reduced to (1.2). Thus Corollary 3.1 and its dual imply the following sufficient conditions for the validity of the hypothesis (g1) and its dual.

Proposition 3.1. *Assume that $g\colon J \times \mathcal{F} \to E$ has the following properties.*

(g2) *$g(t,u)$ is strongly measurable in t for each $u \in \mathcal{F}$ and increasing in u for a.e. $t \in J$, and there exist $h_1 \in L^1_+(J)$ and g_0, g_1 in $L^1(J,E)$ such that $\|g(t,u)\| \le h_1(t)$ and $g_0(t) \le g(t,u) \le g_1(t)$ for all $u \in \mathcal{F}$ and for a.e. $t \in J$.*

Then $g(\cdot,u) \in L^1(J,E)$ for each $u \in \mathcal{F}$, problem (1.2) has for all $q \in L^1(J,E)$ and $\alpha \in C(J_0,E)$ extremal solutions, and they are increasing with respect to q and α.

According to Theorem 1.2.1 of [2] the least fixed point of G is $x_* = \max C = \sup G[C]$, where $C \subset X$ is well-ordered and has the following properties.

$a = \min C$, and if $a < x \in X$, then $x \in C$ iff $x = \sup\{Gy \mid y \in C,\ y < x\}$.

It can be shown when G is increasing that the first elements of C are of the form $x_{n+1} = Gx_n$, $n \in \mathbb{N}$, $x_0 = a = (q_0, \alpha_0)$. In view of definitions (2.1) and (2.2) the least solutions u_n of $Lu = x_n$ can be given as follows: u_0 is the least solution of problem

$$u'(t) = g(t,u) + q_0(t) \text{ a.e. in } J, \quad u(t) = \alpha_0(t),\ t \in J_0, \tag{3.3}$$

and u_{n+1}, $n \in \mathbb{N}$, is the least solution of problem

$$u'(t) = g(t,u) + f(t,u_n,u_n'(t) - g(t,u_n)) \text{ a.e. in } J,\ u(t) = B(t,u_n),\ t \in J_0. \tag{3.4}$$

If there is $n \in \mathbb{N}$ such that $u_{n+1} = u_n$, then u_n is the least solution of problem (1.1). $u_* = \sup_{n\to\infty} u_n$ is the next possible candidate for the least solution of (1.1).

Example 3.1. Choose $E = \mathbb{R}$, $J = [0,1]$ and consider the problem

$$\begin{cases} u'(t) - H(u(t-1) - 2t) = \frac{[u(t-1)-t]}{1+|[u(t-1)-t]|} + \frac{[u'(t)-H(u(t-1)-2t)]}{1+|[u'(t)-H(u(t-1)-2t)]|} \text{ a.e. in } J, \\ u(t) = 1 - \frac{[\int_{-1}^{1} u(s)ds]}{1+|[\int_{-1}^{1} u(s)ds]|}t, \qquad t \in J_0 = [-1,0], \end{cases} \tag{3.5}$$

where H is the Heaviside function: $H(x) = \begin{cases} 1 \text{ if } x \ge 0, \\ 0 \text{ if } x < 0, \end{cases}$ and $[x]$ denotes the greatest integer less than or equal to x. Problem (3.5) is of the form (1.1) with

$$\begin{cases} g(t,u) = H(u(t-1) - 2t), & t \in J,\ u \in \mathcal{F} = C[-1,1], \\ f(t,u,x) = \frac{[u(t-1)-t]}{1+|[u(t-1)-t]|} + \frac{[x]}{1+|[x]|}, & t \in J,\ u \in \mathcal{F},\ x \in \mathbb{R}, \\ B(t,u) = 1 - \frac{[\int_{-1}^{1} u(s)ds]}{1+|[\int_{-1}^{1} u(s)ds]|}t, & t \in J_0,\ u \in \mathcal{F}. \end{cases} \tag{3.6}$$

It is easy to see that the hypotheses of Theorem 2.1 and its dual hold. Thus problem (3.5) has the least solution u_* and the greatest solution u^*. By choosing $q_0(t) \equiv -2$ and $\alpha_0(t) = 1 + t$ in (3.3) and calculating the successive approximations u_n by (3.4), we see that $u_5 = u_4$, whence $u_* = u_4$. The choices $q_0(t) \equiv 2$ and $\alpha_0(t) = 1 - t$ in (3.3) imply that $u_4 = u_3$, whence $u^* = u_3$. Denoting by χ_W the characteristic function of $W \subset \mathbb{R}$, we get the following representations for u_* and u^*.

$$\begin{cases} u_*(t) = (1 - \frac{2}{3}t)\chi_{[-1,0]} + (1 + \frac{3}{2}t)\chi_{[0,\frac{2}{5}]} + (\frac{7}{5} + \frac{1}{2}t)\chi_{[\frac{2}{5},\frac{5}{8}]} + (\frac{81}{40} - \frac{1}{2}t)\chi_{[\frac{5}{8},1]}, \\ u^*(t) = (1 - \frac{3}{4}t)\chi_{[-1,0]} + (1 + 2t)\chi_{[0,\frac{3}{7}]} + (\frac{10}{7} + t)\chi_{[\frac{3}{7},\frac{7}{11}]} + \frac{159}{77}\chi_{[\frac{7}{11},1]}. \end{cases}$$

Moreover, denoting $A = \{(t,x) | t \in [0,1),\ u_*(t) \le x \le u^*(t)\}$, it is easy to show that the points A are bifurcation points for solutions of (3.5). Thus, between u_* and u^* there is a continuum of chaotically behaving solutions of problem (3.5).

Remarks 3.1. Problem

$$F(t,u,u'(t) - g(t,u)) = 0 \ \text{ for a.e. } \ t \in J, \quad u(t) = B(t,u),\ t \in J_0$$

has the same solutions as (1.1) if there is a function $\mu\colon J \times \mathcal{F} \times E \to (0,\infty)$ such that

$$F(t,u,x) = \mu(t,u,x)(x - f(t,u,x)), \quad t \in J,\ u \in \mathcal{F},\ x \in E.$$

By suitable choices of μ the monotonicity condition of f with respect to x can be weakened. The L^1-boundedness of f can also be weakened.

In ordered reflexive Banach spaces all bounded and increasing sequences have weak limits. The spaces $\mathbb{R}^m$, $m = 1,2,\dots$, with maximum norm and coordinatewise partial ordering, the spaces $l^p(\mathbb{R})$, $p \in [1,\infty)$, ordered coordinatewise and normed by p-norm, and spaces $L^p(\Omega,\mathbb{R})$, where $p \in [1,\infty)$ and $\Omega = (\Omega,\mathcal{A},\mu)$ is a measure space, equipped with p-norm and a.e. pointwise ordering, have the same property. In particular, we can choose E to be one of these spaces in the above considerations and get existence, uniqueness and comparison results for finite and infinite systems of differential equations, and also random differential equations when μ above is a probability measure.

The supremum of a well-ordered set $W_1 \subset L^1(J,E)$, a.e. pointwise bounded by an L^1 function, is proved to exist by showing an existence of a sequence of W_1 which converges weakly a.e. in J to $\sup W_1$. (cf. [1]). Because this convergence cannot be topologized, the use of Theorem 1.2.1 of [2] in the proof of Theorem 2.1 is justified.

REFERENCES

[1] Heikkilä, S., *On well-ordered sets in ordered function spaces*, Proceedings of FSDONA-99 (to appear).

[2] Heikkilä, S. and Lakshmikantham, V., *Monotone Iterative Techniques for Discontinuous Differential Equations*, Marcel Dekker, New York, Basel, 1994.

Proceedings of Dynamic Systems and Applications 3 (2001) 319-326

Extrema of nonlocal functionals and boundary value problems for functional differential equations

G.A.Kamenskii, Ju.P.Zabrodina

Department of Differential Equations, Moscow State Aviation Institute, Volokolamskoe shosse 4, Moscow 125871, Russia.

Key words and phrases: Calculus of variations, nonlocal functionals, sufficient conditions.

AMS (MOS) subject classification: 34K10,49K25.

Abstract

The integral depending on a function of two variables, to one of which the operator of differentiation and to the other one the shift operator are applied, is called the nonlocal functional. It is proved a theorem giving sufficient conditions of extrema of nonlocal functionals.

1. Extremum of a nonlocal functional

There is considered the problem of extremum of the functional

$$J(u) = \int_{t_0}^{t_1} dt \int_{s_0}^{s_1} F(t, s, u(t, s - r_1), \dots, u(t, s) \dots, u(t, s + r_2), \quad u'(t, s - r_1), \dots, u'(t, s), \dots, u'(t, s + r_2))ds, \tag{1}$$

where $s_1 - s_0 > r_1 + r_2$,r_1,r_2 are some integers. Here $u'(t, s)$ means the partial derivative of $u(t, s)$ with respect to t.

On the set $E_0 = \{(t, s)|t \in [t_0, t_1], s \in [s_0 - r_1, s_0 + r_2)\}$ there is given the boundary function $\varphi(t, s)$; on the set $E_1 = \{(t, s)|t \in [t_0, t_1], s \in (s_1 - r_1, s_1 + r_2]\}$ there is given the boundary function $\psi(t, s)$. Let $E(z)$ denote the integer part of z. Define sets

$$\mathbb{Q} = [t_0, t_1] \times (s_0 + r_2, s_1 - r_1),$$

$$R_0 = \{(t, s)|t \in [t_0, t_1], s = s_0 + i, i = r_2, r_2 + 1, \dots, E(s_1 - s_0) - r_1\},$$

$$R_1 = \{(t,s)|t \in [t_0,t_1], s = s_1 - i, i = r_1, r_1+1, \dots, E(s_1 - s_0) - r_2\},$$
$$\mathbb{R} = R_0 \cup R_1 \cup R_\mu \cup R_\nu.$$

$$\tilde{\mathbb{R}} = \mathbb{R} \cup \{(t,s) \in \mathbb{Q}|\text{where} \quad u'(t,s) \quad \text{does not exist}\}$$

The problem of extremum of the functional (1) is stated with boundary value conditions

$$\begin{aligned} u(t,s) &= \varphi(t,s), \quad (t,s) \in E_0; \\ u(t,s) &= \psi(t,s), \quad (t,s) \in E_1; \end{aligned} \tag{2}$$

$$\begin{aligned} u(t,s) &= \mu(s), \quad (t,s) \in G_0; \\ u(t,s) &= \nu(s), \quad (t,s) \in G_1; \end{aligned} \tag{3}$$

It is supposed that F is continuous and has continuous second derivatives with respect to all arguments. The function F must satisfy also the growth conditions

$$|F(t,s,[u(t,s)],[v(t,s)])| \le M(t,s)\left(1 + \sum_{i=-r_2}^{r_1} [u^2(t,s+i) + v^2(t,s+i)]\right), \tag{4}$$

$$\sum_{i=-r_2}^{r_1} [|F_{u(t,s+i)}(t,s,[u(t,s)],[v(t,s)])| + |F_{v(t,s+i)}(t,s,[u(t,s)],[v(t,s)])|] \le$$
$$\le N(t,s)\left(1 + \sum_{i=-r_2}^{r_1} [u(t,s+i) + v(t,s+i)]\right), \tag{5}$$

where $M(t,s)$, $N(t,s)$ are some continuous functions. Denote $L_2(\mathbb{Q})$ the space of square integrable functions on $\mathbb{Q}$, and $\mathbb{H}(\mathbb{Q})$ the space of functions that belong together with their partial derivatives with respect to t to $L_2(\mathbb{Q})$. Analogously there are defined the spaces $\mathbb{H}(E_0)$ and $\mathbb{H}(E_1)$.

Assume that $\varphi \in \mathbb{H}(E_0)$, $\psi \in \mathbb{H}(E_1)$, $\mu(s)$ and $\nu(s)$ are some measurable and almost everywhere on G_0 and G_1 bounded functions. Admissible functions $u(t,s)$ belong to $\mathbb{H}(\mathbb{Q})$ and satisfy boundary value conditions (2),(3). Define in $\mathbb{H}(\mathbb{Q})$ the norm

$$\|u\|_H = \max\{\|u\|_{L_2(Q)}, \|u_t\|_{L_2(Q)}\}.$$

From (4),(5) it follows existence of $J(u)$ and $\delta J(u,\delta u)$ for any admissible functions u and δu.

If the problem (1), (2), (3) is considered in the space $\mathbb{H}(\mathbb{Q})$), then we shall say that functional (1) attains a weak extremum.

It was proved in Kamenskii [1] that if $u(t,s)$ supplies a weak extremum to functional (1) with conditions (2), (3), it is necessary that it satisfies almost everywhere on $\mathbb{H}(\mathbb{Q})$ the equation

$$\Phi_{u'(t,s)} = \int_{t_0}^{t} \Phi_{u(t,s)} dt + C(s), \tag{6}$$

where $C(s)$ is an arbitrary function of s and

$$\begin{aligned}\Phi(t, s, u(t, s-r_1-r_2), \dots, u(t,s), \dots, u(t, s+r_1+r_2), \\ u'(t, s-r_1-r_2), \dots, u'(t,s), \dots, u'(t, s+r_1+r_2)) := \\ \sum_{i=-r_2}^{r_1} F(t, s+i, [u(t,s+i)], [u'(t,s+i)]).\end{aligned} \tag{7}$$

Here

$$\begin{aligned}&F(t,s,[u(t,s)],[u'(t,s)]) := \\ &F(t,s,u(t,s-r_1)\dots,u(t,s+r_2),u'(t,s-r_1)\dots,u'(t,s+r_2)).\end{aligned}$$

Differentiating (6) we get

$$\Phi_{u(t,s)} - \frac{d}{dt}\Phi_{u'(t,s)} = 0, \tag{8}$$

which is a generalized Euler equation for this problem.

This equation is a mixed difference-differential equation. The survey of the theory of mixed functional differential equations, which include mixed difference-differential equations, is publised in Kamenskii [2]

Equation (8) is satisfied almost everywhere on $\mathbb{Q}$ in the following generalized sense. The derivative $\frac{d}{dt}\Phi_{u'(t,s)}$ exists almost everywhere on $\mathbb{Q}$ and is almost everywhere equal to the function $\Phi_{u(t,s)}$, but the rule of differentiation of a composite function $\Phi_{u'(t,s)}$ may be not applicable, and it may be that the second derivative $u''(t,s)$ does not exist (see Ivanova & Kamenskii [3]).

2. Sufficient conditions of a weak extremum of a nonlocal functional

We derive here some sufficient conditions of a weak extremum of the nonlocal functional. For simplicity consider here $r_1 = r$, $r_2 = 0$ in (1). Denote $\mathbb{H}_2$ the space of two-dimensional vector-functions $v = (v_1(t,s), v_2(t,s))$, where $v_1, v_2 \in \mathbb{H}(\mathbb{Q})$. In $\mathbb{H}_2$ introduce the norm

$$\|v\|_{H_2} = \max(\|v_1\|_H, \|v_2\|_H)$$

Here $(\cdot,\cdot)$ denotes the scalar product of two-dimensional vectors. Let $P(t,s)$, $Q(t,s)$, $R(t,s)$ be given 2×2 matrices, elements of which are measurable and bounded almost everywhere on $\mathbb{Q}=[t_0,t_1]\times[s_0,s_1-r]$. Suppose that P и R are symmetric and R is positive definite, i.e. there exists such a $k>0$ that $(R(t,s)v,v)\geq k(v,v)$. Therefore there exist almost everywhere on $\mathbb{Q}$ bounded matrices $R^{-1}(t,s)$, $R^{\frac{1}{2}}(t,s)$, $R^{-\frac{1}{2}}(t,s)$.

The set $v\in\mathbb{H}_2$ such that $(Rv'+Qv)\in\mathbb{H}_2$ we denote $\mathbb{S}$. On $\mathbb{S}$ we define the operator A

$$Av=Pv+Q^*v-\frac{d}{dt}(Rv+Qv'), \tag{9}$$

where $*$ denotes the operation of transposition of a matrix.

Let V be a square 2×2 matrix, columns of which belong to $\mathbb{S}$.

Consider the matrix equation

$$PV+Q^*V-\frac{d}{dt}(RV+QV')=0 \tag{10}$$

Equation (10) is a second order equation in t without deviations of argument. For any fixed s it coincides with equation (3.2) from [4] and therefore the initial conditions

$$V(t_0,s)=V_0^0,\quad (RV'+QV)(t_0,s)=V_0^1$$

define the unique solution $V(t,s)$ of equation (10), and the columns of the matrix $V(t,s)$ belong to $\mathbb{S}$.

Definition. The point $\zeta(t_\zeta,s_\zeta)$, $\zeta\in\mathbb{Q}$ we call $\mathbb{S}$-conjugated with the point $\xi(t_\xi,s_\xi)$, $\xi\in\mathbb{Q}$, if the solution $V(t,s)$ of equation (10), satisfying the initial value conditions

$$V(\xi)=0,\quad (RV'+QV)(\xi)=I, \tag{11}$$

where I is the identity 2×2matrix, has the property that the determinant $|V(t,s)|$ is not equal to zero on the interval connecting the points ξ and ζ and $|V(\zeta)|=0$.

Lemma 1 *Let $P(t,s)$, $Q(t,s)$, $R(t,s)$ be 2×2 matrices with bounded almost everywhere on $\mathbb{Q}$ elements, matrices P and R are symmetric, R is positive definite and on $\mathbb{Q}$ there are no points $\mathbb{S}$-conjugated with the point (t_0,s_0). Then*

$$\int\int_Q[(Pv,v)+2(Qv.v')+(Rv',v')]dtds\geq 0 \tag{12}$$

for all v belonging to $\mathbb{H}_2$ and equal to zero on the boundaries of the set $\mathbb{Q}$.

Proof coincides with the proof of the analogous lemma in Kamenskii & Skubachevskii [4]. ∎

In the following theorem the matrices P, Q, R are defined by formulae

$$P = \begin{pmatrix} F_{uu} & F_{uu_r} \\ F_{u_r u} & F_{u_r u_r} \end{pmatrix}, \quad Q = \begin{pmatrix} F_{u'u} & F_{u'u_r} \\ F_{u'_r u} & F_{u'_r u_r} \end{pmatrix}, \quad R = \begin{pmatrix} F_{u'u'} & F_{u'u'_r} \\ F_{u'_r u'} & F_{u'_r u'_r} \end{pmatrix}. \tag{13}$$

Theorem 1 *If the function $u(t,s)$ is a solution of the generalized Euler equation (8) and satisfies the boundary value conditions (2), (3), by substitution of $u(t,s)$ in matrices (13) on the set $\mathbb{Q}$ the matrix $R(t,s)$ is positive definite and on $\mathbb{Q}$ there are no points $\mathbb{S}$-conjugated with the point (t_0, s_0), then the functional (1) attains on $u(t,s)$ a weak local extremum.*

Proof. Introduce the vector $\theta = (\delta u(t,s), \delta u(t, s-r))$. Then the second variation of the functional (1) can be written in the form

$$\delta^2 J = \frac{1}{2} \iint_Q [(P\theta, \theta) + 2(Q\theta, \theta') + (R\theta', \theta')] dt ds.$$

Now we prove that the increment of the functional (1) on the solution $u(t,s)$ of the generalized Euler equation (8) with boundary value conditions (2), (3) can be written in the form

$$\Delta J = \delta^2 J(u, \theta) + \nu \iint_Q (\theta', \theta') dt ds, \tag{14}$$

where $\nu \to 0$ при $\|\delta u\|_H \to 0$.

Using the Taylor formula with the remainder term of the second order and equality to zero of the first variation of the functional (1) along $u(t,s)$, we get

$$\Delta J = \frac{1}{2} \iint_Q (\bar{F}_{uu} \delta u^2 + \bar{F}_{u_r u_r} \delta u_r^2 + \bar{F}_{u'u'} \delta u'^2 + \bar{F}_{u'_r u'_r} \delta {u'_r}^2 +$$
$$+ 2\bar{F}_{uu_r} \delta u \delta u_r + 2\bar{F}_{uu'_r} \delta u \delta u'_r + 2\bar{F}_{uu'} \delta u \delta u' +$$
$$+ 2\bar{F}_{u'u_r} \delta u' \delta u_r + 2\bar{F}_{u'u'_r} \delta u' \delta u'_r + 2\bar{F}_{u'_r u_r} \delta u'_r \delta u_r) dt ds,$$

where

$$\bar{F} = F(t, s, u(t, s-r) + \xi \delta u(t, s-r), u(t,s) + \xi \delta u(t,s),$$
$$u'(t, s-r) + \xi \delta u'(t, s-r), u'(t,s) + \xi \delta u'(t,s)),$$
$$0 \le \xi \le 1.$$

It follows from boundedness of the second partial derivatives of the function F and from $\|\delta u\|_H \to 0$ that

$$\Delta J = \delta^2 J(u,\theta) + b(u,\delta u),$$

where

$$b(u,\delta u) = \iint_Q (b_{11}\delta u^2 + b_{22}\delta u_r^2 + b_{33}\delta u'^2 + b_{44}\delta u_r'^2 +$$
$$+2b_{12}\delta u\delta u_r + 2b_{13}\delta u\delta u' + 2b_{14}\delta u\delta u_r' +$$
$$+2b_{23}\delta u'\delta u_r + 2b_{24}\delta u_r\delta u_r' + 2b_{34}\delta u'\delta u_r')dtds$$

and vrai $\max|b_{ij}(t,s)| \to 0, \quad (i,j = 1,2,3,4)$ as $\|\delta u\|_H \to 0$.

We have $\delta u(t_0,s) = 0$ for any s. Then $\delta u(t,s) = \int_{t_0}^{t} \delta u'(\omega,s)d\omega$, and using the Cauchy inequality, we get

$$\iint_Q \delta u^2(t,s)dtds = \frac{(t_1-t_0)^2}{2}\iint_Q \delta u'^2(t,s)dtds.$$

It follows from boundary value conditions that

$$\iint_Q \delta u^2(t,s-r)dtds = \int_{t_0}^{t_1}\int_{s_0}^{s_1-r} \delta u^2(t,s-r)dtds =$$
$$= \int_{t_0}^{t_1}\int_{s_0-r}^{s_1-2r} \delta u^2(t,z)dtdz \le \iint_Q \delta u^2(t,s)dtds \le$$
$$\le \frac{(t_1-t_0)^2}{2}\iint_Q \delta u'^2(t,s)dtds,$$

Farther, from

$$(\iint_Q |\delta u(t,s)||\delta u'(t,s-r)|dtds)^2 \le$$
$$\le \iint_Q \delta u^2(t,s)dtds \iint_Q \delta u'^2(t,s)dtds \le$$
$$\le \frac{(t_1-t_0)^2}{2}(\iint_Q \delta u'^2(t,s)dtds)^2,$$

it follows

$$|\iint_Q |\delta u(t,s)||\delta u'(t,s-r)|dtds| \le \frac{t_1-t_0}{\sqrt{2}}\iint_Q \delta u'^2(t,s)dtds$$

By the same reasoning we prove that each of integrals of the pairwise products $|\delta u(t,s)|, |\delta u'(t,s)|, |\delta u(t,s-r)|, |\delta u'(t,s-r)|$ does not exceed a constant multiplied by $\iint_Q \delta u'^2(t,s)dtds$.

From inequalities

$$\iint_Q \delta u'^2(t,s)dtds \leq \iint_Q [\delta u'^2(t,s) + \delta u'^2(t,s-r)]dtds \leq \iint_Q (\theta', \theta')dtds,$$

it follows (14).

If we can prove that for any $\delta u \in \mathbb{H}(\mathbb{Q})$, satisfying the boundary value conditions,

$$\delta^2 J(u,\theta) \geq k \iint_Q (\theta', \theta')dtds, \tag{15}$$

then from (14) it follows that the functional (1) attains an extremum on $u(t,s)$ in $\mathbb{H}(\mathbb{Q})$.

Prove now that

$$\iint_Q [(Pv,v) + 2(Qv,v') + (Rv',v')]dtds \geq k \iint_Q (v',v')dtds \tag{16}$$

for any $v \in \mathbb{H}_2$ that are equal to zero on the boundary of the set $\mathbb{Q}$.

Consider the equation

$$Pv + Q^* v' - \frac{d}{dt}[(R-\mu I)v' + Qv] = 0. \tag{17}$$

The matrix R is positive definite and solutions of (17) depend continuously on μ for small μ. There are no points that are $\mathbb{S}$-conjugated with the point (t_0, s_0) on $\mathbb{Q}$. Therefore it is possible to find a so small μ that the matrix $R - \mu I$ is also positive definite. It follows from Lemma 1 that the inequality

$$\iint_Q [(Pv,v) + 2(Qv,v') + ((R-\mu I)v',v')]dtds \geq 0.$$

is true. Then

$$\iint_Q [(Pv,v) + 2(Qv,v') + (Rv',v')]dtds \geq \mu \iint_Q (v',v')dtds$$

that completes the proof of the theorem. ■

References

[1] Kamenskii,G.A, Boundary value problems for differential-difference equations arising from variational problems. Nonlinear Analysis, TMA, Vol. 18, №8, (1992), 801-813.

[2] Kamenskii,G.A, A review of the theory of mixed functional-difference equations. Problems of Nonlinear Analisys in Engineering Systems,(1998), Vol. 2(8), 1-16.

[3] Ivanova, E.P., & Kamenskii G.A. , Variatonal and boundary value problems for differential-difference equations. Moscow, MAI Press, 1993.(In Russian.)

[4] Kamenskii,G.A,& Skubachevskii A.L, Extrema of functionals with deviating arguments. Moscow, MAI Press, 1973. (In Russian.)

Proceedings of Dynamic Systems and Applications 3 (2001) 327-334

Focus of Attention During Image Formation using a Quadtree Approach for Ultra-Wideband Radar *

Lance M. Kaplan

Center for Theoretical Studies of Physical Systems, and
Departemt of Engineering
Clark Atlanta University
Atlanta, GA 30314

ABSTRACT

In this work, we introduce a method to perform focus of attention (FOA) during the image formation processing step for ultra-wideband (UWB) synthetic aperture radar (SAR) systems. The FOA consists of a multiscale detection statistic derived from intermediate radar data generated during the quadtree based backprojection image formation algorithm. As the quadtree algorithm iterates, it is resolving increasingly finer subpatches of the scene. At each stage, the FOA estimates the signal to background ratio of each subpatch in order to detect the subpatches that may contain a target of interest. Whenever a subpatch is not near a detection, the FOA cues the image formation processor to terminate further processing of that subpatch. We demonstrate that the FOA is able to decrease the overall computational load of the image formation process. We also show that the new FOA method produces fewer false alarms than the two-parameter CFAR FOA does over a small database of UWB radar data.

AMS (MOS) subject classification 94A13, 60G35

1.0 INTRODUCTION

The military is interested in using ultra-wideband (UWB) UWB synthetic aperture radar (SAR) systems to detect ground targets that are intentionally concealed under forest and to detect mines at a safe distance from the minefield. The UWB SAR sends and receives a series of low frequency electromagnetic pulses so that the energy can penetrate through foliage and bounce of the hidden target. These systems are referred to as ultra-wideband because the bandwidth to center frequency ratio is large.

While the low frequency nature of UWB radar provides the foliage penetration, it also introduces a number of challenges for a target detection system. This paper addresses one of these problems. At

*Prepared through collaborative participation in the Advanced Sensors Consortium sponsored by the U.S. Army Research Laboratory under Cooperative Agreement DAAL01-96-20-0001.

low frequencies, the SAR system must fly a significantly longer synthetic aperture to achieve equivalent cross-range resolutions as the higher frequency X- and K-band SAR systems. This fact poses a challenge for the image formation processor [1]. Specifically, the distortions and errors due to motion that appear in frequency-based image formers amplify for the UWB SAR case, and more computationally complex image formers such as backprojection are required to form crisp UWB SAR images [2]. In order to lower the complexity of UWB image formation, McCorkle and Rofheart introduced a quadtree backprojection method [3]. The quadtree approach uses a divide-and-conquer strategy with $\log_2 N$ stages similar to the fast Fourier transform (FFT). Unlike the FFT, the quadtree method only approximates the image obtained by standard backprojection. A comparison of images formed by the two methods can be found in [4].

Traditionally, automatic target recognition (ATR) operations such as detection, discrimination, and recognition are performed after image formation [5]. In this work, we integrate a focus of attention (FOA) algorithm into a multistage image formation algorithm. In our approach, we use a detection statistic derived from the intermediate radar data formed during each stage of the quadtree algorithm. The new FOA is able to cue the quadtree to stop processing uninteresting areas of the scene at the very early processing stages. In the end, the hybrid FOA-quadtree algorithm produces imagery of the ROIs that will be passed to the discrimination stage of the ATD system. Obviously, this approach will free up computational resources, which (1) can be used to generate more complex features for later ATR stages or (2) can speed up the end-to-end ATR process. In fact, we show that over a small database, the multistage FOA provides fewer false alarms (at a given probability of detection) than the standard approach of full image formation followed by a two-parameter CFAR FOA. A number of other multiresolution SAR detectors have been presented in the literature [6, 7, 8, 9]. However, these methods are not appropriate as a multistage cuer because they require the availability of all scales to make a detection decision. In addition, they were designed for higher frequency SAR systems whose waveform is a linear frequency modulated chirp.

This paper is organized as follows. Section 2 describes the details for the UWB radar system used to collect the data for this paper, and Section 3 reviews the quadtree backprojection method. A description of the FOA-quadtree algorithm and its theoretical basis is provided in Section 4. Section 5 includes experiments that demonstrate the performance of the new FOA quadtree image formation scheme in terms of false alarms and number of computations. Finally, Section 6 contains some concluding remarks.

2.0 THE ARL BoomSAR DATABASE

In this paper, we use data collected by the Army Research Laboratory (ARL) BoomSAR, which is an experimental radar system [10]. It is composed of an impulse radar sitting atop a 150-ft telescoping boom lift operating at 20-1100 MHz. The system uses a 90° aperture to form imagery with 6 inch resolution both in range and cross-range. In our experiments, we processed raw data generated from the Run 60 data collected at the Aberdeen Proving Grounds (AGP). This data was collected at an elevation angle of 18° and covers an area of 0.42 square kilometers containing eight 230m $\times$ 230m frames. While the BoomSAR is capable of collecting data at all four polarizations, the data used in this study corresponds to an HH polarization. We formed the data into images at a full resolution of 0.075m (or 3in) in both range and cross-range. The scene contains 13 vehicles, a fence, eight trihedrals, and other manmade and natural clutter objects. Ground truth data is available which describes the locations of the vehicles, fence posts, trihedrals and other objects. Fig. 1 shows Frame 2 of the APG data. The resulting image contains two

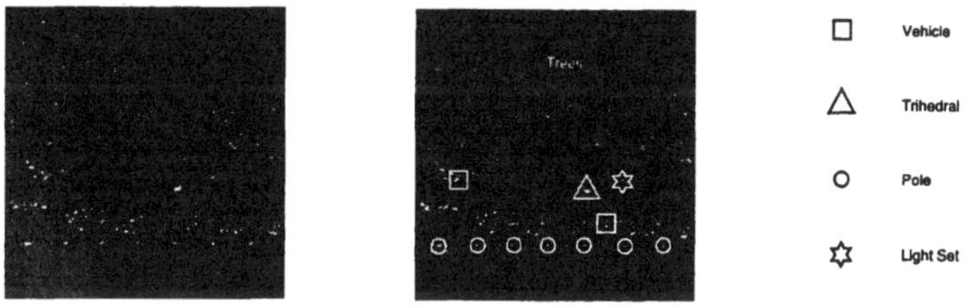

(a) (b) (c)

Figure 1: Frame 1 from the APG database: (a) The image, (b) the image with markings identifying the location of objects listed in the ground truth file, and (c) the ground truth legend.

vehicles. For the ATR problem, we label the 13 vehicles as targets. Any other objects in the scene are considered to be clutter.

3.0 IMAGE FORMATION

The quadtree algorithm for UWB SAR processing uses a divide and conquer decomposition of the backprojection algorithm. A detail description of the algorithm can be found in [3]. For the first stage, the scene is divided into four subpatches. For each subpatch, the echoed pulses are shifted so that a point-like scatterer from the subpatch center lies in the center range, and the algorithm only retains the range bins needed to cover that particular subpatch. Then, adjacent echoed pulses are averaged together in order to reduce the number of pulses by a factor of two. In addition, the phase centers for these integrated pulses are assigned the weighted average of the corresponding phase centers for the original pulses. After the first stage, the intermediate quadtree radar data consists of a set of averaged echo pulse returns for each subpatch. For the next stage, each subpatch is further divided into four smaller subpatches, and the new set of echoed pulses is computed using the corresponding set of pulses generated during the first stage. Then, the division process is repeated until each subpatch is the size of a desired resolution cell. After the final stage, an image is formed by coherently summing up the intermediate quadtree radar for each subpatch (or resolution cell).

The intermediate quadtree radar data can be represented by a four dimension function $v_s(r, a, m, n)$, where s is the quadtree stage, (r, a) represents a sampling in the fast-time/slow-time domain, and (m, n) represents the subpatch location as a sampling in the spatial (or image) domain. A multiresolution measurement scheme can be constructed in the spatial domain by summing the squared data within each subpatch to obtain an *energy image* indexed over (m, n):

$$E_s[m, n] = \sum_r \sum_a |v_s(r, a, m, n)|^2 \tag{1}$$

This image represents the total energy in each subpatch at stage s. The s-th stage provides 4^s subpatch centers, so that the size of the energy image grows with respect to s. As the quadtree progresses, the subpatches cover smaller areas of the ground, and the resolution of the energy images becomes finer.

4.0 MULTISCALE DETECTION

Using information from the energy images, we develop an FOA that can separate regions containing bright objects from benign regions as early as possible in the quadtree image formation process. Similar to detectors that apply the two-parameter CFAR over the formed image, our multistage detector should be able to distinguish bright regions from the darker background. The design of such a detector begins with a model of the raw radar data for a scene containing one coherent reflector in a diffuse background. We model the echoed return from the diffuse background as white noise, and the return from the reflector as a delayed version of the pulse scaled by the reflector's RCS. The raw data for our model is:

$$v(r,a) = Kh(r - r_a) + \eta(r,a), \tag{2}$$

where K is the RCS for the reflector, $h(r)$ is the transmitted pulse, r_a is the distance from the radar to the reflector at each aperture point, and $\eta(r,a)$ is white noise with variance σ_η^2. The detector is a hypothesis test between target, i.e. $K > 0$, and no target, i.e. $K = 0$, given the energy associated to the subpatch data at the available scales. Without loss of generality, we assume that the pulse is normalized so that its energy over all range cells is equal to unity.

Using the scattering model and (1), one can show that the expected energy is:

$$\mathcal{E}\left\{E_s[m_t, n_t]\right\} = 2^s N_a K^2 H(s) + 2^{-s} N_a N_r \sigma_\eta^2 I(s). \tag{3}$$

The $H(s)$ term represents the normalized energy contribution from the impulse response of the radar after each stage:

$$H(s) = \sum_{r=-N_r/2^{s+1}}^{N_r/2^{s+1}-1} h^2(r),$$

and $I(s)$ is a term that accounts for bilinear interpolation. Overall, the expected energy value shows that the energy contribution from a coherent scatterer is geometrically increasing from stage to stage. On the other hand, the energy contribution from an incoherent background is geometrically decreasing from stage to stage. As the quadtree progresses, the overall signal to background ratio is improving during the image formation process, because the reflected energy for the scatterer becomes better localized.

Our multistage detector exploits the fact that energy is growing for a coherent target and decreasing for the background. At stage s, the detector estimates the signal to background ratio. Specifically, our method begins to make a signal hypothesis at stage $s = 2$ by comparing a detection to a threshold, i.e.

$$\frac{E_s[m,n] - \frac{1}{2}\frac{I(s)}{I(s-1)} E_{s-1}[\lfloor m/2 \rfloor, \lfloor n/2 \rfloor]}{2E_1[\lfloor m/2^s \rfloor, \lfloor n/2^s \rfloor] - E_2[\lfloor m/2^{s-1} \rfloor, \lfloor n/2^{s-1} \rfloor]} \underset{\text{Background}}{\overset{\text{Target}}{\gtrless}} 2^s H(s)\tau, \qquad \text{for } s > 2,$$

where τ is a user selected threshold which controls the operating point of the detector, e.g. the probability of detection. The numerator of the statistic is simply the solution for the signal energy at stage s given the present and previous energy measurements. The denominator is the solution for the noise energy at stage $s = 2$ given the energy measurements as stages $s = 1, 2$. The terms in front of the threshold account for the fact that the signal-to-background ratio is increasing as the quadtree progresses. Given the results of the detector at stage s, the FOA quadtree will terminate further processing of a subpatch if that subpatch is not near any detections.

One can show that the detections scheme provides a constant false rate (CFAR) at a given stage in the sense that one can show that the expected false alarm rate is invariant to the incoherent background level σ_η^2. Furthermore, the detector provides approximately a constant probability of detection and decreasing false alarm rate as the quadtree image former improves the signal to background ratio. As the image formation process progresses and the scene is comes into focus, the detector is capable of using the new information to reduce false alarms while maintaining a very high probability of detection.

The termination of formation processing for benign clutter subpatches helps to reduce the overall complexity of the quadtree algorithm. When the raw radar data consists of N_a echo pulses containing N_r range samples, each stage of the full quadtree requires $18 \cdot 2^s N_a + 33 N_r N_a$ floating point operations (flops). In the FOA-quadtree, the extra processing to compute the energy images and the appropriate detection statistics requires $2N_r N_a$ flops. However, at stage s, the quadtree only forms $p(s-1)$ percent of the subpatches, where $p(s-1)$ represents the ratio of subpatches that past the detector from the previous stage over the total number of subpatches. As a result, only $(18 \cdot 2^s N_a + 35 N_r N_a) p(s-1)$ flops are needed to resolve these subpatches into smaller subpatches and compute the corresponding energy values. Obviously, the overall complexity of the FOA quadtree is a function of the number of objects in the scene.

5.0 EXPERIMENTS

All eight frames in the APG dataset were focused using the quadtree image formation algorithm. We used a mosaic of 6×6 patches for each frame to insure that each pixel in the image sees approximately a broadside aperture of about ninety degrees. Fig. 2 shows plots of the energy, detection statistic, and the detection images at stages $s = 2, 4, 6, 8$ of the FOA quadtree image formation algorithm for Frame 2. In the detection images, the white pixels correspond to detections and the grey pixels correspond to neighboring subpatches that are also resolved. The figure does demonstrate that as the quadtree progresses, the signal to background ration increases. Furthermore, the detection statistic is emphasizing the location of the point reflectors. In this example, the threshold was set at a level so that all objects in the ground truth data would be resolved. Fig. 3 shows the final image product of the FOA quadtree and the final detected image with ground truth marking using this threshold. The figure also shows the final detected image when the threshold in raised to a level so that all vehicles are resolved but other objects may not pass the FOA.

Fig. 4 shows plots of the percentage coverage $p(s)$ and the cumulative complexity of both the full and FOA quadtree methods when using the lower threshold over all eight frames. At the earlier stages, the energy image computations make the FOA quadtree a bit more expensive. By the fourth stage, the removal

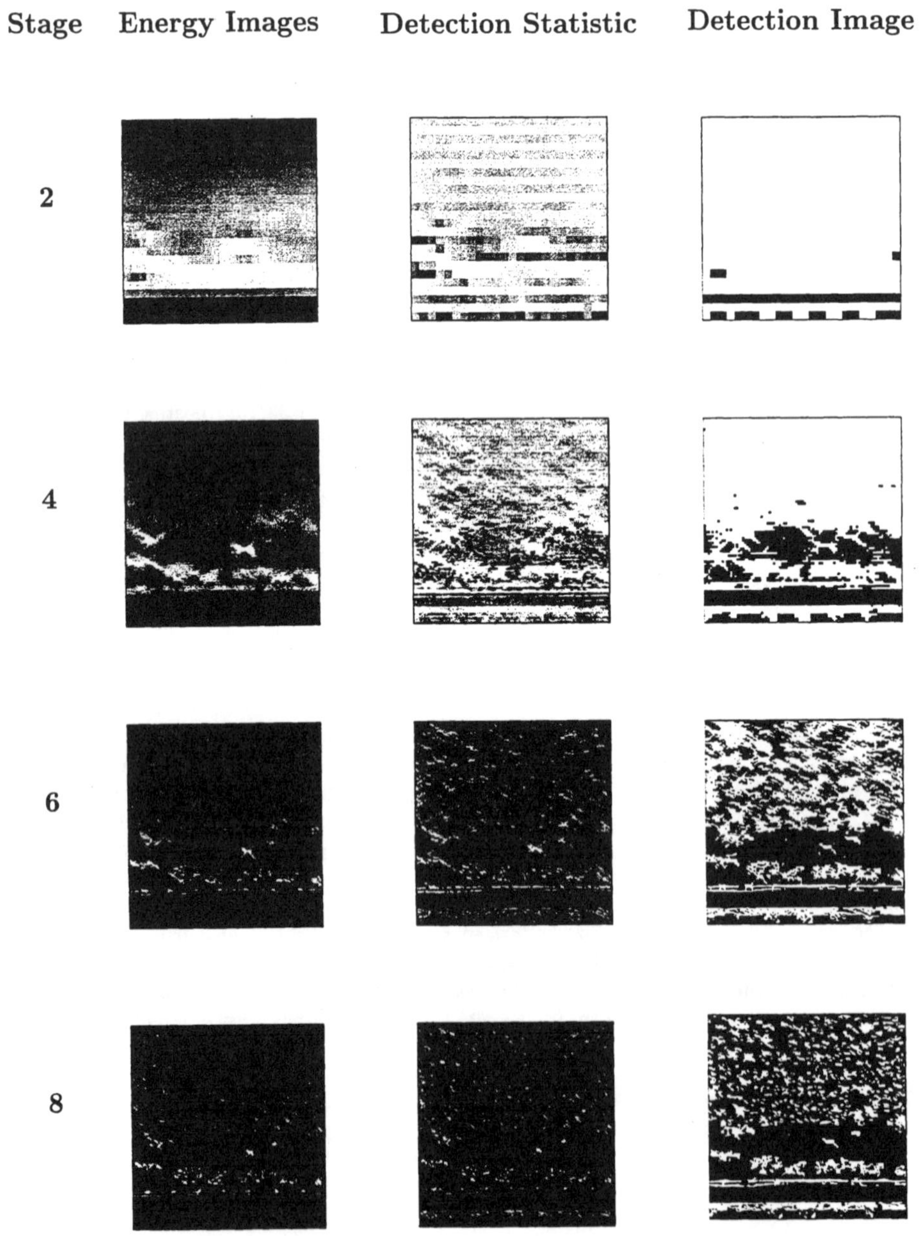

Figure 2: Energy, detection statistics, and detection images during various stages of the FOA quadtree algorithm.

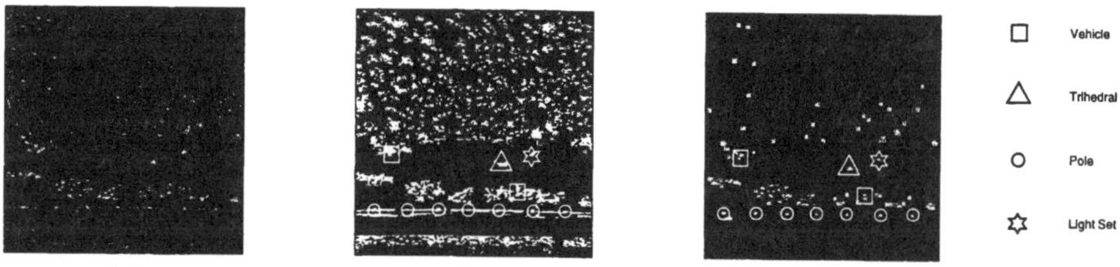

(a) (b) (c) (d)

Figure 3: Results of the FOA quadtree: (a)Final image for $\tau = 0.00015$, (b) detection image for $\tau = 0.00015$, (c) detection image for $\tau = 0.00171$, and (d) ground truth legend.

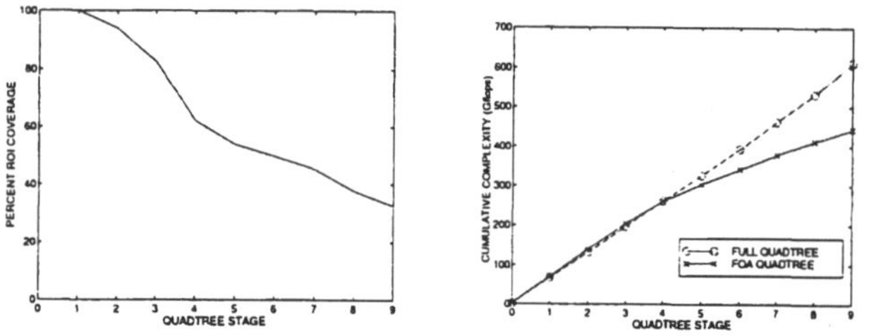

(a) (b)

Figure 4: Complexity results for the FOA quadtree: (a) Percent coverage, and (b) cumulative complexity.

of benign subpatches makes the FOA quadtree more efficient. In the end, the insertion of the FOA reduces the overall complexity by 30 percent.

As mentioned in Section 4, the multistage FOA is looking for bright objects similar to a FOA that uses a two-parameter CFAR over the formed image. By changing the threshold values, we obtained the receiver operating characteristics (ROC) curve for the multistage detector over the APG database. Similarly, we obtained the ROC curve using the two-parameter CFAR with a guard radius of four meters and an annular window width of five pixels. Fig. 5 compares the two ROC curves. For a set probability of detection, the multistage FOA provides a lower false alarm rate. We conjecture that the need to define an annular window for estimating background statistics limits the performance of the two-parameter CFAR FOA.

6.0 CONCLUSIONS

This paper presented a multiresolution FOA that reduces the computational load for UWB SAR image

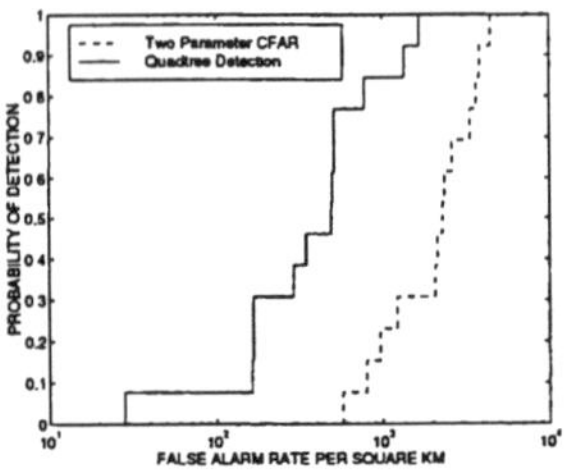

Figure 5: ROC curves comparing the performance of the FOA quadtree and two-parameter CFAR FOA.

formation. Future work will investigate software and hardware implementations of the FOA quadtree. Furthermore, we plan to use information in the intermediate quadtree data do discriminate between scattering due to manmade targets of interest such as vehicle from scattering due to trees an other natural objects.

References

[1] R. Goodman, S. Tummala, and W. Carrara, "Issues in ultra-wideband, widebeam SAR image formation," in *Proceedings of the IEEE International Radar Conference*, pp. 479–485, May 1995.

[2] M. Soumekh, *Synthetic Aperture Radar Signal Processing with MATLAB Algorithms*, J. Wiley, New York, 1999.

[3] J. McCorkle and M. Rofheart, "An order $n^2 \log(n)$ backprojector algorithm for focusing wide-angle wide-bandwidth arbitrary-motion synthetic aperture radar," in *Proc. of SPIE*, vol. 2747, pp. 25–36, 1996.

[4] S.-M. Oh, J. H. McClellan, and M. C. Cobb, "Multi-resolution mixed-radix quadtree SAR image focusing algorithms," in *Proc. of the Third Annual ARL Federated Laboratory Symposium*, pp. 139–143, Feb. 1999.

[5] L. M. Novak, S. D. Halversen, G. J. Owirka, and M. Hiett, "Effects of polarization and resolution on SAR ATR," *IEEE Trans. on Aerospace and Electronic Systems* **33**, pp. 102–116, Jan. 1997.

[6] W. W. Irving, A. S. Willsky, and L. M. Novak, "A multiresolution approach to discriminating targets from clutter in SAR imagery," in *Proc. of SPIE*, vol. 2487, pp. 272–299, 1995.

[7] C. H. Fosgate, H. Krim, W. W. Irving, W. C. Karl, and A. S. Wilsky, "Multiscale segmentation and anomaly enhancement of SAR imagery," *IEEE Trans. on Image Processing* **6**, pp. 7–20, Jan. 1997.

[8] N. S. Subotic, B. J. Thelen, J. D. Gorman, and M. F. Reiley, "Multiresolution detection of coherent radar targets," *IEEE Trans. on Image Processing* **6**, pp. 21–35, Jan. 1997.

[9] A. Kim and H. Krim, "Hierarchial stochastic modeling of SAR imagery for segmentation/compression," *IEEE Trans. on Signal Processing* **47**, pp. 458–468, Feb. 1999.

[10] L. Nguyen, R. Kapoor, and J. Sichina, "Detection algorithms for ultra-wideband foliage penetration radar," in *Proc. of SPIE*, vol. 3066, pp. 165–176, 1997.

*The views and conclusions contained in this document are those of the authors and should not be interpreted as presenting the official policies either express or implied of the Army Research Laboratory or the U. S. Government.

Proceedings of Dynamic Systems and Applications 3 (2001) 335-342

Opial's Type Inequalities On Time Scales And Some Applications

Billur KAYMAKÇALAN and Erol KORALP
Department of Mathematics, Middle East Technical University,
Ankara, 06531, Turkey
billur@metu.edu.tr , erolk@metu.edu.tr

Abstract

Unification of some analogous discrete and continuous Opial's type inequalities , that is certain inequalities involving integrals of functions and their derivatives, are given in the time-scale set-up. Obtained results give way to applications for estimating upper bounds of solutions of some I.V.P.'s.

AMS subject classification: 39A10, 39B72, 26D10, 34C11, 34A40.

1. INTRODUCTION

We are concerned with inequalities of the type

$$\int_0^h |x(\tau)x^\Delta(\tau)|\Delta\tau \leq \frac{h}{4}\int_0^h |x^\Delta(\tau)|^2\Delta\tau \tag{1.1}$$

which unify corresponding similar inequalities in the discrete and continuous cases. To understand the details of this generalization, we first concentrate on the following basics of the concerned unified theory, namely the essentials of the calculus on *measure chains* or *time-scales*. For further details, we refer to [9] and [3].

Definition 1.1 *Let $\mathbb{T}$ be a closed subset of $\mathbb{R}$. For $t \in \mathbb{T}$ which is non-maximal , (i.e. $t < \sup\mathbb{T}$), the mappings $\sigma, \rho : \mathbb{T} \to \mathbb{T}$ given by*

$$\sigma(t) := \inf\{s \in \mathbb{T} : s > t\}, \quad \textit{and} \quad \rho(t) := \sup\{s \in \mathbb{T} : s < t\}$$

are called jump operators, in particular, "forward" and "backward" jump operators, respectively.

In case $\mathbb{T}$ is bounded above, $\sigma(max\mathbb{T}) := max\mathbb{T}$ and accordingly $\rho(min\mathbb{T}) := min\mathbb{T}$, if $\mathbb{T}$ is bounded below. These jump operators enable us to classify the points $\{t\}$ of a time-scale as right-dense, right-scattered, left-dense, left-scattered, respectively, if $\sigma(t) = t$, $\sigma(t) > t$, $\rho(t) = t$, $\rho(t) < t$ for any $t \in \mathbb{T}$.

Maximal element of $\mathbb{T}$ which is left-scattered is called degenerate. Non-degenerate points of $\mathbb{T}$ are denoted by $\mathbb{T}^k$.

Definition 1.2 *The mapping $\mu^*(t) = \sigma(t) - t$ is called grainyness. For fixed s ; $\mu(\sigma(t), s)$ denotes $\sigma(t) - s$ and $\tilde{\mu}([r, s[) := \mu(r, s)$ defines a measure on $\mathbb{T}$.*

In order the speak about integration on time-scales, we have to obtain an appropriate class of functions having anti-derivatives.For this purpose we define the following:

Definition 1.3 *A mapping $g : \mathbb{T} \to X$ is called rd- continous if*
(i) it is continous at each right-dense or maximal $t \in \mathbb{T}$,
(ii) at each left-dense point, left sided limit $g(t^-)$ exists.

We denote by $C_{rd}[\mathbb{T}, X]$ the set of rd-continous mappings from $\mathbb{T}$ to a Banach space X.

Definition 1.4 *Let $\mathbb{T}$ be a time scale and $f : \mathbb{T} \to \mathbb{R}$ be a mapping from $\mathbb{T}$ to $\mathbb{R}$. f is called differentiable at $t_0 \in \mathbb{T}$, if there exists an $a \in \mathbb{R}$ with the following property: For any $\epsilon > 0$, there exists a neighborhood U of t_0, such that*

$$|f(\sigma(t_0)) - f(t) - (\sigma(t_0) - t)a| \leq \epsilon|\sigma(t_0) - t| \text{ for all } t \in U.$$

Remark 1.1 *In the two special cases $\mathbb{R}$ and $\mathbb{Z}$ the "derivative" is uniquely determined. In fact one gets; $a = \frac{df}{dt}(t_0)$ and $a = f(t_0 + 1) - f(t_0)$, respectively.*

We shall denote the derivative of f by $f^\Delta(t)$, which, in general for $f : \mathbb{T} \to X$ may be given by:

$$f^\Delta(t) = \lim_{s \to t} \frac{f(\sigma(s)) - f(t)}{\mu(\sigma(s), t)}, \quad \sigma(s) \neq t.$$

and in the case of t being right-scattered; we write $f^\Delta(t) = \frac{f(\sigma(t)) - f(t)}{(\sigma(t) - t)}$.

Definition 1.5 *Let g and h be mappings such that $g, h : \mathbb{T} \to \mathbb{R}$. Then g is called an anti-derivative of h on $\mathbb{T}$ if it is differentiable on $\mathbb{T}$ and $g^\Delta(t) = h(t)$ for all $t \in \mathbb{T}^k$.*

If $g : \mathbb{T}^k \to X$ has an anti-derivative $f : \mathbb{T}^k \to X$, then for any $r, s \in \mathbb{T}^k, r \leq s$ we define

$$\int_r^s g(t)\Delta t := f(s) - f(r)$$

as a Cauchy integral of g from r to s.

For $\mathbb{T} = \mathbb{R}$ the Cauchy integral coincides with the Riemann integral. For $\mathbb{T} = h\mathbb{Z}$, $h > 0$ the identity

$$\int_r^s g(t)\Delta t = \begin{cases} \sum_{i=\frac{r}{h}}^{\frac{s}{h}-1} g(ih), & \text{if } s > r \\ 0, & \text{if } s = r \\ -\sum_{i=\frac{s}{h}}^{\frac{r}{h}-1} g(ih), & \text{if } s < r \end{cases}$$

can be shown.

In view of the above exposed basics of time scales, our aim now, is to consider the so-called Opial type inequalities , generalized on time-scales and how these results lead their way to applications such as estimating upper bounds of solutions of I.V.P.'s.

In 1960, the Polish Mathematician Zdzidlaw Opial [16] published an inequality involving integrals of functions and their derivatives;

$$\int_0^h |x(t)x'(t)|\, dt \leq \frac{h}{4} \int_0^h |x'(t)|^2\, dt \tag{1.2}$$

where $x \in C^{(1)}[0, h]$, $x(0) = x(h) = 0$ and $x(t) > 0$ in $(0, h)$, and the constant $h/4$ is the best possible.

Inequalities which involve integrals of functions and their derivatives are of great importance in Mathematics with applications in the theory of differential equations, approximations and probability. It has been shown that inequalities of the form (1.2) can be deduced from those of Wirtinger and Hardy type, but the importance of Opial's result is in the establishment of the best possible constant. Recently published monograph [2] is the first book dedicated to the theory of Opial type inequalities .

The positivity requirement of $x(t)$ in the original proof of Opial was shown to be unnecessary later by Olech in [15] where he proved that the inequality (1.2) holds even for functions $x(t)$ which are only absolutely continuous in [0,h]. Moreover, Olech's proof is simpler than that of Opial. Despite the fact that in the historical developement many followers of Opial and Olech were there, for the sake of considering such inequalities in the Time-Scale set-up, we will start with taking into account Olech's result along with its discrete analog given by Lasota in [7] and refer to [2], (Chapters 1 and 5) for further proceedings.

2. UNIFICATIONS OF BASIC OPIAL TYPE INEQUALITIES

Theorem 2.1 *Let $x \in C^1_{rd}\Big([0,h],\mathbb{R}\Big)$ be such that $x(0) = x(h) = 0$. Then the following inequality holds;*

$$\int_0^h |x(\tau)x^\Delta(\tau)|\Delta\tau \le \frac{h}{4}\int_0^h |x^\Delta(\tau)|^2\Delta\tau. \tag{2.1}$$

In (2.1), the constant $\frac{h}{4}$ is the best possible.

Remark 2.1 *Due to the constructive approach of the proof of Theorem 2.1, it can be shown that $h/4$ is indeed the best possible constant for any time scale $\mathbb{T}$.*

Proof: In view of Olech's result [15] and along the lines of Lasota's [7] proof of Opial type discrete inequality , we consider for a given $\epsilon > 0$ the neighborhood

$$N_\epsilon^-(\frac{h}{2}) = \Big\{t \in \mathbb{T}^k : \frac{h}{2} - \epsilon < t \le \frac{h}{2}\Big\}.$$

Take $u = sup\Big\{N_\epsilon^-(\frac{h}{2})\Big\}$.

(i) If $\mathbb{T} = \mathbb{R}$, then $u = \sigma(u) = \frac{h}{2}$.

(ii) If $\mathbb{T} = \mathbb{Z}$ then either $u = \frac{h}{2}$ or u and $\sigma(u)$ cannot be both in $N_\epsilon^-(\frac{h}{2})$.
In either case;

$$u \le \frac{h}{2} \text{ and } h - \sigma(u) \le \frac{h}{2}. \tag{2.2}$$

Starting with the left-hand-side of inequality (2.1) we obtain

$$\begin{aligned}
\int_0^h |x(\tau)x^\Delta(\tau)|\Delta\tau &= \int_0^u |x(\tau)x^\Delta(\tau)|\Delta\tau + \int_{\sigma(u)}^h |x(\tau)x^\Delta(\tau)|\Delta\tau \\
&\le \int_0^u\int_0^\tau |x^\Delta(s)x^\Delta(\tau)|\Delta s\Delta\tau + \int_{\sigma(u)}^h\int_\tau^h |x^\Delta(s)x^\Delta(\tau)|\Delta s\Delta\tau \qquad (2.3)\\
&= \frac{1}{2}\int_0^u\int_0^u |x^\Delta(s)x^\Delta(\tau)|\Delta s\Delta\tau + \frac{1}{2}\int_{\sigma(u)}^h\int_{\sigma(u)}^h |x^\Delta(s)x^\Delta(\tau)|\Delta s\Delta\tau \qquad (2.4)\\
&= \frac{1}{2}\left(\int_0^u |x^\Delta(\tau)|\Delta\tau\right)^2 + \frac{1}{2}\left(\int_{\sigma(u)}^h |x^\Delta(\tau)|\Delta\tau\right)^2 \qquad (2.5)
\end{aligned}$$

where in passing from (2.3) to (2.4), we have made use of the fact that the integral of a symetric integrand over the triangle $0 \leq s \leq t,\ \ 0 \leq \tau \leq a$ is equal to the half of its integral over the square $0 \leq s \leq a,\ \ 0 \leq \tau \leq a$.

Now applying Cauchy-Schwarz inequality to (2.5) we get,

$$\int_0^h |x(\tau)x^\Delta(\tau)|\Delta\tau \leq \frac{1}{2}u\int_0^u |x^\Delta(\tau)|^2\Delta\tau + \frac{1}{2}(h-\sigma(u))\int_{\sigma(u)}^h |x^\Delta(\tau)|^2\Delta\tau$$

which in view of (2.2) becomes,

$$\int_0^h |x(\tau)x^\Delta(\tau)|\Delta\tau \leq \frac{h}{4}\int_0^h |x^\Delta(\tau)|^2\Delta\tau,$$

thereby establishing the desired result (2.1).

Remark 2.2 *The technique used in (2.3)-(2.4) is due to Hua and Pederson who have independently, but both in the year 1965, offered a very simple proof of Opial's inequality (1.1). Their contributions are given in [6] and [13], respectively.*

In view of Hua and Pederson's approach, the following theorem can be given, which is a unification of results similar to discrete and continuous analogs of Theorem 2.1;

Theorem 2.2 *Let $x \in C^1_{rd}\Big([0,a],\mathbb{R}\Big)$ be such that $x(0)=0$. Then the following inequality holds:*

$$\int_0^a |x(\tau)x^\Delta(\tau)|\ \Delta\tau \leq \frac{a}{2}\int_0^a |x^\Delta(\tau)|^2\ \Delta\tau. \tag{2.6}$$

Further, in (2.6) equality holds if and only if $x(t)=ct$.

An enormous literature on Opial-type inequalities has proceeded Beesack's pioneering generalization of Opial's inequality given in [5].

We now give a non-trivial generalization of Theorem 2.2, which is a unified version of the results due to Hua [6] in the continuous case, and the discrete analogues given in [17] by Wong.

Theorem 2.3 (Unified generalization of (1.2) by Hua and Wong)
Let $x \in C^1_{rd}\Big([0,a],\mathbb{R}\Big)$ be such that $x(0)=0$. Further let ℓ be a positive integer. Then the following inequality holds;

$$\int_0^a |x^\ell(t)x^\Delta(t)|\Delta t \leq \frac{a^\ell}{\ell+1}\int_0^a |x^\Delta(t)|^{\ell+1}\Delta t \tag{2.7}$$

with equality being valid in (2.7) if and only if $x(t)=ct$.

As an application of Opial's inequality we will consider upper bounds of solutions of I.V.P.'s, but for the purpose of establishing sharper estimates we need the following observations. We first give a comparison result for linear dynamic inequalities in view of a similar estimate given in [10], which is a weakened form of Theorem 4.2 in [8].

Lemma 2.1 *Let $[a,b] := [a,b]\bigcap \mathbf{T}^k$. Assume that $B(t)$ and $f(t) \in C^1_{rd}\Big([a,b],\mathbb{R}\Big)$ with $B(t) \geq 0$ on $[a,b]$. Let $V(t)$ be a differentiable function on $[a,b]$ and suppose*

$$V^{\Delta}(t) \leq B(t)V(t) + f(t), \qquad t \in [a,b] \tag{2.8}$$

with $V(a) \leq V_0$. Then for $t \geq a$ we have,

$$V(t) \leq V_0\ e_B(t,a) + \int_a^t \Big[f(s)\ e_B(t,\sigma(s))\Big]\Delta s, \quad t \in [a,b], \tag{2.9}$$

where $e_B(t,\tau) = exp(\int_\tau^t B(s)\Delta s)$ is the Fundamental Solution of the linear homogeneous equation corresponding to (2.8).

We end the essentials of the theory that is required for our purpose of demonstrating the powerful and vital usage of Opial inequalities in estimating upper bounds of solutions of some I.V.P.'s, by giving a result which is a generalized version of the Gronwall type inequality on time-scales given in [11].

It should be noted that the above Lemma 2.1 is used in the proof of the following theorem.

Theorem 2.4 *Let u, a, b and k be non-negative, rd-continuous functions on $J := [\alpha,\beta]\bigcap \mathbf{T}^k$, and let $p > 1$ be a constant. Suppose a/b is non-decreasing in J, and*

$$u(t) \leq a(t) + b(t)\int_\alpha^t k(s)u^p(s)\Delta s, \qquad t \in J.$$

Then

$$u(t) \leq a(t)\left[1 - (p-1)\int_{\sigma(\alpha)}^t k(s)b(s)a^{p-1}(s)\Delta s\right]^{\frac{1}{1-p}}, \alpha \leq t \leq \beta_p, \tag{2.10}$$

where

$$\beta_p = \sup\left\{t \in J : (p-1)\int_{\sigma(\alpha)}^t k(s)b(s)a^{p-1}(s)\Delta s < 1\right\}.$$

Remark 2.3 *The above Theorem 2.4 can be obtained in a similar manner; in the* **continuous case**, *observing the results given by Stachurska in [14] and in the* **discrete case**, *analogous inequalities are given in [12] by Patchpatte. These mentioned results can be found in the monograph by Bainov and Simeonov, [4].*

3. APPLICATIONS OF OPIAL'S TYPE INEQUALITIES FOR ESTIMATES OF UPPER BOUNDS OF SOLUTIONS OF I.V.P.'S

We consider the first order I.V.P.

$$y^{\Delta} = f(t,y), \qquad y(0) = 0 \tag{3.1}$$

where the function $f(t,y)$ is defined and rd-continuous on $(0,a] \times \mathbb{R}$, and the inequality

$$|f(t,y)| \leq g(t) + h(t)|y|^{\alpha}, \ \alpha > 1 \tag{3.2}$$

holds, with $g(t)$ and $h(t)$ being defined, nonnegative and rd-continuous on $(0,a]$. Further, we assume that (3.1) has solutions $y(t) \in C^1_{rd}\Big([0,a],\mathbb{R}\Big)$.

Next, we consider the I.V.P.

$$x^{\Delta} = g(t) + h(t)x^{\alpha}, \qquad x(0) = 0 \tag{3.3}$$

and assume that its maximal solution $x_m(t)$ exists on $[0, a]$. Then it follows by (3.2) that $y(t) \leq x_m(t)$, $t \in [0, a]$. But except for some trivial cases such as $\alpha = 1$ or $g(t) = 0$, the I.V.P. (3.3) can hardly be solved, and in this situation to obtain realistic upper bounds for $x_m(t)$, and consequently for $y(t)$, Opial type inequalities play an important role and serve as a very efficient tool.

In our situation, different from the general trend, along with inequality (2.7) we use the estimate (2.10) to obtain sharper results than the one obtained in [2] for the upper bounds of solutions of (3.1).

In this process we note that the integral equation equivalent of the I.V.P. (3.3) can be transformed into the following inequality by making use of (2.7);

$$|x^{\Delta}(t)| \leq g(t) + h(t)\, t^{\alpha-1} \int_0^t |x^{\Delta}(s)|^{\alpha}\, \Delta s. \tag{3.4}$$

This inequality gives way to

$$|x^{\Delta}(t)| \leq G(T) + H(T) \int_0^t |x^{\Delta}(s)|^{\alpha}\, \Delta s. \tag{3.5}$$

for $t \in [0, T]$, where $T \in [0, a]$ is arbitrary but fixed and $sup_{\,0 \leq t \leq T}\; g(t) \leq G(T)$ and $sup_{\,0 \leq t \leq T}\; h(t)t^{\alpha-1} \leq H(T)$.

It can be shown that (3.5) leads to the estimate

$$|y(t)| \leq |x(t)| \leq \int_0^t \left(G^{1-\alpha}(s) + (1-\alpha)H(s)s\right)^{\frac{1}{(1-\alpha)}} \Delta s, \tag{3.6}$$

as long as the right-hand side integral exists.

The above considerations are illustrated in the following example. We note, though, that this example is given in the special case of $\mathbb{T} = \mathbb{R}$.

Example : Lower and upper bounds for the solution $y(t)$ of the I.V.P.

$$y^{\Delta} = t^2 + y^2, \qquad t \in [0, 1] \quad , \; y(0) = 0 \tag{3.7}$$

can be obtained by comparing it with the problems;

$$\begin{cases} y_{\ell}^{\Delta} = t^2 \\ y_{\ell}(0) = 0 \end{cases} \qquad and \qquad \begin{cases} y_u^{\Delta} = 1 + y_u^2 \\ y_u(0) = 0 \end{cases}$$

respectively. It follows that

$$t^3/3 = y_{\ell}(t) \leq y(t) \leq y_u(t) = \tan t, \qquad t \in [0, 1].$$

We note that the explicit known solution of (3.7) in terms of Bessel functions, which is of little practical use, can be found in [1].

In view of the process employed to obtain (3.6) and the Opial's inequality (2.7), the I.V.P. (3.7) can be transformed to the inequality

$$|x^{\Delta}(t)| \leq t^2 + t \int_0^t |x^{\Delta}(s)|^2\, \Delta s. \tag{3.8}$$

From this inequality it is possible to obtain a series of upper bounds for $|x^{\Delta}(t)|$, and hence for $y(t)$.

(i) First noting that $|x^{\Delta}(t)| \leq |x_1^{\Delta}(t)|$, $t \in [0,1]$, where $|x_1^{\Delta}(t)|$ is the solution of the inequality

$$|x_1^{\Delta}(t)| \leq t + \int_0^t |x_1^{\Delta}(s)|^2 \,\Delta s, \tag{3.9}$$

we obtain

$$y(t) \leq |x(t)| \leq \int_0^t \tan s \,\Delta s = \ln(\sec t).$$

Clearly we have that $\ln(\sec t) < \tan t, \quad t \in (0,1]$.

(ii) Next for the inequality (3.8), employing the machinery used to obtain (3.6) it follows that

$$\begin{aligned} y(t) &\leq |x(t)| \leq \int_0^t \frac{s^2}{(1-s^4)} \,\Delta s \\ &= (0.25)\ln(1+t) - (0.25)\ln(1-t) - (0.5)\tan^{-1} t, \qquad t \in [0,1]. \end{aligned}$$

For small t this upper bound of $y(t)$ is very close to the lower bound $t^3/3$, however, as t is close to 1, it diverges. But this can be combined with the previous bound to obtain,

$$\begin{aligned} y(t) &\leq |x(t)| \\ &\leq min\Big\{ \ln(\sec t), 0.25\ln(1+t) - 0.25\ln(1-t) - 0.5\tan^{-1} t \Big\}, \ t \in [0,1]. \end{aligned}$$

(iii) Now let $T \in [0,1]$ be arbitrary, but fixed. Then from (3.8) we have

$$|x^{\Delta}(t)| \leq |x_2^{\Delta}(t)| \leq T\left[t + \int_0^t |x_2^{\Delta}(s)|^2 \,\Delta s\right]. \tag{3.10}$$

By the methods of part (i) , we obtain

$$y(t) \leq |x(t)| \leq \int_0^t \tan s^2 \,\Delta s.$$

This upper bound of $y(t)$ is not only sharper than all the previous bounds, but also it is very close to the lower bound $t^3/3$ for all $t \in [0,1]$.

(iv) Finally for the inequality (3.8) if we don't proceed as in part (ii), but rather apply inequality (2.10) of Theorem 2.4, a generalized form of Gronwall's inequality, using $u(t) = |x^{\Delta}(t)|$, $a(t) = t^2$, $b(t) = t$, $k(s) = 1$ and $p = 2$, we obtain

$$y(t) \leq |x(t)| \leq \frac{1}{\sqrt{2}}\left[\ln(\sqrt{2}+t) - \ln(\sqrt{2}-t) - 2\,\tan^{-1}\left(\frac{t}{\sqrt{2}}\right)\right], t \in [0,1].$$

It can be shown that this estimate is even sharper than the one found in part (iii), and moreover it is closer to but larger than the lower bound $y_\ell = t^3/3$, $t \in [0,1]$. This result not only establishes the usage of Opial type inequality (2.7), but also indicates the strength of it when combined with Gronwall type inequality (2.10).

References

[1] Agarwal, R.P. and Gupta, R.C., Essentials of Ordinary Differential Equations, McGraw-Hill, Singapore,New York, (1991).

[2] Agarwal, R.P. and Pang, P.Y.H., Opial Inequalities with Applications in Differential and Difference Equations, Kluwer, Dordrecht,Netherlands, (1995).

[3] Aulbach, B. and Hilger, S., A Unified Approach to Continuous and Discrete Dynamics,in, Differential Equations: Qualitative Theory, Colloq. Math. Soc. Janos Bolyai, North-Holland, (1988),pp.37-56.

[4] Bainov, D. and Simeonov, P., Integral Inequalities and Applications, Kluwer, Dordrecht, 1992.

[5] Beesack, R.P., On an integral inequality of Z. Opial, Trans. Amer. Math. Soc., 104(1962), pp.470-475.

[6] Hua, L.K., On an inequality of Opial. Scientia Sinica 14(1965), pp.789-790.

[7] Lasota, A., A discrete boundary value problem, Ann. Polon. Math. 20(1968), pp.183-190.

[8] Kaymakçalan, B., Existence and comparison results for dynamic systems on time scale, JMAA, 172(1993), pp.243-255.

[9] Kaymakçalan, B., Lakshmikantham, V. and Sivasundaram, S., Dynamic Systems on on Measure Chains, Kluwer Academic Publishers ,(1996).

[10] Kaymakçalan, B., Özgün, S.A. and Zafer, A., Asymptotic Behavior of Higher Order Equations on Time Scale, Computers and Mathematics with Applications, Special Issue on Difference Equations, 36, no.10-12, (1998), pp.299-306.

[11] Kaymakçalan, B., Özgün, S.A. and Zafer, A., Gronwall and Bihari type inequalities on time scales, Advances in Difference Equations- Proceedings of the Second Int. Conf. on Difference Equations , Veszprem, Hungary, (1997), pp.481-490.

[12] Patchpatte, B.G., Finite-difference inequalities and an extension of Lyapunov's method, Michigan Math. J., 18(1971), pp.385-391.

[13] Pederson, R.N., On an inequality of Opial, Beesack and Levinson, Proc. Amer. Math. Soc. 16(1965), pp.174.

[14] Stachurska, B., On a nonlinear integral inequality, Zeszyty Nauk. Uniw. Jagiellonsk., 252, Prace Math., 15(1971), pp.151-157.

[15] Olech, C., A simple proof of a certain result of Z. Opial, Ann. Polon. Math. 8(1960), pp.61-63,

[16] Opial, Z., Sur une inegalite, Ann. Polon. Math. 8(1960), pp.29-32.

[17] Wong, J.S.W., A discrete analogue of Opial's inequality, Canad. Math. Bull. 10(1967), pp.115-118.

Proceedings of Dynamic Systems and Applications 3 (2001) 343-350

One Method of Liapunov Functionals Construction

V.B. Kolmanovskii and N.P. Kosareva
Cybernetics Department, Moscow Institute of Electronics and Mathematics,
Bolshoy Trekhsvyatitelskiy 3/12, Moscow 109028, Russia

Abstract: We derive some features of the general method of Liapunov functionals construction to investigate stability problems for difference systems. These features allow to use the method more effectively.
AMS (MOS) subject classification. 39A11

1. INTRODUCTION

Let h be a given nonnegative number, i be a discrete time, $i \in Z_0 \bigcup Z$, $Z_0 = \{-h, ..., 0\}$, $Z = \{0, 1, ...\}$, process $x_i \in \mathbf{R}^n$ be a solution of the equation

$$x_{i+1} = F(i, x_{-h}, ..., x_i), \qquad i \in Z, \tag{1.1}$$

with initial function $x_i = \varphi_i, i \in Z_0$. Here $F : Z * S \Leftrightarrow \mathbf{R}^n$, S is a space of sequences with elements from $\mathbf{R}^n$. It is assumed that $F(i, ...)$ is independent on x_j for $j > i$, $F(i, 0, ..., 0) = 0$.

Definition 1. The zero solution of equation (1.1) is called stable if for any $\varepsilon > 0$ there exists a $\delta > 0$ such that $|x_i| < \varepsilon$, $i \in Z$, if $\|\varphi\| = \sup_{i \in Z_0} |\varphi_i| < \delta$. If, besides, $\lim_{i \to \infty} |x_i| = 0$ for all initial functions φ, then the zero solution of equation (1.1) is called asymptotically stable.

Theorem 1. *Let there exist the nonnegative functional $V_i = V(i, x_{-h}, ..., x_i)$, $i \in Z$, which satisfies the conditions*

$$V(0, x_{-h}, ..., x_0) \le c_1 \|\varphi\|^2,$$

$$\Delta V_i \le -c_2 |x_i|^2, \qquad i \in Z,$$

where $\Delta V_i = V_{i+1} - V_i$, $c_k > 0$. Then the zero solution of equation (1.1) is asymptotical stable.

From Theorem 1 it follows that the stability investigation of equations can be reduced to the construction of appropriate Liapunov functionals. The proposed procedure of Liapunov functionals construction consists of four steps.

1. Represent the functional F in equation (1.1) in the form

$$F(i, x_{-h}, ..., x_i) = F_1(i, x_{i-\tau}, ..., x_i) + F_2(i, x_{-h}, ..., x_i) + \Delta F_3(i, x_{-h}, ..., x_i),$$

$$F_1(i, 0, ..., 0) = F_2(i, 0, ..., 0) = F_3(i, 0, ..., 0) = 0, \tag{1.2}$$

Here τ is a given integer.

2. It is supposed that the zero solution of auxiliary difference equation

$$y_{i+1} = F(i, y_{i-\tau}, ..., y_i), \qquad i \in Z, \tag{1.3}$$

is asymptotically stable and there exists the Liapunov function $v_i = v(i, y_{i-\tau}, ..., y_i)$ for this equation which satisfies the conditions of Theorem 1.

3. Liapunov functional V_i is constructed in the form $V_i = V_{1i} + V_{2i}$, where the main component is $V_{1i} = v(i, x_{i-\tau}, ..., x_{i-1}, x_i - F_3(i, x_{-h}, ..., x_i))$.

4. In order to satisfy the conditions of Theorem 1 we need to calculate the value of first difference ΔV_{1i} and to estimate it in a reasonable way. After that the additional component V_{2i} is chosen usually by a standard way.

Consider some pecularities of this procedure.

It is clear that the representation (1.2) in the first step is not unique. Hence for each representation we can get another Liapunov functional and therefore other stability conditions.

In second step for one auxiliary equation (1.3) we can choose different Liapunov functions v_i and therefore to get different Liapunov functionals for equation (1.1).

At last we should like to stress that choosing different ways of estimation of ΔV_{1i} we can construct different Liapunov functionals and as a result to obtain different stability conditions.

Let us demonstrate this procedure for some scalar equations.

2. SYSTEMS WITH FINITE DELAY

2.1. Some conditions of asymptotic stabitity of difference equations

Consider equation

$$\Delta x_i = x_{i+1} - x_i = -\sum_{k=0}^{\tau} b_{i,k} x_{i-k}, \quad i \geq 0, \tag{2.1}$$

where i is a discrete time, τ is a given nonnegative number, $b_{i,k}$ are given constants, $i \geq 0$, $0 \leq k \leq \tau$, $\tau < i$. It is assumed that $b_{i,k} \equiv 0$, $i \geq 0$, $k > \tau$. The initial conditions for equation (2.1) are given by $i = -\tau, \ldots, 0$.
This problem consists of studing asymptotic conditions of equation (2.1) solutions as $i \to \infty$.

Let us use the procedure of Liapunov's functionals construction. For this, represent right hand side of equation (2.1) as a sum

$$\Delta x_i = -q_i x_i + \Delta F_{3i} \tag{2.2}$$

where

$$F_{3i} = \sum_{k=1}^{\tau} \sum_{s=1}^{k} b_{i+k-s,k} x_{i-s}, \tag{2.3}$$

$$q_i = \sum_{k=0}^{\tau} b_{i+k,k}.$$

Equation (2.2) is equivalent to equation (2.1).

2.2. Main result for equation (2.1)

Theorem 2. *If for linear difference equation (2.1): 1) there is a const M, such that*

$$\overline{q}_i = \sum_{k=0}^{\tau} |b_{i+k,k}| \le M q_i, \qquad i \ge 0, \tag{2.4}$$

and also the following conditions are hold

$$q_i = \sum_{k=0}^{\tau} b_{i+k,k} > 0, \quad i \ge 0, \tag{2.5}$$

$$\sup_{i \ge 0} \left[-2 + q_i + \sum_{k=1}^{\tau} \sum_{s=1}^{k} |b_{i+k-s,k}| + M \sum_{j=1}^{\tau} q_{i+j} \right] < 0, \tag{2.6}$$

2)

$$\sum_{i=0}^{\infty} q_i = \sum_{i=0}^{\infty} \sum_{k=0}^{\tau} b_{i+k,k} = \infty. \tag{2.7}$$

3)

$$\sup_{i \ge 0} \left[\sum_{k=1}^{\tau} \sum_{s=1}^{k} |b_{i+k-s,k}| \right] < 1, \tag{2.8}$$

then the zero solution of linear equations (2.1) is asymptotical stable.

For the proof of this Theorem let us consider Liapunov's functional for equation (2.2) in the form $V_i = V_{1i} + V_{2i}$, where $V_{1i} = (x_i - F_{3i})^2$.

Then using condition (2.4) and take into account expression for F_{3i}, we get

$$\Delta V_{1i} \le q_i \left(-2 + q_i + \sum_{k=1}^{\tau} \sum_{s=1}^{k} |b_{i+k-s,k}| \right) x_i^2 + q_i \sum_{s=1}^{\tau} x_{i-s}^2 \sum_{k=s}^{\tau} |b_{i+k-s,k}|.$$

Take the functional V_{2i} in the form

$$V_{2i} = \sum_{k=1}^{\tau} \sum_{j=1}^{k} q_{i+k-j} \sum_{\mu=1}^{j} |b_{i+k-\mu,k}| x_{i-\mu}^2.$$

We get

$$\Delta V_{2i} = \sum_{k=1}^{\tau} \left[\sum_{j=0}^{k-1} q_{i+k-j} |b_{i+k,k}| x_i^2 - q_i \sum_{\mu=1}^{k} |b_{i+k-\mu,k}| x_{i-\mu}^2 \right].$$

Thus for $V_i = V_{1i} + V_{2i}$ we have

$$\Delta V_i \leq q_i \left(-2 + q_i + \sum_{k=1}^{\tau} \sum_{s=1}^{k} |b_{i+k-s,k}| \right) x_i^2 + x_i^2 \sum_{k=1}^{\tau} \sum_{j=0}^{k-1} q_{i+k-j} |b_{i+k,k}|,$$

or

$$\Delta V_i \leq q_i x_i^2 \left(-2 + q_i + \sum_{k=1}^{\tau} \sum_{s=1}^{k} |b_{i+k-s,k}| + M \sum_{j=1}^{\tau} q_{i+j} \right).$$

As a result we get conditions (2.4)-(2.8) of asymptotic stability of nonstationary difference equations (2.1), which depend only on the coefficients of these equations.

2.3. Other stability conditions

Consider the equation

$$x_{i+1} = a x_i + b \sum_{j=1}^{k} (k+1-j) x_{i-j}. \tag{2.9}$$

It is supposed that $k \geq 0$.

Represent equation (2.9) in the form (1.2) with

$$\tau = 0, \quad F_1 = a x_i, \quad F_2 = b \sum_{j=1}^{k} (k+1-j) x_{i-j}, \quad F_3 = 0. \tag{2.10}$$

Choose $V_{1i} = x_i^2$. Calculating ΔV_{1i} we get

$$\Delta V_{1i} \leq (a^2 + |ab| \frac{k(k+1)}{2} - 1) x_i^2 + (|ab| + b^2 \lambda_{k1}) \sum_{j=1}^{k} (k+1-j) x_{i-j}^2,$$

where

$$\lambda_{kl} = \frac{(k+1-l)(k+2-l)}{2}. \tag{2.11}$$

Choosing the functional V_{2i} in the form $V_{2i} = \sum_{l=i-k}^{i-1} A_{i-l} x_l^2$, where

$$A_l = (|ab| + b^2 \lambda_{k1}) \lambda_{kl}, \tag{2.12}$$

we get

$$\Delta V_{2i} = A_1 x_i^2 - (|ab| + b^2 \lambda_{k1}) \sum_{j=1}^{k} (k+1-j) x_{i-j}^2.$$

From (2.11), (2.12) it follows that

$$A_1 = |ab| \frac{k(k+1)}{2} + b^2 \frac{k^2(k+1)^2}{4}.$$

Therefore for $V_i = V_{1i} + V_{2i}$ we have $\Delta V_i \leq [(|a| + |b| \frac{k(k+1)}{2})^2 - 1] x_i^2$. From here and Theorem 1 it follows that the inequality

$$(|a| + |b| \frac{k(k+1)}{2})^2 < 1 \tag{2.13}$$

is a sufficient condition of the zero solution equation (2.9) asymptotic stability.

Using other representations of equation (2.9) in the form (1.2) and making analogous estimations for ΔV_{1i} we can get other stability conditions. For example, using the representation of equation (2.9) in the form (1.2) with

$$\tau = 0, \quad F_1 = (a + b\frac{k(k+1)}{2})x_i, \quad F_2 = 0, \quad F_3 = -\frac{b}{2}\sum_{j=1}^{k}(k+1-j)(k+2-j)x_{i-j},$$

we get sufficient conditions of asymptotic stability of the zero solution of equation (2.9) in the form

$$0 < (1 - a - b\frac{k(k+1)}{2})(1 + a + b\frac{k(k+1)}{2} - |b|\frac{k(k+1)(k+2)}{3}),$$

$$|a + b\frac{k(k+1)}{2}| < 1. \tag{2.14}$$

Using the representation of equation (2.9) in the form (1.2) with

$$\tau = 1, \quad F_1 = ax_i + bkx_{i-1}, \quad F_2 = b\sum_{j=2}^{k}(k+1-j)x_{i-j}, \quad F_3 = 0$$

we get sufficient conditions of asymptotic stability of the zero solution of equation (2.9) in the form

$$\frac{|b|\frac{k(k-1)}{2}[2|a| + (1-bk)|b|\frac{k(k+3)}{2}]}{(1+bk)[(1-bk)^2 - a^2]} < 1,$$

$$|b|k < 1, \qquad |a| < 1 - bk. \tag{2.15}$$

Let us show that using the representation (2.10) for equation (2.9) again but choosing another way of estimation of ΔV_{1i} we can get stability conditions, which differs from conditions (2.13)-(2.15) and gives an additional region of stability.

Calculating ΔV_{1i} and using (2.11) we get as previously

$$\Delta V_{1i} = (a^2-1)x_i^2 + 2ab\sum_{j=1}^{k}(k+1-j)x_ix_{i-j} + b^2(\sum_{j=1}^{k}(k+1-j)x_{i-j})^2 \le$$

$$\le (a^2-1)x_i^2 + 2ab\sum_{j=1}^{k}(k+1-j)x_ix_{i-j} + b^2\lambda_{k1}\sum_{j=1}^{k}(k+1-j)x_{i-j}^2.$$

Let us suppose now that $ab \le 0$ and put

$$V_{2i} = |ab|\sum_{j=1}^{k+1}(\sum_{l=1}^{j}x_{i-l})^2 + \sum_{l=i-k}^{i-1}C_{i-l}x_l^2,$$

where $C_l = b^2\lambda_{k1}\lambda_{kl}$.

Calculating ΔV_{2i} as previously we get

$$\Delta V_{2i} = |ab|\gamma_i + C_1x_i^2 - b^2\lambda_{k1}\sum_{j=1}^{k}(k+1-j)x_{i-j}^2,$$

where

$$\gamma_i = \sum_{j=1}^{k+1}[(\sum_{l=1}^{j} x_{i+1-l})^2 - (\sum_{l=1}^{j} x_{i-l})^2] \leq (k+1)x_i^2 + 2x_i \sum_{l=1}^{k}(k+1-l)x_{i-l}.$$

Therefore for $V_i = V_{1i} + V_{2i}$ we have

$$\Delta V_i \leq (a^2 + |ab|(k+1) + b^2 \frac{k^2(k+1)^2}{4} - 1)x_i^2.$$

Thus the inequalities

$$a^2 + |ab|(k+1) + b^2 \frac{k^2(k+1)^2}{4} < 1, \qquad ab \leq 0,$$

are sufficient condition of asymptotic stability of the zero solution of equation (2.9).

3. VOLTERRA EQUATION

Consider the equation

$$x_{i+1} = ax_i + \sum_{j=1}^{i} b^j x_{i-j}, \qquad i = 0, 1, \dots . \tag{3.1}$$

In [4] were obtained three groups of sufficient conditions of asymptotic stability of the zero solution of equation (3.1). To get new conditions let us transfer this equation for $i \geq 1$ in the following way $x_{i+1} = ax_i + bx_{i-1} + \sum_{j=2}^{i} b^j x_{i-j} = ax_i + bx_{i-1} + b\sum_{j=1}^{i-1} b^j x_{i-1-j} = ax_i + bx_{i-1} + b(x_i - ax_{i-1})$.

As a result the equation (3.1) is reduced to the equivalent form of an equation with finite delays

$$x_1 = ax_0,$$

$$x_{i+1} = Ax_i + Bx_{i-1}, \qquad i = 1, 2, \dots , \tag{3.2}$$

where

$$A = a + b, \qquad B = b(1-a). \tag{3.3}$$

We will construct Liapunov functional for equation (3.2) in the form $V_i = V_{1i} + V_{2i}$, where $V_{1i} = x_i^2$. Then using (3.3) we obtain

$$\Delta V_{1i} \leq [(A^2 + |AB| - 1)x_i^2 + (B^2 + |AB|)x_{i-1}^2.$$

Choosing the functional V_{2i} in the form $V_{2i} = (B^2 + |AB|)x_{i-1}^2$, for the functional V_i we have

$$\Delta V_i \leq [(|A| + |B|)^2 - 1]x_i^2.$$

Using (3.3) we obtain that the inequality

$$(|a+b| + |b(1-a)|)^2 < 1$$

is a sufficient condition of asymptotic stability of the zero solution of equation (3.1).

REFERENCES

1. Kolmanovskii, V.B., Application of the direct Liapunov method to the difference Volterra equations. Avtomatika i Telemekhanika, Vol 11 (1995), 50-64.

2. Lakshmikantham, V., & Trigiante, D., Theory of Difference Equations: Numerical Methods and Applications. N. Y.: Academic Press, 1988.

3. Kolmanovskii, V.B., & Kosareva, N.P., About asymptotical behaviour of linear systems with variable parameters and delays. Differentialniye Uravneniya, Vol 5 (1999), 1-6.

4. Kolmanovskii, V.B., & Shaikhet, L.E., General method of Liapunov functionals construction for stability investigations of stochastic difference equations. Dynamical Systems and Applications, World Scientific Series In Applicable Analysis, Vol 4 (1995), 397-439.

5. Kolmanovskii, V.B., & Rodionov,,A.M., On the stability of some discrete Volterra processes. Avtomatika i Telemekhanika, Vol 2 (1995), 3-13.

Proceedings of Dynamic Systems and Applications 3 (2001) 351-358

DYNAMIC SYSTEMS WITH SINGULAR PERTURBATIONS: SOME PROBLEMS AND APPLICATIONS

Lyudmila K.Kuzmina
Kazan Aviation Institute
Adamuck, 4-6, Kazan-15, 420015
RUSSIA
e-mail: Lyudmila.Kuzmina@ksu.ru

ABSTRACT: The object of investigations is dynamic systems with perturbations of singular class. The work is devoted to the development of general stability methods with reference to the some problems for such systems. Specific interest both from theoretical and from applied view point has the extension of classical traditional statements of stability theory for the cases with non-regular perturbations, for the singular singularly perturbed systems. These special systems arise as mathematical models from engineering applications. Here the singularly perturbed Equations with different types singularities in critical particular cases are considered. Problems of the comparison systems constructing are solved for this class. The reduction principle is formulated for particular critical cases. Besides non-perturbed systems are on the stability domain boundary; the generating systems (comparison ones) are limit systems or asymptotic approximations (s-systems).

The theorems about s-properties (s-stability, s-optimality,...) are obtained. New non-standard approach is developed for singularly perturbed problems.

AMS (MOS) subject classification. 34A20, 35B40, 35K60

1. INTRODUCTION

The object of our study is the complex compound systems, for that the mathematical models may be represented as models with the non-regular perturbations. Here in this work the original systems are modeled by the differential equations with the small parameters before the derivatives. General problem of the qualitative analysis for such systems is investigated in the specific cases (non-Tikhonov's systems, generating systems are singular ones; infinite time interval). With the Lyapunov's theory methods the conditions are determined, under which we can approach to a reduction principle in the theory of the singularly perturbed systems, obtain the generalization of the known results, get the new results for these systems. The research is dedicated to analysis problems and to study of the general qualitative properties for the singular systems. In such systems there are the irregular transitions, and for ones substantiation problems the stability methods [1] are used. Here the ordinary differential equations with the small parameter of type [2,3] are taken as original mathematical model

$$M(\mu)\frac{dy}{dt} = Y(t,\mu,y) \tag{1.1}$$

where μ is small (dimensionless, scalar, positive) parameter; $M(\mu)$ is singular matrix in general case. System (1.1) has the solutions with the very fast changes (when $\mu\to 0$ the system (1.1) has the peculiarities). In accordance with the customary definitions [2,3,4] we shall call (1.1) the perturbed system, and the shortened equations are called the unperturbed system.

$$M(0)\frac{dy}{dt} = Y(t,0,y) \tag{1.2}$$

There is the important problem: how may be established the qualitative conformity (equivalence) between the equations (1.1) and the shortened equations (also between their solutions). Analogous problems were considered by many authors in different statements (beginning from the early works of H.Poincare, A.M.Lyapunov, S.A.Chaplygin, ...).
In the stability theory first rigorous statements and results in reduction of initial system to shortened system were received by A.M.Lyapunov (1892); some general results were obtained by K.P.Persidsky (1952). He devised new statement: when the stability problem for original system

$$\frac{dx}{dt} = X(x,t) \tag{1.3}$$

may be reduced to the solving of stability problems for subsystems

$$\frac{dx_1}{dt} = X_1(x,t) \tag{1.3a}$$

$$\frac{dx_2}{dt} = X_2(x,t) \tag{1.3b}$$

Here $x = \|x_1, x_2\|^{\mathrm{T}}$ is state vector for original system (1.3), $X = \|X_1, X_2\|^{\mathrm{T}}$ is vector-function with the necessary properties. Besides special property of quasi-stability was introduced by K.P.Persidsky [5].
We do not give here the bibliography and the contemporary state of the problem. We note only, that the various qualitative problems with the reduction of original full system to the shortened system in different statements were considered in many early works [3,4,6,7,8,9,10,...]. Also many works were devoted to the investigation of singularly perturbed systems by various methods. But already in early works of I.S.Gradstein (1947, 1949) there is the indication, that all results of singular systems theory are the direct consequence of the Lyapunov's theorems, and the corresponding interesting results were adduced in [4]. The possibility of usage of a stability theory "apparatus" was noted also by N.N.Krasovsky (1953), N.G.Chetayev (1957). Here we study the stability problems and the dynamic problems, that are directly adjoining ones. Following the ideas of N.G.Chetayev [6], we shall take as the comparison system the simplified system, different from (1.2). Moreover, here we shall consider the solving of such problems in critical cases [1]. In these cases the available results of the theory of the singularly perturbed equations can't used (the perturbed systems are near the boundary of stability domain) [11,12,13].

In this paper the qualitative problems in the critical cases are solved for the systems (1.1) with the different powers of small parameter. For considered systems we shall introduce as the generating system the approximate system of s-level (s-system)

$$M^S(\mu)\frac{dy}{dt}=Y^S(y,t,\mu) \tag{1.4}$$

We shall get this s-system from (1.1), when in (1.1) shall take into consideration only the members with $\mu^{\alpha i}$ in power no more than s, ($0\leq \alpha_i \leq s$, $s < r$). Such approach, corresponding to the stability theory methods, allows to get the sequence of the correct comparison systems.

Besides the decomposition conditions of general qualitative properties (stability, operativeness, optimality,...) may be obtained. Also for systems with the peculiarities the problems of the proximity between the solutions of the full and approximate systems on an infinite interval of time are solved. Here we can consider those special cases when the spectrums of corresponding matrices are on imaginary axis.

2. BASIC STATEMENTS

Let the original equations, describing the state of initial object, are led to the form (1.1)

$$M(\mu)\frac{dy}{dt}=Y(t,\mu,y) \tag{2.1}$$

Let the s-system is accepted as the approximate (comparison) system

$$M^S(\mu)\frac{dy}{dt}=Y^S(t,\mu,y) \tag{2.2}$$

It is the simplified system, that is different from traditional one in the singularly perturbed system theory. The system (2.2) has the less order than (2.1). The problem: under what conditions the qualitative investigation of system (2.1) can be reduced to the investigation of system (2.2); when there is the proximity between the corresponding solutions of systems (2.1) and (2.2) on the infinite interval of time; when the dynamical properties possess by the decomposition and when ones can be determined on the shortened system (2.2) (in analysis of stability; operativeness; optimal parameters; ...); how we can estimate the μ-values, that will be permitting such transition.

For the solving of these questions we shall use the stability methods. General approach, based on the methods of Lyapunov's theory, combining the methods of the theory of singular perturbations and the ideas of Chetayev, permits to consider the particular cases, which are interesting both for the theory and for the applications. Following Chetayev [6], we shall introduce in the considering the differential equations for the new variables $b=y-y^S$:

$$\widetilde{M}(\mu)\frac{db}{dt}=B(t,\mu,b) \tag{2.3}$$

Here $y = y(t, \mu)$ is the solution of full system (2.1), $y^S = y^S(t, \mu)$ is the corresponding solution of shortened system (2.2).

Using Lyapunov's methods [1,6] (first and second), analyzing (2.3), we shall be able to determine the required conditions in the solving of singular problems, in the estimates of the parameter values and corresponding domains.

3. SOME SINGULAR PROBLEMS

Here we shall consider the system of type (2.1) for those specific cases, that are inherent in applied problems. Let the original system

$$M(\mu)\frac{dy}{dt}=Y(t,\mu,y) \tag{3.1}$$

where $y = \|y_1, y_2, y_3\|^T$ is vector; $M(\mu) = \|M_{ij}(\mu)\|$ is block matrix, $M_{ii}(\mu) = \mu^{\alpha i} I$; $\alpha_i \geq 0$ $(i=1,2,3)$ are constant numbers; I are identity matrices; $Y(t,\mu,y)$ is holomorphous (on the all variables) vector function with the continuous, limited (on t, μ) coefficients. Assuming the considering the critical cases [1,14], we take in (3.1) $y_1 = x_1$, $y_2 = x_2$, $y_3 = \|x_3, z\|^T$; x_i are n_i -dimensional vectors of basic variables; z is m -dimensional $(m \geq 0)$ vector of critical variables [1]; $Y_i = \|P_i(\mu)x+X_i(t, \mu, z, x)\|$, $(i = 1,2)$;
$Y_3 = \|P_3(\mu)x+X_3(t, \mu, z, x);\ Z(t, \mu, z, x)\|^T$; $P_i(\mu)$ are matrices of the corresponding sizes; X_i, Z are non-linear vector functions; $Y(t, \mu, z, 0)= 0$, $\alpha_1 = 2$, $\alpha_2 = 1$, $\alpha_3 = 0$ (it is characteristic peculiarity of many dynamical systems in mechanics).

3.1. Stability problem

We introduce as a comparison system for such system (3.1) the s-system (let $s=1$), denoting here the members, containing μ in power no more than "s", by "s"-index:

$$M^S(\mu)\frac{dy}{dt}=Y^S(t,\mu,y) \tag{3.2}$$

We shall say, that (3.2) describes the slow motion of "s-level" (with the corresponding characteristic equation $\lambda_m D_S(\lambda,\mu) = 0$). Then the fast motions characteristics are determined by the auxiliary equation

$$d(\beta)=|\beta I - P_{11}(0)| \tag{3.3}$$

The problem: in which conditions the stability property for full system (3.1) is ensuing from the stability property for this s-system.

Theorem 3.1. When $|P(0)| \neq 0$ and the equations $D_S(\lambda,\mu) = 0$, $d(\beta) = (0)$ satisfy the Hurwitz's conditions, then with sufficiently small μ-values the zero-solution stability of system (3.1) follows from the zero-solution stability of system (3.2).
Here we have the asymptotic stability (for $m=0$) or non-asymptotic one (for $m > 0$: the critical case [1]). We shall call this property for full system (3.1) the s-stability property.

Remark 3.1.1. Not dwelling on the proof, we note only, that the system (3.1) with sufficiently small μ-values is the Lyapunov's system [1]. Using the corresponding Lyapunov's manners, we receive this statement **3.1**.

Remark 3.1.2. Let the auxiliary equation (3.3) does not satisfy the theorem conditions, having the imaginary roots (special case) β_0. Then for *s*-stability property we shall have to take the second requirement in theorem **3.1** in form: the equations $D_S(\lambda,\mu) = 0$ and

$$d_S(\beta,\mu) = \begin{vmatrix} \beta I - P^S_{11}; & -P^S_{12} \\ -\mu P_{21}(0); & \beta I - \mu P_{22}(0) \end{vmatrix} = 0 \tag{3.4}$$

satisfy the Hurwitz's conditions, and

$$\frac{\partial}{\partial \mu} d_S(\beta = \beta_0, \mu = 0) \neq 0 \tag{3.5}$$

Here we have the critical spectrums for the matrices of slow and fast variables. This result gives important supplement for critical cases of singular system.

3.2. Proximity problem

The question: in which cases the corresponding solutions of systems (3.1) and (3.2) will remain near ones for all $t>0$?

This problem is directly connected with **3.1** [4].

Let $y=y(t,\mu)$ is the solution of system (3.1) with the initial conditions $y_0 = y\,(t_0,\mu)$; $y^S = y^S\,(t,\ \mu)$ is the solution of system (3.2) with the initial conditions $y^S{}_0 = y^S\,(t_0,\ \mu)$. Here $y^S{}_1 = x^S{}_1 = f_1(t,\ \mu,\ z^S,\ x^S{}_2, x^S{}_3)$ is determined from the equation

$$0 = P^S{}_1 x + X^S{}_1$$

Following to N.G.Chetayev [6] we shall consider the equations of type (2.3) for the deviations. Employing (as in previous cases) stability methods, we shall be able to obtain the results for the particular cases. Not producing the details (and proof), we adduce the general results for the corresponding case.

Theorem 3.2. If $|\,P(0)\,| \neq 0$, equations $D_S(\lambda,\mu) = 0$ and

$$\begin{vmatrix} P_{11}(0); & -P_{12}(0) \\ -P_{21}(0); & \alpha I - P_{22}(0) \end{vmatrix} = 0$$

satisfy Hurwitz's conditions and matrix $P_{11}(\mu)$ is stable matrix (when $\mu=0$), then for sufficiently small μ-values, for the given in advance numbers $\varepsilon > 0,\ \delta > 0,\ \gamma > 0,$ there is such μ_*-value, that in perturbed motion for $0 < \mu \leq \mu_*$, for all $t \geq t_0 + \gamma$ we have

$$\|y - y^S\| < \varepsilon$$

if $\|y_{10} - y^S{}_{10}\| < \delta$ $y_{20} = y^S{}_{20},\ y_{30} = y^S{}_{30}$.

We shall call this property for original system the *s*-proximity property.

Remark 3.2.1. If μ-value is sufficiently small, then $\|y_i - y^S{}_i\| < \varepsilon$, $(i=2,3)$ for all $t \geq t_0$.

Remark 3.2.2. We can receive more general statement analogous to Theorem **3.2** but when all initial conditions for the solutions of systems (3.1), (3.2) do not coincide:

$||y_{i0} - y^S_{i0}|| < \eta$, $(i=2,3)$ where η is the small value.

Remark 3.2.3. Also (as in **3.1**)) there is the corresponding result in critical particular case of imaginary eigen-values of matrix of fast variables.

Remark 3.2.4. Analogous results were obtained also when we as shortened system for (3.1) took the *s*-system (with $s=0$):

$$M(0)\frac{dy}{dt} = Y(t,0,y) \tag{3.6}$$

These results, received by Lyapunov's methods, supplement and generalize the known results for the singular systems in considered here critical cases.
From the point of stability theory it is the analogue of set stability property (on Zubov, [15]).

3.3. Operativeness problem

Now we shall consider the problem: under which conditions we shall be able to determine the quality of full system (the operativeness, the optimality,...) on the shortened *s*-system (let for systems (3.1) and (3.6)).
Here the system operativeness is characterized by the degree of stability [16]: $\delta = \min_j |Re\lambda_j|$,

λ_j is the root of the system characteristic equation. The parameters, with which δ has the maximum value, are called optimal ones (on operativeness).
Let δ is the stability degree for full system (3.1) (for $m=0$);

$$D(\lambda, \mu) = D_0(\lambda) + \mu D_1(\lambda) + \mu^2 D_2(\lambda) + \dots = 0 \tag{3.7}$$

(3.7) is the characteristic equation for (3.1).
Let δ^* is the stability degree for *s*-system (3.6), and the characteristic equation for (3.6) is

$$D_0(\lambda) = D(\lambda, 0) = 0 \tag{3.8}$$

Let $\lambda = \lambda(\mu)$ is the root of equation (3.7); λ_0 is the root of equation (3.8). We have for corresponding roots: $\lambda(\mu) = \lambda_0 + \Delta\,\mu$.
Let δ_* is a given stability degree. Using analogous methods, we can show:

Theorem 3.3. If $|P(0)| \neq 0$, and equations $D(\lambda, 0) = 0$,

$$\begin{vmatrix} P_{11}(0); & P_{12}(0) \\ -P_{21}(0); & \alpha I - P_{22}(0) \end{vmatrix} = 0 \quad , \qquad |\beta E - P_{11}(0)| = 0$$

satisfy Hurwitz's - conditions, and roots λ_0 are simple ones, then the property of stability for the full system (3.1) with the given stability degree δ_* follows from the stability property for the shortened system (3.6) with the stability degree $2\delta_*$, when $0 < \mu \leq \mu^*$.
Here μ^*-value is determined from the solution of equation

$$\max |Re\,\Delta_j(\mu)| = \delta_* \tag{3.9}$$

Here $\Delta = \Delta(\mu)$ is holomorphous function of μ: $\Delta(\mu) = l_1\mu + l_2\mu^2 + \dots$.

We shall call this property for original system the s-operativeness property.

Remark 3.3.1. It is known, that the problem of optimality on operativeness is connected with the case of multiple roots λ_0 [16]. This case corresponds to optimal parameters (here δ has the maximum value). Theorem **3.3** does not include this case. But, using similar approach, we can obtain the corresponding conditions, under which the transition from full system to s-system in the problem of quality is possible also in the case of multiple roots. But here we have:

$\Delta(\mu)= b_1 \mu^{1/N} + b_2 \mu^{2/N}+ \ldots$ (N is the root multiplicity).

In this case the usage of s-system in defining of the system operativeness gives more error than in case of simple roots λ_0. Moreover the domain of admissible μ-values is diminished here. Therefore the every case of optimal parameters requirements the special analysis.

Remark 3.3.2. This new result is very important one at present, because in last years the criteria of maximal stability degree is broadly used [17] for synthesis of control systems.

4. THE ESTIMATION OF μ-VALUE

The stability methods allow to obtain the μ-values estimation, permitting the transition to s-system in concrete problem. For this we can use both first and second method of Lyapunov [1,6]. The example of some estimating by method of the Lyapunov's functions was given in early work of Chetayev (1957).

Here we consider for the concretness, the estimation of permissible μ-values in the stability problem. We shall introduce in the consideration all subsystems, on which the initial system is divided: the system, corresponding to slow variables of s-level ("s-system" in this paper) and the auxiliary subsystems, describing the fast variables. Following Chetayev, defining the conditions, ensuring the stability property for the full system by the conditions on the stability property for the every subsystem, we shall obtain the correlation's for the estimating of μ.

Let the full system is (3.1); let the comparison system is (3.6), corresponding to slow variables y_3. Then, considering in first the subsystem of slow variables, we receive the estimation: $\mu \leq \mu_{3*}$, where μ_{3*} is the solution of the equation-

$$\sup_j |Re\, \Delta_j(\mu)| = \inf_j |Re\, \lambda_{j0}| \tag{4.1}$$

In (4.1) $\lambda_j(\mu)$ is the root of equation (3.7); λ_{j0} is the corresponding root of (3.8) and $\lambda_j(\mu) = \lambda_{j0} + \Delta_j(\mu)$.

Analogous estimations we obtain from the equations, corresponding to components y_2 and y_1 (middle and fast variables): $\mu \leq \mu_{2*}$ and $\mu \leq \mu_{1*}$.

The sought estimate, defining the upper boundary of admissible μ-values domain, is

$$\mu_* = \inf_i \mu_{i*} \tag{4.2}$$

With all $\mu \leq \mu_*$ the transition from the full system to the s-system is possible (in considered here sense: the stability property is conservated and there is the proximity of corresponding solutions on the infinite interval of time).

Remark. The estimation of μ-values, allowing the transition for all kinds problems, is the inexpedient aim (and, generally speaking, practically unsolvable).

5. CONCLUSION

In this research we adduced only some general tenets and results. The different problems of the singularly perturbed systems theory we shall be able to investigate by stability methods. These methods give the possibility to consider the specific critical cases and to obtain the solving of interesting problems both for the theory and for the applications. In the applications these results will enable to come to the rigorous substantiation of the approximate approach and theories in problems of mechanics.

The development of the employed methods is very perspective for the problems of the complex systems dynamics, that are singular systems in general case.

Research is supported by Russian Foundation of Fundamental Investigations (GRANT 99-01-00429).

REFERENCES

1. Lyapunov, A.M., General problem on motion stability. Papers Coll., Vol.2, Moscow, 1956, pp.7-263.
2. Fletto, L., Levinson, N., Periodical solutions of singularly perturbed systems. *Mathematics (Coll.of transl.)*, 2:2, Moscow, 1958, pp.61-68.
3. Tikhonov, A.N., Systems of differential equations with the small parameters before the derivatives. *Math.collect*, Vol.31, No.3, 1952, pp.575-586.
4. Gradstein, I.S., Application of Lyapunov's stability theory to the differential equations theory with small multipliers before derivates. *Math.collect*, Vol.32, No.2, 1953, pp.263-286.
5. Persidsky, K.P., Some specific cases in systems. *News of Kazach. Academy of Sciences, Math.and Mech.*, No.5, 1951, pp.3-24.
6. Chetayev, N.G., On evaluations of approximate integrations. *Appl.Math.and Mech.*, Vol.21, No.6, 1957, pp.419-421.
7. Klimushev, A.I., Krasovsky, N.N., Uniform asymptotic stability of differential equations systems with small parameters. *Appl.Math.and Mech.*, Vol.25, No.4, 1961, pp.680-690.
8. Razumikhin, B.S., On solutions stability of differential equations with small multipliers under derivatives. *Sib. Math.Journal*, Vol.4, No.1, 1963, pp.206-211.
9. Grujic, Lj.T., Uniform asymptotic stability of non-linear singularly perturbed general and large-scale systems. *Int.Journ.Cont.*, 33, No.3, 1981, pp.481-504.
10. Kuzmina, L.K., On some properties of singularly perturbed systems in critical cases. *Appl.Math.and Mech.*, Vol.46, No.3, 1982, pp.382-388.
11. O'Malley, R., Introduction to singular perturbation. Academic Press, NY, 1974.
12. Smith, D., Singular-Perturbation theory. An Introduction with applications. Cambridge Univ.Press, Cambridge, 1985.
13. Vasilyeva, A.B., Butuzov, V.F., Singular perturbed equations in critical cases. Moscow Univ. Press, Moscow, 1978.
14. Kuzmina, L.K., To question about the approximate methods in analysis of singular-perturbed systems. *Appl.Math.and Mech.*, Vol.49, No.6, 1985, pp.909-915.
15. Zubov, V.I., Motion stability. M., High School, 1984, p.230.
16. Tsipkin, I.S., Bromberg, P.V., To the degree stability of linear systems. *News of Sciences Academy USSR*, br. of tech.sciences, No.2, 1945, pp.1163-1167.
17. Zagari, G.I., Shubladze, A.M., Control system synthesis on criteria of maximal stability degree. Moscow, 1988, p.100.

Proceedings of Dynamic Systems and Applications 3 (2001) 359-366

THE INFLUENCE OF A MOVING HEAT SOURCE ON BLOW-UP IN A REACTIVE DIFFUSIVE MEDIUM

C. M. Kirk[1] and W. E. Olmstead[2]
[1]Montclair State University, Upper Montclair, NJ 07043
[2]Northwestern University, Evanston, IL 60208

ABSTRACT: The problem of a moving heat source in a reactive-diffusive medium is modeled by a parabolic partial differential equation that has a nonlinear source term. The source is highly localized and allowed to have rather general one-dimensional motion. Of particular interest is the influence that motion of the source has on the occurrence of blow-up. Criteria for blow-up as well as non blow-up are presented.
AMS subject classification. 35K60, 45D05, 45M05

1. INTRODUCTION

We consider the problem of a concentrated heat source moving in one-dimension through a medium that is both diffusive and reactive. By allowing the source to move, it becomes possible to avoid an explosion that would occur if the source were stationary. When the source is stationary, it gives off an ongoing supply of heat to the same immediate surrounding. This localized in put of thermal energy typically leads to a blow-up. When the source is moving through the medium, it is continually being exposed to new surroundings that are relatively cool. This offers an enhanced ability to diffuse the heat and possibly prevent a blow-up.

Our model for the moving heat source in a reactive-diffusive medium is the nonlinear initial-boundary value problem given by

$$\left(\frac{\partial}{\partial t}-\frac{\partial^2}{\partial x^2}\right)T(x,t)=\delta(x-x_0)F[T(x_0,t)]\,,\quad -\infty<x<\infty\,,\quad t>0 \tag{1}$$

$$T(x,0)=\hat{T}(x)\geq 0\,,\quad -\infty<x<\infty\ , \tag{2}$$

$$T(x,t)\to 0\quad \text{as}\quad |x|\to\infty\,,\quad t>\infty\ . \tag{3}$$

Here $T(x,t)$ is the local temperature of the medium. The initial temperature $\hat{T}(x)$ is taken to be continuous with $\hat{T}(x)\to 0$ as $|x|\to\infty$. The position of the heat source x_0 and its velocity v_0 are given by

$$x_0=x_0(t)\,,\ x_0(0)=0\,;\ \ v_0=v_0(t)=x_0'(t)\,,\ \ t\geq 0$$

A concentration of the source at $x=x_0$ is provided by the singular behavior of the delta function. It is typical of reaction-diffusion models that the nonlinear source function $F(T)$ is smooth and has the properties

$$F(T)>0\,,\ F'(T)>0\,,\ F''(T)>0\ \text{ for }\ T>0\,.$$

It is appropriate to focus on the temperature at $x = x_0(t)$, since any blow-up in the solution of (1)-(3) must occur at the source position. This is evident because (1) represents a homogeneous heat equation for $-\infty < x < x_0$ and $x_0 < x < \infty$, and hence $T(x,t)$ must attain its maximum value at either $t = 0$ or $|x| \to \infty$ or $x = x_0$. Given that $T(x,t)$ is bounded at $t = 0$ and has $|x| \to \infty$, a blow-up could only occur at $x = x_0(t)$.

It is easily shown that (1)-(3) is satisfied by

$$T(x,t) = \int_0^t \int_{-\infty}^{\infty} G(x,t|\,\xi,s)\delta\left[\xi - x_0(s)\right] F\{T\left[x_0(s),s\right]\}d\xi ds + \int_{-\infty}^{\infty} G(x,t|\,\xi,0)\hat{T}(\xi)d\xi\ ,\ -\infty < x < \infty\ ,\ t \geq 0 \tag{4}$$

where Green's function $G(x,t|\,\xi,s)$ is given by

$$G(x,t|\,\xi,s) = \frac{H(t-s)}{2\left[\pi(t-s)\right]^{1/2}} \exp\left[-\frac{(x-\xi^2)}{4(t-s)}\right]\ .$$

An integral equation for the temperature at the source position is obtained from (4) by setting $x = x_0(t)$, while also utilizing the sifting property of the delta function. It follows that

$$T\left[x_0(t),t\right] = \int_0^t k(t,s)F\{T\left[x_0(s),s\right]\}ds + h(t)\ ,\ t \geq 0\ , \tag{5}$$

where

$$k(t,s) = \frac{1}{2\left[\pi(t-s)\right]^{1/2}} \exp\left\{-\frac{\left[x_0(t) - x_0(s)\right]^2}{4(t-s)}\right\}\ ,$$
$$h(t) = \frac{1}{2(\pi t)^{1/2}} \int_{-\infty}^{\infty} \exp\left\{-\frac{\left[x_0(t) - \xi\right]^2}{4t}\right\} \hat{T}(\xi)d\xi\ . \tag{6}$$

By employing the method used in Olmstead [1], Olmstead and Roberts [2], Roberts et al [3], we will investigate (5) for possible blow-up solutions.

2. BASIC RESULTS FOR THE INTEGRAL EQUATION

To put the integral equation (5) into a more convenient form, let

$$u(t) \equiv T\left[x_o(t),t\right] - h(t)\ ,$$

so that $u(t)$ satisfies

$$u(t) = Ku(t) \equiv \int_0^t k(t-s)F\left[u(s) + h(s)\right]ds\ ,\ t \geq 0\ . \tag{7}$$

The properties of $h(t)$ depend upon the initial data $\hat{T}(x)$. Since $\hat{T}(x)$ is non-negative, continuous and approaches zero as $|x| \to \infty$, then $h(t)$ is non-negative, continuous and bounded. We will further assume that

$$0 \le \underline{h} \le h(t) \le \bar{h} < \infty \ .$$

From a physical viewpoint, the occurrence or non-occurrence of an explosion depends upon the outcome of a competition between the thermal energy put into the reactive medium by the source and the ability of the medium to diffuse that energy. A measure of the medium's ability to diffuse heat is given by

$$I(t) \equiv \int_0^t k(t,s)ds \ , \quad t \ge 0 \ .$$

A measure of the strength of the energy source that plays a crucial role in establishing blow-up is given by κ^{-1} where

$$\kappa \equiv \int_{\underline{h}}^{\infty} \frac{dz}{F(z)} < \infty \ .$$

Clearly, the parameter κ becomes smaller as the strength of the nonlinearity is increased.

Our basic result on the existence of a continuous solution to (7) is given by the following:

Theorem 1. Let $x_0(t)$ be continuous for $0 \le t < \infty$. Then (7) has a continuous solution for $0 \le t < t^*$, where $t^* < \infty$ if t^* can be determined as the smallest root of

$$I(t^*) = \Lambda \equiv \max_{0 \le M < \infty} \left[\frac{M}{F(M+\bar{h})} \right] \ , \tag{8}$$

and $t^* = \infty$ if

$$I(t) < \Lambda \ ,$$

for all $t \ge 0$.

Proof. For $x_0(t)$ continuous, it follows from (6) that $k(t,s)$ will be continuous for $0 \le s < t$ and integrable as $s \to t$. If we then consider the space of continuous functions $u(t)$ that satisfy

$$0 \le u(t) \le M < \infty \ , \quad 0 \le t \le t_1 \tag{9}$$

it is clear that the integral operator in (7) will map that space into itself provided that

$$Ku(t) \le M \ , \quad 0 \le t \le t_1 \ . \tag{10}$$

To satisfy (10), it is sufficient that

$$F(M+\bar{h}) \sup_{0\le t\le t_1} I(t) \le M \ . \tag{11}$$

For the operator K to be a contraction in the sup norm, we first note that

$$\begin{aligned} \sup_{0\le t\le t_1} |Ku_1 - Ku_2| &= \sup_{0\le t\le t_1} \left| \int_0^t k(t-s)\{F[u_1(s)+h] - F[u_2(s)+h]\}ds \right| \\ &\le F'(M+\bar{h}) \sup_{0\le t< t_1} I(t) \sup_{0\le t\le t_1} |u_1(t)-u_2(t)| \ . \end{aligned}$$

Then the contraction property holds if

$$F'(M+\bar{h}) \sup_{0\le t\le t_1} I(t) < 1 \ . \tag{12}$$

By the contraction mapping theorem, (7) will have a unique continuous solution satisfying (9) whenever both (11) and (12) are satisfied. To optimize the satisfaction of (11) and (12) for the largest t_1, we require that

$$\sup_{0\le t\le t_1} I(t) < \Lambda \equiv \max_{0\le M<\infty} \left[\frac{M}{F(M+\bar{h})}\right] = \frac{M}{F(M^*+\bar{h})} \ , \tag{13}$$

where M^* is the smallest root of

$$\frac{M^*}{F(M^*+\bar{h})} = \frac{1}{F(M^*+\bar{h})} \ .$$

Thus, if (13) is satisfied, then both (11) and (12) will be satisfied with $M = M^*$. If there exists a $t^* < \infty$ which is the smallest root of (8), then (13) is satisfied for $t_1 < t^*$. If (13) is satisfied for all $t_1 \ge 0$, then $t^* = \infty$.

Our basic result on the nonexistence of a continuous solution to (7) is given by the following:

Theorem 2. Let $x_0(t)$ be continuous and such that $k(t,s)$ is a nonincreasing function of t. Then whenever there exists a $t^{**} < \infty$ such that

$$I(t^{**}) = \kappa \ , \tag{14}$$

it follows that (7) can not have a continuous solution for $t \ge t^{**}$.

Proof. Assume that (7) has a continuous solution for $0 \le t \le t_1$. Since $k(t,s)$ is a nonincreasing function of t, it follows from (7) that

$$u(t) \ge J(t) \equiv \int_0^t k(t_1,s)F[u(s)+h(s)]\,ds \ , \quad 0 \le t \le t_1 \ . \tag{15}$$

Then, from (15) it follows that

$$J'(t) = k(t_1,t)F[u(t)+h(t)] \ge k(t_1,t)F[J(t)+\underline{h}] \ .$$

Integration of this differential inequality yields

$$\int_{b}^{J(t_1)+b} \frac{dz}{F(z)} \geq \int_{0}^{t_1} k(t_1, t)dt = I(t_1) \ . \tag{16}$$

The assumption of a continuous solution of (7) for $0 \leq t \leq t_1$ insures that $J(t_1) < \infty$, and therefore (16) implies that

$$\kappa > I(t_1) \ . \tag{17}$$

If there exists a $t^{**} < \infty$ that satisfies (14), then $t_1 = t^{**}$ violates (17) and contradicts the existence of a continuous solution for $t \geq t^{**}$.

When a blow-up does occur, then the results of Theorems 1 and 2 provide bounds on the blow-up time. That is, if

$$u(t) \rightarrow \infty \ \text{ as } \ t \rightarrow \hat{t} < \infty \ , \quad \text{then} \ \ t^* \leq \hat{t} \leq t^{**} \ .$$

It is useful to be able to apply the results of Theorems 1 and 2 to situations in which the kernel $k(t,s)$ has an appropriate bound. It is straightforward to generalize Theorems 1 and 2 to

Theorem 3. Let $k(t,s) \geq \tilde{k}(t,s)$, $0 \leq s < t < \infty$, where $\tilde{k}(t,s)$ is continuoys for $0 \leq s < t$ and integrable as $s \rightarrow t$. Then Theorem 1 holds with $I(t)$ replaced by $\tilde{I}(t)$ where

$$\tilde{I}(t) \equiv \int \tilde{k}(t,s)ds \ , \quad 0 \leq t < \infty \ .$$

Theorem 4. Let $k(t,s) \geq \tilde{k}(t,s)$, $0 \leq s < t < \infty$, where $\tilde{k}(t,s)$ is a nonincreasing function of t. Then the nonexistence condition (14) of Theorem 2 holds with $I(t)$ replaced by $\tilde{I}(t)$ where

$$\tilde{I}(t) \equiv \int_{0}^{t} \tilde{k}(t,s)ds \ , \quad 0 \leq t < \infty \ .$$

3. BLOW-UP CRITERIA

Here we will present conditionsinvolving the velocity $v_0(t)$ of the moving heat source which guarantee that either a blow-up does occur or that it does not. From a physical viewpoint, it could be argued that if the velocity remains sufficiently high, then a blow-up will be avoided; while if the velocity remains sufficiently low, than a blow-up will occur. This intuitive perception will be sustained by two theorems to follow.

A basic criterion for the avoidance of blow-up is given by the following:

Theorem 5. Let $x_0(t)$ be continuously differentiable for $0 \leq t < \infty$ with a constant $v^* > 0$ such that

$$|x_0'(t)| = |v_0'(t)| \geq v^* \ , \quad 0 \leq t < \infty \ .$$

If v^* is sufficiently large, then (7) will have continuous solution for all $t \geq 0$. In particular this holds if

$$v^* > \frac{1}{\kappa} \, .$$

Proof. By the mean value theorem

$$x_0(t) - x_0(s) = x_0'(\tau)(t-s) = v_0(\tau)(t-s) \, , \; 0 \leq s < \tau < t \, . \tag{18}$$

Then from (6) follows

$$k(t,s) \leq \tilde{k}(t,s) \equiv \frac{1}{2\left[\pi(t-s)\right]^{1/2}} \exp\left[-\frac{(v^*)^2(t-s)}{4}\right] \, . \tag{19}$$

Using Theorem 1 and 3, we have that (7) has a continuous solution for all $t \geq 0$ whenever $t^* = \infty$. This holds if

$$\tilde{I}(t) = \int_0^t \frac{1}{2(\pi s)^{1/2}} \exp\left[-\frac{(v^*)^2 s}{4}\right] ds < \Lambda \, , \;\; 0 \leq t < \infty \, . \tag{20}$$

Evaluation of the integral in (20) yields

$$\frac{1}{v^*} \operatorname{erf}\left(\frac{v^* t^{1/2}}{2}\right) < \Lambda \, , \;\; 0 \leq t < \infty \, . \tag{21}$$

Clearly, there always exists some sufficiently large v^* such that (21) is satisfied. In particular, (21) will hold if $v^* > \Lambda^{-1}$,

A criterion which guarantees that a blow-up does occur is given by the following:

Theorem 6. Let $x_0'(t)$ be continuously differentiable for $0 \leq t < \infty$ with a constant $v^{**} > 0$ such that

$$|x_0'(t)| = |x_0'(t)| \leq v^{**} \, , \;\; 0 \leq t < \infty \, .$$

If v^{**} is sufficiently small, then there exists a $t^{**} < \infty$ such that a continuous solution of (12) can not exist for $t \geq t^{**}$. In particular this holds if

$$v^{**} < \frac{1}{k} \, .$$

Proof. By the mean value theorem, (18) again holds. Then, analogous to (19), we obtain

$$k(t,s) \geq \tilde{k}(t,s) = \frac{1}{2\left[\pi(t-s)\right]^{1/2}} \exp\left[-\frac{(v^{**})^2(t-s)}{4}\right] \, .$$

Using Theorems 2 and 4, we have that if there exists a $t^{**} < \infty$ such that

$$\tilde{I}(t^{**}) = \int_0^{t^{**}} \frac{1}{2(\pi s)^{1/2}} \exp\left[-\frac{(v^{**})^2 s}{4}\right] ds = \kappa \tag{22}$$

then (7) can not have a continuous solution for $t \geq t^{**}$. Evaluation of the integral in (22) yields

$$\frac{1}{v^{**}} \operatorname{erf}\left[-\frac{v^{**}(t^{**})^{1/2}}{2}\right] = \kappa \ . \qquad (23)$$

We note that the left-side of (23) behaves like $(t^{**}/\pi)^{1/2}$ as $v^{**} \to 0$, and hence (23) can always be satisfied for some $t^{**} < \infty$ provided that v^{**} is sufficiently small. In particular (23) will hold for some $t^{**} < \infty$ if $v^{**} < \kappa^{-1}$.

ACKNOWLEDGEMENT

This work was supported in part by ARO Grant DAAH04-96-1-0065.

REFERENCES

1. W. E. Olmstead, Critical speed for the avoidance of blow-up in a reactive-diffusive medium, Z. Angew. Math. Phys. 48 (1997), 1-10.

2. W. E. Olmstead and C. A. roberts, Explosion in a diffusive strip due to a concentrated nonlinear source, Methods Appl. Anal. 1 (1994), 434-445.

3. C. A. Roberts, D. G. Lasseigne and W. E. Olmstead, Volterra equations which model explosion in a diffusive medium, J. Integral. Eqn. Appl. 5 (1993), 531-546.

Proceedings of Dynamic Systems and Applications 3 (2001) 367-374

Smooth Dependence of Solutions of Dynamic Systems On Boundary Matrices*

Bonita A. Lawrence
University of South Carolina Beaufort, Beaufort, SC 29902

Abstract: The goal of this work is to generalize continuous dependence results previously established for both discrete-time and continuous-time dynamic systems. In particular, the dynamic system of interest is

$$x^{\Delta}(t) = f(t, x(t))$$

with multi-point boundary conditions

$$\sum_{m=1}^{k} M_m x(t_m) = r$$

where f is a right-dense continuous function and the solution process, x, is defined on a generalized set known as a time scale. Continuous dependence of solutions of this system on the entries of the boundary matrices, M_m, $1 \le m \le k$, is established.

1. INTRODUCTION

Historically, studies of dynamic systems have been directed by the nature of the mathematical structure within which the solution to the system lives. Both continuous-time and discrete-time processes have been studied quite extensively. In each case a particular domain set and topological space are utilized. A new exciting approach has recently been developed by Hilger [4] utilizing a more generalized domain set, any closed subset of the reals, and an appropriate order topology. The developed structure not only accepts the previously studied systems as special cases but also allows for limitless possibilities.

Hilger's approach gives rise to the question "Are continuous time and discrete time processes merely members of a larger family of dynamic systems with similar properties?" This work addresses, in particular, the concept of continuous dependence of solutions of the dynamic system

$$x^{\Delta}(t) = f(t, x(t)) \,, \tag{1.1}$$

where $f : \mathrm{T}^k \times \Re^n \to \Re^n$, with multi-point boundary conditions

$$\sum_{m=1}^{k} M_m x(t_m) = r \tag{1.2}$$

* This work supported by Research and Productive Scholarship Grant #17220-E21 from the Department of Research and Productive Scholarship, USC, Columbia, South Carolina.

on the entries of the boundary matrices $(a_l)_{i,j} = [M]_{lij}$, $1 \le l \le k$, and $1 \le i, j \le n$. Continuous dependence results of this type have been established for a continuous system by Datta and Lawrence, [1] and for a discrete system by Henderson and Lawrence, [3].

2. PRELIMINARY RESULTS

The domain of the solution of the dynamic system is assumed to be a generalized set called a time-scale and is defined to be any closed subset of the reals that is conditionally complete. Pairing the time-scale, T , with an order relation, $\le$, and a measure, μ, yields measure chain, $(\mathrm{T}, \le, \mu)$. The disconnected nature of T is handled using jump operators defined as follows.

Definition 2.1 – (Jump Operators) Jump operators σ and ρ are mappings from the time-scale to itself defined

$$\sigma(t) = \inf\{s \in \mathrm{T} : s > t\}$$

$$\rho(t) = \sup\{s \in \mathrm{T} : s < t\}$$

The elements of T are classified using these operators. For $t \in \mathrm{T}$, if
$\sigma(t) = t$, then t is right-dense,
$\sigma(t) > t$, then t is right-scattered,
$\rho(t) = t$, then t is left-dense,
$\rho(t) < t$, then t is left-scattered.

A time-scale can be characterized by its graininess, determined by a function constructed from the measure μ.

Definition 2.2 – (Graininess) The graininess of the elements of a time-scale is determined by the function $\mu^* : \mathrm{T}^k \to \Re_0^+$, defined $\mu^*(t) = \mu(\sigma(t), t)$, where T^k is the set of all elements of T that are non-maximal or left-dense. That is, T^k is T less an isolated maximum point.

Within this structure the concept of the derivative is defined. The generalized derivative for a function on a time-scale accepts both the derivative for the continuous case and the difference quotient for the discrete case.

Definition 2.3 - (Generalized Derivative) Let $f : \mathrm{T} \to X$, where X is a Banach space. We say that f has a derivative, f_t^Δ, at $t \in \mathrm{T}$ if and only if for every $\varepsilon > 0$ there exists a neighborhood, U, of t such that for all $s \in U$,

$$\left|f(\sigma(t)) - f(s) - f_t^\Delta \mu(\sigma(t, s))\right| \le \varepsilon \left|\mu(\sigma(t), s)\right|$$

A function f is said to be differentiable at t if it has exactly one derivative at t.

Many important results for functions on time scales have been proved using a modified version of the induction principle. The requirement that T be a conditionally complete chain yields an induction principle adopted from the principle employed by Dieudonne [2] for differential and integral calculus.

Theorem 2.4 - Let $(\mathrm{T},\leq)$ be a chain. The following properties are equivalent.

a) T is conditionally complete.

b) The induction principle holds:
Assume for a family of statements $A(t),\ t\in[\tau,\infty)\subseteq \mathrm{T}$ that the following conditions are satisfied.

1) $A(\tau)$ holds true.
2) For each right-scattered $t\in\mathrm{T}$,
$A(t)\Rightarrow A(\sigma(t))$
3) For each right-dense $t\in\mathrm{T}$, there is a neighborhood, U, such that
$A(t)\Rightarrow A(s)$ for all $s\in U$ with $s>t$.
4) For each left-dense point $t\in\mathrm{T}$
$A(s)\Rightarrow A(t)$ for all $s<t$.

Then $A(t)$ is true for all $t\in[\tau,\infty)\subseteq\mathrm{T}$.

Theorem 2.4 is a portion of a result presented by Lakshmikantham, et al, [5].

3. CONTINUOUS DEPENDENCE RESULTS

The proof of the main result utilizes a continuous dependence result due to Lakshmikantham, Shahzad, and Sivasundaram, [5]. Consider the initial value problem

$$x^{\Delta}(t)=f(t,x(t)) \tag{3.1}$$
$$x(t_0)=x_0 \tag{3.2}$$

where $f:\mathrm{T}^k\times\Re^n\to\Re^n$ is right-dense continuous. The following theorem gives sufficient conditions for continuous dependence of a solution of (3.1),(3.2) on the initial position (as well as the initial time).

Theorem 3.1 - Let $\mathrm{T}=[\tau,s]$ be a compact measurable chain and L be a nonnegative constant with

$$L\mu(\tau,s)<1$$

For dynamic system (3.1) with initial condition

$$x(\tau)=\eta \tag{3.3}$$

assume that f is right-dense continuous and satisfies the Lipschitz condition

$$|f(t,x_1)-f(t,x_2)|\leq L|x_1-x_2|, \quad t\in \mathrm{T}^k, \quad x_1,x_2\in \Re^n$$

Then the solutions of (3.1),(3.3) are unique and continuous with respect to (τ,η).

A sketch of the proof: Define $v(t)=|x(t,\tau,\eta)-x(t,\tau_1,\eta_1)|$. The authors make the claim that $v(t)\leq m(t)$, for $t\geq\tau$, where $m(t)$ is a solution of

$$u^{\Delta}(t)=Lu(t), \quad u(\tau)=v(\tau).$$

They then apply the induction principle of Theorem 2.1 to the statement

$$A(t): v(t)\leq m(t) \quad \text{for all } t\geq\tau.$$

After this is verified they show that one can find a neighborhood of τ (and a neighborhood of η) such that for all elements s

$$|x(t,s,y)-x(t,\tau,\eta)<\varepsilon| \quad \bullet$$

The main result also requires a uniqueness condition for (1.1). The standard uniqueness condition for multi-point boundary value problems is assumed.

Definition 3.2 - (Property $(M_1,\ldots,M_k)$) A solution of (1.1) satisfies Property $(M_1,\ldots,M_k)$ if whenever $x_1(t)$ and $x_2(t)$ are solutions of (1.1) with

$$\sum_{m=1}^{k} M_m x_1(t_m)=\sum_{m=1}^{k} M_m x_2(t_m)$$

it follows that $x_1(t)=x_2(t)$.

And now friends, the main result gives conditions that insure continuous dependence of solutions of (1.1), (1.2) on the entries of the boundary matrices.

Theorem 3.3 - Let $y(t)=y(t_1,M_1,\ldots,M_k,t_1,\ldots,t_k,r)$ be a solution of (3.1) and assume that all assumptions of Theorem 3.1 as well as Property $(M_1,\ldots,M_k)$ are satisfied. Then for every $\varepsilon>0$ there exists a $\delta>0$ such that if

$$\|M_m-\overline{M}_m\|<\delta, \ 1\leq m\leq k, \text{ and } |r_i-\bar{r}_i|<\delta, \ 1\leq i\leq n,$$

then

$$\left\| y(t_1, M_1, \ldots, M_k, t_1, \ldots, t_k, r) - y(t_1, \overline{M}_1, \ldots \overline{M}_k, t_1, \ldots, t_k, \overline{r}) \right\| < \varepsilon$$

Proof: Let $t_1 < t_2 < \ldots < t_k \in \mathrm{T}$ be as defined in (1.2). Define a mapping $\varphi : \Re^{2n} \rightarrow \Re^{2n}$ where

$$\varphi(c_1, \ldots, c_n, r_1, \ldots, r_n) = \left(\left(\sum_{m=1}^{k} M_m v(t_m) \right)_1, \ldots, \left(\sum_{m=1}^{k} M_m v(t_m) \right)_n, r_1, \ldots, r_n \right)$$

where

$$v(t) = v(t, t_1, c)$$

is a solution of initial value problem (3.1) with initial conditions

$$x(t_1) = c \tag{3.4}$$

Claim: The function φ is one-to-one. To verify this fact assume that

$$\varphi(c_1, \ldots, c_n, r_1, \ldots, r_n) = \varphi(c_1', \ldots, c_n', r_1', \ldots, r_n') \ . \tag{3.5}$$

Utilizing the definition of φ, (3.5) yields the expression

$$\left(\left(\sum_{m=1}^{k} M_m v(t_m, t_1, c) \right)_1, \ldots, \left(\sum_{m=1}^{k} M_m v(t_m, t_1, c) \right)_n, r_1, \ldots, r_n \right) = \left(\left(\sum_{m=1}^{k} M_m v(t_m, t_1, c') \right)_1, \ldots, \left(\sum_{m=1}^{k} M_m v(t_m, t_1, c') \right)_n \ldots r_n' \right].$$

Therefore the n-vectors $\sum_{m=1}^{k} M_m v(t_m, t_1, c)$ and $\sum_{m=1}^{k} M_m v(t_m, t_1, c')$ are equal and hence from the uniqueness property of Theorem 3.1

$$v(t, t_1, c) = v(t, t_1, c') \ .$$

Note also that φ is a continuous function. In fact, if the sequence $\{(c_1^{\,l}, \ldots, c_n^{\,l}, r_1^{\,l}, \ldots, r_n^{\,l})$ converges to $\{(c_1, \ldots, c_n, r_1, \ldots, r_n)\}$ as $l \rightarrow \infty$ the continuous dependence of the solutions of (3.1) on the initial data (3.4) implies that $v(t, t_1, c^l)$ converges to $v(t, t_1, c)$ as $l \rightarrow \infty$, where $c^l = (c_1^{\,l}, \ldots, c_n^{\,l})$ and $c = (c_1, \ldots, c_n)$.

An application of the Brouwer Theorem on Invariance of Domain yields the following important point:

1. φ is a homeomorphism of $\Re^{2n}$ onto itself.
2. $\varphi(\Re^{2n})$ is an open subset of $\Re^{2n}$.

Therefore φ^{-1} is continuous on $\varphi(\Re^{2n})$. These facts together with the continuous dependence of (3.1) on the initial values and the fact that

$$\left(\left(\sum_{m=1}^{k} M_m y(t_m)\right)_1, \ldots, \left(\sum_{m=1}^{k} M_m y(t_m)\right)_n, r_1, \ldots, r_n\right) \in \varphi(\Re^{2n})$$

allow the following situation. Given $\eta > 0$ there exists $\delta > 0$ such that if

$$\left\| M_m - \overline{M}_m \right\| < \delta \ , 1 \le m \le k, \text{ and } \left| r_i - \bar{r}_i \right| < \delta \ , 1 \le i \le n,$$

then

$$\left(\left(\sum_{m=1}^{k} \overline{M}_m y(t_m)\right)_1, \ldots, \left(\sum_{m=1}^{k} \overline{M}_m y(t_m)\right)_n, r_1, \ldots, r_n\right) \in \varphi(\Re^{2n}).$$

Also, there is a unique solution

$$y(t) = y\left(t_1, \overline{M}_1, \ldots, \overline{M}_k, t_1, \ldots, t_k, r\right)$$

of the initial value problem (3.1) satisfying multi-point boundary conditions

$$\sum_{m=1}^{k} \overline{M}_m y(t_m) = \bar{r}$$

and

$$\left\| y(t_1, M_1, \ldots, M_k, t_1, \ldots, t_k, r) - y\left(t_1, \overline{M}_1, \ldots \overline{M}_k, t_1, \ldots, t_k, \bar{r}\right) \right\| < \eta$$

Applying Theorem 3.1, given $\varepsilon > 0$ there exists $\eta > 0$ such that

$$\left\| y(t, M_1, \ldots, M_k, t_1, \ldots, t_k, r) - y\left(t, \overline{M}_1, \ldots \overline{M}_k, t_1, \ldots, t_k, \bar{r}\right) \right\| < \varepsilon$$

whenever

$$\left\| y(t_1, M_1, \ldots, M_k, t_1, \ldots, t_k, r) - y\left(t_1, \overline{M}_1, \ldots \overline{M}_k, t_1, \ldots, t_k, \bar{r}\right) \right\| < \eta .$$

Therefore, as $\delta \to 0$

$$y\left(t, \overline{M}_1, \ldots, \overline{M}_k, t_1, \ldots, t_k, \bar{r}\right) \to y(t, M_1, \ldots, M_k, \ldots, t_1, \ldots, t_k, r)$$

The proof is complete. •

A natural extension of this work is differentiability of solutions of (1.1), (1.2) with respect to the entries of the boundary matrices. The author is currently investigating this topic.

REFERENCES

1. A. Datta, B. Lawrence, Smooth Dependence on Boundary Matrices for First Order Differential Equations, Pan Amer. J., in press.

2. J. Dieudonne, Foundations of Modern Analysis, Prentice Hall, New Jersey, 1966.

3. J. Henderson, B. Lawrence, Smooth Dependence on Boundary Matrices. Journal of Difference Equations and Applications, **2**, (1996), 161-166.

4. S. Hilger, Analysis on Measure Chains – A Unified Approach to Continuous and Discrete Calculus. Results Math., **18**, (1990), 19-56.

5. B. Kaymakcalan, V. Lakshmikantham, and s. Sivasundaram, Dynamic Systems on Measure Chains, Kluwer Academic Publishers, Boston, 1996.

Proceedings of Dynamic Systems and Applications 3 (2001) 375-382

Periodic Solutions in Periodic Delayed Gause-Type Predator-Prey Systems

Yongkun Li[1]*and Yang Kuang[2]
[1]Department of Mathematics, Yunnan University,
Kunming, Yunnan 650091, People's Republic of China
[2]Department of Mathematics, Arizona State University,
Tempe, AZ 85287-1804, USA

ABSTRACT: Reasonable sufficient conditions are obtained for the existence of positive periodic solutions in periodic delayed Gause-type predator-prey systems. Our approach involves the application of coincidence degree theorem and estimations of uniform upper bounds on solutions. This method imposes minimum restrictions on the form and magnitude of time delays. Indeed, our results are applicable to discrete, distributed and state-dependent delays. Our results indicate that seasonal effects on population models often lead to synchronous solutions. In addition, we may conclude that when both seasonality and time delay are present, the seasonality is often the generating force for the often observed fluctuations in population densities, including the inherently oscillatory predator-prey dynamics.

Keywords: Coincidence degree, Periodic solution, delay equation, state-dependent delay, predator-prey model.

AMS (MOS) subject classification: Primary 34K15; secondary 34K20, 92A15.

1 INTRODUCTION

It is often observed that populations in the real world tend to fluctuate. This is especially so for predator-prey interactions. There are three typical approaches for modeling such behavior: (i) introduce more species into the model, and consider the higher dimensional systems (like predator-prey interactions, May [6]); (ii) assume that the per capita growth function is time dependent and periodic in time; (iii) take into account the time delay effect in the population dynamics (Smith and Kuang [7], Zhao et al. [10]). Although all of them are good mechanisms of generating periodic solutions (and therefore offer some explanations to the often observed oscillatory behavior in population densities), it does not give us any insight as which is the real generating or dominating force behind the oscillatory behavior if only one of such mechanism is considered. Naturally, more realistic and interesting models of population interactions

*Currently, visiting at Department of Mathematics, Arizona State University.

should take into account both the seasonality of the changing environment and the effects of time delays. Therefore, it is interesting and important to study the following general nonlinear nonautonomous delayed Gause-type predator-prey system:

$$\begin{cases} \dot{x}(t) = x(t)[g(t, x(t)) - p(x(t))y(t)], \\ \dot{y}(t) = y(t)\left[-\alpha + h\left(\int_{-\gamma}^{-\gamma_0} x(t+\theta)\, d\eta(\theta)\right)\right], \end{cases} \tag{1.1}$$

where

(A1) $p \in C^1([0,+\infty), R)$, and $p(x) > 0$ for $x \geq 0$, $p'(x) \leq 0$;

(A2) $h(x) \in C^1([0,+\infty), R)$ and $h'(x) > 0$ for $x \geq 0$; $h(0) = 0$ and there exists $x_0 > 0$ such that $h(x_0) = \alpha$;

(A3) $g(t,x) \in C^1([0,+\infty) \times R, R)$, $g(t+\omega, x) = g(t,x)$ for $(t,x) \in R \times [0,+\infty)$, where ω is a nonnegative constant. There exists a constant $H > 0$ such that $g(t,x) \leq H$ for $(t,x) \in R \times [0,+\infty)$. In addition, $\int_0^\omega g(t,x_0)\, dt > 0$.

(A4) $\gamma_0, \gamma > 0$ are constants, η is a nondecreasing function satisfying $\eta(-\gamma_0^+) - \eta(-\gamma^-) = 1$.

Readers familiar with predator-prey models may notice that the above assumptions are needed to make the system a standard Gause-type predator-prey model(see Kuang(1993)) and to ensure the existence of a positive steady state when the system is reduced to an autonomous ODE model. To facilitate this observation, one can let

$$g(t,x(t)) = r(t)(1 - x(t)/K(t)), \quad p(x) = a/(1+\alpha x), \quad h(x) = bx/(1+\alpha x),$$

where $r(t), K(t)$ are periodic functions of period ω, and a, b, α are constants.

At the end of this paper, we will point out that our results for the above system can be easily extended to the one with infinite delay or a state-dependent delay.

Existing results on the existence of periodic solutions in periodic systems (population models, in particular) often fall into one of these three categories: (1) the results of the applications of contraction principle or fluctuation principle, which establish both the existence and attractivity of the periodic solutions in periodic equations with time delay (Kuang [4], p. 181); (2) the existence simply follow the observation that the periodic solution exist when there is no time delay and this periodic solution remains so when time delay is a multiple of the period of the periodic equation (Gopalsamy et al. [3], Zhang and Gopalsamy [9]); (3) the results of the application of Horn's asymptotic fixed point theorem (Freedman and Wu [1], Tang and Kuang [8]). While these methods often allow the investigator to address the stability issues of the periodic solutions, the conditions for the existence part are often unnecessary numerous, tedious, stringent, and difficult to satisfy. Specifically, all the above methods are ill suited to problems with state-dependent delay equations.

By employing the powerful and effective coincidence degree method, we found that the existence of periodic solutions in periodic predator-prey systems with or without time delays require only a set of natural and easily verifiable conditions. Such approach was first adopted in Li(1999) for a more specific predator-prey model, where $g(t,x(t)) = r(t)(1 - x(t)/K(t)), p(x) = x$. Our conditions are readily satisfied in many realistic population models that can be found in the literature. This strongly suggests that seasonal effects on population models indeed often lead to synchronous solutions. In addition, we may conclude that when both seasonality and time delay are present and deserve consideration, the seasonality is often the main generating force for the often observed oscillatory behavior in population densities.

2 MAIN RESULTS

The method to be used in this paper involves the application of the continuous theorem of coincidence degree (Gaines and Mawhin [2], p. 40). This requires us to introduce a few notations.

Let X, Y be real Banach spaces, $L : \text{dom } L \subset X \to Y$ a Fredholm mapping of index zero (Index $L = \dim \ker L - \text{codim Im } L$) and $P : X \to X$, $Q : Y \to Y$ are continuous projectors such that Im P = Ker L, Ker Q = Im L and $X = \text{Ker } L \oplus \text{Ker } P$, $Y = \text{Im } L \oplus \text{Im } Q$. Consequently, the restriction L_P of L to dom $L \cap \text{Ker } P$ is one-to-one and onto map to Im L, so that its (algebraic) inverse $K_P : \text{Im } L \to \text{dom } L \cap \text{Ker } P$ is well defined. Denote by $J : \text{Im } Q \to \text{Ker } L$ an isomorphism of Im Q onto Ker L.For convenience we cite the continuous theorem from Gaines and Mawhin ([2], p. 40):

Theorem A *Let $\Omega \subset X$ be an open bounded set and $N : X \to Y$ be a continuous operator which is L-compact on $\bar{\Omega}$ (i.e., $QN : \bar{\Omega} \to Y$ and $K_P(I - Q)N : \bar{\Omega} \to X$ are compact). Assume*

(a) *for each $\lambda \in (0, 1)$, every solution x of $Lx = \lambda Nx$ is such that $x \notin \partial\Omega$,*

(b) *$QNx \neq 0$ for each $x \in \text{Ker } L \cap \partial\Omega$,*

(c) *the Brouwer degree $\deg [JQN, \Omega \cap \text{Ker } L, 0] \neq 0$.*

Then the operator equation $Lx = Nx$ has at least one solution in dom $L \cap \bar{\Omega}$.

Consider the following system

$$\begin{cases} \dot{u}(t) = g(t, e^{u(t)}) - p(e^{u(t)})e^{v(t)}, \\ \dot{v}(t) = -\alpha + h\left(\int_{-\gamma}^{-\gamma_0} e^{u(t+\theta)}\, d\eta(\theta)\right). \end{cases} \tag{2.1}$$

It is easy to see that if system (2.1) has an ω-periodic solution $(u^*(t), v^*(t))$, then $(e^{u^*(t)}, e^{v^*(t)})$ is a positive ω-periodic solution of system (1.1).

In order to use continuation theorem of coincidence degree theory to establish the existence of an ω-periodic solution of system (2.1), we take $X = \{(u, v)^T \in C(R, R^2) : u(t + \omega) = u(t), v(t + \omega) = v(t)\}$ and

$$\|(u, v)^T\| = \max_{t\in[0,\omega]} |u(t)| + \max_{t\in[0,\omega]} |v(t)|.$$

With this norm, X is a Banach space. Let

$$L : \text{dom } L \to X, \quad (u, v)^T \mapsto (\dot{u}(t), \dot{v}(t))^T,$$

where dom $L = \{(u, v)^T \in X : (u, v)^T \in C^1(R, R^2)\}$. Obviously, Ker $L = \{(u, v)^T \in X : (u, v)^T = (c_1, c_2)^T \in R^2\}$, Im $L = \{(u, v)^T \in X : \int_0^\omega u(t)\, dt = \int_0^\omega v(t)\, dt = 0\}$ is closed in X and dim Ker L = codim Im $L = 2$. Therefore, L is a Fredholm mapping of index 0.

The following is our main result.

Theorem 2.1 *Suppose that* (A1)-(A4) *hold, then system* (1.1) *has a positive ω-periodic solution.*

Proof. To make use of Theorem A, we need to define a continuous operator N,

$$N: X \to X, \quad \begin{bmatrix} u \\ v \end{bmatrix} \mapsto \begin{bmatrix} g(t, e^{u(t)}) - p(e^{u(t)})e^{v(t)} \\ -\alpha + h\left(\int_{-\gamma}^{-\gamma_0} e^{u(t+\theta)}\right) d\eta(\theta) \end{bmatrix}.$$

We define the two continuous projectors as follows:

$$P : X \to X, \quad P((u,v)^T) = \left(\frac{1}{\omega}\int_0^\omega u(t)\,dt, \frac{1}{\omega}\int_0^\omega u(t)\,dt\right)^T,$$

$$Q : X \to X, \quad Q((u,v)^T) = \left(\frac{1}{\omega}\int_0^\omega u(t)\,dt, \frac{1}{\omega}\int_0^\omega v(t)\,dt\right)^T.$$

Then, through an elementary computation, we can obtain the explicit form of the corresponding inverse $K_P : \text{Im } L \to \text{Ker } P \cap \text{ dom } L$ of L_P as:

$$\begin{bmatrix} u \\ v \end{bmatrix} \mapsto \begin{bmatrix} \int_0^t u(s)\,ds - \frac{1}{\omega}\int_0^\omega \int_0^\mu u(s)\,ds\,d\mu \\ \int_0^t v(s)\,ds - \frac{1}{\omega}\int_0^\omega \int_0^\mu v(s)\,ds\,d\mu \end{bmatrix}.$$

Notice that $QN : X \to X$ takes the form of

$$\begin{bmatrix} u \\ v \end{bmatrix} \mapsto \begin{bmatrix} \frac{1}{\omega}\int_0^\omega [g(t, e^{u(t)}) - p(e^{u(t)})e^{v(t)}]\,dt \\ \frac{1}{\omega}\int_0^\omega \left[-\alpha + h\left(\int_{-\gamma}^{-\gamma_0} e^{u(t+\theta)}\,d\eta(\theta)\right)\right] dt \end{bmatrix}.$$

Hence, we have $K_P(I-Q)N : X \to X$ takes the form of

$$\begin{bmatrix} u \\ v \end{bmatrix} \mapsto \begin{bmatrix} \int_0^t [g(s, e^{u(s)}) - p(e^{u(s)})e^{v(s)}]\,ds \\ \int_0^t \left[-\alpha + h\left(\int_{-\gamma}^{\gamma_0} e^{u(s+\theta)}\,d\eta(\theta)\right)\right] ds \end{bmatrix}$$

$$- \begin{bmatrix} \frac{1}{\omega}\int_0^\omega \int_0^\mu [\rho(s, e^{u(s)}) - p(e^{u(s)})e^{v(s)}]\,ds\,d\mu \\ \frac{1}{\omega}\int_0^\omega \int_0^\mu \left[-\alpha + h\left(\int_{-\gamma}^{\gamma_0} e^{u(s+\theta)}\,d\eta(\theta)\right)\right] ds\,d\mu \end{bmatrix}$$

$$+ \begin{bmatrix} \left(\frac{1}{2} - \frac{t}{\omega}\right)\int_0^\omega [g(s, e^{u(s)}) - p(e^{u(s)})e^{v(s)}]\,ds \\ \left(\frac{1}{2} - \frac{t}{\omega}\right)\int_0^\omega \left[-\alpha + h\left(\int_{-\gamma}^{-\gamma_0} e^{u(s+\theta)}\,d\eta(\theta)\right)\right] ds \end{bmatrix}.$$

Clearly, QN and $K_P(I-Q)N$ are continuous by the Lebesgue theorem and, furthermore, $QN(\overline{\Omega})$ and $K_P(1-Q)N(\overline{\Omega})$ are relatively compact for any open bounded set $\Omega \subset X$. Hence, N is L-compact on $\overline{\Omega}$ for any open bounded set $\Omega \subset X$. The corresponding operator equation $Lx = \lambda Nx$, $\lambda \in (0,1)$, takes the form of

$$\begin{cases} \dot{u}(t) = \lambda[g(t, e^{u(t)}) - p(e^{u(t)})e^{v(t)}], \\ \dot{v}(t) = \lambda\left[-\alpha + h\left(\int_{-\gamma}^{-\gamma_0} e^{u(t+\theta)}\, d\eta(\theta)\right)\right], \end{cases} \quad \lambda \in (0,1). \tag{2.2}$$

Suppose that $(u(t), v(t))^T \in X$ is a solution of system (2.2) for a certain $\lambda \in (0,1)$. Integrating system (2.2) over the interval $[0,\omega]$ we obtain

$$\int_0^\omega g(t, e^{u(t)})\, dt = \int_0^\omega p(e^{u(t)})e^{v(t)}\, dt \tag{2.3}$$

and

$$\int_0^\omega h\left(\int_{-\gamma}^{-\gamma_0} e^{u(t+\theta)}\, d\eta(\theta)\right) dt = \alpha\omega. \tag{2.4}$$

It follows from (A3) and (2.3) that

$$\int_0^\omega p(e^{u(t)})e^{v(t)}\, dt \le H\omega. \tag{2.5}$$

Denote $\Delta_1 = \{t \in [0,\omega] : g(t, e^{u(t)}) \ge 0\}$, $\Delta_2 = \{t \in [0,\omega] : g(t, e^{u(t)}) < 0\}$, then

$$\int_0^\omega g(t, e^{u(t)})\, dt = \int_{\Delta_1} g(t, e^{u(t)})\, dt + \int_{\Delta_2} g(t, e^{u(t)})\, dt.$$

¿From this, (A3) and (2.3), we find

$$\begin{aligned} -\int_{\Delta_2} g(t, e^{u(t)})\, dt &= \int_{\Delta_1} g(t, e^{u(t)})\, dt - \int_0^\omega p(e^{u(t)})e^{v(t)}\, dt \\ &< \int_{\Delta_1} g(t, e^{u(t)})\, dt \le H\omega. \end{aligned}$$

Hence

$$\begin{aligned} \int_0^\omega |g(t, e^{u(t)})|\, dt &= \int_{\Delta_1} |g(t, e^{u(t)})|\, dt + \int_{\Delta_2} |g(t, e^{u(t)})|\, dt \\ &= \int_{\Delta_1} g(t, e^{u(t)})\, dt - \int_{\Delta_2} g(t, e^{u(t)})\, dt < 2H\omega. \end{aligned} \tag{2.6}$$

By (2.2), (2.5) and (2.6), we obtain

$$\begin{aligned} \int_0^\omega |\dot{u}(t)|\, dt &\le \int_0^\omega |g(t, e^{u(t)}) - p(e^{u(t)})e^{v(t)}|\, dt \\ &< \int_0^\omega |g(t, e^{u(t)})|\, dt + \int_0^\omega p(e^{u(t)})e^{v(t)}\, dt \\ &< 3H\omega. \end{aligned} \tag{2.7}$$

Moreover, from (2.2) and (2.4) it follows that

$$\begin{aligned}\int_0^\omega |\dot v(t)|\,dt &\le \int_0^\omega \left| -\alpha + h\Big(\int_{-\gamma}^{-\gamma_0} e^{u(t+\theta)}\,d\eta(\theta)\Big)\right| dt \\ &\le \int_0^\omega \alpha\,dt + \int_0^\omega h\Big(\int_{-\gamma}^{-\gamma_0} e^{u(t+\theta)}\,d\eta(\theta)\Big)\,dt \\ &= 2\alpha\omega. \end{aligned} \tag{2.8}$$

¿From (A2) and (2.4), it is easy to see that there exists a constant $B_1 > 0$, independent of initial conditions, and a point $t^* \in [-\gamma, \omega - \gamma_0]$ such that $|u(t^*)| < B_1$. Since $u(t)$ is periodic of period ω, we see that there is a $t_1 \in [0,\omega]$, such that $|u(t_1)| < B_1$. It follows from this and (2.7) that

$$\begin{aligned}|u(t)| &\le |u(t_1)| + \int_0^\omega |\dot u(t)|\,dt \\ &< B_1 + 3H\omega \stackrel{\text{def}}{=} B_2. \end{aligned} \tag{2.9}$$

In view of (A1), (A2), (2.3), (2.5) and (2.9), we have

$$p(e^{B_2})\int_0^\omega e^{v(t)}\,dt \le \int_0^\omega p(e^{u(t)})e^{v(t)}\,dt \le H\omega \tag{2.10}$$

and

$$\begin{aligned}p(0)\int_0^\omega e^{v(t)}\,dt &\ge \int_0^\omega p(e^{u(t)})e^{v(t)}\,dt \\ &= \int_0^\omega g(t, e^{u(t)})\,dt \ge \underline{g}\omega, \end{aligned} \tag{2.11}$$

where

$$\underline{g} = \min_{\substack{t\in[0,\omega] \\ u\in[-B_2,B_2]}} \{g(t,e^u)\}.h$$

By (2.10), (2.11) and the continuity of $v(t)$, one can easily see that there exists a constant B_3, independent of initial conditions, and a point $t_2 \in [0,\omega]$ such that

$$|v(t_2)| < B_3.$$

¿From this and (2.8) we obtain that

$$\begin{aligned}|v(t)| &\le |v(t_2)| + \int_0^\omega |\dot v(t)|\,dt \\ &< 2\alpha\omega + B_3 \stackrel{\text{def}}{=} B_4. \end{aligned}$$

Let $B = B_2 + B_4 + \ln x_0 + \ln(\omega^{-1}\int_0^\omega g(t,x_0)\,dt/p(x_0))$ and take $\Omega = \{(u(t),v(t))^T \in X : \|(u,v)^T\| < B\}$. It is clear that Ω verifies the requirement (a) in Lemma A.

When $(u,v) \in \partial\Omega \cap \text{Ker } L = \partial\Omega \cap R^2$, (u,v) is a constant vector in R^2 with $|u|+|v| = B$. Then

$$QN\begin{bmatrix} u \\ v \end{bmatrix} = \begin{bmatrix} \frac{1}{\omega}\int_0^\omega g(t,e^u)\,dt - p(e^u)e^v \\ -\alpha + h(e^u) \end{bmatrix} \neq 0.$$

Moreover, take $J = I$: Im $Q \to$ Ker L, $x \mapsto x$. Then in view of (A2), it is easy to see that

$$\begin{aligned} &\deg\{JQN((u,v)^T), \Omega \cap \text{ Ker } L, 0\} \\ &= \deg\{QN((u,v)^T), \Omega \cap R^2, 0\} \\ &= \text{sgn}\{h'(x_0)p(x_0)x_0e^{\bar{v}}\} = 1 \neq 0, \end{aligned}$$

where $e^{\bar{v}} = \omega^{-1}\int_0^\omega g(t,x_0)\,dt/p(x_0)$.

By now we have proved that Ω verifies all the requirements of Lemma A and hence system (2.1) has an ω-periodic solution. The proof is complete.

By Theorem 2.1, one can easily see that

Corollary 2.1 *Suppose that* (A1)–(A3) *hold. Then the following system*

$$\begin{cases} \dot{x}(t) = x(t)[y(t,x(t)) - p(x(t))y(t)], \\ \dot{y}(t) = y(t)[-\alpha + h(x(t-\tau))] \end{cases}$$

has a positive ω-periodic solution, where $\tau > 0$ is a constant.

Next consider the following infinite delayed Gause-type prey-predator systems

$$\begin{cases} \dot{x}(t) = x(t)[g(t,x(t)) - p(x(t))y(t)], \\ \dot{y}(t) = y(t)\left[-\alpha + h\left(\int_{-\infty}^0 x(t+\theta)\,d\eta(\theta)\right)\right], \end{cases} \tag{2.12}$$

where η is a nondecreasing function satisfying $\eta(0^+) - \eta(-\infty) = 1$.

Theorem 2.2 *Suppose that* (A1)–(A3) *hold. Then system* (2.12) *has a positive ω-periodic solution.*

Proof. The proof is similar to the proof of Theorem 2.1 and hence is omitted here.

Finally, consider the following state dependent delay Gause-type prey-predator system.

$$\begin{cases} \dot{x}(t) = x(t)[g(t,x(t)) - p(x(t))y(t)], \\ \dot{y}(t) = y(t)[-\alpha + h(x(t-\tau(x(t))))] \end{cases} \tag{2.13}$$

where $\tau \in C(R, R^+)$.

Theorem 2.3 *Suppose that* (A1)–(A3) *hold. Then the system* (2.13) *has a positive ω-periodic solution.*

Proof. Again, the proof is similar to the proof of Theorem 2.1 and is thus omitted.

REFERENCES

1. Freedman, H.I., & Wu, J. Periodic solutions of single-species models with periodic delay. *SIAM J. Math. Anal.*, Vol. 23(1992). 689–701.

2. Gaines, R.E., & Mawhin, J.L. *Coincidence Degree and Nonlinear Differential Equations*, Springer-Verlag, Berlin, 1977.

3. Gopalsamy, K., Kulenovic, M.R.S., & Ladas, G. Environmental periodicity and time delays in a "food-limited" population model, *J. Math. Anal. Appl.*, Vol. 147(1990). 545–555.

4. Kuang, Y. (1993). *Delay Differential Equations with Applications in Population Dynamics*, Academic Press, Boston.

5. Li, Y. Periodic solutions of a periodic delay predator-prey system, *Proc. Amer. Math. Soc.*, Vol. 127(1999). 1331–1335.

6. May, R.M. *Stability and Complexity in Model Ecosystems.* Princeton University Press, Princeton, 1974.

7. Smith, H.L., & Kuang, Y. (1992). Periodic solutions of delay differential equations of threshold-type delays, in: *Oscillation and Dynamics in Delay Equations*, Graef and Hale ed. Contemporary Mathematics, Vol. 129 (1992), AMS, Providence, 153–176.

8. Tang, B.R., & Kuang, Y. Existence, uniqueness and asymptotic stability of periodic solutions of periodic functional differential systems, *Tohoku Mathematical Journal*, Vol. 49 (1997). 217–239.

9. Zhang, B.G., & Gopalsamy, K. Global attractivity and oscillations in a periodic delay-logistic equation, *J. Math. Anal. Appl.*, Vol. 150(1990). 274–283.

10. Zhao, T., Kuang, Y., & Smith, H.L. (1997), Global existence of periodic solutions in a class of delayed Gause-type predator-prey systems, *Nonlinear Analysis, TMA*, Vol. 28(1997). 1373–1394.

Proceedings of Dynamic Systems and Applications 3 (2001) 383-390

APPROXIMATION METHODS FOR INTEGRODIFFERENTIAL EQUATIONS

J. Liu, E. Parker, J. Sochacki, A. Knutsen

Department of Mathematics, James Madison University, Harrisonburg, VA 22807

ABSTRACT : Consider the following integrodifferential equation

$$u'(t) = Au(t) + \int_0^t b(t-s)Au(s)ds + f(t), \quad t \geq 0, \tag{0.1}$$

in a general Banach space X, with A an unbounded linear operator which generates a strongly continuous semigroup, and $b(\cdot)$ a scalar function. We propose an approximation method that will reduce the integrodifferential equation (0.1) to an ordinary differential equation in the product space X^k, where k is determined by the accuracy of the approximation. Therefore, the numerical solutions for the ordinary differential equation in X^k can be used to approximate the solutions of the integrodifferential equation (0.1). An application in viscoelasticity is given.

1 INTRODUCTION.

Consider the following integrodifferential equation

$$u'(t) = Au(t) + \int_0^t b(t-s)Au(s)ds + f(t), \quad t \geq 0, \tag{1.1}$$

$$u(0) = u_0, \tag{1.2}$$

in a general Banach space X with A an unbounded linear operator which generates a strongly continuous semigroup, and $b(\cdot)$ a scalar function. In most studies, Eq.(1.1)-(1.2) is first reformulated as an abstract ordinary differential equation in some product space $X \times Y$. Then the abstract ordinary differential equation is shown to be wellposed. See, e.g., [3]. Due to the fact that Eq.(1.1) involves an integral, the coordinate Y of the product space is given by some function space after the reformulation, such as $Y = L^2([0,\infty), X)$, so that great effort must be made in order to prove the wellposedness and to obtain numerical solutions using the approximation techniques in the function spaces.

In this note, we want to propose the following approach to approximate the solutions of Eq.(1.1)-(1.2): In the first step, we invert $Au(\cdot)$ so as to transform Eq.(1.1)-(1.2) into a form with a continuous kernel,

$$u'(t) = \Big(A - B(0)I\Big)u(t) - \int_0^t B'(t-s)u(s)ds + \hat{f}(t), \quad t \geq 0, \tag{1.3}$$

$$u(0) = u_0, \tag{1.4}$$

where I is the identity operator on X, $B(\cdot)$ is a scalar determined by $b(\cdot)$, and

$$\hat{f}(t) = f(t) + B(t)u_0 + \int_0^t B(t-s)f(s)ds. \tag{1.5}$$

In the second step, we approximate the continuous scalar function $-B'(\cdot)$ by a polynomial $P_n(\cdot)$ of degree n. In the third step, we show that on any finite time interval $[0, T_0]$, the solutions $v_n(\cdot)$ of

$$v_n'(t) = \Big(A - B(0)I\Big)v_n(t) + \int_0^t P_n(t-s)v_n(s)ds + \hat{f}(t), \quad t \geq 0, \tag{1.6}$$

$$v_n(0) = u_0, \tag{1.7}$$

can be used to approximate the solution u of Eq.(1.3)-(1.4). The degree n of the polynomial $P_n(\cdot)$ is determined by how accurate an approximation v_n is to be of u. In the fourth step, we show that Eq.(1.6)-(1.7) is equivalent to a wellposed ordinary differential equation in the product space X^{n+2}. Now, in the fourth step, the product space is X^{n+2} and no function spaces are involved, thus the proof of the wellposedness is straightforward. We will also see that the leading operator in the ordinary differential equation in X^{n+2} has many zero entries. This will lead to a simple computational procedure.

When $X = L^2[0,1]$, $A = \frac{\partial^2}{\partial x^2}$ with (domain) $D(A) = W_0^{1,2}[0,1] \cap W^{2,2}[0,1]$, which is a special and important case for equations in viscoelasticity, we are able to reduce the ordinary differential equation in the product space X^{n+2} into a system of simple partial differential equations, and then use Maclaurin polynomials and the finite difference method to approximate solutions of the system of partial differential equations.

2 APPROXIMATION METHODS.

Here, we assume that the operator A generates a strongly continuous semigroup. We also assume that $b(\cdot) \in C^2([0,\infty), R)$ and $f(\cdot) \in C^1([0,\infty), X)$. We first transform Eq.(1.1)-(1.2) into Eq.(1.3)-(1.4).

Theorem 2.1. Eq.(1.1)-(1.2) has a unique (classical) solution for every $u_0 \in D(A)$ (the domain of A); and Eq.(1.1)-(1.2) is equivalent to Eq.(1.3)-(1.4).

Proof. The existence and the uniqueness are known (see e.g., [3]). To transform the equation, we define

$$R * H(t) = \int_0^t R(t-s)H(s)ds \text{ and } \delta * H = H.$$

Then we find a C^1 solution B of $B + b + B * b = 0$. (See, e.g., [2, 4, 5].) So that

$$(\delta + B) * (\delta + b) = \delta. \tag{2.1}$$

Now, write equation (1.1)-(1.2) as

$$u' = (\delta + b) * Au + f. \tag{2.2}$$

Then we have

$$(\delta + B) * u' = Au + (\delta + B) * f. \tag{2.3}$$

Hence

$$u' = Au + (\delta + B) * f - B * u'. \tag{2.4}$$

The result is true by observing that integration by parts yields

$$B * u'(t) = \int_0^t B'(t-s)u(s)ds + B(0)u(t) - B(t)u_0. \quad \square \tag{2.5}$$

For Eq.(1.3)-(1.4), we approximate the continuous scalar function $-B'(\cdot)$ by a polynomial $P_n(\cdot)$ of degree n, and then use Eq.(1.6)-(1.7) to approximate Eq.(1.3)-(1.4). To do this, we need the following inequality.

Lemma 2.1. [1] Let $u(t)$ be a nonnegative continuous function for $t \geq \alpha$, and suppose

$$u(t) \leq C + \int_\alpha^t k(t,s)u(s)ds + \int_\alpha^t \Big(\int_\alpha^s h(t,s,r)u(r)dr\Big)ds, \quad t \geq \alpha,$$

where $C \geq 0$ is a constant and $k(t,s)$ and $h(t,s,r)$ are nonnegative continuous functions for $\alpha \leq r \leq s \leq t$. If the functions $k(t,s)$ and $h(t,s,r)$ are nondecreasing in t for fixed s, r, then

$$u(t) \leq C \exp\Big\{\int_\alpha^t k(t,s)ds + \int_\alpha^t \Big(\int_\alpha^s h(t,s,r)dr\Big)ds\Big\}, \quad t \geq \alpha.$$

Now we can prove the following

Theorem 2.2. Let $T_0 > 0$ be fixed. For any $\varepsilon > 0$, there is a polynomial $P_n(\cdot)$ of degree n such that for the solution $v_n(\cdot)$ of Eq.(1.6)-(1.7) and the solution $u(\cdot)$ of Eq.(1.3)-(1.4) (or Eq.(1.1)-(1.2)), one has

$$\max_{t\in[0,T_0]} \|u(t) - v_n(t)\| \leq \varepsilon. \tag{2.6}$$

Proof. Note that $B(0)I$ is a bounded linear operator, then from the semigroup theory ([7]), $A - B(0)I$ generates a strongly continuous semigroup, which we denote by $T(\cdot)$. Also from the semigroup theory, there are $M > 1$ and $a > 0$ such that $\|T(t)\| \leq Me^{at}$, $t \geq 0$. We also know that the solutions $u(\cdot)$ and $v_n(\cdot)$ satisfy

$$u(t) = T(t)u_0 + \int_0^t T(t-s)\Big[\int_0^s -B'(s-r)u(r)dr + \hat{f}(s)\Big]ds, \tag{2.7}$$

$$v_n(t) = T(t)u_0 + \int_0^t T(t-s)\Big[\int_0^s P_n(s-r)v_n(r)dr + \hat{f}(s)\Big]ds. \tag{2.8}$$

Using Lemma 2.1, we can verify that there is a constant C independent of n such that $\|v_n(t)\| < C$ for $n = 1, 2, \ldots$ and $t \in [0, T_0]$. (See the treatment of $u(t) - v_n(t)$ below for details.) Therefore,

$$\begin{aligned}
\|u(t) - v_n(t)\| &\leq \Big\|\int_0^t T(t-s)\Big[\int_0^s \big(-B'(s-r)u(r) - P_n(s-r)v_n(r)\big)dr\Big]ds\Big\| \\
&\leq \Big\|\int_0^t T(t-s)\Big[\int_0^s \Big\{-B'(s-r)[u(r) - v_n(r)] \\
&\quad +[-B'(s-r) - P_n(s-r)]v_n(r)\Big\}dr\Big]ds\Big\| \qquad (2.9) \\
&\leq \int_0^t\int_0^s Me^{a(t-s)}|-B'(s-r) - P_n(s-r)|\|v_n(r)\|drds \\
&\quad + \int_0^t\int_0^s Me^{a(t-s)}|B'(s-r)|\|u(r) - v_n(r)\|drds \qquad (2.10) \\
&\leq Me^{aT_0}T_0^2C \max_{s\in[0,T_0]} |-B'(s) - P_n(s)| \\
&\quad + \int_0^t\int_0^s Me^{a(t-s)}|B'(s-r)|\|u(r) - v_n(r)\|drds. \qquad (2.11)
\end{aligned}$$

Now, we can apply Lemma 2.1 with $k(t,s) = 0$ and $h(t,s,r) = Me^{a(t-s)}|B'(s-r)|$ to obtain

$$\|u(t) - v_n(t)\| \leq$$
$$\Big(\max_{s\in[0,T_0]} |B'(s) + P_n(s)|\Big)Me^{aT_0}T_0^2C \exp\Big\{\int_0^t\int_0^s Me^{a(t-s)}|B'(s-r)|drds\Big\}.$$

Therefore we can find P_n to approximate $-B'$ to complete the proof. □

For this polynomial $P_n(\cdot)$ of degree n, the $(n+1)$th derivative is zero. So we have

Theorem 2.3. Eq.(1.6)-(1.7) is equivalent to an ordinary differential equation

$$Y'(t) = GY(t) + F(t), \quad t \geq 0, \tag{2.12}$$

$$Y(0) = Y_0, \tag{2.13}$$

on X^{n+2}, where

$$G = \begin{bmatrix} A - B(0)I & I & 0 & . & . & . & 0 & 0 \\ P_n(0)I & 0 & I & . & . & . & 0 & 0 \\ P_n'(0)I & 0 & 0 & . & . & . & 0 & 0 \\ . & . & . & . & . & . & 0 & 0 \\ . & . & . & . & . & . & 0 & 0 \\ . & . & . & . & . & . & I & 0 \\ P_n^{(n-1)}(0)I & 0 & 0 & . & . & . & 0 & I \\ P_n^{(n)}(0)I & 0 & 0 & . & . & . & 0 & 0 \end{bmatrix}, \quad F(t) = \begin{bmatrix} \hat{f}(t) \\ 0 \\ 0 \\ . \\ . \\ . \\ 0 \\ 0 \end{bmatrix}, \quad Y_0 = \begin{bmatrix} u_0 \\ 0 \\ 0 \\ . \\ . \\ . \\ 0 \\ 0 \end{bmatrix}. \tag{2.14}$$

And G generates a strongly continuous semigroup on X^{n+2}. That is, Eq.(2.12)-(2.13) is wellposed on X^{n+2}.

Proof. In Eq.(1.6), define

$$y_1(t) = v_n(t), \tag{2.15}$$

$$y_2(t) = \int_0^t P_n(t-s)v_n(s)ds = \int_0^t P_n(t-s)y_1(s)ds. \tag{2.16}$$

Then

$$y_1'(t) = [A - B(0)I]y_1(t) + y_2(t) + \hat{f}(t), \tag{2.17}$$

$$y_2'(t) = P_n(0)y_1(t) + \int_0^t P_n'(t-s)y_1(s)ds. \tag{2.18}$$

Next, define

$$y_3(t) = \int_0^t P_n'(t-s)y_1(s)ds, \tag{2.19}$$

to get

$$y_3'(t) = P_n'(0)y_1(t) + \int_0^t P_n''(t-s)y_1(s)ds. \tag{2.20}$$

Continue in this way to obtain

$$y_k(t) = \int_0^t P_n^{(k-2)}(t-s)y_1(s)ds, \tag{2.21}$$

$$y_k'(t) = P_n^{(k-2)}(0)y_1(t) + \int_0^t P_n^{(k-1)}(t-s)y_1(s)ds. \tag{2.22}$$

Finally we have

$$y_{n+2}(t) = \int_0^t P_n^{(n)}(t-s)y_1(s)ds, \tag{2.23}$$

$$y_{n+2}'(t) = P_n^{(n)}(0)y_1(t). \tag{2.24}$$

Thus, Eq.(1.6)-(1.7) is equivalent to Eq.(2.12)-(2.13). Next, G generates a strongly continuous semigroup on X^{n+2} by the perturbation results of the semigroup theory. □

Now, we see that by applying Theorems 2.1 – 2.3, the solutions of Eq.(1.1)-(1.2) can be approximated by the first component of the solutions of Eq.(2.12)-(2.13).

3 WHEN $X = L^2[0,1]$ AND $A = \frac{\partial^2}{\partial x^2}$.

Let us apply the above procedures to the following equation in viscoelasticity,

$$u_t(t,x) = u_{xx}(t,x) + \int_0^t b(t-s)u_{xx}(s,x)ds + f(t,x), \quad t \geq 0, \ x \in [0,1], \tag{3.1}$$

$$u(t,0) = u(t,1) = 0, \ t \geq 0, \quad u(0,x) = g(x), \ x \in [0,1]. \tag{3.2}$$

If we let $X = L^2[0,1]$, $A = \frac{\partial^2}{\partial x^2}$, $D(A) = W_0^{1,2}[0,1] \cap W^{2,2}[0,1]$, then Eq.(3.1)-(3.2) is represented by Eq.(1.1)-(1.2). From Theorem 2.1, Eq.(3.1)-(3.2) can be replaced by Eq.(1.3)-(1.4). That is, Eq.(3.1)-(3.2) is equivalent to an integrodifferential equation with a continuous kernel. Thus, without loss of generality, and also for computational reasons, we only consider the following equation with a continuous kernel $E(\cdot)$,

$$u_t(t,x) = u_{xx}(t,x) + \int_0^t E(t-s)u(s,x)ds + f(t,x), \quad t \geq 0, \ x \in [0,1], \tag{3.3}$$

$$u(t,0) = u(t,1) = 0, \ t \geq 0, \quad u(0,x) = g(x), \ x \in [0,1]. \tag{3.4}$$

Now, applying Theorems 2.2 - 2.3 to Eq.(3.3)-(3.4), (that is, now let $B(0) = 0$, $-B' = E$, $\hat{f} = f$ in Eq.(1.3)-(1.4)), we see that Eq.(3.3)-(3.4) can be approximated by Eq.(2.12)-(2.13) (with $B(0) = 0$, $\hat{f} = f$, and $P_n(\cdot)$ the approximation of $E(\cdot)$). Therefore, for $Y(t) = (y_1(t,\cdot), y_2(t,\cdot), ..., y_{n+2}(t,\cdot))$ in X^{n+2}, the first component y_1 of the solution of

$$\begin{cases} \frac{\partial}{\partial t}y_1(t,x) &= \frac{\partial^2}{\partial x^2}y_1(t,x) + y_2(t,x) + f(t,x), \\ \frac{\partial}{\partial t}y_2(t,x) &= P_n(0)y_1(t,x) + y_3(t,x), \\ \frac{\partial}{\partial t}y_3(t,x) &= P_n'(0)y_1(t,x) + y_4(t,x), \\ \quad \cdot & \\ \quad \cdot & \\ \quad \cdot & \\ \frac{\partial}{\partial t}y_{n+1}(t,x) &= P_n^{(n-1)}(0)y_1(t,x) + y_{n+2}(t,x), \\ \frac{\partial}{\partial t}y_{n+2}(t,x) &= P_n^{(n)}(0)y_1(t,x), \end{cases} \tag{3.5}$$

with the corresponding initial and boundary conditions can be used to approximate the solution of Eq.(3.3)-(3.4).

Example. Let $E(t) = \sin t$ in (3.3) and consider

$$u_t(t,x) = u_{xx}(t,x) + \int_0^t \sin(t-s)u(s,x)ds, \quad t \geq 0, \ x \in [0,1], \tag{3.6}$$

$$u(t,0) = u(t,1) = 0, \ t \geq 0, \quad u(0,x) = x(1-x), \ x \in [0,1]. \tag{3.7}$$

We use $P_n(t) = \sum_{i=0}^{(n-1)/2}(-1)^i \frac{t^{2i+1}}{(2i+1)!}$ (when n is odd) for the approximation of $\sin t$.

So that now Eq.(3.5) becomes

$$\begin{cases} \frac{\partial}{\partial t}y_1(t,x) &= \frac{\partial^2}{\partial x^2}y_1(t,x) + y_2(t,x), \\ \frac{\partial}{\partial t}y_2(t,x) &= y_3(t,x), \\ \frac{\partial}{\partial t}y_3(t,x) &= y_1(t,x) + y_4(t,x), \\ \quad\cdot \\ \quad\cdot \\ \quad\cdot \\ \frac{\partial}{\partial t}y_{n+1}(t,x) &= y_{n+2}(t,x), \\ \frac{\partial}{\partial t}y_{n+2}(t,x) &= (-1)^{(n-1)/2}y_1(t,x), \end{cases} \tag{3.8}$$

with the initial conditions

$$y_1(0,x) = x(1-x), \quad y_j(0,x) = 0, \quad j = 2,3,...,n+2, \tag{3.9}$$

and the boundary conditions $y_j(t,0) = y_j(t,1) = 0$, $j = 1,...,n+2$.

We solved this equation numerically using Forward Euler in t and centered differences in x. The stability conditions restricted testing n and letting $t \to \infty$. We, therefore, also used the modified Picard method presented in [6]. Since (3.5) is a polynomial system in the y_i's, the method presented in [6] is directly applicable if $\sin t$ is approximated by a Maclaurin polynomial. A polynomial system is easily integrated and produces polynomials in t. These polynomials are easily evaluated for any t. In [6] the authors show that iterating on the system using Picard's method of successive approximations produces higher order Maclaurin polynomial approximations to the solution of the system. Applying the method to (3.8) gives

$$\begin{cases} y_{1,k}(t_{j+1},x_i) &= y_{1,k}(t_j,x_i) + \int_{t_j}^{t_{j+1}} \Big[\frac{y_{1,k-1}(s,x_{i+1}) - 2y_{1,k-1}(s,x_i) + y_{1,k-1}(s,x_{i-1})}{(\Delta x)^2} \\ &\quad + y_{2,k-1}(s,x_i)\Big]ds, \\ y_{2,k}(t_{j+1},x_i) &= y_{2,k}(t_j,x_i) + \int_{t_j}^{t_{j+1}} y_{3,k-1}(s,x_i)ds, \\ \quad\cdot \\ \quad\cdot \\ \quad\cdot \\ y_{n+1,k}(t_{j+1},x_i) &= y_{n+1,k}(t_j,x_i) + \int_{t_j}^{t_{j+1}} y_{n+2,k-1}(s,x_i)ds, \\ y_{n+2,k}(t_{j+1},x_i) &= y_{n+2,k}(t_j,x_i) + \int_{t_j}^{t_{j+1}} (-1)^{(n-1)/2}y_{1,k-1}(s,x_i)ds. \end{cases}$$

The Picard iterate is designated by k and is easily iterated on. This increases the degree of the Maclaurin polynomial and the accuracy of the solution.

In the figure below we present a solution to (3.8) using the above Picard iteration with $n = 5$, $x_i = i\Delta x$, $i = 1,...,20$, $\Delta x = 0.05$, $t_{j+1} - t_j = \Delta t = 0.00125$ and let $t \to 5$.

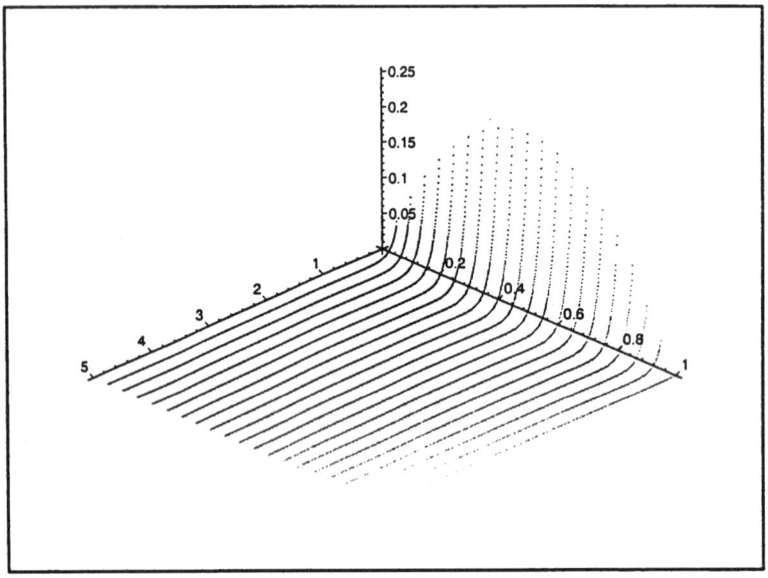

References

[1] D. Bainov and P. Simeonov, *Integral inequalities and applications*, Kluwer Academic Publishers, Boston, 1992.

[2] W. Desch, R. Grimmer, *Propagation of singularities for integrodifferential equations*, J. Diff. Eq., **65**(1986), 411-426.

[3] W. Desch, R. Grimmer and W. Schappacher, *Some considerations for linear integrodifferential equations*, J. Math. Anal. & Appl., **104**(1984), 219-234.

[4] G. Gripenberg, S-O. Londen and O. Staffans, *Volterra integral and functional equations*, Cambridge University Press, Cambridge, 1990, 12-13.

[5] R. MacCamy, *An integro - differential equation with application in heat flow*, Q. Appl. Math., **35**(1977), 1-19.

[6] E. Parker and J. Sochacki, *Implementing the Picard iteration*, Neural, Parallel & Scientific Computations, **4**(1996), 97-112.

[7] A. Pazy, *Semigroups of linear operators and applications to partial differential equations*, Springer - Verlag, New York, 1983.

Proceedings of Dynamic Systems and Applications 3 (2001) 391-398

MULTI-TIME SCALES IN SINGULARLY PERTURBED FORWARD EQUATIONS FOR CONTINUOUS-TIME MARKOV CHAINS

Q. G. LIU[1], G. YIN[1], and Q. ZHANG[2]

[1]Department of Mathematics, Wayne State University, Detroit, MI 48202.
[2]Department of Mathematics, University of Georgia, Athens, GA 30602.

Abstract

Stemming from the application of a cost-minimizing manufacturing firm involving production planning and capacity expansion and a two-level hierarchical decomposition, this work develops asymptotic properties of solutions of forward equations for singularly perturbed Markov chains with two small parameters. We obtain asymptotic properties of the solution of the forward equation, and present an example to illustrate the asymptotic theorem.

AMS (MOS) subject classifications. 34D15, 35B25, 60J27

1. INTRODUCTION

Recently, much effort has been shown in studying singularly perturbed Markov chains in continuous time; see [1, 13, 14, 15, 16] among others. Much of the current interest stems from the motivation of various applications in manufacturing systems and production planning. In many applications, one frequently needs to deal with large-scale systems. For example, one may wish to find an optimal control of a manufacturing system involving a large number of states and subject to random break downs and repairs. Even for a seemingly not so complex system, the optimal control can be very difficult to obtain (see [16, Section 5.2]), not to mention the added difficulty of very large dimensional state space involved. In fact, not only is it an impossible task to obtain analytic solutions, but also the direct numerical approximation can be intractable, and the straight forward use of numerical methods may not be adequate. Naturally, one seeks to split the large job into small pieces with the hope that certain decompositions and aggregations will lead to a simplification of the formidable task–treating a large dimensional systems. This idea may be traced back to the work of Simon and Ando [17]. Dealing with manufacturing systems, it has been found that an approach known as hierarchical decomposition and *aggregation* of states is a feasible alternative. Such an idea has also been used in large-scale dynamical systems. A viable alternative calls for taking advantages of the high contrast rates of changes in the physical systems, in which some states vary an order of magnitude faster than the rest, and using singular perturbation approach as a tool to reduce the complexity of the underlying systems. To reduce the complexity, it is crucial to many important applications that we have a thorough understanding of the structures and asymptotic properties of the underlying Markov chains. In [8], Khasminskii, Yin and Zhang use matched asymptotic expansion to establish the convergence of a sequence of the probability distribution vectors. Singularly perturbed Markov chains with recurrent states, naturally divisible into a number of classes are then treated in [9]. This line of work has been continued in [18], in which we have further derived asymptotic expansions for Markov chains with the inclusion of transient states and absorbing states and asymptotic distributions for Markov chains having recurrent states (see also the related papers [4, 6, 7, 19, 20, 21, 22]).

This work continues our investigation on the asymptotics of singularly perturbed Markov chains. We examine a continuous-time model with three-time scales. The motivation stems from the production-marketing systems treated in Sethi and Zhang [16, Chapter 11]. The underlying problem is concerned with a cost-minimizing manufacturing firm involving production planning and capacity expansion and a two-level hierarchical decomposition. In the aforementioned reference, Sethi and Zhang have found the asymptotic optimal strategies by using the limit distribution of the Markov chains. Our main effort in this paper is devoted to the asymptotic properties of the solution of the forward equation satisfied by the probability vector. Using singular perturbation techniques, we derive the leading term and the next term in the asymptotic expansion. Quite different from the case with one small parameter that we have considered before, the construction of the asymptotic expansion is based upon solutions of partial differential equations and albegraic-differential equations. The problem is thus more difficult and complex and the results are useful in the subsequent studies on Markov decision processes and controlled Markovian systems involving three-time scales. In the current paper, in addition to obtaining the first a couple of terms in the expansion, we give an illustrative example to explain the utility of the theorem. In a subsequent work [12], we will obtain the asymptotic expansion up to order n.

The rest of the paper is organized as follows. Section 2 presents the formulation of the problem. Section 3 gives an account on the asymptotic properties of the solutions of the forward equation. The main results are presented. Section 4 then gives an example to illustrate the results. Section 5 concludes the paper with a few more remarks.

2. PROBLEM FORMULATION

Given $\varepsilon > 0$ and $\delta > 0$, let $\alpha^{\varepsilon,\delta}(t)$ be a finite state Markov chain with state space $\mathcal{M} = \{1, \ldots, m\}$ generated by $Q^{\varepsilon,\delta}(t)$. This work concentrates on weak and strong interaction model with the generator of the underlying Markov chain given by

$$Q^{\varepsilon,\delta}(t) = \frac{1}{\varepsilon}A(t) + \frac{1}{\delta}B(t) + \widehat{Q}(t),$$

where $\varepsilon > 0$ and $\delta > 0$ are two small parameters and $Q^{\varepsilon,\delta}(t) \in \mathbb{R}^m$.

Let the row vector

$$p^{\varepsilon,\delta}(t) = \Big(P(\alpha^{\varepsilon,\delta}(t) = 1), \ldots, P(\alpha^{\varepsilon,\delta}(t) = m)\Big) \in \mathbb{R}^{1\times m}$$

denote the probability distribution of the Markov chain at time t. Then $p^{\varepsilon,\delta}(\cdot)$ is a solution of the forward equation

$$\frac{dp^{\varepsilon,\delta}(t)}{dt} = p^{\varepsilon,\delta}(t)Q^{\varepsilon,\delta}(t), \quad p^{\varepsilon,\delta}(0) = p^0, \tag{1}$$

where $p_i^0 \geq 0$ for each i, $\sum_{i=1}^m p_i^0 = 1$, and $p^0 = (p_1^0, \ldots, p_m^0)$ with p_i^0 being the ith component of p^0.

Note that three-time scales are involved. This extends our previous consideration of two-time scale singularly perturbed Markov chains. The motivation of the current study and the selection of the model come the modeling of a profit-maximizing production-marketing firm that needs to decide over time on the rate of promotional expenditures that creates additional demand for the products and the rate of production to meet the demand. The two parameters ε and δ signify the order of magnitudes corresponding to the rates of variation of the capacity and demand processes; see [16].

In the subsequent study, motivated by the manufacturing models studied in [16, Chapter 11], we take $A(t)$ and $B(t)$ to be of the following form. For some positive integer $m = lm_0$,

$$A(t) = \begin{pmatrix} \tilde{Q}(t) & & \\ & \ddots & \\ & & \tilde{Q}(t) \end{pmatrix}, \tag{2}$$

where $\tilde{Q}(t) = (\tilde{q}_{ij}(t)) \in \mathbb{R}^{m_0 \times m_0}$ is a generator of a Markov chain with state space $\{a_1, \ldots, a_{m_0}\}$, and

$$B(t) = \begin{pmatrix} \overline{q}_{11}(t)I_{m_0} & \cdots & \overline{q}_{1l}(t)I_{m_0} \\ \vdots & & \vdots \\ \overline{q}_{l1}(t)I_{m_0} & \cdots & \overline{q}_{ll}(t)I_{m_0} \end{pmatrix}, \tag{3}$$

where $\overline{Q}(t) = (\overline{q}_{ij}(t)) \in \mathbb{R}^{l \times l}$ is a generator of a Markov chain with state space $\{1, \ldots, l\}$ and $I_{m_0} \in \mathbb{R}^{m_0 \times m_0}$ is an identity matrix. The slowly changing component is described by the generator

$$\hat{Q}(t) = \begin{pmatrix} \hat{q}_{11}(t)I_{m_0} & \cdots & \hat{q}_{1l}(t)I_{m_0} \\ \vdots & & \vdots \\ \hat{q}_{l1}(t)I_{m_0} & \cdots & \hat{q}_{ll}(t)I_{m_0} \end{pmatrix}. \tag{4}$$

Our effort in what follows is to develop asymptotic properties of the solutions of the forward equation with $A(t)$, $B(t)$, and $\hat{Q}(t)$ given above.

3. ASYMPTOTIC RESULTS

We begin this section by stating a lemma.

Lemma 1. *The solution $p^{\varepsilon,\delta}(t)$ of (1) satisfies the condition $0 \le p_i^{\varepsilon,\delta}(t) \le 1$ and $\sum_{i=1}^{m} p_i^{\varepsilon,\delta}(t) = 1$.*

In view of this lemma the solutions of (1) are bounded uniformly since they are probabilities. In what follows, we use $\mathbb{1}_j = (1, \ldots, 1)' \in \mathbb{R}^{j \times 1}$ to denote a column j vector with all components being one.

To proceed, we make the following assumptions:

(A1) Given $0 \le T < \infty$, for each $t \in [0,T]$, $\widetilde{Q}(t)$ and $\overline{Q}(t) = (\overline{q}_{ij}(t))$ are weakly irreducible (see [18]).

(A2) For some $n \ge 0$, $A(\cdot)$, $B(\cdot)$ and $\widehat{Q}(\cdot)$ are $(n+1)$-time continuously differentiable on [0,T]. In addition, $(d^{n+1}/dt^{n+1})A(\cdot)$, $(d^{n+1}/dt^{n+1})B(\cdot)$ and $(d^{n+1}/dt^{n+1})\widehat{Q}(\cdot)$ are Lipschitz on [0,T].

Define an operator $L^{\varepsilon,\delta}$ by

$$L^{\varepsilon,\delta} f = \frac{df}{dt} - f\left(\frac{1}{\varepsilon}A(t) + \frac{1}{\delta}B(t) + \widehat{Q}(t)\right) \tag{5}$$

for any smooth vector-valued function $f(\cdot)$. Then $L^{\varepsilon,\delta} f = 0$ if and only if f is a solution to the differential equation (1). We seek the asymptotic expansion of the form

$$p^{\varepsilon,\delta}(t) = \varphi^{\varepsilon,\delta,n}(t) + \psi^{\varepsilon,\delta,n}\left(\frac{t}{\varepsilon}, \frac{t}{\delta}\right) + e^{\varepsilon,\delta,n}(t).$$

where $e^{\varepsilon,\delta,n}(t)$ is the remainder,

$$\varphi^{\varepsilon,\delta,n}(t) = \sum_{i+j=0}^{n} \varepsilon^i \delta^j \varphi^{(i,j)}(t), \tag{6}$$

and

$$\psi^{\varepsilon,\delta,n}\left(\frac{t}{\varepsilon}, \frac{t}{\delta}\right) = \sum_{i+j=0}^{n} \varepsilon^i \delta^j \psi^{(i,j)}\left(\frac{t}{\varepsilon}, \frac{t}{\delta}\right), \tag{7}$$

with the function $\varphi^{(i,j)}(\cdot)$ and $\psi^{(i,j)}(\cdot,\cdot)$ to be determined via the direct construction.

Theorem 2. *Assume* (A1) *and* (A2). *For small parameters $\varepsilon > 0$ and $\delta > 0$, denote the unique solution of* (1) *by $p^{\varepsilon,\delta}(\cdot)$. Then for $0 \le i+j \le 1$, the functions $\varphi^{(i,j)}(\cdot)$ and $\psi^{(i,j)}(\cdot,\cdot)$ can be constructed such that:*

(a) *$\varphi^{(i,j)}(\cdot)$ is twice differentiable on $[0,T]$.*

(b) *For each i and j, there are $\hat{\kappa}_1 > 0$ and $\hat{\kappa}_2 > 0$ such that*

$$\left|\psi^{(i,j)}\left(\frac{t}{\varepsilon}, \frac{\varepsilon}{\delta}\right)\right| \le K_1 \exp\left(-\frac{\hat{\kappa}_1 t}{\varepsilon}\right) + K_2 \exp\left(-\frac{\hat{\kappa}_2 t}{\delta}\right).$$

(c) *Suppose there exist constants $h_1 > 0$ and $h_2 > 0$ such that $h_1\varepsilon \le \delta \le h_2\varepsilon$. Then the following estimate holds:*

$$\sup_{t\in[0,T]} \left|p^{\varepsilon,\delta}(t) - \varphi^{\varepsilon,\delta,1}(t) - \psi^{\varepsilon,\delta,1}\left(\frac{t}{\varepsilon}, \frac{t}{\delta}\right)\right| \le K(\varepsilon + \delta). \tag{8}$$

Remark. In the above result, we only obtain the leading term and the next term in the asymptotic expansion. Actually, an asymptotic series can be constructed. Regarding to the smoothness of the generators, if the generators are $(n+1)$-times continuously differentiable with Lipschitz continuous $(n+1)$st derivatives. Then an approximation series of order n can be constructed. Moreover, it can be established that the approximation error holds uniformly in t.

Remark. It should be emphasized that unlike previous result in this direction, where to construct the asymptotic expansions, one solves a number of algebraic-differential equations (ordinary differential equations), the construction of the expansion for the current model require the solutions of a number of partial differential equations in addition to the algebraic-differential systems. Nevertheless, the main tools that we use are still Fredholm alternative, and notion of orthogonality in an appropriate sense. As a consequence of Theorem 2, we obtain:

Corollary 3. *Suppose $Q^{\varepsilon,\delta}(\cdot)$ is continuously differentiable on $[0,T]$, which satisfies (A1), and $(d/dt)Q(\cdot)$ is Lipschitz on $[0,T]$. Then for all $t>0$, $\lim_{\varepsilon,\delta\to 0} p^{\varepsilon,\delta}(t) = \varphi^{(0,0)}(t)$.*

We present the construction of the leading term in of the expansion. Consider

$$L^{\varepsilon,\delta}\varphi^{\varepsilon,\delta,1}(t) = 0.$$

That is,

$$\begin{aligned}
&\frac{d\varphi^{(0,0)}(t)}{dt} + \varepsilon\frac{d\varphi^{(1,0)}(t)}{dt} + \delta\frac{d\varphi^{(0,1)}(t)}{dt} \\
&= \left(\varepsilon^{-1}\varphi^{(0,0)}(t)A(t) + \delta^{-1}\varphi^{(0,0)}(t)B(t) + \varphi^{(0,0)}(t)\widehat{Q}(t)\right) \\
&\quad + \left(\varphi^{(1,0)}(t)A(t) + \varepsilon\delta^{-1}\varphi^{(1,0)}(t)B(t) + \varepsilon\varphi^{(1,0)}(t)\widehat{Q}(t)\right) \\
&\quad + \left(\varepsilon^{-1}\delta\varphi^{(0,1)}(t)A(t) + \varphi^{(0,1)}(t)B(t) + \delta\varphi^{(0,1)}(t)\widehat{Q}(t)\right).
\end{aligned}$$

Using the proposed asymptotic expansion, equating the coefficients of the powers of $\varepsilon^i\delta^j$ with $i+j\le 1$,

$$\begin{aligned}
&\varepsilon^{-1}\delta^0 : 0 = \varphi^{(0,0)}(t)A(t), \\
&\varepsilon^0\delta^{-1} : 0 = \varphi^{(0,0)}(t)B(t), \\
&\varepsilon^0\delta^0 : \frac{d\varphi^{(0,0)}(t)}{dt} = \varphi^{(1,0)}(t)A(t) + \varphi^{(0,1)}(t)B(t) + \varphi^{(0,0)}(t)\widehat{Q}(t), \\
&\varepsilon^1\delta^{-1} : 0 = \varphi^{(1,0)}(t)B(t), \\
&\varepsilon^{-1}\delta^1 : 0 = \varphi^{(0,1)}(t)A(t).
\end{aligned} \tag{9}$$

Based on the assumption, $\widetilde{Q}(t)$ and $\overline{Q}(t)$ are irreducible. Let $\pi(t)$ and $\lambda(t)$ denote their stationary distributions respectively. Then $\pi(t) = (\pi_1(t),\ldots,\pi_{m_0}(t))$ is the unique solution to the equations:

$$\pi(t)\widetilde{Q}(t) = 0, \text{ and } \sum_{i=1}^{m_0}\pi_i(t) = 1,$$

and $\lambda(t) = (\lambda_1(t),\ldots,\lambda_l(t))$ is the unique solution to the equations:

$$\lambda(t)\overline{Q}(t) = 0, \text{ and } \sum_{i=1}^{l}\lambda_i(t) = 1.$$

Denote

$$\pi_\lambda(t) = (\lambda_1(t)\pi(t), \ldots, \lambda_l(t)\pi(t)) \in \mathbb{R}^m.$$

It then can be shown that $\varphi^{(0,0)}(t) = \pi_\lambda(t)$.

4. AN EXAMPLE

Let $\alpha^{\varepsilon,\delta}$ be a Markov chain of 4 states with a constant generator $Q^{\varepsilon,\delta} = \frac{1}{\varepsilon}A + \frac{1}{\delta}B + \widehat{Q}$. The constant matrices A, B and $\widehat{Q}$ is defined as follows

$$A = \begin{pmatrix} -a & a & 0 & 0 \\ a & -a & 0 & 0 \\ 0 & 0 & -a & a \\ 0 & 0 & a & -a \end{pmatrix}, \quad B = \begin{pmatrix} -b & 0 & b & 0 \\ 0 & -b & 0 & b \\ b & 0 & -b & 0 \\ 0 & b & 0 & -b \end{pmatrix}, \quad \widehat{Q} = \begin{pmatrix} -c & 0 & c & 0 \\ 0 & -c & 0 & c \\ c & 0 & -c & 0 \\ 0 & c & 0 & -c \end{pmatrix},$$

where a, b and c are positive constants. Denote

$$\widetilde{Q} = \begin{pmatrix} -a & a \\ a & -a \end{pmatrix}, \quad \overline{Q} = \begin{pmatrix} -b & b \\ b & -b \end{pmatrix}, \quad Q^h = \begin{pmatrix} -c & c \\ c & -c \end{pmatrix}.$$

The solution to the forward equation (1) can be expressed as

$$\begin{aligned} p^{\varepsilon,\delta}(t) &= C_1 + C_2 \exp\left(-\frac{2at}{\varepsilon}\right) + C_3 \exp\left(-\frac{2bt}{\delta} - 2ct\right) \\ &\quad + C_4 \exp\left(-\frac{2at}{\varepsilon} - \frac{2bt}{\delta} - 2ct\right), \end{aligned}$$

where

$$C_1 = \frac{1}{4}(1,1,1,1), \; C_2 = -\frac{p_2^0 + p_4^0}{4}(1,-1,1,-1),$$
$$C_3 = -\frac{p_3^0 + p_4^0}{4}(1,1,-1,-1), \; C_4 = -\frac{p_1^0 + p_4^0}{4}(1,-1,-1,1).$$

It is easy to see that the stationary distribution of $\widetilde{Q}$ is $\pi = \left(\frac{1}{2}, \frac{1}{2}\right)$ and the stationary distribution of $\overline{Q}$ is $\lambda = \left(\frac{1}{2}, \frac{1}{2}\right)$, thus

$$\varphi^{(0,0)}(t) = \pi_\lambda = \left(\frac{1}{2}\lambda, \frac{1}{2}\lambda\right) = \left(\frac{1}{4}, \frac{1}{4}, \frac{1}{4}, \frac{1}{4}\right).$$

and

$$\varphi^{(1,0)}(t) = 0, \text{ and } \varphi^{(0,1)}(t) = 0.$$

The boundary layer terms are given by

$$\begin{aligned}\psi^{(0,0)}(\tau,\mu) &= \left(p^0-\varphi^{(0,0)}(0)\right)\exp(A\tau+B\mu),\\ \psi^{(1,0)}(\tau,\mu) &= 0,\\ \psi^{(0,1)}(\tau,\mu) &= \mu\left(p^0-\varphi^{(0,0)}(0)\right)\widehat{Q}\exp(A\tau+B\mu).\end{aligned}$$

Note that

$$\overline{P}_A=\begin{pmatrix}1\!\!1_2\pi & \\ & 1\!\!1_2\pi\end{pmatrix},\quad \overline{P}_B=\begin{pmatrix}\frac{1}{2}I_2 & \frac{1}{2}I_2\\ \frac{1}{2}I_2 & \frac{1}{2}I_2\end{pmatrix}.$$

Hence

$$\overline{P}_A\overline{P}_B=\frac{1}{4}1\!\!1_4 1\!\!1_4'=1\!\!1_4\pi_\lambda.$$

Now the limits $\lim_{\varepsilon,\delta\to 0}\psi^{(0,0)}(\tau,\mu)=0$ and $\lim_{\varepsilon,\delta\to 0}\psi^{(0,1)}(\tau,\mu)=0$ can be obtained by the fact that $\left(p^0-\varphi^{(0,0)}(0)\right)$ is orthogonal to $1\!\!1_4$ and $\widehat{Q}$ is orthogonal to $1\!\!1_4$. It is also easy to check that

$$\sup_{t\in[0,T]}\left|e^{\varepsilon,\delta,0}(t)\right|=O(\varepsilon+\delta),\quad\text{and}\quad \sup_{t\in[0,T]}\left|e^{\varepsilon,\delta,1}(t)\right|=O(\varepsilon+\delta).$$

5. FURTHER REMARKS

This paper has dealt with singularly perturbed Markov chains with three-time scales. The approach we are using is the matched asymptotic expansions. We have found the first a couple of terms in the asymptotic expansion. This approach may be adopted to obtain the full asymptotic expansion of the solutions of the forward equation. Due to the use of multi-timing parameters, the number of partial differential equations needed to be solved increases as the order of the expansion increases. For future study, one may proceed to the investigation on unscaled and scaled occupation measures as defined in [18]. A number of asymptotic properties such as asymptotic normality, exponential bounds will be of interest from both analysis and application point of views.

ACKNOWLEDGMENT

The research of G. Yin was supported in part by the National Science Foundation under grants DMS-9877090. The research of Q. Zhang was supported in part by the Office of Naval Research Grant N00014-96-1-0263.

References

[1] F. Delebecque and J. Quadrat, Optimal control for Markov chains admitting strong and weak interactions, *Automatica* **17** (1981), 281-296.

[2] S. N. Ethier and T. G. Kurtz, *Markov Processes: Characterization and Convergence*, J. Wiley, New York, 1986.

[3] A. Il'in, R.Z. Khasminskii, and G. Yin, Singularly perturbed switching diffusions: Rapid switchings and fast diffusions, *J. Optim. Theory Appl.* **102** (1999), 555-591.

[4] A. Il'in, R.Z. Khasminskii, and G. Yin, G. Asymptotic expansions of solutions of integro-differential equations for transition densities of singularly perturbed switching diffusions: Rapid switchings, *J. Math. Anal. Appl.* **238** (1999), 516-539.

[5] M. Iosifescu, *Finite Markov Processes and Their Applications*, Wiley, Chichester, 1980.

[6] R.Z. Khasminskii and G. Yin, Asymptotic series for singularly perturbed Kolmogorov-Fokker-Planck equations, *SIAM J. Appl. Math.* **56** (1996), 1766-1793.

[7] R.Z. Khasminskii and G. Yin, On transition densities of singularly perturbed diffusions with fast and slow components, *SIAM J. Appl. Math.* **56** (1996), 1794-1819.

[8] R. Z. Khasminskii, G. Yin and Q. Zhang, Asymptotic expansions of singularly perturbed systems involving rapidly fluctuating Markov chains, *SIAM J. Appl. Math.* **56** (1996), 277-293.

[9] R. Z. Khasminskii, G. Yin and Q. Zhang, Constructing asymptotic series for probability distribution of Markov chains with weak and strong interactions, *Quart. Appl. Math.* LV (1997), 177-200.

[10] H. J. Kushner, *Approximation and Weak Convergence Methods for Random Processes, with Applications to Stochastic Systems Theory*, MIT Press, Cambridge, MA, 1984.

[11] H. J. Kushner, *Weak Convergence Methods and Singularly Perturbed Stochastic Control and Filtering Problems*, Birkhäuser, Boston, 1990.

[12] Q.G. Liu, G. Yin, and Q. Zhang, Singularly perturbed Markov chains with two small parameters: A matched asymptotic expansion, preprint, 1999.

[13] Z. G. Pan and T. Başar, H^∞-control of Markovian jump linear systems and solutions to associated piecewise-deterministic differential games, in *New Trends in Dynamic Games and Applications*, G. J. Olsder (Ed.), 61-94, Birkhäuser, Boston, 1995.

[14] A. A. Pervozvanskii and V. G. Gaitsgori, *Theory of Suboptimal Decisions: Decomposition and Aggregation*, Kluwer, Dordrecht, 1988.

[15] R. G. Phillips and P. V. Kokotovic, A singular perturbation approach to modelling and control of Markov chains, *IEEE Trans. Automat. Control* **26** (1981), 1087-1094.

[16] S. P. Sethi and Q. Zhang, *Hierarchical Decision Making in Stochastic Manufacturing Systems*, Birkhäuser, Boston, 1994.

[17] H. A. Simon and A. Ando, Aggregation of variables in dynamic systems, *Econometrica* **29** (1961), 111-138.

[18] G. Yin and Q. Zhang, *Continuous-time Markov Chains and Applications: A Singular Perturbation Approach*, Springer-Verlag, New York, 1998.

[19] G. Yin and Q. Zhang, Singularly perturbed discrete-time Markov chains, to appear in *SIAM J. Appl. Math.*.

[20] G. Yin, Q. Zhang, and G. Badowski, Asymptotic properties of a singularly perturbed Markov chain with inclusion of transient states, to appear in *Ann. Appl. Probab.*

[21] G. Yin, Q. Zhang, and G. Badowski, Singularly perturbed Markov chains: Convergence and aggregation, *J. Mult. Anal.* **72** (2000), 208-229.

[22] Q. Zhang and G. Yin, Central limit theorems for singular perturbations of nonstationary finite state Markov chains, *Ann. Appl. Probab.* **6** (1996), 650-670.

Proceedings of Dynamic Systems and Applications 3 (2001) 399-408

Lyapunov method in the pole assignment problem

Maximov Y.M., Brusin A.V.
Nizhny Novgorod State Technical University

Abstract. A new approach to solving the problem of pole assignment by gain output feedback is presented. This approach consists of forming discrete procedure of transformation of the closed-loop spectrum by its small shifts on every step of this process. After this procedure either the desirable closed-loop spectrum is achieved (if spectrum controlability conditions are valid) or the Euclidean measure of deviation between desirable and current closed-loop spectra acquires minimum value.
Keywords: pole assignment problem, closed-loop systems, output feedback, spectrum evolution.

Introduction.

The problem of pole assignment by output feedback has been extensively studied in the last decades. However, it has not got exhaustive solution yet. H. Kimura [1] relates this fact with essentially nonlinear nature of this problem. Matrix algebra loses its power in face of nonlinear nature and various structural properties obtained in linear theory become irrelevant.

In this paper the nonlinear theory of pole assignment by output feedback is presented. The basic idea consists of creating the sequence of small shifts of closed-loop spectrum $\Lambda(i) = \{p_1(i), p_2(i), \dots, p_n(i)\}$ towards the desirable spectrum $\eta = (\eta_1, \eta_2, \dots, \eta_n)$. Thus, the closed-loop poles describe some trajectories on complex plane. Here open-loop spectrum $\Lambda(0)$ is an initial condition for these trajectories. The spectrum evolution is described by the nonlinear equations. Using Lyapunov-type function permits to design output feedback matrix so that either the desirable closed-loop spectrum (if spectrum controllability conditions are valid) is achieved or the

Euclidean measure of deviation between desirable and current spectra acquires minimum value.

1. Problem statement.

Let's consider a linear controllable and observable dynamic plant

$$\dot{x}(t) = Ax(t) + Bu(t), \quad y(t) = Cx(t), \tag{1.1}$$

where $x(t) \in R^n$ is the state vector, $u(t) \in R^m$ is the control vector, $y(t) \in R^l$ is the vector of the observed output; A, B and C are system matrices of appropriate dimensions. Modal control of the plant (1.1) is built by iteration method leading to generation of sequence of linear systems (LS). The process of generating LS is described within the following scheme

number of step	Control	Equation of LS
1.	$u_1(t) \equiv u(t) = \varepsilon_0 G(0)y(t) + u_2(t)$	$\dot{x}(t) = A(1)x(t) + Bu_2(t),\ y(t) = Cx(t)$ $A(1) = A(0) + \varepsilon_0 BG(0)C$
2.	$u_2(t) = \varepsilon_1 G(1)y(t) + u_3(t)$	$\dot{x}(t) = A(2)x(t) + Bu_3(t),\ y(t) = Cx(t)$ $A(2) = A(1) + \varepsilon_1 BG(1)C$
....	....	....

ets. Here, ε_{i-1} and $G(i-1)$ are, accordingly, a small parameter and matrix of regulator at the i-th step of modal control, $u_{i+1}(t) \in R^m$ is the control vector of the i-th LS. The matrices of these LS satisfy the following recurrent equation

$$A(i+1) = A(i) + \varepsilon_i BG(i)C, \quad i = 0,1,2,\ldots$$

where $A(0) = A$. The spectra of this sequence of LS are interrelated in the following manner:

$$\Lambda(i+1) = \Lambda(i) + \delta\Lambda(i),$$

where $\Lambda(i) = \{p_1(i), p_2(i), \ldots, p_n(i)\}$ is the set of eigenvalues (spectrum) of the i-th LS, $\delta\Lambda(i) = \{\delta p_1(i), \delta p_2(i), \ldots, \delta p_n(i)\}$ is their small (because ε_i is small) perturbation and $\Lambda(0)$ is the spectrum of the plant (1.1). The procedure of the small shifts of spectrum (procedure of construction of new LS) is completed if the goal of modal control is achieved. Let the N-th step correspond to this moment, then $u_{i+1}(t) \equiv 0 \ \forall t$.

Thus, the solution of the problem of transformation of the spectrum $\Lambda(0)$ to desirable closed-loop spectrum η is connected with the solution of the two problems:

1. The problem of obtaining the equations describing the evolution of the closed-loop spectrum.
2. The problem of obtaining such a sequence of matrices $G(0), G(1),\ldots$ that $\lim_{i\to N}\Lambda(i) = \eta$ occurs.

Basing on the above stated, the controller solving the problem of spectrum transformation has the form: $u(t) = Ky(t)$ where $K = \sum_{i=0}^{N}\varepsilon_i G(i)$.

2. The recurrent equation for spectrum

Consider the $(i+1)$-th step of modal control. At this step the matrix $A(i)$ is perturbed by matrix $\delta A(i) = \varepsilon_i BG(i)C$, spectrum $\Lambda_n(i)$ is perturbed by $\delta\Lambda_n(i)$. The following theorem defines the value of the spectrum perturbation, as well as left and right eigenvectors perturbations.

Theorem 1. Let all the eigenvalues of matrix $A(i)$ be different. Then, there exists a small number $\tilde{\varepsilon}$ that the following formulas are valid if $\varepsilon_i \in (0,\tilde{\varepsilon})$:

$$
\begin{aligned}
&p_k(i+1) = p_k(i) + \varepsilon_i l_k^T(i)BG(i)Ce_k(i) + o(\varepsilon^2) \\
&e_k(i+1) = \left(I + \varepsilon_i E_k(i)BG(i)C\right)e_k(i) + o(\varepsilon^2) \\
&l_k^T(i+1) = l_k^T(i)\left(I + \varepsilon_i(i)BG(i)CE_k(i)\right) + o(\varepsilon^2)
\end{aligned}
\tag{2.1}
$$

where $\delta e_k(i)$ and $\delta l_k(i)$ are, accordingly, perturbations of the right $e_k(i)$ and left $l_k(i)$ eigenvectors of matrix $A(i)$ corresponding to the eigenvalue $p_k(i)$, the matrix $E_k(i) = \sum_{j=1, j\neq k}^{n}\left(p_k(i) - p_j(i)\right)^{-1} e_j(i) l_j^T(i) \in \mathrm{E}^n$, E^n denotes the n- dimensional linear complex space and T is the transposition sign.

Proof: See the Appendix.

The theorem 1 may be used for the formulating the following recurrent equations describing the evolution of closed-loop spectrum:

$$p_k(i+1) = p_k(i) + \varepsilon_i l_k^T(i)BG(i)Ce_k(i),$$
$$e_k(i+1) = \left(I + \varepsilon_i E_k(i)BG(i)C\right)e_k(i), \qquad (k = 1,\dots,n) \qquad (2.2)$$
$$l_k^T(i+1) = l_k^T(i)\left(I + \varepsilon_i(i)BG(i)CE_k(i)\right)$$

where $I \in R^{n\times n}$ is the unit matrix, $o(\varepsilon^2)$ denotes members of higher order of smallness. The set of eigenvalues and eigenvectors of matrix $A(0)$ are the initial conditions for the recurrent equations (2.2).

3. Design of the closed-loop matrices.

Let $\eta = (\eta_1, \eta_2, \dots, \eta_n)$ be desirable spectrum to which the spectrum $\Lambda(0)$ is necessary to transform. Without loss of generality let us assume that the desirable spectrum η has been taken such that complex-conjugate poles of the spectrum $\Lambda(0)$ correspond to complex-conjugate poles of spectrum η and real poles of the spectrum $\Lambda(0)$ match to real poles of the spectrum η. Let us introduce an Euclidean measure of the deviation between current and the desirable spectra:

$$V(i) = \sum_{k=1}^{n} (p_k(i) - \eta_k)(p_k(i) - \eta_k)^*, \qquad (3.1)$$

where $*$ denotes the complex-conjugate value. The closed-loop matrix $G(i)$ will be designed on every step i so that the function $V(i)$ satisfies the Lyapunov's theorem for discrete systems [2]. Let's calculate $\Delta V(i) = V(i+1) - V(i)$ on the strength of equation (2.2). Using this equation we obtain:

$$\Delta V(i) = \varepsilon\left(\sum_{k=1}^{n}\left[\Delta p_k(i)(p_k(i) - \eta_k)^* + \Delta p_k^*(i)(p_k(i) - \eta_k)\right]\right) + \varepsilon^2\left(\sum_{k=1}^{n}\Delta p_k(i)\Delta p_k^*(i)\right) + o(\varepsilon^2) \quad (3.2)$$

Let us introduce the vector $g(i) \in R^{ml}$ composed of elements g_{jk} of the matrix $G(i)$, i.e. $g(i) = col(g_{11}, \dots, g_{1m}, \dots g_{l1}, \dots g_{lm})$. Obviously, determination of vector-function $g(i)$ is equivalent to determination of matrix function $G(i)$ and vice versa. Using Kroneker's product properties it is easy to obtain the following relation:

$$\Delta p_k(i) = \left(l_k^T(i) \otimes e_k^T(i)\right)\left(B \otimes C^T\right)g(i), \qquad (3.3)$$

where $\otimes$ - stands for Kroneker product. Taking into account the relation (3.3) the expression (3.2) is reduced to the form

$$\Delta V(i) = \varepsilon_i\left(s^T(i) + s^{*T}(i)\right)g(i) + \varepsilon_i^2 g^T(i)r(i)r^T(i)g(i) + o(\varepsilon_i^2) \qquad (3.4)$$

where $s^T(i) = \left[\sum_{k=1}^{n} \left(l_k^T(i) \otimes e_k^T(i) \right) \left(p_k(i) - \eta_k \right)^* \right] \left(B \otimes C^T \right)$, $r^T(i) = \left(B \otimes C^T \right) \sum_{k=1}^{n} \left(l_k^T(i) \otimes e_k^T(i) \right)$

Since the complex-conjugate poles of the spectrum $\Lambda(0)$ must be transformed to the complex-conjugate poles of spectrum η and real poles of the spectrum $\Lambda(0)$ - to real poles of the spectrum η it is easy to conclude that the vectors $r(i)$ and $s^T(i)$ are the real vectors: $r(i) \in R^{ml}$, $s^T(i) \in R^{ml}$. Hence it follows

$$\Delta V(i) = 2\varepsilon_i s^T(i) g(i) + \varepsilon_i^2 g^T(i) R(i) g(i) + o(\varepsilon_i^2), \tag{3.5}$$

where matrix $R(i) = r(i) r^T(i) \in R^{ml \times ml}$. Let us take the vector-function $g(i)$ in two forms:

$$g_1(i) = -\gamma_i s(i), \tag{3.6}$$

$$g_2(i) = -\gamma_i \frac{s(i)}{\|s(i)\|^2}, \tag{3.7}$$

where $\gamma_i > 0$ is the amplification coefficient. Substituting the (3.6) and (3.7) into (3.5) the following are obtained accordingly

$$\Delta V_1(i) = -2\varepsilon_i \gamma_i \|s(i)\|^2 + \varepsilon_i^2 \gamma_i^2 s^T(i) R(i) s(i) + o(\varepsilon_i^2) \tag{3.8}$$

$$\Delta V_1(i) = -2\varepsilon_i \gamma_i + \varepsilon_i^2 \gamma_i^2 \frac{s^T(i) R(i) s(i)}{\|s(i)\|^2} + o(\varepsilon_i^2) \tag{3.9}$$

Determine conditions by which the sign of the $\Delta V(i)$ is defined by sign of the main member of the series (3.8) and (3.9). . Since the matrix $R(i)$ is real and symmetric the following unequation holds [3]

$$g^T(i) R(i) g(i) \le \hat{\lambda}(i) \|g(i)\|^2,$$

where $\hat{\lambda}(i) = \max_{k=1,ml} |\lambda_k(i)|$, λ_k, $k = 1, \ldots, ml$ are eigenvalues of the matrix $R(i)$. Then it is evident that if the unequation

$$\varepsilon_i \gamma_i \hat{\lambda}(i) << 2 \tag{3.10}$$

is valid, the members of the ε^2-order and higher order of smallness in (3.8) and (3.9) can be ignored.

Consider possible variants of choosing the parameters of ε_i and amplification coefficient γ_i so that the unequality (3.10) is performed. There exist two simple enough variants.

1. Let $\gamma_i = \gamma = const$. In this case the choice of parameter ε_i satisfying the condition (3.10) is connected with estimation of $\hat{\lambda}(i)$ at every step. However, this variant is uncomfortable for realization.

2. Let $\varepsilon_i = \varepsilon = const$. In this case the amplification coefficient must be selected so that the condition (3.10) is satisfied. This variant does not require the determination of $\hat{\lambda}(i)$ value at every step, if the law of variation of the γ_i is taken in

the form $\gamma_i = \frac{1}{i}$. It is easy to see that in this case there exists a number i^* so that the condition (3.10) will be obtained for all $i \geq i^*$.

4. On the spectrum controllability.

From (2.7) and (2.8) it follows that the spectrum η is attainable from the spectrum $\Lambda(0)$ under above modal control law if the vector $s(i) \neq 0 \; \forall i \in [0, N]$. If $\exists \hat{i} \in [0, N]$: $s(\hat{i}) = 0$, then the recurrent procedure is stopped. Thereat $V(\hat{i}) \neq 0$, and therefore $V(\hat{i}) \neq \eta$. So that current spectrum does not achieve the desirable spectrum. Since the function $V(i)$ is decreased monotonically then Euclidean measure of deviation attains on the step $\hat{i}$ its minimum.

After some calculations the function $s(i)$ may be presented in the form:

$$s(i) = \sum_{k=1}^{n} \left[\left(B^T l_k(i)\right) \otimes \left(Ce_k(i)\right) \right] \left(p_k(i) - \eta_k\right)^{\bullet}$$

In its turn this relation may be reduced to :

$$s(i) = \Psi(i) \delta p^*(i) \qquad (4.1)$$

where vector $\delta p(i) = col(p_1 - \eta_1, \ldots, p_n - \eta_n)$, matrix $\Psi(i) \in E^{ml \times n}$ is determined as

$$\Psi(i) = \begin{bmatrix} c_1^T Z_1 b_1 & c_1^T Z_2 b_1 & \cdots & c_1^T Z_n b_1 \\ c_2^T Z_1 b_1 & c_1^T Z_2 b_1 & \cdots & c_2^T Z_n b_1 \\ \vdots & \vdots & & \\ c_l^T Z_1 b_1 & c_l^T Z_2 b_1 & \cdots & c_l^T Z_n b_1 \\ \hdashline c_1^T Z_1 b_2 & c_1^T Z_2 b_2 & \cdots & c_1^T Z_n b_2 \\ \vdots & \vdots & & \vdots \\ c_l^T Z_1 b_2 & c_l^T Z_2 b_2 & \cdots & c_l^T Z_n b_2 \\ \hdashline \vdots & \vdots & & \vdots \\ \hdashline c_1^T Z_1 b_m & c_1^T Z_2 b_m & \cdots & c_1^T Z_n b_m \\ \vdots & \vdots & & \vdots \\ c_l^T Z_1 b_m & c_l^T Z_2 b_m & \cdots & c_l^T Z_n b_m \end{bmatrix}$$

The vector c_μ is composed of elements of the μ-th row of the matrix C, the vector b_ν is composed of elements of the ν-th column of the matrix B.

Using the relation (4.1) it is easy to prove the next theorem.

Theorem 2.

Let the plant (1.1) be closed by output feedback and the desirable spectrum η of the closed-loop system is given. Then, if

$$rank \Psi(i) = n \quad \forall i, \qquad (4.2)$$

then the spectrum η is attainable from the spectrum $\Lambda(0)$.

Proof: See the Appendix.

The condition (4.2) implies the well-known requirements of full-controllability and full-observability of the closed-loop system at all values $i \in [0, \hat{i}]$. Obviously if this requirement is not satisfied then such right $e_k(i)$ or left $l_k(i)$ eigenvectors of matrix $A(i)$ are found then $Ce_k(i) = 0$ or $l_k^T(i)B = 0$ are valid. In this case the corresponding row or column of matrix $A(i)$ becomes null-vector and $rank\Psi(i) < n$.

The condition (4.2) implies also the unequality

$$ml > n \qquad (4.3)$$

Otherwise it automatically follows $rank\Psi(i) \le ml < n$.

But the realization of (4.2) is not reduced to satisfaction of the above two requirements. For full controlled and full observed closed-loop system with $ml > n$ there exists some set of eigenvectors converting all minors of matrix $\Psi(i)$ with $rang = n$ to zero.

The next example shows that this set of eigenvalues is not empty. Let $n = 4$, $m = 3$, $l = 2$ and eigenvalues satisfy the following conditions: $c_1^T e_1(i) = 0$, $c_1^T e_2(i) = 0$, $c_2^T e_3(i) = 0$, $c_2^T e_4(i) = 0$, $l_3^T b_1(i) = 0$, $l_4^T b_1(i) = 0$, $l_3^T b_2(i) = 0$, $l_4^T b_2(i) = 0$, $l_1^T b_3(i) = 0$, $l_2^T b_3(i) = 0$. Then rewriting the matrix $\Psi(i)$ for this case we obtain $rank\,\Psi(i) = 3$

Conclusion.

We have considered the new approach to pole assignment problem solution by output feedback using the Lyapunov-type function. The following fundamental results are obtained:

1. the nonlinear recurrent equations for closed-loop spectrum evolution;
2. using the Lyapunov-type function the algorithms of modal control were designed. These algorithms provide for the minimum Euclidean measure of deviation between desirable and current closed-loop spectra;
3. the conditions of spectrum controlability.

Appendix.

Proof of theorem 1.

Consider the matrix $A(i)$ corresponding to the i-th step of the modal control. Let the solution of the eigenvalues problem be known

$$A(i)e_k(i) = p_k(i)e_k(i),\; l_k^T(i)A(i) = p_k(i)l_k^T(i) \qquad (A1)$$

If the matrix $A(i)$ is perturbed by the $\delta A(i) = \varepsilon_i U(i)$, where $U(i) = BG(i)C$, then this problem is represented as

$$\left(A(i)+\delta A(i)\right)\left(e_k(i)+\delta e_k(i)\right)=\left(p_k(i)+\delta p_k(i)\right)\left(e_k(i)+\delta e_k(i)\right) \qquad \text{(A2)}$$

Opening the brackets in (A2), taking into account (A1) and smallness of the value ε, and neglecting the terms of the second order we have

$$\left(A(i)-p_k(i)I\right)\delta\hat{e}_k(i)=-\left(\delta A(i)-\delta\hat{p}_k(i)\,I\right)e_k(i) \qquad \text{(A3)}$$

where $\delta\hat{p}_k(i)$ и $\delta\hat{e}_k(i)$ are, accordingly, the estimates of perturbations of the eigenvalue $p_k(i)$ and the eigenvector $e_k(i)$, which are proportional to the first order ε_i. Let's multiply (A3) by $l_k^T(i)$ on the left. With the help (A1) and taking into account that $e_k(i)$ and $l_k(i)$ always be determined in the same manner as $l_k^T(i)e_k(i)=1$, we obtain

$$\delta\hat{p}_k(i)=\varepsilon_i l_k^T(i)U(i)e_k(i) \qquad \text{(A4)}$$

The value $\delta p_k(i)=\delta\hat{p}_k(i)+o(\varepsilon_i)$ is continuously dependent upon ε_i. Thus, we can always determine $\tilde{\varepsilon}$, such that for $\varepsilon_i\in(0,\tilde{\varepsilon})$, the second order terms of ε_i may be neglected. Considering $\delta p_k(i)\approx\delta\hat{p}_k(i)$ we obtain the upper part of (2.1).

It is known [3] that for the matrix A and its resolvent $L(p)=(A-pI)^{-1}$ the following spectral decomposition takes place

$$A=\sum_{k=1}^{n}p_k Z_k\,, \qquad L(p)=\sum_{k=1}^{n}(p-p_k)^{-1}Z_k\,,$$

where $Z_k=e_k l_k^T$ is the spectral projector corresponding to the eigenvalue p_k. Introduce $L(p)$ as $L(p)=(p-p_k)^{-1}z_k+E_k(p)$, where $E_k(p)=\sum_{j=1,\,j\neq k}^{n}(p-p_j)^{-1}Z_j(i)$. It is obvious that $E_k(p)$ is analytical in the region of p_k. Taking into account that $Z_kZ_j=\delta_{kj}Z_k$ we have

$$E_k(p_k(i))(p_k(i)\cdot I-A(i))=I-Z_k(i) \qquad \text{(A5)}$$

Multiplying (A3) by $E_k(p_k(i))$ leftward and taking into account (A5) we obtain

$$(I-Z_k(i))\delta\hat{e}_k(i)=E_k(p_k(i))\left(U(i)-\varepsilon_i l_k^T(i)U(i)e_k(i)\right)e_k(i) \qquad \text{(A6)}$$

As the vector $\tilde{e}_k(i)=e_k(i)+\delta\hat{e}_k(i)$ is determined to the accuracy of nonzero multiplicative constant (which can be varied together with ε_i) we may find $\tilde{e}_k(i)$ such that $l_k^T(i)\tilde{e}_k(i) = l_k^T(i)(e_k(i) + \delta\hat{e}_k(i)) = 1$. By virtue of $l_k^T(i)e_k(i) = 1$, we have $l_k^T(i)\delta\hat{e}_k(i) = 0$. In this case

$$Z_k(i)\delta\hat{e}_k(i) = e_k(i)l_k^T(i)\delta\hat{e}_k(i) = 0 \qquad \text{(A7)}$$

Since $l_j^T(i)e_k(i) = 0$ under $j \neq k$ then

$$E_k(p_k(i))\delta\hat{p}_k(i)\cdot I\cdot e_k(i) = \sum_{j=1,\ j\neq k}^{n}\left(p_k^{\prime} - p_j^{\prime}\right)^{-1}\delta\hat{p}_k(i)e_j(i)l_j^T(i)e_k(i) = 0.$$

Taking this as well as (A7) into consideration we find $\delta\hat{e}_k(i)$.

The vector $\delta\hat{l}_k(i)$ may be found in the same way. □

Proof of theorem 2.

Let $rank\Psi(i) = n\,\forall i$. In this case null space $N(\Psi(i))$ of the matrix $\Psi(i)$ consists of null-vector $\delta p(i) = 0$ and the process of modal control is completed when the spectrum η is attained.

If $rank\,\Psi(i) < n$ then there exists set of non-zero $\delta p(i) \in N(\Psi(i))$. It means that for given current spectrum $\Lambda(i)$ there exists such a manifold $H(i)$ that if $\eta \in H(i)$ then $s(i) = 0$ is valid. Hence the spectrum η is not attainable from spectrum $\Lambda(0)$ by the above-mentioned method. □

References

1. H. Kimura Pole Assignment by Output Feedback: A Longstanding Open Problem. Proc. of the 33rd Conf. Of Decision and Control, Lake Buena Vista, pp. 2101-2105, 1994.
2. V.N Fomin. Control methods for the linear discrete plants. St. Peterburg: Len. University, 1985.(In Russia)
3. Lankaster P. Theory of matrices. Academic Press, New York – London, 1969.

Proceedings of Dynamic Systems and Applications 3 (2001) 409-416

Active and Passive Damping in a Beam

N. G. Medhin
Department of Mathematics and
Center for Theoretical Studies
of Physical Systems
Clark Atlanta University
Atlanta, GA 30314

M. Sambandham
Department of Mathematics
Morehouse College
Atlanta, GA 30314

ABSTRACT: The purpose of this article is to indicate certain mechanisms of passive and active damping in a beam. We consider a layered beam with viscoelastic layers and embedded piezo ceramic patches. The viscoelastic layers effect passive damping while active damping is effected through the smart material components.

1. INTRODUCTION

There has been a growing interest in the engineering and industrial community over several decades in the application of composite materials in structures such as beams, plates, and other complex structures [3], [6], [4]. In numerous designs viscoelastic materials are used to effect/enhance damping. Recently there has been a growing interest in the use of smart materials as actuators and sensors [1], [5].

Often, in the literature, there is a gap in the rigorous mathematical foundation of the damping devices mentioned above. In this article, we will discuss damping mechanisms based on a model whose dynamics is rigorously established [2]. We will also indicate how one can proceed to develop a mathematically sound numerical scheme. We will present a variational problem that one can use for the design of controls to effect/enhance suppression in the beam. Finally, we present the basis for passive damping through the viscoelastic layers by considering a flattened beam and solutions involving exponential families.

2. THE PHYSICAL PROBLEM

The physical problem that we consider deals with the vibration of a layered beam. The beam consists of an elastic core of thickness h. This layer is sandwiched between two viscoelastic layers of thickness Δ each. Smart material patches, piezo ceramic patches in this case, are then bonded to the viscoelastic layers symmetrically on the viscoelastic layers. Each component of this layered beam is configured to have a finite radius of curvature, R. Each layer has the same thickness which is negligible compared to the circumferential length of the layers. See figure below.

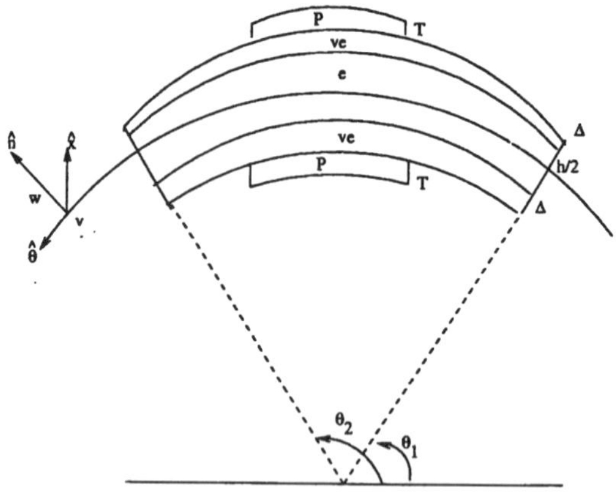

e: elastic core
P: piezo patch
ve: viscoelastic layer
T = thickness of piezo patch

The longitudinal and transverse vibrations are coupled due to the curvature. However, the transverse vibration is our primary concern. Any motion in the direction of the width of the beam is assumed to be negligible. Certain simplifying assumptions have to be made. For example, separation of layers, slippage of one layer past another are not allowed. Transverse normal stresses are negligible, and a cross section which is originally normal to the beam reference surface will remain normal to the deformed reference surface and will remain unstrained. We assume the deformation is small enough for a linear model.

A dynamical model which includes these basic simplifying assumptions as a starting point, given in [2], is

$$\bar{\rho}\begin{pmatrix}\ddot{v}\\ \ddot{w}\end{pmatrix} = -\hat{A}\begin{pmatrix}v\\ w\end{pmatrix} + \int_{-r}^{0} g(s)B\begin{pmatrix}v(t)-v_t(s)\\ w(t)-w_t(s)\end{pmatrix}ds + f, \tag{2.1}$$

where

$$\hat{A} = \begin{bmatrix}\partial\circ\xi_1^M\circ\partial + \alpha\partial\circ\beta_1^M\circ\partial & \partial\circ\xi_1^M + \alpha\partial^0\circ\beta_1^M\\ -\xi_1^M\circ\partial - \alpha\beta_1^M\circ\partial & -\xi_1^M + \partial^2\circ\xi_2^M\circ\partial^2 - \alpha\beta_1^M\\ & -\alpha\partial^2\circ\beta_2^M\circ\partial^2\end{bmatrix}, \tag{2.2}$$

$$B = \begin{bmatrix}\partial\circ\beta_1^M\circ\partial & \partial\circ\beta_1^M\\ -\beta_1^M\circ\partial & -\beta_1^M - \partial^2\circ\beta_2^M\circ\partial^2\end{bmatrix}, \tag{2.3}$$

$$\alpha = \int_{-r}^{0} g(s)ds, \tag{2.4}$$

and ξ_k^M,β_k^M denote multiplication operators by ξ_k and β_k respectively, $v_t(s) = v(t+s)$, $w_t(s) = w(t+s)$, $g > 0$, $g' \geq 0$, $g \in L_1([-r,0])$. We assume that $-(\xi_1 + \alpha\beta_1)$, and $\xi_2 - \alpha\beta_2$ are positive functions.

The functions ξ_k and β_k depend on physical parameters of the composite material such as Young's modulus, Poisson ratio and thicknesses of the various layers, radius of curvature and width of the beam. In (2.1) v represents the longitudinal vibration, w represents the transverse vibration, $\bar{\rho}$ represents the linear mass density of the composite beam. Finally, the input f represents external load, part of which is due to the piezo patches. The piezo patches can generate significant external forces and moments. For further details see [2].

We supplement (2.1) with the boundary conditions

$$v(t,\theta_1) = v(t,\theta_2) = w(t,\theta_1) = w(t,\theta_2) = 0$$

$$\partial_\theta w(t,\theta_1) = \partial_\theta w(t,\theta_2) = 0$$

Let

$$H = L_2(\theta_1,\theta_2) \times L_2(\theta_1,\theta_2), \qquad |\psi|_H^2 = \langle \bar{\rho}\psi_1, \psi_1\rangle + \langle \bar{\rho}\psi_2, \psi_2\rangle,$$

where $\langle \cdot,\cdot\rangle$ denotes L_2-inner product.

$$V = H_0^1(\theta_1,\theta_2) \times H_0^2(\theta_1,\theta_2),$$

$$|\varphi|_V^2 = -\langle(\xi_1 + \alpha\beta_1)\partial\varphi_1, \partial\varphi_1\rangle + \langle(\xi_2 - \alpha\beta_2)\partial^2\varphi_2, \partial^2\varphi_2\rangle$$

$$W = L_G^2(-r,0;V)$$

$$|\eta|_W^2 = \int_{-r}^0 g\left[\beta_1\langle\partial\eta_1,\partial\eta_1\rangle + \beta_2\langle\partial^2\eta_2,\partial^2\eta_2\rangle\right] ds$$

Finally, define $\hat{K} : W \to V^*$ so that for $\zeta \in V$,

$$\langle \hat{K}\eta, \zeta\rangle = -\langle \eta,\zeta\rangle_W - \int_{-r}^0 g(s)\beta_1\left[\langle \eta_2, \partial\zeta_1\rangle + \langle\partial\eta_1,\zeta_2\rangle + \langle\eta_2,\zeta_2\rangle\right] ds.$$

Define

$$D\eta = \frac{d}{ds}\eta, \qquad D : \text{dom } D \to W,$$

$$\text{dom } D = \{\eta \in H^1(-r,0;V) \mid \eta(0) = 0\}$$

Setting

$$\varphi = \begin{pmatrix} v \\ w \end{pmatrix} \in V, \qquad \psi = \begin{pmatrix} \dot{v} \\ \dot{w} \end{pmatrix} \in H, \qquad \bar{\gamma}(t) = \begin{pmatrix} v(t) - v_t(\cdot) \\ w(t) - w_t(\cdot) \end{pmatrix} \in W,$$

we obtain from (2.1)

$$M\begin{pmatrix} \dot{\varphi} \\ \dot{\psi} \\ \dot{\bar{\gamma}} \end{pmatrix} = \begin{pmatrix} 0 & I & 0 \\ -\hat{A} & 0 & \hat{K} \\ 0 & I & 0 \end{pmatrix}\begin{pmatrix} \varphi \\ \psi \\ \bar{\gamma} \end{pmatrix} + \begin{bmatrix} 0 \\ f \\ 0 \end{bmatrix} = A\begin{pmatrix} \varphi \\ \psi \\ \bar{\gamma} \end{pmatrix} + \begin{pmatrix} 0 \\ f \\ 0 \end{pmatrix}$$

$$M = \begin{pmatrix} I & 0 & 0 \\ 0 & \bar{\rho} I & 0 \\ 0 & 0 & I \end{pmatrix}, \qquad I = \text{identity } 2 \times 2\text{-matrix}$$

3. SEMI-DISCRETE GALERKAIN APPROXIMATION

Let

$$Z = V \times H \times L^2_G(-r, 0; V),$$

$$D^* = -g' g^{-1} I - D,$$

$$\text{dom } D^* = \{\eta \in H^1(-r, 0; V) \mid \eta(-r) = 0, -g' g^{-1}\eta \in W, D\eta \in W\},$$

Let

$$\Delta \hat{A} = \begin{bmatrix} \partial \circ (\xi_1 + \alpha\beta_1)^M \circ \partial & 0 \\ 0 & \partial^2 \circ (\xi_2 + \alpha\beta_2)^M \circ \partial^2 \end{bmatrix},$$

$$\Delta B = \begin{bmatrix} \partial \circ \beta_1^M \circ \partial & 0 \\ 0 & \partial^2 \circ \beta_2^M \circ \partial^2 \end{bmatrix}$$

We define $\Delta \hat{K} : W \to V^*$ by

$$\Delta \hat{K}(\eta) = \int_{-r}^{0} g(s) \Delta B \eta(s) ds$$

Then,

$$\mathcal{A}^* = \begin{pmatrix} 0 & -(\Delta \hat{A})^{-1} \hat{A} & 0 \\ \Delta \hat{A} & 0 & \Delta \hat{K} \\ 0 & (\Delta B)^{-1} B & D^* \end{pmatrix},$$

$$\text{dom } \mathcal{A}^* = \left\{ \begin{pmatrix} \varphi \\ \psi \\ \eta \end{pmatrix} : \psi \in V, \eta \in \text{dom } D^*, \Delta \hat{A}(\varphi) + \Delta \hat{K}(\eta) \in H \right\}$$

We note that if $\eta \in \text{dom } D^* \cap \text{dom } D$ then $\langle D^*\eta, \eta \rangle \leq 0$. Writing

$$\mathcal{A}^* = \mathcal{A}_1^* + Q,$$

we can verify that $\mathcal{A}_1^*$ is dissipative and

$$\|Qz\|_Z \leq \frac{c(h, \Delta, T)}{R^2} \|z\|_Z,$$

$$c(h, \Delta, T) \to 0 \quad \text{if } (h + \Delta + T) \to 0.$$

Letting

$$\mathcal{Y} = V \times V \times W \hookrightarrow V \times H \times W \to \mathcal{Y}^*$$

$$\mathcal{V} = \text{dom}(\mathcal{A}^*) \subset \mathcal{Y}$$

we extend $\mathcal{A}$ to $\tilde{\mathcal{A}} : Z \to \mathcal{V}^*$ by setting

$$(\tilde{\mathcal{A}}h)(v) = \langle h, \mathcal{A}^* v\rangle_Z, \qquad v \in \mathcal{V}.$$

We associate a continuous semigroup $\tilde{T}(t)$ to $\tilde{\mathcal{A}}$. This forms a basis to implement a semi-discrete Galerkin approximation for (2.1) using weak formulation.

In the next section we look at a related variational problem based on the weak formulation. One can use the variational problem to design controls, i.e., obtain information as to where to locate the piezo patches and the level of voltage input into them.

4. ASSOCIATED VARIATIONAL PROBLEM

Once we form a semi-discrete Galerkin approximation scheme we are led to consider the following variational problem

$$\begin{aligned} \varphi'(t) &= A\varphi(t) + \int_{-r}^{0} g(s)K\varphi(t+s)ds + Bu(t) + f(t), \\ \varphi(t) &= \tilde{\varphi}_0(t), \qquad -r \le t \le 0, \end{aligned} \tag{4.1}$$

where A, B and K are constant matrices, and $\tilde{\varphi}_0 \in C([-r,0])$. Here $C([-r,0])$ denotes the set of continuous functions on $[-r,0]$. In order to deal with the idea of suprressing the transverse vibration of the beam we consider the quadratic cost criterion

$$J(u) = \frac{1}{2}\int_0^{t_1} \left[\varphi^T(t)X(t)\varphi(t) + u^T(t)R(t)u(t)\right] dt, \tag{4.2}$$

where $X(t)$ is symmetric positive semidefinite, $R(t)$ is symmetric positive definite. We also assume that $X(t)$ and $R(t)$ are bounded and continuous functions for $t \ge 0$. The function g has the properties $g \in L_1(-\infty, 0)$, $g \ge 0$, $g' \ge 0$, $g(s) = 0$ for $s < -r$. We also extend f by setting $f(t) = 0$, $t \ge t_1$. Also we set $g(s) = 0$ for $s > 0$.

Let

$$\mathcal{U}_{ad} = \{u : [0, t_1 + r] \to R \mid u \text{ measurable}, \ u(t) = 0, \ t_1 \le t \le t_1 + r, \ \|u\|_\infty < \infty\},$$

$$H(t) = \begin{cases} 1 & t > 0 \\ 0 & t \le 0 \end{cases}$$

and x^T means transpose of x.

Theorem 4.1. *Let (φ_0, ψ) be the solution of the system*

$$\begin{aligned} \varphi_0'(t) &= A\varphi_0(t) + BR^{-1}B^t\psi^T(t) + \int_{t-r}^{t} g(s-t)K\varphi(s)ds + f(t) \\ \psi'(t) &= -\psi(t)A - \int_t^{t_1} g(t-\tau)H(t-\tau+r)\psi(\tau)K\, d\tau + \varphi_0^T(t)X \\ \varphi_0(t) &= \tilde{\varphi}_0, \qquad -r \le t \le 0 \\ \psi(t) &= 0, \qquad t_1 \le t \le t_1 + r \end{aligned}$$

Let

$$u_0(t) = R^{-1}B^T\psi^T(t).$$

Then,

$$J(u_0) \le J(u), \qquad u \in \mathcal{U}_{ad}.$$

Proof. Let φ_ϵ be the state corresponding to $u_0 + \epsilon u$ in (4.1). Let $\delta\varphi_0(t) = \lim_{\epsilon \to 0+} (\varphi_\epsilon(t) - \varphi_0(t))/\epsilon$. Then, use the fact that

$$\int_0^{t_1} \delta\varphi_0(t)X(t)\varphi_0(t)dt + \int_0^{t_1} u_0^T(t)R(t)u(t)dt = 0.$$

5. DAMPING EFFECT OF THE VISCOELASTIC LAYERS

To get insight into the passive damping mechanism of the viscoelastic layers we remove the pieze patches from our composite beam and consider a flat beam. Then, the transverse and longitudinal, horizontal in this case, decouple. The transverse vibration is governed by the equation

$$\begin{aligned} \bar{\rho}hb\frac{\partial^2 w}{\partial t^2} &+ b\left(\frac{h^3}{12}\beta_1 + \frac{2}{3}\mu_1\beta_2\right)\partial_y^4 w \\ &- \int_{-r}^{0} \frac{2}{3}\mu_1\beta_2 g(s)\partial_y^4 w(t+s)ds = f_2, \end{aligned} \tag{5.1}$$

$$\mu_1 = \left(\frac{h}{2} + \Delta\right)^3 - \left(\frac{h}{2}\right)^3,$$

$$b = \text{width of beam}.$$

The parameters β_1, β_2 are physical parameters involving Poisson ratio and Young's modulus of the elastic core and viscoelastic layers.

If we look for solutions of the homogeneous equation in (5.1) (i.e., set $f_2 = 0$) in the form $e^{\lambda t}\varphi_n(x)$, $\varphi_n'' = -\ell_n\varphi_n$, $\ell_n > 0$ we are led to consider the transcendental equation

$$\lambda^2 + \beta - \int_{-r}^{0} g(s)e^{\lambda s}ds = 0, \tag{5.2}$$

where $\beta > 0$, $g > 0$, $g' \ge 0$, $g \in L_1(-r, 0)$. To investigate the roots of the equation in (5.2) we follow our colleague, E. Wilkins.

Letting $t = -s/r$, $h(t) = r^3 g(s)$, $\mu = \lambda r$, $\gamma = \beta r^2$ we consider the roots of the equation

$$\varphi(z) = z^2 + \gamma - \int_0^1 h(t)e^{-zt}dt = 0, \tag{5.3}$$

where $h \in L_1(0,1)$, $h > 0$, on $(0,1)$, $h' \le 0$ a.e. on $(0,1)$.

We assume that $\gamma > \int_0^1 h(t)dt$. This is consistent with assumptions made earlier following (2.1)–(2.4). One can verify that $\lim_{t\to 0+} th(t) = 0$. Now, if $z = x + iy$ is a root of the equation $\varphi(z) = 0$ we integrate by parts and note that

$$x\left[2y^2 + \int_0^1 h(t)e^{-xt}(1-\cos yt)dt\right]$$
$$= -h(1)e^{-x}(1-\cos y) + \int_0^1 h'(t)e^{-xt}(1-\cos t)dt.$$

Since $h > 0$, $h' \le 0$ on $(0,1)$, $y \ne 0$ implies $x \le 0$. Further, $x = 0$ iff $h' = 0$ a.e. on $(0,1)$, i.e., $h(t) = h(1)$ on $(0,1)$, and if $\cos y = 1$, i.e., $y = 2n\pi$, n a positive integer. Then, $\gamma = 4n^2\pi^2 > 0$. Thus, we see that, if $\varphi(z) = 0$ then Re $z < 0$ unless h is constant and $\gamma = 4n^2\pi^2$ for some nonzero integer n, in which case $z = \pm 2n\pi i$. In the case $y = 0$, the roots must be negative as a result of the assumption $\gamma > \int_0^1 h(t)dt$.

REFERENCES

1. B. Azvine, G. R. Tomlinson and R. J. Wynne, Initial studies in the use of active constrained layer damping for controlled resonant vibration, *Proc. SPIE Conf. on Smart Structures and Materials*, Orlando, 1994; *Jour. of Smart Material and Structures*, submitted.

2. H. T. Banks, N. G. Medhin, Y. Zhang, A Mathematical Framework for Curved Active Constrained Layer Structures: Well-Posedness and Approximation, *Numerical Func. Anal. and Optim.*, Vol. 17, 1996, pp. 1–22.

3. D. J. Mead and S. Markus, The forced vibration of a three-layer, damped sandwiched beam with arbitrary boundary conditions, *Jour. of Sound and Vibration,* **10** (1969), pp. 163–175.

4. D. K. Rao, Frequency and loss factors of sandwich beams under various boundary conditions, *Jour. of Mechanical Engineering Science,* **20** (1978), pp. 271–282.

5. J. A. Rongong, J. R. Wright, G. R. Tomlinson and R. J. Wynne, Modeling of hybrid constrained layer/piezoceramic approach to active damping, 1995, preprint.

6. M.-J. Yan and E. H. Dowell, Governing equations for vibration constrained layer damping sandwich plates and beams, *Journal of Applied Mechanics,* **39** (1972), pp. 1041–1047.

Proceedings of Dynamic Systems and Applications 3 (2001) 417-424

BOUNDEDNESS OF PERTURBED VOLTERRA DISCRETE EQUATIONS

RIGOBERTO MEDINA
Departamento de Ciencias Exactas, Universidad de Los Lagos
Casilla 933, Osorno, CHILE

ABSTRACT. Volterra difference equations with a linear part are considered. Results on the boundedness and stability of solutions of these equations based on growth properties of the resolvent matrix of a linear Volterra difference equation, are proven. The results obtained give good estimates and explicit radius of attraction for the solutions.

AMS subject classification: 39A11, 65Q05.

1. INTRODUCTION

Volterra difference equations (VDE) mainly arise in the modeling process of some real phenomena or by applying a numerical method to a Volterra integral equation. Actually much of their general quantitative and specially their qualitative theory remains to be developed. To be more precise there are some results on the existence of solutions of VDE and their stability [1-6], but there are only a few papers dealing with their asymptotic behavior. Stability and boundedness are among the most important properties of the solutions of Volterra difference equations. See Crisci et al [7]. For applications of the VDE in Combinatorics see Babai & Lengyel [8], in Epidemics see Lauwerier [9], and in Numerical Methods for Volterra integro-differential equations see Crisci et al [7].

Consider the following system of Volterra difference equations

$$x(n+1) = A(n)x(n) + \sum_{j=0}^{n} B(n,j)x(j), \quad x(0) = x_0, \tag{1}$$

and its perturbation

$$y(n+1) = A(n)y(n) + \sum_{j=0}^{n} B(n,j)y(j) + f(n, y(n)), \quad y(0) = y_0, \tag{2}$$

where $A(n)$, $B(n,j)$ are $q \times q$ matrix functions on $\mathbf{Z}^+$ and $\mathbf{Z}^+ \times \mathbf{Z}^+$, respectively, and $f(n,x)$ is a continuous $\mathbb{R}^q$-valued function defined on $\mathbf{Z}^+ \times \mathbb{R}^q$. In this paper, we will discuss the question under what conditions equation (2) preserves some properties of equation (1) related to the asymptotic behavior of the solutions. We will investigate the properties: boundedness, stability and uniform asymptotic stability. To be more precise, based on certain growth properties of the resolvent matrix of Eq.(2), we obtain some boundedness results for the solutions of Eq.(2) (See Theorems 1-2, Corollaries 1-3) and assuming that the perturbation term $f(n,x)$

satisfies $||f(n,x)|| \leq \sum_{i=1}^{p} \lambda_i(n)\omega_i(||x||)$, $p \in \mathbf{Z}^+$, with $\lambda_i : \mathbf{Z}^+ \longrightarrow [0,\infty)$ properly summable functions and $\omega_i : [0,\infty) \longrightarrow [0,\infty)$ suitable monotone functions, then we can prove that there exists a ball $B(0,\rho) \subseteq \mathbb{R}^q$ such that all solution $y(n)$ of Eq.(2), with initial conditions $y(n_0) \in B(0,\rho)$, is bounded. (See Theorems 3-4, and Corollary 4). Furthermore, in this case, the radius ρ can be explicitly computed (see Medina & Pinto [10]).

Finally, we want to point out that in the existing literature about the boundedness of VDE the corresponding bounds are not explicitly shown.

In the present paper, we continue the research initiated in [11-13] addressing our study to establish new boundedness and stability results for Volterra difference equations.

2. PRELIMINARIES

We need the following results for our discussion:
Consider a linear difference system of the form

$$y(n+1) = A(n)y(n) + \sum_{j=0}^{n} B(n,j)y(j) + g(n), \quad y(0) = y_0, \tag{3}$$

where $y(n) \in \mathbb{R}^q$, $A(n)$, $B(n,j)$ are $q \times q$ matrices defined on $\mathbf{Z}^+$ and $\mathbf{Z}^+ \times \mathbf{Z}^+$ respectively, and $g(n)$ is a vector function defined on $\mathbf{Z}^+$.

We define the resolvent matrix $R(n,m)$ of Eq.(3) as the unique solution of the matrix equation

$$R(n+1,m) = A(n)R(n,m) + \sum_{j=0}^{n} B(n,j)R(j,m), \quad n \geq m, \tag{4}$$

with $R(m,m) = I$ for $0 \leq m \leq n$.

Elaydi [14] proved that Eq. (3) has a unique solution $y(n)$ which can be expressed as

$$y(n) = y(n,0,y_0) = R(n,0)y_0 + \sum_{j=0}^{n-1} R(n,j+1)g(j). \tag{5}$$

In the case $A(n) = A$ constant matrix (with A nonsingular) and $B(n,j) = B(n-j)$, equation (4) reduces to

$$R(n+1) = AR(n) + \sum_{j=0}^{n} B(n-j)R(j). \tag{6}$$

The next Theorem of Bihari-type will be fundamental in the application of the comparison principle.

Consider the discrete inequality:

$$u(n) \leq d + \sum_{i=1}^{p} \{ \sum_{j=0}^{n-1} \lambda_i(j) \omega_i(u(j)) \} \tag{7}$$

under the following Condition (D) :

D_1) $d \in \mathbb{R}$ nonnegative, $u(n) \geq 0$ for all $n \geq 0$;

D_2) $0 \leq \lambda_i \in \ell_1(\mathbf{Z}^+)$; $i = 1, 2, \cdots, p$, and

D_3) the functions ω_i $(i = 1, 2, \cdots, p)$ are continuous and nondecreasing on $[0, \infty)$ and positive on $(0, \infty)$ such that ω_{i+1}/ω_i $(i = 1, 2, \cdots, p)$ are nondecreasing on $(0, \infty)$.

To solve inequality (7), we define:

I) the functions

$$W_k(u) = \int_{u_k}^{u} \frac{ds}{\omega_k(s)}, \quad u > 0, \ u_k > 0, \ k = 1, 2, \cdots, p \tag{8}$$

and W_k^{-1} represents their inverse functions.

II) the functions $\varphi_0(u) = u$ and

$$\varphi_k = \Psi_k \circ \Psi_{k-1} \circ \cdots \circ \Psi_1, \Psi_k(u) = W_k^{-1}[W_k(u) + \alpha_k], \tag{9}$$

where $\alpha_k = \sum_{n=0}^{\infty} \lambda_k(n)$.

We remark that any φ_i $(i = 1, 2, \cdots, p)$ is a continuous, positive and nondecreasing function on its domain.

Thus, we can establish the following theorem:

Theorem A[15, Medina & Pinto]. Under Condition (D), if $d < \varphi_p^{-1}(\infty)$ then for any $n \in \mathbf{Z}^+$, we have

$$u(n) \leq \varphi_p(d). \tag{10}$$

We remark that if

$$\int_1^{\infty} \frac{ds}{\omega_i(s)} = \infty \quad (i = 1, 2, \cdots, p), \tag{11}$$

then φ_k (and Ψ_k) is defined for all u. Consequently, (10) is valid for all $d > 0$. The dual condition to (11), namely

$$\int_{0+}^{1} \frac{ds}{\omega_i(s)} = \infty \quad (i = 1, 2, \cdots, p) \tag{12}$$

implies that φ_k (and Ψ_k) is defined for all u small enough. Thus (10) is valid if d is small enough.

On the other hand, if $\int_0^\infty \frac{ds}{\omega_i(s)} < \infty$, $i = 1, 2, \cdots, p$ then the inequality $\sum_{j=0}^{\infty} \lambda_i(j) \geq \int_0^\infty \frac{ds}{\omega_i(s)}$, for some $i = 1, 2, \cdots, p$ implies that there is not $d > 0$ satisfying Theorem A. In any case, the biggest $d \in \mathbb{R}$ satisfying the conditions of Theorem A is $d = \varphi_p^{-1}(\infty)$.

We want to point out that the concepts of stability, uniform stability, and uniform asymptotic stability for the Volterra difference equation (2), considered here, are analogues of the corresponding concepts in the continuous case. (See Crisci et al [1] and Elaydi [14]).

We introduce the following assumption:

Assumption A. For the linear Volterra discrete equation (1) we assume that there exist positive constants $\eta \geq 1$ and $\gamma > 0$ which satisfies the estimate

$$||R(i,j)|| \leq \eta\gamma^{i-j} \ , \ i \geq j \ , \tag{13}$$

where $R(i,j)$ is the resolvent matrix of Eq. (2).

3. MAIN RESULTS

We are now in a position to establish our main results dealing with the boundedness and stability of Eq. (2).

Theorem 1 Assume that

i) $||f(n,x)|| \leq F(n, ||x||)$, for all $(n,x) \in \mathbf{Z}^+ \times \mathbb{R}^q$, where $F : \mathbf{Z}^+ \times \mathbb{R} \longrightarrow \mathbb{R}^+$ is a monotone nondecreasing continuous function with respect to the second variable for each fixed $n \in \mathbf{Z}^+$. Furthermore, suppose that one of the following conditions are satisfied:

ii) Assumption A holds for a constant $\gamma > 1$ and $\alpha F(n,u) \leq F(n, \alpha u)$, with $0 < \alpha < 1$, for all $(n,x) \in \mathbf{Z}^+ \times \mathbb{R}^+$, or

iii) Assumption A holds for a constant γ, $0 < \gamma < 1$ and $\beta F(n,u) \leq F(n, \beta u)$, $\beta > 1$ for all $(n,u) \in \mathbf{Z}^+ \times \mathbb{R}^+$, or

iv) Assumption A holds for a constant $\gamma = 1$.

Then, all solution $y(n)$ of Eq. (2) such that $||y(n_0)|| < \frac{\gamma^{n_0}}{\eta} z(n_0)$ satisfies $||y(n)|| < \gamma^n z(n)$, for all $n \in \mathbf{Z}^+$, where $z(n)$ is a solution of the difference equation $z(n) = z(n_0) + \sum_{j=n_0}^{n-1} \frac{\eta}{\gamma} F(j, z(j))$, $z(n_0) = z_0$.

Proof. Let condition (ii) holds. From (5) the solution $y(n)$ of Eq. (2) satisfies $y(n) = R(n,n_0)y_0 + \sum_{j=n_0}^{n-1} R(n,j+1)f(j,y(j)),\ \ n \in \mathbf{Z}^+$.

Thus, from condition (i) it follows that

$$\begin{aligned}||y(n)|| &\le ||R(n,n_0)||\ ||y(n_0)|| + \sum_{j=n_0}^{n-1} ||R(n,j+1)||\ ||f(j,y(j))|| \\ &\le \eta\gamma^{n-n_0}||y(n_0)|| + \sum_{j=n_0}^{n-1} \eta\gamma^{n-j-1}\cdot F(j,||y(j)||).\end{aligned}$$

Hence

$$||y(n)||\gamma^{-n} \le \eta||y(n_0)||\gamma^{-n_0} + \sum_{j=n_0}^{n-1} \frac{\eta}{\gamma} F(j,||y(j)||\gamma^{-j}).$$

By the assumption $\eta||y(n_0)|| < z(n_0)\gamma^{n_0}$, we infer that

$$||y(n)||\gamma^{-n} - \sum_{j=n_0}^{n-1} \frac{\eta}{\gamma} F(j,||y(j)||\gamma^{-j}) < z(n) - \sum_{j=n_0}^{n-1} \frac{\eta}{\gamma} F(j,z(j)),$$

from which, it follows that $||y(n)|| < z(n)\gamma^n, \qquad n \in \mathbf{Z}^+$, because $||y(n_0)||\gamma^{-n_0} < z(n_0)$ and $F(r,u)$ is a monotone nondecreasing continuous function with respect to u for each $r \in \mathbf{Z}^+$.

In the case that (iii) or (iv) holds, the proof is carried out similarly, which completes the proof of the theorem. ■

The next corollary of Theorem 1 shows an explicit bound for the solutions of Eq. (2).

Corollary 1. Assume that the perturbation term of Eq.(2) satisfies

v) $||f(n,x)|| \le (g(n)+L)||x|| + h(n), g(n)+L \ge 0,\ \ h(n) \ge 0$ for all $n \in \mathbf{Z}^+$ and $x \in \mathbb{R}^q$.

If Assumption A is satisfied for all $n \in \mathbf{Z}^+$, then every solution $y(n) = y(n;n_0,y(n_0))$ of Eq.(2) such that $||y(n_0)|| < \frac{\gamma^{n_0}}{\eta}z(n_0)$ satisfies

$$\begin{aligned}||y(n)|| \le & \eta||y(n_0)|| \prod_{j=n_0}^{n-1} \{\gamma + \eta(g(j)+L)\} + \\ & + \sum_{j=n_0}^{n-1} \eta h(j) \prod_{s=j+1}^{n-1} \{\gamma+\eta(g(s)+L)\}, \text{ for all } \ n \in \mathbf{Z}^+.\end{aligned}$$

As a consequence of Corollary 1 we get:

Corollary 2. Under the assumptions of Corollary 1, suppose that the Assumption A is satisfied for γ, $0 < \gamma < 1$ and all $n \in \mathbf{Z}^+$. If $0 < \gamma + \eta(g(n) + L) < 1$, for all $n \in \mathbf{Z}^+$ then every solution $y(n) = y(n, n_0, y_0)$ of Eq. (2) such that $||y(n_0)|| < \frac{\gamma^{n_0}}{\eta} z(n_0)$ satisfies $||y(n)|| \leq \eta||y(n_0)|| + \sum_{j=n_0}^{n-1} \eta h(j)$.

If in Corollary 1 $h(n) = 0$ for all $n \in \mathbf{Z}^+$, then we obtain the next corollary:

Corollary 3. Assume that the perturbation term of Eq.(2) satisfies

vi) $||f(n, x)|| \leq (g(n) + L)||x||$, for all $n \in \mathbf{Z}^+$ and $x \in \mathbb{R}^q$.

If Assumption A is satisfied, for all $n \in \mathbf{Z}^+$ then, every solution $y(n) = y(n, n_0, y(n_0))$ of Eq.(2) such that $||y(n_0)|| < \frac{\gamma^{n_0}}{\eta} z(n_0)$ satisfies

$$||y(n)|| \leq \eta||y(n_0)|| \prod_{j=n_0}^{n-1} \{\gamma + \eta(g(j) + L)\}, \quad n \in \mathbf{Z}^+.$$

Now, we will discuss the question under what conditions equation (2) preserves some stability properties of equation (1).

Thus, in the next theorem, we will give a uniform asymptotic stability theorem to the zero solution of equation (2).

Theorem 2. Suppose condition (vi) holds for the perturbed Eq. (2) and let Assumption A be satisfied for all $n \in \mathbf{Z}^+$. Suppose that $f(n, 0) = 0$ for all $n \in \mathbf{Z}^+$. If $\limsup_{n\to\infty} \eta g(n) < 1 - (\gamma + \eta L)$, then the zero solution of Eq. (2) is uniformly asymptotically stable, with respect to n_0.

Proof. To prove this theorem we may simply mimic the proof of Taniguchi's Theorem 2 [16] for difference equations, and therefore we will omit it. ■

Theorem 3. Assume that

a) $||f(n, x)|| \leq \sum_{i=1}^{p} \lambda_i(n)\omega_i(||x||)$, $p \in \mathbf{Z}^+$, where ω_i $(i = 1, 2, \cdots, p)$ satisfy Condition (D.3) and they are "r_i-multiplicative", that is, for each $i = 1, 2, \cdots, p$ there exists a nonnegative function r_i defined on $(0, \infty)$ such that

$$\omega_i(\alpha u) \leq r_i(\alpha)\omega_i(u), u \geq 0, \quad \alpha > 0 \tag{14}$$

b) $\sum_{j=0}^{\infty} \lambda_i(j) r_i(j) < \infty$, $i = 1, 2, \cdots, p$.

c) Assumption A holds.

Then, every solution $y(n)$ of Eq. (2) such that $||y(n_0)|| < \frac{\gamma^{n_0}}{\eta} \cdot \varphi_p^{-1}(\infty)$, satisfies

$$||y(n)|| \leq \gamma^n \cdot \varphi_p(\eta||y(n_0)||\gamma^{-n_0}), \text{ for all } n \in \mathbf{Z}^+. \tag{15}$$

Proof. From (5), the solution $y(n)$ of Eq. (2) satisfies $y(n) = R(n,n_0)y_0 + \sum_{j=n_0}^{n-1} R(n,j+1)f(j,y(j)), \quad n \in \mathbf{Z}^+$.

Thus, we obtain that from conditions (a) and (c), $||y(n)|| \leq \eta\gamma^{n-n_0} \cdot ||y_0|| + \sum_{i=1}^{p}\sum_{j=n_0}^{n-1} \frac{\eta}{\gamma}\lambda_i(j)r_i(\gamma^j)\omega_i(||y(j)||\gamma^{-j})$.

Introducing the abbreviation $v(j) = \gamma^{-j}||y(j)||$, we then get $v(n) \leq \eta v(n_0) + \sum_{i=1}^{p}\sum_{j=n_0}^{n-1} \frac{\eta}{\gamma}\lambda_i(j)r_i(\gamma^j)\omega_i(v(j))$.

If $\eta||v(n_0)|| < \varphi_p^{-1}(\infty)$ then, by Conditions (a) and (b), Theorem A can be applied and the result follows at once. ∎

Corollary 4.

A) If (11) holds, then $\varphi_p^{-1}(\infty) = \infty$ and the results of Theorem 3 are valid for all the solutions of Eq. (2).

B) If (12) holds, then the results of Theorem 3 are valid only for solutions of Eq. (2) with initial conditions small enough, namely $\eta||y_0|| < \gamma^{n_0} \cdot \varphi_p^{-1}(\infty)$.

Proof. To prove this Corollary it is enough to apply remarks (11) and (12), respectively. ∎

Theorem 4. Suppose conditions (a) and (b) of Theorem 3 hold, and the functions $\omega_i \quad (i = 1, 2, \cdots, p)$ satisfy (12). Furthermore, assume that the zero solution of Eq.(1) is asymptotically stable. If $f(n,0) = 0$ for all $n \in \mathbf{Z}^+$ and the initial conditions $y(n_0)$ are small enough, then the zero solution of Eq.(2) is asymptotically stable.

Proof. To prove this theorem is sufficient to apply (15) and Corollary 4-(B). ∎

ACKNOWLEDGEMENT

This research was partially supported by Fondap de Matemáticas Aplicadas and Fondecyt under Grant N° 1970427.

REFERENCES

[1] M.R. Crisci, Z. Jackiewicz, E. Russo and A. Vecchio, Stability analysis of discrete recurrence equations of Volterra type degenerate kernel, J. Math. Anal. 162, 49-62 (1991).

[2] S.N. Elaydi and V.L. Kocic, Global stability of nonlinear Volterra difference system, Differential Equations and Dynamical Systems 2(4), 337-345 (1994).

[3] S. Zhang, Stability of infinite delay difference systems, Nonlinear Anal., Theory, Methods & Applications 22, 1121-1129 (1994).

[4] M. Murakami, Representation of solutions of linear functional difference equations in phase space, Nonlinear Anal., Theory, Methods & Applications 30, 1153-1164 (1997).

[5] R.P. Agarwal and P.Y. Pang, On a generalized difference system, Nonlinear Anal., Theory, Methods & Applications 30(1), 365-376 (1997).

[6] X. Fu and L. Zhang, Criteria on boundedness in terms of two measures for Volterra type discrete systems, Nonlinear Anal., Theory, Methods & Applications 30(5), 2673-2681 (1997).

[7] M.R. Crisci, V.B. Kolmanovskii, E. Russo and A. Vecchio, Boundedness of discrete Volterra equations, J. Math. Anal., Appl. 211, 106-130 (1997).

[8] L. Babai and T. Lengyel, A convergence for recurrent sequences with applications to the partition lattice, Analysis 12, 109-119 (1992).

[9] H. Lauwerier, Mathematical Models of Epidemcs, Math. Centrum, Amsterdam (1981).

[10] R. Medina and M. Pinto, Asymptotic behavior of solutions of second order nonlinear difference equations, Nonlinear Anal., Theory, Methods & Appl. 19(2) 187-195 (1992).

[11] R. Medina, The asymptotic behavior of the solutions of a Volterra difference equation, Computers Math. Applic. 34, 19-26 (1997).

[12] R. Medina, Difference equations of the advanced type, Nonlinear Studies 3, 193-201 (1996).

[13] R. Medina, Solvability of discrete Volterra equations in weighted spaces, Dynamic Systems and Appl. 5, 407-422 (1996).

[14] S.N. Elaydi, Periodicity and stability of linear Volterra difference systems, J. Math. Anal. Appl. 181, 483-492 (1994).

[15] R. Medina and M. Pinto, Nonlinear discrete inequalities and stability of difference equations, World Sci. Ser. Appl. Anal. 3, 467-482 (1994).

[16] T. Taniguchi, On the estimate of solutions of perturbed linear difference equations, J. Math. Anal. Appl. 149, 599-610 (1990).

[17] M. Zouyousefain, Difference equations of Volterra type and extension of Lyapunov's method, J. Applied Mathematics and Stochastic Anal. 3(3), 193-202 (1990).

Proceedings of Dynamic Systems and Applications 3 (2001) 425-431

POSITIVE SOLUTIONS FOR FREDHOLM MULTIVALUED EQUATIONS

Maria Meehan[1] and Donal O'Regan[2]

[1]Department Of Mathematics, National University of Ireland, Dublin, Ireland
[2]Department Of Mathematics, National University of Ireland, Galway, Ireland.

ABSTRACT: We discuss the existence of positive solutions to the integral inclusion $y(t) \in \int_0^1 k(t,s)\,F(s,y(s))\,ds$ for $t \in [0,1]$ where $k : [0,1] \times [0,1] \to \mathbf{R}$ and $F : [0,1] \times \mathbf{R} \to 2^{\mathbf{R}}$. Our analysis relies on a multivalued version of Krasnoselskii's fixed point theorem in a cone.

AMS(MOS) Subject Classification. 45D05, 45G10

1. INTRODUCTION

In [1] Agarwal and O'Regan studied the existence of nonnegative solutions to the Fredholm integral inclusion

$$y(t) \in \int_0^1 k(t,s)\,F(s,y(s))\,ds \quad \text{for } t \in [0,1]; \tag{1.1}$$

here $k : [0,1]\times[0,1] \to \mathbf{R}$ and $F : [0,1]\times\mathbf{R} \to 2^{\mathbf{R}}$ (note $2^{\mathbf{R}}$ denotes the family of nonempty subsets of $\mathbf{R}$). Using the multivalued version of Krasnoselskii's fixed point theorem in a cone (see [1, 7]) Agarwal and O'Regan were able in [1] to guarantee the existence of a nonnegative solution to (1.1) provided some strong conditions were assumed on the kernel k. In particular it was assumed that:

$$\begin{cases} \exists M,\ 0 < M < 1,\ \kappa \in C[0,1] \text{ and an interval } [a,b] \subseteq [0,1],\ a < b \\ \text{such that } k(t,s) \geq M\,\kappa(s) \text{ for } t \in [a,b] \text{ and a.e. } s \in [0,1], \\ \text{together with } k(t,s) \leq \kappa(s) \text{ for } t \in [0,1] \text{ and a.e. } s \in [0,1]. \end{cases} \tag{1.2}$$

Condition (1.2) was motivated from the Green's function which arises in Sturm Liouville, Hermite, focal etc. differential equations (single and multivalued [2, 3]). However for the general integral inclusion (1.1) condition (1.2) may be too restrictive. In this paper by strengthening the conditions on the nonlinearity F we are able to reduce the restrictions on the kernel k (in fact (1.2) is not assumed) and again guarantee the existence of a positive solution to (1.1). Some of the ideas in this paper were motivated by a recent paper of the authors [5] in the single valued case.

Finally in this introduction we state the multivalued analogue of Krasnoselskii's fixed point theorem in a cone (see [1, 7]). Let $E = (E, \|\cdot\|)$ be a Banach space and $C \subseteq E$ a cone. Let $\rho > 0$ and we let

$$\Omega_\rho = \{x \in E : \ \|x\| < \rho\} \quad \text{and} \quad \partial\Omega_\rho = \{x \in E : \ \|x\| = \rho\}.$$

Theorem 1.1. *Let $E = (E, \|.\|)$ be a Banach space, $C \subseteq E$ a cone and let $\|.\|$ be increasing with respect to C. Also r, R are constants with $0 < r < R$. Suppose $F : \overline{\Omega_R} \cap C \to K(C)$ (here $K(C)$ denotes the family of nonempty, convex, compact subsets of C) is an upper semicontinuous, compact map and assume one of the following conditions*

(A). $\|y\| \leq \|x\|$ for all $y \in F\,x$ and $x \in \partial\Omega_R \cap C$ and $\|y\| > \|x\|$ for all $y \in F\,x$ and $x \in \partial\Omega_r \cap C$

or

(B). $\|y\| > \|x\|$ for all $y \in F\,x$ and $x \in \partial\Omega_R \cap C$ and $\|y\| \leq \|x\|$ for all $y \in F\,x$ and $x \in \partial\Omega_r \cap C$

is true. Then F has a fixed point in $C \cap \left(\overline{\Omega_R} \setminus \Omega_r\right)$.

2. FREDHOLM INCLUSIONS

In this section we use Theorem 1.1 to show the existence of a positive solution to

$$y(t) \in \int_0^1 k(t,s)\, F(s, y(s))\, ds \quad \text{for } t \in [0,1] \tag{2.1}$$

where $k : [0,1] \times [0,1] \to \mathbf{R}$ and $F : [0,1] \times \mathbf{R} \to 2^{\mathbf{R}}$. The following conditions will be assumed throughout:

$$F : [0,1] \times \mathbf{R} \to K([0,\infty)) \tag{2.2}$$

$$t \mapsto F(t,x) \text{ is measurable for every } x \in \mathbf{R} \tag{2.3}$$

$$x \mapsto F(t,x) \text{ is upper semicontinuous for a.e. } t \in [0,1] \tag{2.4}$$

$$\begin{cases} \text{for each } b > 0 \text{ there exists a function } h_b \in L^1[0,1] \text{ such that} \\ |F(t,x)| \leq h_b(t) \text{ for a.e. } t \in [0,1] \text{ and every } x \in \mathbf{R} \text{ with } |x| \leq b \end{cases} \tag{2.5}$$

$$\begin{cases} \text{for each } t \in [0,1],\ k(t,s) \text{ is measurable on } [0,1] \text{ and} \\ k(t) = ess\ sup\,|k(t,s)|,\ 0 \leq s \leq 1, \text{ is bounded on } [0,1] \end{cases} \tag{2.6}$$

$$\text{the map } t \mapsto k_t \text{ is continuous from } [0,1] \text{ to } L^\infty[0,1]; \text{ here } k_t(s) = k(t,s) \tag{2.7}$$

$$k(t,s) \geq 0 \text{ for a.e. } (t,s) \in [0,1] \times [0,1] \tag{2.8}$$

$$\begin{cases} \text{there exists a } f : [0,\infty) \to [0,\infty) \text{ continuous and nondecreasing, a} \\ \text{constant } A_0,\ 0 < A_0 \leq 1, \text{ and a function } q \in L^1[0,1] \text{ with} \\ A_0\, q(s)\, f(y) \leq u \leq q(s)\, f(y) \text{ for any } u \in F(s,y) \text{ with } (s,y) \in [0,1] \times [0,\infty) \end{cases} \tag{2.9}$$

$$K_2 \equiv \inf_{t \in [0,1]} \int_0^1 q(s)\, k(t,s)\, ds > 0 \tag{2.10}$$

$$\begin{cases} \text{there exists a continuous function } \psi : (0,1) \to (0,\infty) \text{ such that} \\ \text{for any } 0 < m < 1 \text{ and } u \geq 0 \text{ we have } f(m u) \geq \psi(m)\, f(u) \end{cases} \tag{2.11}$$

$$\exists\, 0 < M < 1 \text{ with } \frac{M}{\psi(M)} \leq \frac{A_0\, K_2}{K_1} \text{ where } K_1 = \sup_{t \in [0,1]} \int_0^1 k(t,s)\, q(s)\, ds \tag{2.12}$$

$$\exists\, \alpha > 0 \text{ with } \alpha > K_1\, f(\alpha) \tag{2.13}$$

and

$$\exists\, \beta > 0,\ \beta \neq \alpha \text{ with } \beta < A_0\, K_2\, f(M\beta). \tag{2.14}$$

Theorem 2.1. *Suppose* (2.2) − −(2.14) *hold. Then* (2.1) *has at least one positive solution* $y \in C[0,1]$ *and either*

(A). $0 < \alpha < |y|_0 = \sup_{t\in[0,1]} |y(t)| < \beta$ *and* $y(t) \geq M\,\alpha,\ t \in [0,1]$ *if* $\alpha < \beta$

or

(B). $0 < \beta < |y|_0 < \alpha$ *and* $y(t) \geq M\,\beta,\ t \in [0,1]$ *if* $\beta < \alpha$

hold.

PROOF: Let

$$E = (C[0,1], |\,.\,|_0) \quad \text{and} \quad C = \{u \in C[0,1] :\ u(t) \geq M\,|u|_0 \ \text{ for } \ t \in [0,1]\}$$

where M is as defined in (2.12). Let $A : C \to 2^E$ be defined by

$$A\,x = \left\{ v \in C[0,1] :\ \exists\, w \in \mathcal{F}\,x \ \text{ such that } \ v(t) = \int_0^1 k(t,s)\,w(s)\,ds \ \text{ for all } \ t \in [0,1] \right\}$$

with

$$\mathcal{F}\,x = \{u \in L^1[0,1] :\ u(t) \in F(t, x(t)) \ \text{ for a.e. } \ t \in [0,1]\},$$

here $x \in C$.

Remark 2.1. Note A is well defined since if $x \in C$ then (2.2) − −(2.4) and [3] guarantee that $\mathcal{F}\,x \neq \emptyset$.

We first show $A : C \to 2^C$. To see this let $x \in C$ and $y \in A\,x$. Then $M\,|x|_0 \leq x(t)$ for $t \in [0,1]$ and there exists a $v \in \mathcal{F}\,x$ with

$$y(t) \ = \ \int_0^1 k(t,s)\,v(s)\,ds \quad \text{for } \ t \in [0,1].$$

In addition (2.9) guarantees that $v(s) \leq q(s)\,f(x(s))$ for $s \in [0,1]$ and so for $t \in [0,1]$ we have

$$|y(t)| \ \leq \ f(|x|_0) \int_0^1 k(t,s)\,q(s)\,ds \ \leq \ f(|x|_0)\,K_1.$$

Consequently

$$|y|_0 \leq K_1\,f(|x|_0). \tag{2.15}$$

On the other hand (2.9) guarantees that $v(s) \geq A_0\,q(s)\,f(x(s))$ for $s \in [0,1]$. Now $x \in C$ (so $x(t) \geq M\,|x|_0$ for $t \in [0,1]$) together with (2.11), (2.15) and (2.12) gives for $t \in [0,1]$,

$$\begin{aligned} y(t) \ &\geq \ \int_0^1 k(t,s)\,A_0\,q(s)\,f(x(s))\,ds \geq A_0\,f(M\,|x|_0) \int_0^1 k(t,s)\,q(s)\,ds \\ &\geq \ A_0\,\psi(M)\,f(|x|_0) \int_0^1 k(t,s)\,q(s)\,ds \geq A_0\,\psi(M)\,K_2\,f(|x|_0) \\ &\geq \ \frac{K_2}{K_1}\,A_0\,\psi(M)\,|y|_0 \ \geq \ M\,|y|_0, \end{aligned}$$

and so $y \in C$. Consequently $A : C \to 2^C$. Also notice (2.2) − −(2.7) and a standard result [6] guarantees that

$$A : C \to K(C) \ \text{ is upper semicontinuous and completely continuous.}$$

Let

$$\Omega_\alpha = \{u \in C[0,1] : \ |u|_0 < \alpha\} \quad \text{and} \quad \Omega_\beta = \{u \in C[0,1] : \ |u|_0 < \beta\}.$$

We wish to apply Theorem 1.1. We can do so if we show

$$|y|_0 < |x|_0 \quad \text{for all } y \in A\,x \text{ and } x \in C \cap \partial\Omega_\alpha \tag{2.16}$$

and

$$|y|_0 > |x|_0 \quad \text{for all } y \in A\,x \text{ and } x \in C \cap \partial\Omega_\beta \tag{2.17}$$

hold. Let $x \in C \cap \partial\Omega_\alpha$ (so $|x|_0 = \alpha$ and $M\,\alpha \le x(t) \le \alpha$ for $t \in [0,1]$) and $y \in A\,x$. Then there exists a $v \in \mathcal{F}\,x$ with

$$y(t) = \int_0^1 k(t,s)\,v(s)\,ds \quad \text{for } t \in [0,1].$$

In addition (2.9) guarantees that $v(s) \le q(s)\,f(x(s))$ for $s \in [0,1]$ and so for $t \in [0,1]$ we have

$$|y(t)| \le f(|x|_0) \int_0^1 k(t,s)\,q(s)\,ds \le f(\alpha) \int_0^1 k(t,s)\,q(s)\,ds \le f(\alpha)\,K_1$$

This together with (2.13) yields

$$|y|_0 \le f(\alpha)\,K_1 < \alpha = |x|_0, \tag{2.18}$$

so (2.16) is true. Next let $x \in C \cap \partial\Omega_\beta$ (so $|x|_0 = \beta$ and $M\,\beta \le x(t) \le \beta$ for $t \in [0,1]$) and $y \in A\,x$. Then there exists a $v \in \mathcal{F}\,x$ with

$$y(t) = \int_0^1 k(t,s)\,v(s)\,ds \quad \text{for } t \in [0,1].$$

In addition (2.9) guarantees that $v(s) \ge A_0\,q(s)\,f(x(s))$ for $s \in [0,1]$ and so for $t \in [0,1]$ we have

$$y(t) \ge A_0 \int_0^1 k(t,s)\,q(s)\,f(x(s))\,ds \ge A_0\,f(M\,\beta) \int_0^1 k(t,s)\,q(s)\,ds \ge A_0\,f(M\,\beta)\,K_2.$$

This inequality together with (2.14) yields

$$y(t) \ge A_0\,f(M\,\beta)\,K_2 > \beta = |x|_0 \quad \text{for } t \in [0,1],$$

and so $|y|_0 > |x|_0$ i.e. (2.17) is true.

Theorem 1.1 guarantees that there is a positive solution $x \in C[0,1]$ to (2.1) with $x \in C \cap \left(\overline{\Omega_\alpha} \setminus \Omega_\beta\right)$ if $\beta < \alpha$ whereas $x \in C \cap \left(\overline{\Omega_\beta} \setminus \Omega_\alpha\right)$ if $\alpha < \beta$. Finally note $|x|_0 \ne \alpha$ and $|x|_0 \ne \beta$. To see this suppose $|x|_0 = \alpha$. Then since $x \in A\,x$ we have (follow the ideas used to prove (2.18)),

$$\alpha = |x|_0 \le f(\alpha)\,K_1 < \alpha = |x|_0,$$

a contradiction. A similar argument shows $|x|_0 \ne \beta$. □

Remark 2.2. It is easy to use the results of Case (A) and Case (B) to obtain a multiplicity result for (2.1). We leave the details to the reader.

REFERENCES

[1]. R.P. Agarwal and D. O'Regan, A note on the existence of multiple fixed points for multivalued maps with applications, *Jour. Differential Equations*, to appear.

[2]. R.P. Agarwal, D. O'Regan and P.J.Y. Wong, Positive solutions of differential, difference and integral equations, *Kluwer*, Dordrecht, 1999.

[3]. K. Deimling, Multivalued differential equations, *Walter de Gruyter*, Berlin, 1992.

[4]. L.H. Erbe, S. Hu and H. Wang, Multiple positive solutions of some boundary value problems, *J. Math. Anal. Appl.*, **184**(1994), 640–648.

[5]. M. Meehan and D. O'Regan, Positive solutions of singular and nonsingular Fredholm integral equations, to appear.

[6]. D. O'Regan, Integral inclusions of upper semicontinuous or lower semicontinuous type, *Proc. Amer. Math. Soc.*, **124**(1996), 2391–2399.

[7]. W.V. Petryshyn, Existence of fixed points of positive k–set contractive maps as consequences of suitable boundary conditions, *J. London Math. Soc.*, **38**(1988), 503–512.

Proceedings of Dynamic Systems and Applications 3 (2001) 431-436

Spurious Limit-Cycles

Ronald E. Mickens
Physics Department, Clark Atlanta University, Atlanta, GA 30314

ABSTRACT: A number of important dynamic systems are modeled by second order ODE's having limit-cycle solutions. However, the application of averaging methods, such as that derived by Krylov-Bogoliubov-Mitropolsky, can lead to the prediction of spurious limit-cycles. These are periodic solutions that are not actual behaviors of the original systems. We investigate why this phenomena occurs and suggest possibilities for eliminating them. A modified van der Pol equation is used to study these issues.

AMS (MOS) subject classification: 34A45, 34C15, 34E10

1. INTRODUCTION

The second order ODE

$$\ddot{x} + x = \epsilon f(x, \dot{x}), \qquad 0 < \epsilon \ll 1, \tag{1}$$

where ϵ is a small positive parameter, provides a model of 1-dim oscillatory systems [3, 4, 5]. For many of these systems $f(x, \dot{x})$ takes the form

$$f(x, \dot{x}) = f_1(x) + f_2(x, \dot{x}^2)\dot{x} \tag{2}$$

where

$$f_1(-x) = -f_1(x), \qquad f_2(-x, \dot{x}^2) = f_2(x, \dot{x}^2). \tag{3}$$

In general, Eq. (1) has limit-cycle solutions. These are solutions that correspond to isolated closed curves [4, 5] in the (x, y) phase space, where $y = \dot{x}$. Note that the system equations for Eq. (1) are

$$\frac{dx}{dt} = y, \tag{4a}$$

$$\frac{dy}{dt} = -x + \epsilon f(x, y), \tag{4b}$$

while the equation determining the phase space trajectories is

$$\frac{dy}{dx} = \frac{-x + \epsilon f(x, y)}{y}. \tag{5}$$

Limit-cycles are characterized by an "amplitude" and "frequency," and, for small values of ϵ, these can be calculated using one of the standard perturbation procedures [4]: Lindstédt-Poincaré, multi-times, and averaging methods. Our purpose here is to highlight a difficulty that arises in using standard averaging methods to determine the number and properties of limit-cycles for Eq. (1). In particular, we examine these issues within the framework of the Krylov-Bogoliubov-Mitropolsky (KBM) procedure [1, 4, 5] for calculating analytic perturbation solutions for the limit-cycles of Eq. (1). The major problem is that standard KBM can predict the existence of "spurious" limit-cycles in higher orders of the calculation. Spurious limit-cycles do not actually exist, but are artifacts of the perturbation procedure. Our task is to examine this situation by considering the example of a modified van der Pol equation for which it is known that such spurious limit-cycles do not exist and show why they occur in the KBM method. Finally, we suggest several ways for changing the KBM method such that spurious limit-cycles may be eliminated.

The work reported here is preliminary and extends that given in the paper by Burnette and Mickens [2].

2. BACKGROUND

The averaging method of KBM represents the solution to Eq. (1) as [1, 4]

$$x(t,\epsilon) = a\cos\psi + \epsilon u_1(a,\psi) + \epsilon^2 u_2(a,\psi) + \cdots, \tag{6}$$

where the amplitude, $a(t,\epsilon)$, and phase, $\psi(t,\epsilon)$, functions are determined by the equations

$$\frac{da}{dt} = \epsilon A_1(a) + \epsilon^2 A_2(a) + \cdots, \tag{7a}$$

$$\frac{d\psi}{dt} = 1 + \epsilon B_1(a) + \epsilon^2 B_2(a) + \cdots, \tag{7b}$$

and the $u_i(a,\psi)$ satisfy

$$u_i(a,\psi+2\pi) = u_i(a,\psi). \tag{7c}$$

The following conditions are also imposed on the $u_i(a,\psi)$:

$$\int_0^{2\pi} u_i(a,\psi)\begin{pmatrix}\cos\psi\\ \sin\psi\end{pmatrix} d\psi = 0. \tag{8}$$

The prime justification for this requirement is to eliminate possible secular terms that may arise as the calculation proceeds [1, 4]. Under these assumptions, the functions $\{A_1(a), A_2(a), \ldots; B_1(a), B_2(a), \ldots; u_1(a,\psi), \ldots\}$ are calculated by substituting Eq. (6) into Eq. (1), and using Eqs. (7) and (8). Thus, to order ϵ, we obtain

$$u_1(a,\psi) = g_0(a) + \sum_{n=2}^{\infty} \frac{g_n(a)\cos(n\psi) + h_n(a)\sin(n\psi)}{1-n^2}, \tag{9a}$$

$$A_1(a) = -\left(\frac{1}{2\pi}\right)\int_0^{2\pi} f(a\cos\psi, -a\sin\psi)\sin\psi\, d\psi, \tag{9b}$$

$$B_1(a) = -\left(\frac{1}{2\pi a}\right)\int_0^{2\pi} f(a\cos\psi, -a\sin\psi)\cos\psi\, d\psi, \tag{9c}$$

$$g_n(a) = \left(\frac{1}{\pi}\right)\int_0^{2\pi} f(\cdots,\cdots)\cos(n\psi)d\psi, \tag{9d}$$

$$h_n(a) = \left(\frac{1}{\pi}\right)\int_0^{2\pi} f(\cdots,\cdots)\sin(n\psi)d\psi. \tag{9e}$$

Similar, but more complicated relations exist for $A_2(a)$, $B_2(a)$, etc.

Within the context of the KBM method, to order ϵ^n, the "amplitude" and frequency of the limit-cycles are given from the solutions of the equations [1, 4].

$$\epsilon A_1(\bar{a}) + \epsilon^2 A_2(\bar{a}) + \cdots + \epsilon^n A_n(\bar{a}) = 0, \tag{10a}$$

$$\omega(\bar{a}) = 1 + \epsilon B_1(\bar{a}) + \cdots + \epsilon^n B_n(\bar{a}). \tag{10b}$$

The actual procedure requires solving Eq. (10a) for the amplitudes, $\{\bar{a}_j;\ j = 1, 2, \ldots, J\}$, and substituting their values into Eq. (10b) to obtain the corresponding frequencies. (In most cases, Eq. (10a) is a function of $\bar{a}^2$, and only the positive roots need to be selected.)

An examination of Eq. (10a) indicates the genesis of spurious limit-cycles. If $f(x, \dot{x})$ is a polynomial function of its arguments, then the $A_i(a)$ will also be polynomials. Two cases arise. Suppose that

$$\text{degree}\,[A_m(\bar{a})] \le \text{degree}\,[A_1(\bar{a})], \qquad m \ge 2, \tag{11}$$

and denote the solutions to $A_1(\bar{a}) = 0$ by $\left\{\bar{a}_j^{(0)}; j = 1, 2, \ldots, J\right\}$. For this case, the inclusion of higher order in ϵ terms will only make a small change in the amplitude values [1, 4, 5], i.e.,

$$\bar{a}_j = \bar{a}_j^{(0)} + \epsilon\bar{a}_j^{(1)} + \cdots + \epsilon^n \bar{a}_j^{(n)}, \tag{12}$$

where the various coefficients of powers of ϵ can be calculated from Eq. (10a). Thus, higher order calculations do not introduce any change in the number of limit-cycles as determined from the lowest order result, $A_1(\bar{a}) = 0$.

Now let the following situation occur

$$\text{degree}\,[A_m(\bar{a})] > \text{degree}\,[A_{m-1}(\bar{a})] > \cdots > \text{degree}\,[A_1(\bar{a})]. \tag{13}$$

This case leads to two types of limit-cycle amplitudes as determined by Eq. (10a). For small ϵ, the first type of limit-cycles satisfy a relation such as that given by Eq. (12). However, the second type of limit-cycles have amplitudes (for small ϵ) that take the form [2]

$$a_i = \frac{K_i}{\epsilon^{\gamma_i}} + \cdots, \tag{14}$$

where K_i and γ_i are positive constants with $i > J$. Note that the actual limit-cycles correspond to roots of Eq. (10a) that are bounded as $\epsilon \to 0$, while the spurious limit-cycles correspond to roots of Eq. (10a) that become unbounded as $\epsilon \to 0$.

The above results can be observed in the application of the KBM method to the following modified van der Pol equation

$$\ddot{x} + x = \epsilon\left[-\beta x^3 + (1 - x^2)\dot{x}\right], \qquad \beta = O(1), \tag{15}$$

where β is a positive parameter. A direct calculation gives

$$A_1(a) = \left(\frac{a}{2}\right)\left(1 - \frac{a^2}{4}\right), \qquad B_1(a) = \epsilon\left(\frac{3\beta a^2}{8}\right), \tag{16a}$$

$$A_2(a) = \left(\frac{\beta a^3}{16}\right)\left[3 + \left(\frac{19}{8}\right)a^2\right], \tag{16b}$$

$$B_2(a) = -\left(\frac{1}{8}\right)\left[1 - a^2 + \left(\frac{15\beta^2 + 7}{32}\right)a^4\right], \tag{16c}$$

$$u_1(a, \psi) = -\left(\frac{a^3}{32}\right)(\beta\cos 3\psi - \sin 3\psi). \tag{16d}$$

For $\beta = 0$, the van der Pol equation is obtained and $A_2(a) \equiv 0$. Thus, to terms of order ϵ^2, there is a single stable limit-cycle with amplitude and phase

$$\beta = 0 : \bar{a} = 2, \qquad \omega(2) = 1 - \frac{\epsilon^2}{16}. \tag{17}$$

This is in agreement with Lindstedt-Poincaré theory and other mathematical results which prove that the van der Pol equation has a unique (stable) limit-cycle for all $\epsilon > 1$ [4, 5].

In Eq. (10a), let $n = 2$, substitute Eqs. (16), and take $\beta > 0$. If we let $z = a^2$, then the limit-cycles, to order ϵ^2, satisfy the equation

$$\bar{z}\left[\left(\frac{19\epsilon\beta}{64}\right)\bar{z}^2 - \left(\frac{1}{4}\right)\left(1 - \frac{3\epsilon\beta}{2}\right)\bar{z} + 1\right] = 0, \tag{18}$$

with solutions

$$\bar{z}_1 = 0, \tag{19a}$$

$$\bar{z}_2 = 4 + O(\epsilon), \tag{19b}$$

$$\bar{z}_3 = \left(\frac{16}{19\beta}\right)\left(\frac{1}{\epsilon}\right) - \frac{100}{19} + O(\epsilon). \tag{19c}$$

The root $\bar{z}_1$ corresponds to an unstable fixed-point; the root $\bar{z}_2$ is a stable limit-cycle; while $\bar{z}_3$ is an unstable limit-cycle. Thus, based on the usual formulation of

KBM theory, the modified van der Pol equation has, to order ϵ^2, *two* limit-cycles. However, application of the Liénard-Levinson-Smith theorem [4] to Eq. (15) shows that the modified van der Pol equation has a unique stable limit-cycle and one unstable fixed-point. This means that $\bar{z}_3$ corresponds to a spurious limit-cycle, while $\bar{z}_1$ and $\bar{z}_2$ give, to order ϵ^2, the correct location of the fixed-point and the actual limit-cycle. Comparison of Eqs. (14) and (19c) shows that

$$K = \frac{16}{19\beta}, \qquad \gamma = 1. \tag{20}$$

A more general discussion of the nature of the spurious limit-cycles is given in the paper by Burnette and Mickens [2]. In summary, regular (actual) limit-cycles have amplitudes that go to finite values when $\epsilon \to 0$, while for spurious limit-cycles, the amplitudes become unbounded when $\epsilon \to 0$.

In the next section, we discuss a possibility for eliminating spurious limit-cycles in the KBM averaging method.

3. POSSIBLE RESOLUTION

A direct way to eliminate spurious limit-cycles in the KBM method is to replace Eq. (7a) by

$$\frac{da}{dt} = \epsilon A_1(a), \tag{21}$$

i.e., formulate the perturbation expansion such that

$$A_i(a) \equiv 0, \qquad \text{for } i = 2, 3, 4, \ldots. \tag{22}$$

This means that the limit-cycle amplitudes are independent of ϵ and do not become "corrected" in higher order calculations. To do this, the requirement of Eq. (8) will have to be dropped! Note that at the order of ϵ^i calculation, two constants will be introduced, the coefficients of the Fourier expansion involving the $\cos\psi$ and $\sin\psi$ terms in $u_i(a, \psi)$. This means that the lowest harmonic terms in $u_i(a, \psi)$ are to be determined in the next level of calculation. In other words, $u_i(a, \psi)$ is only fully determined by eliminating secular terms at the order ϵ^{i+1} level of the calculation. Such a procedure already occurs in the Lindstédt-Poincaré perturbation method for nonlinear ODE's having limit-cycles [4]. The major question is whether this procedure can be consistently done to arbitrary orders of the calculation in terms of the ϵ expansion.

A related issue is that the "steady-state" limit of Eq. (6), as determined by the solutions of Eqs. (7a) and (7b) with condition (7c), do not reduce to the solution given by the Lindstédt-Poincaré method. This indicates another crisis in the KBM method and is almost certainly related to the existence of spurious limit-cycles. We are currently working to resolve these problems.

ACKNOWLEDGMENT

The work reported in this paper was supported in part by grants from DOE and the MBRS-Program at Clark Atlanta University.

REFERENCES

1. Bogoliubov, N. N. and Mitropolsky, J. A., Asymptotical Methods in the Theory of Nonlinear Oscillations. Hindustan Publishing Company; Delhi, India; 1963.
2. Burnette, J. E. and Mickens, R. E., Spurious limit-cycles arising in higher order averaging methods. Journal of Sound and Vibration, Vol. 193 (1996), 743–746.
3. Krylov, N. and Bogoliubov, N., Introduction to Nonlinear Mechanics. Princeton University Press; Princeton, NJ; 1943.
4. Mickens, R. E., Introduction to Nonlinear Oscillations. Cambridge University Press, New York, 1981.
5. Minorsky, N., Introduction to Non-linear Mechanics. J. W. Edwards; Ann Arbor, MI; 1947. See pps. 205–206.

Proceedings of Dynamic Systems and Applications 3 (2001) 437-444

OPTIMAL CONTROL AND DOMAIN IDENTIFICATION FOR HEMIVARIATIONAL INEQUALITIES

Stanislaw Migórski

Jagiellonian University, Faculty of Mathematics and Physics
Institute of Computer Science, ul. Nawojki 11, 30072 Cracow, Poland

ABSTRACT: The paper deals with two optimal control problems. First one is an optimal control problem for systems governed by a parabolic hemivariational inequality. In the second control problem the system is described by an elliptic hemivariational inequality and the control set is a set of domains. The existence results for both problems are provided.

AMS (MOS) subject classification. 49J27, 49J20, 34G20, 35J85.

1. INTRODUCTION

In this note we present results on the existence of solutions to two problems: an optimal control problem and a domain identification one. Both problems are considered for systems which are described by hemivariational inequalities. The latter are variational expressions which have been introduced by P. D. Panagiotopoulos (cf. [14], [12]) as the mathematical models for some mechanical problems (in the theory of nonlinear elasticity, plasticity, fluid flow, heat transfer, etc.) with nonconvex and not everywhere differentiable energy functions (nonconvex superpotentials). Since the superpotentials are supposed to be locally Lipschitz functions, the hemivariational inequalities are derived with the help of the generalized subdifferential in the sense of Clarke [5]. The last two decades show that the notion of hemivariational inequality is a proper and useful tool to describe large classes of important problems with nonmonotone and multivalued laws (for details see Panagiotopoulos [14]).

In the control problem we consider the evolution hemivariational inequality as "a state equation". Since the state of the system is not uniquely determined we deal with a double minimization problem. The cost functional to be minimized is of Bolza type and the control constraints are time dependent. It should be noted that the theory of evolution hemivariational inequalities is still in its development (see references in Migórski [10]) while the control problems for such inequalities are of great importance from both theoretical and practical viewpoint (see Haslinger and Panagiotopoulos [8] and Migórski and Ochal [11] for the models for distributed parameter systems).

The second problem under consideration is the optimal shape design one in which the role of controls is played by a class of admissible domains. This class consists of all subsets of $\mathbb{R}^N$ satisfying the cone property. Our aim is to give a result on the existence of optimal shapes. We restrict ourselves to systems governed by the elliptic hemivariational inequality with a linear differential operator, Neumann boundary condition and the one dimensional nonmonotone law. The generalization of Theorem 2 to the evolution case seems to be an open problem.

2. AN OPTIMAL CONTROL PROBLEM

Let V and X be reflexive separable Banach spaces and let H be a Hilbert space such that $V \subset X \subset H$ continuously and densely, and $V \subset X$ compactly. Let $A(t): V \to V'$ for $t \in (0,T)$ and $\mathcal{B}: L^2(0,T;U) \to \mathcal{X}'$ be given operators (U is a Hilbert space of controls) and $L: [0,T] \times X \times U \to \mathbb{R}$, $l: H \to \mathbb{R}$ be given mappings.

We consider the following optimal control problem:

$$\min\left\{\int_0^T L\left(t, y(t), u(t)\right) dt + l(y(T))\right\} \tag{1}$$

subject to

$$\begin{cases} y' + \mathcal{A}y + \partial\mathcal{J}(y) \ni f + \mathcal{B}u \\ y(0) = y_0, \quad u(t) \in C(t) \text{ a.e. } t \in (0,T), \ u(\cdot) \text{ measurable,} \end{cases} \tag{2}$$

where $\mathcal{A}: \mathcal{V} \to \mathcal{V}'$ is the Nemitsky operator corresponding to $A(\cdot)$, i.e. $(\mathcal{A}v)(t) = A(t)v(t)$, $\mathcal{V} = L^p(0,T;V)$, $\mathcal{X} = L^p(0,T;X)$, $\mathcal{H} = L^2(0,T;H)$, $\mathcal{X}' = L^q(0,T;X')$, $\mathcal{V}' = L^q(0,T;V')$, $2 \le p < \infty$, $1/p + 1/q = 1$, $0 < T < +\infty$, $\mathcal{W} = \{w \in \mathcal{V} : w' \in \mathcal{V}'\}$ and $\mathcal{J}: \mathcal{X} \to \mathbb{R}$ is a locally Lipschitz function.

Recall (see Clarke [5]) that the generalized directional derivative of a locally Lipschitz function $\mathcal{J}$ at y in the direction v is defined by

$$\mathcal{J}^0(y;v) = \limsup_{z \to y,\ \lambda \downarrow 0} \frac{1}{\lambda}\left(\mathcal{J}(z + \lambda v) - \mathcal{J}(z)\right)$$

and the generalized gradient of $\mathcal{J}$ at y is given by

$$\partial\mathcal{J}(y) = \{\zeta \in \mathcal{X}' : \mathcal{J}^0(y;v) \ge \langle \zeta, v\rangle_{\mathcal{X}\times\mathcal{X}'} \text{ for all } v \in \mathcal{X}\}.$$

Recall that a mapping $T: Z \to 2^{Z'}$ (Z being a reflexive Banach space with dual Z') is said to be pseudomonotone (Browder and Hess [3]) if it satisfies

(a) for every $z \in Z$, Tz is a nonempty, convex and weakly compact set in Z',
(b) T is usc from every finite dimensional subspace F of Z into *weak* $- Z'$,
(c) if $z_n \to z$ weakly in Z, $z_n^* \in T(z_n)$ and $\limsup \langle z_n^*, z_n - z\rangle_Z \le 0$, then for each $y \in Z$ there exists $z^*(y) \in T(z)$ such that $\langle z^*(y), z - y\rangle_Z \le \liminf \langle z_n^*, z_n - y\rangle_Z$.

Let $L: D(L) \subset Z \to Z'$ be a linear densely defined maximal monotone operator. We say that T is generalized pseudomonotone with respect to $D(L)$ if the conditions (a) and (b) hold and

(d) if $\{z_n\} \subset D(T) \cap D(L)$ is such that $z_n \to z$ weakly in Z, $Lz_n \to Lz$ weakly in Z', $z_n^* \in T(z_n)$, $z_n^* \to z^*$ weakly in Z' and $\limsup \langle z_n^*, z_n\rangle_Z \le \langle z^*, z\rangle_Z$, then $z^* \in T(z)$ and $\langle z_n^*, z_n\rangle_Z \to \langle z^*, z\rangle_Z$.

We admit the following hypotheses:

$\underline{H(A)}$: $A(t): V \to V'$ is demicontinuous and pseudomonotone such that
(1) $t \mapsto A(t,v)$ is weakly measurable on $(0,T)$ for all $v \in V$,
(2) $\|A(t)v\|_{V'} \le \alpha(t) + c_1 \|v\|_V^{p-1}$, a.e. $t \in (0,T)$ with $\alpha \in L^q$, $c_1 > 0$,
(3) $\langle A(t)v, v\rangle_V \ge c_2 \|v\|_V^p$ for $v \in V$ with $c_2 > 0$.

$\underline{H(\mathcal{J})}$: $\mathcal{J}: \mathcal{X} \to \mathbb{R}$ is Lipschitz continuous on each bounded subset of $\mathcal{X}$ and

$$\mathcal{J}^0(v; -v) \le k\,(1 + \|v\|_{\mathcal{X}}^\sigma) \quad \text{for all } v \in \mathcal{X}$$

with $k \ge 0$ and $1 \le \sigma < p$.

$\underline{H(\mathcal{B})}$: $\mathcal{B}: L^2(0,T;U) \to \mathcal{X}'$ is a linear continuous operator.

$\underline{H(C)}$: $C: [0,T] \to 2^U$ is a measurable map with nonempty, weakly compact and convex values, and $C(t) \subseteq C_0$ a.e. $t \in (0,T)$, where C_0 is a fixed convex weakly compact subset of U.

$\underline{H(L)}$: $L: [0,T] \times X \times U \to \mathbb{R}$ is an integrand such that
(1) $(t,y,u) \mapsto L(t,y,u)$ is measurable,
(2) $(y,u) \mapsto L(t,y,u)$ is sequentially lower semicontinuous, for a.e. t,
(3) $L(t,y,\cdot)$ is convex on U,
(4) $\beta(t) - M(\|y\|_X + |u|_U) \le L(t,y,u)$ for a.e. t with $\beta \in L^1$, $M > 0$.

$\underline{H(l)}$: $l: H \to \mathbb{R}$ is weakly lower semicontinuous.

Denoting by $S(u) \subset \mathcal{W}$ the solution set to (2) for every $u \in L^2(0,T;U)$ and by F the expression under the minimum in (1), we have

Theorem 1. *Under the hypothees $H(A)$, $H(\mathcal{J})$, $H(\mathcal{B})$, $H(C)$, $H(L)$, $H(l)$ and $f \in \mathcal{V}'$, $y_0 \in V$, the problem (1) has a solution, that is, there exists an admissible pair $(\widehat{y}, \widehat{u}) \in \mathcal{W} \times L^2(0,T;U)$ such that $F(\widehat{y},\widehat{u}) = \inf\{F(y,u) : u \in L^2(0,T;U), y \in S(u)\}$.*

Proof. The proof is based on the direct method of the calculus of variations. Let $\{y_n, u_n\}_{n \in \mathbb{N}}$ be a minimizing sequence for F. Following Migórski [10], we know that $S(u_n) \ne \emptyset$ for all $n \in \mathbb{N}$. From $H(C)$ it follows that $\{u_n\}$ remains in a sequentially weakly compact subset of $L^2(0,T;U)$. Therefore we may assume that $u_n \to \widehat{u}$ weakly in $L^2(0,T;U)$ and $\widehat{u}(t) \in C(t)$ a.e. $t \in (0,T)$. Next by (2) we have

$$y_n' + \mathcal{A}y_n + w_n^* = f + \mathcal{B}u_n \quad \text{in } \mathcal{V}'$$

with $w_n^* \in \partial\mathcal{J}(y_n)$. Hence by using the integration by parts formula, $H(A)$ and $H(\mathcal{J})$, we obtain

$$|y_n(T)|_H^2 - |y_n(0)|_H^2 + 2c_2\|y_n\|_{\mathcal{V}}^p - 2k(1 + \|y_n\|_{\mathcal{X}}^\sigma) \le 2\|y_n\|_{\mathcal{V}}\|f + \mathcal{B}u_n\|_{\mathcal{V}'}.$$

Since $\{u_n\}$ is bounded in $L^2(0,T;U)$ independently of n, from $H(\mathcal{B})$, we get

$$\gamma\,(\|y_n\|_{\mathcal{V}})\,\|y_n\|_{\mathcal{V}} - k \le c_3 + \|f\|_{\mathcal{V}'} + \frac{1}{2}c_3\|y_0\|_V^2$$

where $c_3 > 0$ and $\gamma: \mathbb{R}_+ \to \mathbb{R}_+$ is a function such that $\gamma(s) \to \infty$, as $s \to \infty$. Hence we deduce that $\{y_n\}$ is bounded in $\mathcal{V}$ uniformly with respect to n. Thus, since $\mathcal{A}$, $\partial\mathcal{J}$ and $\mathcal{B}$ are bounded operators, we have $\{y_n'\}$ is bounded in $\mathcal{V}'$. So we have shown that $\{y_n\}$ lies in a bounded subset of $\mathcal{W}$. Therefore, by passing to a subsequence if necessary, we may assume that $y_n \to \widehat{y}$ weakly in $\mathcal{W}$ with $\widehat{y} \in \mathcal{W}$. Since $\mathcal{W} \subset \mathcal{X}$ compactly and $\mathcal{W} \subset C(0,T;H)$ continuously (cf. Lions [9], Zeidler [15]), we have

$$y_n \to \widehat{y} \text{ weakly in } \mathcal{X},$$

$$y_n(t) \to \widehat{y}(t) \text{ weakly in } H \text{ for all } t \in [0,T].$$

By hypothesis $H(l)$, we have

$$l(\widehat{y}(T)) \le \liminf_n l(y_n(T)).$$

Applying Theorem 2.1 of Balder [1], we obtain

$$\int_0^T L(t,\widehat{y}(t),\widehat{u}(t))\,dt \le \liminf_n \int_0^T L(t,y_n(t),u_n(t))\,dt.$$

Hence $F(\widehat{y},\widehat{u}) \le \inf\{F(y,u) : u \in L^2(0,T;U), y \in S(u)\}$. It remains to prove that the pair $(\widehat{y},\widehat{u})$ is admissible for the problem (1). Let $z_n(t) = y_n(t) - y_0$. Then $z_n \in \mathcal{W}$ and it is a solution to the following problem

$$z_n' + \widetilde{\mathcal{A}}z_n + h_n^* \ni f + \mathcal{B}u_n, \quad h_n^* \in \widetilde{\partial\mathcal{J}}(z_n), \quad z_n(0) = 0, \tag{3}$$

where $(\widetilde{\mathcal{A}}z)(t) = \mathcal{A}(z(t) + y_0)$ and $\widetilde{\partial\mathcal{J}}(z)(t) = \partial\mathcal{J}(z(t) + y_0)$. We may assume that $z_n \to \widehat{z}$ weakly in $\mathcal{W}$ and also in $\mathcal{X}$, and $z_n(t) \to \widehat{z}(t)$ weakly in H, for each $t \in [0,T]$, where $\widehat{z}(t) = \widehat{y}(t) - y_0$. From the fact that both $\widetilde{\mathcal{A}}$ and $\widetilde{\partial\mathcal{J}}$ are bounded operators, we can suppose further that

$$\widetilde{\mathcal{A}}(z_n) \to z^* \text{ weakly in } \mathcal{V}'$$

$$h_n^* \to h^* \text{ weakly in } \mathcal{X}'.$$

Passing to the limit in the first inclusion in (3), as $n \to \infty$, we have $\widehat{z}\,' + z^* + h^* = f + \mathcal{B}\widehat{u}$. Moreover, we have $h^* \in \widetilde{\partial\mathcal{J}}(\widehat{z})$, since the graph of $\widetilde{\partial\mathcal{J}}$ is closed in $\mathcal{X} \times (weak - \mathcal{X}')$ topology. Using the weak lower semicontinuity of the norm, from (3), we get

$$\limsup_n \langle \widetilde{\mathcal{A}}(z_n), z_n\rangle_{\mathcal{V}} \le \langle f + \mathcal{B}\widehat{u}, \widehat{z}\rangle_{\mathcal{V}} - \langle h^*, \widehat{z}\rangle_{\mathcal{X}} - \frac{1}{2}|\widehat{z}(T)|_H^2 = \langle z^*, \widehat{z}\rangle_{\mathcal{V}}.$$

Applying Theorem 2(b) of Berkovits and Mustonen [2], we know that $\widetilde{\mathcal{A}}$ is generalized pseudomonotone with respect to $D(d/dt) = \{v \in \mathcal{V} : v' \in \mathcal{V}', v(0) = 0\}$. Thus we deduce that $z^* = \widetilde{\mathcal{A}}\widehat{z}$. So $\widehat{z}$ solves the problem

$$\begin{cases} \widehat{z}\,' + \widetilde{\mathcal{A}}\widehat{z} + \widetilde{\partial\mathcal{J}}(\widehat{z}) \ni f + \mathcal{B}\widehat{u}, \\ \widehat{z}(0) = 0 \end{cases}$$

and this means that $\widehat{y}(t) = \widehat{z}(t) + y_0$ is a solution to $\widehat{y}\,' + \mathcal{A}\widehat{y} + \partial\mathcal{J}(\widehat{y}) \ni f + \mathcal{B}\widehat{u}$, $\widehat{y}(0) = y_0$ with $\widehat{u}(t) \in C(t)$ a.e. t. The pair $(\widehat{y},\widehat{u})$ is admissible and the proof is complete.

3. A DOMAIN IDENTIFICATION PROBLEM

In this section we consider an optimal control problem for the stationary hemivariational inequality, where controls are geometric domains. For such problem we present a result on the existence of optimal shapes.

Let D be a given bounded open subset of $\mathbb{R}^N$ and let θ, h and r be three constants such that $\theta \in (0, \pi/2)$, $h, r > 0$ and $2r \leq h$. Let $\Pi = \Pi(\theta, h, r)$ be a family of all open subsets of D which satisfy the cone property with constants θ, h and r. Recall (cf. Chenais [4]) that a subset Ω of $\mathbb{R}^N$ is said to satisfy the cone property with constants θ, h and r iff

$$\forall\, z \in \partial\Omega \;\; \exists\, C_z = C(\xi_z, \theta, h) \;\text{ such that }\; y + C_z \subset \Omega \text{ for all } y \in \Omega \cap B(z, r),$$

where $C(\xi_z, \theta, h) = \{x \in \mathbb{R}^N : \langle x, \xi_z \rangle > |x| \cos\theta, |x| < h\}$ is the cone of angle θ, height h and axis ξ_z ($|\xi_z| = 1$) and $B(z, r)$ denotes the open ball of radius r and center at z.

For $\Omega \in \Pi$ we denote by 1_Ω the characteristic function of Ω in D. On Π we consider the following topology. We say that a sequence $\{\Omega_n\} \subset \Pi$ converges to Ω in Π iff $1_{\Omega_n} \to 1_\Omega$ in $L^2(D)$. It is well known (see Chenais [4]) that the set Π is closed and relatively compact in such topology. Moreover, the family Π satisfies the so-called uniform extension property, i.e. we have

Lemma 1. *There exists a positive constant $K = K(\theta, h, r)$ depending on $\Omega \in \Pi(\theta, h, r)$ through θ, h and r only, and such that for any $\Omega \in \Pi$ there exists a linear continuous extension operator $p_\Omega : H^1(\Omega) \to H^1(\mathbb{R}^N)$ such that $\|p_\Omega\| \leq K$.*

Let now Ω_0 be a given subset of D. By $\Pi_0 = \Pi_0(\theta, h, r)$ we denote the sets of $\Pi(\theta, h, r)$ which contain Ω_0. We assume that θ, h and r are chosen uniformly for all the domains of Π.

The domain identification problem is as follows: find $(\Omega^*, y^*) \in \bigcup_{\Omega \in \Pi_0} (\Omega \times S(\Omega))$ such that

$$G(\Omega^*, y^*) = \min_{\Omega \in \Pi_0} \min_{y \in S(\Omega)} G(\Omega, y), \tag{4}$$

where the control is the set Ω changing in the family Π_0, $G(\Omega, y) = \int_\Omega g(x, y(x))\, dx$ is the cost functional and $S(\Omega)$ denotes the solution set of the following hemivariational inequality: find $y \in H^1(\Omega)$ such that there exists $\chi \in L^2(\Omega)$ and

$$\begin{cases} a_\Omega(y, v) + (\chi, v)_{L^2(\Omega)} = (f, v)_{L^2(\Omega)}, \quad \forall\, v \in H^1(\Omega) \\ \chi(x) \in \partial j(x, y(x)) \text{ for a.e. } x \in \Omega. \end{cases} \tag{5}$$

Hypotheses:

$\underline{H(a)}$: $a_\Omega : H^1(\Omega) \times H^1(\Omega) \to \mathbb{R}$ is a form such that

$$a_\Omega(u, v) = \int_\Omega \left(\sum_{i,j=1}^N a_{ij}(x) D_i u D_j v + a_0(x) uv \right) dx,$$

where $a_{ij}, a_0 \in L^\infty(D)$, $a_0 \geq \alpha_1 > 0$, $\sum_{ij} a_{ij}(x)\xi_i\xi_j \geq \alpha_2|\xi|^2$ for $\xi \in \mathbb{R}^N$.

$\underline{H(j)}$: $j: \Omega \times \mathbb{R} \to \mathbb{R}$ is a function such that
(1) $x \mapsto j(x, \xi)$ is measurable on Ω, for all $\xi \in \mathbb{R}$,
(2) $\xi \mapsto j(x, \xi)$ is locally Lipschitz on $\mathbb{R}$ for a.e. x and $j(\cdot, 0) \in L^1(\Omega)$,
(3) for any $\xi \in \mathbb{R}$ and a.e. $x \in \Omega$ and for any $\eta \in \partial j(x, \xi)$ we have $|\eta| \leq c_4(1 + |\xi|)$ with $c_4 \geq 0$,
(4) $j^0(x, \xi; -\xi) \leq \rho(x)|\xi|$ for all $\xi \in \mathbb{R}$ with a nonnegative $\rho \in L^2(\Omega)$.

$\underline{H(g)}$: $g: \mathbb{R}^n \times \mathbb{R} \to \mathbb{R}$ is a nonnegative normal integrand such that

$$g(x, s) \leq k(x)(1 + |s|^r) \text{ for } s \in \mathbb{R}$$

with $r \geq 1$ and $k \in L^1$.

Theorem 2. *Assume that $H(a)$, $H(j)$, $H(g)$ hold and $f \in L^2(D)$. Then the problem (4) has at least one solution.*

Theorem 2 extends the main result of Chenais [4]. The proof is a consequence of the following result on the continuity of the solution set to (5) with respect to the variations of the domain.

Lemma 2. *Assume that $H(a)$, $H(j)$ hold and $f \in L^2(D)$. Then the mapping $\Pi_0 \ni \Omega \mapsto S(\Omega) \subset H^1(\Omega)$ has a closed graph in the following sense: if $\Omega_n, \Omega \in \Pi_0$ and $\Omega_n \to \Omega$ in Π, as $n \to \infty$, then for every solution $(y_n, \chi_n) \in S(\Omega_n) \times L^2(\Omega_n)$ of the problem*

$$\begin{cases} a_{\Omega_n}(y_n, v) + (\chi_n, v)_{L^2(\Omega_n)} = (f, v)_{L^2(\Omega_n)}, & \forall\, v \in H^1(\Omega_n) \\ \chi_n(x) \in \partial j(x, y_n(x)) \text{ for a.e. } x \in \Omega_n, \end{cases}$$

there exists a subsequence $\{n_k\} \subset \{n\}$ such that

$$p_{n_k} y_{n_k} \to \xi \text{ weakly in } H^1(D)$$

$$\overline{\chi}_{n_k} \to \zeta \text{ weakly in } L^2(D),$$

and the pair $(y, \chi) = (\xi|_\Omega, \zeta|_\Omega)$ is a solution to the problem (5) (here $p_n = p_{\Omega_n}$ denotes the extension operator from Lemma 1 and $\overline{w} \in L^2(D)$ is the extension of a function $w \in L^2(\Omega)$ by zero outside Ω).

For the proof of Lemma 2 and more details in this direction see Ochal [13]. The domain identification problems for elliptic hemivariational inequalities with the cost criterion of the form $\int_\Omega g(x, y(x), Dy(x))\, dx$ were treated by using the mapping method in Denkowski and Migórski [6] and [7]. For more papers on the subject we refer to the above mentioned works.

ACKNOWLEDGEMENT

The research has been supported by the State Committee for Scientific Research of the Republic of Poland (KBN) under Research Grant No. 2 P03A 040 15.

REFERENCES

1. E. Balder, Necessary and sufficient conditions for L^1-strong-weak lower semicontinuity of integral functionals, *Nonlinear Analysis*, 11 (1987) 1399–1404.

2. J. Berkovits and V. Mustonen, Monotonicity methods for nonlinear evolution equations, *Nonlinear Analysis*, 27 (1996) 1397–1405.

3. F. E. Browder and P. Hess, Nonlinear mappings of monotone type in Banach spaces, *J. Funct. Anal.*, 11 (1972) 251–294.

4. D. Chenais, On the existence of a solution in a domain identification problem, *J. Math. Anal. Appl.*, 52 (1975) 189–219.

5. F. H. Clarke, *Optimization and Nonsmooth Analysis*, John Wiley & Sons, New York, 1983.

6. Z. Denkowski and S. Migórski, *Optimal Shape Design for Elliptic Hemivariational Inequalities in Nonlinear Elasticity,* Proceedings of International Conference Variational Calculus, Optimal Control and Applications, Trassenheide, Germany, September 23–27, 1996, W. H. Schmidt et al. (Eds.), International Series of Numerical Mathematics, vol. 124, 31–40, Birkhäuser Verlag, Basel-Boston-Berlin, 1998.

7. Z. Denkowski, S. Migórski, Optimal shape design problems for a class of systems described by hemivariational inequalities, *J. Global Optim.*, 12 (1) (1998), 37–59.

8. J. Haslinger, P. D. Panagiotopoulos, Optimal Control of Systems Governed by Hemivariational Inequalities. Existence and Approximation Results, *Nonlinear Analysis*, 24 (1995) 105–119.

9. J. L. Lions, *Quelques méthodes de résolution des problémes aux limites non linéaires*, Dunod, Paris, 1969.

10. S. Migórski, On the Existence of Solutions for Parabolic Hemivariational Inequalities, *J. Comput. and Appl. Math.*, submitted.

11. S. Migórski and A. Ochal, Optimal control of parabolic hemivariational inequalities, *J. Global Optim.*, submitted.

12. Z. Naniewicz and P. D. Panagiotopoulos, *Mathematical Theory of Hemivariational Inequalities*, Dekker, New York, 1995.

13. A. Ochal, Domain Identification Problem for Elliptic Hemivariational Inequalities, *Universitatis Iagellonicae Acta Mat.*, submitted.

14. P. D. Panagiotopoulos, *Hemivariational Inequalities; Applications in Mechanics and Engineering*, Springer-Verlag, Berlin, 1993.

15. E. Zeidler, *Nonlinear Functional Analysis and Applications, Vol. II*, Springer, New York, 1990.

Proceedings of Dynamic Systems and Applications 3 (2001) 445-452

Efficient 3-D Filter Banks for Motion Estimation

Fernando Mujica[1,2], Sang Park[1], Mark J. T. Smith[1], and Romain Murenzi[2]

Center for Signal and Image Processing[1]
School of Electrical Engineering
Georgia Institute of Technology

Center for Theoretical Studies of Physical Systems[2]
Department of Physics
Clark Atlanta University

Abstract

This paper introduces an efficient approach to object tracking. It is based on previous work in which the continuous wavelet transform (CWT) was used to estimate motion parameters corresponding to moving objects in a 3-D sequence. The CWT algorithm was shown to perform well in noisy environments and with multiple objects present. However, the CWT algorithm relies on an oversampled representation of the wavelets, which is computationally inefficient. The spatio-temporal filter banks discussed in this paper remove that barrier and provide additional flexibility and adaptivity.

1 Introduction

Many applications require the estimation of object motion parameters (i.e. position and velocity) from video signals, and furthermore, the ability to follow the time evolution of these parameters. We refer to this process as *object tracking*. Two applications in which object tracking are of interest are missile interception and traffic monitoring. In the former, the goal is for a missile interceptor, equipped with on-board imaging sensors, to track and neutralize a missile threat. In the traffic monitoring scenario, the aim is to detect and monitor speeding, reckless driving, improper line changes, and the like, using imagery obtained from highway cameras. These applications typically involve imagery with low SNRs, complex motion, and multiple objects, making the tracking problem a challenging one.

In this paper, we consider a velocity selective filtering approach based on *spatio-temporal filter banks*. These filter banks represent an extension of the two-dimensional directional filter bank approach [7, 8] with the goal of emulating the behavior of the spatio-temporal continuous wavelet transform (CWT), which has proven useful for motion estimation [5, 6]. The directional filter bank approach offers a computational advantage over direct implementation of the CWT-based filters, while exploiting the temporal information, which is much needed in real-world scenarios.

Techniques relying on spatial information from the current and previous frames are, in general, highly sensitive to noise and fail to resolve temporary occlusions. Two popular approaches that fall into this category are block matching [1, 10] and optical flow [11, 12]. Other techniques combine a spatial processing stage with a temporal one [13, 14]. Another class of approaches attack the problem from a velocity selective filtering point of view. Three-dimensional matched filtering is the most popular technique in this category as it corresponds to the optimum filter for SNR improvement of linearly moving objects in noise. The technique presented in this paper belongs to this category as well, except that it also embodies adaptivity

In the following section we briefly describe the object tracking algorithm in terms of the CWT filters and their interpretation as motion selective filters. Following this, we introduce efficent spatio-temporal directional filter banks that can be used in lieu of the CWT for efficient implementation.

2 The Spatio-temporal CWT Filters

The spatio-temporal CWT filters have the following form in the wavenumber-frequency domain

$$\hat{\psi}_{\vec{b},\tau,a,\theta,c}(\vec{k},\omega) = a^{3/2}\hat{\psi}\left(ac^{1/3}r^{-\theta}\vec{k}, \frac{a}{c^{2/3}}\omega\right) e^{-j(\vec{k}\cdot\vec{b}+\omega\tau)} , \tag{1}$$

where the parameters $\vec{b}$, τ, a, c, and θ correspond to the spatial translation, temporal translation, scale, velocity magnitude, and velocity orientation respectively. The filter prototype, or mother wavelet, used in this work is the Morlet wavelet which is given, in the wavenumber-frequency, domain by

$$\hat{\psi}(\vec{k},\omega) = \left(e^{-\frac{1}{2}|\vec{k}-\vec{k}_o|^2} - e^{-\frac{1}{2}(|\vec{k}|^2+|\vec{k}_o|^2)}\right)\left(e^{-\frac{1}{2}(\omega-\omega_o)^2} - e^{-\frac{1}{2}(\omega^2+\omega_o^2)}\right) . \tag{2}$$

This Morlet wavelet is shown in Figure 1 (a). Note that an object moving linearly with velocity $\vec{v}_o = [1,\ 0]^T$ has its energy concentrated around the plane $\omega = -\vec{k}\cdot\vec{v}_o$ passing through the center of the filter. Therefore, this filter is selective to velocity $\vec{v}_o$.

Filters tuned to other velocities and scales are obtained by appropriately choosing the CWT parameters in equation (1). The effect of three of the CWT parameters on the energy distribution of the filters is shown on Figure 1 (b). The scale parameter a distributes the energy of the filter along a conic volume without changing its associated velocity and it is associated with the size selectivity of the filter. The speed tuning parameter c distributes the energy along a $1/x$-like path and is directly related with speed selectivity. Finally the orientation parameter rotates the filter with respect to the frequency axes, therefore, changing its direction selectivity. In [15] the CWT, which is defined as the inner product of the signal of interest and the filters in equation (1), is used to extract motion parameters. More explicitly, three CWT cost functions depending on spatial location, velocity, and scale, are sequentially optimized on a frame-by-frame basis. This procedure permits a piecewise approximation

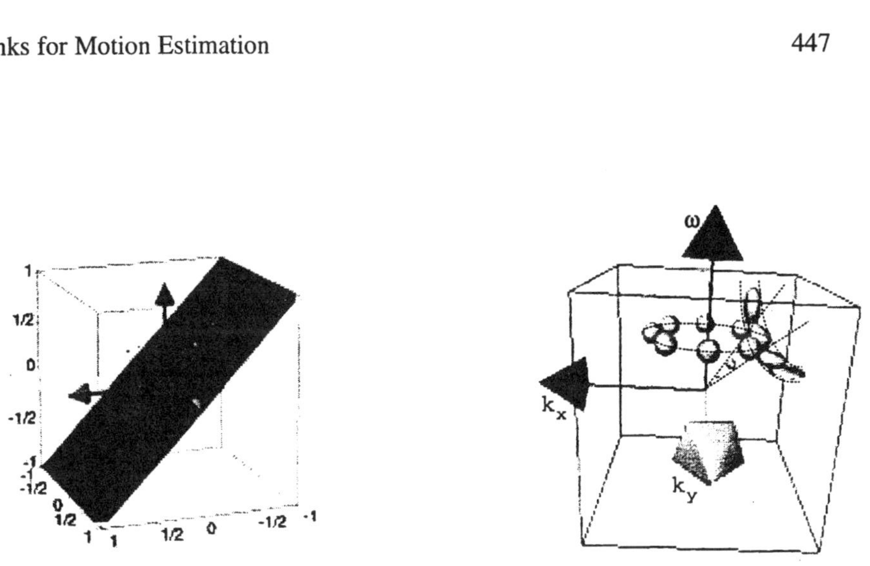

(a) (b)

Figure 1: The 3-D plot of Morlet wevelet. (a) The Morlet wavelet in the wavenumber-frequency domain with $\vec{k}_o = [-6,\ 0]^T$ $\vec{k}_o = [-6,\ 0]^T$ and $\omega_o = 6$. (2+1)D isosurface at value $0.5\,\psi_{\max}$ with $\psi_{\max}$ the maximum of ψ. (b) Effect of the DDM transformations on the Morlet wavelet tuned to velocity $\vec{v}_o = [1,\ 0]^T$. Isosurfaces of half absolute value in the wavenumber-frequency domain are shown for different values of the scale a, orientation θ, and speed tuning c parameters.

of non-linear motion and results in robust estimates in the presence of noise, temporary occlusions, and image coding artifacts [16].

3 Digital Spatio-Temporal Filter Banks

The CWT is effective in serving as an adaptive filter for motion estimation, as demonstrated previously [15]. However, it is a continuous-time filter represented in discrete-time for simulation purposes by oversampling. As such, its discrete-time implementation is not efficient by digital filtering standards. In the next subsections, we will discuss separable filter banks that can be used in place of CWT as part of the motion tracking algorithm, that provide adaptivity and spatio-temporal selectivity.

The new spatio-temporal decomposition is composed of two component filter banks:a conventional subband tree, which produces 3-D cubes in the Fourier domain; and a directional filter bank, producing 3-D Fourier wedges and cones. When applied separably in space and time and in cascade, spatio-temporal frequency components can be extracted with higher precision, compared to the CWT approach discussed in [15]. Moreover, it is shown that the computational complexity of the trees can be reduced by a factor a two over that of conventional QMFs applied in the same separable manner.

3.1 Spatio-Temporal Subband Cubes

Two-band filter banks have a long history, dating back to 1976. Over the years, many two-band filter banks have been designed, one of the more popular sets being

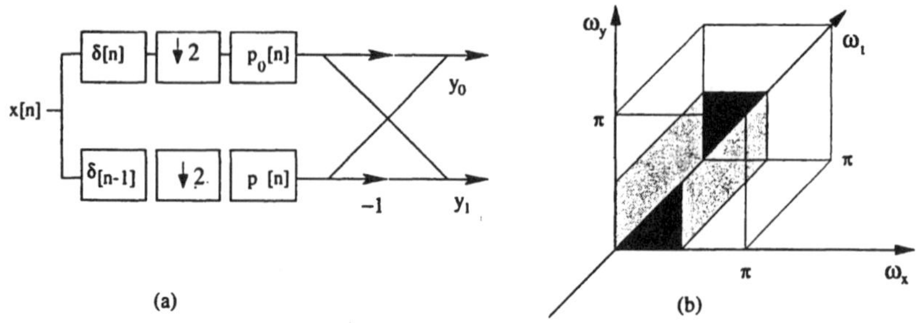

Figure 2: (a) Polyphase filter bank structure. (b) Illustration of 3-D filter bank passbands.

the QMFs of Johnston [9].

In this work, we employ a polyphase representation, as in Figure 2 (a), whereby the analysis filters $h_0[n]$ and $h_1[n]$ are represented by the polyphase filters $p_0[n]$ and $p_1[n]$ such that

$$H_0[z] = P_0[z^2] + z^{-1}P_1[z^2]$$

and

$$\begin{aligned} p_0[n] &= h_0[2n] \\ p_1[n] &= h_0[2n+1]. \end{aligned}$$

Unique to the work presented in our paper is the use of odd length two-band lowpass prototype filters. These filters are designed to have a cutoff frequency of $\pi/2$ and equal frequency deviation in the passband and the stopband. An example filter with seven coefficients is shown below:

$$h_0[n] = [-0.042, 0, 0.29, 0.5, 0.29, 0, -0.042].$$

Using such filters, the computational complexity is almost cut in half, compared to using QMFs in their most efficient polyphase form.

By applying these filter banks separably in the spatial and temporal dimensions of a video sequence, they can produce frequency selective cubes in the 3-D Fourier domain as illustrated in Figure 2 (b).

3.2 Spatio-Temporal Subband Wedges

Rectangular and cubic geometries in the 3-D Fourier domain, do not generally match the motion signatures of moving objects. Typically, the energy associated with linear motion resides in a plane in the 3-D Fourier space, as illustrated in Figure ??. For more complex motion, more refined divisions of the 3-D Fourier space are appropriate.

Toward that end, we consider an efficient class of directional filter banks. The directional filter bank is a tree of two-band filter banks with resampling/downsampling matrices in lieu of 1-D downsamplers. The first two stages in the tree are identical and use quincunx downsampling matrices, while the remaining stages of all of the same

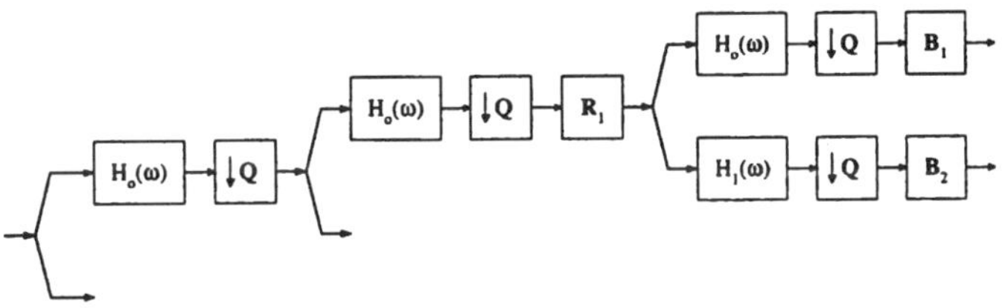

Figure 3: Structure of directional filter banks.

form and employ additional matrices, which will be considered in the next section. The structure is illustrated in Figure 3, where the filters effective have an hour-glass shaped passband. Such filters can also be converted to polyphase form for efficient implementation [7]. The remaining stages are the tree employ filters with parallelogram shaped passbands. All stages employ the two-band filters of the type mentioned in the early section. The passband shapes are controlled by the resampling/downsampling matrices. In particular, we employ the quincunx matrices

$$q_1 = \begin{bmatrix} 1 & 1 \\ -1 & 1 \end{bmatrix} \quad q_2 = \begin{bmatrix} 1 & -1 \\ 1 & 1 \end{bmatrix} \tag{3}$$

using q_1 for the first two stages. Thereafter, a pair of quincunx matrices is used for 2^n subbands in the following way:

$$Q_i = \begin{cases} q_2 & \text{if } i \text{ is 1 or 4} \\ q_1 & \text{if } i \text{ is 2 or 3} \end{cases}$$

where i is the remainder when 2^n is divided by 4.

3.3 Resampling Matrices

The definition of a resampling matrix is a matrix whose determinant should be 1 with integer matrix entries so that its inverse matrix is also a resampling matrix. Among the infinitely many resampling matrices,

$$r_1 = \begin{bmatrix} 1 & 1 \\ 0 & 1 \end{bmatrix} \quad r_2 = \begin{bmatrix} 1 & -1 \\ 0 & 1 \end{bmatrix}$$

$$r_3 = \begin{bmatrix} 1 & 0 \\ 1 & 1 \end{bmatrix} \quad r_4 = \begin{bmatrix} 1 & 0 \\ -1 & 1 \end{bmatrix}$$

can be used to convert filters with diamond passbands to ones with parallelograms passbands. At any stage after the second stage, $R_i = r_i$.

3.4 Backsampling Matrices

In the new directional filter bank as in [8], the backsampling matrix, B_i^n, makes the overall sampling for each of 2^n ($n > 2$) subbands either simple vertical or horizontal sampling so that each of the subbands can contain spatial information.

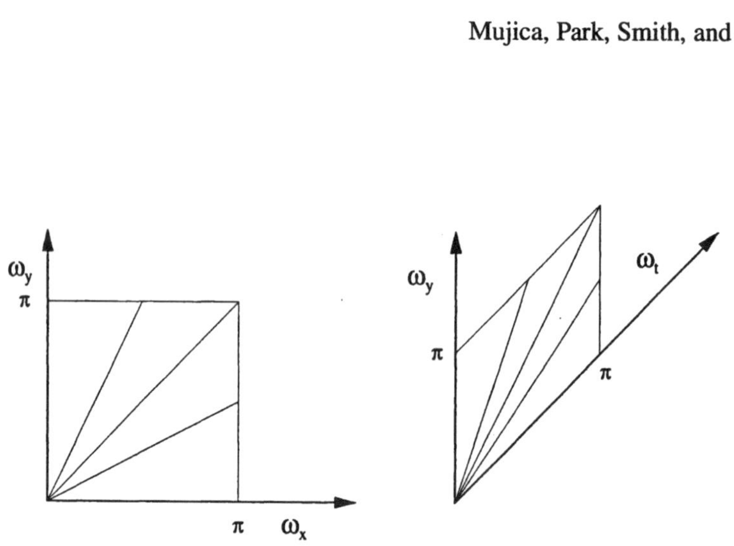

Figure 4: Illustration of wedge shaped passbands for 2-D filter bank.

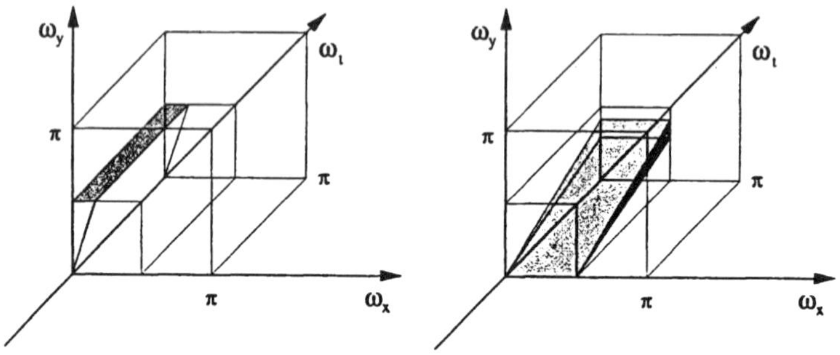

Figure 5: Illustration of wedge passbands arising from separable filter bank cascades.

$$
\begin{aligned}
B_i^n &= \left[\left[B^{T(n-1)} \right]^{-1} R_i * Q_i \right]^{-1} * D_i, \text{ where} \\
D_i &= \begin{cases} \left[\begin{smallmatrix} 1 & 0 \\ 0 & 2 \end{smallmatrix}\right] & \text{if } i = 1 \text{ or } 2 \\ \left[\begin{smallmatrix} 2 & 0 \\ 0 & 1 \end{smallmatrix}\right] & \text{if } i = 3 \text{ or } 4 \end{cases} \\
B_i^3 &= R_{5-i}, \text{ where } i = 1, 2, 3, 4
\end{aligned}
$$

In 2-D, these filters provide passbands as shown in Figure 4. By implementing these filter banks separably, one can achieve 3-D wedges as illustrated in Figure 5. Such passband geometries can provide greater flexibility to capture motion signatures of moving objects than can the CWT.

The computational complexity of these spatio-temporal filter banks istwice as efficient as QMF decompositions and orders of magnitude more efficient than the CWT implementations considered previously in [15, 16].

In the next phase of this work, the discrete spatio-temporal filter banks will be employed in simulation tracking studies to evaluate the performance gains over CWT implementations.

Acknowledgments

This paper is based upon work supported in part by the U.S. Army Research Office under grant numbers DAAH-04-96-1-0161.

References

[1] R. Bamberger and M. Smith, "Filter Banks for the Directional Decomposition of Images: Theory and Design," IEEE Transactions on Acoustics, Speech, and Signal Processing". 40, No. 4, pp. 882-892, April 1992.

[2] Sang-il Park, Mark J. T. Smith and Russell M. Mersereau, "A New Directional Filter Bank for Image Analysis and Classification," IEEE, ICASSP99, Phoenix, AZ, March 16-19, 1999

[3] Bernd Jähne, "Digital Image Processing," Springer-Verlag, 1997

[4] A. Grossmann, J. Morlet and T. Paul "Transforms Associated to Square Intregrable Representation. II." Ann. Inst. Henri Poincare, 45 (1986) 293.

[5] M. Duval-Destin and Romain Murenzi "Spatio-Temporal Wavelet: Application to the Analysis of Moving Patters" in "Wavelets and Applications", (Proc. Toulouse 1992), Y. Meyer, S. Roques (eds).

[6] Romain Murenzi (and J-P Antoine, Carrette, B. Piette) Image Analysis with Two-dimensional Continuous Wavelet Transform,"'Signal Proc. 31 (1993).

[7] Fernando Mujica, Romain Murenzi, Jean-Pierre Leduc and Mark J. T. Smith, "Robust Object Tracking in Compressed Image Sequence," Journal of Electronic Imaging, Vol. 7, No. 4, pp. 746-754, October 1998

[8] F. Mujica, J.P. Leduc, R. Murenzi, and M. Smith, "Spatio-temporal Continuous Wavelets Applied to Missile Warhead Detection and Tracking," Proceedings of SPIE Conference on Visual Communications and Image Processing, 1997, February 1997.

[9] J. Johnston, " A Filter Family Designed for use in QMF Banks," IEEE, ICASSP, April 1980, pp.291-294

[10] A. Murat Tekalp, "Digital Video Processing," Prentice Hall, 1995

[11] David J. Heeger, "Optical Flow Using Spatiotemporal Filters," International Journal of Computer Vision, 1988, pp. 279-302

[12] Gilad Adiv, "Determining Three-Dimensional Motion and Structure from Optical Flow Generated by Several Moving Objects," ieeePAMI, 1985, PAMI-7, 4, Jul, pp. 384–401

[13] Thomas J. Burns and Steven K. Rogers and Dennis W. Ruck and Mark E. Oxley, "Computing Optical Flow Using a Discrete, Spatio-Temporal, Wavelet Multiresolution Analysis," spieWA, vol. 2242,1994, pp. 549–560

[14] I. S. Reed and R. M. Gagliardi and L. Stotts, A Recursive "Moving-Target-Indication Algorithm for Optical Image Sequences," ieeeAES, 1990, vol. 26, no. 3, May, pp. 434–439

[15] Fernando Mujica and Jean-Pierre Leduc and Romain Murenzi and Mark J. T. Smith, "A New.Motion Parameter Estimation Algorithm Based on the Continuous Wavelet Transform," Submitted to IEEE Transactions On Image Processing 1998, Jan

[16] Fernando Mujica and Romain Murenzi and Jean-Pierre Leduc and Mark J. T. Smith, "Robust Object Tracking in Compressed Image Sequence," SPIE Journal of Electronic Imaging, 1998, vol. 7, no. 4, Oct, pp. 746–754

Proceedings of Dynamic Systems and Applications 3 (2001) 453-466

NAMBU DYNAMICAL SYSTEMS

Ramón Navarro[1], S. Codriansky[2] and C. González[2]

[1]Universidad Rómulo Gallegos
Decanato de Postgrado
San Juan de los Morros, Venezuela

[2]Departamento de Matemáticas y Física
Instituto Pedagógico de Caracas
Caracas 1010, Venezuela

ABSTRACT: To construct a geometric formulation of Nambu Dynamical Systems with more than one 3-tuples, a modified exterior calculus which incorporates a symmetric part in the exterior product is proposed. This extended exterior calculus reduces to the classical antisymmetric one in the particular case of a single tuple. A redefinition of the differential, the contraction and the Lie derivative implies that all powers of the basic 3-form for m 3-tuples are invariants of the flow associated to the Nambu vector field.

AMS subject classification: 53B30, 53B50, 58A10

1. INTRODUCTION

Nambu dynamical systems [see 10] is presented as a modification of the classical Hamiltonian dynamical systems. In this approach a three-dimensional phase space with coordinates $\mathbf{r} = (x, y, z)$ is introduced, with the dynamics given by the system of first-order differential equations:

$$\frac{dx}{dt} = \frac{\partial(H,G)}{\partial(y,z)}, \ \frac{dy}{dt} = \frac{\partial(H,G)}{\partial(z,x)}, \ \frac{dz}{dt} = \frac{\partial(H,G)}{\partial(x,y)} \tag{1}$$

where the functions H and G of (x, y, z) serve as a pair of Hamiltonians to determine the motion of points in phase space as opposed to a single one, H, in the Hamiltonian case and $\frac{\partial(\cdots)}{\partial(\cdots)}$ is a Jacobian of order 2.

In vector notation

$$\frac{d\mathbf{r}}{dt} = \nabla H \times \nabla G \tag{2}$$

For any function $F(x, y, z)$, we have

$$\frac{dF}{dt} = \frac{\partial(F, G, H)}{\partial(x, y, z)} = \nabla F \circ (\nabla H \times \nabla G) \tag{3}$$

Since both H and G are constants of the motion, the orbit of a system in phase space is the intersection of two surfaces, $H = \text{const}$ and $G = \text{const}$. Moreover, the velocity field $\frac{d\mathbf{r}}{dt}$ is divergenceless: $\nabla \circ (\nabla H \times \nabla G) = 0$. This is a Liouville theorem in our phase space.

The Euler equations for a free rigid rotator takes just this form if we identify $\mathbf{r} = (x, y, z)$ with the angular momentum components L_x, L_y, L_z in the body fixed frame on the one hand, and G, H, respectively, with the total kinetic energy and the square of the angular momentum in that frame, $G = \frac{1}{2}(\frac{L_x^2}{I_x} + \frac{L_y^2}{I_y} + \frac{L_z^2}{I_z})$, $H = \frac{1}{2}(L_x^2 + L_y^2 + L_z^2)$. It is obvious that this dynamical system can be extended to a phase space of any dimensionality (odd or even): In $\mathbb{R}^n$ with coordinates $\mathbf{x} = (x_1, x_2, \ldots, x_n)$, for any real function $F(\mathbf{x})$, its time evolution is specified by $(n-1)$ functions $H_i(\mathbf{x})$, $i = 1, 2, \ldots, n-1$, called the Hamiltonians of the dynamical system, through

$$\frac{dF(\mathbf{x})}{dt} = \frac{\partial(F, H_1, \ldots, H_{n-1})}{\partial(x_1, x_2, \ldots, x_n)} \tag{4}$$

where $\frac{\partial(\cdots)}{\partial(\cdots)}$ is a Jacobian of order n. It is obvious that all $H_i(\mathbf{x})$ are constants of the motion and they correspond to the integral surfaces of the system of differential equations that describe the time evolution of the coordinates x_i:

$$\frac{dx_i}{dt} = \frac{\partial(x_i, H_1, H_2, \ldots, H_{n-1})}{\partial(x_1, x_2, \ldots, x_n)} \tag{5}$$

We can justify the physical relevance of such generalizations because (see [4]) in addition to the rigid rotator, other physical systems such as a free particle, a harmonic oscillator and its application on the simple Hubbard model for superconductivity and electromagnetic systems can be in Nambu form.

Introducing a geometrical framework for the description of Nambu dynamical system (see[9]), we consider the autonomous ordinary differential equation

$$\frac{d\mathbf{x}}{dt} = \mathbf{f}(\mathbf{x}) \tag{6}$$

Here $\mathbf{x} \in M$, whith M a smooth manifold embedded in $\mathbb{R}^n$, $\mathbf{x} = (x_1, x_2, \ldots, x_n)$, $\mathbf{f} : M \longrightarrow \mathrm{T}M$ and $\mathbf{f}(\mathbf{x}) = (f_1(\mathbf{x}), f_2(\mathbf{x}), \ldots, f_n(\mathbf{x}))$

We shall say that (6) is a Nambu dynamical system if there exists functions $H_1, H_2, \ldots, H_{n-1}$ (the Hamiltonians of the system), $H_j : M \longrightarrow \mathbb{R}$, $j = 1, 2, \ldots, n-1$ such that

$$\frac{dx_i}{dt} = f_i(\mathbf{x}) = \frac{\partial(x_i, H_1, \ldots, H_{n-1})}{\partial(x_1, x_2, \ldots, x_n)} \tag{7}$$

$i = 1, 2, \ldots, n$.

In this case, in the $\mathbf{x}$ coordinates, given the vector field χ by $\chi = f_i(\mathbf{x})\dfrac{\partial}{\partial x_i}$, the volume n-form $\Omega = \mathrm{d}\,x_1 \wedge \mathrm{d}\,x_2 \wedge \cdots \wedge \mathrm{d}\,x_n$ and the form ω by the contraction $\omega = i_\chi(\Omega)$, it is clear, using (7), that $\mathrm{d}\,\omega = \left[\sum_{i=1}^{n} \dfrac{\partial f_i}{\partial x_i}\right] \Omega = 0$ and $\omega = \mathrm{d}\,H_1 \wedge \mathrm{d}\,H_2 \wedge \cdots \wedge \mathrm{d}\,H_{n-1}$

If $n = m \cdot s$, then the n variables $x_1, x_2, \ldots, x_n$ are grouped in m sets, each one containing s variables and we say that the Nambu dynamical system has m s-tuples and $s-1$ Hamiltonians $H_1(\mathbf{x}), H_2(\mathbf{x}), \ldots, H_{s-1}(\mathbf{x})$ are needed in order to determine the dynamical evolution of a given function $F(\mathbf{x})$ according to

$$\frac{dF(\mathbf{x})}{dt} = \sum_{\alpha=1}^{m} \frac{\partial(F, H_1, \ldots, H_{s-1})}{\partial(x_1^\alpha, x_2^\alpha, \ldots, x_s^\alpha)} \tag{8}$$

where the superindex labels a tuple and the subindex the place of the variable within that particular tuple.

If $s = 2$, then it is a Hamiltonian dynamical system since $\frac{dx_1^\beta}{dt} = \sum_{\sigma=1}^{m} \frac{\partial(x_1^\beta, H)}{\partial(x_1^\sigma, x_2^\sigma)} = \frac{\partial(x_1^\beta, H)}{\partial(x_1^\beta, x_2^\beta)} + \sum_{\substack{\sigma \neq \beta \\ \sigma=1}}^{m} \frac{\partial(x_1^\beta, H)}{\partial(x_1^\sigma, x_2^\sigma)} = \frac{\partial H}{\partial x_2^\beta}$ and $\frac{dx_2^\beta}{dt} = -\frac{\partial H}{\partial x_1^\beta}$ for any $\beta = 1, 2, \ldots, m = n/2$.

It is possible to conceive Nambu dynamical systems with tuples of different dimensionality but this case will not be considered here. Attempts to construct a geometric formulation of a Nambu dynamical system when there are m s-tuples $(m > 1)$, s odd $(n = m \cdot s)$, have met with difficulties related to the antisymmetry of the exterior product of forms. For example, in the case $n = 3m$ (see[6]), construct the following set of forms similarly to Hamiltonian structure:

$$\mathrm{d}\,H \wedge \mathrm{d}\,G; \quad \omega = \sum_{\alpha=1}^{m-1} \mathrm{d}\,x_1^\alpha \wedge \mathrm{d}\,x_2^\alpha \wedge \mathrm{d}\,x_3^\alpha - \mathrm{d}\,H \wedge \mathrm{d}\,G \wedge \mathrm{d}\,t$$

so with respect to the Nambu field χ is

$$i_\chi(\omega) = \mathrm{d}\,H \wedge \mathrm{d}\,G, \ \mathrm{d}\omega = 0 \ \text{ and } \ \mathrm{L}_\chi \omega = 0 \tag{9}$$

We have, from (9), that the following are invariant integrals (Poincare-Cartan's invariant integrals) with respect to χ:

$$\int \omega, \int \omega \wedge \omega, \cdots, \int (\omega \wedge \cdots \wedge \omega)$$

where the last integrand consist of $n/3$ factors. Since ω is a 3-form, however, only the first of these is nonvanishing. Therefore, the $3m$ dimensional volume of a region of phase space vanishes, devoiding of any meaning to the Liouville condition, which states that the volume of a region of phase space is preserved during evolution of the system. Then, only in the case $n = 3$ for this approach does there seem to be a generalization of the Liouville theorem.

On the other hand, let us define a set of "partial" exterior derivative operators (see[7], $n = 3m$) d_α, $\alpha = 1, 2, \ldots, m$, such that for any p-form $\sigma = \sigma_{j_1 \cdots j_p} \, \mathrm{d}x_{j_1} \wedge \cdots \wedge \mathrm{d}x_{j_p} \equiv \boldsymbol{\sigma}_\mathbf{j} \, \mathrm{d}\mathbf{x}_\mathbf{j}$ is $\mathrm{d}_\alpha \sigma = \sum_{i=1}^{3} \frac{\partial \sigma_\mathbf{j}}{\partial x_i^\alpha} \, \mathrm{d}x_i^\alpha \wedge \mathrm{d}\mathbf{x}_\mathbf{j}$

Define now, m three-forms as $\omega_\alpha = \mathrm{d}x_1^\alpha \wedge \mathrm{d}x_2^\alpha \wedge \mathrm{d}x_3^\alpha$. Then, using that $\mathrm{d}_\alpha \mathrm{d}_\beta = -\mathrm{d}_\beta \mathrm{d}_\alpha$, $\mathrm{d}_\alpha^2 = 0$ and $\mathrm{d} = \mathrm{d}_1 + \mathrm{d}_2 + \cdots + \mathrm{d}_m$, a simple calculation shows that for a Nambu vector field χ is $i_\chi \omega_\alpha = \mathrm{d}_\alpha H \wedge \mathrm{d}_\alpha G$, $\alpha = 1, 2, \ldots, m$.

Also, $\mathrm{L}_\chi \omega_\alpha = i_\chi(\mathrm{d}\omega_\alpha) + \mathrm{d}(i_\chi \omega_\alpha) = \mathrm{d}(\mathrm{d}_\alpha H \wedge \mathrm{d}_\alpha G) \neq 0$ since $\mathrm{d}\,\mathrm{d}_\alpha \neq 0$ and for $\omega = \sum_{\alpha=1}^{m} \omega_\alpha$ is $\mathrm{L}_\chi \omega = \sum_{\alpha=1}^{m} \mathrm{L}_\chi \omega_\alpha \neq 0$. Therefore, neither $\int \omega_\alpha$ nor $\int \omega$ are invariants. In the same manner, none of the products $\omega_\alpha \wedge \omega_\beta$, $\omega_\alpha \wedge \omega_\beta \wedge \omega_\gamma, \ldots$ is χ invariant except for the last one, the form of volume $\Omega = \omega_1 \wedge \cdots \wedge \omega_m = \mathrm{d}x_1^1 \wedge \mathrm{d}x_2^1 \wedge \mathrm{d}x_3^1 \wedge \cdots \wedge \mathrm{d}x_3^m$, since $\mathrm{L}_\chi \Omega = 0$ (Ω is up to constant multiple, the only candidate which can be constructed from ω_α).

Thus, no invariant integral exists but the highest one $\int \Omega$ and this result is just the opposite to that given in (9).

In an attempt to clarify the situation in $n = 3m$ dimensional phase space of Nambu dynamical systems, we propose a modification of the exterior calculus of Cartan which allows that all powers of the basic 3-form become invariant integrals up to the volume $3m$-form: The new exterior product (denoted by $\bar{\wedge}$ and called extended exterior product or just e-product) distinguishes when two variables belong to different tuples. The extended exterior product of two 1-forms is symmetric if they belongs to different tuples and is antisymmetric if they belong to the same tuple. The forms built upon the $\bar{\wedge}$ product will be referred to as extended exterior forms or e-forms.

Notation and Conventions: In $\mathbb{R}^n$, $n = ms$, $\mathbf{x} = (x_1^\alpha, x_2^\alpha, \ldots, x_s^\alpha)$ denotes an s-tuple, $\alpha = 1, 2, \ldots, m$. We denote by $\mathrm{d}x_j^\alpha$ (1-form) the dual to the natural basis $\partial_\alpha^j = \dfrac{\partial}{\partial x_j^\alpha}$. Also θ_j^α denotes an 1-form and for each multi-index $\mathbf{j} = (j_1, \ldots, j_k)$, $\boldsymbol{\phi} = (\phi_1, \ldots, \phi_k)$, $\boldsymbol{\omega}_\mathbf{j}^{\boldsymbol{\phi}}$, $\boldsymbol{\mu}_\mathbf{j}^{\boldsymbol{\phi}}$ are k-eforms, i.e., $\boldsymbol{\omega}_\mathbf{j}^{\boldsymbol{\phi}} = \omega_{j_1}^{\phi_1} \bar{\wedge} \omega_{j_2}^{\phi_2} \bar{\wedge} \cdots \bar{\wedge} \omega_{j_k}^{\phi_k}$. Furthermore, $\boldsymbol{\omega}_I^{\boldsymbol{\phi}} = \boldsymbol{\omega}_{I_1}^{\phi_1} \bar{\wedge} \boldsymbol{\omega}_{I_2}^{\phi_2} \bar{\wedge} \cdots \bar{\wedge} \boldsymbol{\omega}_{I_k}^{\phi_k}$, if $I, I_1, \ldots, I_k$ are sets and $\phi_i \neq \phi_j$; $\boldsymbol{\omega}_I^\alpha = \omega_{i_1}^\alpha \bar{\wedge} \omega_{i_2}^\alpha \bar{\wedge} \omega_{i_3}^\alpha$ whenever $I = \{i_1, i_2, i_3\}$. The summation convention is adopted.

2. MODIFIED RULES

With linear forms, cotangent spaces at points of an n-manifold and its tensor products (space of n-fold covariant tensors), symmetrization, antisymmetrization, we recall that, given two covariant tensors $z_p \in \mathbf{T}_p$ and $z_q \in \mathbf{T}_q$, their exterior product, denoted by $z_p \wedge z_q$, is defined as the antisymmetric tensor of order $p+q$ given by the formula

$$z_p \wedge z_q = \frac{1}{p!\,q!}\, a(z_p \otimes z_q)$$

while the symmetric product $z_p \wedge_s z_q$ is defined by

$$z_p \wedge_s z_q = \frac{1}{p!\,q!}\, s(z_p \otimes z_q)$$

2. 1. THE EXTENDED EXTERIOR PRODUCT

Consider $\mathbb{R}^n$, with $n = 3m$. Group the variables in m 3-tuples by labeling them with two indexes: the superindex for the 3-tuple (there are m 3-tuples) and the subindex to specify its place inside the 3-tuple. Thus $\mathbf{x} = (x_j^\alpha)$, where $j = 1, 2, 3$ and $\alpha = 1, 2, \ldots, m$.

Definition 1 *A product on 1-forms, denoted $\bar{\wedge}$ and called extended exterior product or eproduct is introduced by $\bar{\wedge} = \delta_{\alpha\beta} \wedge + (1-\delta_{\alpha\beta})\wedge_s$; $\alpha, \beta = 1, 2, \ldots, m$.*

With this new product, we get for two 1-forms θ_i^α, θ_j^β that

$$\theta_i^\alpha \,\bar{\wedge}\, \theta_j^\alpha = \theta_i^\alpha \wedge \theta_j^\alpha = \theta_i^\alpha \otimes \theta_j^\alpha - \theta_j^\alpha \otimes \theta_i^\alpha, \quad \textit{and}$$
$$\theta_i^\alpha \,\bar{\wedge}\, \theta_j^\beta = \theta_i^\alpha \wedge_s \theta_j^\beta = \theta_i^\alpha \otimes \theta_j^\beta + \theta_j^\beta \otimes \theta_i^\alpha, \ \textit{for } \alpha \neq \beta$$

Thus $\theta_i^\alpha \,\bar{\wedge}\, \theta_j^\alpha = -\theta_j^\alpha \,\bar{\wedge}\, \theta_i^\alpha$ and $\theta_i^\alpha \,\bar{\wedge}\, \theta_j^\beta = \theta_j^\beta \,\bar{\wedge}\, \theta_i^\alpha$ whenever $\alpha \neq \beta$, i.e. $\theta_i^\alpha \wedge \theta_j^\beta = (-1)^{\delta_{\alpha\beta}}\, \theta_j^\beta \,\bar{\wedge}\, \theta_i^\alpha$.

Definition 2 *The extended product $\bar{\wedge}$ of an 1-eform, θ_i^α, with a k-eform, $\omega_I^{\boldsymbol\beta}$ (being I a set and $\boldsymbol\beta = (\beta_1, \beta_2, \ldots, \beta_s)$ with $\beta_i \neq \beta_j$) is defined as*

$$\theta_i^\alpha \,\bar{\wedge}\, \omega_I^{\boldsymbol\beta} = \theta_i^\alpha \,\bar{\wedge}\, (\omega_{I_1}^{\beta_1} \,\bar{\wedge}\, \omega_{I_2}^{\beta_2} \,\bar{\wedge} \cdots \bar{\wedge}\, \omega_{I_s}^{\beta_s}) = \frac{1}{k!}\, s(\theta_i^\alpha \otimes \omega_I^{\boldsymbol\beta})$$

if α is different to each of the components $\beta_1, \beta_2, \ldots, \beta_s$ of the multindex $\boldsymbol\beta$. In this case, $\theta_i^\alpha \,\bar{\wedge}\, \omega_I^{\boldsymbol\beta} = \omega_I^{\boldsymbol\beta} \,\bar{\wedge}\, \theta_i^\alpha$.

If $\alpha = \beta_j$ for some j, then it is defined by

$$\begin{aligned} \theta_i^\alpha \,\bar{\wedge}\, \omega_I^{\boldsymbol\beta} &= \theta_{I_\alpha}^\alpha \,\bar{\wedge}\, (\omega_{I_1}^{\beta_1} \,\bar{\wedge}\, \omega_{I_2}^{\beta_2} \,\bar{\wedge} \cdots \bar{\wedge}\, \overset{\circ}{\omega}{}_{I_j}^{\alpha} \,\bar{\wedge} \cdots \bar{\wedge}\, \omega_{I_s}^{\beta_s}) \\ &= \frac{1}{o(I_\alpha)!\,(k - o(I_j))!}\, s(\theta_{I_\alpha}^\alpha \otimes \overset{\circ}{\omega}{}_I^{\boldsymbol\beta}) \end{aligned}$$

where $I_\alpha = \{i\} \cup I_j$ and $\overset{\circ}{\omega}{}_{I_j}^{\alpha}$ means that all $\omega_{i_j}^\alpha$, $i_j \in I_j$ were deleted. Note that $\theta_{I_\alpha}^\alpha$ is antisymmetric. In this case, $\theta_i^\alpha \,\bar{\wedge}\, \omega_I^{\boldsymbol\beta} = (-1)^{\delta_{\alpha\beta}}\, \omega_I^{\boldsymbol\beta} \,\bar{\wedge}\, \theta_i^\alpha$, where $\delta_{\alpha\beta} = o(I_j)$.

We can generalize the eproduct for arbitrary forms: On an arbitrary p-form, order the factors in such a way that all belonging to a same tuple, appears grouped; inside each group, a strict order is defined and $\wedge_s$ will replace $\bar{\wedge}$ in those places where two neighboring factors have different superindex. Moreover, by the commutative property of $\bar{\wedge}_s$ and bilinearity, they results grouped in partial factors, each one with his own superindex (each factor is at most a 3-eform). To make the product of a p-form with a q-form, we use the notation ω_I^α. We can write uniquely the eproduct of a p-form $\omega_{\mathrm{j}}^{\phi}$ with a q-form $\mu_{\mathrm{k}}^{\varphi}$ as $\wedge_s$ products of at most 3-forms ω_I^α, where I is $\{1\}$, $\{1,2\}$ or $\{1,2,3\}$.

Formally, $\omega_j{}^{\phi}\bar{\wedge}\mu_{\mathrm{k}}^{\varphi} = \omega_{I_1}^{\alpha_1}\wedge_s\cdots\wedge_s\omega_{I_l}^{\alpha_l}$ (assuming that $\alpha_1, \alpha_2, \ldots, \alpha_l$ are different between them) is a $(p+q)$-eform if $\mathrm{o}(I_1)+\cdots+\mathrm{o}(I_l) = p+q$. Now, we can use the associative and commutative properties of $\wedge_s$. Note that for each α, $\wedge$ appears inside ω_I^α, i.e. $\omega_I^\alpha = \omega_1^\alpha\wedge\omega_2^\alpha\wedge\omega_3^\alpha$ (at most).

2. 2. THE EXTENDED DIFFERENTIAL

Let M be a $3m$-manifold (whenever we write $3m$-manifold, we are assuming it is also smooth, i.e., a C^k manifold where k is large enough for the statement to be meaningful, eventually $k=\infty$). For every point $a\in M$, there is an admissible local chart (U,φ) with $a\in U$ and $\varphi(U)\subset \mathbb{R}^{3m}$. Let $f: M\longrightarrow\mathbb{R}$ be C^∞ so $\mathrm{T}f:\mathrm{T}M\longrightarrow \mathrm{T}\mathbb{R}=\mathbb{R}\times\mathbb{R}$. Recall that the tangent of f is defined by $\mathrm{T}f(u,e) = (f(u), Df(u)\cdot e)$, where $Df(u)$ is the derivative of f at u. For each $m\in M$, we can think of $\mathrm{T}f$ acting on a vector $v\in\mathrm{T}_m M$ in the following way: $\mathrm{T}f(v) = (f(m), \mathrm{d}f(m)\cdot v)$. This defines the element $\mathrm{d}f(m)\in\mathrm{T}_m^* M$. Thus $\mathrm{d}f: M\longrightarrow \mathrm{T}^*M$ is a section of T^*M, a covector field or a form and it is called the differential of f. If local coordinates (x_j^α); $j=1,2,3$; $\alpha=1,2,\ldots,m$ are introduced, then the local components of $\mathrm{d}f$ are $(\mathrm{d}f)_j^\alpha = \frac{\partial f}{\partial x_j^\alpha}$ and we can write $\mathrm{d}f = \frac{\partial f}{\partial x_j^\alpha}\,\mathrm{d}x_j^\alpha$.

Now we can define partial derivatives subordinated to the superindex (tuple) α:

$$\hat{\partial}_\alpha^i f = p^\alpha \partial_\alpha^i f \quad\text{or}\quad \frac{\hat{\partial} f}{\partial x_i^\alpha} = p^\alpha\frac{\partial f}{\partial x_i^\alpha},\ i=1,2,3$$

for each superindex α, with p^α an invertible algebraic object, independent of x, depending on the 3-tuple determined by α and to be specified soon, used to introduce non-commutativeness.

For the second partial derivatives, we have

$$\begin{gathered}\hat{\partial}_\beta^j\hat{\partial}_\alpha^i f = \hat{\partial}_\beta^j(p^\alpha\partial_\alpha^i f) = p^\beta p^\alpha \partial_\beta^j\partial_\alpha^i f,\\ \hat{\partial}_\alpha^i\hat{\partial}_\beta^j f = \hat{\partial}_\alpha^i(p^\beta\partial_\beta^j f) = p^\alpha p^\beta\partial_\alpha^i\partial_\beta^j f \quad\text{and}\\ (p^\beta)^{-1}(p^\alpha)^{-1}\hat{\partial}_\alpha^i\hat{\partial}_\beta^j f = \partial_\alpha^i\partial_\beta^j f\end{gathered}$$

Since $\partial_\beta^j\partial_\alpha^i f = \partial_\alpha^i\partial_\beta^j f$, it follows that $\hat{\partial}_\beta^j\hat{\partial}_\alpha^i f = p^\beta p^\alpha (p^\beta)^{-1}(p^\alpha)^{-1}\hat{\partial}_\alpha^i\hat{\partial}_\beta^j f$. Note that

i) If $\alpha = \beta$ then $\hat{\partial}^j_\beta \hat{\partial}^i_\alpha f = \hat{\partial}^i_\alpha \hat{\partial}^j_\beta f$, i.e. the second partial derivatives commute since $p^\beta p^\alpha (p^\beta)^{-1}(p^\alpha)^{-1} = 1$

ii) If $\alpha \neq \beta$, we take $p^\beta p^\alpha (p^\beta)^{-1}(p^\alpha)^{-1} = -1$, so $\hat{\partial}^j_\beta \hat{\partial}^i_\alpha f = -\hat{\partial}^i_\alpha \hat{\partial}^j_\beta f$, i.e. the second partial derivatives anticommute if $p^\beta p^\alpha = -p^\alpha p^\beta$. Thus, we have the following

Proposition 1 *The partial derivatives operators $\hat{\partial}$ satisfies*

$$\hat{\partial}^j_\beta \hat{\partial}^i_\alpha f = (-1)^{1-\delta_{\alpha\beta}} \hat{\partial}^i_\alpha \hat{\partial}^j_\beta f$$

Remark 1 *If $f \in \mathcal{F}(M)$, the space $C^\infty(M, I\!R)$, then f is a 0-form*

Remark 2 *The differential $\mathrm{d}\, f$, $f \in \mathcal{F}(M)$, in local coordinates is $\mathrm{d}\, f = \frac{\partial f}{\partial x^\alpha_i} \mathrm{d}\, x^\alpha_i = (p^\alpha)^{-1} \frac{\hat{\partial} f}{\partial x^\alpha_i} \mathrm{d}\, x^\alpha_i = \frac{\hat{\partial} f}{\partial x^\alpha_i} (p^\alpha)^{-1} \mathrm{d}\, x^\alpha_i = \frac{\hat{\partial} f}{\partial x^\alpha_i} \hat{\mathrm{d}} x^\alpha_i$, where $\hat{\mathrm{d}} x^\alpha_i = (p^\alpha)^{-1} \mathrm{d}\, x^\alpha_i$.*

Clearly $\hat{\mathrm{d}} x^\alpha_i$ is dual to $\hat{\partial}^i_\alpha = p^\alpha \partial^i_\alpha$.

The extended differential of $f \in \mathcal{F}(M)$ in local coordinates is defined by $\hat{\mathrm{d}} f = \frac{\hat{\partial} f}{\partial x^\alpha_i} \hat{\mathrm{d}} x^\alpha_i$. This formula, applied to the coordinate functions x^α_i shows that the differential of x^α_i is the 1-form $\hat{\mathrm{d}} x^\alpha_i$

Lemma 1 *If f is a 0-form, then $\hat{\mathrm{d}}^2 f = 0$*

Proof. $\hat{\mathrm{d}}^2 f = \hat{\mathrm{d}}(\hat{\mathrm{d}} f) = \hat{\mathrm{d}}(\hat{\partial}^i_\alpha f\, \hat{\mathrm{d}} x^\alpha_i) = \hat{\partial}^j_\beta \hat{\partial}^i_\alpha f \hat{\mathrm{d}} x^\beta_j \bar{\wedge} \hat{\mathrm{d}} x^\alpha_i$

Also $\hat{\mathrm{d}}^2 f = \hat{\partial}^i_\alpha \hat{\partial}^j_\beta f\, \hat{\mathrm{d}} x^\alpha_i \bar{\wedge} \hat{\mathrm{d}} x^\beta_j$. Thus $\hat{\partial}^i_\alpha \hat{\partial}^j_\beta f\, \hat{\mathrm{d}} x^\alpha_i \bar{\wedge} \hat{\mathrm{d}} x^\beta_j = \hat{\partial}^j_\beta \hat{\partial}^i_\alpha f\, \hat{\mathrm{d}} x^\beta_j \bar{\wedge} \hat{\mathrm{d}} x^\alpha_i = (-1)^{1-\delta_{\alpha\beta}} \hat{\partial}^i_\alpha \hat{\partial}^j_\beta f\, (-1)^{\delta_{\alpha\beta}}\, \hat{\mathrm{d}} x^\alpha_i \bar{\wedge} \hat{\mathrm{d}} x^\beta_j = -\hat{\partial}^i_\alpha \hat{\partial}^j_\beta f\, \hat{\mathrm{d}} x^\alpha_i \bar{\wedge} \hat{\mathrm{d}} x^\beta_j$.

Hence $\hat{\mathrm{d}}^2 f = \hat{\partial}^i_\alpha \hat{\partial}^j_\beta f\, \hat{\mathrm{d}} x^\alpha_i \bar{\wedge} \hat{\mathrm{d}} x^\beta_j = 0$. ∎

Since $\hat{\mathrm{d}}(\hat{\mathrm{d}} x_i) = 0$, define $\hat{\mathrm{d}}$ on 1-forms by $\hat{\mathrm{d}}(f(x)\, \hat{\mathrm{d}} x^\alpha_i) = \hat{\mathrm{d}} f(x) \bar{\wedge} \hat{\mathrm{d}} x^\alpha_i$, i.e.

$$\hat{\mathrm{d}}(f(x)\, \hat{\mathrm{d}} x^\alpha_i) = \hat{\partial}^j_\beta f\, \hat{\mathrm{d}} x^\beta_j \bar{\wedge} \hat{\mathrm{d}} x^\alpha_i$$

By analogy, define $\hat{\mathrm{d}}(f(x)\, \hat{\mathrm{d}} \mathrm{x_j}^\phi) = \hat{\mathrm{d}} f \bar{\wedge} \hat{\mathrm{d}} \mathrm{x_j}^\phi = \hat{\mathrm{d}} f \bar{\wedge} \hat{\mathrm{d}} x^{\phi_1}_{j_1} \bar{\wedge} \hat{\mathrm{d}} x^{\phi_2}_{j_2} \bar{\wedge} \cdots \bar{\wedge} \hat{\mathrm{d}} x^{\phi_k}_{j_k}$

Lemma 2 *Let $\theta^\alpha_1 = f(x)\, \hat{\mathrm{d}} x^\alpha_i$ and $\theta^\beta_2 = g(x)\, \hat{\mathrm{d}} x^\beta_j$ be 1-eforms, then $\hat{\mathrm{d}}(\theta^\alpha_1 \bar{\wedge} \theta^\beta_2) = \hat{\mathrm{d}}\theta^\alpha_1 \bar{\wedge} \theta^\beta_2 + (-1)^{\delta_{\alpha\beta}}\, \hat{\mathrm{d}}\theta^\beta_2 \bar{\wedge} \theta^\alpha_1$*

Proof. We have $\hat{\mathrm{d}}\theta^\alpha_1 = \hat{\mathrm{d}} f \bar{\wedge} \hat{\mathrm{d}} x^\alpha_i$, $\hat{\mathrm{d}}\theta^\beta_2 = \hat{\mathrm{d}} g \bar{\wedge} \hat{\mathrm{d}} x^\beta_j$. Thus

$$\begin{aligned}
\hat{\mathrm{d}}\theta_1 \bar{\wedge} \theta_2 &= \hat{\mathrm{d}} f \bar{\wedge} \hat{\mathrm{d}} x^\alpha_i \bar{\wedge} g \hat{\mathrm{d}} x^\beta_j = g\, \hat{\mathrm{d}} f \bar{\wedge} \hat{\mathrm{d}} x^\alpha_i \bar{\wedge} \hat{\mathrm{d}} x^\beta_j \\
\hat{\mathrm{d}}\theta_2 \bar{\wedge} \theta_1 &= \hat{\mathrm{d}} g \bar{\wedge} \hat{\mathrm{d}} x^\beta_j \bar{\wedge} f \hat{\mathrm{d}} x^\alpha_i = f\, \hat{\mathrm{d}} g \bar{\wedge} \hat{\mathrm{d}} x^\beta_j \bar{\wedge} \hat{\mathrm{d}} x^\alpha_i \quad \text{and} \\
\hat{\mathrm{d}}(\theta_1 \bar{\wedge} \theta_2) &= \hat{\mathrm{d}}(fg\, \hat{\mathrm{d}} x^\alpha_i \bar{\wedge} \hat{\mathrm{d}} x^\beta_j) = \hat{\mathrm{d}}(fg) \bar{\wedge} \hat{\mathrm{d}} x^\alpha_i \bar{\wedge} \hat{\mathrm{d}} x^\beta_j \\
&= (\hat{\mathrm{d}} f\, g + f\, \hat{\mathrm{d}} g) \bar{\wedge} \hat{\mathrm{d}} x^\alpha_i \bar{\wedge} \hat{\mathrm{d}} x^\beta_j \\
&= g\, \hat{\mathrm{d}} f \bar{\wedge} \hat{\mathrm{d}} x^\alpha_i \bar{\wedge} \hat{\mathrm{d}} x^\beta_j + f\, \hat{\mathrm{d}} g \bar{\wedge} \hat{\mathrm{d}} x^\alpha_i \bar{\wedge} \hat{\mathrm{d}} x^\beta_j \\
&= \hat{\mathrm{d}}\theta_1 \bar{\wedge} \theta_2 + f\, \hat{\mathrm{d}} g \bar{\wedge} \hat{\mathrm{d}} x^\alpha_i \bar{\wedge} \hat{\mathrm{d}} x^\beta_j \\
&= \hat{\mathrm{d}}\theta_1 \bar{\wedge} \theta_2 + (-1)^{\delta_{\alpha\beta}} f\, \hat{\mathrm{d}} g \bar{\wedge} \hat{\mathrm{d}} x^\beta_j \bar{\wedge} \hat{\mathrm{d}} x^\alpha_i \\
&= \hat{\mathrm{d}}\theta_1 \bar{\wedge} \theta_2 + (-1)^{\delta_{\alpha\beta}}\, \hat{\mathrm{d}}\theta_2 \bar{\wedge} \theta_1
\end{aligned}$$

■

Similarly, we can demonstrate

Lemma 3 *Let* $\omega = f(x)\hat{\mathrm{d}}\mathbf{x}^{\alpha}_{\mathbf{j}}$ *be a p-eform and* $\mu = g(x)\hat{\mathrm{d}}\mathbf{x}^{\beta}_{\mathbf{k}}$ *a q-eform with* α, β, $\mathbf{j}$, $\mathbf{k}$ *multiindexes and* $f, g \in \mathcal{F}(M)$. *Then*

$$\hat{\mathrm{d}}(\omega \mathbin{\bar{\wedge}} \mu) = \hat{\mathrm{d}}\omega \mathbin{\bar{\wedge}} \mu + (-1)^{\delta_{\alpha\beta}} \hat{\mathrm{d}}\mu \mathbin{\bar{\wedge}} \omega$$

where $\delta_{\alpha\beta}$ *is a natural number depending heavily on* α *and* β

Let ω be a differential p-eform. The exterior differentiation operator $\hat{\mathrm{d}}$ maps ω into a differential $(p+1)$-eform called the extended exterior derivative of ω. Still better, if $\bar{\Omega}^k(M)$ denotes C^∞ sections of the exterior k-eforms bundle on the tangent spaces of M (M is, from here to the end, a smooth manifold), then we extended the differential of functions to a map $\hat{\mathrm{d}}$, $\hat{\mathrm{d}} : \bar{\Omega}^k(M) \longrightarrow \bar{\Omega}^{k+1}(M)$ for any k, so

Theorem 2 *Let* M *be a* $3m$*-dimensional manifold. There is a unique family of mappings* $\hat{\mathrm{d}}^k$, $\hat{\mathrm{d}}^k : \bar{\Omega}^k(U) \longrightarrow \bar{\Omega}^{k+1}(U)$ ($k = 0, 1, \ldots, 3m$ *and* U *is open in* M), *which we denote by* $\hat{\mathrm{d}}$, *called the extended exterior derivative on* M, *such that*

(i) $\hat{\mathrm{d}}$ *is linear:*
$$\begin{aligned} \hat{\mathrm{d}}(\omega + \mu) &= \hat{\mathrm{d}}\omega + \hat{\mathrm{d}}\mu \\ \hat{\mathrm{d}}(\lambda\omega) &= \lambda\,\hat{\mathrm{d}}\omega \quad \text{for} \quad \lambda \in I\!R \end{aligned}$$

(ii) For $\omega \in \bar{\Omega}^k(U)$ *and* $\mu \in \bar{\Omega}^q(U)$, *we have*

$$\hat{\mathrm{d}}(\omega \mathbin{\bar{\wedge}} \mu) = \hat{\mathrm{d}}\omega \mathbin{\bar{\wedge}} \mu + (-1)^{\delta_{\alpha\beta}} \hat{\mathrm{d}}\mu \mathbin{\bar{\wedge}} \omega$$

where $\delta_{\alpha\beta}$ *is a natural number depending on* α, β

(iii) If $f \in \mathcal{F}(U)$, *then* $\hat{\mathrm{d}}f = \hat{\partial}^i_\alpha f \,\hat{\mathrm{d}}x^\alpha_i$

(iv) $\hat{\mathrm{d}}^2 = \hat{\mathrm{d}} \circ \hat{\mathrm{d}} = 0$

(v) $\hat{\mathrm{d}}$ *is a local operator, that is, if* $U \subset V \subset M$ *are open and* $\omega \in \bar{\Omega}^k(V)$ *then* $\hat{\mathrm{d}}(\omega|_U) = (\hat{\mathrm{d}}\omega)|_U$ *[here,* $\omega|_U$ *means* ω *restricted to* U*]*

Proof. The proof is close to theorem 6.4.1 in [2].

3. THE CANONICAL 3-EFORM

In $I\!R^n$, $n = 3m$, a system is described in terms of the following 3-eform (which will be called the canonical 3-eform)

$$\omega = \sum_{\alpha=1}^{m} \hat{\mathrm{d}}x^\alpha_1 \mathbin{\bar{\wedge}} \hat{\mathrm{d}}x^\alpha_2 \mathbin{\bar{\wedge}} \hat{\mathrm{d}}x^\alpha_3 = \sum_{\alpha=1}^{n} \hat{\mathrm{d}}x^\alpha_1 \wedge \hat{\mathrm{d}}x^\alpha_2 \wedge \hat{\mathrm{d}}x^\alpha_3$$

Proposition 3 *In $\mathbb{R}^{3m}$, k-eforms can be constructed with $k \leq 3m$ only*

Proof. It is easy to check that with the rules for the eproduct, its square

$$\begin{aligned}(\bar{\wedge}\omega)^2 &= \omega \bar{\wedge} \omega = \sum_{\alpha,\beta=1}^{m} \hat{d}x_1^\alpha \bar{\wedge} \hat{d}x_2^\alpha \bar{\wedge} \hat{d}x_3^\alpha \;\bar{\wedge}\; \hat{d}x_1^\beta \bar{\wedge} \hat{d}x_2^\beta \bar{\wedge} \hat{d}x_3^\beta \\ &= \sum_{\alpha,\beta=1}^{m} (-1)^{\delta_{\alpha\beta}} \hat{d}x_1^\beta \bar{\wedge} \hat{d}x_2^\beta \bar{\wedge} \hat{d}x_3^\beta \;\bar{\wedge}\; \hat{d}x_1^\alpha \bar{\wedge} \hat{d}x_2^\alpha \bar{\wedge} \hat{d}x_3^\alpha\end{aligned}$$

does not vanish as occurs when the antisymmetric product $(\wedge\omega)^2$ is computed. An easy computation shows that $(\bar{\wedge}\omega)^{3m+1} = 0$ ■

4. THE EXTENDED INTERIOR PRODUCT

We recall that the usual interior product of a vector $v \in \mathrm{T}_x$ with a tensor $\tau \in \mathrm{T}^n_m$ is the $(m, n-1)$ tensor defined by

$$(i_v(\tau))(\theta^1, \theta^2, \ldots, \theta^m, v_1, v_2, \ldots, v_{n-1}) = \tau(\theta^1, \ldots, \theta^m, v, v_1, v_2, \ldots, v_{n-1})$$

If e_k (resp θ^k) denotes the k-th basis (resp dual basis) element of T_x, we have

$$\begin{aligned}&i_{e_k}(e_{i_1} \otimes \cdots \otimes e_{i_r} \otimes \theta^{j_1} \otimes \cdots \otimes \theta^{j_s}) = \\ &= \delta_k^{j_1}\, e_{i_1} \otimes \cdots \otimes e_{i_r} \otimes \theta^{j_2} \otimes \cdots \otimes \theta^{j_s} \quad \text{and}\end{aligned}$$

$$i_{e_k}(\theta^{j_1} \otimes \theta^{j_2} \otimes \cdots \otimes \theta^{j_s}) = \delta_k^{j_1}\, \theta^{j_2} \otimes \cdots \otimes \theta^{j_s}$$

These formulae and linearity enable us to compute any interior product.

In $\mathbb{R}^{3n}$, let v be a vector field denoted by $v = v_i^\alpha \hat{\partial}_\alpha^i$

Since $f \in \mathcal{F}(M)$ is an 0-eform, we define $\hat{\imath}_v(f) = 0$. Furthermore $\hat{\imath}_v\, \hat{d}x_j^\alpha = v_j^\alpha$ since $\hat{d}x_j^\alpha \circ \hat{\partial}_\beta^i = 0$ and $\hat{d}x_j^\alpha \circ \hat{\partial}_\alpha^j = 1$. The extended interior product or e-interior product of a vector $v \in \mathrm{T}_x$ with the eproduct of two 1-eforms is

$$\begin{aligned}\hat{\imath}_v(\hat{d}x_i^\alpha \bar{\wedge} \hat{d}x_j^\alpha) &= \hat{\imath}_v(\hat{d}x_i^\alpha \otimes \hat{d}x_j^\alpha - \hat{d}x_j^\alpha \otimes \hat{d}x_i^\alpha) \\ &= v_i^\alpha \hat{d}x_j^\alpha - v_j^\alpha \hat{d}x_i^\alpha \quad \text{and for } \alpha \neq \beta \text{ it is} \\ \hat{\imath}_v(\hat{d}x_i^\alpha \bar{\wedge} \hat{d}x_j^\beta) &= \hat{\imath}_v(\hat{d}x_i^\alpha \otimes \hat{d}x_j^\beta + \hat{d}x_j^\beta \otimes \hat{d}x_i^\alpha) \\ &= v_i^\alpha \hat{d}x_j^\beta + v_j^\beta \hat{d}x_i^\alpha = (\hat{\imath}_v\, \hat{d}x_i^\alpha) \bar{\wedge} \hat{d}x_j^\beta + \hat{\imath}_v(\hat{d}x_j^\beta) \bar{\wedge} \hat{d}x_i^\alpha\end{aligned}$$

Thus $\hat{\imath}_v(\hat{d}x_i^\alpha \bar{\wedge} \hat{d}x_j^\beta) = \hat{\imath}_v(\hat{d}x_i^\alpha) \bar{\wedge} \hat{d}x_j^\beta + (-1)^{\delta_{\alpha\beta}}\, \hat{\imath}_v(\hat{d}x_j^\beta) \bar{\wedge} \hat{d}x_i^\alpha$

This means that the e-interior product is either a derivation or antiderivation as the result of its action on a 2-eform.

Definition 3 *Let* T *be a linear operator on forms;* T *is said to be a* $\overline{\wedge}$*-antiderivation if it is a* $\wedge_s$*-derivation and a* $\wedge$*-antiderivation.*

Definition 4 *The extended interior product or contracted multiplication of a vector* $v \in \mathrm{T}_x$ *and a e-form* ω*, denoted* $\hat{\imath}_v\ \omega$ *is defined as follows:*

i) $\hat{\imath}_v f = 0,\ f \in \mathcal{F}(M)$

ii) $\hat{\imath}_v \hat{\mathrm{d}} x_i^\alpha = v_i^\alpha$

iii) $\hat{\imath}_v$ *is a* $\overline{\wedge}$*-antiderivation*

5. THE LIE DERIVATIVE

The Lie derivative is defined in the usual way as a derivation. If M is a $3m$-manifold, a vector field on M is a section of the tangent bundle $\mathrm{T}\,M$ of M and we will suppose that it is C^∞.

For $f \in \mathcal{F}(M)$, define the Lie derivative of f along the vector field $\mathbf{v}$ by

$$\mathrm{L}_\mathbf{v} f(m) = \mathbf{v}(f)(m) = \hat{\mathrm{d}} f(m) \circ \mathbf{v}(m) \quad \text{for any } m \in M.$$

In local coordinates (x_i^α), $i = 1,2,3$; $\alpha = 1,2,\ldots,m$

$$\mathrm{L}_\mathbf{v} f = \mathbf{v}(f) = \hat{\partial}^i_\alpha f \hat{\mathrm{d}} x_i^\alpha \cdot v_i^\alpha \hat{\partial}^i_\alpha = v_i^\alpha \hat{\partial}^i_\alpha f$$

This properties and the tensor derivation property can be used to compute L_X on forms.

For tensor τ and θ, it is $\mathrm{L}_\mathbf{v}(\tau + \theta) = \mathrm{L}_\mathbf{v}(\tau) + \mathrm{L}_\mathbf{v}(\theta)$ and $\mathrm{L}_\mathbf{v}(\tau \otimes \theta) = \mathrm{L}_\mathbf{v}(\tau) \otimes \theta + \tau \otimes \mathrm{L}_\mathbf{v}(\theta)$ and for forms ω, μ we have $\mathrm{L}_\mathbf{v}(\omega \mathbin{\overline{\wedge}} \mu) = \mathrm{L}_\mathbf{v}(\omega) \mathbin{\overline{\wedge}} \mu + \omega \mathbin{\overline{\wedge}} \mathrm{L}_\mathbf{v}(\mu)$

We recall that $\mathrm{L}_\mathbf{v} \hat{\partial}^i_\alpha = -\hat{\partial}^i_\alpha v_k^\beta \hat{\partial}^k_\beta$ and $\mathrm{L}_\mathbf{v} \hat{\mathrm{d}} x_i^\alpha = \hat{\partial}^k_\beta v_i^\alpha \hat{\mathrm{d}} x_k^\beta = \hat{\mathrm{d}} v_i^\alpha = \hat{\mathrm{d}}\, \mathrm{L}_\mathbf{v} x_i^\alpha$

Thus, the Lie derivative of the vector field $\mathbf{u}$ with respect to $\mathbf{v}$ is

$$\begin{aligned}
\mathrm{L}_\mathbf{v} \mathbf{u} &= \mathrm{L}_\mathbf{v}(u_i^\alpha \hat{\partial}^i_\alpha) = \mathrm{L}_\mathbf{v}(u_i^\alpha) \hat{\partial}^i_\alpha + u_i^\alpha \mathrm{L}_\mathbf{v}(\hat{\partial}^i_\alpha) \\
&= v_k^\beta \hat{\partial}^k_\beta u_i^\alpha \hat{\partial}^i_\alpha - u_i^\alpha \hat{\partial}^i_\alpha v_k^\beta \hat{\partial}^k_\beta \\
&= (v_k^\beta \hat{\partial}^k_\beta u_i^\alpha - u_k^\beta \hat{\partial}^k_\beta v_i^\alpha) \hat{\partial}^i_\alpha
\end{aligned}$$

Remark 3 *The famous formula of Cartan* $\mathrm{L}_\mathbf{v} = \mathrm{i}_v\ \mathrm{d} + \mathrm{d}\ \mathrm{i}_v$ *is not true. Furthermore,* $\mathrm{L}_\mathbf{v}$ *and* $\hat{\mathrm{d}}$ *conmute only on the coordinate functions* x_i^α*.*

6. NAMBU DYNAMICAL SYSTEMS

Let M be a $3m$-manifold and $\mathbf{v} : M \longrightarrow \mathrm{T}\,M$ a vector field.

Definition 5 *The vector field* $\mathbf{v}$ *is said to define a Nambu vector field if there exists two real functions (Hamiltonians)* $H(x)$, $G(x)$ *such that in the* (x_i^α), $\alpha = 1, 2, \ldots, m$; $i = 1, 2, 3$ *local coordinates, it can be expressed as* $\mathbf{v} = v_i^\alpha(x)\hat{\partial}_\alpha^i$ *where* $v_1^\alpha = \frac{\partial(H,G)}{\partial(x_2^\alpha, x_3^\alpha)}$, $v_2^\alpha = \frac{\partial(H,G)}{\partial(x_3^\alpha, x_1^\alpha)}$, $v_3^\alpha = \frac{\partial(H,G)}{\partial(x_1^\alpha, x_2^\alpha)}$, *for each* $\alpha = 1, 2, \ldots, m$.

Now consider on M a dynamical system defined by $\frac{\hat{d}x}{dt} = f(x)$ or $\frac{\hat{d}x_i^\alpha}{dt} = f_i^\alpha(x)$ in the (x_i^α) local coordinates. This function identifies a vector field $\mathbf{v}$, $\mathbf{v} : M \longrightarrow \mathrm{T}\,M$, given by $\mathbf{v} = f_i^\alpha(x)\hat{\partial}_\alpha^i$

Definition 6 *A dynamical system* $\dfrac{\hat{d}x}{dt} = f(x)$ *is said to be a Nambu dynamical system if there exists two real functions (Hamiltonians)* $H(x)$, $G(x)$ *such that in the* (x_i^α) *local coordinates, for each* $\alpha = 1, 2, \ldots, m$ *it is*

$$f_1^\alpha(x) = \frac{\partial(H,G)}{\partial(x_2^\alpha, x_3^\alpha)},\ f_2^\alpha(x) = \frac{\partial(H,G)}{\partial(x_3^\alpha, x_1^\alpha)},\ f_3^\alpha(x) = \frac{\partial(H,G)}{\partial(x_1^\alpha, x_2^\alpha)}$$

i.e., if the associated vector field $\mathbf{v}$ *is a Nambu vector field.*

In this case, for any function $F : M \longrightarrow \mathbb{R}$, we have

$$\frac{\hat{d}F}{dt} = \sum_{\alpha=1}^{m} \frac{\partial(F, H, G)}{\partial(x_1^\alpha, x_2^\alpha, x_3^\alpha)}$$

which determines the dynamical evolution of $F(x)$.

For each 3-tuple $(x_1^\alpha, x_2^\alpha, x_3^\alpha)$, $\alpha = 1, 2, \ldots, m$ we have

$$\frac{\hat{d}x_i^\alpha}{dt} = \frac{\hat{\partial}(H,G)}{\partial(x_2^\alpha, x_3^\alpha)},\ \frac{\hat{d}x_2^\alpha}{dt} = \frac{\hat{\partial}(H,G)}{\partial(x_3^\alpha, x_1^\alpha)},\ \frac{\hat{d}x_3^\alpha}{dt} = \frac{\hat{\partial}(H,G)}{\partial(x_1^\alpha, x_2^\alpha)}$$

and with velocity field given by

$$\frac{\hat{d}r(t)}{dt} = \left(\frac{\hat{d}x_1^\alpha}{dt}, \frac{\hat{d}x_2^\alpha}{dt}, \frac{\hat{d}x_3^\alpha}{dt}\right)$$

We will use the notation $\mathbf{v} = \sum_{\alpha=1}^m \mathbf{v}^\alpha$ with $\mathbf{v}^\alpha = v_1^\alpha\hat{\partial}_\alpha^1 + v_2^\alpha\hat{\partial}_\alpha^2 + v_3^\alpha\hat{\partial}_\alpha^3$.

Thus $\frac{\hat{d}r(t)}{dt} = \nabla H \times \nabla G$ and since $\nabla \cdot (\nabla H \times \nabla G) = 0$, it is

$$\hat{\partial}_\alpha^1 \frac{\hat{d}x_1^\alpha}{dt} + \hat{\partial}_\alpha^2 \frac{\hat{d}x_2^\alpha}{dt} + \hat{\partial}_\alpha^3 \frac{\hat{d}x_3^\alpha}{dt} = 0 \quad \text{(Liouville Theorem)}$$

Then $\hat{\partial}_\alpha^1 v_1^\alpha + \hat{\partial}_\alpha^2 v_2^\alpha + \hat{\partial}_\alpha^3 v_3^\alpha = 0$ and $\sum_{\alpha=1}^m \hat{\partial}_\alpha^1 v_1^\alpha + \hat{\partial}_\alpha^2 v_2^\alpha + \hat{\partial}_\alpha^3 v_3^\alpha = 0$, i.e., the divergence of a Nambu vector field $\mathbf{v}$ is zero.

Let ω be the canonical 3-eform, i.e. $\omega = \sum_{\alpha=1}^m \hat{d}x_1^\alpha \mathbin{\hat{\wedge}} \hat{d}x_2^\alpha \mathbin{\hat{\wedge}} \hat{d}x_3^\alpha$, then $\omega = \sum_{\alpha=1}^m \hat{d}x_1^\alpha \wedge \hat{d}x_2^\alpha \wedge \hat{d}x_3^\alpha$ is antisymmetric.

For each α, the Lie derivative of ω is

$$\begin{aligned} \mathrm{L}_{\mathbf{v}^\alpha}\,\omega &= \mathrm{L}_{\mathbf{v}^\alpha} \sum_{\alpha=1}^{m} \mathrm{d}\,x_1^\alpha \wedge \mathrm{d}\,x_2^\alpha \wedge \mathrm{d}\,x_3^\alpha = \mathbf{v}^\alpha (\sum_{\alpha=1}^{m} \hat{\mathrm{d}}x_1^\alpha \wedge \hat{\mathrm{d}}x_2^\alpha \wedge \hat{\mathrm{d}}x_3^\alpha) \\ &= \hat{\mathrm{d}}v_1^\alpha \wedge \hat{\mathrm{d}}x_2^\alpha \wedge \hat{\mathrm{d}}x_3^\alpha + \hat{\mathrm{d}}x_1^\alpha \wedge \hat{\mathrm{d}}v_2^\alpha \wedge \hat{\mathrm{d}}x_3^\alpha + \hat{\mathrm{d}}x_1^\alpha \wedge \hat{\mathrm{d}}x_2^\alpha \wedge \hat{\mathrm{d}}v_3^\alpha \\ &= \hat{\partial}_\alpha^1 v_1^\alpha \hat{\mathrm{d}}x_1^\alpha \wedge \hat{\mathrm{d}}x_2^\alpha \wedge \hat{\mathrm{d}}x_3^\alpha + \hat{\partial}_\alpha^2 v_2^\alpha \hat{\mathrm{d}}x_1^\alpha \wedge \hat{\mathrm{d}}x_2^\alpha \wedge \hat{\mathrm{d}}x_3^\alpha + \hat{\partial}_\alpha^3 v_3^\alpha \hat{\mathrm{d}}x_1^\alpha \wedge \hat{\mathrm{d}}x_2^\alpha \wedge \hat{\mathrm{d}}x_3^\alpha \\ &= (\hat{\partial}_\alpha^1 v_1^\alpha + \hat{\partial}_\alpha^2 v_2^\alpha + \hat{\partial}_\alpha^3 v_3^\alpha)\hat{\mathrm{d}}x_1^\alpha \wedge \hat{\mathrm{d}}x_2^\alpha \wedge \hat{\mathrm{d}}x_3^\alpha = 0 \end{aligned}$$

Thus $\mathrm{L}_{\mathbf{v}}\,\omega = \sum \mathrm{L}_{\mathbf{v}^\alpha}(\omega) = 0$.

On the other hand, the result of computing the extended contraction on the canonical 3-eform is

$$\begin{aligned} \hat{\imath}_v\,\omega &= \hat{\imath}_v\,(\sum_{\alpha=1}^{m} \hat{\mathrm{d}}x_1^\alpha \wedge \hat{\mathrm{d}}x_2^\alpha \wedge \hat{\mathrm{d}}x_3^\alpha) \\ &= \sum_{\alpha=1}^{m} v_i^\alpha \hat{\mathrm{d}}x_2^\alpha \wedge \hat{\mathrm{d}}x_3^\alpha - v_2^\alpha \hat{\mathrm{d}}x_1^\alpha \wedge \hat{\mathrm{d}}x_3^\alpha + v_3^\alpha \hat{\mathrm{d}}x_1^\alpha \wedge \hat{\mathrm{d}}x_2^\alpha \\ &= \sum_{\alpha=1}^{m} (\hat{\partial}_\alpha^2 H \hat{\partial}_\alpha^3 G - \hat{\partial}_\alpha^3 H \hat{\partial}_\alpha^2 G)\hat{\mathrm{d}}x_2^\alpha \wedge \hat{\mathrm{d}}x_3^\alpha + \\ &\quad -(\hat{\partial}_\alpha^3 H \hat{\partial}_\alpha^1 G - \hat{\partial}_\alpha^1 H \hat{\partial}_\alpha^3 G)\hat{\mathrm{d}}x_1^\alpha \wedge \hat{\mathrm{d}}x_3^\alpha + \\ &\quad +(\hat{\partial}_\alpha^1 H \hat{\partial}_\alpha^2 G - \hat{\partial}_\alpha^2 H \hat{\partial}_\alpha^1 G)\hat{\mathrm{d}}x_1^\alpha \wedge \hat{\mathrm{d}}x_2^\alpha \\ &= \sum_{\alpha=1}^{m} (\hat{\partial}_\alpha^1 H \hat{\mathrm{d}}x_1^\alpha + \hat{\partial}_\alpha^2 H \hat{\mathrm{d}}x_2^\alpha + \hat{\partial}_\alpha^3 H \hat{\mathrm{d}}x_3^\alpha) \,\bar{\wedge}\, (\hat{\partial}_\alpha^1 G \hat{\mathrm{d}}x_1^\alpha + \hat{\partial}_\alpha^2 G \hat{\mathrm{d}}x_2^\alpha + \hat{\partial}_\alpha^3 G \hat{\mathrm{d}}x_3^\alpha) \end{aligned}$$

But

$$\begin{aligned} \hat{\mathrm{d}}H \,\bar{\wedge}\, \hat{\mathrm{d}}G &= \sum_{\alpha=1}^{m} \hat{\partial}_\alpha^1 H \hat{\mathrm{d}}x_1^\alpha + \hat{\partial}_\alpha^2 H \hat{\mathrm{d}}x_2^\alpha + \hat{\partial}_\alpha^3 H \hat{\mathrm{d}}x_3^\alpha \,\bar{\wedge} \sum_{\beta=1}^{m} \hat{\partial}_\beta^1 G \hat{\mathrm{d}}x_1^\beta + \hat{\partial}_\beta^2 G \hat{\mathrm{d}}x_2^\beta + \\ &\quad + \hat{\partial}_\beta^3 G \hat{\mathrm{d}}x_3^\beta \\ &= \sum_{\alpha=1}^{m} \hat{\partial}_\alpha^i H \hat{\partial}_\alpha^k G \hat{\mathrm{d}}x_i^\alpha \wedge \hat{\mathrm{d}}x_k^\alpha + 2\sum_{\alpha\neq\beta} \hat{\partial}_\alpha^i H \hat{\partial}_\beta^j G \hat{\mathrm{d}}x_i^\alpha \wedge_s \hat{\mathrm{d}}x_k^\beta \end{aligned}$$

and by linearity and anticommutativity it is

$$\begin{aligned} \tfrac{\hat{\mathrm{d}}H \wedge \hat{\mathrm{d}}G - \hat{\mathrm{d}}G \bar{\wedge} \hat{\mathrm{d}}H}{2} &= \textstyle\sum_{\alpha=1}^{m} \hat{\partial}_\alpha^i H \hat{\partial}_\alpha^k G \hat{\mathrm{d}}x_i^\alpha \wedge x_k^\alpha \\ &= \textstyle\sum_{\alpha=1}^{m} (\hat{\partial}_\alpha^1 H \hat{\mathrm{d}}x_1^\alpha + \hat{\partial}_\alpha^2 H \hat{\mathrm{d}}x_2^\alpha + \hat{\partial}_\alpha^3 H \hat{\mathrm{d}}x_3^\alpha) \,\bar{\wedge}\, (\hat{\partial}_\alpha^1 G \hat{\mathrm{d}}x_1^\alpha + \\ &\quad + \hat{\partial}_\alpha^2 G \hat{\mathrm{d}}x_2^\alpha + \hat{\partial}_\alpha^3 G \hat{\mathrm{d}}x_3^\alpha), \quad \text{so} \\ &\qquad \hat{\imath}_{\mathbf{v}}\,\omega = \tfrac{\hat{\mathrm{d}}H \bar{\wedge} \hat{\mathrm{d}}G - \hat{\mathrm{d}}G \bar{\wedge} \hat{\mathrm{d}}H}{2} \end{aligned}$$

We have just proved

Theorem 4 *Let M be a $3m$-manifold. Consider a Nambu dynamical system on M*

$$\frac{\hat{\mathrm{d}}x}{dt} = f(x)$$

Let $\mathbf{v}$ be the vector field associated to f, i.e., $\mathbf{v} : M \longrightarrow \mathrm{T}\,M$ with $v_i^\alpha = f_i^\alpha(x)\hat{\partial}_\alpha^i$, $\alpha = 1, 2, \ldots, m$; $i = 1, 2, 3$ in local coordinates. Then there exists a 3-eform $\boldsymbol{\omega}$ (non-volume form) and a form $\boldsymbol{\alpha}$ associated to $\mathbf{v}$ such that

i) $\operatorname{div} \mathbf{v} = 0$

ii) $\alpha = \dfrac{\hat{\mathrm{d}}H \,\bar{\wedge}\, \hat{\mathrm{d}}G - \hat{\mathrm{d}}G \,\bar{\wedge}\, \hat{\mathrm{d}}H}{2}$, *where $H(x)$, $G(x)$ are the Hamiltonians.*

iii) $\hat{\mathrm{d}}\alpha = 0$

iv) $\mathrm{L}_\mathbf{v}\, \boldsymbol{\omega} = 0$

Remark 4 *(i), (iii) and (iv) are equivalent since $\hat{\mathrm{d}}\hat{\imath}_\mathbf{v}\,(\boldsymbol{\omega}) = \hat{\mathrm{d}}\alpha = \sum_\alpha (\operatorname{div} \mathbf{v}^\alpha) \hat{\mathrm{d}}x_1^\alpha \wedge \hat{\mathrm{d}}x_2^\alpha \wedge x_3^\alpha = \mathrm{L}_\mathbf{v}\, \boldsymbol{\omega}$. Furthermore,*

Theorem 5 *Let M be a $3m$-manifold. The dynamical system $\frac{\hat{\mathrm{d}}x}{dt} = f(x)$ on M (being v the associated vector field) is a Nambu dynamical system iff there exists two Hamiltonians, H and G, such that $\hat{\imath}_v\, \omega = \frac{\hat{\mathrm{d}}H \bar{\wedge} \hat{\mathrm{d}}G - \hat{\mathrm{d}}G \bar{\wedge} \hat{\mathrm{d}}H}{2}$ where ω is the canonical 3-eform.*

Proof. $(\Leftarrow)$

Since $\hat{\imath}_v\, \omega = \sum_\alpha v_1^\alpha \hat{\mathrm{d}}x_2^\alpha \wedge \hat{\mathrm{d}}x_3^\alpha - v_2^\alpha \hat{\mathrm{d}}x_1^\alpha \wedge \hat{\mathrm{d}}x_3^\alpha + v_3^\alpha \hat{\mathrm{d}}x_1^\alpha \wedge \hat{\mathrm{d}}x_2^\alpha$ and

$$\frac{\hat{\mathrm{d}}H \,\bar{\wedge}\, \hat{\mathrm{d}}G - \hat{\mathrm{d}}G \,\bar{\wedge}\, \hat{\mathrm{d}}H}{2} = \sum \hat{\partial}_\alpha^1 H \hat{\mathrm{d}}x_1^\alpha + \hat{\partial}_\alpha^2 H \hat{\mathrm{d}}x_2^\alpha + \hat{\partial}_\alpha^3 H \hat{\mathrm{d}}x_3^\alpha \,\bar{\wedge}\, \hat{\partial}_\alpha^1 G \hat{\mathrm{d}}x_1^\alpha + \\ + \hat{\partial}_\alpha^2 G \hat{\mathrm{d}}x_3^\alpha + \hat{\partial}_\alpha^3 G \hat{\mathrm{d}}x_3^\alpha$$

it is clear that

$$\begin{aligned} v_1^\alpha &= \hat{\partial}_\alpha^2 H \hat{\partial}_\alpha^3 G - \hat{\partial}_\alpha^3 H \hat{\partial}_\alpha^2 G &= \tfrac{\hat{\partial}(H,G)}{\partial(x_2^\alpha, x_3^\alpha)} \\ v_2^\alpha &= \hat{\partial}_\alpha^3 H \hat{\partial}_\alpha^1 G - \hat{\partial}_\alpha^1 H \hat{\partial}_\alpha^3 G &= \tfrac{\hat{\partial}(H,G)}{\partial(x_3^\alpha, x_1^\alpha)} \\ v_3^\alpha &= \hat{\partial}_\alpha^1 H \hat{\partial}_\alpha^2 G - \hat{\partial}_\alpha^2 H \hat{\partial}_\alpha^3 G &= \tfrac{\hat{\partial}(H,G)}{\partial(x_1^\alpha, x_2^\alpha)} \end{aligned}$$

■

Thus $\boldsymbol{\omega}$ is an invariant 3-eform of $\mathbf{v}$, i.e., $\boldsymbol{\omega}$ is constant along the integral curves of the Nambu dynamical system. Then we have

Theorem 6 *A Nambu dynamical system on a $3m$-manifold has the m integral invariants $\int \omega^k$ with $k \leq 3m$ and ω a 3-eform*

Proof. Since L_v is a $\bar{\wedge}$-derivation then $L_v(\omega \bar{\wedge} \omega) = 0$. Thus all m $\bar{\wedge}$-powers of ω are constant along the integral curves of the Nambu dynamical system. ■

REFERENCES

1. **Abraham, R.**, *Foundations of Mechanics.* W. A. Benjamin (1967)

2. **Abraham, R., Marsden, J. and Ratiu, T.**, *Manifolds, Tensor Analysis and Applications.* 2° ed, Springer Verlag (1988)

3. **Codriansky, S. and Gonzalez-Bernardo, C. A.**, *A singlet in generalized dynamics.* Il Nuovo Cimento 97B (1987), 64-

4. **Codriansky, S. and Gonzalez-Bernardo, C. A.**, *Developments in Nambu mechanics.* J. Phys. A.: Math. Gen. 27 (1994), 2565-

5. **Codriansky, S., Navarro, R., Pedroza, M.**, *The Liouville condition and Nambu mechanics.* J. Phys. A.: Math. Gen. 29 (1996), 1037-1044.

6. **Estabrook, F.B.**, *Comments on generalized Hamiltonian dynamics.* Phys. Rev. D8 (1973), 2740-2743.

7. **Fecko, M.**, *On a geometric formulation of the Nambu dynamics.* J. Math. Phys. 33 (1992), 926-929.

8. **Marmo, G., Saletan, E. J., Simoni, A., Zaccaria, F.**, *Liouville dynamics and Poisson brackets.* J. Math. Phys. 22 (1981), 835-842.

9. **Morando, P.**, *Liouville condition, Nambu mechanics and differential forms.* J. Phys. A.: Math. Gen. 29 (1996), L329-L331.

10. **Nambu, Y.**, *Generalized Hamiltonian mechanics.* Phys. Rev. D 7, (1973), 2405-2412.

Proceedings of Dynamic Systems and Applications 3 (2001) 467-474

Ergodic linear Hamiltonian systems with absolutely continuous dynamics

Carmen Núñez and Ana M. Sanz

Dpto. de Matemática Aplicada a la Ingeniería
Escuela Técnica Superior de Ingenieros Industriales
Universidad de Valladolid, Paseo del Cauce s/n, 47011 Valladolid, Spain

ABSTRACT: The authors consider the flow induced by an ergodic family of linear Hamiltonian systems and establish a sufficient condition to ensure the absolutely continuous character of the dynamics. Such a condition is formulated in terms of certain regularity properties of the Weyl M-matrices. These properties are equivalent to the existence of a square-integrable symplectic matrix-valued function which defines a change of variables that takes the initial systems to skew-symmetric form and preserves the character of the dynamics.

AMS (MOS) subject classification. 34B20, 58F11, 28D05

1. INTRODUCTION

This paper is concerned with the ergodic structure of random linear Hamiltonian systems. More precisely, we establish conditions which are sufficient for the presence of *absolutely continuous dynamics*. Such conditions are formulated in terms of the behavior of the Weyl M-functions in the direction of a suitable Atkinson's perturbation.

We consider a family of linear Hamiltonian systems

$$\mathbf{z}' = H(\omega{\cdot}t)\,\mathbf{z}\,, \qquad \omega \in \Omega\,, \tag{1.1}$$

where Ω is a compact metric space with a continuous flow $\mathbb{R}\times\Omega\to\Omega$, $(t,\omega)\mapsto\omega{\cdot}t$, and H is a continuous real $2n\times 2n$ matrix-valued function on Ω which belongs to the algebra of the symplectic matrices $\mathfrak{sp}(n,\mathbb{R})$; i.e. $H^T J + J\,H = 0$, with $J = \left[\begin{smallmatrix} 0 & -I_n \\ I_n & 0 \end{smallmatrix}\right]$.

This family of systems induces in a natural way a skew-product flow on the trivial bundle $\Omega\times\mathbb{C}^{2n}$ and on its invariant subset $\Omega\times\mathbb{R}^{2n}$. If $U(t,\omega)$ represents the fundamental matrix solution of the system (1.1) for $\omega\in\Omega$ with $U(0,\omega)=I_{2n}$, the orbit of $(\omega,\mathbf{z}_0)$ is $\{(\omega{\cdot}t, U(t,\omega)\,\mathbf{z}_0)\mid t\in\mathbb{R}\}$. The absolutely continuous character of the dynamics of the family means, roughly speaking, that its orbits are nearly bounded. See below for a precise definition.

Several of the structures related to the qualitative description of this kind of flows are given in terms of real or complex Lagrange planes. In fact, the family (1.1) induces flows on the complex and real Lagrange bundles, whose definitions and main characteristics we describe briefly in what follows.

Recall that an n-dimensional vector subspace $F\subset\mathbb{C}^{2n}$ is called a *Lagrange plane* if $\mathbf{v}^T J\mathbf{w}=0$ for all $\mathbf{v},\mathbf{w}\in F$. One of this subspaces can be represented by a $2n\times n$ matrix $F\equiv\left[\begin{smallmatrix} F_1 \\ F_2 \end{smallmatrix}\right]$ of range n with $F_1^T F_2 = F_2^T F_1$. The column vectors form a basis of the Lagrange plane; so, two matrices $\left[\begin{smallmatrix} F_1 \\ F_2 \end{smallmatrix}\right]$ and $\left[\begin{smallmatrix} G_1 \\ G_2 \end{smallmatrix}\right]$ represent the same Lagrange plane if and only if there is a non-singular $n\times n$ matrix P such that $F_1=G_1P$ and $F_2=G_2P$. Consequently, the space $L_{\mathbb{C}}$ of all Lagrange planes of $\mathbb{C}^{2n}$ is an orientable compact manifold which can be identified with the homogeneous space $\widetilde{G}/\widetilde{H}$, where $\widetilde{G} = Sp(n,\mathbb{C}) \cap SO(2n,\mathbb{C})$ and $\widetilde{H} = O(n,\mathbb{C})$ (see Mishchenko et al [13]).

However, the set of all Lagrange planes of $\mathbb{R}^{2n}$ is orientable if and only if n is odd (see Matsushima [12]). In order to work with an orientable manifold we consider the set $L_\mathbb{R}$ of real Lagrange planes with an assigned orientation, which can be identified with G/H for $G = Sp(n,\mathbb{R}) \cap SO(2n,\mathbb{R})$ and $H = SO(n,\mathbb{R})$. The space $L_\mathbb{R}$ admits a G-invariant volume measure, denoted by l.

It is immediate to check that $U(t,\omega)$ lies in the symplectic group $Sp(n,\mathbb{R})$, and hence $U(t,\omega)\,F$ belongs to $L_\mathbb{C}$ (resp. $L_\mathbb{R}$) whenever F lies in $L_\mathbb{C}$ (resp. $L_\mathbb{R}$). This shows that the map $\tau(t,\omega,F) = (\omega{\cdot}t, U(t,\omega)\,F)$ defines a continuous skew-product flow τ on $K_\mathbb{C} = \Omega \times L_\mathbb{C}$ and $K_\mathbb{R} = \Omega \times L_\mathbb{R}$. The equations describing the evolution of these flows in adequate coordinates can be found in Novo et al [15] (see also Barret [1] and Reid [17, 18]).

We fix an ergodic measure m_0 on Ω and define $m_1 = m_0 \otimes l$ on a complete σ-algebra of $K_\mathbb{R}$. The dynamics of $(K_\mathbb{R}, \tau)$ is said to be *absolutely continuous* if there exists a τ-invariant measure on $K_\mathbb{R}$ equivalent to m_1; otherwise, it is said to be *singular*.

One can prove that the existence of this kind of measures assures that all the Lyapunov exponents of the family (1.1) are zero. At the same time, the appearance of positive Lyapunov exponents is a generic property in several different contexts (see Johnson [7], Knill [10], Fabbri [2] and Nerurkar [14] for examples of this situation in the two-dimensional case), which explains the interest in finding theoretical conditions to guarantee the absolutely continuous character of the dynamics.

Absolutely continuous dynamics occurs, for instance, whenever all the solutions of the systems (1.1) are bounded (see Ellis and Johnson [3]), in certain cases of non-bounded orbits (see Novo and Obaya [16]) and, in general, when Kotani theory applies (see Kotani and Simon [11] and Novo et al [15]). In all these cases the following condition is satisfied:

Property (c). *There exists a symplectic real matrix-valued function $C \in L^2(\Omega, m_0)$ such that the change of variables $\widetilde{\mathbf{z}} = C(\omega{\cdot}t)\,\mathbf{z}$ takes the systems* (1.1) *to*

$$\widetilde{\mathbf{z}}' = \widetilde{H}(\omega{\cdot}t)\,\widetilde{\mathbf{z}}\,, \qquad \omega \in \Omega\,, \tag{1.2}$$

where $\widetilde{H}$ belongs to $\mathfrak{sp}(n,\mathbb{R})$ and is skew-symmetric.

The skew-symmetric character of the transformed systems assures that the corresponding fundamental matrix solution $\widetilde{U}(t,\omega)$ (with $\widetilde{U}(0,\omega) = I_{2n}$) belongs to the group G, and hence m_1 is an invariant measure for the new flow. Consequently, (1.2) is in the absolutely continuous case and, since the change of variables preserves this property, also (1.1) is.

Our goal in this paper is to provide an equivalent formulation for Property (c) -and hence a sufficient condition for the absolutely continuous dynamics- in terms of the behavior of the Weyl M-matrices. In Section 2 we recall the definition and main properties of these functions before stating the main result.

2. A CONDITION FOR ABSOLUTELY CONTINUOUS DYNAMICS

Let us consider the perturbed random Hamiltonian systems

$$\mathbf{z}' = (H(\omega{\cdot}t) + EJ^{-1}\Gamma(\omega{\cdot}t))\,\mathbf{z}\,, \qquad \omega \in \Omega\,, \tag{2.1}$$

where $E \in \mathbb{C}$ is a complex parameter and $\Gamma \geq 0$ is a continuous positive semidefinite symmetric real $2n{\times}2n$ matrix-valued function on Ω. Note that the family of perturbed

n-dimensional random Schrödinger equations

$$-\mathbf{x}'' + V(\omega\cdot t)\,\mathbf{x} = E\,\Gamma_1(\omega\cdot t)\,\mathbf{x}\,, \qquad \omega\in\Omega\,,$$

with V and Γ_1 continuous symmetric real $n\times n$ matrix-valued functions on Ω and $\Gamma_1 \geq 0$, is included in the general formulation (2.1) by setting $\mathbf{z} = \left[\begin{smallmatrix}\mathbf{x}\\ \mathbf{x}'\end{smallmatrix}\right]$, $H = \left[\begin{smallmatrix}0 & I_n\\ V & 0\end{smallmatrix}\right]$ and $\Gamma = \left[\begin{smallmatrix}\Gamma_1 & 0\\ 0 & 0\end{smallmatrix}\right]$.

We denote by τ_E the flow induced by systems (2.1) in $K_{\mathbb{C}}$ and $K_{\mathbb{R}}$ and by $U_E(t,\omega)$ the fundamental matrix solution satisfying $U_E(0,\omega) = I_{2n}$. $U_0(t,\omega)$ will be denoted, as in the previous section, by $U(t,\omega)$.

We also assume that the perturbation matrix Γ satisfies the following Atkinson's condition: each minimal subset of Ω contains at least a point ω such that

$$\int_{-\infty}^{\infty} \mathbf{v}^T U^T(t,\omega)\,\Gamma^2(\omega\cdot t)\,U(t,\omega)\,\mathbf{v}\,dt > 0 \qquad \forall\,\mathbf{v}\in\mathbb{C}^{2n}-\{0\}\,. \tag{2.2}$$

According to the results of Johnson and Nerurkar [8], this property guarantees the *exponential dichotomy* (see Sacker and Sell [19]) of the systems (2.1) for $\Im E \neq 0$. That is, the existence of a splitting of the complex bundle into two τ_E-invariant n-dimensional closed subbundles, $\Omega\times\mathbb{C}^{2n} = F^+_{\Gamma,E}\oplus F^-_{\Gamma,E}$, and positive constants C, β such that

(i) $\|U_E(t,\omega)\,\mathbf{z}\| \leq Ce^{-\beta t}\|\mathbf{z}\|$ for every $t\geq 0$ and $(\omega,\mathbf{z})\in F^+_{\Gamma,E}$,
(ii) $\|U_E(t,\omega)\,\mathbf{z}\| \leq Ce^{\beta t}\|\mathbf{z}\|$ for every $t\leq 0$ and $(\omega,\mathbf{z})\in F^-_{\Gamma,E}$.

Moreover, for every $\omega\in\Omega$, the sections $F^{\pm}_{\Gamma,E}(\omega) = \{\mathbf{z}\in\mathbb{C}^{2n}\,|\,(\omega,\mathbf{z})\in F^{\pm}_{\Gamma,E}\}$ are complex Lagrange planes and can be represented in terms of the *Weyl M-matrices* by $\left[\begin{smallmatrix}I_n\\ M^{\pm}_{\Gamma}(\omega,E)\end{smallmatrix}\right]$ (see also Hinton and Shaw [4, 5] and Johnson [6]).

The functions $M^{\pm}_{\Gamma}(\omega,E)$, defined for $\Im E\neq 0$ and $\omega\in\Omega$, are symmetric complex $n\times n$ matrix-valued functions, jointly continuous in both variables and analytic on E outside the real axis for each $\omega\in\Omega$ fixed. Besides, $\pm\Im E\,\Im M^{\pm}_{\Gamma}(\omega,E) > 0$ and $M^{\pm}_{\Gamma}(\omega,\overline{E}) = (M^{\pm}_{\Gamma})^*(\omega,E)$.

Recall that our aim is to reformulate Property (c) in terms of the behavior of the Weyl matrices. Thus, assume first that the initial family of systems (1.1) satisfies this property. Then, according to the results of Novo et al [15], the limits

$$A_\Gamma(\omega) = \lim_{t\to\infty}\frac{1}{2t}\int_{-t}^{t} U^T(s,\omega)\,\Gamma(\omega\cdot s)\,U(s,\omega)\,ds$$

exist on an invariant subset $\Omega_0\subset\Omega$, with $m_0(\Omega_0)=1$. In addition, the matrix-valued function A_Γ belongs to $L^1(\Omega,m_0)$ and is a positive definite solution along the flow of the equation

$$Z' = -H^T(\omega\cdot t)\,Z - Z\,H(\omega\cdot t)\,, \qquad \omega\in\Omega\,. \tag{2.3}$$

That is, for each $\omega\in\Omega_0$ the map $t\mapsto A_\Gamma(\omega\cdot t)$ is differentiable and (2.3) is satisfied if we substitute Z by $A_\Gamma(\omega\cdot t)$. Moreover,

(i) there exists an invariant subset $\Omega_1\subset\Omega$ with $m_0(\Omega_1)=1$ and real positive numbers $\lambda_{\Gamma,1},\ldots,\lambda_{\Gamma,n}$ such that, for every $\omega\in\Omega_1$, the eigenvalues of $JA_\Gamma(\omega)$ are $-i\lambda_{\Gamma,1},\ldots,-i\lambda_{\Gamma,n}, i\lambda_{\Gamma,1},\ldots,i\lambda_{\Gamma,n}$.
(ii) If $\omega\in\Omega_1$, the n-dimensional linear subspaces $F^+_\Gamma(\omega)$ and $F^-_\Gamma(\omega)$ of $\mathbb{C}^{2n}$, respectively generated by the eigenvectors associated with the eigenvalues with negative and positive imaginary part, are complex Lagrange planes, and the sets $\{(\omega,F^{\pm}_\Gamma(\omega))\,|\,\omega\in\Omega_1\}\subset K_{\mathbb{C}}$ are τ-invariant.

(iii) The planes $F_\Gamma^\pm(\omega)$ can be represented by $\begin{bmatrix} I_n \\ N_\Gamma^\pm(\omega) \end{bmatrix}$, with $N_\Gamma^-(\omega) = \overline{N_\Gamma^+(\omega)}$ and $\pm\Im N_\Gamma^\pm(\omega) > 0$.

(iv) If $\omega \in \Omega_1$, the matrix-valued function

$$C_\Gamma(\omega) = \begin{bmatrix} \Im^{1/2} N_\Gamma^+(\omega) & 0 \\ -\Im^{-1/2} N_\Gamma^+(\omega)\, \Re N_\Gamma^+(\omega) & \Im^{-1/2} N_\Gamma^+(\omega) \end{bmatrix} \tag{2.4}$$

is a symplectic matrix and C_Γ, $C_\Gamma^{-1} \in L^2(\Omega, m_0)$.

Furthermore, the change of variables $\widetilde{\mathbf{z}} = C_\Gamma(\omega{\cdot}t)\,\mathbf{z}$ takes the family (1.1) to skew-symmetric form. That is, from the matrix A_Γ (and hence from the non-perturbed equation (1.1) and the matrix Γ) we obtain the expression of one of the changes of variables whose existence is theoretically assured by Property (c).

The expression of the new systems and mainly the relation between the matrices A_Γ and C_Γ, allow us to show the existence of the derivative of the *rotation number* (see [6] and [15] for its precise definition) in the direction of the matrix Γ at the point $E = 0$, and to obtain its value: $\alpha'_\Gamma(0) = \lambda_{\Gamma,1} + \cdots + \lambda_{\Gamma,n}$. And this property is used in the proof of the radial convergence of the Weyl M-matrices from the upper-half plane when $\Gamma > 0$ (or when $\Gamma_1 > 0$ in the Schrödinger case): there exist

$$\lim_{\varepsilon \to 0^+} M_\Gamma^\pm(\omega, i\varepsilon) = N_\Gamma^\pm(\omega)$$

in measure. These results show the importance of the matrices $N_\Gamma^\pm$ in the ergodic description of the flow on the Lagrange bundles.

Taking these results of [15] as starting point, Johnson et al [9] establish the above convergence in the $L^1(\Omega, m_0)$-topology. In fact, they prove that the matrix-valued functions

$$B_{\Gamma,\varepsilon}(\omega) = C_{\Gamma,\varepsilon}^T(\omega)\, C_{\Gamma,\varepsilon}(\omega),$$

where

$$C_{\Gamma,\varepsilon}(\omega) = \begin{bmatrix} \Im^{1/2} M_\Gamma^+(\omega, i\varepsilon) & 0 \\ -\Im^{-1/2} M_\Gamma^+(\omega, i\varepsilon)\, \Re M_\Gamma^+(\omega, i\varepsilon) & \Im^{-1/2} M_\Gamma^+(\omega, i\varepsilon) \end{bmatrix},$$

converge as $\varepsilon \to 0^+$ in the $L^1(\Omega, m_0)$-topology to the matrix-valued function

$$B_\Gamma(\omega) = C_\Gamma^T(\omega)\, C_\Gamma(\omega),$$

where C_Γ is given by (2.4). From here they deduce that $C_\Gamma(\omega) = \lim_{\varepsilon \to 0^+} C_{\Gamma,\varepsilon}(\omega)$ in the $L^2(\Omega, m_0)$-topology, which provides the desired convergence for the Weyl M-matrices.

In particular, Property (c) guarantees that the family

$$\{B_{\Gamma,\varepsilon}(\omega) \mid 0 < \varepsilon \le \varepsilon_0\} \tag{2.5}$$

is uniformly integrable. That is, for every $\tau > 0$ there exists $\delta > 0$ such that if $\Delta \subset \Omega$ is a measurable subset with $m_0(\Delta) < \delta$ then $\int_\Delta \operatorname{tr} B_{\Gamma,\varepsilon}(\omega)\, dm_0 < \tau$ for every $0 < \varepsilon \le \varepsilon_0$.

Now we establish the converse of this result. We will also assume that $m_0(\{\omega\}) = 0$ for every $\omega \in \Omega$, which means that the flow on the base Ω does not have critical points.

Theorem 1. *Assume that the family* (2.5) *is uniformly integrable for an Atkinson's perturbation* Γ. *Then Property* (c) *holds.*

This result establishes conditions on the behavior of the Weyl M-matrices in the direction of a suitable perturbation which are sufficient to ensure the presence of absolutely continuous dynamics, which was our objective. Its proof requires the following technical lemma. Given $\Delta \subset \Omega$, we will denote $\Delta_T = \cup_{s\in[-T,T]}\Delta\cdot s$ where $\Delta\cdot s = \{\omega\cdot s \,|\, \omega \in \Delta\}$.

Lemma 2. *Let $S_\varepsilon : \Omega \to M_n(\mathbb{R})$, $0 \le \varepsilon \le \varepsilon_0$ be a family of measurable matrix-valued functions satisfying*

(1) *S_0 is continuous;*
(2) *there is $h \in L^1(\Omega, m_0)$ with $\|S_\varepsilon(\omega)\| \le h(\omega)$ for every $0 < \varepsilon \le \varepsilon_0$ and m_0-almost every $\omega \in \Omega$.*

For each $0 \le \varepsilon \le \varepsilon_0$ we represent by $V_\varepsilon(t,\omega)$ the fundamental matrix solution of the linear systems

$$\mathbf{x}' = (S_0(\omega\cdot t) + \varepsilon\, S_\varepsilon(\omega\cdot t))\,\mathbf{x}\,, \qquad \omega \in \Omega \tag{2.6}$$

with $V_\varepsilon(0,\omega) = I_n$. Then, there exists a compact subset $\Delta \subset \Omega$ with $m_0(\Delta) > 0$ such that, as $\varepsilon \to 0^+$, $V_\varepsilon(t,\omega)$ converges to $V_0(t,\omega)$ uniformly on the compact sets $[-T_1, T_1] \times \Delta_{T_2}$ for all $T_1, T_2 > 0$.

Proof. Let us fix $T_1, T_2 > 0$. Birkhoff's ergodic theorem guarantees that

$$\lim_{|t|\to\infty} \frac{1}{t}\int_0^t h(\omega\cdot s)\,ds = \int_\Omega h(\omega)\,dm_0 \tag{2.7}$$

almost everywhere on Ω. We take a compact subset $\Delta \subset \Omega$ with $m_0(\Delta) > 0$ where the above convergence is uniform, and find a real positive constant $\rho_1 > 0$ such that

$$\left|\int_0^t h(\omega\cdot s)\,ds\right| \le \rho_1 \tag{2.8}$$

for every $t \in [-T_1, T_1]$ and $\omega \in \Delta$. We also define

$$\rho_2 = \sup\left\{\|V_0(t,\omega)\|,\ \|V_0^{-1}(t,\omega)\| \mid t \in [-T_1,T_1]\,,\ \omega \in \Delta\right\}\,.$$

The linear change of variables $\mathbf{x} = V_0(t,\omega)\,\mathbf{y}$ takes the linear systems (2.6) to

$$\mathbf{y}' = \varepsilon V_0^{-1}(t,\omega)\, S_\varepsilon(\omega\cdot t)\, V_0(t,\omega)\,\mathbf{y}\,, \qquad \omega \in \Omega\,. \tag{2.9}$$

The fundamental matrix solution $W_\varepsilon(t,\omega)$ of (2.9) with $W_\varepsilon(0,\omega) = I_n$ satisfies

$$W_\varepsilon(t,\omega) = I_n + \varepsilon\int_0^t V_0^{-1}(s,\omega)\, S_\varepsilon(\omega\cdot s)\, V_0(s,\omega)\, W_\varepsilon(s,\omega)\,ds\,,$$

which can be written as

$$\begin{aligned} W_\varepsilon(t,\omega) - I_n = \varepsilon &\int_0^t V_0^{-1}(s,\omega)\, S_\varepsilon(\omega\cdot s)\, V_0(s,\omega)\,(W_\varepsilon(s,\omega) - I_n)\,ds \\ + \varepsilon &\int_0^t V_0^{-1}(s,\omega)\, S_\varepsilon(\omega\cdot s)\, V_0(s,\omega)\,ds\,. \end{aligned} \tag{2.10}$$

From (2.8), (2.10) and hypothesis (ii) we deduce that

$$\rho(\varepsilon) = \sup\{\|W_\varepsilon(t,\omega) - I_n\| \mid t \in [-T_1,T_1]\,,\ \omega \in \Delta\}$$

is finite and satisfies $\rho(\varepsilon) \le \varepsilon\,\rho_2^2\rho_1\,\rho(\varepsilon) + \varepsilon\,\rho_2^2\rho_1$. Consequently, for ε sufficiently small, $0 \le \rho(\varepsilon) \le \varepsilon\,\rho_2^2\rho_1/(1 - \varepsilon\,\rho_2^2\rho_1)$, which provides $\lim_{\varepsilon\to 0^+}\rho(\varepsilon) = 0$. Moreover, since $V_\varepsilon(t,\omega) - V_0(t,\omega) = V_0(t,\omega)\,(W_\varepsilon(t,\omega) - I_n)$, we conclude that

$$\rho^*(\varepsilon) = \sup\{\|V_\varepsilon(t,\omega) - V_0(t,\omega)\| \mid t \in [-T_1,T_1]\,,\ \omega \in \Delta\} \le \rho_2\,\rho(\varepsilon)\,.$$

Hence, $\lim_{\varepsilon\to 0^+}\rho^*(\varepsilon) = 0$ and $V_\varepsilon(t,\omega)$ converges to $V_0(t,\omega)$ as $\varepsilon \to 0^+$ uniformly on $[-T_1,T_1]\times\Delta$. Finally, the uniform convergence of (2.7) also on Δ_{T_2}, combined with the previous arguments, shows that $V_\varepsilon(t,\omega)$ converges to $V_0(t,\omega)$ uniformly on $[-T_1,T_1]\times\Delta_{T_2}$, as stated. □

Proof of Theorem 1. From the uniform integrability and the non-existence of critical points for the measure m_0, it is easy to deduce that the family (2.5) is bounded in the $L^1(\Omega, m_0)$-topology.

Represent $B_{\Gamma,\varepsilon}(\omega) = (b_\varepsilon^{kj}(\omega))_{k,j=1,\dots,2n}$ and consider the Borel measures $d\nu_\varepsilon = b_\varepsilon^{11}(\omega)\, dm_0$. There exists a sequence $(\varepsilon_j)_{j\in\mathbb{N}}$ with $\lim_{j\to\infty}\varepsilon_j = 0$ such that $(\nu_{\varepsilon_j})_{j\in\mathbb{N}}$ converges to a Borel measure ν in the weak topology $\sigma(\mathcal{M}(\Omega), C(\Omega))$. Given $\tau > 0$, there exists $\delta > 0$ such that if $\Delta\subset\Omega$ is a measurable subset with $m_0(\Delta) < \delta$ then $\nu_{\varepsilon_j}(\Delta) < \tau$ for every $j\in\mathbb{N}$. Besides, if $\Delta\subset\Omega$ and $m_0(\Delta)<\delta$ we can find an open set O with $\Delta\subset O$ and $m_0(O)<\delta$. Hence, $\nu(\Delta)\le\nu(O)\le\liminf_{j\to\infty}\nu_{\varepsilon_j}(O)\le\tau$. This shows that the measure ν is absolutely continuous with respect to m_0; thus it can be written as $d\nu = b^{11}(\omega)\, dm_0$, and $b_\varepsilon^{11}(\omega)$ converges to $b^{11}(\omega)$ in the $\sigma(L^1(\Omega,m_0), C(\Omega))$-topology.

From this, Lusin's theorem and again the uniform integrable character of the family, one can show that $b_{\varepsilon_j}^{11}(\omega)$ converges to $b^{11}(\omega)$ in the $\sigma(L^1(\Omega,m_0), L^\infty(\Omega,m_0))$-topology. Applying successively the above arguments to each component of the matrix $B_{\Gamma,\varepsilon}(\omega)$, we obtain a decreasing sequence of positive numbers $(\varepsilon_j)_{j\in\mathbb{N}}$ with $\sum_{j=1}^\infty\sqrt{\varepsilon_j}<\infty$ and a symmetric matrix $B(\omega)\in L^1(\Omega,m_0)$ such that $B_{\Gamma,\varepsilon_j}(\omega)$ converges to $B(\Omega)$ in the $\sigma(L^1(\Omega,m_0), L^\infty(\Omega,m_0))$-topology.

For $\omega\in\Omega$, $0<\varepsilon\le\varepsilon_0$ and $X = \begin{bmatrix} X_1 & X_2\\ X_2^T & X_3\end{bmatrix}\in M_{\mathbb{R}}(2n)$, we define $G_\varepsilon^k(\omega,X)$ ($k = 1,2,3$) by

$$\begin{aligned}
\frac{1}{\sqrt{\varepsilon}}G_\varepsilon^1(\omega,X) ={}& X_1\left(\Gamma_2^T\,\Im^{-1}M_\Gamma^+\,\Re M_\Gamma^+ + \Gamma_3\left(\Im M_\Gamma^+ + \Re M_\Gamma^+\,\Im^{-1}M_\Gamma^+\,\Re M_\Gamma^+\right)\right)\\
&- X_2\left(\Gamma_1\,\Im^{-1}M_\Gamma^+\,\Re M_\Gamma^+ + \Gamma_2\left(\Im M_\Gamma^+ + \Re M_\Gamma^+\,\Im^{-1}M_\Gamma^+\,\Re M_\Gamma^+\right)\right)\\
&- X_3\left(\Im M_\Gamma^+\,\Gamma_1\right),\\
\frac{1}{\sqrt{\varepsilon}}G_\varepsilon^2(\omega,X) ={}& -X_1\left(\Gamma_2^T\,\Im^{-1}M_\Gamma^+ + \Gamma_3\,\Re M_\Gamma^+\,\Im^{-1}M_\Gamma^+\right)\\
&+ X_2\left(\Gamma_1\,\Im^{-1}M_\Gamma^+ + \Gamma_2\,\Re M_\Gamma^+\,\Im^{-1}M_\Gamma^+\right) - X_3\left(\Im M_\Gamma^+\,\Gamma_2\right),\\
\frac{1}{\sqrt{\varepsilon}}G_\varepsilon^3(\omega,X)) ={}& -X_2^T\left(\Gamma_2^T\,\Im^{-1}M_\Gamma^+ + \Gamma_3\,\Re M_\Gamma^+\,\Im^{-1}M_\Gamma^+\right)\\
&+ X_3\left(\Gamma_1\,\Im^{-1}M_\Gamma^+ + \Gamma_2\,\Re M_\Gamma^+\,\Im^{-1}M_\Gamma^+ - \Im M_\Gamma^+\,\Gamma_3\right)
\end{aligned}$$

(where $\Gamma = \begin{bmatrix}\Gamma_1 & \Gamma_2\\ \Gamma_2^T & \Gamma_3\end{bmatrix}$ and M_Γ^+ represent $\Gamma(\omega)$ and $M_\Gamma^+(\omega, i\varepsilon)$ respectively), and

$$G_\varepsilon(\omega,X) = \begin{bmatrix} G_\varepsilon^1(\omega,X) & G_\varepsilon^2(\omega,X)\\ (G_\varepsilon^2)^T(\omega,X) & G_\varepsilon^3(\omega,X)\end{bmatrix}.$$

The matrix $B_{\Gamma,\varepsilon}(\omega)$ is then a solution along the flow of the systems

$$X' = -H^T(\omega\cdot t)X - XH(\omega\cdot t) + \sqrt{\varepsilon}\,G_\varepsilon(\omega,X), \qquad \omega\in\Omega. \tag{2.11}$$

Writing the components of X as a column vector $\mathbf{x}$, we obtain the $4n^2$-dimensional linear systems

$$\mathbf{x}' = \big(S_0(\omega\cdot t) + \sqrt{\varepsilon}\, S_\varepsilon(\omega\cdot t)\big)\,\mathbf{x}\,, \qquad \omega\in\Omega\,, \tag{2.12}$$

equivalent to (2.11). The components of S_ε are obtained as a linear combination of a finite number of terms which are the product of a continuous function and a component of the matrix $\sqrt{\varepsilon}B_{\Gamma,\varepsilon}(\omega)$. Consequently, there exists a positive constant $C>0$ such that $\|S_{\varepsilon_j}(\omega)\| \le C\sqrt{\varepsilon_j}\,\mathrm{tr}\, B_{\Gamma,\varepsilon_j}(\omega)$ for every $j\ge 1$ and $\omega\in\Omega$. Hence, denoting by K an upper bound of $\int_\Omega \mathrm{tr}\, B_{\Gamma,\varepsilon_j}(\omega)\, dm_0$, we conclude that

$$\sum_{j=1}^{\infty}\int_\Omega \|S_{\varepsilon_j}(\omega)\|\, dm_0 \le CK\sum_{j=1}^{\infty}\sqrt{\varepsilon_j} < \infty\,.$$

This implies that the measurable function $h(\omega)=\sum_{j=1}^\infty \|S_{\varepsilon_j}(\omega)\|$ belongs to the space $L^1(\Omega, m_0)$, and $\|S_{\varepsilon_j}(\omega)\|\le h(\omega)$ for every $\omega\in\Omega$ and $j\ge 1$.

We represent by $V_\varepsilon(t,\omega)$ the fundamental matrix of (2.12) satisfying $V_\varepsilon(0,\omega)=I_{4n^2}$. Note that if we denote as $\mathbf{b}_\varepsilon(\omega)$ the $4n^2$-dimensional column vector corresponding to the matrix $B_{\Gamma,\varepsilon}(\omega)$, then $\mathbf{b}_\varepsilon(\omega)$ is a solution along the flow of this equation. According to Lemma 2 (applied to the sequence $(\varepsilon_j)_{j\in\mathbb{N}}$), there exists a compact subset $\Delta\subset\Omega$ of positive measure $m_0(\Delta)>0$ such that when $j\to\infty$, $V_{\varepsilon_j}(t,\omega)$ converges to $V_0(t,\omega)$ uniformly on the compact sets $[-T_1,T_1]\times\Delta_{T_2}$ for every $T_1,T_2>0$. Let us fix $T_1,T_2>0$. Then

$$\mathbf{b}_{\varepsilon_j}(\omega\cdot T_1) = V_{\varepsilon_j}(T_1,\omega)\,\mathbf{b}_{\varepsilon_j}(\omega)$$

for every $j\in\mathbb{N}$. Thus, $\mathbf{b}_{\varepsilon_j}(\omega\cdot T_1)\,\chi_{\Delta_{T_2}}(\omega) = V_{\varepsilon_j}(T_1,\omega)\,\mathbf{b}_{\varepsilon_j}(\omega)\,\chi_{\Delta_{T_2}}(\omega)$, and taking limits as $j\to\infty$ in the $\sigma(L^1(\Omega,m_0), L^\infty(\Omega,m_0))$-topology we deduce that

$$\mathbf{b}(\omega\cdot T_1)\,\chi_{\Delta_{T_2}}(\omega) = V_0(T_1,\omega)\,\mathbf{b}(\omega)\,\chi_{\Delta_{T_2}}(\omega)\,,$$

which provides, when $T_2\to\infty$,

$$\mathbf{b}(\omega\cdot T_1) = V_0(T_1,\omega)\,\mathbf{b}(\omega) \tag{2.13}$$

m_0-almost everywhere on Ω and for every $T_1\in\mathbb{R}$. Modifying $\mathbf{b}(\omega)$ in a null subset of Ω we can assume that equality (2.13) holds for every $T_1\in\mathbb{R}$ and $\omega\in\Omega$. Thus, $\mathbf{b}(\omega)$ is a solution along the flow of the systems (2.12) for $\varepsilon=0$, and $B(\omega)$ is a solution along the flow of (2.11) for $\varepsilon=0$.

Let $\Omega_0 = \{\omega\in\Omega \mid B(\omega)>0\}$. Since $B(\omega\cdot t) = (U^{-1})^T(t,\omega)\,B(\omega)\,U^{-1}(t,\omega)$ we deduce that the set Ω_0 is invariant. We first assume that $m_0(\Omega_0)=0$. It is easy to construct a measurable map $\Omega\to\mathbb{R}^{2n}$, $\omega\mapsto\mathbf{v}(\omega)$ with $\|\mathbf{v}(\omega)\|=1$ and $\mathbf{v}^T(\omega)\,B(\omega)\,\mathbf{v}(\omega)\le 0$ for almost every $\omega\in\Omega$. For each ω we represent by $\lambda_j(\omega)$ the smallest eigenvalue of $B_{\Gamma,\varepsilon_j}(\omega)$; then $\Omega\to\mathbb{R}$, $\omega\to\lambda_j(\omega)$ is upper-semicontinuous. We have

$$\lim_{j\to\infty}\int_\Omega \mathbf{v}^T(\omega)\,B_{\Gamma,\varepsilon_j}(\omega)\,\mathbf{v}(\omega)\,dm_0 = \int_\Omega \mathbf{v}^T(\omega)\,B(\omega)\,\mathbf{v}(\omega)\,dm_0 = 0\,,$$

and $\mathbf{v}^T(\omega)\,B_{\Gamma,\varepsilon_j}(\omega)\,\mathbf{v}(\omega)\ge\lambda_j(\omega)>0$, hence, $\lim_{j\to\infty}\int_\Omega\lambda_j(\omega)\,dm_0=0$. In addition, since B_Γ is a symplectic matrix

$$\limsup_{j\to\infty}\int_\Omega \mathrm{tr}\, B_{\Gamma,\varepsilon_j}(\omega)\,dm_0 \ge \limsup_{j\to\infty}\int_\Omega\frac{1}{\lambda_j(\omega)}dm_0 \ge \lim_{j\to\infty}\frac{1}{\int_\Omega\lambda_j(\omega)\,dm_0} = \infty\,,$$

which contradicts our hypothesis and proves that $m_0(\Omega_0)=1$. We conclude that $B(\omega)$ is a positive definite solution along the flow of the systems (2.3). Finally, an argument similar to the one given in Theorem 4.5 of [15] assures the construction from B of a symplectic positive definite matrix $\widetilde{B} \in L^1(\Omega, m_0)$, solution along the flow of the same equation, which in its turn can be written as $\widetilde{B} = \widetilde{C}^T\widetilde{C}$ for a new matrix $\widetilde{C}$ satisfying the conditions of Property (c). This completes the proof. □

ACKNOWLEDGEMENT

Partially supported by Junta de Castilla y León under project VA01/97 and by Ministerio de Educación y Cultura. Authors would also like to thank Drs. R. Obaya and S. Novo for helpful discussions.

REFERENCES

[1] Barret, J.H., A Prüfer transformation for matrix differential equations. Proc. Amer. Math. Soc., Vol 8 (1957) pp. 510–518.

[2] Fabbri, R., Genericità dell'Iperbolicità nei sistemi differenziali lineari di dimensione due. Ph. Dissertation, Universitá degli Studi de Firenze, 1997.

[3] Ellis, R., & Johnson, R., Topological dynamics and linear differential systems. J. Differential Equations, Vol 44 (1982) pp. 21–39.

[4] Hinton, D., & Shaw, J., On Titchmarsh-Weyl m-functions for linear Hamiltonian systems. J. Differential Equations, Vol 40 (1981) pp. 316–342.

[5] Hinton, D., & Shaw, J., Hamiltonian systems of limit point or limit circle type with both endpoints singular. J. Differential Equations, Vol 50 (1983) pp. 444–464.

[6] Johnson, R., m-functions and Floquet exponents for linear differential systems. Ann. Mat. Pura Appl., Vol 147 (1987) pp. 211–248.

[7] Johnson, R., Hopf bifurcation for non-periodic solutions of differential equations I: Linear theory. *J. Dynamics Differential Equations*, Vol 1 (1989) pp. 179–198.

[8] Johnson, R., & Nerurkar, M., Exponential dichotomy and rotation number for linear Hamiltonian systems. J. Differential Equations, Vol 108 (1994) pp. 201–216.

[9] Johnson, R., Novo, S., & Obaya, R., Ergodic properties and Weyl M-functions for linear Hamiltonian systems. Preprint, 1999.

[10] Knill, O., Positive Lyapunov exponents for a dense set of bounded measurable $SL(2,\mathbb{R})$-cocycles. Ergod. Th. & Dynam. Sys., Vol 12 (1992) pp. 319–331.

[11] Kotani, S., & Simon, B., Stochastic Schrödinger operators and Jacobi matrix on the strip. Commun. Math. Phys., Vol 119 (1988) pp. 403–429.

[12] Matsushima, Y., Differentiable Manifolds. Marcel Dekker, New York, 1972.

[13] Mishchenko, A.S., Shatalov, V.E., & Sternin, B.Yu., Lagrangian Manifolds and the Maslov Operator. Springer-Verlag, New York, 1990.

[14] Nerurkar, M., Positive exponents for a dense set of continuous $SL(2,\mathbb{R})$ valued cocycles which arise as solutions to strongly accesible linear differential systems. Contemporary Mathematics, Vol 215 Amer. Math. Soc., Providence, Rhode Island, 1998 pp. 265–278.

[15] Novo, S., Núñez, C., & Obaya, R., Ergodic properties and rotation number for linear Hamiltonian systems. J. Differential Equations, Vol 148 (1998) pp. 148–185.

[16] Novo, S., & Obaya, R., On the dynamical behaviour of an almost periodic linear system with an ergodic 2-sheet studied by R. Johnson. Israel J. Math., Vol 105 (1998) pp. 235–249.

[17] Reid, W.T., A Prüfer transformation for differential systems. Pac. J. Math., Vol 8 (1958) pp. 575–584.

[18] Reid, W. T., Sturmian Theory for Ordinary Differential Equations. Applied Mathematical Sciences, Vol 31, Springer-Verlag, New York, 1980.

[19] Sacker, R.J., & Sell, G.R., A spectral theory for linear differential systems. J. Differential Equations, Vol 27 (1978) pp. 320–358.

Proceedings of Dynamic Systems and Applications 3 (2001) 475-482

SINGULAR PERTURBATION PROBLEMS FOR STOCHASTIC FUNCTIONAL DIFFERENTIAL EQUATIONS WITH CAUSAL OPERATORS

ZEPHYRUNUS C. OKONKWO
DEPARTMENT OF MATHEMATICS AND COMPUTER SCIENCE
ALBANY STATE UNIVERSITY
504 COLLEGE STREET ALBANY, GA 31705, USA

Abstract: This paper deals with the construction of approximation solutions for abstract Itô-Volterra functional equations by means of singularly perturbed Itô-Volterra functional differential equations. Existence theorems are proven. An LQ-optimal control problem described by a singularly perturbed Itô-Volterra functional differential equation is briefly discussed.

AMS (MOS) SUBJECT CLASSIFICATION: 34F05, 34K15, 34K30, 60H20

1. INTRODUCTION

In this paper, we construct approximate solutions to stochastic functional equations of the form

$$x(t,\omega) = (\mathcal{U}x)(t,\omega) \tag{1.1}$$

by means of Itô-Volterra functional differential equation

$$d(\epsilon x_\epsilon(t,\omega)) = -x_\epsilon(t,\omega)dt + (Vx_\epsilon)(t,\omega)dt + g(t, x_\epsilon(t,\omega))dz(t,\omega) \tag{1.2}$$

with the random initial condition

$$x(0,\omega) = x^0 \in R^n. \tag{1.3}$$

V is a Volterra operator acting on the function space $E([0,T] \times \Omega, R^n)$ which may be the $L^2([0,T] \times \Omega, R^n)$ or $C([0,T] \times \Omega, R^n)$. g is a matrix-valued random function satisfying certain regularity properties, $\epsilon>0$, is a small parameter, and Ω is the sample space.

For completeness, it is essential to characterize the underlying space $L^2([0,T] \times \Omega, R^n)$, (the function space $C([0,T] \times \Omega, R^n)$ can be characterized analogously). $L^2([0,T] \times \Omega, R^n)$, $T < \infty$, is the space of all product measurable random functions x: $[0,T] \times \Omega \to R^n$ with square integrable sample paths. V is a causal operator acting on the function space $L^2([0,T] \times \Omega, R^n)$ and satisfying certain regularity properties to be specified in the sequel. g is a nonlinear map belonging to the space $\mathcal{M}_2([0,T] \times R^n, R^{n \times q})$, the space of all $n \times q$ product measurable matrix-valued random functions satisfying the inequality

$$\mathcal{P}\left(\omega \backslash \int_0^T \|g(t, x(t,\omega))\|^2 dt \ < \infty\right) = 1, \tag{1.4}$$

with the integral in (1.4) being in the ordinary Lebesgue sense.

Throughout this paper, we will assume that $\left(\Omega, \mathcal{F}, \mathcal{P}\right)$ is a complete probability space. $\mathcal{F}$ is the σ-algebra of subsets of Ω; $\mathcal{F}_t$ is the underlying filtration and $\mathcal{P}$ is the probability measure. Ω is the sample space, with $\omega \in \Omega$. Furthermore, $z(t,\omega)$ is assumed to be a normalized R^q valued Wiener process, x^0 is a Gaussian random variable with a positive definite covariance matrix Q_0 and independent of the Wiener process $z(t,\omega)$.

Singularly perturbed differential equations arise as modeling tools in various areas of science and engineering, for example, in solid mechanics, and hydrodynamics. Some applications of singular perturbation in solid mechanics can be found in Olunloyo [9] where the effect of viscous damping on the response of a thick membrane to dynamic loading was considered. Cole [1] dealt with various singular perturbation problems in boundary layer theory. See also Van-Dyke [10] for applications of singular perturbation techniques in hydrodynamic flows. Golec & Ladde [4] applied averaging principle in the solution of singularly perturbed stochastic functional differential equations. Using auxiliary results (in Golec & Ladde [4]), a modified version of the stochastic averaging principle was developed to study systems consisting of fast and slow phenomena. Corduneanu [2] dealt with approximation techniques to solutions of a class of functional equations using singularly perturbed functional differential equations.

In this paper, the singular perturbation technique should be viewed as a regularization/ approximation procedure for solutions of equation (1.1). We will however obtain an alternative proof of the basic existence results in $C([0, T] \times \Omega, R^n)$ as well as $L^2([0, T] \times \Omega, R^n)$. In Section 2, we construct approximate solutions for the abstract Volterra equation in the case the underlying space $C([0, T] \times \Omega, R^n)$, which for each $\omega \in \Omega$ is the space of continuous functions. The case where the underlying function space is $L^2([0, T] \times \Omega, R^n)$ is discussed in Section 3, in particular, the case of singularly perturbed Itô-Volterra functional differential equation is considered. Section 4 deals with the formulation of a possible LQ- optimal control problem for singularly perturbed Itô-Volterra functional differential equation.

2. PRELIMINARY RESULTS

Consider the abstract Volterra equation

$$x(t,\omega) = (Vx)(t,\omega) \tag{2.1}$$

where V is continuous and compact on $C([0, T] \times \Omega, R^n)$. V may be of the form

$$(Vx)(t,\omega) = f(t,\omega) + \int_0^t k(t,s)x(s,\omega)ds, \; t \leq T, \tag{2.2}$$

or of the form

$$(Vx)(t,\omega) = f(t,\omega) + \int_0^t k(t,s,x(s,\omega))ds, \; t \leq T. \tag{2.3}$$

In equation (2.3), V is a nonlinear Volterra operator. Hence, (2.2) can be viewed as a stochastic integral equation of Volterra type. It is also possible for $(Vx)(t,\omega)$ to be of the form (see Corduneanu [2])

$$(Vx)(t,\omega) = \sum_{j=0}^{\infty} A(t)x(t-t_j,\omega) + \int_0^t B(t,s)x(s,\omega)ds, \; t_j \leq t. \tag{2.4}$$

Results pertaining to such classical equations are abound in the literature.

Let us recall that the solution of the first order linear initial value problem

$$\epsilon\, \dot{y}(t) + y(t) = q(t), \; y(0) = y^0, \tag{2.5}$$

is

$$y(t) = y^0 \cdot exp(-\tfrac{t}{\epsilon}) + \tfrac{1}{\epsilon}\int_0^t q(s) \cdot \{exp[-(\tfrac{t-s}{\epsilon})]\}\, ds. \tag{2.6}$$

We shall extend this notion to functional differential equation (1.2).

Let us assume that V is acting in the space $C([0, T] \times \Omega, R^n)$, which for each $\omega \in \Omega$, is the space of continuous maps. V therefore has the initial value property, that is, there exists $x^0 \in R^n$ such that $(Vx)(0, \omega) = x^0$ for any $x \in C([0, T] \times \Omega, R^n)$ and the solution of (1.2) if any will satisfy the initial value

$$x(0, \omega) = x^0.$$

We therefore impose on $x_\epsilon(t, \omega)$ the initial condition

$$x_\epsilon(0, \omega) = x^0 = \text{fixed initial value of } V, \tag{2.7}$$

(see Corduneanu [2]).

Using (2.5) therefore, we have that (1.2) can be put in the form

$$x_\epsilon(t, \omega) = x^0 \cdot exp(-\tfrac{t}{\epsilon}) + \tfrac{1}{\epsilon} \cdot \int_0^t \{exp[-(\tfrac{t-s}{\epsilon})]\} \cdot (Vx_\epsilon)(s, \omega)ds$$
$$+ \tfrac{1}{\epsilon} \cdot \int_0^t \{exp[-(\tfrac{t-s}{\epsilon})]\} \cdot g(s, x_\epsilon(s, \omega))\, dz(s,\omega). \tag{2.8}$$

Let us introduce the following notions.

Let

$$\Psi(t,\epsilon) = \begin{cases} 0, & t<0 \\ \frac{1}{\epsilon} e^{-\frac{t}{\epsilon}}, & t \in R_+ \end{cases}, \tag{2.9}$$

then (2.8) becomes

$$x_\epsilon(t, \omega) = x^0 \cdot exp(-\tfrac{t}{\epsilon}) + \int_0^t \Psi(t-s, \epsilon) \cdot (Vx_\epsilon)(s, \omega)ds$$
$$+ \int_0^t \Psi(t-s, \epsilon) \cdot g(s, x_\epsilon(s, \omega))\, dz(s,\omega). \tag{2.10}$$

For completeness we notice that the scalar function $\Psi(t, \epsilon)$ satisfies the following properties:

(i) $\int_0^t \Psi(s, \epsilon)ds = 1 - exp(-\tfrac{t}{\epsilon}),\ t \in R_+$

(ii) $\lim_{\epsilon \to 0} \int_0^a \Psi(s, \epsilon)ds = 1,\ a > 0,$

(i) $\lim_{\epsilon \to 0} \int_\delta^T \Psi(s, \epsilon)ds = 0,\ \delta>0, T>0.$

Properties (i) – (iii) establish the fact that the family of functions $\{\Psi(t, \epsilon), \epsilon>0\}$ provides an approximate identity for the convolution for any interval $[0, T]$, $T<\infty$. Hence for any $\{\epsilon_k\}$ such that $\epsilon_k \to 0$ as $k \to 0$, we have

$$\lim_{k \to \infty} \int_0^t \Psi(t-s, \epsilon_k) l(s, \omega)ds = l(t, \omega), \quad l(t, \omega) \in C([0, T] \times \Omega, R^n) \tag{2.11}$$

uniformly on $[0, T]$.

$$\lim_{k \to \infty} \int_0^t \Psi(t-s, \epsilon_k) l(s, \cdot)ds = l(t, \cdot), \quad l(t, \cdot) \in L^2([0, T], R^n) \tag{2.12}$$

in convergence of $L^2([0, T] \times \Omega, R^n)$.

Let us make the following observations which are essential for the subsequent theorem. Instead of considering equation (1.2) and the random initial condition (1.3), let us consider the following differential equation

$$d(\epsilon x_\epsilon(t,\omega)) = -x_{x\epsilon}(t,\omega)dt + (Vx_\epsilon)(t,\omega)dt \tag{2.13}$$

and the initial condition (1.2). The following theorem deals with the construction of the approximate solutions to equation (1.1) by means of (2.13), (1.3).

Theorem (2.1) Consider the stochastic functional differential equation (2.13) with the random initial condition (1.3). Assume that the following conditions are satisfied:

(*i*) V is an operator of Volterra type;
(*ii*) V is continuous on $C([0,T]\times\Omega,R^n)$;
(*iii*) V is compact on $C([0,T]\times\Omega,R^n)$;

Then there exists a unique solution process $x(t,\omega)\in C([0,\ T]\times\Omega,R^n)$ satisfying equation (2.13) with $x(0,\omega)=x^0$. Moreover, the solutions of (2.13) can be regarded as regularized solutions for (2.10) with $g=0$.

Proof: First, let us observe that equation (2.13), (1.3) can be rewritten in the form

$$x_\epsilon(t,\omega) - x^0 = \int_0^t \Psi(t-s,\epsilon)\cdot[(Vx_\epsilon)(s,\omega)-x^0]\,ds. \tag{2.14}$$

Using Schauder fixed point theorem, we will obtain a local existence of the solution process of (2.13), (1.3). Observe that, for any $\epsilon>0$, the operator on the left -hand side of (2.14) is almost surely compact on $C([0,a]\times\Omega,R^n)$. Let us define $(\mathfrak{J}u_\epsilon)(t,\omega)$ by

$$(\mathfrak{J}u_\epsilon)(t,\omega) = x^0 + \int_0^t \Psi(t-s,\epsilon)\cdot[(Vu_\epsilon)(s,\omega)-x^0]\,ds. \tag{2.15}$$

For any $u\in C([0,a]\times\Omega,R^n)$, the integrand on the right hand side of (2.15) is a product of two continuous functions on $C([0,a]\times\Omega,R^n)$, with $(Vu_\epsilon)(s,\omega)-x^0$ being compact. Hence $\mathfrak{J}_\epsilon$ is continuous and almost surely compact operator on $C([0,a]\times\Omega,R^n)$. It remains to show that $\mathfrak{J}_\epsilon$ takes the set $\mathcal{M}_r$ into itself where $\mathcal{M}_r$ is a closed convex set in $C([0,a]\times\Omega,R^n)$. Indeed, consider the ball

$$\mathcal{M}_r = \{x\colon x\in C([0,a]\times\Omega,R^n),\ \mathcal{E}|x(t,\omega)-x^0|\le r\}.$$

Since the operator $V-x^0$ is compact on $\mathcal{M}_r$, there exists a positive number $\xi(r)>0$ such that $\mathcal{E}|(Vu)(t,\omega)-(Vu)(s,\omega)|<r$ for each $u\in\mathcal{M}_r$, provided $|t-s|<\xi(r)$, $t,s\in[0,a]$, $\omega\in\Omega$. V has the initial value property, that is, $(Vx)(0,\omega)=x^0$, and hence $\mathcal{E}|(Vu)(s,\omega)-x^0|<r(1-e^{-\frac{\xi}{\epsilon}})<r$ for $0\le s\le\xi$, which shows that $\mathfrak{J}_\epsilon\mathcal{M}_r\subset\mathcal{M}_r$, where $\mathcal{M}_r$ consists of the functions restricted to $[0,\xi]$. Since $\mathcal{M}_r$ is closed and convex in $C([0,\xi]\times\Omega,R^n)$, by Schauder fixed point theorem, there exists at least one solution process $x_\epsilon(t,\omega)$ of (2.14) such that $x_\epsilon(t,\omega)\in\mathcal{M}_r$.

Let $\{\epsilon_k\}$ be a sequence of positive numbers such that $\lim_{k\to\infty}\epsilon_k=0$. For each k, let $\{x_{\epsilon_k}(t,\omega)\}$ be the solution of (2.14), and such that $x_{\epsilon_k}\in\mathcal{M}_r$, $t\in[0,\xi]$.

$$x_{\epsilon_k}(t,\omega) - x^0 = \int_0^t \Psi(t-s,\epsilon)\cdot[(Vx_{\epsilon_k})(s,\omega)-x^0]\,ds. \tag{2.16}$$

Since the operator V is compact, $\{Vx_{\epsilon_k}\}$ is convergent in $C([0,\xi]\times\Omega,R^n)$. From the property of approximate identity of the convolution, the limit of the right hand side of

(2.16) exists. Existence of such a limit means the convergence of $\{x_{\epsilon_k}(t,\omega)\}$ to some $x(t,\omega) \in \mathcal{M}_r$ and by letting $k \to \infty$ in (2.16), we have $x_\epsilon(t,\omega) - x^0 = (Vx_\epsilon)(t,\omega) - x^0$ which is the same as (1.1).

In the discussions presented above, we have been able to construct approximate continuous solutions for the Volterra equation (1.1). This completes the proof.

Remark (2.1): Assume that in (1.1), $(\mathcal{U}x)(t,\omega)$ is given by

$$(\mathcal{U}x)(t,\omega) = f(t,\omega) + \int_0^t k(t,s)x(s,\omega)ds,\ t \le T. \tag{2.17}$$

Then equation (1.1) can be written as

$$x(t,\omega) = f(t,\omega) + \int_0^t k(t,s)x(s,\omega)ds, \tag{2.18}$$

where $f \in C([0,T] \times \Omega, R^n)$ and $k(t,s)$ is the kernel of the integral equation. (2.18) is a Volterra integral equation of the second kind whose solution can be obtained by the method of successive approximations, see for example Corduneanu [2]. Similarly, equation (2.13), (1.3) can be written in the form

$$x(t,\omega) = x^0 + \int_0^t \Psi(t-s,\epsilon) f(s,\omega)ds + \int_0^t \Psi(t-s,\epsilon)\, k(t,s)(Vx)(s,\omega)ds, \tag{2.19}$$

where $\Psi(t,\epsilon)$ satisfies (2.9). One can easily verify therefore that the approximate solutions to (2.18) can be obtained using (2.19).

Remark (2.2): We have carefully dealt with the construction of the approximate identity for (1.1) using (2.13), (1.3) instead of (1.2), (1.3). The latter case will be dealt with in the next section where we shall assume that the abstract causal operator $(\mathcal{U}x)(t,\omega)$ is of Itô-Volterra type, and the underlying space will be $L^2([0,T] \times \Omega, R^n)$.

3. CONSTRUCTION OF APPROXIMATE SOLUTIONS IN L^2-SPACE

Let us consider the abstract equation (1.1) in the case $x \in L^2([0,T] \times \Omega, R^n)$. We will construct approximate solutions of (1.1) by means of the solutions of (1.2), (1.3). Since we are dealing with the space of measurable functions, without loss of generality we shall choose the initial condition $x(0,\omega) = \theta \in R^n$ (the zero vector in R^n). Notice that $|e^{-\frac{t}{\epsilon}}| \to 0$ as $\epsilon \to 0$. Hence, we shall deal with the integral equation

$$x_\epsilon(t,\omega) = \int_0^t \Psi(t-s,\epsilon)\,(Vx_\epsilon)(s,\omega)ds + \int_0^t \Psi(t-s,\epsilon)\, g(s, x_\epsilon(s,\omega))dz(s,\omega). \tag{3.1}$$

The Itô-Volterra equation (3.1) can be shown to possess local solution $x_\epsilon(t,\omega) \in L^2([0,T] \times \Omega, R^n)$ provided adequate assumptions are made on the operator V and the nonlinear random matrix valued function g. Indeed, g must satisfy the following properties:

$$\mathcal{E}\left\{ \int_0^t g(s, x(s,\omega))dz(s,\omega) \right\} = 0, \tag{3.2}$$

and

$$\mathcal{E}\,|\int_0^t g(s,x(s,\omega))dz(s,\omega)|^2 = \int_0^t \mathcal{E}|g(s,x(s,\omega))|^2 ds < \infty. \tag{3.3}$$

Here and in the sequel $\mathcal{E}\{h\}$ shall mean the mathematical expectation of h.

Assume that V is a continuous compact operator on $L^2([0, T] \times \Omega, R^n)$ with $\epsilon>0$ fixed. Consider the operator

$$(\mathcal{K}u_\epsilon)(t,\omega) = \int_0^t \Psi(t-s,\epsilon)\,(Vx_\epsilon)(s,\omega)ds + \int_0^t \Psi(t-s,\epsilon)\, g(s, x_\epsilon(s,\omega))dz(s,\omega) \quad (3.4)$$

We recall the following well known properties of convolutions: for $A \in L^1$ and $B \in L^p$, $1 \le p \le \infty$, one has that

$$A*B \in L^p \text{ and } |A*B|_{L^p} \le |A|_{L^1}\,|B|_{L^p}. \quad (3.5)$$

Consider the operator $\mathcal{K}_\epsilon$ in (3.4) on the ball

$$\mathcal{B}_\rho = \{ x\colon\ x \in L^2([0, a] \times \Omega, R^n),\ \mathcal{E}|x|_2 \le \rho\}.$$

We will prove that $\mathcal{K}_\epsilon$, or its restriction to some $L^2([0, \delta] \times \Omega, R^n)$, $\delta \le a$, has a fixed point. From inequality (3.5), we see that $\mathcal{K}_\epsilon$ is continuous and almost surely compact on $L^2([0, a] \times \Omega, R^n)$. We obtain from (3.4) on $[0, t]$;

$$\mathcal{E}\left(\int_0^t |(\mathcal{K}_\epsilon u)(s,\omega)|^2 ds\right) \le 2\left(1 - e^{-\frac{t}{\epsilon}}\right) \mathcal{E}\int_0^t |(Vu_\epsilon)(s,\omega)|^2 ds$$
$$+ 2\left(1 - e^{-\frac{t}{\epsilon}}\right) \mathcal{E}\int_0^t |g(s, u_\epsilon(s,\omega))|^2 ds, \quad (3.6)$$

for any $t \in [0, a]$, and $u \in \mathcal{B}_\rho$.

It is obvious that the inclusion $\mathcal{K}_\epsilon\, \mathcal{B}_\rho \subset \mathcal{B}_\rho$ will follow from the fact that

$$\left(\mathcal{E}\int_0^t |(Vu_\epsilon)(s,\omega)|^2 ds\right) \le \frac{\rho^2}{4}, \quad \left(\mathcal{E}\int_0^t |g(s, u_\epsilon(s,\omega))|^2 ds\right) \le \phi(\rho). \quad (3.7)$$

Notice that from (3.6), we have

$$\left(\int_0^t \mathcal{E}|(\mathcal{K}_\epsilon u)(s,\omega)|^2 ds\right)^{\frac{1}{2}} \le \sqrt{2\left(1 - e^{-\frac{t}{\epsilon}}\right)\left(\frac{\rho^2}{4} + \phi(\rho)\right)} \quad \le \rho. \quad (3.8)$$

(3.8) implies that $\phi(\rho) \le \frac{\rho^2}{4}\dfrac{\left(1+e^{-\frac{t}{\epsilon}}\right)}{\left(1-e^{-\frac{t}{\epsilon}}\right)}$, $t \in (0, a)$. (3.9)

One can validate inequalities (3.7) on some interval $[0,\delta]$ based on the compactness of the set of images $\{\mathcal{K}_\epsilon u\colon\ u \in \mathcal{B}_\rho \in L^2([0, a] \times \Omega, R^n)\}$. Indeed, since Vu is compact and g is continuous on $[0,T] \times R^n$, $\{\mathcal{K}_\epsilon u\colon u \in \mathcal{B}_\rho\}$ is relatively compact in $L^2([0, a] \times \Omega, R^n)$. Hence, it admits a finite α-net for every $\alpha>0$. Also if $\{u_k\}$, $k = 1,2,3, \cdots, m$ is the finite α-net, them (3.7) holds for every $u = u_k$, $k = 1,2,3, \cdots, m$ for some $\delta>0$, $\delta \le a$. For every arbitrary $u \in \mathcal{B}_\rho$, we have

$$\mathcal{E}\int_0^t |(Vu_\epsilon)(s,\omega)|^2 ds \le 2\alpha,\ 0 \le t \le \delta;$$
$$\mathcal{E}\int_0^t |g(s, u_\epsilon(s,\omega))|^2 ds \le 2\alpha,\ 0 \le t \le \delta; \quad (3.10)$$

It suffices to choose $4\alpha \le \rho^2$ in order to establish inequality (3.7). The Schauder fixed point theorem can be applied to the operator $\mathcal{K}_\epsilon$ and the ball $\mathcal{B}_\rho$, on the interval $[0,\delta]$. Hence, the existence in $L^2([0, \delta] \times \Omega, R^n)$ for (3.1) is established for every $\epsilon>0$. We must notice that δ does not depend on ϵ.

Now, let $\{\epsilon_k\}$ be a sequence of positive numbers such that $\epsilon_k \to 0$. For each k, let $x_{\epsilon_k}(t,\omega)$ be a solution of (3.1) defined on $[0, \delta]$, for each ω, and such that $x_{\epsilon_k}(t,\omega) \in \mathcal{B}_\rho$, $t \in [0,\delta]$,

$$x_{\epsilon_k}(t,\omega)=\int_0^t \Psi(t-s,\epsilon)\,(Vx_{\epsilon_k})(s,\omega)ds+\int_0^t \Psi(t-s,\epsilon)\,g(s,x_{\epsilon_k}(s,\omega))dz(s,\omega) \quad (3.11)$$

Since $\mathcal{K}_\epsilon$ is compact in $L^2([0,\ t]\times\Omega,R^n)$, $\{\mathcal{K}_{\epsilon_k}x\}$ is convergent (in the L^2-sense). Because of property (2.12) for the approximate identity for the convolution, the existence of the limit to the right-hand of (3.11) implies the convergence of $\{x_{\epsilon_k}(t,\omega)\}$ to some $x(t,\omega)\in\mathcal{B}_\rho$ for every ω. By letting $k\to\infty$ in (3.11), we obtain $x(t,\omega)=(\mathcal{U}x)(t,\omega)$ which is the same as (1.1). We summarize the following the following discussions with the following theorem.

Theorem (3.1) Assume that the operator $\mathcal{U}$ (in (1.1) is of Itô-Volterra type, continuous and compact from $L^2([0,\ a]\times\Omega,R^n)$ into itself. Then, there exists at least one solution of (1.1) that belongs to the space $L^2([0,\ a]\times\Omega,R^n)$ (possibly to some restriction to some interval $[0.\delta]$, $\delta\le a$). In addition, the singularly perturbed equation (1.2) is solvable in $L^2([0,\ a]\times\Omega,R^n)$ for some $\epsilon>0$, and its solutions are regarded as regularized solutions for (3.11).

4. OPTIMAL CONTROL AND OTHER RELATED PROBLEMS

Consider the following stochastic functional equation

$$x(t,\omega)=(\mathcal{L}x)(t,\omega)+(Bu)(t,\omega)+G(t,\omega) \quad (4.1)$$

where $\mathcal{L}:L^2([0,\ T]\times\Omega,R^n)\to L^2([0,\ T]\times\Omega,R^n)$ is a linear causal operator and $B:L^2([0,\ T]\times\Omega,R^m)\to L^2([0,\ T]\times\Omega,R^n)$ is also a linear causal operator taking elements of the control space into the space of the trajectories. $G(t,\omega)$ is an Itô-integral, that is

$$G(t,\omega)=\int_0^t g(s,\omega)dz(s,\omega). \quad (4.2)$$

Our problem here is to minimize the cost functional

$$\mathcal{C}(x,u)=\int_0^T \mathrm{Tr}\,\mathcal{E}[P(x(t,\omega))(x(t,\omega)^T]dt+\int_0^T \mathrm{Tr}\,\mathcal{E}[\mathcal{Q}(u(t,\omega))(u(t,\omega)^T]dt, \quad (4.3)$$

subject to (4.1). In (4.3), P is a self-adjoint positive semi-definite operator on $L^2([0,\ T]\times\Omega,R^n)$, and $\mathcal{Q}$ is a self-adjoint positive definite operator on the control space $L^2([0,T]\times\Omega,R^m)$. Under suitable conditions, the existence and uniqueness of the solution process of (4.1) can be proven, (see for example Okonkwo & Ladde [8]).

Now, consider the singularly perturbed stochastic functional differential equation

$$d(\epsilon x(t,\omega))=-x(t,\omega)\,dt+(\mathcal{L}x)(t,\omega)dt+(Bu)(t,\omega)dt+g(t,\omega)dz(t,\omega) \quad (4.4)$$

with the initial condition

$$x(0,\omega)=\theta\in R^n\ (\ \theta \text{ is the zero element in } R^n). \quad (4.5)$$

Of course, equation (4.4) is equivalent to

$$x_\epsilon(t,\omega)=\int_0^t \Psi(t-s,\epsilon)\,(\mathcal{L}x_\epsilon)(s,\omega)ds+\int_0^t \Psi(t-s,\epsilon)\,(Bu)(s,\omega)ds$$
$$+\int_0^t \Psi(t-s,\epsilon)\,g(s,x_\epsilon(s,\omega))dz(s,\omega). \quad (4.6)$$

Since $\mathcal{L}$ and B are linear causal operators, we can without loss of generality assume that $\mathcal{L}_1=\Psi\mathcal{L}$ and $B_1=\Psi B$ are causal operators. Let $g_1=\Psi g$.

One can write that

$$\int_0^t (\mathcal{L}_1x)(s,\omega)ds=\int_0^t K(t,s)x(s,\omega)ds$$

which is usually established by Riesz representation theorem where $K(t,s)$ is the kernel associated with the operator $\mathcal{L}_1$. The solution process can be written in the form

$$x_\epsilon(t,\omega) = \int_0^t \mathcal{X}(t,s)(B_1 u)(s,\omega)ds + \int_0^t \mathcal{X}(t,s)\, g_1(s, x_\epsilon(s,\omega))dz(s,\omega) \tag{4.7}$$

or

$$x_\epsilon(t,\omega) = \mathcal{T} u_\epsilon + \mathcal{G}_\epsilon \tag{4.8}$$

$$(\mathcal{T}_\epsilon u)(t,\omega) = \int_0^t \mathcal{X}(t,s)(B_1 u)(s,\omega)ds$$

and

$$\mathcal{G}_\epsilon = \int_0^t \mathcal{X}(t,s)\, g_1(s, x_\epsilon(s,\omega))dz(s,\omega).$$

$\mathcal{X}(t,s)$ is the Cauchy matrix associated with the operator $\mathcal{L}_1$. Let $\widetilde{K}(t,s)$ be the resolvent kernel associated with $K(t,s)$, then, the Cauchy kernel is related to the resolvent kernel as follows (see Corduneanu [2])

$$\mathcal{X}(t,s) = I_{n\times n} + \int_s^t \widetilde{K}(t,v)dv. \tag{4.9}$$

One can deal with the following problem:
Minimize (4.3) subject to (4.7) with $u \in U \subset L^2([0,T]\times\Omega, R^m)$ where U is closed and convex in $L^2([0,T]\times\Omega, R^m)$. The following question is relevant: How is the solution of the LQ – optimal control problem for the singularly perturbed system related to the solution of the LQ-optimal control problem for the unperturbed system ? The complete answer to this problem will be dealt with in a separate paper.

REFERENCES

[1] J. Cole. *Perturbation methods in applied mathematics*, Ginn (Blaisdel) Boston, MA, 1968.

[2] C. Corduneanu. *Integral Equations and Applications.* Cambridge University Press, 1991

[3] C. Corduneanu. Neutral Functional Differential Equations with Abstract Volterra Operators, *Advances in Nonlinear Dynamics*, V. 5 Gordon and Breach, 1995.

[4] J. Golec, & G. S. Ladde. Averaging Principle and Systems of Singularly Perturbed Stochastic Differential Equations. J. Math. Phys. 31 (5), 1990, 1116-1123.

[5] P. A. Lagerstrom. Matched asymptotic expansions. Springer-Verlag, 1988

[6] Yizeng Li. *Global Existence and Stability of Functional Differential Equations with Abstract Volterra Operators*. Ph.D. Thesis, University of Texas at Arlington, 1993.

[7] Z. C. Okonkwo. Stochastic Functional Differential Equations With Abstract Volterra Operators 1: Existence of Solutions. *Volterra Equations and Applications*, Gordon and Breach, 1999.

[8] Z. Okonkwo, & G.S. Ladde, Itô- Type Stochastic Differential Equations with Abstract Volterra Operators and Their Control, *Dynamic Systems and Applications 6* (1997) 461 – 468.

[9] V. O. S. Olunloyo. The effect of viscous damping on the response of a thick rectangular membrane to dynamic loading. Iranian Journal of science and technology, Vol. 6, 177-192, 1977.

[10] M. Van Dyke. *Perturbation Methods in Fluid Mechanics*, parabolic press Stanford, CA, 1975.

[11] J. Yeh. *Martingales and Stochastic Analysis.* World Scientific Publishing Co., Singapore, 1995.

Proceedings of Dynamic Systems and Applications 3 (2001) 483-490

Symplectic Multiple Time Step Integration of Hamiltonian Dynamical Systems

Daniel Okunbor[1] and David Hardy[2]

[1]Department of Mathematics and Computer Science, University of Maryland Eastern Shore, Princess Anne, MD 21801

[2]Theoretical Biophysics Group, Beckman Institute of Science and Technology, University of Illinois at Champaign-Urbana, Urbana, Ill 61801

ABSTRACT: In this paper, we provide the theoretical foundation for multiple time scale integration of Hamiltonian dynamical systems applied to symplectic integration algorithms. Using the Lie group approach for the construction of symplectic integration methods, we demonstrate that symplectic multiple time stepping implementation of these methods can be obtained by compositions. In particular, we investigate the symplectic multiple time step implementation of leap/frog scheme, otherwise called Stórmer-Verlet integration method. The symplectic multiple time step Stórmer-Verlet was tested on the linear spring model and was found to be significantly more efficient than the standard approach. An analysis indicated that the optimal partitioning of the potential function is proportionally related to the maximum spring constant.

AMS(MOS) Subject Classifications: 65L05

1. Introduction

We consider the computer solution of dynamical Hamiltonian systems

$$\frac{d}{dt}z = J\nabla H(z),\ z = \begin{pmatrix} q \\ p \end{pmatrix} \in \mathcal{R}^{2d}$$

where $J = \begin{pmatrix} 0_d & I_d \\ -I_d & 0_d \end{pmatrix}$, I_d is an identity matrix of dimension d, H is the scalar-valued function. The function H is called the Hamiltonian and it is assumed to be sufficiently smooth with respect to z for existence and uniqueness of solution. Dynamical Hamiltonian systems is of considerable importance to areas of scientific research, such as modeling biomolecular structures and planetary motion. We consider the autonomous separable Hamiltonian for an N-body system,

$$H(\mathbf{q}(t), \mathbf{p}(t)) = T(\mathbf{p}) + V(\mathbf{q}) = \frac{1}{2}\mathbf{p}^{\mathrm{T}}M^{-1}\mathbf{p} + V(\mathbf{q}),$$

where $\mathbf{q}(t), \mathbf{p}(t) \in \mathbb{R}^{\nu}$ are the positions and momenta of the system, respectively, with $\nu\,(= 3N)$ degrees of freedom, and M is the diagonal mass matrix (with each mass replicated three times). The quadratic term $T(\mathbf{p})$ is the kinetic energy, and $V(\mathbf{q})$ is the potential energy. The corresponding equations of motion are

$$\dot{\mathbf{q}} = M^{-1}\mathbf{p}, \quad \dot{\mathbf{p}} = -\nabla V(\mathbf{q}) = \mathbf{f}(\mathbf{q}),$$

where $\mathbf{f}$ is the force of the system.

The motivation for symplectic multiple time step integration comes from analysis by Skeel & Gear [1] showing that a fixed stepsize symplectic integrator will lose symplecticness, in the traditional sense, when used with a variable stepsize. These results were also verified empirically by Calvo & Sanz-Serna [2], although experimentation in [4] suggests otherwise for second-order integrators. More recently, Hairer and Stoffer[5] describe an approach to symplectic variable stepsize integration by perturbing the Hamiltonian to one with an identical solution. However, his methods are implicit and do not make use of multiple time step techniques.

Multiple time stepping is described in [6] as a method for making force computation more efficient. This idea was later applied to Verlet integration [7, 9] in a manner that retains symplecticness, time-reversibility, and second-order accuracy. We present a similar approach in which the potential energy function is smoothly partitioned into the sum of potential energy terms, allowing different parts of the system to be integrated using multiples of a smallest stepsize. Multiple time stepping seems more natural than variable stepsize for molecular dynamics, since faster forces will always be present.

This paper shows techniques for partitioning both bonded and nonbonded potential energy functions. Integration for both cases is implemented efficiently by assigning interactions to classes, with each class represented by its own potential energy term. The experiments performed here for the bond-length potential provide empirical verification of analytic results from Littell, Skeel, & Zhang [12].

2. Symplectic Integration

The time derivatives of $\mathbf{z}(t) = \left(\mathbf{q}(t)^{\mathrm{T}}, \mathbf{p}(t)^{\mathrm{T}}\right)^{\mathrm{T}} \in \mathbb{R}^{2\nu}$ are expressed as

$$\frac{d^k}{dt^k}\mathbf{z}(t) = \mathcal{D}_H^k \mathbf{z}(t)$$

for the differential operator $\mathcal{D}_H$ defined by

$$\mathcal{D}_H = \sum_{i=1}^{\nu} \left(\frac{\partial H}{\partial p_i}\frac{\partial}{\partial q_i} - \frac{\partial H}{\partial q_i}\frac{\partial}{\partial p_i} \right).$$

Since

$$\mathbf{z}(t) = \sum_{k=0}^{\infty} \frac{t^k}{k!}\frac{d^k}{dt^k}\mathbf{z}(0) = e^{t\mathcal{D}_H}(\mathbf{z}(0)), \tag{1}$$

the exponential operator maps $\mathbf{z}(0)$ to $\mathbf{z}(t)$. For separable Hamiltonians, the differential operator is expressed as $\mathcal{D}_H = \mathcal{D}_T + \mathcal{D}_V$, and the exponential operator (1) applied to some $\mathbf{z} = \left(\mathbf{q}^{\mathrm{T}}, \mathbf{p}^{\mathrm{T}}\right)^{\mathrm{T}}$ can be approximated by alternate applications of

$$e^{t\mathcal{D}_T}(\mathbf{z}) = \begin{bmatrix} \mathbf{q} + t\nabla T(\mathbf{p}) \\ \mathbf{p} \end{bmatrix}, \quad e^{t\mathcal{D}_V}(\mathbf{z}) = \begin{bmatrix} \mathbf{q} \\ \mathbf{p} - t\nabla V(\mathbf{q}) \end{bmatrix}. \tag{2}$$

The transformations in (2) are both symplectic since they satisfy the necessary and sufficient condition for a mapping ψ to be symplectic, given by

$$(\nabla\psi)^{\mathrm{T}} J (\nabla\psi) = J, \tag{3}$$

where $\nabla\psi$ is the Jacobian matrix of ψ and J is the skew-symmetric matrix

$$J = \begin{bmatrix} 0_\nu & I_\nu \\ -I_\nu & 0_\nu \end{bmatrix}.$$

It follows from (3) that the composition of symplectic transformations is also symplectic.

A single step of Verlet is expressed as the composition

$$\mathbf{z}_{n+1} = \phi_h(\mathbf{z}_n) = e^{\frac{h}{2}\mathcal{D}_V} e^{h\mathcal{D}_T} e^{\frac{h}{2}\mathcal{D}_V}(\mathbf{z}_n),$$

showing that it is symplectic. Using the Baker-Campbell-Hausdorff (BCH) formula from Lie algebra [8, 3, 13], the product of exponentials of operators can be expanded as

$$e^{\mathcal{A}} e^{\mathcal{B}} = e^{\mathcal{A}+\mathcal{B}+\frac{1}{2}[\mathcal{A},\mathcal{B}]+\frac{1}{12}([[\mathcal{B},\mathcal{A}],\mathcal{A}]+[\mathcal{B},[\mathcal{B},\mathcal{A}]])-\frac{1}{24}[\mathcal{B},[[\mathcal{B},\mathcal{A}],\mathcal{A}]]+\cdots},$$

with the Poisson bracket notation $[\mathcal{A}, \mathcal{B}] = \mathcal{A}\mathcal{B} - \mathcal{B}\mathcal{A}$. This can be used to show that Verlet is second-order accurate, since its operator approximates $e^{h\mathcal{D}_H}$ up to the h^3 terms

$$e^{\frac{h}{2}\mathcal{D}_V} e^{h\mathcal{D}_T} e^{\frac{h}{2}\mathcal{D}_V} = e^{h\mathcal{D}_H + \frac{h^3}{24}(2[\mathcal{D}_T,[\mathcal{D}_T,\mathcal{D}_V]]-[[\mathcal{D}_T,\mathcal{D}_V],\mathcal{D}_V])+\mathcal{O}(h^5)}. \tag{4}$$

3. Symplectic Multiple Time Step Stórmer-Verlet

A single step of multiple time step Verlet can be expressed as

$$\begin{aligned} \mathbf{z}_{n+1} &= \Phi_h(\mathbf{z}_n) \\ &= e^{\frac{h}{2}w_{\mu-1}(n+1)\mathcal{D}_V^{[\mu-1]}} e^{\frac{h}{2}w_{\mu-2}(n+1)\mathcal{D}_V^{[\mu-2]}} \cdots e^{\frac{h}{2}w_0(n+1)\mathcal{D}_V^{[0]}} \\ &\cdot e^{h\mathcal{D}_T} e^{\frac{h}{2}w_0(n)\mathcal{D}_V^{[0]}} e^{\frac{h}{2}w_1(n)\mathcal{D}_V^{[1]}} \cdots e^{\frac{h}{2}w_{\mu-1}(n)\mathcal{D}_V^{[\mu-1]}}(\mathbf{z}_n), \end{aligned} \tag{5}$$

where

$$e^{t\mathcal{D}_V^{[i]}}(\mathbf{z}_n) = \begin{bmatrix} \mathbf{q}_n \\ \mathbf{p}_n - t\nabla V^{[i]}(\mathbf{q}_n) \end{bmatrix}.$$

This transformation is obviously symplectic, since it is the composition of symplectic mappings. The local truncation error is computed over a sequence of $\pi_{\mu-1}$ steps $n, n+1, \ldots, n+\pi_{\mu-1}-1$, where $n \equiv 0 \pmod{\pi_{\mu-1}}$. This can be written as the recurrence

$$(\Phi_h)^{\pi_{\mu-1}} = \Phi^{[\mu-1]}_{\pi_{\mu-1}h},$$

with

$$\Phi^{[i]}_{\pi_i h} = e^{\frac{\pi_i h}{2}\mathcal{D}_V^{[i]}} \left(\Phi^{[i-1]}_{\pi_{i-1}h}\right)^{\frac{\pi_i}{\pi_{i-1}}} e^{\frac{\pi_i h}{2}\mathcal{D}_V^{[i]}},$$

for $i = 1, \ldots, \mu-1$, and

$$\Phi^{[0]}_{\pi_0 h} = e^{\frac{\pi_0 h}{2}\mathcal{D}_V^{[0]}} \left(e^{h\mathcal{D}_T}\right)^{\pi_0} e^{\frac{\pi_0 h}{2}\mathcal{D}_V^{[0]}} = e^{\frac{h}{2}\mathcal{D}_V^{[0]}} e^{h\mathcal{D}_T} e^{\frac{h}{2}\mathcal{D}_V^{[0]}}.$$

The multiple time step Verlet integrator is shown to have second-order accuracy by applying (4) to the recurrence. The base case is

$$\Phi^{[0]}_{\pi_0 h} = e^{\pi_0 h(\mathcal{D}_T + \mathcal{D}_V^{[0]}) + \frac{\pi_0 h^3}{24}(\pi_0^2 C_0) + \mathcal{O}(h^5)},$$

and the inductive step is

$$\begin{aligned}
\Phi^{[i]}_{\pi_i h} &= e^{\frac{\pi_i h}{2}\mathcal{D}_V^{[i]}} \left(\Phi^{[i-1]}_{\pi_{i-1}h}\right)^{\frac{\pi_i}{\pi_{i-1}}} e^{\frac{\pi_i h}{2}\mathcal{D}_V^{[i]}} \\
&= e^{\frac{\pi_i h}{2}\mathcal{D}_V^{[i]}} \left(e^{\pi_{i-1}h\left(\mathcal{D}_T + \sum_{j=0}^{i-1}\mathcal{D}_V^{[j]}\right) + \frac{\pi_{i-1}h^3}{24}\left(\sum_{j=0}^{i-1}\pi_j^2 C_j\right) + \mathcal{O}(h^5)}\right)^{\frac{\pi_i}{\pi_{i-1}}} e^{\frac{\pi_i h}{2}\mathcal{D}_V^{[i]}} \\
&= e^{\pi_i h\left(\mathcal{D}_T + \sum_{j=0}^{i}\mathcal{D}_V^{[j]}\right) + \frac{\pi_i h^3}{24}\left(\sum_{j=0}^{i}\pi_j^2 C_j\right) + \mathcal{O}(h^5)},
\end{aligned} \tag{6}$$

for $i = 1, \ldots, \mu-1$, showing that the operator expansion over $\pi_{\mu-1}$ steps yields

$$\begin{aligned}
(\Phi_h)^{\pi_{\mu-1}} &= e^{\pi_{\mu-1}h\left(\mathcal{D}_T + \sum_{i=0}^{\mu-1}\mathcal{D}_V^{[i]}\right) + \frac{\pi_{\mu-1}h^3}{24}\left(\sum_{i=0}^{\mu-1}\pi_i^2 C_i\right) + \mathcal{O}(h^5)} \\
&= e^{\pi_{\mu-1}h\mathcal{D}_H + \frac{\pi_{\mu-1}h^3}{24}\left(\sum_{i=0}^{\mu-1}\pi_i^2 C_i\right) + \mathcal{O}(h^5)},
\end{aligned} \tag{7}$$

where

$$C_i = 2\left[\mathcal{D}_T + \sum_{j=0}^{i-1}\mathcal{D}_V^{[j]}, \left[\mathcal{D}_T + \sum_{j=0}^{i-1}\mathcal{D}_V^{[j]}, \mathcal{D}_V^{[i]}\right]\right] - \left[\left[\mathcal{D}_T + \sum_{j=0}^{i-1}\mathcal{D}_V^{[j]}, \mathcal{D}_V^{[i]}\right], \mathcal{D}_V^{[i]}\right].$$

4. Application to bond-length force

4.1 Partitioning the potential energy

The bond-length force joining two particles is modeled as an ideal spring. The potential energy for the spring that joins particles i and j is given by Hooke's Law

$$V_{ij}(\|\mathbf{q}_j - \mathbf{q}_i\|) = \frac{\kappa_{ij}}{2}(\|\mathbf{q}_j - \mathbf{q}_i\| - \ell_{ij})^2,$$

where κ_{ij} is the spring constant and ℓ_{ij} is the equilibrium length. For notational convenience, we define $\kappa_{ij} = 0$ if no spring joins particles i and j. The potential energy for the system is given by

$$V(\mathbf{q}) = \sum_{i<j} V_{ij}(\|\mathbf{q}_j - \mathbf{q}_i\|). \tag{8}$$

We partition V by assigning each spring to a class. The following restrictions are placed on class assignment.

1. Every spring is assigned to exactly one class.

2. Class 0 has a spring.

3. Class $\mu - 1$ has a spring.

The first restriction ensures that the spring potentials sum as required by (8). The second restriction makes sense, for if class 0 had no springs, then the effective microstepsize would be $\pi_k h$ for the smallest numbered class k having a spring. The third restriction ensures that no more than μ classes are necessary. The class assignment of springs can be effectively determined by the spring constants. Since the integration method is second-order, the error from a spring in class k with spring constant κ is proportional to $(\pi_k h)^2\kappa$. Classes can be assigned so as to balance the error contributed by each spring. Let κ_{max} be the largest spring constant, and assume we wish to determine the class c of a spring with constant κ. We assign the stiffest spring to class 0 and select the largest c possible so that the proportionality of error introduced for the spring is no greater than that for the stiffest spring. This is done by selecting the largest c such that

$$(\pi_c h)^2\kappa \le h^2\kappa_{max},$$

or equivalently,

$$\pi_c \le \sqrt{\frac{\kappa_{max}}{\kappa}}. \tag{9}$$

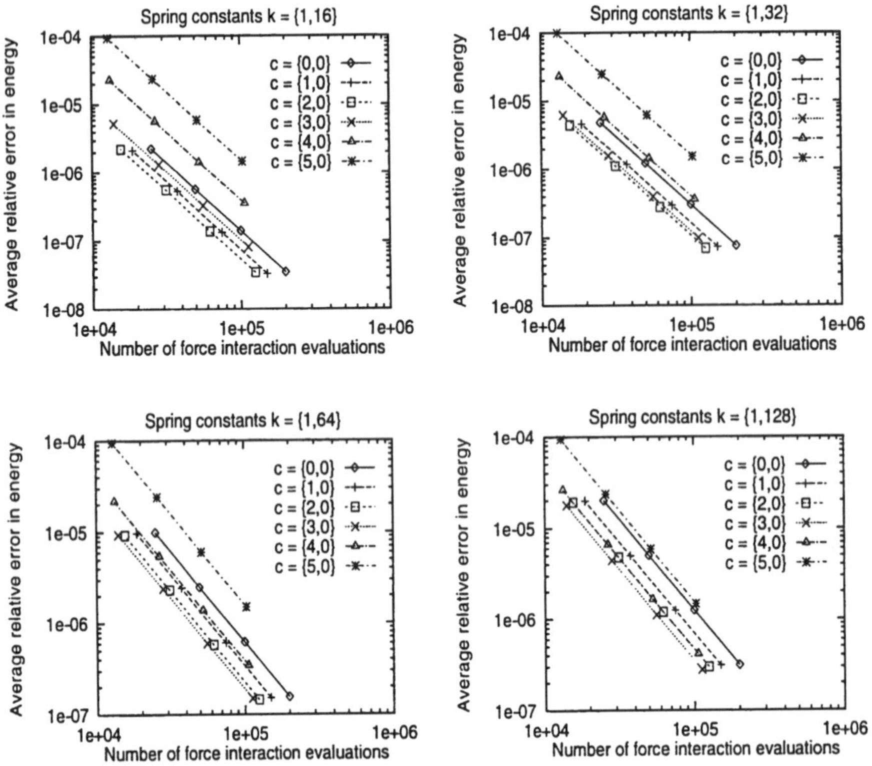

Figure 1: Experimental results for the two spring system.

We choose the number of classes to be $\mu = c+1$ for the largest such c over all springs. Doing this should make the integrator more efficient by minimizing the amount of computation without increasing the magnitude of the error.

The integrator is implemented most efficiently by keeping a list L_k of springs for each class k. Each list element is a 4-tuple (i, j, κ, ℓ) representing a spring, identifying the spring constant and equilibrium length as well as the pair of particles joined. For each force evaluation, the divisibility of the step number is first determined, establishing the highest class to be evaluated. The list for each of these classes is traversed, and the contribution to the force by each spring is computed.

4.2 Numerical experiments

For the numerical experiments performed in this paper, the number of force interaction evaluations, which is equal to the number of square roots, are plotted against the average relative error in energy $\bar{E}$ for different stepsizes. The line formed when plotted on a logarithmic scale should have slope -2, indicating a second-order method. The value $\bar{E}$ is given by

$$\bar{E} = \frac{1}{s}\sum_{i=1}^{s}\left|\frac{E_i - E_0}{E_0}\right|,$$

where s is the number of energy samples, E_0 is the initial total energy, and E_i is ith total energy sample. The local truncation error analysis indicates that energy sampling should be restricted to steps divisible by $\pi_{\mu-1}$. For a time τ simulation using microstepsize h, we satisfy this divisibility condition by sampling energy on steps numbered

$$n_k = \pi_{\mu-1}\left\lfloor\frac{k}{s}\left\lfloor\frac{\tau}{\pi_{\mu-1}h}\right\rfloor\right\rfloor \qquad \text{for } k = 1, 2, \ldots, s.$$

Numerical experiments have been performed on the simple linear spring system with three masses connected by two springs. The masses are $m_1 = m_2 = m_3 = 1$, and the spring equilibrium lengths are $\ell_1 = \ell_2 = 6$. The system is started with initial conditions $q_1(0) = -7$, $q_2(0) = 0$, $q_3(0) = 7$, $p_1(0) = p_2(0) = p_3(0) = 0$ and integrated for $\tau = 10$ seconds using microstepsizes $h = 1\times10^{-4}, 2\times10^{-4}, 4\times10^{-4}, 8\times10^{-4}$. The first spring constant $\kappa_1 = 1$ is fixed while the second is varied to define four different systems $\kappa_2 = 16, 32, 64, 128$. For each system, we use the multiple time step Verlet method with the second spring assigned to class $c_2 = 0$ and the first spring assigned to classes $c_1 = 0, 1, 2, 3, 4, 5$. The resulting four plots for $s = 100$ energy samples are shown in Figure 1. Notice that the class assignment strategy given by (9) yields the most efficient integration for each system.

Acknowledgment

This work is partly supported by National Science Foundation Grant CCR-9408973.

References

[1] R. D. Skeel & C. W. Gear, Does variable step size ruin a symplectic integrator?, *Physica D*, **60**, 311, 1992.

[2] M. P. Calvo & J. M. Sanz-Serna, The development of variable-step symplectic integrators, with applications to the two-body problem, *SIAM J. Sci. Statist. Comput.* **14**, 936, 1993.

[3] D. I. Okunbor, Canonical integration methods for Hamiltonian dynamical systems, Doctoral Thesis, University of Illinois at Urbana-Champaign, Illinois, 1992.

[4] D. I. Okunbor, Variable step size does not harm second-order integrators for Hamiltonian systems, *J. Comput. Appl. Math.* **47**, 273, 1993.

[5] E. Hairer, Variable time step integration with symplectic methods, *Appl. Numer. Math.*, **25**, 219, 1997.

[6] W. B. Streett, D. J. Tildesley, & G. Saville, Multiple time-step methods in molecular dynamics, *Molec. Phys.* **35**, 639, 1978.

[7] H. Grubmüller, H. Heller, A. Windemuth, & K. Schulten, Generalized Verlet algorithm for efficient molecular dynamics simulations with long-range interactions, *Molec. Sim.* **6**, 121, 1991.

[8] D. H. Stattinger & O. L. Weaver, *Lie Groups and Algebras with Applications to Physics, Geometry, and Mechanics*, Berlin: Springer-Verlag, 1986.

[9] M. Tuckerman, B. J. Berne, & G. J. Martyna, Reversible multiple time scale molecular dynamics, *J. Chem. Phys.* **97**, 1990, 1992.

[10] J. J. Biesiadecki & R. D. Skeel, Dangers of multiple time step methods, *J. Comput. Phys.* **109**(2), 318, 1993.

[11] R. D. Skeel & J. J. Biesiadecki, Symplectic Integration with Variable Stepsize, *Annals of Numer. Math.*, 191, 1994.

[12] T. R. Littell, R. D. Skeel, & M. Zhang, Error analysis of symplectic multiple time stepping, *SIAM J. Numer. Anal.* **34**(5), 1792, 1997.

[13] V. I. Arnold, Mathematical Methods of Classical Mechanics, 2nd ed., Springer-Verlag, New York, 1989.

Proceedings of Dynamic Systems and Applications 3 (2001) 491-496

ON OSCILLATION OF HALF-LINEAR FUNCTIONAL DIFFERENTIAL EQUATIONS

Hiroshi Onose
Department of Mathematics, Faculty of Sciences, Ibaraki University,
Mito 310-8512, Japan
E-mail onose@mito.ipc.ibaraki.ac.jp

ABSTRACT: Some oscillation results are proposed mainly for half-linear functional differential equations with deviating arguments.
AMS(MOS) Subject classification: 34K15

1. INTRODUCTION

We consider the following nonlinear functional differential equations

$$(r(t)|x'(t)|^{\alpha-1}x'(t))' + a(t)|x(g(t))|^{\beta-1}x(g(t)) = f(t), \quad t \geq t_0, \tag{A}$$

$$(r(t)\phi(x'(t)))' + a(t)|x(g(t))|^{\beta-1}x(g(t)) = f(t), \quad t \geq t_0, \tag{B}$$

and

$$(r(t)\phi(x'(t)))' + F(t, x(g(t))) = f(t), \quad t \geq t_0, \tag{C}$$

where $\alpha > 0$ and $\beta > 0, r \in C([t_0, \infty); (0, \infty)), a, f \in C([t_0, \infty); R), g \in C([t_0, \infty); (0, \infty))$ and $\lim_{t\to\infty} g(t) = \infty, \phi \in C(R; R), \phi$ is strictry monotone increasing, $\mathrm{sgn}\phi(u) = \mathrm{sgn}u, \phi(0) = 0, \phi(R) = R$ and $F \in C([t_0, \infty) \times R; R)$. We call equation (A) as half-linear. Our main pourpose of this paper is to analize the oscillatory properties on (A), (B) and (C). A nontrivial solution $x(t)$ of (A) (or (B), (C)) is said to be *oscillatory* if there exists a sequence $\{t_k\}_{k=1}^{\infty}$ such that $\lim_{k\to\infty} t_k = \infty$ and $x(t_k) = 0$ for all k. Otherwise, a solution is said to be *nonoscillatory*. Since Elbert[1], many authors study for the oscillatory properties and the more essential fundamental theory of (A) in the case $f(t) = 0$. These are interesting and have good applications to the partial differential equations. Recently the oscillation of half-linear equation with forcing term is considered by Jaroš and Kusano[3], Kusano and Lalli[4], and Usami and Tokuhira[7]. Here, we will analize them to more general equations, especialy with deviating argument $g(t)$ in (C). Some new simple equations are presented as examples to illustrate the results.

2. THE RESULTS

Theorem 1. *Suppose that* $\int_{t_0}^{\infty} |a(s)|ds < \infty$, *and that either one of*

$$\lim_{t\to\infty} \int_{t_0}^{t} f(s)ds = \infty \tag{1}$$

or

$$\lim_{t\to\infty} \int_{t_0}^{t} f(s)ds = -\infty. \tag{2}$$

holds. Then every bounded solution $x(t)$ *of* (A) *is monotonically increasing for supposing* (1), *and monotonically decreasing for suposing* (2).

Proof. Let $x(t)$ be a bounded solution of (A). Then, we may suppose $|x(t)| \le N$ and $|x(g(t))| \le N$, where N is a positive constant, for $t \ge t_1 \ge t_0$. By integrating (A) from t_1 to t, we obtain

$$r(t)|x'(t)|^{\alpha-1}x'(t) - r(t_1)|x'(t_1)|^{\alpha-1}x'(t_1) +$$

$$\int_{t_1}^{t} a(s)|x(g(s))|^{\beta-1}x(g(s))ds = \int_{t_1}^{t} f(s)ds, \quad t \ge t_1. \tag{3}$$

As the first integral in (3), we obtain

$$|\int_{t_1}^{t} a(s)|x(g(s))|^{\beta-1}x(g(s))ds| \le N^{\beta} \int_{t_1}^{\infty} |a(s)|ds < \infty, \quad t \ge t_1. \tag{4}$$

From (3) and (4), we obtain $x'(t) > 0$ for $t \ge T_1 \ge t_1$ in case (1) holding, and $x'(t) < 0$ for $t \ge T_2 \ge t_1$ in case (2) holding. Hence the assertions of our theorem are satisfied.

Q. E. D.

Corollary 1. *Suppose that* $\int_{t_0}^{\infty} |a(s)|ds < \infty$, *and that either one of*

$$\lim_{t\to\infty} \int_{t_0}^{t} f(s)ds = \infty$$

or

$$\lim_{t\to\infty} \int_{t_0}^{t} f(s)ds = -\infty$$

holds. Then (A) *has no bounded oscillatory solutions.*

Proof. The proof is contained in theorem 1.

Theorem 2. *Suppose that the following conditions hold;*

$$\limsup_{m\to\infty} \left| \int_{t_0}^{t_m} f(s)ds \right| = \infty \tag{5}$$

for any sequence $\{t_m\}_{m=1,2,\ldots}$, $t_m \in [T,\infty)$, t_m *is increasing as* m *and* $\lim_{m\to\infty} t_m = \infty$, *for any positive sufficiently large* $T \geq t_0$ *and* $\int_{t_0}^{\infty} |a(s)|ds < \infty$. *Then* (B) *has no bounded oscillatory solution.*

Proof. Let $x(t)$ be a bounded oscillatory solution of (B). Then we may assume that there exists a sequence $\{t_n\}_{n=1,2,\ldots}, t_n \in [T,\infty), t_n$ is increasing as n and $\lim_{n\to\infty} t_n = \infty$ and $x'(t_n) = 0$ for $n = 1, 2, \ldots$, and that $|x(t)| \leq N, |x(g(t))| \leq N$ for $t \geq T_1 \geq T$. Integrating (B) from t_1 to t_n, we obtain

$$r(t_n)\phi(x'(t_n)) - r(t_1)\phi(x'(t_1)) - \int_{t_1}^{t_n} a(s)|x(g(s))|^{\beta-1}x(g(s))ds = \int_{t_1}^{t_n} f(s)ds,$$

for this we obtain

$$-\int_{t_1}^{t_n} a(s)|x(g(s))|^{\beta-1}x(g(s))ds = \int_{t_1}^{t_n} f(s)ds. \tag{6}$$

We find that

$$\left| \int_{t_1}^{t_n} a(s)|x(g(s))|^{\beta-1}x(g(s))ds \right| \leq N^{\beta} \int_{t_1}^{\infty} |a(s)|ds < \infty. \tag{7}$$

From (6), (7) we see that $\limsup_{n\to\infty} \left| \int_{t_1}^{t_n} f(s)ds \right| < \infty$, which contradicts to (5).

Q. E. D.

Theorem 3. *Assume* $uF(t,u) > 0 (u \neq 0)$. *Suppose that for any sufficiently large* $T \geq t_0$, *and any constants* $k > 0$ *and* ℓ,

$$\liminf_{t\to\infty}\{k + \int_T^t \phi^{-1}[\frac{1}{r(s)}(\ell + \int_T^s f(\sigma)d\sigma)]ds\} < 0, \tag{8}$$

and

$$\limsup_{t\to\infty}\{-k + \int_T^t \phi^{-1}[\frac{1}{r(s)}(\ell + \int_T^s f(\sigma)d\sigma]ds\} > 0, \tag{9}$$

where ϕ^{-1} *is an inverse function of* ϕ. *Then, every solution of* (C) *is oscillatory.*

Proof. The proof is proceeded as Jaroš and Kusano[3]. Let $x(t)$ be a nonoscillatory solution of (C). Then, we may suppose $x(t) > 0$ (or $x(t) < 0$) for $t \in [t_1,\infty), t_1 \geq t_0, t_1$ being sufficiently large. First we suppose $x(t) > 0$ and $x(g(t)) > 0$ for $t \geq t_2 \geq t_1$. From (C) we see

$$(r(t)\phi(x'(t)))' = f(t) - F(t, x(g(t))) \leq f(t), \quad t \geq t_2.$$

By integrating this implies

$$r(t)\phi(x'(t)) \leq c_1 + \int_{t_2}^{t} f(s)ds, \quad \text{where } c_1 = r(t_2)\phi(x'(t_2)),$$

and dividing this by $r(t)$,

$$\phi(x'(t)) \leq \frac{1}{r(t)}(c_1 + \int_{t_2}^{t} f(s)ds)$$

and from this we see

$$x'(t) \leq \phi^{-1}[\frac{1}{r(t)}(c_1 + \int_{t_2}^{t} f(s)ds),$$

where ϕ^{-1} exists as our assumption, integrating it again from t_2 to t, we obtain

$$x(t) \leq c_2 + \int_{t_2}^{t} \phi^{-1}[\frac{1}{r(s)}(c_1 + \int_{t_1}^{s} f(\sigma)d\sigma)]ds, \quad t \geq t_2, \tag{10}$$

where $c_2 = x(t_2)$.

As we take lim inf of (10) as $t \to \infty$, we see $\liminf_{t\to\infty} x(t) < 0$, by (8). But, this is a contradiction to $x(t) > 0, t \geq t_1$. The next, if we suppose $x(t) < 0$ and $x(g(t)) < 0$ for $t \geq t_3 \geq t_1$. From (C) we see

$$(r(t)\phi(x'(t)))' = f(t) - F(t, x(g(t))) \geq f(t), \quad t \geq t_3.$$

By integrating this implies

$$r(t)\phi(x'(t)) \geq c_3 + \int_{t_3}^{t} f(s)ds, \quad \text{where } c_3 = r(t_3)\phi(x'(t_3))$$

and dividing this by $r(t)$, integrating it again from t_3 to t, we obtain

$$x(t) \geq c_4 + \int_{t_3}^{t} \phi^{-1}[\frac{1}{r(s)}(\ell + \int_{t_3}^{s} f(\sigma)d\sigma)]ds, \quad t \geq t_3, \tag{11}$$

where $c_4 = x(t_3)$.

As we take lim sup of (11) as $t \to \infty$, we see $\limsup_{t\to\infty} x(t) > 0$ by (9). This is a contradiction.

Q. E. D.

Theorem 4. *Assume*
(H1); $|F(t,x)|$ *is bounded foy any* $t \in [t_0, \infty)$ *and* $|x| \le N$, *where* N *is any positive constant.*
Suppose that for any sufficiently large $T \ge t_0$ *any constants* $k > 0$, $,\ell$ *and* k',

$$\liminf_{t\to\infty}\{k + \int_T^t \phi^{-1}[\frac{1}{r(s)}(\ell + \int_T^s (f(\sigma)+k')d\sigma)]ds\} = -\infty$$

and

$$\limsup_{t\to\infty}\{-k + \int_T^t \phi^{-1}[\frac{1}{r(s)}(\ell + \int_T^s (f(\sigma)+k')d\sigma)]ds\} = +\infty,$$

where ϕ^{-1} *is an inverse function of* ϕ. *Then every bounded solution of* (C) *is oscillatory.*

Proof. Let $x(t)$ be a bounded nonoscillatory solution of (C). Then we may suppose $x(t) > 0$ (or $x(t) < 0$) for sufficiently large $t_1, t \ge t_1 \ge t_0$. We may also assume $|F(t, x(g(t))| \le K$ for $|x(t)| \le N, |x(g(t))| \le N, t \ge t_2 \ge t_1$, where K and$N$ are sufficiently large positive numbers. From (C) and (H1), we see

$$(r(t)\phi(x'(t)))' = f(t) - F(t, x(g(t))) \le f(t) + K, \quad t \ge t_2.$$

By integrating this, we obtain

$$r(t)\phi(x'(t)) \le c_1 + \int_{t_2}^t (f(s)+K)ds, \text{where } \ c_1 = r(t_2)\phi(x'(t_2)),$$

dividing this by $r(t)$ and integrating again from t_2 to t, we obtain

$$x(t) \le c_2 + \int_{t_2}^t \phi^{-1}[\frac{1}{r(s)}(c_1 + \int_{t_2}^s (f(\sigma)+K)d\sigma)]ds, \quad t \ge t_2, \tag{12}$$

where $c_2 = x(t_2)$.
As we take lim inf of (12) as $t \to \infty$, we see $\liminf_{t\to\infty} x(t) < 0$. This is a contradiction to $x(t) > 0, t \ge t_1$. The next, if we suppose $x(t) < 0$ then we have also a contradiction by the same argument as above.

Q. E. D.

Example 1. Consider the equation

$$(e^{4t}|x'(t)|^2 x'(t))' + e^{-t}x(t) = e^{-2t} - e^t, \quad t \ge 1. \tag{13}$$

Every bounded solution of (13) is nonoscillatory by theorem 2. In fact, $x(t) = e^{-t}$ is a such solution.

Example 2. Consider the equation

$$((x'(t))^{\frac{1}{3}})' + t(x(t-\pi))^{\frac{1}{3}} = t\sin t, \quad t \geq \pi. \tag{14}$$

Every solution of (14) is oscillatory by theorem 3.

Example 3. Consider the equations

$$((x'(t))^{\frac{1}{3}})' - \frac{1}{t^2}(x(t-\pi))^{\frac{1}{3}} = t\sin t, \quad t \geq \pi \tag{15}$$

and

$$((x'(t))^{\frac{1}{3}})' + \sin t(x(t+\pi))^{\frac{1}{3}} = t\sin t, \quad t \geq \pi. \tag{16}$$

Every bounded solution of (15) and (16) is oscillatory by theorem 4.

Example 4. Consider the equation

$$(|x'(t)|^2 x'(t))' + t(x(t-2\pi))^3 = t\sin^3 t - 3\sin t\cos^2 t, \quad t \geq 4\pi. \tag{17}$$

Every solution of (17) is oscillatory by theorem 3. In fact $x(t) = \sin t$ is a such solution.

Remark. Examples 2-4 are not examined by Jaroš and Kusano[3] which is considered only the case without deviating argument.

REFERENCES

1. Elbelt, Á., Oscillation and nonoscillation theorems for some nonlinear ordinary differential equations, Lecture Notes in Mathematics Vol. 964(1982): Ordinary and Partial Differential Equations. 187-212.
2. Elbert, Árpád, Integral characterization of principal solution of half-linear differential equations, Abstract of third international conference on dynamic systems & applications(1999), Morehous Colleage. 15-16.
3. Jaroš, Jaroslav & Kusano, Takaŝi, Second-order semilinear differential equations with external forcing terms(Japanese). RIMS Kokyuroku No. 984(1997). 191-197.
4. Kusano, T. & Lalli, B. S., On oscillation of half-linear functional differential equations with deviating arguments. Vol 24(1994), Hiroshima Math. J.. 549-563.
5. Kusano, Takaŝi & Onose, Hiroshi, On oscillation of functional differential equations with forcing terms(to appear).
6. Ladde, G. S., Lakshmikantham, V. & Zhang, B. G., Oscillation theory of differential equations with deviating arguments. Marcel Dekker, INC., New York and Basel (1987).
7. Usami, H. & Tokuhira, T., Asymptotic behaviour of oscillatory solutions of quasilinear ordinary differential equations. Seminar abstract at Fukuoka University (1998). 1-4.
8. Wong, P. J. Y. & Agarwal, R. P., Oscillatory behaviour of solutions of certain second order nonlinear differential equations.Vol 198 (1996), J. Math. Anal. Appl.. 337-354.

Proceedings of Dynamic Systems and Applications 3 (2001) 497-504

TWIN SOLUTIONS TO RIGHT FOCAL $(1, n-1)$ SINGULAR BOUNDARY VALUE PROBLEMS

Donal O'Regan

Department of Mathematics
National University of Ireland, Galway
Ireland.

ABSTRACT: The existence of two nonnegative solutions to the $(1, n-1)$ focal boundary value problem is established in this paper. Our nonlinearity may be singular at $y = 0$, $t = 0$ and/or $t = 1$.

AMS(MOS) Subject Classification. 34B15

1. INTRODUCTION

This paper discusses the existence of two nonnegative solutions to the $(1, n-1)$ focal boundary value problem (here $n \geq 2$),

$$\begin{cases} (-1)^{n-1} y^{(n)}(t) = \phi(t)\,[g(y(t)) + h(y(t))], \quad 0 < t < 1 \\ y(0) = 0 \\ y^{(i)}(1) = 0, \quad 1 \leq i \leq n-1. \end{cases} \tag{1.1}$$

Here our nonlinear term $g + h$ may be singular at $y = 0$. Some multiple solution results were presented in [4] using Krasnoselskii's fixed point theorem in a cone. However in [4] a strong integrability assumption had to be assumed on ϕ and g, namely we assumed

$$\int_0^1 \phi(s)\, g(s)\, ds < \infty. \tag{1.2}$$

In this paper by using a more general fixed point theorem than that of Krasnoselskii (which was established in [5] using degree theory and in [8] using the essential map approach) we are able to guarantee the existence of two solutions to (1.1) under very general assumptions (in particular (1.2) is not assumed). In [3] we established the existence of one solution to (1.1) using a Leray–Schauder alternative and in this paper we note by adding one extra condition we are able to guarantee the existence of two solutions .

For the remainder of this section we present some results from the literature which will be needed in Section 2.

Theorem 1.1. [2,5,8] *Let $E = (E, \|.\|)$ be a Banach space, $K \subset E$ a cone and let $\|.\|$ be increasing with respect to K. Also r, R are constants with $0 < r < R$. Suppose*

$A : \overline{\Omega_R} \cap K \to K$ (here $\Omega_R = \{x \in E : \|x\| < R\}$) is a continuous, compact map and assume the following conditions hold:

$$x \neq \lambda A(x) \quad for \quad \lambda \in [0,1) \quad and \quad x \in \partial_E \Omega_r \cap K \tag{1.3}$$

and

$$\|Ax\| > \|x\| \quad for \quad x \in \partial_E \Omega_R \cap K. \tag{1.4}$$

Then A has a fixed point in $K \cap \{x \in E : r \leq \|x\| \leq R\}$.

Remark 1.1. In Theorem 1.1 if (1.3) and (1.4) are replaced by

$$x \neq \lambda A(x) \quad \text{for} \quad \lambda \in [0,1) \quad \text{and} \quad x \in \partial_E \Omega_R \cap K \tag{1.5}$$

and

$$\|Ax\| > \|x\| \quad \text{for} \quad x \in \partial_E \Omega_r \cap K. \tag{1.6}$$

then A has a fixed point in $K \cap \{x \in E : r \leq \|x\| \leq R\}$.

In [3] we proved the following lower type inequality.

Theorem 1.2. *Suppose $y \in C^{n-1}[0,1] \cap C^n(0,1)$ satisfies*

$$\begin{cases} (-1)^{n-1}\, y^{(n)}(t) > 0 \quad for \quad t \in (0,1) \\ y(0) = a \geq 0 \\ y^{(i)}(1) = 0, \quad 1 \leq i \leq n-1. \end{cases}$$

Then

$$y(t) \geq t\,|y|_0 = t\,y(1) \quad for \quad t \in [0,1].$$

2. TWIN SOLUTIONS

First we recall the following existence result for (1.1) which was established in [3].

Theorem 2.1. *Suppose the following conditions are satisfied:*

$$\phi \in C(0,1) \quad with \quad \phi > 0 \quad on \quad (0,1) \quad and \quad \phi \in L^1[0,1] \tag{2.1}$$

$$g > 0 \quad is\ continuous\ and\ nonincreasing\ on \quad (0,\infty) \tag{2.2}$$

$$h \geq 0 \quad continuous\ on \quad [0,\infty) \quad with \quad \frac{h}{g} \quad nondecreasing\ on \quad (0,\infty) \tag{2.3}$$

and

$$\exists\, r > 0 \quad with \quad \frac{1}{\left\{1 + \frac{h(r)}{g(r)}\right\}} \int_0^r \frac{du}{g(u)} > a_0; \tag{2.4}$$

here

$$a_0 = \int_0^1 \int_x^1 \frac{(s-x)^{n-2}}{(n-2)!}\, \phi(s)\, ds\, dx. \tag{2.5}$$

Then (1.1) has a solution $y \in C^{n-1}[0,1] \cap C^n(0,1)$ with $y > 0$ on $(0,1]$ and $|y|_0 = \sup_{t \in [0,1]} |y(t)| < r$.

Our next result guarantees that there exists a solution y to (1.1) with $|y|_0 > r$. We will use Theorem 1.1 to achieve this.

Theorem 2.2. *Suppose* (2.1) – –(2.4) *hold. Choose* $a \in \left(0, \frac{1}{2}\right)$ *and fix it and suppose there exists*

$$R > r \quad with \quad \frac{R\, g(a\, R)}{g(R)\, g(a\, R) + g(R)\, h(a\, R)} < \int_a^1 (-1)^{n-1}\, G(\sigma, s)\, \phi(s)\, ds; \tag{2.6}$$

here $0 \leq \sigma \leq 1$ *is such that*

$$\int_a^1 (-1)^{n-1}\, G(\sigma, s)\, \phi(s)\, ds = \sup_{t\in[0,1]} \int_a^1 (-1)^{n-1}\, G(t, s)\, \phi(s)\, ds \tag{2.7}$$

with

$$G(t, s) = \frac{(-s)^{n-1}}{(n-1)!} \quad if \quad 0 \leq s \leq t,$$

whereas

$$G(t, s) = -\frac{1}{(n-1)!} \sum_{i=1}^{n-1} \binom{n-1}{i}\, t^i\, (-s)^{n-i-1} \quad if \quad t \leq s \leq 1.$$

Then (1.1) *has a solution* $y \in C^{n-1}[0,1] \cap C^n(0,1)$ *with* $y > 0$ *on* $(0,1]$ *and* $r < |y|_0 \leq R$.

PROOF: To show existence we apply Theorem 1.1. Choose $\epsilon > 0$ and $\epsilon < r$ with

$$\frac{1}{\left\{1 + \frac{h(r)}{g(r)}\right\}} \int_\epsilon^r \frac{du}{g(u)} > a_0. \tag{2.8}$$

Let $m_0 \in \{1, 2, ...\}$ be chosen so that $\frac{1}{m_0} < \epsilon$ and $\frac{1}{m_0} < a\, R$ and let $N_0 = \{m_0, m_0 + 1, ...\}$. We first show that

$$\begin{cases} (-1)^{n-1}\, y^{(n)}(t) = \phi(t)\, [g(y(t)) + h(y(t))], \quad 0 < t < 1 \\ y(0) = \frac{1}{m} \\ y^{(i)}(1) = 0, \quad 1 \leq i \leq n-1 \end{cases} \tag{2.9^m}$$

has a solution y_m for each $m \in N_0$ with $y_m(t) > \frac{1}{m}$ for $t \in (0,1]$ and $r \leq |y_m|_0 \leq R$. To show $(2.9)^m$ has such a solution for each $m \in N_0$, we look at

$$\begin{cases} (-1)^{n-1}\, y^{(n)}(t) = \phi(t)\, [g^\star(y(t)) + h(y(t))], \quad 0 < t < 1 \\ y(0) = \frac{1}{m} \\ y^{(i)}(1) = 0, \quad 1 \leq i \leq n-1 \end{cases} \tag{2.10^m}$$

with

$$g^\star(u) = \begin{cases} g(u), & u \geq \frac{1}{m} \\ g\left(\frac{1}{m}\right), & 0 \leq u \leq \frac{1}{m}. \end{cases}$$

Remark 2.1. Notice $g^\star(u) \leq g(u)$ for $u > 0$.

Fix $m \in N_0$. Let $E = (C[0,1], |\,.\,|_0)$ and

$$K = \{u \in C[0,1] : \ u(t) \geq 0 \ \text{ for } \ t \in [0,1] \ \text{ and } \ u(t) \geq t\, |u|_0 \ \text{ for } \ t \in [0,1]\}.$$

Note K is a cone of E. Let $A : K \to C[0,1]$ be defined by

$$A\, y(t) = \frac{1}{m} + \int_0^1 (-1)^{n-1}\, G(t, s)\, \phi(s)\, [g^\star(y(s)) + h(y(s))]\, ds \quad \text{for} \quad y \in K$$

A standard argument [9] implies $A : K \to C[0,1]$ is continuous and completely continuous. Next we show $A : K \to K$. If $u \in K$ then $A\,u(t) \geq 0$ for $t \in [0,1]$ since [1] for $(t,s) \in [0,1] \times [0,1]$ we have $(-1)^{n-1}\,G(t,s) \geq 0$. Also notice that

$$\begin{cases} (-1)^{n-1}\,(A\,u)^{(n)}(t) > 0 \quad \text{on} \quad (0,1) \\ A\,u(0) = \frac{1}{m} \\ (A\,u)^{(i)}(1) = 0, \quad 1 \leq i \leq n-1 \end{cases}$$

so Theorem 1.2 guarantees that $A\,u(t) \geq t\,|A\,u|_0$ for $t \in [0,1]$. Consequently $A\,u \in K$ so $A : K \to K$. Let

$$\Omega_1 = \{u \in C[0,1] : \ |u|_0 < r\} \quad \text{and} \quad \Omega_2 = \{u \in C[0,1] : \ |u|_0 < R\}.$$

We first show

$$y \neq \lambda\,A\,y \quad \text{for} \quad \lambda \in [0,1) \quad \text{and} \quad y \in K \cap \partial\Omega_1. \tag{2.11}$$

Suppose this is false i.e. suppose there exists $y \in K \cap \partial\Omega_1$ and $\lambda \in [0,1)$ with $y = \lambda\,A\,y$. We can assume $\lambda \neq 0$. Now since $y = \lambda\,A\,y$ we have

$$\begin{cases} (-1)^{n-1}\,y^{(n)}(t) = \phi(t)\,[g^\star(y(t)) + h(y(t))], \quad 0 < t < 1 \\ y(0) = \frac{1}{m} \\ y^{(i)}(1) = 0, \quad 1 \leq i \leq n-1. \end{cases} \tag{2.12}$$

Recall [1] for $(t,s) \in [0,1] \times [0,1]$ that $(-1)^{n-1}\,G^{(i)}(t,s) \geq 0$ for $1 \leq i \leq n-1$, so $y^{(i)} \geq 0$ on $(0,1)$ if $i \in \{1,....,n-1\}$ is odd, and $y^{(i)} \leq 0$ on $(0,1)$ if $i \in \{2,....,n-1\}$ is even. Also notice

$$g^\star(y(t)) + h(y(t)) \leq g(y(t)) + h(y(t)) \quad \text{for} \quad t \in (0,1)$$

so for $x \in (0,1)$ we have

$$|(-1)^{n-1}\,y^{(n)}(x)| \leq \phi(x)\,g(y(x))\left\{1 + \frac{h(y(1))}{g(y(1))}\right\}. \tag{2.13}$$

Note $y \in K \cap \partial\Omega_1$ so $y(1) = r$ and

$$|(-1)^{n-1}\,y^{(n)}(x)| \leq \phi(x)\,g(y(x))\left\{1 + \frac{h(r)}{g(r)}\right\} \quad \text{for} \quad x \in (0,1).$$

Integrate from t (here $t > 0$) to 1 to obtain

$$|y^{(n-1)}(t)| \leq g(y(t))\left\{1 + \frac{h(r)}{g(r)}\right\}\int_t^1 \phi(s)\,ds.$$

Another integration from x (here $x > 0$) to 1 yields

$$|y^{(n-2)}(x)| \leq g(y(x))\left\{1 + \frac{h(r)}{g(r)}\right\}\int_x^1 (s-x)\,\phi(s)\,ds.$$

Successive integrations will give for $x \in (0,1)$,

$$y'(x) = |y'(x)| \leq g(y(x))\left\{1 + \frac{h(r)}{g(r)}\right\}\int_x^1 \frac{(s-x)^{n-2}}{(n-2)!}\,\phi(s)\,ds$$

and so

$$\frac{y'(x)}{g(y(x))} \leq \left\{1 + \frac{h(r)}{g(r)}\right\} \int_x^1 \frac{(s-x)^{n-2}}{(n-2)!} \phi(s)\, ds. \tag{2.14}$$

Integration from 0 to 1 yields

$$\int_{\frac{1}{m}}^{r} \frac{du}{g(u)} \leq \left\{1 + \frac{h(r)}{g(r)}\right\} \int_0^1 \int_x^1 \frac{(s-x)^{n-2}}{(n-2)!} \phi(s)\, ds\, dx$$

and consequently

$$\int_{\epsilon}^{r} \frac{du}{g(u)} \leq \left\{1 + \frac{h(r)}{g(r)}\right\} \int_0^1 \int_x^1 \frac{(s-x)^{n-2}}{(n-2)!} \phi(s)\, ds\, dx. \tag{2.15}$$

This contradicts (2.8) and so (2.11) is true.

Next we show

$$|A\,y|_0 > |y|_0 \quad \text{for} \quad y \in K \cap \partial\Omega_2. \tag{2.16}$$

To see this let $y \in K \cap \partial\Omega_2$ so $|y|_0 = R$. Also since $y \in K$ we have $y(t) \geq t\,|y|_0 = t\,R$ for $t \in [0,1]$. Now for $s \in [a,1]$ we have

$$g^\star(y(s)) + h(y(s)) = g(y(s)) + h(y(s))$$

since $y(s) \geq a\,R > \frac{1}{m_0}$ for $s \in [a,1]$. Note in particular that

$$y(s) \in [a\,R,\, R] \quad \text{for} \quad s \in [a,1]. \tag{2.17}$$

With σ as defined in (2.7) we have using (2.17) and (2.6) that

$$\begin{aligned} A\,y(\sigma) &= \frac{1}{m} + \int_0^1 (-1)^{n-1}\, G(\sigma,s)\, \phi(s)\, [g^\star(y(s)) + h(y(s))]\, ds \\ &\geq \int_a^1 (-1)^{n-1}\, G(\sigma,s)\, \phi(s)\, [g^\star(y(s)) + h(y(s))]\, ds \\ &= \int_a^1 (-1)^{n-1}\, G(\sigma,s)\, \phi(s)\, g(y(s)) \left\{1 + \frac{h(y(s))}{g(y(s))}\right\} ds \\ &\geq g(R) \left\{1 + \frac{h(a\,R)}{g(a\,R)}\right\} \int_a^1 (-1)^{n-1}\, G(\sigma,s)\, \phi(s)\, ds \\ &> R = |y|_0, \end{aligned}$$

and so $|A\,y|_0 > |y|_0$. Hence (2.16) is true.

Now Theorem 1.1 implies A has a fixed point $y_m \in K \cap \left(\overline{\Omega_2} \setminus \Omega_1\right)$ i.e. $r \leq |y_m|_0 \leq R$. In fact $|y_m|_0 > r$ (note if $|y_m|_0 = r$ then following essentially the same argument from (2.13)–(2.15) will yield a contradiction). Consequently $(2.10)^m$ (and so $(2.9)^m$) has a solution $y_m \in C^{n-1}[0,1] \cap C^n(0,1)$, $y_m \in K$, with

$$\frac{1}{m} \leq y_m(t) \quad \text{for} \quad t \in [0,1], \quad r < |y_m|_0 \leq R \tag{2.18}$$

and (note $y_m \in K$)

$$y_m(t) \geq t\,r \quad \text{for} \quad t \in [0,1]. \tag{2.19}$$

Consider

$$I(z) = \int_0^z \frac{du}{g(u)}.$$

Now I is an increasing map from $[0,\infty)$ onto $[0,\infty)$ (notice $I(\infty)=\infty$ since $g>0$ is nonincreasing on $(0,\infty)$) with I continuous on $[0,A]$ for any $A>0$. In fact

$$\{I(y_m)\}_{m\in N_0} \text{ is a bounded, equicontinuous family on } [0,1]. \tag{2.20}$$

To see this return to (2.13) (with y replaced by y_m) so

$$|(-1)^{n-1}\, y_m^{(n)}(x)| \le \phi(x)\, g(y_m(x)) \left\{1+\frac{h(R)}{g(R)}\right\} \quad \text{for } x\in(0,1).$$

Successive integration gives

$$\frac{y_m'(x)}{g(y_m(x))} \le \left\{1+\frac{h(R)}{g(R)}\right\} v(x) \quad \text{for } x\in(0,1);$$

here

$$v(x)=\int_x^1 \frac{(s-x)^{n-2}}{(n-2)!}\,\phi(s)\,ds.$$

Consequently for $t_1,\, t_2\in[0,1]$ we have

$$\begin{aligned} |I(y_m(t_2))-I(y_m(t_1))| &= \left|\int_{y_m(t_1)}^{y_m(t_2)} \frac{du}{g(u)}\right| = \left|\int_{t_1}^{t_2} \frac{y_m'(x)}{g(y_m(x))}\,dx\right| \\ &\le \left\{1+\frac{h(R)}{g(R)}\right\} \left|\int_{t_1}^{t_2} v(x)\,dx\right|, \end{aligned}$$

and so $\{I(y_m)\}_{m\in N_0}$ is an equicontinuous family on $[0,1]$. In addition the previous inequality, the uniform continuity of I^{-1} on $[0,I(R)]$, and

$$|y_m(t_2)-y_m(t_1)| = |I^{-1}(I(y_m(t_2))) - I^{-1}(I(y_m(t_1)))|$$

imply that

$$\{y_m\}_{m\in N_0} \text{ is a bounded, equicontinuous family on } [0,1]. \tag{2.21}$$

The Arzela–Ascoli Theorem guarantees the existence of a subsequence N of N_0 and a function $y\in C[0,1]$ with y_m converging uniformly on $[0,1]$ to y as $m\to\infty$ through N. Also $y(0)=0$, $r\le|y|_0\le R$ and $y(t)\ge t\,r$ for $t\in[0,1]$. In particular $y>0$ on $(0,1]$. Fix $t\in(0,1)$. Now y_m, $m\in N$, satisfies the integral equation

$$y_m(t)=y_m(1)-\int_t^1 \frac{(s-t)^{n-1}}{(n-1)!}\,\phi(s)\,[g(y_m(s))+h(y_m(s))]\,ds.$$

Let $m\to\infty$ through N to obtain

$$y(t)=y(1)-\int_t^1 \frac{(s-t)^{n-1}}{(n-1)!}\,\phi(s)\,[g(y(s))+h(y(s))]\,ds.$$

We can do the above argument for each $t\in(0,1)$. Thus $(-1)^{n-1}\,y^{(n)}(t)=\phi(t)\,[g(y(t))+h(y(t))]$ for $0<t<1$ and $y^{(i)}(1)=0$ for $1\le i\le n-1$. Finally it is easy to see that $|y|_0>r$ (note if $|y|_0=r$ then following essentially the argument from (2.13)–(2.15) will yield a contradiction). □

Remark 2.2. If in (2.6) we have $R<r$ then (1.1) has a solution $y\in C^{n-1}[0,1]\cap C^n(0,1)$ with $y>0$ on $(0,1]$ and $R\le|y|_0<r$. The argument is similar to that in Theorem 2.2 except here we use Remark 1.1.

Theorem 2.3. *Suppose* (2.1) – –(2.4) *and* (2.6) *hold. Then* (1.1) *has two solutions* $y_1, y_2 \in C^{n-1}[0,1] \cap C^n(0,1)$ *with* $y_1 > 0,\ y_2 > 0$ *on* $(0,1]$ *and* $|y_1|_0 < r < |y_2|_0 \leq R$.

PROOF: The existence of y_1 follows from Theorem 2.1 and the existence of y_2 follows from Theorem 2.2. □

Example 2.1. The singular boundary value problem

$$\begin{cases} y'' + \frac{1}{2(\alpha+1)}\left(y^{-\alpha} + y^{\beta} + 1\right) = 0 \quad \text{on} \quad (0,1) \\ y(0) = y'(1) = 0, \quad \alpha > 0, \ \beta > 1 \end{cases} \tag{2.22}$$

has two solutions $y_1, y_2 \in C^{n-1}[0,1] \cap C^n(0,1)$ with $y_1 > 0,\ y_2 > 0$ on $(0,1]$ and $|y_1|_0 < 1 < |y_2|_0$.

To see this we will apply Theorem 2.3 with $n = 2$, $\phi = \frac{1}{2(\alpha+1)}$, $g(u) = u^{-\alpha}$ and $h(u) = u^{\beta} + 1$. Clearly (2.1), (2.2) and (2.3) hold. Note also that

$$a_0 = \int_0^1 \int_x^1 \frac{1}{2(\alpha+1)}\, ds\, dx = \frac{1}{4(\alpha+1)},$$

and so (2.4) holds (with $r = 1$) since

$$\frac{1}{\left\{1 + \frac{h(r)}{g(r)}\right\}} \int_0^r \frac{du}{g(u)} = \frac{1}{(1 + r^{\alpha+\beta} + r^{\alpha})}\left(\frac{r^{\alpha+1}}{\alpha+1}\right) = \frac{1}{3(\alpha+1)} > \frac{1}{4(\alpha+1)} = a_0.$$

Finally note (since $\beta > 1$), take $a = \frac{1}{4}$, that

$$\lim_{R\to\infty} \frac{R\, g\left(\frac{R}{4}\right)}{g(R)\, g\left(\frac{R}{4}\right) + g(R)\, h\left(\frac{R}{4}\right)} = \lim_{R\to\infty}\left(\frac{R^{\alpha+1}\left(\frac{1}{4}\right)^{-\alpha}}{\left(\frac{1}{4}\right)^{-\alpha} + \left(\frac{1}{4}\right)^{\beta} R^{\alpha+\beta} + R^{\alpha}}\right) = 0$$

so there exists $R > 1$ with (2.6) holding. The result now follows from Theorem 2.3.

REFERENCES

[1]. R.P. Agarwal, D. O'Regan and P.J.Y. Wong, Positive solutions of differential, difference and integral equations, *Kluwer Academic Publishers*, Dordrecht, 1999.

[2]. R.P. Agarwal and D. O'Regan, Twin solutions to singular boundary value problems, *Proc. Amer. Math. Soc.*, to appear.

[3]. R.P. Agarwal and D. O'Regan, Right focal singular boundary value problems, *Zeitschrift für Angewandte Mathematik und Mechanik*, to appear.

[4]. R.P. Agarwal and D. O'Regan, Multiplicity results for singular conjugate, focal and (n,p) problems, to appear.

[5]. K. Deimling, Nonlinear functional analysis, *Springer*, New York, 1985.

[6]. P.W. Eloe and J. Henderson, Singular nonlinear boundary value problems for higher order ordinary differential equations, *Nonlinear Analysis*, **17**(1991), 1–10.

[7]. L.H. Erbe, S. Hu and H. Wang, Multiple positive solutions of some boundary value problems, *Jour. Math. Anal. Appl.*, **184**(1994), 640–648.

[8]. D. O'Regan, Existence of nonnegative solutions to superlinear non–positone problems via a fixed point theorem in cones of Banach spaces, *Dynamics of Continuous, Discrete and Impulsive Systems*, **3**(1997), 517–530.

[9]. D. O'Regan, Existence Theory for Nonlinear Ordinary Differential Equations, *Kluwer Academic Publishers*, Dordrecht, 1997.

Proceedings of Dynamic Systems and Applications 3 (2001) 505-510

Destroyable Restrictive Intervals and Two Unimodal Map Models Arbitrarily Close to Homoclinic Tangencies

Steven M. Pederson
Department of Mathematics
Morehouse College
Atlanta, Georgia 30314

ABSTRACT: First we show for a class of unimodal maps that there are only three ways for a map to have a destroyable restrictive interval. Second we discuss two unimodal map models with destroyable restrictive intervals and show through the bifurcation of a restrictive interval that if f is either of these map models, then f is arbitrarily close to a map with a homoclinic tangency.
AMS (MOS) subject classification. 58F03

1. INTRODUCTION

For an interval map f, a restrictive interval M may be thought of as a renormalization domain where $f^n : M \to M$ for some $n \geq 1$. In this paper we first show for a class of unimodal maps that there are only three ways for a map to have a destroyable restrictive interval. Second we discuss two qualitatively different unimodal map models with destroyable restrictive intervals and show that if f is either of these map models, then f itself does not necessarily have a homoclinic tangency, but there is a unimodal map g arbitrarily close to f with a homoclinic tangency. In either map model, f has no homoclinic tangency to a periodic point outside of M, but g has a homoclinic tangency to a periodic point outside of M. Exhibition of g involves the bifurcation of the restrictive interval M.

Next we review some background definitions and results. A continuous map f of a compact real interval I into itself is *piecewise monotone* if I can be partitioned into a finite number of subintervals with disjoint interiors, on each of which f is strictly monotone. Also, for a piecewise monotone map f, the *turning points* of f are the points in the interior of I at which f has a local extremum. If f has exactly one turning point, then f is *unimodal*. Let $C^r([0,1])$ denote the C^r maps of $[0,1]$ into itself, $U^r([0,1])$ denote the unimodal maps in $C^r([0,1])$, $C^{s,r}([0,1])$ denote C^s families of maps in $C^r([0,1])$, $P(f)$ denote the set of periodic points of f, and $P_k(f)$ denote the set of periodic points of f of period k. Let f be piecewise monotone. If for some $p \in P(f)$ there exist a $q \neq p$ arbitrarily close to p and a $k = k(q) \in \mathbb{Z}^+$ such that $f^k(q) = p$, then f has a *homoclinic orbit* to p. If a homoclinic orbit contains a turning point c of f, meaning that $f^l(q) = c$ for some $l = l(q) \in \mathbb{Z}^+$, $l < k$, then f has a *homoclinic tangency* (HCT).

Let $f \in C^0([0,1])$ be piecewise monotone. As in Melo & Strien [5], a closed proper subinterval M of $[0,1]$ is called a *restrictive interval* (RI) with *period* $n \geq 1$ for f if

1. the interiors of $M, \ldots, f^{n-1}(M)$ are disjoint,
2. $f^n(M) \subset M$, $f^n(\partial M) \subset \partial M$,
3. at least one of the intervals $M, \ldots, f^{n-1}(M)$ contains a turning point, and
4. M is maximal with respect to these properties.

For an S-unimodal map f, if M is a RI of period n and $f^n(M) = M$, then f has a HCT to a periodic point in ∂M. However, there exist unimodal maps f such that f has a RI M of period n with $f^n(M) = M$ but f has no HCT. See Figure 1. In the two unimodal map models to be presented, this is one reason to look for a HCT outside of M.

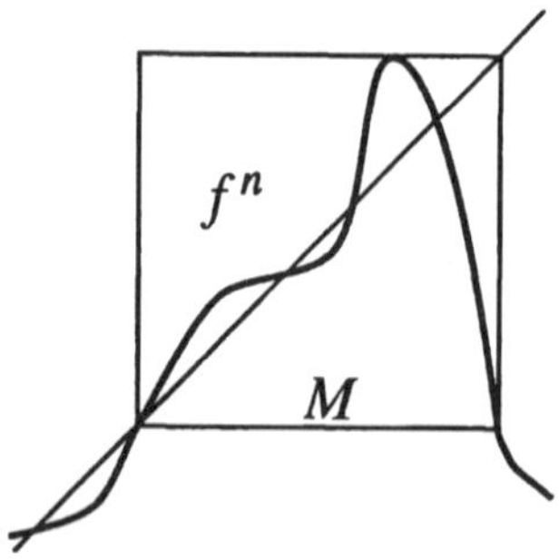

FIGURE 1. Even if $f^n(M) = M$ for an $f \in U^r([0,1])$ with a RI M of period n, it is possible that f has no HCT.

In Section 2, after defining C^r-destroyable RI, we show for a class of maps in $U^r([0,1])$, $r \geq 2$, that there are only three ways for a map to have a destroyable RI. In one of these three scenarios, a RI can be destroyed by loss of the maximality property (4) in the definition of RI without affecting the requirements (1), (2), and (3) in the definition of RI. From this point of view, there are only two ways for a RI to be destroyed without loss of the maximality property (4).

Let $\Omega(f)$ denote the nonwandering set of f. For $f \in U^0([0,1])$ such that $f : \partial[0,1] \to \partial[0,1]$, Jonker & Rand [3] used RIs to obtain a canonical decomposition of $\Omega(f)$ into a countable number of sets,

$$\Omega(f) = \bigcup_{i=0}^{j} \Omega_i(f).$$

In the Jonker & Rand decomposition, $\Omega_0(f) = \Omega(f) \cap \partial[0,1]$. Let $\text{orb}_f(x)$ denote $\{x, f(x), \dots\}$ and $\text{int}(A)$ denote the interior of a set A. If f has a RI, then

$$\Omega_1(f) = \Omega(f) \setminus (\text{int}(\text{orb}_f(M)) \cup \Omega_0(f)),$$

where M is the largest RI of f containing the turning point c. If f has no RI, then $\Omega_1(f) = \Omega(f) \setminus \Omega_0(f)$.

In Section 3, we discuss two particular maps in $U^r([0,1])$, $r \geq 2$, with RIs. If f is either of these maps, then f itself does not neccesarily have a HCT to a periodic point of $\Omega_1(f)$, but there is a sequence $\{g_i\}_{i=1}^{\infty} \subset U^r([0,1])$ converging to f in the C^r topology such that each g_i has a HCT to a periodic point of $\Omega_1(f)$. Exhibition of $\{g_i\}_{i=1}^{\infty}$ involves the bifurcation of a RI of f, allowing $\text{orb}_{g_i}(c)$ to reach $\Omega_1(f)$.

2. DESTROYABLE RESTRICTIVE INTERVALS

Let $CS[0,1]$ denote the nonempty closed subintervals of $[0,1]$ with the Hausdorff metric.

Definition 2.1. A RI M of period n of f is *C^r-destroyable* if for all $\epsilon > 0$ there is a $g \in C^r([0,1])$ such that $d_r(g, f) < \epsilon$ and either

1. g has no RI of period n, or
2. g has a RI of period n and there exists at least one $\{g_\lambda\}_{\lambda\in[0,1]} \in C^{0,r}([0,1])$ such that $g_0 = g$, $g_1 = f$, and $d_r(g_\lambda, f) < \epsilon$ for all $\lambda \in [0,1]$, but there is no continuous function $M : [0,1] \to \mathcal{CS}[0,1]$ such that $M(1) = M$ and $M(\lambda)$ is a RI of period n of g_λ.

Lemma 2.2. *Let $f \in U^r([0,1])$, $r \geq 2$, be a map with a RI M of period n, $c \in M$, $c \notin P(f)$, $p = \partial M \cap P_n(f)$, and $p' = \partial M \setminus \{p\}$. If c is the only critical point of f, and it is nondegenerate, and p is not a limit point of $P_n(f)$, then if M is C^r-destroyable, one of the following occurs:*

1. *$f^n(M) = M$ and the forward orbit of c can be perturbed out of M,*
2. *p is destroyable through a saddle-node bifurcation,*
3. *only the maximality condition on M as a RI is lost.*

Proof. It is not hard to show $c \in \text{int}(M)$. Assume (1) and (3) do not happen. Suppose for all g sufficiently C^r-close to f, g has a periodic point $p(g)$, which is the perturbation of p, and there is a point $p'(g)$, which is the perturbation of p', such that $g^n(p'(g)) = p'(g)$. Also for the closed interval N_g with endpoints $p(g)$ and $p'(g)$, suppose $g^n(N_g) \subset N_g$, $g^n(\partial N_g) \subset \partial N_g$, and $c \in N_g$. Therefore for some map g arbitrarily C^r-close to f, for all such N_g, the interiors of $N_g, \ldots, g^{n-1}(N_g)$ are not disjoint. Then at least two of the intervals $M, \ldots, f^{n-1}(M)$ intersect at their boundaries. Let $0 \leq i < j < n$ and suppose $f^i(M) \cap f^j(M) \neq \emptyset$. $f^i(M)$ has endpoints $f^i(p)$ and $f^i(c)$. $f^j(M)$ has endpoints $f^j(p)$ and $f^j(c)$. If $f^i(c) = f^j(p)$, then $f^n(c) = f^{n-i+j}(p)$. Since $n < n-i+j < 2n$,

$$f^{n-i+j}(p) = f^{j-i}(p) \in f^{j-i}(M).$$

But $f^n(c) \in \text{int}(M)$. This contradicts that $\text{int}(M)$ and $\text{int}(f^{j-i}(M))$ are disjoint. A similar contradiction arises if either $f^j(c) = f^i(p)$ or $f^j(c) = f^i(c)$. Therefore $f^j(p) = f^i(p)$. Then $f^{n-i+j}(p) = f^n(p)$. Since $f^{j-i}(p) \in P_n(f)$,

$$f^{j-i}(p) = f^n(f^{j-i}(p)) = f^n(p) = p.$$

Since $0 < j-i < n$, $f^{j-i}(M) \cap M \neq \emptyset$ and we may assume $i = 0$. Since $f^{n-j}(f^j(M)) \cap f^{n-j}(M) \neq \emptyset$, $f^n(M) \cap f^{n-j}(M) \neq \emptyset$. $f^n(M)$ and $f^{n-j}(M)$ intersect at p. Therefore $f^{n-j}(M) = f^j(M)$ and $n - j = j$. So $j = n/2$. Then because p and p' are not critical points of f^n, M is not C^r-destroyable. See Figure 2.

Therefore whether or not two of the intervals $M, \ldots, f^{n-1}(M)$ intersect at their boundaries, to destroy M as a RI of f through an arbitrarily small C^r perturbation of f requires bifurcation of p as a periodic point. Suppose for some ϵ that for all δ between 0 and ϵ that $f^n(p+\delta) < p+\delta$, $f^n(p-\delta) > p-\delta$. Then it is not hard to show that M is not C^r-destroyable. Therefore for a sufficiently small neighborhood U_p of p, either $f^n(x) < x$ for all $x \neq p$ in U_p, or $f^n(x) > x$ for all $x \neq p$ in U_p, and p must be destroyed as a periodic point through a saddle-node bifurcation. □

3. TWO UNIMODAL MAP MODELS AND HCTS

Let $\omega_f(x)$ denote the omega limit set of x under f.

Proposition 3.1. (First Map) *Suppose $f \in U^r([0,1])$, $r \geq 2$, is such that:*

1. *f has only one critical point, which is the nondegenerate turning point c,*
2. *f has a unique RI M_2 which contains c,*
3. *the first return map $f^{n_1}|_{M_2} : M_2 \to M_2$ is onto,*

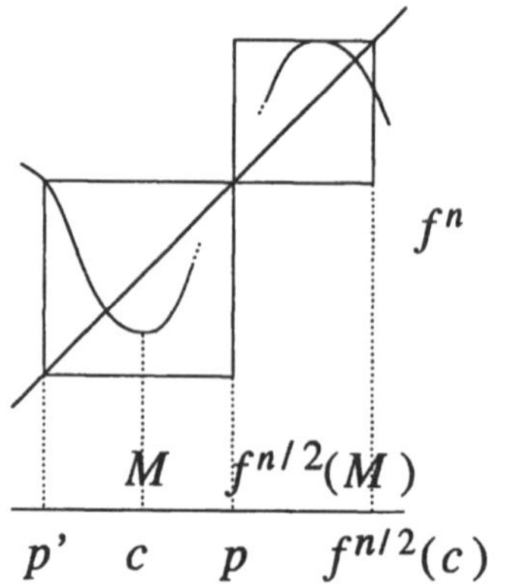

FIGURE 2. f^n if at least two of the intervals $M, \ldots, f^{n-1}(M)$ intersect at their boundaries. Then $M \cap f^{n/2}(M) = p$.

4. *if* $M_1 = [0,1]$ *and* $p \in P(f)$ *is the left endpoint of* M_2*, then* p *is repelling on* $M_1 \setminus M_2$ *and isolated in* $\Omega_1(f)$*, and for all* $x \in \text{int}(M_2)$ *sufficiently near* p*,* $f^{n_1}(x) > x$*,*
5. *if* $y \in P_k(f)$*, then* y *is not a limit point of* $P_k(f)$*.*

Then f *is arbitrarily* C^r*-close to* $g \in U^r([0,1])$ *with a HCT to a periodic point of* $\Omega_1(f)$*.*

Proof. Assume for all $q \in M_1 \setminus M_2$ sufficiently close to p that $\omega_f(q) \subset \Omega_2(f)$ or is isolated in $\Omega_1(f)$. Without loss of generality, suppose $M_2 = [p, p']$. Let

$$a = \sup_{x<p}\{x : \omega_f(x) \subset \Omega_1(f) \setminus \partial\text{orb}_f(M_2), \omega_f(x) \text{ not isolated in } \Omega_1(f) \setminus \partial\text{orb}_f(M_2)\}$$

and $W = [a,p]$. Let $W' = [p', b]$ be the smallest closed subinterval of $[0,1]$ with left endpoint p' and $f(W') = f(W)$. So $f(a) = f(b)$. Let $N_2 = W \cup M_2 \cup W'$. Then for all $y \in \text{int}(N_2)$, $\omega_f(y) \subset \Omega_2(f)$ or is isolated in $\Omega_1(f)$. Clearly $f^{n_1}(\text{int}(N_2)) \subset N_2$. By continuity of f^{n_1}, $f^{n_1}\partial(N_2) \subset \partial(N_2)$ and $f^{n_1}(N_2) \subset (N_2)$. By a similar argument, $\text{int}(N_2), f(\text{int}(N_2)), \ldots, f^{n_1-1}(\text{int}(N_2))$ are disjoint. But this contradicts that M_2 is a RI of f. Therefore for some $q \in M_1 \setminus M_2$ arbitrarily close to p, $\omega_f(q)$ is not isolated in $\Omega_1(f)$. Then $\omega_f(q)$ can not be a two-sided periodic attractor.

Suppose $\omega_f(q)$ is not a one-sided periodic attractor. Then partially by the results of Blokh [1, 2] and Melo & Strien [5], either (i) $\omega_f(q)$ is a periodic orbit that is not strictly contained in the ω-limit set of any point, or (ii) $\omega_f(q)$ is contained in a basic Cantor set. In both (i) and (ii), $f^k(q) \in \Omega_1(f)$ for some $k \in \mathbb{Z}^+$ since $\omega_f(q)$ is not a periodic attractor.

If $\omega_f(q)$ is not a one-sided periodic attractor, let $d = f^k(q)$, otherwise let d be a point in $\omega_f(q)$. There is a sequence $\{x_i\}_{i=1}^{\infty} \subset \Omega_1(f)$ such that $\lim_{i\to\infty} x_i = d$. By Young [6], $\Omega_1(f) \subset \overline{P}(f)$. If $x_i \notin P(f)$ for some i, then $x_i \in \overline{P}(f)$. Therefore we may assume there is a sequence of periodic points $\{p_i\}_{i=1}^{\infty} \subset \Omega_1(f)$ and $\lim_{i\to\infty} p_i = d$. Suppose for an infinite number of i's that p_i is attracting. By Theorem C of Mañé [4], the periods of all attracting periodic points in $\Omega_1(f)$ are bounded. An infinite number of the p_i's have the same period j. By continuity of f^j, $f^j(f^k(q)) \in P_j(f)$. But this contradicts (5) in Proposition 3.1. Therefore we may assume that each p_i is repelling (on both sides). As in Young [6], let

$$\Gamma_-(y) = \text{cl}\{z : f^n(z) = y \text{ for some } n > 0\}.$$

An interval $K \subset [0,1]$ is a *homterval* if $f^n : K \to f^n(K)$ is a homeomorphism for all $n \geq 0$. Suppose for N sufficiently large that for all $i > N$, $p_i \notin \Gamma_-(c)$. Then for each $i > N$, there is a neighborhood U_{p_i} of p_i which is a homterval. By Theorem A and Lemma II.3.1 of Melo & Strien [5], f maps U_{p_i} into an interval L_i such that $f^{k_i} : L_i \to L_i$ is monotone. Since $p_i \in P(f)$, we may assume that $p_i \in L_i$. Since $f^{k_i} : L_i \to L_i$ is monotone, if $f^{k_i}(p_i) \neq p_i$, then p_i is in the basin of some attracting fixed point of f^{k_i}. Therefore $f^{k_i}(p_i) = p_i$. Since f^{k_i} is monotone on L_i, L_i must contain at least one attracting fixed point a_i of f^{k_i}. Also for $j \neq i$, $p_j \notin L_i$. By Theorem C of Mañé [4], the periods of the a_i's are bounded and an infinite number of the a_i's have the same period a and $\lim_{i\to\infty} a_i = d$. By continuity of f^a, $f^a(d) \in P_a(f)$, contradicting (5) in Proposition 3.1. For all N, there is an $i > N$ such that $p_i \in \Gamma_-(c)$. We may assume that $p_i \in \Gamma_-(c)$ for all i. For each p_i, there is a $q_i \in M_1 \setminus M_2$ near q so that $f^k(q_i) = p_i$. Arbitrarily close to p_i is r_i such that $f^{j(i)}(r_i) = c$.

Next fix an i. Let U_c be a neighborhood of c strictly contained in M_2 such that $f(r_i), \ldots, f^{j(i)-1}(r_i) \notin U_c$. If $\omega_f(q)$ is a one-sided periodic attractor with period l, let U_d be a neighborhood of d such that $r_i, f(r_i), \ldots, f^{j(i)-1}(r_i) \notin U_d$ and f^l is monotone on U_d. If $\omega_f(q)$ is not a one-sided periodic attractor, make an arbitrarily small C^r perturbation of f on U_c to a map g such that $g^{n_1}(c) \notin M_2$ and for some i, $g^{2n_1}(c) = q_i^*$, such that for some $\alpha \in \mathbb{Z}^+$, $g^{\alpha n_1}(q_i^*) = f^{\alpha n_1}(q_i^*) = q_i$. It suffices to perturb f only on U_c because $f^{n_1 - 1} : f(U_c) \to f^{n_1}(U_c)$ is a homeomorphism. Since g maps q_i to p_i, g has a HCT to a periodic point of $\Omega_1(f)$. If $\omega_f(q)$ is a one-sided periodic attractor, make a similar perturbation of f on both U_c and U_d to get g. □

The proof of the next proposition is similar to that of Proposition 3.1.

Proposition 3.2. (Second Map) *Suppose $f \in U^r([0,1])$, $r \geq 2$, is such that:*

1. *f has only one critical point, which is the nondegenerate turning point c,*
2. *f has a unique RI M_2 which contains c and $\omega_f(c) = \mathrm{orb}_f(p)$,*
3. *the first return map $f^{n_1}|_{M_2} : M_2 \to M_2$ is not onto,*
4. *$p \in P(f) \cap \partial M_2$ is an isolated periodic point of $\Omega_1(f)$ and is destroyable through a saddle-node bifurcation,*
5. *if $y \in P_k(f)$, then y is not a limit point of $P_k(f)$.*

Then f is arbitrarily C^r-close to $g \in U^r([0,1])$ with a HCT to a periodic point of $\Omega_1(f)$.

ACKNOWLEDGEMENTS

I thank Profs. Zhihong Xia and Thomas Morley for their support and advice during the writing of my Ph.D. thesis which has led to this paper.

REFERENCES

1. A. Blokh, *Decomposition of Dynamical Systems of an Interval*, Russ. Math. Surv. **38** (1983), 133–134.
2. A. Blokh, *The "Spectral" Decomposition for One-Dimensional Maps*, in Dynamics Reported, vol. 4 (New Series), Springer-Verlag, Berlin, 1995, pp. 1–59.
3. L. Jonker & D. Rand, *Bifurcations in One Dimension I. The Nonwandering Set*, Invent. Math. **62** (1981), 347–365.
4. R. Mañé, *Hyperbolicity, Sinks and Measure in One Dimensional Dynamics*, Comm. Math. Phys. **100** (1985), 495–524.
5. W. de Melo & S. van Strien, *One-Dimensional Dynamics*, Springer-Verlag, Berlin, 1993.
6. L.-S. Young, *A Closing Lemma on the Interval*, Invent. Math. **54** (1979), 179–187.

Proceedings of Dynamic Systems and Applications 3 (2001) 511-520

Lossless Compression of Images and Video using the EZW Framework

Veeru N. Ramaswamy

Advance Multimedia Comm.
Bell Laboratories
Lucent Technologies
Holmdel, NJ - 07733

Kamesh R. Namuduri

Center for Theoretical Studies
of Physical Systems
Clark Atlanta University
Atlanta, GA - 30314

Abstract

Embedded Zerotree Wavelet (EZW) coding proposed by Shapiro is a recent research advancement in wavelet and subband coding. EZW is a framework that exploits correlation across different subbands unlike the traditional wavelet coding which exploits correlation of wavelet coefficients only within a subband. The EZW-based lossless image coding framework consists of three stages: (i) reversible discrete wavelet transform of an image, (ii) ordering and selection of wavelet coefficients using the EZW data structure and (iii) context modeling based arithmetic coding of wavelet coefficients. The performance of the compression algorithm depends on the choice of various parameters and the implementation strategies employed in all the three stages.

The set partitioning scheme proposed by Said and Pearlman is an enhancement of the EZW technique for lossy and lossless image compression. In this paper, we modify the set partitioning scheme to study various parameters involved in each stage and present several experiments to optimize the performance of the EZW based lossless compression algorithm for images and video.

1 Introduction

The present and future multimedia applications are focusing on the problem of optimizing storage and transmission bandwidth for different applications such as HDTV, internet, video conferencing and telemedicine. Hence, a very high interest exists among researchers on the problem of compressing still images and video. In this paper, we concentrate on lossless intraframe compression of video sequences using a set partitioning based EZW framework. Each frame of the video sequence is compressed individually as a single image.

Image and video compression is an important task in multimedia and teleconferencing applications. The EZW based lossless image coding framework consists of three stages: (i) reversible discrete wavelet transform (ii) ordering of wavelet coefficients using the EZW data structure and (iii) context modeling based arithmetic coding. The choice of wavelets plays an important role in decorrelating and compacting the image data and thereby improving the compression efficiency. The EZW data structure orders the wavelet coefficient to exploit the interband correlation. Finally, in the third stage, the arithmetic coding exploits the skewness existing in the distribution of wavelet coefficients using different context models. The research work presented in this chapter consists of several experiments to investigate (i) different choices of wavelet filters that can be used in lossless image coding, (ii) different methods of ordering the wavelet coefficients using the EZW data structure and (iii) different context models that can be used in conjunction with arithmetic coding [1]. A combination of a good wavelet filter that can produce maximum data decorrelation, a proper data structure to order and group the wavelet coefficients and an appropriate context model for arithmetic coding will help in improving the compression efficiency. Context modeling of wavelet coefficients is being recently studied for lossy and lossless image coding applications [6][7]. In this work, several experiments were conducted to find appropriate context modeling of wavelet coefficients in a lossless compression scheme. In this paper, various experiments that were conducted in all three stages of the framework to obtain maximum compression efficiency are presented.

2 Abbreviations and Terminology

The terminology used in this paper is similar to that used by Shapiro [2], and Said *et al* [3] and presented here for convenience. Let W, H represent the width and height of image and let $T = 2^n$ represent the current threshold.

- $S_n(i,j)$: Significance of a coefficient w.r.t T
- $S_n(D(i,j))$: Significance of descendants w.r.t T
- $S_n(L(i,j))$: Significance of grand-descendants w.r.t T
- *S Bitmap*: This corresponds to the set of $S_n(i,j)$ values for all coordinates (i,j) in the transformed image with respect to one threshold say 2^n or the n-th bit plane. Then S Bitmap corresponds to $\{S_n(i,j) : 0 \leq i < W, 0 \leq j < H, T = 2^n\}$
- *SD Bitmap*: This corresponds to the set of $S_n(D(i,j))$ values for all coordinates (i,j) in the transformed image with respect to one threshold say 2^n or the n-th bit plane. Then SD Bitmap corresponds to $\{S_n(D(i,j)) : 0 \leq i < W, 0 \leq j < H, T = 2^n\}$
- *SL Bitmap*: This corresponds to the set of $S_n(L(i,j))$ values for all coordinates (i,j) in the transformed image with respect to one threshold say 2^n or the n-th bit plane. Then SL Bitmap corresponds to $\{S_n(L(i,j)) : 0 \leq i < W, 0 \leq j < H, T = 2^n\}$
- LIP: *List of Insignificant Pixels*
- LSP: *List of Significant Pixels*
- LIS: *List of Insignificant Sets*
- *Significance Bit:* The bit corresponding to $S_n(i,j)$ needed to update the wavelet coefficients in the decoder.
- *Message Bit:* The bit corresponding to $S_n(D(i,j))$ or $S_n(L(i,j))$ that is needed for ordering information.

3 The Framework

The framework used for wavelet based lossless image coding is shown in Figure 1. The outline of the algorithm is given below. Each step in the algorithm is explained in the following sections.
The Algorithm:
Begin

- Perform the reversible discrete wavelet transform.
- Split the coefficients into:
  - significance map containing the first few bit planes of the wavelet coefficients.

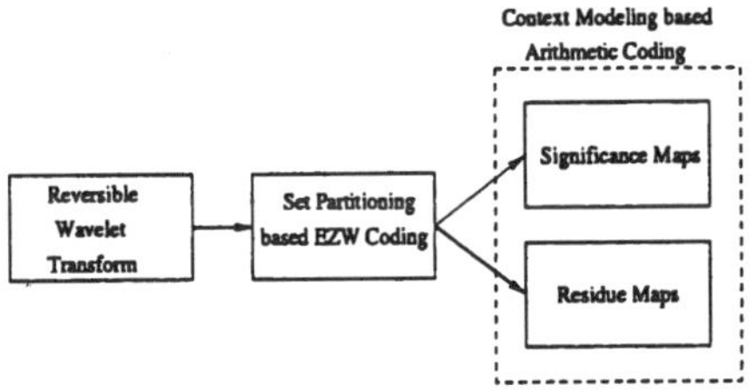

Figure 1: Framework for EZW based Lossless Coding

  - residue map containing the remaining bits of the wavelet coefficients

- Encode significance map (i) without grouping or (ii) with grouping and context model based arithmetic coding.
- Encode residue map using context model based arithmetic coding.

End

In the first stage, a reversible wavelet transform is performed. An appropriate choice of wavelet filter will help in improving data decorrelation and compaction which intern enhances the compression efficiency to a large extent. A performance analysis of different wavelets for lossless coding can be found in [8]. In the second stage, a set partitioning based EZW coding (refer Chapter 2 for the algorithm) is performed. The set partitioning consists of two passes namely the sorting and the refinement pass. The sorting pass sorts the coefficients in LIP and moves them to LSP when they become significant. In other words, the wavelet coefficients remain in LIP till the most significant bit (MSB) that is set to 1 is encountered in the sorting pass. The refinement pass outputs the remaining bits following the first significant bit,namely the first bit that is set to 1 in the wavelet coefficients. The algorithm is iterated by decrementing the threshold (threshold usually being a power of two) each time and terminated when the minimum threshold is reached. The set partitioning based EZW coding scheme, in effect, splits the wavelet coefficients into (1) significance map and (2) the residue map (the coefficients ranging from -2^L to 2^L, where 2^L is the last threshold). The significance map consisting of first few bit planes are encoded as the significance and message bits. The significance bit is used to update the wavelet coefficient in the decoder and the message bits are used for ordering significance information. In the sorting pass, the S, SD and SL bitmaps are used to compute the significance information. The S, SD and SL bitmaps are computed for each threshold using recursive functions 'Create_SMaps', 'Create_SDMaps' and 'Create_SLMaps'. These functions can be found in appendix section of [9].

Finally, in the third stage, the significance map and the residue map are encoded using context model based arithmetic coding. The coefficients in the significance map were kept in a group of 2×2 to exploit higher order entropy. In this stage, context modeling and context selection play an important role in achieving high compression efficiency. Several experiments were conducted on context models and the results will be discussed in the section to follow.

4 Ordering of Wavelet Coefficients in the EZW coding

As discussed in section 2, the wavelet coefficients are split into significance and residue map. Grouping of significance helps to exploit higher order entropy. The significance information consists of the significance and message bits. The sorting and refinement passes in the set partitioning algorithm produces significance and message bits to allow proper decoding of symbols. The significance bits in the sorting pass are encoded as either groups of 2×2 or as individual bits. Similarly the message bits namely the $S_n(D(i,j))$ and $S_n(L(i,j))$ can also be encoded as a 2×2 group or individual bits. the encoding of individual bits and grouping of 2×2 bits in the sorting pass is discussed with the aid of above example.

4.1 Encoding of Individual Significance Bits

The individual $S_n(i,j)$, $S_n(D(i,j))$ and $S_n(L(i,j))$ bits are packed into a double precision integer and are output into a file. A brief algorithm is presented below. This is a variation of the algorithm in [3]
Algorithm A:
Begin

- Perform the reversible forward wavelet transform.
- Perform set partitioning:
  - (†) For each current threshold (2^n) $\geq$ minimum threshold (2^L) do
    - *Sorting Pass:* Output $S_n(i,j)$, $S_n(D(i,j))$ and $S_n(L(i,j))$ as individual bits depending on the coordinate (i,j) in LIS as Type A or Type B. Pack the significance information into a double precision integer.
    - *Refinement Pass:* Output the bits corresponding to the current threshold (2^n or n-th bit) for all coordinates in LSP except those added in the current pass.
  - Perform adaptive arithmetic coding without any modeling for the packed integers.
  - Divide the threshold by 2 and go to (†).
- Classify the residue symbols into categories. Perform context modeling based arithmetic coding on the categorized residue symbols and adaptive arithmetic coding (without any modeling) for offset within a category.

End

In the sorting pass of the above algorithm, the encoder outputs the significance information as individual bits to the decoder. The bits are packed into double precision integers irrespective of the bits being significance or message bits. Then, adaptive arithmetic coding is used to encode the packed integers. In the refinement pass, the n-th MSB (corresponding to threshold 2^n) of all the coordinates in LSP, except those that have been added in the current pass are output.

4.2 Encoding 2×2 Groups of Significance Information

As discussed in the earlier section, the significance information can be either encoded as individual bits or can be encoded as symbols formed by a 2×2 set. In this section, the significance and the message bits can be encoded as a 2×2 group. For this reason, the coordinates are always added as a 2×2 set in the LIS list. The 2×2 set at coordinate (i,j) is represented by the set

$$\{(i,j),(i,j+1),(i+1,j),(i+1,j+1)\}$$

Hence only the left-top coordinate, namely (i,j) is added into the LIS list. Group is straight forward to implement. Every addition into the LIS list will be added as a 2×2 set. This is due to the fact that either all the four descendants are added into the list or none of them depending on the $S_n(D(i,j))$ or $S_n(L(i,j))$ value. Bits from the LSP during the refinement pass are encoded individually. A modification of Algorithm A to group significance and message bits is given below:
Algorithm B:
Begin

- Perform the reversible forward wavelet transform.
- Perform set partitioning:
  - (†) For each current threshold (2^n) $\geq$ minimum threshold (2^L) do
    - * *Sorting Pass:* Output $S_n(i,j)$, $S_n(D(i,j))$ and $S_n(L(i,j))$ information as follows:
      - Form a symbol from a 2×2 group of $S_n(D(i,j))$ at the coordinate (i,j).
      - For each bit in $S_n(D(i,j))$ that is set to 1, form a symbol from a 2×2 group of $S_n((k,l))$ at the coordinate (k,l) with (k,l) in the set $\{(2i,2j),(2i,2j+2),(2i+2,2j),(2i+2,2j+2)\}$
      - Form a symbol from a 2×2 group of $S_n(D(i,j))$ at the coordinate (i,j)
    - * *Refinement Pass:* Output the bits corresponding to the current threshold (2^n or n-th bit) for all coordinates in LSP except those added in the current pass.
  - Perform context modeling based adaptive arithmetic coding to encode the symbols.
  - Divide the threshold by 2 and go to (†).
- Classify the residue symbols into categories. Perform context modeling based arithmetic coding on the categorized residue symbols and adaptive arithmetic coding (without any modeling) for offset within a category.

End

It needs to be noted that the symbols formed from the significance information is coded according to the conditions in set partitioning. Depending on the bit that is set to 1 in the symbol formed by 2×2 grouping of $S_n(D(i,j))$, the significance of four children corresponding to that bit alone (which represents a coordinate in the 2×2 set) is again coded as one symbol with a grouping performed at a coordinate in $\{(2i,2j),(2i,2j+2),(2i+2,(2j),(2i+2,2j+2)\}$.

The data packet to be encoded has the following form:

- A symbol formed by a 2×2 grouping of $S_n(D(i,j))$.
- Depending on the bit that is set to 1 in $S_n(D(i,j))$, a symbol is formed by a 2×2 grouping of $S_n(i,j)$. Hence a maximum of four $S_n(i,j)$ symbols needs to be coded.
- A symbol formed by a 2×2 grouping of $S_n(L(i,j))$ if the grandchildren exist.

The grouping of significance information allows us to exploit up to m-th order entropy, m being {1, 2, 3, 4}. Better context models can be formed to encode the symbols rather than individual bits and hence the context models can exploit the correlation between the symbols formed by $S_n(i,j)$ and $S_n(D(i,j))$. In the following sections, the experiments conducted on different context modeling techniques to encode significance and residue values are discussed.

5 Context Modeling of Significance Information and Residue

In most of the source modeling techniques, a neighborhood is used to form the contexts or otherwise known as conditioning events in a Markov model. In case of zerotree based coding of wavelet coefficients a good context selection involves the parent coefficient and the coefficients from the previous thresholds apart from the neighborhood in current or previous scans. The source modeling and source coding of alphabets can be viewed as a two dimensional list, where one dimension contains the model numbers and the other the symbol set. Generating an appropriate context such that the skewness in the data can be exploited by the arithmetic coding in an optimum fashion is a challenging task.

5.1 Context Modeling of Significance Information

In our experiments, it was observed that one common set of context models for the significance information cannot exploit the correlation between the significance bits and the message bits namely $S_n(i,j)$ and $S_n(D(i,j))$ (or $S_n(L(i,j))$). Hence a set of 16 different models each for $S_n(D(i,j))$ and $S_n(L(i,j))$ and a set of 4 models for $S_n(i,j)$ were used. Initially, the contexts were a polynomial function of the types of the coordinates (Type A, B, or C). Type C is our modification to the set partitioning algorithm to convey the message that the coordinates have already been processed as Type A and Type B and does not need any further processing. Then contexts based on the previously encoded symbols formed by 2 × 2 block were used. Both were unsuccessful attempts to code the significance information efficiently. Finally, after extensive experiments with the context models, it was found that the only correlation in the wavelet coefficients is between two successive thresholds $T = 2^n$ and $T = 2^{n-1}$ This is due to the fact that once a coordinate becomes significant with respect to a threshold, that coordinate is significant with respect to all lower thresholds. It was observed that this context modeling improved the compression efficiency to a small extent only. The symbol encoded formed from the significance information in a 2 × 2 group with respect to previous threshold was stored as a context information in the LIS list structure.

5.2 Context Modeling of Residue

The last step in the two algorithms A and B is context modeling of the residue. The residue coefficients are formed after the set partitioning based EZW coding has been performed up to a last threshold (say 2^L, usually L=3). The range of residue is between ($-2^L - 1$ to $2^L - 1$) As mentioned in [8], since the significance of the coefficient was found by checking whether a coefficient is $\geq$ to the current threshold, there could be a conflict in decoding the symbol zero. This problem was avoided by mapping the coefficients that are equal to current threshold to an escape symbol and mapping the symbols back to 2^L. A major contribution to the storage cost is from the residue since the entropy content is high. A vast number of experiments were conducted to find an optimal coding of the residue coefficients. One experiment was to adapt the CALIC (Context based Adaptive Lossless Image Coding) encoding to exploit the edges and the texture patterns in the residue, if there are any [5]. A brief discussion of the actual CALIC encoding and the experiments on adapting it to the residue coefficients can be found in [9]. Another experiment was to improve the context modeling proposed in [4].

6 Conclusion

The Compression Efficiencies for 10 different images and for 2 different thresholds (32 and 8) are provided in tables 1 and 2. The compression efficiencies for every fifth frame of the football video sequence is tabulated in Table 5.2. Each frame of the football video sequence is of size 740 × 486 and 8 bpp. It was observed from our experiments while context modeling of residue based on SPIHT helped in improving the compression efficiency, context modeling of significance did not enhance the efficiency. Also the (5,3) pair performed the best among all the filters. The results for other experiments can be found in [9].

References

[1] I. H. Witten, R. M. Neal and J. G. Cleary, "Arithmetic Coding for Data Compression", *Comm. of the ACM*, Vol. 30, No. 6, pp 520-540, 1987.

Table 1: Compression Efficiencies of S+P and (5,3) filters with Last Threshold = 32

	Compression Efficiency					
	S+P (Type C)			(5,3)		
Image	M1	M2	M3	M1	M2	M3
Challenger	43.34	43.24	42.25	43.79	43.60	42.65
Coral	38.22	38.09	37.08	38.32	38.10	37.12
F16	57.03	56.94	56.21	58.37	58.25	57.41
House	44.85	44.63	44.39	45.96	45.65	45.48
Planets	78.41	78.29	77.99	78.83	78.60	78.26
LAX	37.49	37.48	36.52	39.18	39.17	38.44
Lenna	34.51	34.24	33.38	35.31	34.97	34.45
Man	63.09	63.05	62.54	61.23	61.15	60.67
Shuttle	49.35	49.27	48.12	50.82	50.67	49.89
Sphere	46.28	46.19	45.67	45.16	45.10	44.37

M1: With Inidividual Bit Encoding of Signifiance

M2: With 2 × 2 Grouping of Significance and Context Modeling

M3: With CALIC Encoding of Residue Coefficients.

Table 2: Compression Efficiencies of S+P and (5,3) filters with Last Threshold = 8

	Compression Efficiency					
	S+P (Type C)			(5,3)		
Image	M1	M2	M3	M1	M2	M3
Challenger	41.21	40.66	40.73	42.17	41.15	41.54
Coral	36.03	35.33	35.57	36.62	35.28	35.99
F16	56.06	55.56	55.47	57.74	57.01	56.89
House	43.26	42.16	43.07	44.67	43.18	44.40
Planets	77.46	76.88	77.11	78.02	77.10	77.55
LAX	33.95	33.61	33.80	36.28	35.90	36.24
Lenna	32.04	31.15	31.85	33.42	32.18	33.27
Man	62.74	62.43	62.03	60.66	60.10	60.00
Shuttle	47.57	47.11	46.95	49.82	49.09	49.28
Sphere	44.68	44.10	44.10	43.60	42.68	42.73

M1: With Inidividual Bit Encoding of Signifiance

M2: With 2 × 2 Grouping of Significance and Context Modeling

M3: With CALIC Encoding of Residue Coefficients.

Table 3: Results for Football Video Sequence (in bpp)

Frame #	S+P[B]	S+P[C]	(5,3)
00	4.54	4.61	4.48
05	4.56	4.56	4.57
10	4.52	4.59	4.55
15	4.56	4.56	4.53
20	4.64	4.63	4.63
25	4.63	4.64	4.64
30	4.61	4.62	4.61
35	4.64	4.64	4.62
40	4.68	4.67	4.66
45	4.65	4.65	4.65
50	4.62	4.66	4.65
55	4.64	4.64	4.62

[2] J. M. Shapiro, "Embedded image coding using zerotrees of wavelet coefficients", *IEEE Trans. on Signal Processing*, Vol. 41, No. 12, Dec 1993.

[3] A. Said and W. A. Pearlman, "A New, Fast, and Efficient Image Coded Based on Set Partitioning in Hierarchichal Trees" *IEEE Trans. on Circuits and Systems for Video Technology*, Vol. 6, No. 3, pp 243-249, June 1996.

[4] A. Said and W. A. Pearlman, "An image multiresolution representation for lossless and lossy image compression", *IEEE Trans. on Image Processing*, Vol. 5, No. 9, pp 1303-1310, September 1996.

[5] X. Wu and N. Memon, "Context-based, Adaptive, Lossless Image Codec", *IEEE Trans. on Communications*, vol. 45, no. 4, pp 437-444, April 1997.

[6] S. M. LoPresto, K. Ramchandran, and M. T. Orchard, "Image Coding based on Mixture Modeling of Wavelet Coefficients and a fast Estimation-Quantization Framework", *Proc. of 1997 Data Compression Conference*, pp 221-230, March 1997.

[7] C. Chrysafis, and A. Ortega, "Efficient context-based entropy Coding for Lossy Wavelet Coefficients Image Compression", *Proc. of 1997 Data Compression Conference*, pp 241-250, March 1997.

[8] V. N. Ramaswamy, K. R. Namuduri, and N. Ranganathan, "Performance Analysis of Wavelets in Embedded Zerotree based Lossless Image Coding Schemes", to appear in *IEEE Trans. on Signal Processing*, 1998.

[9] V. N. Ramaswamy, "Lossless Image Compression using Wavelet Decomposition", *Ph.D Dissertation*, Dept. of Computer Science and Engg., Univ. of South Florida, Tampa, Aug. 1998.

Proceedings of Dynamic Systems and Applications 3 (2001) 519-526

Global Existence Theorem and Asymptotic Behavior via Semigroup Theory

By
V. Raghavendra and Varsha Gupta*
Department of Mathematics
Indian Institute of Technology
KANPUR-208016, INDIA

Abstract. In this study, we establish the existence of a unique solution of a nonlinear parabolic equation subject to Initial Neumann Boundary Value Problem with an additional constraint of impulses at fixed moments. The above problem is first reduced to the study of nonlinear equation with impulses in abstract spaces. A modification of the classical Hille-Yosida theorem is used to take into account of the impulses.

---

* presently at Bombay

Global Existence Theorem and Asymptotic Behavior via Semigroup Theory

By
V. Raghavendra and Varsha Gupta*
Department of Mathematics
Indian Institute of Technology
Kanpur-208016, INDIA

1. Introduction :

We first study the existence of solutions of a nonlinear problem

$$\frac{du}{dt} = Au + f(t,u) \tag{1.1}$$

$$u(0) = u_o \tag{1.2}$$

$$u(t_k^+) = au(t_k) \tag{1.3}$$

where u is a left-continuous function taking values in a Banach space X, $\{t_k\}^\infty$ is a sequence of positive real numbers such that $t_k \to \infty$ as $k \to \infty$, A is the infinitesimal generator of a C –semigroup. We investigate sufficient conditions for the3 global existence and asymptotic behavior of solutions of (1.1) – (1.3) by using a modified version of the Hille-Yosida theorem. Finally we show the usefulness of the result which is related a class of nonlinear parabolic equations with Initial-Neumann boundary conditions under additional constraints of impulses at fixed moments.

2. Preliminaries :

Let (X, $\| \, \|$) be a real or a complex Banach space. $\Re$ and $\Re^+$ denotes the real line and $[0, \infty)$ respectively. Let $\{t_k\}$ be a real sequence such that $0 < t_1 < t_2 < \ldots < t_k \ldots$ with $t_k \to \infty$ as $k \to \infty$ and $I_k = [t_{k-1}, t_k)$.

2.1 Definition :

A family $\{S(t): t \in \Re\}$ of continuous linear operators from X to X is said to form a semigroup of class C_0^a (or a C_0^a –semigroup) if

(i) $S(0) = I$, the identity operator,

(ii) $S(t+s) = a^{n+1-(i+j)} S(t)\, S(s)$ for $t \in I_j$, $s \in I_i$, $t + s \in I_n$;

(iii) $t \in S(t)u$ is continuous on $\Re^+\backslash\{t_k\}^\infty_{k=1}$ for each $u \in X$ and $S(t^+_k) = a\, S(t_k)$. $S(t^-_k) = S(t_k)$ for $k = 1,2,....$

C^a_0-semigroup reduces to a C_o-semigroup when $a = 1$. The generator A of a C^a_0-semigroup is defined in a similar way as in the case of a C_o-semigroup. For details we refer to Pazy [7]. In the sequel, we assume that A is the generator of a C^a_0-semigroup S(t).

2,2Definition (Mild and Strong solution):

Let $u_o \in X$, $f \in C[0,T]\backslash\{t_k\}_{k=1} Xx, X)$ (where $t_n < T$). The function $u \in C([0,T]\backslash\{t_k\}_{k=1}, X)$ given by

$$u(t) = S(t)u_0 + \int_0^t S(t-s)f(s)ds, \quad 0 \le t < T$$

is called a mild solution of the IVP (1.1) – (1.3) on [0,T]. A map u: $[0,T] \to X$ is called a strong solution of (1.1) to (1.3) if u is continuously differentiable on $[0,T]\backslash\{t_k\}_{k=1}$, $u(t) \in D(A)$ for $0 < t \le T$.

The following is an existence theore , which uses the Banach contraction mapping principle.

2.2 Theorem:

Let $\{S(t)\}_{t\geq o}$ be a C_o-semigroup generated by A on X. Let f: $\Re^+ x\, X \to X$ be continuous in its first argument t on $\Re^+\backslash\{t_k\}_{u=1}$ and be uniformly Lipschitz continuous in second argument (with Lipschitz constant L) on X. Then, for every u_o in X the IVP (1.1) – (1.13) has a unique mild solution u, which is continuous.

2.3 Theorem:

Let $S(t)_{t\geq o}$ be a C_o-semigroup generated by A on X. Let f: $\Re^+ \, x\, X \to X$ be a c^1-function on $\Re^+\backslash(\{t_k\}_{k=1}.) x\, X$ Let $u_o \in D(A)$. Then a function u: $\Re^+ \to X$ is a strong solution of (1.1) – (1.3) if and only if u satisfies (2.1).

A similar theorem is proved in [3] and hence omitted here. We adopt the following definition from [1].

2.4 Definition:

A strong solution u of (1.1) – (1.3) (on many occasions referred to equilibrium solution) is called D(A)-stable if for any $\varepsilon > 0$ there exists a $\delta(\varepsilon) > 0$ such that for every strong solution u(t) in D(A) of (1.1) – (1.3) satisfies

$$\|u(t) - u(t)\| < \varepsilon \text{ for } t \in [0, T_{max}] \text{ if } \|u - u_0\| < \delta$$

where $[0,T_{max}]$ is the maximum interval of existence of u. In addition, we say that u is asymptotically stable in $K \subset D(A)$ if u is D(A)-stable and for any $u_o \in K$ we have $T_{max} = \infty$ and $u(t) - u(t) \to 0$ as $t \to \infty$.

3. Asymptotic Behavior:

We now study the asymptotic behavior of (both mild and strong) solutions of a class of IVP

$$\frac{du}{dt} = Au + f(t,u) \tag{3.1}$$

$$u(0) = u_o \tag{3.2}$$

$$u(t_k^+) = au(t_k) \tag{3.3}$$

The following result is important in itself and is crucially employed later.

Theorem 3.1:

Let A be the infinitesimal generator of a C_o^a-semigroup S(t) for $1 \le p < \infty$, let

$$\int_0^t \|S(t)u\|^p \, dt < \infty \text{ for every } u \in X \tag{3.5}$$

Then there exists constants $M \ge 1$ and $\mu > 0$ such that

$$\|S(t)\| \le M a^{n-i} e^{-\mu t}, t \in I_n, \; n \ge 1. \tag{3.6}$$

The proof is similar to the proof as given in Pazy [3] and hence omitted. A consequence of Theorem 3.1 is the following result on asymptotic behavior.

Theorem 3.2:

Let S(t) satisfy the condition of Theorem 3.1. Let f(t,u) satisfy the Lipschitz condition with respect to x on X with Lipschitz constant L and $f(t,0) \equiv 0$. If $\mu \ge ML$, $|a| < 1$, then for every $u_o \in X$ the mild solution u(t) of (3.1) – (3.3) tends to 0 as $t \to \infty$.

Proof: By Theorem 2.3, we have

$$u(t) = S(t)u_0 + \int_0^t S(t-s)f(s)ds,$$

For $t \in I_1$ and by an application of Gronwall's Inequality we get

$$\|u(t)\| \le M \|u_0\| \, e^{(ML-\mu)t}$$

By iteration, for $t \in I_n$, we have

$$\|u(t)\| \leq M^n \, |a|^{n-1} \, \|u_0\| \; e^{(ML-\mu t)}$$

Which shows that $\|u(t)\| \to 0$ as $t \to \infty$.

Theorem 3.3:

Let S(t) satisfy the conditions of Theorem 3.1 and $u_o \in D(A)$. Let $f: \Re^+ \times X \to X$ be continuously differentiable on $\Re^+ \backslash \{t_k\}_{k=1}$ and $f(t,0) \equiv 0$. Let $\left\|\frac{\partial f}{\partial u}\right\| \leq L$. If $\mu \geq ML$, then the strong solution u(t) of (1.1) – (1.3) is asymptotically stable in $K(\subset D(A))$.

The proof is similar to that of Theorem 3.2. We conclude with an example.

Example: Consider the IBVP

$$\frac{\partial u}{\partial t} = D\,\frac{\partial^2 u}{\partial x^2} - u \;,(t,x) \in \Re^+ \times \bar{I},\, t \neq t_k \tag{3.5}$$

$$u(0,x) = u_0(x)\;,\quad x \in \bar{I} \tag{3.6}$$

$$u_x(t_k^+,-b) = u_x(t,b) = 0\,,\quad t \neq t_k \tag{3.7}$$

$$u(t_k^=,x) = u\,(t_k,x)\;,\; k = 1,2,\ldots . \tag{3.8}$$

where $u: \Re^+ \to X = L^2(I)$, $I = (-b,b)$, $D > 0$ is a constant. By the standard method, we have

$$u(t,x) = \sum_{m=0}^{\infty} \left\{\frac{1}{2b}\int_{-b}^{b} \sin\left(\frac{(2m+1)}{b} y u_0(y) dy\right\} a^{n-i}\; e^{-\lambda_m^2 t} \sin\left(\frac{(2m+1)}{b}x\right)$$

for any $u_o \in L^2(I)$ we define, for $t \in \Re^+$,

$$(S(t)u_o)(x) = \text{RHS of the above expression.}$$

Then we can easily verify that S(t) is a semigroup on $L^2(I)$ and moreover we have, for $u_o \in L^2(I)$,

$$\int_0^\infty \|S(t)u_0\|_2^2 dt \;\leq\; \frac{3b^3}{8D}|a|^{n-1}\;\|u_0\|_2^2 \;<\infty$$

which shows that $S(t)u_o \in L_2(I)$, for $t \in \Re^+$. Also (see Page 155, [2]) we have, for $u_o \in L^2(I)$

$$\|S(t)u_0\|_2 \leq \|u_0\|_2 .$$

Thus S(t), for $t \in \Re^+$, is a C_o^a-semigroup of contractions. Thus all the conditions of Theorem 3.1 are satisfied which shows that these constants $\mu \leq 1$ and $\mu > 0$ such that

$$\|S(t)\| \leq a^{n-1} \ \mu \ e^{\mu t} .$$

REFERENCES

[1] I.V. Capasso and D. Fortunato, "Stability results for semilinear evolution equations and there applications to some reaction-diffusion problems," Siam. J. Appl. Math, Vol. 39, No. 1, 1980, 37-47.

[2] T.W. Korner, "Fourier Analysis," Cambridge University Press, New York, 1990.

[3] A. Pazy, "Semigroups of linear operators and applications to P.D.E.," Applied Math Series, Vol. 44, Springer-Verlag, New York.

[4] V. Raghavendra and V. Gupta, "Hille-Yosida theorem for left-continuous functions," (To appear).

Proceedings of Dynamic Systems and Applications 3 (2001) 527-532

Stabilization/Regulation of a Multirobotic system modeled with Petri nets using vector Lyapunov and Comparison methods

Zvi Retchkiman
Centro de Investigacion en Computo
Lab. de Automatizacion
Instituto Politecnico Nacional
Apartado Postal 75-476
C.P. 07738 Mexico, D.F.
Col. Lindavista
Zacatenco
e-mail: mzvi@cic.ipn.mx

ABSTRACT: This paper studies the stabilitzation/regulation problem of a multirobotic system modeled by Petri nets using vector Lyapunov methods. After reviewing some preliminaries about stability of discrete event systems modeled with Petri nets, the stabilization/regulation problem is recalled. A multirobotic system helps to illustrate how the methodology is applied in order to get some desired performance. The proposed approach shows to be an easy and convenient way to deal with the systems stabilization/regulation problem.
AMS (MOS) Subject Classification. 93D30, 93A30, 93D05.

1. Introduction

A discrete event system, is a dynamical system which evolves in time by the occurrence of events at possibly irregular time intervals. Some examples include Flexible manufacturing systems, Computer networks, Logic circuits. Petri nets are a graphical and mathematical modeling tool applicable to Discrete event systems in order to represent and analyze the systems properties. Petri nets are known to be useful for describing and studying information processing systems that are characterized as being: concurrent, asynchronous, distributed, parallel, no determinists, and-or stochastic. For a detailed discussion of Petri net theory see Murata or Peterson [3, 4] and the references quoted therein.

A flexible manufacturing system is an efficient production line with versatile machines, an automatic transport system and a sophisticated decision making system (Silva [5]). Flexible manufacturing systems can be formed by subsystems that work concurrently and, therefore are suitable to be modeled by Petri nets.

In this paper, the stabilitzation/regulation theory philosophy proposed in Retchkiman [2] is applied to a multirobotic system, (a particular example of a flexible manufacturing system), in order to get some preassigned behavior.

The paper is organized as follows. In Section 2, some stability preliminaries of discrete event systems modeled with Petri nets are recalled (Retchkiman [1, 2]) . In Section 3, the stabilization/regulation problem introduced in Retchkiman [2] is addressed. Finally in section 4, the multirobotic stabilization/regulation problem is solved.

NOTATION: $N = \{0,1,2,...\}$, $R_+ = [0,\infty)$, $N_{n0}^+ = \{n_0, n_0+1, ..., n_0+k, ...\}, n_0 \geq 0$. Given $x, y \in R^n$, we usually denote the relation "$\leq$" to mean componentwise inequalities with the same relation, i.e., $x \leq y$ is equivalent to $x_i \leq y_i, \forall i$. A function $f(n,x)$, $f : N_{n0}^+ \times R^n \rightarrow R^n$ is called nondecreasing in x if given $x, y \in R^n$ such that $x \geq y$ and $n \in N_{n0}^+$ then, $f(n,x) \geq f(n,y)$.

2. Preliminaries

Consider systems of first ordinary difference equations given by

$$x(n+1) = f[n, x(n)],\; x(n_o) = x_0,\; n \in N_{n0}^+ \tag{1}$$

where $n \in N_{n0}^+$, $x(n) \in R^n$ and $f : N_{n0}^+ \times R^n \rightarrow R^n$ is continuous in $x(n)$.

Definition 1 *The n vector valued function $\Phi(n, n_0, x_0)$ is said to be a solution of (1) if $\Phi(n_0, n_0, x_0) = x_0$ and $\Phi(n+1, n_0, x_0) = f(n, \Phi(n, n_0, x_0))$ for all $n \in N_{n0}^+$.*

Definition 2 *The system (1) is said to be*
i). Practically stable, if given (λ, A) with $0 < \lambda < A$, then

$$|x_0| < \lambda \Rightarrow |x(n, n_0, x_0)| < A,\; \forall n \in N_{n0}^+,\; n_0 \geq 0;$$

ii). Uniformly practically stable, if it is practically stable for every $n_0 \geq 0$.

The following class of function is defined.

Definition 3 *A continuous function $\alpha : [0,\infty) \rightarrow [0,\infty)$ is said to belong to class $\mathcal{K}$ if $\alpha(0) = 0$ and it is strictly increasing.*

2.1. Comparison principles and Lyapunov methods for Practical stability

Consider a vector Lyapunov function $v(n, x(n))$, $v : N_{n0}^+ \times R^n \rightarrow R_+^p$ and define the variation of v relative to (1) by

$$\Delta v = v(n+1, x(n+1)) - v(n, x(n)) \tag{2}$$

Then, the following result concerns the practical stability of (1).

Theorem 1 *Let $v : N_{n0}^+ \times R^n \rightarrow R_+^p$ be a continuous function in x, define the function $v_0(n, x(n)) = \sum_{i=1}^{p} v_i(n, x(n))$ such that satisfies the estimates*

$$b(|x|) \leq v_0\left(n, x\left(n\right)\right) \leq a(|x|) \text{for } a, b \in \mathcal{K} \text{and}$$

$$\Delta v(n, x(n)) \leq w(n, v(n, x(n)))$$

for $n \in N_{n0}^+$, $x(n) \in R^n$, *where* $w : N_{n0}^+ \times R_+^p \rightarrow R^p$ *is a continuous function in the second argument.*

Assume that : $g(n,e) \triangleq e + w(n,e)$ *is nondecreasing in* e, $0 < \lambda < A$ *are given and finally that* $a(\lambda) < b(A)$ *is satisfied. Then, the practical stability properties of*

$$e(n+1) = g(n, e(n)),\ e(n_0) = e_0 \geq 0. \tag{3}$$

imply the corresponding practical stability properties of system (1).

Proof. From the continuity of v with respect to the second argument, it is always possible to make $v_0(n_0, x_0) < a(\lambda) \Rightarrow v_0(n, x(n)) \leq \sum_{i=1}^{p} e_i(n, n_0, e_0) < b(A)\ \forall n \in N_{n0}^+$, we want to prove that $|x(n, n_0, x_0)| < A$ for $n \geq n_0$. If not, then there exists an $n_1 \geq n_0$ and a solution $x(n, n_0, x_0)$ such that $x(n_1) \geq A$ and $x(n) < A$ for $n_0 \leq n < n_1$. We then have that $b(A) \leq b(x(n_1)) \leq v_0(n_1, x(n_1)) \leq \sum_{i=1}^{p} e_i(n_1, n_0, e_0) < b(A)$ which is a contradiction and therefore our claim is proved ∎.

Corollary 2 *In Theorem 1:*

i). If $w(n,e) \equiv 0$ *we get uniform practical stability of (1) which implies structural stability.*
ii). If $w(n,e) = -c(e)$, *for* $c \in \mathcal{K}$, *we get uniform practical asymptotic stability of (1).*

2.2. Petri Nets and Practical stability of Discrete event systems

A Petri net is a 5-tuple, $PN = \{P, T, F, W, M_0\}$ where: $P = \{p_1, p_2, ..., p_m\}$is a finite set of places, $T = \{t_1, t_2, ..., t_n\}$ is a finite set of transitions, $F \subset (P \times T) \cup (T \times P)$ is a set of arcs, $W : F \rightarrow N_1^+$ is a weight function, M_0: $P \rightarrow N$ is the initial marking, $P \cap T = \varnothing$ and $P \cup T \neq \varnothing$.

A Petri net structure without any specific initial marking is denoted by N. A Petri net with the given initial marking is denoted by (N, M_0). Notice that if $W(p,t) = \alpha$ (or $W(t,p) = \beta$) then, this is often represented graphically by α, (β) arcs from p to t (t to p) each with no numeric label.

Let $M_k(p_i)$ denote the marking (i.e., the number of tokens) at place $p_i \in P$ at time k and let $M_k = [M_k(p_1), ..., M_k(p_m)]^T$ denote the marking (state) of PN at time k. A transition $t_j \in T$ is said to be enabled at time k if $M_k(p_i) \geq W(p_i, t_j)$ for all $p_i \in P$ such that $(p_i, t_j) \in F$. It is assumed that at each time k there exists at least one transition to fire. If a transition is enabled then, it can fire. If an enabled transition $t_j \in T$ fires at time k then, the next marking for $p_i \in P$ is given by

$$M_{k+1}(p_i) = M_k(p_i) + W(t_j, p_i) - W(p_i, t_j).$$

Let $A = [a_{ij}]$ denote an $n \times m$ matrix of integers (the incidence matrix) where $a_{ij} = a_{ij}^+ - a_{ij}^-$ with $a_{ij}^+ = W(t_i, p_j)$ and $a_{ij}^- = W(p_j, t_i)$. Let $u_k \in \{0,1\}^n$ denote a firing vector where if $t_j \in T$ is fired then, its corresponding firing vector is $u_k = [0, ..., 0, 1, 0, ..., 0]^T$ with the one in the jth position in the vector and zeros everywhere else. The matrix equation (nonlinear difference equation) describing the dynamical behavior represented by a Petri net is:

$$M_{k+1} = M_k + A^T u_k \tag{4}$$

where if at step k, $a_{ij}^- < M_k(p_j)$ for all $p_i \in P$ then, $t_i \in T$ is enabled and if this $t_i \in T$ fires then, its corresponding firing vector u_k is utilized in the difference equation to generate the next step. Notice that if M' can be reached from some other marking M and, if we fire some sequence of d transitions with corresponding firing vectors $u_0, u_1, ..., u_{d-1}$ we obtain that

$$M' = M + A^T u, \; u = \sum_{k=0}^{d-1} u_k. \tag{5}$$

Let (N_{n0}^+, d) be a metric space where $d : N_{n0}^+ \times N_{n0}^+ \rightarrow R_+$ is defined by

$$d(M_1, M_2) = \sum_{i=1}^{m} \zeta_i \mid M_1(p_i) - M_2(p_i) \mid ; \zeta_i > 0, \; i = 1, ..., m.$$

and consider the matrix difference equation which describes the dynamical behavior of the discrete event system modeled by a Petri net (5) then we have,

Proposition 3 *Let N be a Petri net. N is uniform practical stable if there exists a Φ strictly positive m vector such that*

$$\Delta v = u^T A \Phi \leq 0 \Leftrightarrow A\Phi \leq 0 \tag{6}$$

Moreover, N is uniform practical asymptotic stability if the following equation holds

$$\Delta v = u^T A \Phi \leq -c(e) \Leftrightarrow A\Phi \leq -c(e) \text{for } c \in \mathcal{K}. \tag{7}$$

3. The Stabilization Problem

Consider the matrix difference equation which describes the dynamical behavior of the discrete event system modeled by a Petri net

$$M' = M + A^T u \tag{8}$$

We are interested in finding a firing sequence vector, control law, such that system (8) remains bounded.

Definition 4 *Let N be a Petri net. N is said to be stabilizable if there exists a firing transition sequence with transition count vector u such that system (8) remains bounded.*

Proposition 4 *Let N be a Petri net. N is stabilizable if there exists a firing transition sequence with transition count vector u such that the following equation holds*

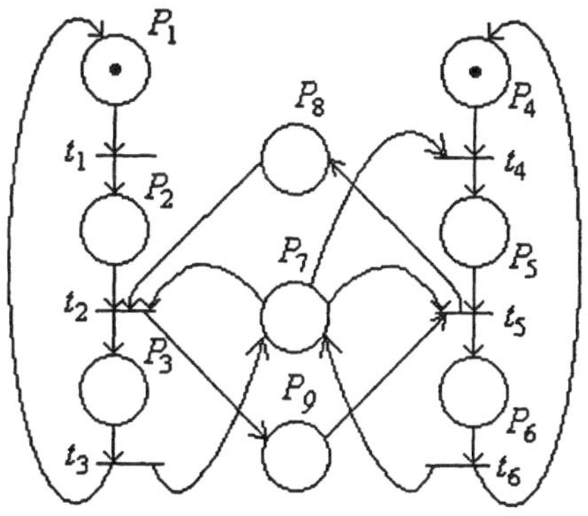

Figure 1:

$$\Delta v = A^T u \leq 0. \tag{9}$$

Remark 1 *It is important to underline that by fixing a particular u, which satisfies (9), we restrict the coverability tree to those markings (states) that are finite. The technique can be utilized to get some type of regulation and/or eliminate some undesirable events (transitions). Notice that in general (6)$\not\Rightarrow$(9) and that (9)$\not\Rightarrow$* (6).

4. The Multirobotic System

The multirobotic system consists of two robot arms which perform pick and place operations accessing a common workspace at times to obtain or transfer parts. In order to avoid collision, it is assumed that only one robot can access the workspace at a time. In addition, it is assumed that the common workspace has a buffer with a limited space for products. the process represents the operation of the two robots serving two different machining tools, with one robot arm transferring products from one machining tool to the buffer, and the other robot arm transferring semiproducts from the buffer to the other machining tool. The Petri net which represents the system is shown in Fig.1, where the interpretation of places and transitions is as follows:

Place	Interpretation
$P_1(P_4)$	Robot $R_1(R_2)$ performs tasks outside the common workspace
$P_2(P_5)$	Robot $R_1(R_2)$ waits for access to the common workspace
$P_3(P_6)$	Robot $R_1(R_2)$ performs in the common workspace
P_7	mutual exclusion
$P_8(P_9)$	number of empty (full) positions in buffer
Transition	Interpretation
$t_1(t_4)$	Robot $R_1(R_2)$ requests access to the common workspace
$t_2(t_5)$	Robot $R_1(R_2)$ enters the common workspace
$t_3(t_6)$	Robot $R_1(R_2)$ leaves the common workspace

We are interested in finding a firing transition sequence with transition count vector u such that the system will process, exactly k semiproducts. The incidence matrix of the Petri net model is

$$A = \begin{bmatrix} -1 & 1 & 0 & 0 & 0 & 0 & 0 & 0 & 0 \\ 0 & -1 & 1 & 0 & 0 & 0 & -1 & -1 & 1 \\ 1 & 0 & -1 & 0 & 0 & 0 & 1 & 0 & 0 \\ 0 & 0 & 0 & -1 & 1 & 0 & 0 & 0 & 0 \\ 0 & 0 & 0 & 0 & -1 & 1 & -1 & 1 & -1 \\ 0 & 0 & 0 & 1 & 0 & -1 & 1 & 0 & 0 \end{bmatrix}$$

Then, picking $\Phi = [1,1,2,1,1,2,1,1,1] \Rightarrow A\Phi \leq 0$ therefore, the multirobotic system is uniform practical stable. Even more, choosing $u = [k,k,k,k,k,k]$ we get that $A^T u \leq 0$ and since the transitions fire exactly k times, the objective of processing k semiproducts is fulfilled.

Acknowledgment

I would like to thank the Department of Mathematics of Morehouse College for providing part of the financial support needed to attend the conference.

References

[1] Zvi Retchkiman, " Practical Stability of Discrete event systems using Lyapunov and comparison methods", Proc. ACC '98, Philadelphia, Pennsylvania 1998.

[2] Zvi Retchkiman, "A Vector Lyapunov function approach for the Stabilization of Discrete evenrt systems", International Journal of Applied Mathematics, vol.2, No.7, 2000.

[3] T.Murata, "Petr nets: Properties, analysis, and applications", in Proc.IEEE, 1989, vol. 77, no 4, pp. 541-580.

[4] J.L.Peterson, *Petri net modeling and the Modeling of systems.* Englewood Cliffs, NJ: Prentice Hall, 1981.

[5] Silva, *Las redes de Petri en la Automatica y la Informatica.* Ed. AC, Madrid.

Proceedings of Dynamic Systems and Applications 3 (2001) 533-544

Variational Lyapunov Method and Stability Theory of Hybrid Systems

Rebecca L. Rizzo
Florida Institute of Technology
Applied Mathematics Program
Melbourne, FL 32901, USA

Abstract: Employing theVariational Lyapunov method to hybrid systems, the stability criteria is investigated.

Key words and phrases: Variational Lyapunov method, hybrid system, stability theory.

1. Introduction

Recently [4,5,6], the method of variation of parameters and the Lyapunov method have been combined to provide a unified technique, known as the variational Lyapunov method. This unification offers a comparison result via Lyapunov-like functions and consequently provides a flexible mechanism to preserve the nature of perturbations. This method proves extremely useful and effective in the investigation of quantitative properties of nonlinear problems.

The problem of stabilizing a continuous plant governed by differential equations through the interaction with a discrete time controller has recently been investigated. This investigation has lead to the study of hybrid systems [2,3].

In this paper, we will extend the known variational Lyapunov theory to hybrid differential systems. Section 2 shows the use of this unification method via example, Section 3 provides proof of the variational comparison theorem for hybrid systems in terms of Lyapunov functions, finally, in Section 4, the variational comparison result is used in the stability analysis of hybrid differential system.

2. Unification of Nonlinear Variation of Parameters and Lyapunov Method

Consider the differential system

$$y'(t) = F(t,y);\ y(t_0) = x_0 \tag{2.1}$$

where $F \in C[R_+ \times R^n,\ R^n]$. Let $y(t,t_0,x_0)$ be the unique solution existing for $t \geq t_0$ and $0 \leq t_0 < t_1 < t_2 < t_3 < \ldots < t_k < \ldots$ such that $t_k \to \infty$ as $k \to \infty$.

Now, consider the hybrid differential system given by

$$x'(t) = f(t,x,\lambda_k(x_k));\ x_k = x(t_k), \quad t \in [t_k,\ t_{k+1}] \tag{2.2}$$

where $f \in C[R_+ \times R^n \times R^m,\ R^n]$; $\lambda_k \in C[R^n, R^m]$, $k = 0, 1, 2, 3, \ldots$

We assume existence and uniqueness of solutions of the systems

$$x'(t) = f(t, x, \lambda_k(z)), \ x(t_k) = z \tag{2.3}$$

on $t \in [t_k, t_{k+1}]$ for any fixed $z \in R^n$ and $k = 0, 1, 2, 3, 4...$

To be specific, the system will look like the following

$$x'(t) = \begin{bmatrix} x_0\,'(t) = f(t, x_0(t), \lambda_0(x_0))\,;\, x_0(t_0) = x_0\,;\, t_0 \le t \le t_1 \\ x_1\,'(t) = f(t, x_1(t), \lambda_1(x_1));\, x_1(t_1) = x_1\,;\, t_1 \le t \le t_2 \\ x_2\,'(t) = f(t, x_2(t), \lambda_2(x_2));\, x_2(t_2) = x_2\,;\, t_2 \le t \le t_3 \\ x_3\,'(t) = f(t, x_3(t), \lambda_3(x_3));\, x_3(t_3) = x_3\,;\, t_3 \le t \le t_4 \\ \cdot \\ \cdot \\ \cdot \\ x_k\,'(t) = f(t, x_k(t), \lambda_k(x_k));\, x_k(t_k) = x_k\,;\, t_k \le t \le t_{k+1} \\ \cdot \\ \cdot \\ \cdot \end{bmatrix}$$

By a solution of (2.2) we mean the following

$$x(t) = \begin{bmatrix} x_0(t), & t_0 \le t \le t_1 \\ x_1(t), & t_1 \le t \le t_2 \\ x_2(t), & t_2 \le t \le t_3 \\ x_3(t), & t_3 \le t \le t_4 \\ \cdot & \cdot \\ \cdot & \cdot \\ \cdot & \cdot \\ x_k(t), & t_k \le t \le t_{k+1} \\ \cdot & \cdot \\ \cdot & \cdot \\ \cdot & \cdot \end{bmatrix}$$

We should note that solutions of (2.2) are piecewise differentiable in each interval of $t \in [t_k, t_{k+1}]$ for any fixed $x_k \in R^n$ and for all $k = 0, 1, 2, \ldots$

We introduce the known definitions for convenience.

Definition 2.1
Let $V \in C[R^n, R_+]$, $t \in (t_k, t_{k+1})$, $x, z \in R^n$. Then we define

$$D^+V(x,z) = \limsup_{h\to 0^+} \tfrac{1}{h}[V(y(t,s+h,x+hf(s,x,\lambda_k(z)))) - V(y(t,s,x))]. \tag{2.4}$$

If $V \in C^1[R^n, R_+]$ and $y(t,s,x)$ is differentiable with respect to the initial values,

then

$$D^+V(x,z) = \tfrac{\partial V}{\partial x}(y(t,s,x))\left[\tfrac{\partial y}{\partial t_0}(t,s,x) + \tfrac{\partial y}{\partial x_0}y(t,s,x)f(s,x,\lambda_k(z))\right].$$

Definition 2.2
A function $a \in C[R_+, R_+]$ is said to belong to the class $\mathcal{K}$ if $a(u)$ is strictly increasing and $a(0) = 0$

To see the unification of the method of variation of parameters and Lyapuov method, consider the system

$$x' = G(t,x), \;\; x(t_0) = x_0 \tag{2.5}$$

where $G \in C[R_+ \times R^n, R^n]$. Assume that the solution $y(t,t_0,x_0)$ of (2.1) exists and is differentiable with respect to the initial values for $t \geq t_0$ and $V \in C^1[R_+ \times R^n, R_+]$. Then, setting $p(s) = V(s, y(t,s,x(s))$, where $x(t) = x(t,t_0,x_0)$ is any solution of (2.5) existing on $[t_0, \infty)$, we have

$$p'(s) = \tfrac{\partial V}{\partial t}(s, y(t,s,x(s))) + \tfrac{\partial V}{\partial x}(t,s,x(s))\Big[\tfrac{\partial y}{\partial t_0}(t,s,x(s)) + \tfrac{\partial y}{\partial x_0}(t,s,x(s))G(s,x(s))\Big]$$

Now, let this expression for $p'(s) \equiv \widetilde{G}\,(t,s,x(s))$ and integrate from t_0 to t. This yields

$$V(t,x(t)) = V(t_0, y(t,t_0,x_0)) + \int_{t_0}^{t} \widetilde{G}\,(t,s,x(s))\,ds, \; t \geq t_0. \tag{2.6}$$

Example 1: If $V(t,x) = |\,x\,|^2$, then this reduces to

$$|x(t)|^2 = |y(t)|^2 + \int_{t_0}^{t} 2y(t,s,x(s))\Big[\tfrac{\partial y}{\partial t_0}(t,s,x(s)) + \tfrac{\partial y}{\partial x_0}(t,s,x(s))G(t,s,x(s))\Big]ds.$$

Example 2: If $V(t,x) = x$, then this simplifies to

$$x(t) = y(t) + \int_{t_0}^{t}\Big[\tfrac{\partial y}{\partial t_0}(t,s,x(s)) + \tfrac{\partial y}{\partial x_0}(t,s,x(s))G(s,x(s))\Big]\,ds.$$

Moreover, when $G(t,x) = F(t,x) + R(t,x)$, where $R(t,x)$ is the perturbation term and we assume $\frac{\partial F}{\partial y}(t,y)$ exists and is continuous in $R_+ \times R^n$, then we arrive at the relation

$$\frac{\partial y}{\partial t_0}(t,t_0,x_0) + \frac{\partial y}{\partial x_0}(t,t_0,x_0)F(t_0,x_0) \equiv 0. \tag{2.7}$$

Using (2.6) and (2.7), we obtain the following result

$$x(t) = y(t) + \int_{t_0}^{t} \frac{\partial y}{\partial x_0}(t,s,x(s))R(s,x(s))\,ds,\ t \geq t_0, \tag{2.8}$$

which is Alexseev's nonlinear variation of parameters formula [1].

Using (2.8) to estimate $|x(t)|$ leads to estimating both, $\left|\frac{\partial y}{\partial x_0}(t,s,x(s))\right|$ and $\left|R(s,x(s))\right|$, destroying any good behavior of $R(s,x(s))$, if any exists [4]. Now, if we use the results from Example 1, it is easy to see that the expression $\frac{\partial y}{\partial x_0}(t,s,x(s))y(t.s.x(s))R(s,x(s))$ does not need to be nonnegative. Therefore, the behavior of $R(s,x(s))$ plays an important role in improving the properties of the unperturbed system [3].

3. Variational comparison result

We now prove the following variational comparison result for hybrid systems in terms of Lyapunov functions. This is a useful tool in describing the stability criteria of the hybrid system and furthermore relating the solutions of (2.2) to the solutions of (2.1).

Theorem 3.1: *Assume that*

(i) $V \in C[R^n, R_+]$, *$V(x)$ and $|y(t,s,x)|$ are locally Lipschitzian in x for each (t,s) in $R_+ \times R^n$;*

(ii) $D^+V(x,z) \leq g(s,V(x),\sigma_k(V(z)))$ *for $t \in (t_k,t_{k+1})$ and $s \leq t$, where $g \in C[R_+^3, R]$ and $\sigma_k \in C[R_+, R_+]$, $x,z \in R^n$, $k = 0,1,2,3...$;*

(iii) *$r(t,t_0,u_0)$ is the maximal solution of the hybrid comparison equation*

$$\begin{bmatrix} \frac{du}{dt} = u_k'(t) = g(t,u_k(t),\sigma_k(u_k)),\ t \in [t_k,t_{k+1}] \\ u(t_k) = u_k \end{bmatrix} \tag{2.8}$$

existing for $t \in [t_0,\infty)$.

Then for any solution $x(t) = x(t,t_0,x_0)$ of (2.2) such that $V(y(t,t_0,x_0)) \leq u_0$, we have $V(x(t)) \leq r(t,t_0,V(y(t)))$, $t \geq t_0$ where $r(t,t_0,u_0)$ and $y(t) = y(t,t_0,x_0)$ are the solutions of (2.8) and (2.1) respectively.

Proof: Let $x(t) = x(t, t_0, x_0)$ be any solution of (2.2) existing on $[t_0, \infty)$ and set $m(s) = y(t, s, x)$. Using (i) and (ii), we get the differential inequality

$$D^+m(s) \le g(t, m(s), \sigma_k(m_k)), \text{ for } t_k < s < t_{k+1},$$

where $m_k = V(y(t, t_k, x(t_k)))$.

For $t \in [t_0, t_1]$, since $m(t_0) = V(y(t, t_0, x_0)) \le u_0$, by Theorem 4.3.1 [6] we get

$$V(x_0(t)) \le r_0(t, t_0, u_0),\ t_0 \le t \le t_1 .$$

We note that $r_0(t, t_0, u_0)$ is the maximal solution of

$$u_0'(t) = g(t, u_0(t), \sigma_0(u_0)),\ \ u_0(t_0) = u_0 \ge 0,\ t \in [t_0, t_1]$$

and $x_0(t)$ is the solution of

$$x_0{}'(t) = f(t, x_0(t), \lambda_0(x_0)),\ x_0(t_0) = x_0,\ \ t \in [t_0, t_1].$$

Similarly, for $t \in [t_1, t_2]$, we get

$$V(x_1(t)) \le r_1(t, t_1, u_1),\ \ t_1 \le t \le t_2,$$

where $u_1 = r_0(t_1, t_0, V(y(t, t_0, x_0)), r_1(t, t_1, u_1)$ is the maximal solution of

$$u_1{}'(t) = g(t, u_1(t), \sigma_1(u_1)),\ \ u_1(t_1) = u_1 \ge 0,\ t \in [t_1, t_2]$$

and $x_1(t)$ is the solution of

$$x_1{}'(t) = f(t, x_1(t), \lambda_1(x_1)),\ x_1(t_1) = x_1,\ t \in [t_1, t_2].$$

Proceeding similarly for $t \in [t_k, t_{k+1}]$, we get

$$V(x_k(t)) \le r_k(t, t_k, u_k),\ \ t_k \le t \le t_{k+1},$$

where $x_k(t)$ is the solution of the differential hybrid system

$$x_k{}'(t) = f(t, x_k(t), \lambda_k(x_k));\ x_k(t_k)) = x_k \text{ for } t \in [t_k, t_{k+1}],$$

and $r_k(t, t_k, u_k)$ is the maximal solution of the comparison equation

$$u_k{}'(t) = g(t, u_k(t), \sigma_k(u_k));\ u(t_k) = u_k \text{ for } t \in [t_k, t_{k+1}],$$

where $u_k = r_{k-1}[t_k, t_{k-1}, r_{k-2}(t_{k-1}, t_{k-2}, u_{k-1})]$.

Thus, defining $r(t, t_0, u_0)$ as the maximal solution of the comparison hybrid system (2.8) as

$$r(t) = \begin{bmatrix} r_0(t,t_0,u_0); & t_0 \le t \le t_1 \\ r_1(t,t_1,u_1); & t_1 \le t \le t_2 \\ r_2(t,t_2,u_2); & t_2 \le t \le t_3 \\ r_3(t,t_3,u_3); & t_3 \le t \le t_4 \\ \cdot & \cdot \\ \cdot & \cdot \\ \cdot & \cdot \\ r_k(t,t_k,u_k); & t_k \le t \le t_{k+1} \\ \cdot & \cdot \\ \cdot & \cdot \\ \cdot & \cdot \end{bmatrix}$$

Taking $V(y(t,t_0,x_0)) = u_0$, we conclude with the desired result,

$V(x(t)) \le r(t,t_0,u_0)$. The proof is complete.
Throughout this paper, Theorem 3.1 will be referred to as the Variational Comparison Theorem for hybrid systems.

The following corollaries give a few interesting and useful special cases of Theorem 3.1 [2].

Corollary 3.1 : *If we choose* $g(t,u,\sigma_k(u_k)) = \sigma_k u_k$, $\sigma_k \ge 0$ *for each* k, *then Theorem* 3.1 *yields*

$$V(x(t)) \le [1 - \sigma_k(t - t_k)] \prod_{i=1}^{k} [1 - \sigma_{i-1}(t_i - t_{i-1})] V(y(t,t_0,x_0)), \text{ for } t \in [t_k, t_{k+1}]$$

Corollary 3.2 : *If we choose* $g(t,u,\sigma_k(u_k)) = -\alpha u + \sigma_k u_k$; $\alpha > 0$, $\sigma_k \ge 0$ *for each* k, *then Theorem* 3.1 *yields*

$$V(x(t)) \le [e^{-\alpha(t-t_k)} - \tfrac{\sigma_k}{\alpha}(e^{-\alpha(t-t_k)} - 1) \prod_{i=1}^{k} [e^{-\alpha(t_i - t_{i-1})} - \cdot \tfrac{\sigma_{i-1}}{\alpha}(e^{-\alpha(t_i - t_{i-1})} - 1)] V(y(t,t_0,x_0)),$$

for $t \in [t_k, t_{k+1}]$

4. Stability Criteria

We shall discuss the stability properties of the trivial solution of the hybrid system (2.2). Since we developed the variational comparison theorem 3.1 in terms of the hybrid comparison equation, it is easy to investigate stability criteria. For this purpose, let us suppose

$$f(t,0,\lambda_k(0)) \equiv 0 \text{ and } g(t,0,\sigma_k(0)) = 0,$$

so that we have the trivial solutions of both the hybrid and comparison hybrid systems, respectively. We can then prove the following result.

First, we recall the known definitions for convenience [6].

Definition 4.1 : The trivial solution $y \equiv 0$ of (2.1) is said to be

(i) <u>*uniformly stable*</u> , if for each $\epsilon > 0$ and $t_0 \in R_+$, there exists a positive function $\delta = \delta(\epsilon)$ such that $| x_0| < \delta$ implies $| x(t) | < \epsilon, t \geq t_0$;

(ii) <u>*quasi-uniformly asymptotically stable*</u> , if for each $\epsilon > 0$ and $t_0 \in R_+$, there exists $\delta_0 > 0$ and $T = T(\epsilon)$ such that $|x_0| < \delta_0$ implies $|x(t)| < \epsilon$ for $t \geq t_0 + T$.

Theorem 4.1: *Assume that all the conditions from Theorem 3.1 hold. Suppose further that*

$b(| x |)| \leq V(x) \leq a(| x |)$ *for* $a, b \in \mathcal{K}$ *and the trivial solution of* (2.1) *is uniformly stable.* (4.1)

Then the uniform stability properties of the trivial solution of the comparison hybrid system (2.8) *imply the corresponding uniform stability properties of the trivial solution of the hybrid system* (2.2).

The proof consists of two parts. We will first prove the uniform stability of (2.2), and then prove uniform asymptotic stability of (2.2)

Proof: Uniform stability of the trivial solution of (2.2)

Let $\epsilon > 0$ and $t_0 \in R_+$ be given. Suppose the trivial solution $u = 0$ of (2.8) is uniformly stable. Then, given $b(\epsilon) > 0$ and $t_0 \in R_+$, there exists a $\delta_1(\epsilon) = \delta_1 > 0$ such that

$$0 \leq u_0 < \delta_1 \text{ implies } u(t,t_0,u_0) < b(\epsilon), \quad t \geq t_0, \tag{4.2}$$

for $u(t,t_0,x_0)$ being any solution of (2.8).

Now since the trivial solution (2.1) is uniformly stable, given $\eta(\epsilon) = \eta > 0$ and $t_0 \in R_+$, there exists $\delta(\epsilon) = \delta > 0$ such that

$$|x_0| < \delta \text{ implies } |y(t, t_0, x_0)| < \eta, \quad t \geq t_0. \tag{4.3}$$

Choosing $\eta = a^{-1}(\delta_1)$, we claim that the trivial solution of (2.2) is uniformly stable. Thus, given $\epsilon > 0$, there exists $\delta(\epsilon) > 0$ such that

$$|x_0| < \delta \text{ implies } |x(t)| < \epsilon, \quad t \geq t_0.$$

If this is not true, there would exist a solution $x(t) = x(t, t_0, x_0)$ of (2.2) with $|x_0| < \delta$ and $t^* > t_0$ satisfying

$$|x(t^*)| = \epsilon \text{ and } |x(t)| < \epsilon \text{ for } t_0 \leq t \leq t^*. \tag{4.4}$$

Then by Theorem 3.1, we get

$$V(x(t)) \leq r(t, t_0, u_0), \quad t_0 \leq t \leq t^*,$$

where $r(t, t_0, u_0)$ is the maximal solution of (2.8).

Now, (4.1- 4.4) from above yield

$$\begin{aligned} b(\epsilon) = b|x(t^*)| \leq V(x(t^*)) &\leq r(t^*, t_0, V(y(t^*, t_0, x_0))) \\ &\leq r(t^*, t_0, a(|y(t^*, t_0, x_0)|)) \\ &\leq r(t^*, t_0, \delta_1) < b(\epsilon). \end{aligned}$$

This contradiction proves $x \equiv 0$ of (2.2) is uniformly stable.

We will now prove uniform asymptotic stability of the trivial solution of (2.2)

Let us suppose next that $u \equiv 0$ of (2.8) is uniformly asymptotically stable. Then this trivial solution is also uniformly stable. Thus, letting $\epsilon = \rho$ and $\delta_0 = \delta(\rho)$, uniform stability of $u \equiv 0$ of (2.8) yields

$$|x_0| < \delta_0 \text{ implies } |x(t)| < \rho, \quad t \geq t_0.$$

In addition to uniform stability of $u \equiv 0$ of (2.8), solution is uniformly attractive. Thus, given $\epsilon > 0$ and $t_0 \in R_+$, there exists $\delta^* > 0$ and $T = T(\epsilon) > 0$ such that

(i) $\quad 0 \leq u_0 < \delta^*$ implies $u(t, t_0, u_0) < b(\epsilon), \quad t \geq t_0 + T$

By Theorem 3.1, we have

(ii) $\quad V(x(t)) \leq r(t, t_0, V(y(t, t_0, x_0)), \ t \geq t_0.$

Since the trivial solution of (2.1) is uniformly stable,

$$(iii)\quad |x_0| < \delta_0 \text{ implies } |y(t, t_0, x_0)| < \eta_0, \quad t \geq t_0$$

for $\delta_0 = \delta(\rho)$ and $\eta_0 = a^{-1}(\delta_1(\rho)) = a^{-1}(\delta^*)$ for $\delta^* = \delta_1(\rho)$.

Now suppose that there exists a sequence $\{\tau_n\}$ such that $\tau_n \to \infty$ as $n \to \infty$, $\tau_n \geq t_0 + T$ and $|x(\tau_n)| \geq \epsilon$ for any solution $x(t)$ of (2.2) with $|x_0| < \delta_0$.

Then $(i) - (iii)$ from above yields

$$b(\epsilon) \leq b(|x(\tau_n)|) \leq V(x(\tau_n)) \leq r(\tau_n, t_0, a(|y(\tau_n, t_0, x_0)|)) \leq r(\tau_n, t_0, \delta^*) < b(\epsilon)$$

for $\tau_n \geq t_0 + T$. This contradiction proves the uniform asymptotical stability of the trivial solution of (2.2). The proof is complete.

The following corollaries correspond to the original Lyapunov theorems on stability in the present hybrid system framework and give a few interesting and useful special cases of Theorem 4.1 [2].

Corollary 4.1 : If we choose $g(t, u, \sigma(u_k)) = \sigma_k u_k$; $\sigma_k \geq 0$ for each k, then Theorem 4.1 yields uniform stability of $x \equiv 0$ of (2.2) provided the infinite series $\sum_{i=1}^{\infty} [\sigma_{i-1}(t_i - t_{i-1})]$ converges.

Proof: If the series $\sum_{i=1}^{\infty}[\sigma_{i-1}(t_i - t_{i-1})]$ converges, then the infinite product $\prod_{i=1}^{\infty}[1 + \sigma_{i-1}(t_i - t_{i-1})]$ converges as well. Let $\prod_{i=1}^{\infty}[1 + \sigma_{i-1}(t_i - t_{i-1})] = N$.

By Corollary 3.1, we get

$$u(t, t_0, V(y(t, t_0, x_0)) = [1 + \sigma_k(t_i - t_k)] \prod_{i=1}^{k}[1 + \sigma_{i-1}(t_i - t_{i-1})] \; V(y(t, t_0, x_0))$$

for $t \in [t_k, t_{k+1}]$, $k = 0, 1, 2, \ldots$ It then follows that

$$u(t, t_0, V(y(t, t_0, x_0)) \leq N\, V(y(t, t_0, x_0) \leq N\, a(y(t, t_0, x_0)) \leq N\delta_1 \leq b(\epsilon),$$

which implies uniform stability of $x \equiv 0$ of (2.2) and the proof is complete.

Corollary 4.2 : If we choose $g(t, u, \sigma(u_k)) = -C(u(t)) + \sigma_k u_k$; $C \in \mathcal{K}$, $\sigma_k \geq 0$ for each k, then Theorem 4.1 yields uniform asymptotic stability of $x \equiv 0$ of (2.2) provided the infinite series $\sum_{i=1}^{\infty} [\sigma_i(t_i - t_{i-1})]$ converges.

Proof: Let us assume the series $\sum_{i=1}^{\infty} [\sigma_i(t_i - t_{i-1})]$ converges. Thus, we will let
$M = \sum_{i=1}^{\infty} [\sigma_i(t_i - t_{i-1})]$.
From the conditions in Theorem 3.1,

$$\begin{aligned} D^+V(t, x, z) &\leq g(t, V(x(t)), \sigma_k(V(z)) \\ &\leq -C(V(x(t)) + \sigma_k V(z) \\ &\leq \sigma_k(V(z)). \end{aligned}$$

By Corollary 4.1, we have uniform stability of the trivial solution $u \equiv 0$ of the hybrid system

$$\begin{bmatrix} u'(t) = -C(u(t)) + \sigma_k u_{k,} t \in [t_k, t_{k+1}] \\ u_k = u(t_k),\ u(t_0) = u_0 \end{bmatrix} \tag{4.5}$$

Thus, for a fixed $\rho > 0$ there exists an $\eta = \eta(\epsilon) > 0$
such that

$$0 \leq u_0 < \eta \quad \text{implies } u(t, t_0, u_0) < \rho,\ t \geq t_0, \tag{4.6}$$

where $u(t, t_0, u_0)$ is any solution of (4.5) satisfying (4.6). Given $\epsilon \in (0, \rho)$, there exists a $\delta = \delta(\epsilon)$ such that, with this delta, we have uniform stability.

To prove uniform asymptotic stability, it is enough to show uniform attractivity. That is, whenever $u_0 < \eta$ and $u(t, t_0, u_0)$ any solution of (4.5), there exists a $t^* \in [t_0, t_0 + T]$ where $T = T(\epsilon) = \frac{M+\eta}{C(\delta(\epsilon))} + 1$ such that

$$|u(t^*, t_0, u_0)| < \delta\,,\ t^* \in [t_0, t_0 + T].$$

If not, then

$$|u(t,t_0, u_0)| \geq \delta\,,\ t \in [t_0, t_0 + T].$$

Integrating from t_0 to $t_0 + T$ yields

$$\begin{aligned} u(t_0 + T) &= u(t_0) - \int_{t_0}^{t_0+T} C(u(s))ds + \sum_{i=1}^{\infty} [\sigma_i(t_i - t_{i-1})] \\ &\leq \eta - C(\delta)T + M. \end{aligned}$$

Now, our choice of T leads to a contradiction. Thus $u(t^*, t_0, u_0) < \delta(\epsilon)$ implies uniform asymptotic stability of $u \equiv 0$ and then Theorem 4.1 yields uniform asymptotic stability of $x \equiv 0$ of (2.2). The proof is complete.

References

[1] Lakshmikantham, V. and Leela, S., *Differential and Integral Inequalities* Vol. I, Academic Press, New York 1969.

[2] Lakshmikantham, V. and Liu, X. *Impulsive Hybrid Systems and Stability Theory* (to appear)

[3] Clarke, F.H., Ledyaev, Yu.S., Sontag, E.D. and Subbotin, A.I., *Asymptotic controllability implies feedback stabilization.* (to appear).

[4] Lakshmikantham, V., Liu, X., and Leela, S., *Variational Lyapunov Method and Stability Theory* (to appear)

[5] Lakshmikantham, V. and Deo, S.G., *Method of Variation of Parameters for Dynamic Systems*, Gordon and Breach Science Publishers, Canada 1998.

[6] LakshmikanthamV., Leele, S., and Martynyuk, A.A., *Stability Analysis of Nonlinear Systems*, Marcel Dekker, INC., New York 1989.

Proceedings of Dynamic Systems and Applications 3 (2001) 545-552

A Time-Dependent Model for Pumping Dynamics in Solid State Lasers

Lila F. Roberts
Georgia Southern University, Statesboro, GA 30458-8093

ABSTRACT: The temporal output of solid state lasers has been investigated using a set of nonlinear ordinary differential equations that relate the spatially averaged population densities of the electronic levels and the averaged photon density within the laser cavity. Previous models do not directly account for the dynamics induced by depletion of the pumping photons as they move through the active material. To investigate the evolution of the dynamic quantities under these conditions, the model is reformulated to account for changes in the pumping photons due to absorption. In this study, several qualitative results are presented as well as a numerical investigation of the qualitative behavior of solutions.

AMS(MOS) subject classification. 34C60, 37N20

1. INTRODUCTION

Over the past decade, lasers have become increasingly important in science and medical applications. Solid state lasers have potential for meeting performance requirements for various applications [1]-[2]. Operational parameters are often controlled using various intracavity elements and injection seeding. To design a stable and efficient laser and to understand the effect on laser output of these intracavity elements, it is necessary to gain an understanding of the development and long-term behavior of the dynamical processes in the laser.

Investigations of the evolution of the spatially averaged dynamic variables in solid state lasers have been studied in [3]-[7] using systems of coupled nonlinear ordinary differential equations. These equations describe the development of the population densities at the electronic levels of the active material and the photon density within the laser cavity. Previous models do not directly account for absorption of the pumping energy as the pump pulse propagates through the active material. In this study the mathematical model developed by Buoncristiani et al [5] is reformulated to account for the spatially averaged dynamics of the pump pulse.

In the next section, the equations describing the evolution of the population and photon densities in a typical straight laser cavity are discussed. We assume that the laser material is end-pumped and we include the temporal evolution of the pumping photons to complete the model. Several qualitative results are established using parameters consistent with a Titanium-doped sapphire prototype. A numerical study of parameters leading to local stability of solutions is presented.

2. RATE EQUATIONS

2.1 Electronic Populations

In a four-level laser, ions at the ground state ($i = 0$) are excited to the highest energy level ($i = 3$) at a rate $W_{03}(t)$ (time^{-1}). Let $n_i(t)$, $i = 0, 1, 2, 3$ denote the normalized density of active ions at the ith energy level and let $n(t)$ denote the normalized population inversion, $n_2(t) - \frac{g_2}{g_1} n_1(t)$. The rate equations that govern the evolution of spatially averaged densities of active ions in the an idealized four level laser system are

$$\begin{aligned} \frac{dn(t)}{dt} &= c_{11}n(t) + c_{12}n_1(t) - \gamma\beta n(t)\phi(t) + W(t)n_0(t) \\ \frac{dn_1(t)}{dt} &= c_{21}n(t) + c_{22}n_1(t) + \beta n(t)\phi(t) \end{aligned} \tag{1}$$

The normalized photon density in the material is denoted by $\phi(t)$. $W(t)$ is the rate at which ions are excited to the upper state.

The parameters in (1) are algebraic combinations of the laser system parameters:

$$\gamma = 1 + \frac{g_2}{g_1} \qquad \beta = \sigma v N_T$$

$$c_{11} = -\left(\frac{1}{\tau_2} + \frac{\gamma - 1}{\tau_{fl}}\right) \qquad c_{12} = (1-\gamma)\left(-c_{11} - \frac{1}{\tau_1}\right)$$

$$c_{21} = \frac{1}{\tau_{fl}} \qquad c_{22} = \frac{\gamma - 1}{\tau_{fl}} - \frac{1}{\tau_1}$$

where τ_i, $i = 1, 2, 3$, represents the lifetime (ns) of an ion at the ith energy level and τ_{fl} is the fluorescence lifetime of the material. The cross section for stimulated emission (cm^2) at the peak of the emission spectrum is denoted by σ and the degeneracy at the ith energy level by g_i. The speed of light in the material is represented by $v = \frac{c}{\eta}$, where c is the speed of light in a vacuum (cm/sec) and η is the index of refraction of the material. N_T represents the total concentration of active ions per cm^3.

2.2 Photon Density

The normalized photon density, $\phi(t)$, at the peak emission wavelength evolves due to spontaneous and stimulated emission; losses to ϕ occur as a result of absorption, scattering, and output to the cavity. The governing equation for photon density in the active medium is

$$\frac{d\phi(t)}{dt} = -\frac{1}{\tau_c}\phi(t) + \beta_\ell n(t)\phi(t) \tag{2}$$

where $\beta_\ell = \frac{\ell}{L}\beta$. Here, ℓ and L represent the lengths of the active material and the cavity, respectively. The parameter τ_c (time^{-1}) is the lifetime of a photon in the cavity, computed [5] for a straight laser cavity,

$$\tau_c = \frac{2L}{c}\left(2\alpha\ell - \ln R_1 R_2 T_s^2\right)^{-1} \tag{3}$$

where α (cm^{-1}) is the absorption coefficient at the peak of the emission spectrum. R_1 and R_2 are the reflectivities of the mirrors and T_s is the Fresnel coefficient for the material. In this study we assume that $\ell = L$, i.e. the material completely fills the cavity.

We assume that initially a small number of photons is present in the cavity, $\phi(t) > 0$, so that the photon density may evolve.

2.3 Pumping Dynamics

In this study we assume the laser is end-pumped. To account for propagation of the pump flux through the material, we assume that the pumping photons are available for only one pass through the material. We assume that pumping photons propagate along the optic axis (z-axis) and interactions transverse to the z axis are neglected. The equation that describes the pump flux, $W_f(t,z)$ at time t and position z is (Ganiel [3])

$$\frac{\partial W_f(t,z)}{\partial t} + v\frac{\partial W_f(t,z)}{\partial z} = -v\sigma_{ab}N_T n_0(t,z)W_f(t,z) \tag{4}$$

where σ_{ab} (cm^2) is the cross section for stimulated absorption.

The pump density, W_p, is related to the flux by $W_f = vW_p$ and the rate at which ions at time t and position z are excited is given by $W(t,z) = \sigma_{ab}W_p(t,z)$. Thus (4) can be written as

$$\frac{\partial W(t,z)}{\partial t} + v\frac{\partial W(t,z)}{\partial z} = -v\sigma_{ab}N_T n_0(t,z)W(t,z) \tag{5}$$

The material is end-pumped. Let $W_E(t)$ represent the rate at which pumping photons enter the material so that $W(t,0) = W_E(t)$.

To average (5) over the length of the active material, let $W(t) = \frac{1}{\ell}\int_0^\ell W(t,z)dz$.

Making the usual assumption (Tarasov [8]) that

$$\frac{1}{\ell}\int_0^\ell n_0(t,z)W(t,z)dz \approx \left(\frac{1}{\ell}\int_0^\ell n_0(t,z)dz\right)\left(\frac{1}{\ell}\int_0^\ell W(t,z)dz\right) \tag{6}$$

we obtain the ordinary differential equation

$$\frac{dW(t)}{dt} = -v\sigma_{ab}N_T n_0(t)W(t) + \frac{v}{\ell}\left(W_E(t) - W(t,l)\right). \tag{7}$$

We approximate $W(t, \ell)$ using the fact that pump intensity decreases exponentially with z (Lalanne et al [9]) so that

$$W(t,z) \approx W(t,0)\exp(-\sigma_{ab}N_T z)$$
$$= W_E(t)\exp(-\sigma_{ab}N_T z)$$

and at $z = \ell$, $W(t,\ell) = W_E(t)$.

Thus, (7) can be now be written as

$$\frac{dW(t)}{dt} = -\beta_p n_0(t)W(t) + kW_E(t) \tag{8}$$

where $k = \frac{v}{\ell}(1-\exp(-\sigma_{ab}N_T\ell))$ and $\beta_p = v\sigma_{ab}N_T$.

Equations (1), (2), and (8) constitute the model.

3. QUALITATIVE RESULTS

In this section we establish qualitative behavior predicted by the model. Theorem 1 establishes that under conditions consistent with those present in solid state lasers, the physical variables remain nonnegative as long as they are initially nonnegative, thus demonstrating that the model predicts physically reasonable behavior.

We assume that the physical parameters satisfy

$$\begin{gathered} 0 < \tau_1 << \tau_2 < \tau_{fl} \\ 0 < \tau_c \\ 1 < \gamma \\ 0 < \beta_p \end{gathered} \tag{9}$$

and that initially,

$$n(0) \geq 0,\ n_1(0) \geq 0, n_0(0) \geq 0,\ \phi(0) \geq 0. \tag{10}$$

Note that since $n_0(0) = 1 - n(0) - \gamma n_1(0)$, $n_0(0) \geq 0$ implies that $0 \leq n(0) + \gamma n_1(0) \leq 1$.

In Roberts [4], the model consisting of (1) and (2) was studied in detail. In particular, nonnegativity of the dynamic variables was established when the pumping rate was continuous and nonnegative on $[0, \infty)$. To apply the result to the model consisting of (1), (2) and (8), it is sufficient to demonstrate nonnegativity of W.

Lemma 1. Suppose that the physical parameters and initial conditions for the model (1), (2), and (8) satisfy (9) and (10), respectively. Suppose $W_E = W_E(t)$ is continuous and positive on $[0, \infty)$ and suppose that $W(0) \geq 0$. Then $W = W(t) \geq 0$ on $[0, \infty)$.

Proof. We write (8) as

$$\frac{dW}{dt} + \beta_p n_0 W = kW_E. \tag{11}$$

On $[0, \infty)$, (11) has solution

$$\begin{aligned} W(t) &= \exp\left[-\int_0^t \beta_p n_0(\tau)d\tau\right]W(0) + k\int_0^t \exp\left[-\int_s^t \beta_p n_0(\tau)d\tau\right]W_E(s)ds \\ &\geq 0. \end{aligned} \tag{12}$$

This proves the lemma. ■

The proof of Theorem 1 now follows from Roberts [4].

Theorem 1. Suppose that the physical parameters and initial conditions for the model (1), (2), and (8) satisfy (9) and (10), respectively. Suppose $W_E = W_E(t)$ is continuous and positive on $[0, \infty)$ and suppose that $W(0) \geq 0$. Then, $n(t) \geq 0, n_1(t) \geq 0, n_0(t) \geq 0, \text{and } \phi(t) \geq 0$ for $t > 0$.

We now prove that if the input pumping rate is integrable or bounded, so is the rate at which pumping photons are available for excitation. This result will enable us to utilize the proof in Roberts [4] to show that if the input pumping rate is integrable or bounded, so are the physical variables in the laser system. Integrability guarantees that if the pumping rate dies off for large t, so do the dynamics in the laser. Boundedness of the input guarantees that output cannot become unbounded. These properties demonstrate that the model predicts that the solutions are consistent with the physical behavior of an actual laser system under similar conditions.

Lemma 2. Assume that conditions (9) and (10) hold. Suppose that $W_E(t)$ is continuous and positive on $[0, \infty)$.

(a) Suppose $W_E(t)$ is integrable on $[0, \infty)$. Then, if $W(0) \geq 0$, $W(t)$ is integrable on $[0, \infty)$.

(b) Suppose $W_E(t)$ is bounded on $[0, \infty)$. Then, if $0 < W(0) < \infty$, $W(t)$ is bounded on $[0, \infty)$.

Proof. Theorem 1 guarantees that if (9) and (10) hold, then $0 \leq n + \gamma n_1 \leq 1$ for all t.

(a) From (8),

$$\begin{aligned} \frac{dW}{dt} &= -\beta_p n_0 W + kW_E \\ &= -\beta_p(1-(n+\gamma n_1))W + kW_E \\ &\leq -\beta_p W + kW_E \end{aligned} \tag{13}$$

Thus,

$$W(t) \leq e^{-\beta_p t}W(0) + k\int_0^t e^{-\beta_p(t-\tau)}W_E(\tau)d\tau \tag{14}$$

Integrating,

$$\begin{aligned} \int_0^s W(t)dt &\leq W(0)\int_0^s e^{-\beta_p t}dt + k\int_0^s \int_0^t e^{-\beta_p(t-\tau)}W_E(\tau)d\tau dt \\ &\leq \frac{1}{\beta_p}W(0)(1-e^{-\beta_p s}) + k\int_0^s \int_\tau^s e^{-\beta_p(t-\tau)}W_E(\tau)dtd\tau \\ &= \frac{1}{\beta_p}W(0)(1-e^{-\beta_p s}) + \frac{k}{\beta_p}\int_0^s (1-e^{-\beta_p s})W_E(\tau)d\tau \\ &\leq \frac{1}{\beta_p}W(0) + \frac{k}{\beta_p}\int_0^s W_E(\tau)d\tau \end{aligned} \tag{15}$$

Integrability of W on $[0, \infty)$ now follows from the integrability of W_E.

(b) From (14), and boundedness of $W_E(t)$, i.e. $W_E(t) \leq M$ for all t, it follows that

$$
\begin{aligned}
W(t) &\le e^{-\beta_p t}W(0) + k\int_0^t e^{-\beta_p(t-\tau)}W_E(\tau)d\tau \\
&\le W(0) + \frac{kM}{\beta_p}\left(1-e^{-\beta_p t}\right) \\
&\le W(0) + \frac{kM}{\beta_p}
\end{aligned}
$$

This shows that W is bounded on $[0, \infty)$. ■

The proof of Theorem 2 now follows from Roberts [4].

Theorem 2. Suppose that the physical parameters and initial conditions for the model (1), (2), and (8) satisfy (9) and (10), respectively. Suppose $W_E = W_E(t)$ is continuous and positive on $[0, \infty)$ and suppose that $0 \le W(0) \le \infty$. Then,

(a) if $W_E(t)$ is integrable on $[0, \infty)$, so are $n(t)$, $n_1(t)$, and $\phi(t)$.

(b) if $W_E(t)$ is bounded on $[0, \infty)$, so are $n(t)$, $n_1(t)$, and $\phi(t)$.

4. LOCAL STABILITY RESULTS

The case where W_E is constant corresponds to continuous wave (cw) operation of the laser system. In this case, the system (1), (2), and (8) has two equilibrium solutions, given in Table 1.

$$n^1 = \frac{-c_{22}kW_E}{\beta_p\left(c_{11}c_{22} - c_{12}c_{21}\right)} \qquad n^2 = \frac{-c_{33}}{\beta}$$

$$n_1^1 = -\frac{c_{21}}{c_{22}}n^1 \qquad n_1^2 = \frac{c_{33}\beta_p(c_{11} + \gamma c_{21}) - k\beta W_E}{\beta\beta_p(c_{12} + \gamma c_{22})}$$

$$\phi^1 = 0 \qquad \phi^2 = \frac{c_{21}c_{33} - \beta c_{22}n_1^2}{-c_{33}}$$

$$W^1 = \frac{kW_E}{\beta_p\left(1 - n^1 - \gamma n_1^1\right)} \qquad W^2 = \frac{kW_E}{\beta_p(1 - n^2 - \gamma n_1^2)}$$

where $k = \frac{v}{\ell}\left(1 - \exp(-\sigma_{ab}N_T\ell)\right)$ and $\beta_p = v\sigma_{ab}N_T$.

Table 1. Equilibrium points for system (1), (2), (8).

Utilizing the Routh-Hurwitz criteria for linearized systems, it follows that a necessary and sufficient condition for asymptotic stability of the first equilibrium solution is

$$\beta n^1 - \frac{1}{\tau_c} < 0 \tag{16}$$

Laser action cannot commence unless photon density increases from zero, i.e. the first equilibrium point must be unstable. Moreover, to achieve laser oscillation, we expect the second equilibrium point to be asymptotically stable.

Investigating the stability conditions for the second equilibrium point by invoking the Routh-Hurwitz criteria is extremely difficult, however, as in Wangler et al [6], we utilize a numerical investigation to estimate a minimum pumping rate required for the active variables to reach a nonzero steady state.

In the numerical investigation, we used physical parameter values for titanium doped sapphire (Ti^{3+}:Al_2O_3) given in Buoncristiani et al [5]. In Fig. 1, we investigate the stability condition (16) as a function of W_E. We see that as W_E increases from zero, $\beta n^1 - \frac{1}{\tau_c}$ is negative, but increasing. Denote by W^* the pumping rate for which $\beta n^1 - \frac{1}{\tau_c} = 0$. For $W_E < W^*$, the first equilibrium point ($E1$) is asymptotically stable and the second ($E2$) is unstable. At $W_E = W^*$, the two equilibrium points coincide. This is the threshold pumping rate. For $W_E > W^*$, there is an interchange in the stability of the equilibrium points. In the numerical experiments, with $\tau_c = 0.15779\text{ns}$, the threshold condition $W^* \approx 5.9\times10^{-5}$ (ns^{-1}).

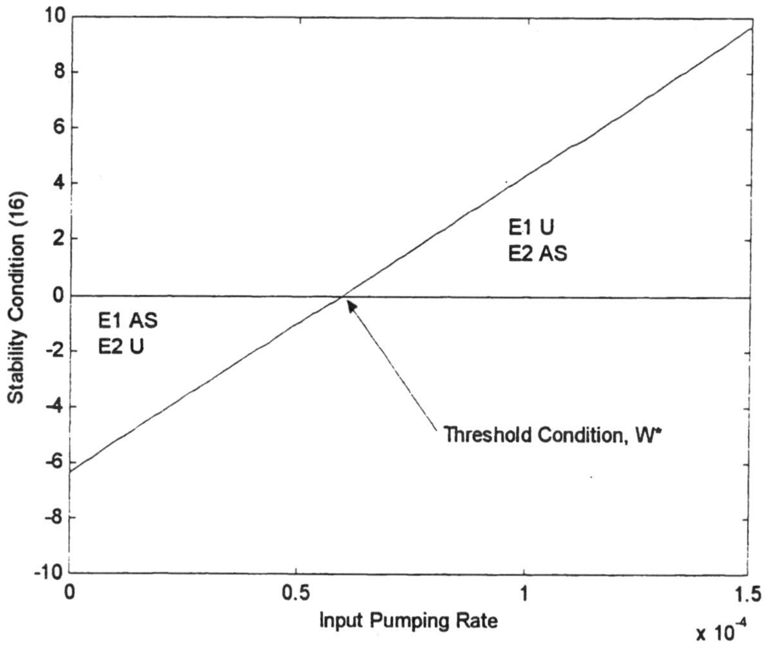

Figure 1. Stability condition (16) as a function of input pumping rate W_E.

It is also of interest to compare the predicted values of the steady states in Table 1 to the computed values. In Table 2, numerical values for the second equilibrium point for $W_E \approx 0.01088$ ns^{-1}, corresponding to a pumping energy of 30 mJ, are compared to the steady state values as computed using a standard ODE solver in MATLAB. Agreement in the predicted values and the computed values is quite good, indicating that the ODE solver effectively predicts long-term behavior.

Coordinate	Predicted Value	Computed Value
Population Inversion	0.07704160021707	0.07704155876471
Lower Level	0.00447980808146	0.00447980874685
Photon	0.00070363402682	0.00070363445004
Pumping Rate	0.00489849662767	0.00489912509983

Table 2. Comparison between predicted and computed steady states. In this computation, $W_E \approx 0.01088\ ns^{-1}$.

5. CONCLUSIONS AND FUTURE INVESTIGATIONS

In this study, we have developed a model for the dynamics of a solid state laser that accounts for propagation of the pumping photons across the active medium. We have demonstrated that the model predicts results that are physically reasonable and in qualitative agreement with previous models. We estimated a pumping threshold for parameters consistent with Ti^{3+}:Al_2O_3 lasers, however, the study is sufficiently general to have application to other four-level solid state laser systems. We have also demonstrated that the numerical procedure for computing solutions effectively captures long-term behavior for a range of input pumping values above threshold.

The strategy for estimating the pumping rate available for excitation was based on assuming that pumping photons (1) traveled in only one direction and (2) were not available for excitation after exiting the material. Future study will involve development and analysis of a model that accounts for left and right traveling pumping photons and an investigation of impulsive behavior of the pumping photons.

REFERENCES

1. Apolonskii, A. A., Baraulya, V. I., Zinnatov, E I Zinin, 'Actively mode-locked Ti:sapphire laser,' **Quantum Electronics**, Vol 27, No 9 (1997), 756-758.
2. Sarukura, N. & Ishida, Y., 'Ultrashort pulse generation from a passively mode-locked Ti:sapphire laser based system,' IEEE Journal of Quantum Electronics, Vol 28, No 10 (1992), 2134-2141.
3. Ganiel, U., Hardy, A., & Neumann, G., 'Amplified spontaneous emission and signal amplification in dye-laser systems,' IEEE Journal of Quantum Electronics, Vol QE-11, No 11 (1975), 881-892.
4. Roberts, Lila F., **A Mathematical Model of the Dynamics in an Optically Pumped Four-Level Solid State Laser System**, Doctoral Dissertation , Old Dominion University, 1988.
5. Buoncristiani, A. M., Roberts, L. F., & Swetits, J. J., 'Model of an end-pumped, injection seeded solid state laser,' **Mathematical and Compter Modelling**, Vol 12, No 3 (1989), 303-312.
6. Wangler, T. G., Swetits, J. J., & Buoncristiani, A. M., 'Temporal model of an optically pumped co-doped solid state laser,' **Mathematical and Computer Modelling**, Vol 17, No 6 (1993), 67-82.
7. Moulton, Peter F., 'An investigation of the Co:MgF_2 laser system,' IEEE Journal of Quantum Electronics, Vol QE-21, No 10 (1985), 1982-1595.
8. Tarasov, L. V., **Laser Physics**, MIR Publishers, Moscow, 1983.
9. Lalanne, J. R., Ducasse, A., & Kielich, S., ***Laser-Molecule Interaction***, Wiley, New York, 1996.

Proceedings of Dynamic Systems and Applications 3 (2001) 553-560

Mathematical modelling and analysis for low frequency design of transformers, in cables, transmission lines

Kumud Singh-Altmayer[1] and Michèle Vanmaele[2]
[1]Georgia Southern University, Math. and Comp. Sc. Dept., Statesboro, GA 30460
[2]University of Ghent, Dept. of Applied Math. and Computer Sc., Ghent, Belgium

ABSTRACT: Using numerical and analytical methods, magnetic field intensity as well as electric field intensity are calculated in terms of the vector potential for the configuration of a shell containing three conductors. The current density distribution may be uniform or non-uniform. Eventually the field flux may change which in turn creates power losses inside the conducting media. We are interested in the exact losses or an approximation of these losses and their effect on the performance of the designed model for a transformer.
AMS (MOS) subject classification. 45L10, 65R20, 78A25

1. INTRODUCTION

This paper will deal with the so-called eddy currents, which are of great interest to electrical engineers since their appearance in power systems is undesirable. In fact, eddy currents are one of the main problems encountered in designing electrical apparatus. In particular as the energy density of power transmission is increasing, the eddy current loss problems are becoming more and more important. However for some applications eddy currents are desired and are consciously generated, as for example for induction heating. In either case the determination of the eddy current density distribution, and from this the loss produced, is the main subject of research. To stress the importance of this research we note that not more than 40% of heat (or sometimes even less) is converted into electrical energy. Thus if one can save one joule of energy in an electrical load, one saves in fact on average three times this energy content in the form of fossil fuels.

2. NATURE OF EDDY CURRENTS

Eddy currents have been distinguished as a separate entity in the field of electromagnetism since Jean Bernard Leon Foucault (1819 - 1868) (French physicist) first discovered their nature. In fact, they can be deduced from Faraday's law as well as from Maxwell's equations (see Tegopoulos & Kriezis [11]), although Maxwell himself did not focus his attention on this problem. Nevertheless, the eddy current problem can be considered as a result of electromagnetic induction but with the electromotive force induced in a massive conductor instead of a circuit made of wires or filaments. More explicitly, Faraday's law of induction applied to linear circuits gives rise to induced voltages, and these voltages cause currents which flow in the corresponding circuits. Eddy currents are produced by a similar process but in extended conductors.

Eddy currents manifest themselves as skin and proximity effects. In the case of the skin effect, the eddy current produced or redistribution of the time varying current flowing through the cross-section of a certain conductor, as compared with the flow of direct current, is due to its own field. The current tends to flow near the surface of the conductor, and the loss is higher than the case where direct current flows in the same conductor. The proximity effect takes place when a conductor is subjected to a time-varying field existing in its neighbourhood. Usually this field comes from an adjacent time-varying current-carrying conductor. This field induces eddy currents regardless of whether the first conductor carries current or not. If it does, then skin effect eddy currents and proximity effect eddy currents superimpose to form the total eddy current distribution.

3. THE MATHEMATICAL MODEL

3.1. Configuration and assumptions

We consider three tubular conductors inside a big shell composing an equilateral configuration as in figure 1. In Kawasaki et al. [3] the cradle and the trefoil configurations are studied.

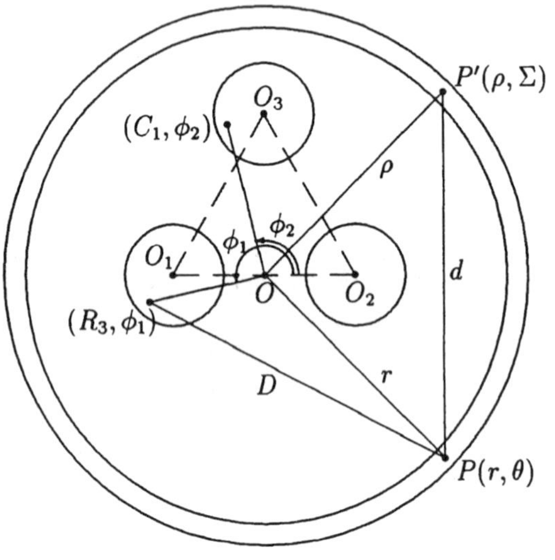

Figure 1: Configuration of a conductor/shell containing three conductors.

The centres O_i, $i = 1,2,3$, of the conductors G_i, $i = 1,2,3$ are the edges of an equilateral triangle with side length R_2. The centre O of the outer shell lies on the side O_1O_2 at a distance R_1 from the edges O_1 and O_2 and at a distance C_2 from O_3. We assume that the electrical excitation is sinusoidal with angular frequency $\omega = 2\pi f$, so as to obtain a steady-state electromagnetic problem. We restrict ourselves to the case of low frequency problems, for example $f = 50 - 60$ Hz. This means that the frequency has a wavelength which is large compared with the dimensions of the apparatus. For low frequency problems it is valid to assume that the displacement current is neglected. This assumption is equivalent to the assumption that electric and magnetic fields are propagated instantaneously. As a consequence the charge distribution on the conducting surface of the material is absent.

Further we assume that the conducting media are magnetic with constant permeability $\mu > 1$, as for example steel and copper. Also, the conducting media within a conductor are homogeneous and isotropic. The conducting objects are treated macroscopically, in other words no atomic effects are considered.
The length of the shell as well as of the conductors inside of it is taken to be infinite, so as to obtain a two-dimensional problem. The current density will have in that case only one component in the z-direction.

3.2. Integral formulation and notations

Let's denote by $\vec{E}_O$ the electric field at the centre O of the outer shell. $\vec{A}_O$ represents the magnetic vector potential at O and $\vec{A}_P$ stands for the total sum of magnetic potential at a point P in the outer shell chosen as depicted in figure 1.
The current density vectors $\vec{J}_1$, $\vec{J}_2$ and $\vec{J}_3$ are respectively due to the presence of conductors inside the outer shell's field, to the field current on the inner conductors and to the interaction of the inner conductors affecting each other's fields.
We will use the integral form of Faraday's law to find these current density vectors. For the case of time-harmonic fields this law is written as follows (see Sadiku [8]):

$$\oint \vec{E} \cdot d\vec{\ell} = -j\omega \oint \vec{A} \cdot d\vec{\ell} \qquad \text{with } j = \sqrt{-1}\,.$$

For the sake of simplicity we consider G_i, $i = 1, 2, 3$, as conductors rather than as shells, because in case of shells fields both inside the outer boundaries as well as outside the shells should be considered.
Further we have symmetry of the current density and of the vector potential in view of azimuthal symmetry which is common for this type of problem when conductors and shells have infinite length (see Krawczyck & Tegopoulos [4], Kriezis [5], Tegopoulos & Kriezis [11]). This means

$$\vec{J}_i(r, \theta + \pi) = -\vec{J}_i(r, \theta)\,, \qquad i = 1, 2, 3\,.$$

To simplify the notations in what follows, we introduce $\vec{J}_4 := \vec{J}_2 + \vec{J}_3$. Moreover,

$$\vec{J}_2 = \vec{J}_{G1} + \vec{J}_{G2} + \vec{J}_{G3}, \quad \vec{J}_{G1} = \vec{J}_{G2} = \vec{J}_{G3},$$

where $\vec{J}_{G_i}$ is the current density due to the field current on conductor G_i. Next we define the vector potentials as follows:

$$\vec{A}_P(r, \theta) = \frac{\mu_0}{2\pi} \int_{S_O} \vec{J}_1(\rho, \Sigma) \ln(d)\, \mathrm{d}S(\rho, \Sigma) - \frac{3\mu_0}{2\pi} \int_{S_G} \vec{J}_2(R_3, \phi_1) \ln(D)\, \mathrm{d}S(R_3, \phi_1) \quad (1)$$

$$\begin{aligned} \vec{A}_O(r, \theta) \;=\; & \frac{\mu_0}{2\pi} \int_{S_O} \vec{J}_1(\rho, \Sigma) \ln(\rho)\, \mathrm{d}S(\rho, \Sigma) - \frac{3\mu_0}{2\pi} \int_{S_G} \vec{J}_2(R_3, \phi_1) \ln\left(\frac{D}{R_3}\right) \mathrm{d}S(R_3, \phi_1) \\ & - \frac{3\mu_0}{2\pi} \int_{S_G} \vec{J}_3(C_1, \phi_2) \ln\left(\frac{D}{C_1}\right) \mathrm{d}S(C_1, \phi_2), \end{aligned} \quad (2)$$

with d and D the distances as depicted in figure 1 and μ_O the permitivity of free space. S_O is the surface area of the outer shell while S_G denotes the surface area of any inner conductor G_i, $i = 1, 2, 3$, since in this case all three inner conductors have the same surface area and size.
The factor three in front of the integrals comes in due to the fact that all $\vec{J}_2$'s will be the same for the inner conductors. The mutual effect of the three conductors produces the third term, i.e. the last term in relation (2), represented by $\vec{J}_3$.
The points $P'(\rho, \Sigma)$ and $P(r, \theta)$ are chosen on the inner boundary of the outer shell since there are no currents outside the shell due to insulation. (R_3, ϕ_1) are the polar-coordinates of a point on any of the three inner conductors. The point with coordinates (C_1, ϕ_2) is chosen to calculate the mutual interaction of the three conductors G_i at any point of their surfaces.

We use a Cartesian coordinate system with the centre O of the outer shell as origin to formulate the following cosine relationships (see also figure 1):

$$r_1 = (r^2 + R_1^2 + 2rR_1 \sin\theta)^{1/2} \qquad r_2 = (r^2 + C_1^2 + 2rC_1 \sin\theta)^{1/2},$$

with R_1 and C_1 defined above. For the rotation of fields with respect to the outer shell and for the mutual effects inside this shell we define

$$\rho_1 = (R_1^2 + R_3^2 + 2R_1R_3 \sin\phi_1)^{1/2} \qquad \rho_2 = (R_1^2 + C_1^2 + 2R_1C_1 \sin\phi_2)^{1/2}.$$

For the calculation of the effects of the fields at $P(r, \theta)$ and $P'(\rho, \Sigma)$ we need the following total sum of distances:

$$d_i = d_i(r, \theta; \rho, \Sigma) = [r^2 + R_i^2 - 2rR_i \cos(\theta - \Sigma)]^{1/2} \qquad i = 1, 2,$$

and

$$D_i = D_i(r, \theta; \rho, \Sigma) = [r^2 + r_i^2 - 2rr_i cos(\theta - \phi_i)]^{1/2} \qquad i = 1, 2.$$

3.3. Calculation of current densities

Introducing the notations

$$\begin{aligned} h_{O1} &= \text{radius of inner boundary of the shell with centre } O, \\ h_{O2} &= \text{radius of outer boundary of the shell with centre } O, \\ h_G &= \text{radius of the conductors } G_1, G_2 \text{ and } G_3, \end{aligned}$$

we define the kernels of the integrals for the current densities as follows where the factor n comes in through the frequency of the wave number and where $C := j\omega\sigma\frac{\mu_O}{2\pi}$ with σ denoting the conductivity:

$$\begin{aligned}
\mathcal{F}_1(r, \theta; \rho, \Sigma) &= C\left[\ln\left(\frac{2d_1}{\rho} + \frac{d_2}{\rho}\right) - 3n\frac{h_{O1}^2}{S_O}\ln\left(\frac{h_{O2}}{\rho}\right) + \frac{3}{2S_O}(h_{O2}^2 - \rho^2)\right] \\
\mathcal{F}_2(r, \theta{:}\ R_3, \phi_1) &= C\left[n\frac{h_{O2}^2}{S_O}\left(\ln\left(\frac{2h_{O1}}{R_3}\right) + \ln\left(\frac{h_{O1}}{C_1}\right)\right) - \ln\left(\frac{2D_1}{R_3}\right) + \ln\left(\frac{D_2}{C_1}\right)\right] \\
\mathcal{F}_3(r, \theta; \rho, \Sigma) &= -C\left[\ln\left(\frac{2D_1}{r_1}\right) + \ln\left(\frac{D_2}{r_2}\right)\right] \\
\mathcal{F}_4(r, \theta; R_3, \phi_1) &= C\left[\ln\left(\frac{2d_1}{h_G}\right) + \ln\left(\frac{d_2}{h_G}\right) + \frac{3}{2} - \frac{\rho_1^2}{h_G} - \frac{\rho_2^2}{2h_G}\right]
\end{aligned}$$

$$\mathcal{F}_5(r,\theta;C_1,\phi_2) = C\left[3\ln\left(\frac{R_2}{h_G}\right)+\frac{3}{2}-\frac{\rho_1^2}{h_G}-\frac{\rho_2^2}{2h_G}\right]$$

By these formulas for the kernels one can write the total current distribution as follows when denoting by $\vec{I}$ the total current related to $\vec{E}_O$:

$$\begin{aligned}\vec{J}_1(r,\theta) &= 3\frac{\vec{I}}{S_O}+\int_{S_O}\mathcal{F}_1(r,\theta;\rho,\Sigma)3\vec{J}_1(\rho,\Sigma)\,\mathrm{d}S(\rho,\Sigma)\\ &\quad+\int_{S_G}\mathcal{F}_2(r,\theta;R_3,\phi_1)3\vec{J}_2(R_3,\phi_1)\,\mathrm{d}S(R_3,\phi_1)\end{aligned} \tag{3}$$

$$\begin{aligned}\vec{J}_4(r,\theta) &= 3\frac{\vec{I}}{S_G}+\int_{S_O}\mathcal{F}_3(r,\theta;\rho,\Sigma)3\vec{J}_1(\rho,\Sigma)\,\mathrm{d}S(\rho,\Sigma)\\ &\quad+\int_{S_G}\mathcal{F}_4(r,\theta;R_3,\phi_1)3\vec{J}_2(R_3,\phi_1)\,\mathrm{d}S(R_3,\phi_1)\\ &\quad+\int_{S_G}\mathcal{F}_5(r,\theta;C_1,\phi_2)3\vec{J}_3(C_1,\phi_2)\,\mathrm{d}S(C_1,\phi_2)\end{aligned} \tag{4}$$

After calculating the current densities over the cross section of shells and conductors one can find the *force*. The force will have two components, one in the x-direction and one in the y-direction, denoted as F_x and F_y. In other words we have $F(r,\theta) = \sqrt{F_x^2(r,\theta)+F_y^2(r,\theta)}$, with

$$F_x = 3\frac{\mu_0}{2\pi}\int_{S_G}J_4(r,\theta)\left(\int_{S_O}\frac{J_1(\rho,\Sigma)\sin(\beta)}{d_3}\,\mathrm{d}S(\rho,\Sigma)\right)\mathrm{d}S(r,\theta), \tag{5}$$

where d_3 stands for the distance between a point inside the outer shell and a point chosen on the surfaces of the inner conductors. The angle β is the angle between θ and $\theta-\phi_1$ or between θ and $\theta-\phi_2$ minus a rotation of $3\frac{\pi}{2}$. Similarly,

$$F_y = -3\frac{\mu_0}{2\pi}\int_{S_G}J_4(r,\theta)\left(\int_{S_O}\frac{J_1(\rho,\Sigma)\cos(\beta)}{d_3}\,\mathrm{d}S(\rho,\Sigma)\right)\mathrm{d}S(r,\theta). \tag{6}$$

4. DISCRETIZATION OF THE MODEL

Now we would like to find numerically the values of forces and current densities. We can plug in then numerical values for a particular material, e.g. copper or steel, to find the losses in the respective materials.

To perform the discretization we subdivide the cross-sectional areas of the outer shell, respectively of the three inner conductors, into N_O, respectively N_G, subsections. This is done in such a way that each layer of a subsection passes through the common centre or an axis of centres parallel to each other. In other words such that all axis of rotation are colinear.

In order to apply the method of moments (see Sadiku [8]) we expand the scalar components of the current density distributions $\vec{J}_i$, $i = 1, 2, 3, 4$, in terms of basisfunctions. We will show the method for one scalar component of the vectors and therefore omit the subscript for the component. We have:

$$J_1 = \sum_{k=1}^{N_O} \alpha_{k1} p_k^O, \qquad J_2 = \sum_{j=1}^{N_G} \alpha_{j2} p_j^G, \qquad J_3 = \sum_{j=1}^{N_G} \alpha_{j3} p_j^G, \qquad J_4 = \sum_{k=1}^{N_O} \alpha_{k4} p_k^O,$$

where p_k^O, resp. p_j^G, are piecewise constant basisfunctions which take the value 1 in the kth subsection of the surface S_O, resp. in the jth subsection of S_G, and zero otherwise.
After introducing the short hand notations $B_O = 3\frac{I}{S_O}$ and

$$\begin{aligned} L_1(J_1) &= J_1(r,\theta) - \int_{S_O} \mathcal{F}_1(r,\theta;\rho,\Sigma) 3 J_1(\rho,\Sigma)\, \mathrm{d}S(\rho,\Sigma) \\ L_2(J_2) &= -\int_{S_G} \mathcal{F}_2(r,\theta;R_3,\phi_1) 3 J_2(R_3,\phi_1)\, \mathrm{d}S(R_3,\phi_1) \end{aligned}$$

and substituting the expansions for J_i, $i = 1, 2$, into these linear functions, we weight the resulting equation (3) for one component with functions w_ℓ and integrate both sides over S_O:

$$\sum_{k=1}^{N_O} \alpha_{k1} \langle L_1(p_k^O), w_\ell\rangle + \sum_{j=1}^{N_G} \alpha_{j2} \langle L_2(p_j^G), w_\ell\rangle = \langle B_O, w_\ell\rangle, \tag{7}$$

where $\langle\cdot,\cdot\rangle$ denotes the innerproduct. Similarly, we introduce $B_G = 3\frac{I}{S_G}$ and

$$\begin{aligned} L_3(J_1) &= -\int_{S_O} \mathcal{F}_3(r,\theta;\rho,\Sigma) 3 J_1(\rho,\Sigma)\, \mathrm{d}S(\rho,\Sigma) \\ L_4(J_2) &= -\int_{S_G} \mathcal{F}_4(R_3,\phi_1) 3 J_2(R_3,\phi_1)\, \mathrm{d}S(R_3,\phi_1) \\ L_5(J_3) &= -\int_{S_G} \mathcal{F}_5(C_1,\phi_2) 3 J_3(C_1,\phi_2)\, \mathrm{d}S(C_1,\phi_2) \end{aligned}$$

and substitute the expansions for J_i, $i = 1, 2, 3, 4$, into these linear functions which compose equation (4) for one component. Weighting the resulting equation with the functions w_ℓ and integrating both sides over the surface S_O, we obtain

$$\begin{aligned} \sum_{k=1}^{N_O} \alpha_{k1} \langle L_3(p_k^O), w_\ell\rangle &+ \sum_{j=1}^{N_G} \alpha_{j2} \langle L_4(p_j^G), w_\ell\rangle \\ &+ \sum_{j=1}^{N_G} \alpha_{j3} \langle L_5(p_j^G), w_\ell\rangle + \sum_{k=1}^{N_O} \alpha_{k4} \langle p_k^O, w_\ell\rangle = \langle B_G, w_\ell\rangle. \end{aligned} \tag{8}$$

We have to choose $N_O + N_G$ weightfunctions w_ℓ such that the system of equations (7)-(8) can be solved for the $2(N_O + N_G)$ unknowns α_{k1}, α_{j2}, α_{j3} and α_{k4}.
The above system of equations can be expressed in matrix form

$$[L]\,[\alpha] = [B] \tag{9}$$

where $[L]$ is a $2(N_O + N_G) \times 2(N_O + N_G)$ matrix and $[\alpha]$ and $[B]$ are $2(N_O + N_G)$ column matrices. In fact these matrices have the following bloc structure:

$$\begin{bmatrix} [L_{11}] & [L_{12}] & [O_G] & [O_O] \\ [L_{21}] & [L_{22}] & [L_{23}] & [L_{24}] \end{bmatrix} \begin{bmatrix} [\alpha_1] \\ [\alpha_2] \\ [\alpha_3] \\ [\alpha_4] \end{bmatrix} = \begin{bmatrix} [B_1] \\ [B_2] \end{bmatrix},$$

where $[O_G]$, respectively $[O_O]$, is the zero matrix of dimension $(N_O + N_G) \times N_G$, respectively $(N_O + N_G) \times N_O$, and with

$$[L_{11}] = \begin{bmatrix} \langle L_1(p_1^O), w_1 \rangle & \dots & \langle L_1(p_{N_O}^O), w_1 \rangle \\ \vdots & & \vdots \\ \langle L_1(p_1^O), w_{N_O+N_G} \rangle & \dots & \langle L_1(p_{N_O}^O), w_{N_O+N_G} \rangle \end{bmatrix}.$$

We omit the explicit expression of the matrices $[L_{12}]$, $[L_{21}]$, $[L_{22}]$ and $[L_{23}]$ since they are similar to the one given above, apart from $[L_{24}]$ which has the following form

$$[L_{24}] = \begin{bmatrix} \langle p_1^O, w_1 \rangle & \dots & \langle p_{N_O}^O, w_1 \rangle \\ \vdots & & \vdots \\ \langle p_1^O, w_{N_O+N_G} \rangle & \dots & \langle p_{N_O}^O, w_{N_O+N_G} \rangle \end{bmatrix}.$$

Futher, we have

$$[\alpha_s] = {}^t\left[\alpha_{1s} \;\; \dots \;\; \alpha_{N_O,s}\right] \quad \text{for } s = 1,4, \qquad [\alpha_t] = {}^t\left[\alpha_{1t} \;\; \dots \;\; \alpha_{N_G,t}\right] \quad \text{for } t = 2,3$$

and

$$[B_1] = {}^t\left[\langle B_O, w_1 \rangle \;\; \dots \;\; \langle B_O, w_{N_O+N_G} \rangle\right] \qquad [B_2] = {}^t\left[\langle B_G, w_1 \rangle \;\; \dots \;\; \langle B_G, w_{N_O+N_G} \rangle\right]$$

The matrix $[B]$ is known, while the elements of the matrix $[L]$ are to be calculated. Once these elements are evaluated the matrix equation (9) can be solved for $[\alpha]$ by inverting $[L]$ or by employing a Gauss-Jordan elimination method. If matrix inversion is used we obtain

$$[\alpha] = [L]^{-1}[B].$$

The elements of $[\alpha]$ eventually provide the current distributions J_1 and J_4 which in turn are used to calculate the force components F_x and F_y in (5)-(6). The forces will determine the rotation of fields according to Faraday's law.

5. CONCLUSION

In eddy current analysis the real size and distribution of current densities within the conductors is an important factor. One must also give thought that whether material used is of high or low resistivity. In other words we must see how the dimensions of conductors/shells in use are in proportion with the skin depth. The method used above has similarity to references Dokopoulos & Tampakis [1], Shepperd [9], Snelling [10] and other citations there. This is a general method which can be used for any combination of conductors/shells inside a shell with proper insulating materials and shielding.

Also, the method of conformal transformation has been used by Higgins [2]. In his case an eccentric-annular cross-section has been considered. In the present case the conductors can be placed anywhere with respect to the outer conductor in that the centres need not to coincide. However, the axis of all conductors and shells passing through their centre are parallel to each other. This provides part of eccentricity for concentric case (see Perry [6]). In the work of Higgins [2] as well as in other old papers, steel and copper were the main materials for conductors. For more details on material properties we refer to Russell [7].
The analysis and design for different materials and combination of materials have been considered in Tegopoulos & Kriezis [11].

In future work we will test the numerical discretization via the construction of numerical examples and its implementation for real-world design problems of good conductors or shells, where the aim is to prevent eddy current losses .

ACKNOWLEDGEMENTS

The second author thanks the Faculty of Economics and Business Administration, University of Ghent, for the financial support to attend the conference.
Both authors would like to express their gratitude to conference organizor M. Sambandham for his invitation to represent their work in the DSA meeting, May 1999.

REFERENCES

1. Dokopoulos, P.,& Tampakis, D., Eddy currents in a system of tubular conductors, IEEE Trans. on Magnetics, Vol 20 (1984), no. 5, 365-369.
2. Higgins, T.J., The induction of tubular conductors of eccentric annular region, J. Math. & Phy., Vol 20 (1942), no. 2, 390-398.
3. Kawasaki, A., Inami, M., & Ishikawa, T., Theoretical considerations on eddy current losses in non-magnetic and magnetic pipes for power transmission systems, IEEE Trans. on Power Apparatus and Systems, Vol 100 (1981), no. 2, 474-484.
4. Krawczyk, A., & Tegopoulos, J.A., Numerical Modelling of Eddy Currents, Oxford University Press, Oxford, 1993.
5. Kriezis, E.E., Calculation of field and forces in a cylinderical shell, IEEE Trans. on Magnetics, Mag. 20, vol. 5 (1984).
6. Perry, M.P., Low Frequency Electromagnetic Design, Marcel Dekker Inc., New York, 1985.
7. Russell, A., The induction coefficients of a circuit and their applications., J. IEEE, Vol 69 (1931), 270-282.
8. Sadiku, M.N.O., Numerical Techniques in Electromagnetics, CRC Press, Boca Raton, 1992.
9. Shepperd J. Salon., Finite Element Analysis of Electrical Machines, Kluwer Academic Publishers, 1995.
10. Snelling, E.C., Soft Ferrites: Properties and Applications, Buttersworth & Co. Ltd., London, 1988.
11. Tegopoulos, J.A., & Kriezis, E.E., Eddy Currents in Linear Conducting Media, Elsevier, Amsterdam, 1985.

Proceedings of Dynamic Systems and Applications 3 (2001) 561-568

Another Extension of the Method of Generalized Quasilinearization to Integro-Differential Equations

Sonya A. F. Stephens
Florida A&M University, Mathematics Department, Tallahassee, FL 32307

Abstract. The method of quasilinearization is generalized for integro-differential equations. Conditions on the integro-differential equation are relaxed by allowing the addition of a convex function instead of requiring that the integro-differential equation be convex. In other words, $f_{uu} + \phi_{uu} \geq 0$ and $K_{uu} + L_{uu} \geq 0$ where $\phi_{uu} > 0$ and $L_{uu} > 0$ as opposed to $f_{uu} \geq 0$ and $K_{uu} \geq 0$.

AMS(MOS) subject classification. 35K60, 35K57

1. INTRODUCTION

The method of quasilinearization has been generalized and extended by requiring less restrictive assumptions in Deo & Knoll [3], Lakshmikantham [5], Lakshmikantham & Malek [6], Lakshmikantham et al [8], Lakshmikantham et al [9], and Stephens [10]. In Lakshmikantham et al [9], the first exposure to a multistage process was presented when adding a convex function to the original differential equation. Although this process follows the desired methodology, it is very complicated. It was shown in Lakshmikantham et al [5], that the same results could be achieved using a simpler algorithm that offers the construction of monotone sequences that are solutions of linear differential equations. Here we use this simpler approach on a specific nonlinear differential equation, the integro-differential equation.

2. MAIN RESULT

Consider the IVP

$$u'(t) = f(t, u(t)) + \int_{t_0}^{t} K(t, s, u(s))\, ds, \quad u(t_0) = u_0 \tag{2.1}$$

with $t \in J = [t_0, t_0 + \mathrm{T}]$, $t_0 \geq 0$, $T > 0$, and assume suitable conditions imposed on f and K.

We now prove our main result.

Theorem 2.1 Assume that

(A1) $f \in C[J \times \mathfrak{R}, \mathfrak{R}]$, and $f_u(t,u)$, $f_{uu}(t,u)$ exist and are continuous satisfying $f_{uu}(t,u) + \phi_{uu}(t,u) \geq 0$ for all $t \in J$ where $\phi \in C[J \times \mathfrak{R}, \mathfrak{R}]$ and $\phi_u(t,u)$, $\phi_{uu}(t,u)$ exist and are continuous and $\phi_{uu}(t,u) > 0$, $t \in J$;

(A2) $K \in C[J \times J \times \mathfrak{R}, \mathfrak{R}]$ and $K_u(t,s,u)$, $K_{uu}(t,s,u)$ exist and are continuous satisfying $K_{uu}(t,s,u) + L_{uu}(t,s,u) \geq 0$ for all $(t,s) \in J \times J$

where $L \in C[J \times J \times \mathfrak{R}, \mathfrak{R}]$ and $L_u(t,s,u)$, $L_{uu}(t,s,u)$ exist and are continuous and $L_{uu}(t,s,u) > 0$, $(t,s) \in J \times J$ and $K(t,s,u)$ and $L(t,s,u)$ are monotone nondecreasing;

(A3) $\alpha_0, \beta_0 \in C^1[J, \mathfrak{R}]$, are lower and upper solutions of (2.1) such that $\alpha_0(t) \le \beta_0(t)$ on $J = [t_0, t_0 + T]$, $t_0 \ge 0$, $T > 0$.

Then there exist monotone sequences $\{\alpha_n(t)\}$, $\{\beta_n(t)\}$, which converge uniformly and quadratically to a unique solution of (2.1).

Proof: In view of (A1) and (A2), it is clear that for $u \ge v$

$$f(t,u) \ge f(t,v) + \big[f_u(t,v) + \phi_u(t,v)\big](u-v) - \big[\phi(t,u) - \phi(t,v)\big] \tag{2.2}$$
$$K(t,s,u) \ge K(t,s,v) + \big[K_u(t,s,v) + L_u(t,s,v)\big](u-v) - \big[L(t,s,u) - L(t,s,v)\big] \tag{2.3}$$

Also, both functions, $f(t,u)$ and $K(t,s,u)$, satisfy the Lipschitz condition since $f_u(t,u)$ and $K_u(t,s,u)$ exist. This yields

$$f(t,u_1) - f(t,u_2) \le M(u_1 - u_2), \quad M > 0, \text{ for all } t \in J \tag{2.4}$$
$$K(t,s,u_1) - K(t,s,u_2) \le N(u_1 - u_2), \quad N > 0, \text{ for all } (t,s) \in J \times J \tag{2.5}$$

for any $\alpha_0(t) \le u_2 \le u_1 \le \beta_0(t)$. Also, due to the Lipschitzian condition of f and K, there exists a unique solution to (2.1) using Theorem 1.2.1, Lakshmikanthm & Rao [7] repeatedly.

Let α_1 and β_1 be solutions of the following IVPs

$$\begin{aligned}\alpha_1' = {} & f(t,\alpha_0) + \textstyle\int_{t_0}^t K(t,s,\alpha_0(s))\,ds + \Big[f_u(t,\alpha_0) + \phi_u(t,\alpha_0) \\ & - \phi_u(t,\beta_0)\Big](\alpha_1 - \alpha_0) + \textstyle\int_{t_0}^t \Big\{\big[K_u(t,s,\alpha_0(s)) + L_u(t,s,\alpha_0(s)) \\ & - L_u(t,s,\beta_0(s))\big](\alpha_1(s) - \alpha_0(s))\Big\}\,ds; \quad \alpha_1(t_0) = u_0,\end{aligned} \tag{2.6}$$

$$\begin{aligned}\beta_1' = {} & f(t,\beta_0) + \textstyle\int_{t_0}^t K(t,s,\beta_0(s))\,ds + \big[f_u(t,\alpha_0) + \phi_u(t,\alpha_0) \\ & - \phi_u(t,\beta_0)\big](\beta_1 - \beta_0) + \textstyle\int_{t_0}^t \Big\{\big[K_u(t,s,\alpha_0(s)) + L_u(t,s,\alpha_0(s)) \\ & - L_u(t,s,\beta_0(s))\big](\beta_1(s) - \beta_0(s))\Big\}\,ds; \quad \beta_1(t_0) = u_0.\end{aligned} \tag{2.7}$$

Set $p = \alpha_0 - \alpha_1$. Then $p(t_0) \le 0$. So using (A3) and (2.6),

$$\begin{aligned}p' &= \alpha_0' - \alpha_1' \\ &\le f_u(t,\alpha_0) + \phi_u(t,\alpha_0) - \phi_u(t,\beta_0))p \\ &\quad + \textstyle\int_{t_0}^t \Big\{\big[K_u(t,s,\alpha_0(s)) + L_u(t,s,\alpha_0(s)) - L_u(t,s,\beta_0(s))\big]p(s)\Big\}ds\end{aligned}$$

Using this and a variation of Theorem 1.2.1, Lakshmikantham & Rao [7] yields $p(t) \le 0$, $t \ge t_0$, and therefore proves that $\alpha_0(t) \le \alpha_1(t)$, $t \in J$.

Next, set $p = \beta_1 - \beta_0$ which yields $p(t_0) \leq 0$. Again, using (A3) and (2.7)

$$\begin{aligned} p' &= \beta_1' - \beta_0' \\ &\leq \Big(f_u(t,\alpha_0) + \phi_u(t,\alpha_0) - \phi_u(t,\beta_0)\Big)p \\ &\quad + \int_{t_0}^t \Big\{ [K_u(t,s,\alpha_0(s)) + L_u(t,s,\alpha_0(s)) - L_u(t,s,\beta_0(s))]p(s) \Big\}\, ds \end{aligned}$$

As before, this and a special case of Theorem 1.2.1, Lakshmikantham & Rao [7] proves that $\beta_1(t) \leq \beta_0(t)$, $t \in J$.

We now set $p = \alpha_1 - \beta_0$ noting that $p(t_0) \leq 0$. Then using (2.6)

$$\begin{aligned} p' &= \alpha_1' - \beta_0' \\ &= f(t,\alpha_0) + \int_{t_0}^t K(t,s,\alpha_0(s))\, ds \\ &\quad + \Big[f_u(t,\alpha_0) + \phi_u(t,\alpha_0) - \phi_u(t,\beta_0)\Big](\alpha_1 - \alpha_0) \\ &\quad + \int_{t_0}^t \Big\{ \big[K_u(t,s,\alpha_0(s)) + L_u(t,s,\alpha_0(s)) \\ &\quad - L_u(t,s,\beta_0(s))\big](\alpha_1(s) - \alpha_0(s)) \Big\}\, ds \\ &\quad - f(t,\beta_0) - \int_{t_0}^t K(t,s,\beta_0(s))\, ds \end{aligned}$$

From (2.2), (2.3) and the Mean Value Theorem, we can see that

$$\begin{aligned} f(t,\beta_0) + \int_{t_0}^t K(t,s,\beta_0(s))\, ds \geq\ & f(t,\alpha_0) + \int_{t_0}^t K(t,s,\alpha_0(s)) \\ & + \Big[f_u(t,\alpha_0) + \phi_u(t,\alpha_0)\Big](\beta_0 - \alpha_0) \\ & + \int_{t_0}^t \Big\{ [K_u(t,s,\alpha_0(s)) \\ & + L_u(t,s,\alpha_0(s))](\beta_0(s) - \alpha_0(s)) \Big\}\, ds \\ & - \phi_u(t,\beta_0)(\beta_0 - \alpha_0) \\ & - \int_{t_0}^t \Big[L_u(t,s,\beta_0(s))(\beta_0(s) - \alpha_0(s))\Big]\, ds. \end{aligned}$$

This in turn yields

$$\begin{aligned} p' \leq\ & \Big[f_u(t,\alpha_0) + \phi_u(t,\alpha_0) - \phi_u(t,\beta_0)\Big]p + \int_{t_0}^t \Big\{ [K_u(t,s,\alpha_0(s)) + L_u(t,s,\alpha_0(s)) \\ & - \mathrm{L}_u(t,s,\beta_0(s))]p(s) \Big\}\, ds \end{aligned}$$

This implies that $\alpha_1(t) \leq \beta_0(t)$, $t \in J$ using a variation of Theorem 1.2.1, Lakshmikantham & Rao [7] which yields $\alpha_0(t) \leq \alpha_1(t) \leq \beta_0(t)$. To prove $\alpha_0(t) \leq \beta_1(t) \leq \beta_0(t)$, $t \in J$, a similar approach is used. Next we need to show that $\alpha_1(t) \leq \beta_1(t)$, $t \in J$ using (2.2), (2.3), (2.6), (2.7), and the Mean Value Theorem.

$$\alpha_1' \leq f(t,\alpha_1) + \int_{t_0}^t K(t,s,\alpha_1(s))\, ds - \Big[f_u(t,\alpha_0) + \phi_u(t,\alpha_0)\Big](\alpha_1 - \alpha_0)$$

$$
\begin{aligned}
&- \int_{t_0}^t \Big\{ \big[K_u(t,s,\alpha_0(s)) + L_u(t,s,\alpha_0(s))\big](\alpha_1(s) - \alpha_0(s)) \Big\} \, ds \\
&+ \phi_u(t,\beta_0)(\alpha_1 - \alpha_0) + \int_{t_0}^t \Big[L_u(t,s,\beta_0(s))(\alpha_1(s)) - \alpha_0(s))\big] \Big] ds \\
&+ \Big[f_u(t,\alpha_0) + \phi_u(t,\alpha_0) - \phi_u(t,\beta_0) \Big] (\alpha_1 - \alpha_0) \\
&+ \int_{t_0}^t \Big\{ \big[K_u(t,s,\alpha_0(s)) + L_u(t,s,\alpha_0(s)) \\
&- L_u(t,s,\beta_0(s))\big](\alpha_1(s) - \alpha_0(s)) \Big\} \, ds \\
= {}& f(t,\alpha_1) + \int_{t_0}^t K(t,s,\alpha_1(s))
\end{aligned}
$$

and

$$
\begin{aligned}
\beta_1' \geq {}& f(t,\beta_1) + \int_{t_0}^t K(t,s,\beta_1(s)) \, ds - \Big[f_u(t,\alpha_0) + \phi_u(t,\alpha_0) \Big] (\beta_0 - \beta_1) \\
&- \int_{t_0}^t \Big\{ \big[K_u(t,s,\alpha_0(s)) + L_u(t,s,\alpha_0(s))\big](\beta_0(s) - \beta_1(s)) \Big\} \, ds \\
&- \phi_u(t,\beta_0)(\beta_0 - \beta_1) - \int_{t_0}^t \Big[L_u(t,s,\beta_0(s))(\beta_0(s) - \beta_1(s)) \Big] ds \\
&+ \Big[f_u(t,\alpha_0) + \phi_u(t,\alpha_0) - \phi_u(t,\beta_0) \Big] (\beta_1 - \beta_0) \\
&+ \int_{t_0}^t \Big\{ \big[K_u(t,s,\alpha_0(s)) + L_u(t,s,\alpha_0(s)) - L_u(t,s,\beta_0(s))\big](\beta_1(s) - \beta_0(s)) \Big\} ds \\
= {}& f(t,\beta_1) + \int_{t_0}^t K(t,s,\beta_1(s)) \, ds.
\end{aligned}
$$

Using Theorem 1.2.1, Lakshmikantham & Rao [7], we get $\alpha_1(t) \leq \beta_1(t)$, $t \in J$. We have now proven $\alpha_0(t) \leq \alpha_1(t) \leq \beta_1(t) \leq \beta_0(t)$, $t \in J$.

Now we will use induction to prove

$$
\alpha_0(t) \leq \alpha_1(t) \leq \ldots \leq \alpha_n(t) \leq \beta_n(t) \leq \ldots \leq \beta_1(t) \leq \beta_0(t), \; t \in J. \tag{2.8}
$$

Assume that for $k > 1$,

$$
\begin{aligned}
\alpha_k' &\leq f(t,\alpha_k) + \int_{t_0}^t K(t,s,\alpha_k(s)) \, ds \\
\beta_k' &\geq f(t,\beta_k) + \int_{t_0}^t K(t,s,\beta_k(s)) \, ds.
\end{aligned}
$$

Let α_{k+1} and β_{k+1} be the solutions to the following IVPs:

$$
\begin{aligned}
\alpha_{k+1}' = {}& f(t,\alpha_k) + \int_{t_0}^t K(t,s,\alpha_k(s)) \, ds + \Big[f_u(t,\alpha_k) + \phi_u(t,\alpha_k) \\
&- \phi_u(t,\beta_k) \Big] (\alpha_{k+1} - \alpha_k) + \int_{t_0}^t \Big\{ \big[K_u(t,s,\alpha_k(s)) + L_u(t,s,\alpha_k(s)) \\
&- L_u(t,s,\, \beta_k(s))\big](\alpha_{k+1}(s) - \alpha_k(s)) \Big\} \, ds; \quad \alpha_{k+1}(t_0) = u_0,
\end{aligned}
$$

and

$$
\beta_{k+1}' = f(t,\beta_k) + \int_{t_0}^t K(t,s,\beta_k(s)) \, ds + \Big[f_u(t,\alpha_k) + \phi_u(t,\alpha_k)
$$

$$- \phi_u(t,\beta_k)\Big](\beta_{k+1}-\beta_k) + \int_{t_0}^t \Big\{ [K_u(t,s,\alpha_k(s)) + L_u(t,s,\alpha_k(s))$$
$$- L_u(t,s,\beta_k(s))](\beta_{k+1}(s)-\beta_k(s))\Big\}\, ds; \quad \beta_{k+1}(t_0) = u_0.$$

Following the previous work, set $p = \alpha_k - \alpha_{k+1}$, $p(t_0) \le 0$. This yields $\alpha_k(t) \le \alpha_{k+1}(t)$, $t \in J$ where

$$p' \le \Big[f_u(t,\alpha_k) + \phi_u(t,\alpha_k) - \phi_u(t,\beta_k)\Big]p$$
$$+ \int_{t_0}^t \Big\{ [K_u(t,s,\alpha_k(s)) + L_u(t,s,\alpha_k(s)) - L(t,s,\beta_k(s))]p(s)\Big\}\, ds.$$

Now, setting $p = \alpha_{k+1} - \beta_k$, $p(t_0) \le 0$, yields $\alpha_{k+1} \le \beta_k(t)$, $t \in J$, where

$$p' \le f(t,\alpha_k) + \int_{t_0}^t K(t,s,\alpha_k(s))\, ds$$
$$+ \Big[f_u(t,\alpha_k) + \phi_u(t,\alpha_k) - \phi_u(t,\beta_k)\Big](\alpha_{k+1}-\alpha_k)$$
$$+ \int_{t_0}^t \Big\{ [K_u(t,s,\alpha_k(s)) + L_u(t,s,\alpha_k(s))$$
$$- L_u(t,s,\beta_k(s))](\alpha_{k+1}(s)-\alpha_k(s))\Big\} ds - f(t,\beta_k) - \int_{t_0}^t K(t,s,\beta_k(s))\, ds$$

From (2.2), (2.3) and the Mean Value Theorem, we can see that

$$f(t,\beta_k) + \int_{t_0}^t K(t,s,\beta_k(s))\, ds \ge f(t,\alpha_k) + \int_{t_0}^t K(t,s,\alpha_k(s))$$
$$+ \Big[f_u(t,\alpha_k) + \phi_u(t,\alpha_k)\Big](\beta_k - \alpha_k)$$
$$+ \int_{t_0}^t \Big\{ [K_u(t,s,\alpha_k(s))$$
$$+ L_u(t,s,\alpha_k(s))](\beta_k(s)-\alpha_k(s))\Big\}\, ds$$
$$- \phi_u(t,\beta_k)(\beta_k - \alpha_k)$$
$$- \int_{t_0}^t \Big[L_u(t,s,\beta_k(s))(\beta_k(s)-\alpha_k(s))\Big] ds.$$

This in turn yields

$$p' \le \Big[f_u(t,\alpha_k) + \phi_u(t,\alpha_k) - \phi_u(t,\beta_k)\Big]p + \int_{t_0}^t \Big\{ [K_u(t,s,\alpha_k(s)) + L_u(t,s,\alpha_k(s))$$
$$- L_u(t,s,\beta_k(s))]p(s)\Big\}\, ds$$

Similarly, $\alpha_k(t) \le \beta_{k+1}(t) \le \beta_k(t)$, $t \in J$, can be shown. It is left for us to show $\alpha_{k+1}(t) \le \beta_{k+1}(t)$, $t \in J$.

$$\alpha'_{k+1} \le f(t,\alpha_{k+1}) + \int_{t_0}^t K(t,s,\alpha_{k+1}(s))\, ds - \Big[f_u(t,\alpha_k) + \phi_u(t,\alpha_k)\Big](\alpha_{k+1}-\alpha_k)$$
$$- \int_{t_0}^t \Big\{ [K_u(t,s,\alpha_k(s)) + L_u(t,s,\alpha_k(s))](\alpha_{k+1}(s)-\alpha_k(s))\Big\} ds$$
$$+ \phi_u(t,\beta_k)(\alpha_{k+1}-\alpha_k) + \int_{t_0}^t \Big[L_u(t,s,\beta_k(s))(\alpha_{k+1}(s)) - \alpha_k(s))\Big] ds$$

$$
\begin{aligned}
&+ \Big[f_u(t,\alpha_k) + \phi_u(t,\alpha_k) - \phi_u(t,\beta_k)\Big](\alpha_{k+1}(s) - \alpha_k(s)) \\
&+ \textstyle\int_{t_0}^t \Big\{ [K_u(t,s,\alpha_k(s)) + L_u(t,s,\alpha_k(s)) \\
&- L_u(t,s,\beta_k(s))](\alpha_{k+1}(s) - \alpha_k(s))\Big\}\, ds \\
&= f(t,\alpha_{k+1}) + \textstyle\int_{t_0}^t K(t,s,\alpha_{k+1}(s))
\end{aligned}
$$

again using (2.2), (2.3), (2.6), and the Mean Value Theorem. Similar arguments yield $\beta'_{k+1} \geq f(t,\beta_{k+1}) + \int_{t_0}^t K(t,s,\beta_{k+1}(s))\, ds$ which leads to $\alpha_{k+1}(t) \leq \beta_{k+1}(t)$, $t \in J$ using Theorem 1.2.1, Lakshmikantham & Rao [7]. This proves (2.8) for all n. Using standard arguments from Ladde et al [4], we get that the sequences $\{\alpha_n(t)\}$, $\{\beta_n(t)\}$ converge uniformly to the unique solution of (2.1) on J.

Now we show that $\{\alpha_n(t)\}$ and $\{\beta_n(t)\}$ converge quadratically. Let

$$p_n(t) = u(t) - \alpha_n(t) \geq 0 \tag{2.9}$$

$$q_n(t) = \beta_n(t) - u(t) \geq 0. \tag{2.10}$$

Then

$$p_n(t_0) = 0 \quad \text{and} \quad q_n(t_0) = 0.$$

So,

$$
\begin{aligned}
p_n'(t) &= f(t,u) + \textstyle\int_{t_0}^t K(t,s,u(s))\, ds - \Big[f_u(t,\alpha_{n-1}) + \textstyle\int_{t_0}^t K_u(t,s,\alpha_{n-1}(s))\, ds \\
&+ \{f_u(t,\alpha_{n-1}) + \phi_u(t,\alpha_{n-1}) - \phi_u(t,\beta_{n-1})\}(\alpha_n - \alpha_{n-1}) \\
&+ \textstyle\int_{t_0}^t \Big\{ [K_u(t,s,\alpha_{n-1}(s)) + L_u(t,s,\alpha_{n-1}(s)) \\
&- L_u(t,s,\beta_{n-1}(s))](\alpha_n(s) - \alpha_{n-1}(s))\Big\}\, ds\Big] \\
&= f(t,u) - f(t,\alpha_{n-1}) + \textstyle\int_{t_0}^t \Big[K(t,s,u(s)) - K(t,s,\alpha_{n-1}(s))\Big]\, ds \\
&+ \phi(t,u) - \phi(t,\alpha_{n-1}) + \textstyle\int_{t_0}^t \Big[L(t,s,u(s)) - L(t,s,\alpha_{n-1}(s))\Big]\, ds \\
&- \Big[\{f_u(t,\alpha_{n-1}) + \phi_u(t,\alpha_{n-1}) - \phi_u(t,\beta_{n-1})\}(\alpha_n - u + u - \alpha_{n-1}) \\
&+ \textstyle\int_{t_0}^t \Big\{ [K_u(t,s,\alpha_{n-1}(s)) + L_u(t,s,\alpha_{n-1}(s)) \\
&- L_u(t,s,\beta_{n-1}(s))](\alpha_n(s) - u(s) + u(s) - \alpha_{n-1}(s))\Big\}\, ds\Big] \\
&- \Big[\phi(t,u) - \phi(t,\alpha_{n-1})\Big] - \textstyle\int_{t_0}^t \Big[L(t,s,u(s)) - L(t,s,\alpha_{n-1}(s))\Big]\, ds \\
&\leq f_{uu}(t,\xi_1)p_{n-1}^2 + \textstyle\int_{t_0}^t \Big[K_{uu}(t,s,\xi_1(s))p_{n-1}^2(s)\Big]\, ds + \phi_{uu}(t,\xi_1)p_{n-1}^2 \\
&+ \textstyle\int_{t_0}^t \Big[L_{uu}(t,s,\xi_1(s))p_{n-1}^2(s)\Big]\, ds + \Big[f_u(t,\alpha_{n-1}) + \phi_u(t,\alpha_{n-1}) \\
&- \phi_u(t,\beta_{n-1})\Big]p_n + \textstyle\int_{t_0}^t \Big\{ [K_u(t,s,\alpha_{n-1}(s)) + L_u(t,s,\alpha_{n-1}(s)) \\
&- L_u(t,s,\beta_{n-1}(s))]p_n(s)\Big\}\, ds + \phi_{uu}(t,\xi_2)(q_{n-1} + p_{n-1})p_{n-1} \\
&+ \textstyle\int_{t_0}^t \Big[L_{uu}(t,s,\xi_2(s))\Big(q_{n-1}(s) + p_{n-1}(s)\Big)p_{n-1}\Big]\, ds
\end{aligned}
$$

where $\alpha_{n-1} \le \xi_1 \le u$ and $\alpha_{n-1} \le \xi_2 \le \beta_{n-1}$. Hence,

$$p_n'(t) \le (M+3N)p_{n-1}^2 + (M_1+3N_1)\int_{t_0}^t p_{n-1}^2(s)\,ds + Rp_n + R_1\int_{t_0}^t p_n(s)\,ds + Nq_{n-1}^2 + N_1\int_{t_0}^t q_{n-1}^2(s)\,ds.$$

where $|f_u(t,u)| \le R$, $|K_u(t,s,u)| \le R_1$, $|f_{uu}(t,u)| \le M$, $|K_{uu}(t,s,u)| \le M_1$, $|\phi_{uu}(t,u)| \le N$, and $|L_{uu}(t,s,u)| \le N_1$. Theorem 1.4.1 Lakshmikantham & Rao [7] yields $p_n(t) \le r(t)$, $t \in J$, where $r(t)$ is the solution to the related linear integro-differential equations

$$r'(t) = Rr(t) + R_1\int_{t_0}^t r(s)\,ds + F(t), \;\; p_n(t_0) = r(t_0) = 0 \tag{2.11}$$

where $F(t) = (M+3N)p_{n-1}^2 + (M_1+3N_1)\int_{t_0}^t p_{n-1}^2(s)\,ds + Nq_{n-1}^2 + N_1\int_{t_0}^t q_{n-1}^2(s)\,ds$. Note that

$$F(t) \le (M+3N+M_1T+3N_1T)\max_{t\in J} p_{n-1}^2(t) + (N+N_1T)\max_{t\in J} q_{n-1}^2(t). \tag{2.12}$$

We now find an estimate for $r(t)$ given by (2.11). Let

$$z(t) = \int_{t_0}^t r(s)\,ds \tag{2.13}$$

Then (2.11) becomes $\quad z''(t) = Rz'(t) + R_1 z(t) + F(t), \;\; z'(t_0) = z(t_0) = 0$

which is a second order linear nonhomogeneous differential equation with constant coefficients which will be solved using variation of parameters in view of (2.11) and (2.12). This yields

$$p_n(t) \le r(t) \le \frac{2e^{RT}}{\sqrt{R^2+4R_1}}F(t)$$

and therefore,

$$\max_{t\in J} p_n(t) \le \frac{2e^{RT}}{\sqrt{R^2+4R_1}}\Big[(M+3N+M_1T+3N_1T)\max_{t\in J} p_{n-1}^2(t) + (N+N_1T)\max_{t\in J} q_{n-1}^2(t)\Big], \; t \in J.$$

Hence, $\{\alpha_n(t)\}$ converges quadratically to $u(t)$ which is the unique solution to (2.1). The proof of the quadratic convergence of $\{\beta_n(t)\}$ is done is a similar fashion. Here

$$\max_{t\in J} q_n(t) \le \frac{2e^{RT}}{\sqrt{R^2+4R_1}}\Big[(2M+3N+2M_1T+3N_1T)\max_{t\in J} q_{n-1}^2(t) + (M+N+M_1T+N_1T)\max_{t\in J} p_{n-1}^2(t)\Big], \; t \in J.$$

The proof is complete.

REFERENCES

[1] Bellman, R. Methods of Nonlinear Analysis, Vol. II. Academic Press, New York, 1973.

[2] Bellman, R., & Kalaba, R. Quasilinearization and Nonlinear Boundary Value Problems. American Elsevier, New York, 1965.

[3] Deo, S. G., & Knoll, C. McGloin. Extension of the Method of Quasilinearization to Integro-Differential Equations. Nonlinear World, 4, (1997), 43 - 52.

[4] Ladde, G. S., Lakshmikantham, V., & Vatsala, A. S. Monotone Iterative Techniques for Nonlinear Differential Equations. Pitman, Boston, 1985.

[5] Lakshmikantham, V., Further Improvement of Generalized Quasilinearization Method. Nonlinear Analysis, 27, (1996), 223 - 227.

[6] Lakshmikantham, V., & Malek, S., Generalized Quasilinearization. Nonlinear World, 1, (1994), 59 - 63.

[7] Lakshmikantham, V., & Rao, M. Rama Mohana Theory of Integro-Differential Equation, Stability and Control: Theory , Methods, and Applications Volume 1. Gordon and Breach Science Publishers, Switzerland, 1995.

[8] Lakshmikantham, V., Leela, S., & Sivasundaram, S., Extensions of the Method of Quasilinearization. Journal of Optimization Theory and Applications 87, (1995), 379 - 401.

[9] Lakshmikantham, V., Leela, S., & McRae, F. A., Improved Generalized Quasilinearization Method. Nonlinear Analysis, 24, (1995), 1627 - 1637.

[10] Stephens, S., The Method of Generalized Quasilinearization and Integro-Differential Equations. Engineering Simulation, Vol. 16, (1999), 439 - 450.

Proceedings of Dynamic Systems and Applications 3 (2001) 569-576

Mesoscopic Resonance of Self-Excited Defect Oscillations.

Lev G. Steinberg
Department of Mathematics
University of Puerto Rico, Mayagüez, PR 00681-9018, USA

ABSTRACT: We approach the problem within the framework of continuum mechanics which describes the defective body in terms of dislocation and disclination fields. It is shown that this continuum model produces massless defect-waves under specific conductivity type conditions. It is shown that interactions of the defect waves produce some resonances, which have been experimentally detected.

AMS(MOS) subject classification. 35R99, 35Q72, 74C99, 74J05, 74M99

1 Introduction

Contemporary studies of defect materials naturally develope at several levels: (a) microscopic, which includes studies atomic microclusters and single imperfections in lattice structure,single dislocation generation, motion and interaction with other defects, etc.(b) mesoscopic level, at an intermediate scale between microscopic and macroscopic approaches;(c) the macroscopic level, which includes studies of inhomogeneous elasticity, plasticity, viscoelasticity, etc.

The history of dislocation theory started in 1907 with works by Volterra and Love that consider both line and point defects as singularities in an elastic continuum. It was proposed by (1934) Taylor, Orowan, and Polyani to consider plastic deformation in crystal as collective motion of the dislocations. Later, (1983) based on gauge ideas, the theory of dislocation and disclinations at macro level was developed by Edelen and Kadić.This theory contains several input parameters: moduli of elasticity, velocity of dislocations and disclinations that are assumed to be known in the development of a complete theory. It is also assumed that the reference state is free of defects. Eshelby attempted to reveal the macroscopic behavior from microscopic considerations. His study leads to the description of configuration forces [3] related to lattice imperfections. Later these forces were considered by Noll, Truesdell,Toupin,Gurtin [6], Maugin.

This paper to an attempt to focus on defect forces at mesoscopic level and show their role in mesoscopic resonance phenomena.

2 Continuum Model

By the continuum point of view at the mesoscopic level, given portion of material is treated as a collection of particles with defect tangles that can be placed in one-to one correspondence with any continuum closed region of three dimensional Euclidean space. This approach is increasingly important when the dislocation density reaches $10^{11} - 10^{12}$ dislocation/cm^2 or more.

At the mesoscopic level we chose the phenomenological description of dislocation and disclinations. Our basic framework at this step is the continuum theory which we describe in terms differential forms.

2.1 The kinematics of defects

Equations of the defect kinematics in differential form

Based on the disclination 3-forms

$$\Omega^i = -S^i\wedge dT + \theta^i = -S^{Ai}\mu_A\wedge dT + \theta^i\mu$$

and the dislocation 2-forms

$$D^i = J^i\wedge dT + \alpha^i = J^i_A dX^A\wedge dT + \alpha^{Ai}\mu_A$$

the general system of defect kinematics is the following [2]:

$$dD^i = \Omega^i \qquad d\Omega^i = 0$$

where
$\alpha^i = \alpha^{Ai}\mu_A$ represents the 2-forms of dislocation density,
$J^i = J^i_A dX^A$ represents the 1-forms of dislocation current,
$S^i = S^{Ai}\mu_A$ represents the 2-forms of disclination current, and
$\Theta^i = \theta^i\mu$ represents the 3-forms of disclination density,
The first integrals

$$D^i = dB^i + K^i$$

are admitted,
where
$B^i = V^i dT + \beta^i = V^i dT + \beta^i_A dX^A$
$K^i = -\omega^i\wedge dT + k^i = -\omega^i_A dX^A\wedge dT + k^{A,i}\mu_A$
$k^i = k^{Ai}\mu_A$ is the notation for the bend-twist 2-forms,
$\omega^i = \omega^i_A dX^A$ is the notation for the spin 2-forms,
$\beta^i = \beta^i_A dX^A$ is the notation for the distortion 1-forms, and
V^i is the notation for the velocity 0-forms.

Defect dynamics in 4-dimensional space-time

The equations of balance of linear momentum and total energy are in the following form:

$$\begin{aligned} \partial_4 \Im_i &= dm_i \\ \partial_4 E &= dW \end{aligned}$$

or in terms of the components:

$$\begin{aligned} \partial_4 \jmath_i &= \partial_A m_i^A \\ \partial_4 E &= \partial_A W^A \end{aligned}$$

where
$m_i = m_i^A \mu_A$ are the 2-forms of mechanical surface torque,
$\Im_i = \jmath_i \mu$ are the 3-forms of angular momentum density,
$W = W^A \mu_A$ is the 2-form of total energy flux, and
$E = e\mu$ is the 3-form of total energy density.
The equations of balance of linear momentum and total energy have general solution in the form [2]

$$\begin{aligned} \nabla \times \vec{H}_i - \partial_4 \vec{D}_i &= \vec{m}_i \quad \nabla \cdot \vec{D}_i = -\jmath_i \\ \nabla \times \vec{H}_4 - \partial_4 \vec{D}_4 &= \vec{W} \quad \nabla \cdot \vec{D}_4 = -e \end{aligned} \tag{1}$$

which will be considered with the defect kinematics equation

$$-\nabla \times \vec{J}^i - \partial_4 \vec{\alpha}^i = \vec{S}^i \quad \nabla \cdot \vec{\alpha}^i = -\theta^i \tag{2}$$

To complete the system 1 - 2 we should add constitutive equations. Our main assumption is that at mesoscopic level the defective stresses depend on corresponding dislocation fields.

Particularly we offer to use the following hypotheses:

Hypothesis 1. Dislocation density $\vec{\alpha}^i$ and stress current fields $\vec{H}_i$ are linearly dependent.

Hypothesis 2. Stress current $\vec{D}_i$ and dislocation current fields $\vec{J}^i$ are linearly dependent.

These means that general constitutive relations are in the following form:

$$\vec{D}_i = \nu_{ij} \vec{J}^j \qquad \vec{H}_i = \mu_{ij}^{-1} \vec{\alpha}^j,$$

where $\nu, \mu, \varepsilon, \delta$ are defect conductivity tensors with positive components components. We assume that properties of these tensor components satisfy requirements of the hyperbolicity (wave properties) of the system 1-2.

For the isotropic case $\mu_{ij} = \mu\delta_{ij}$ and $\nu_{ij} = \nu\delta_{ij}$ and the corresponding constitutive relations

$$\vec{D}_i = \nu \vec{J}^i \qquad \vec{H}_i = \mu^{-1} \vec{\alpha}^i$$

produce traveling defect-waves.

2.1.1 Interactions defect waves

Hereafter we assume that at the mesoscopic level the energy of the defect waves dissipates due to defect friction stress. We propose that for this case

(a) the defect friction is linear dependent on the momentum

$$\overrightarrow{m}_i^{fr} = -\eta_1^{ij} \jmath_j$$

and

(b) the inertia forces are linear dependent on dislocation currents

$$f_{in}^i = -\partial_4 \jmath_i = \eta_2^j J_1^{ij}$$

where

$\jmath_i = \partial_t (I\omega_i)$ is the momentum.

We also introduce the Hamiltonian, which is in high frequency approximation could be the following

$$H(\jmath_i, \vec{J}_1^i) = -\sum_i \left(\frac{1}{2}\kappa_{11}^i \jmath_i^2 + \kappa_{12}^i \,\vec{\jmath} \cdot \vec{J}_1^i + \frac{1}{2}\kappa_{22}^i \vec{J}_1^{i2} \right) \tag{3}$$

Taking into account our previous assumption we obtain

$$\begin{aligned}
\partial_4 \jmath_i &= \frac{\partial \mathcal{H}(\jmath_i, \vec{J}_1^j)}{\partial_i J_1^j} = -\kappa_{12}^i \jmath_i - \kappa_{22}^j J_1^{ij} \\
\partial_4 \overrightarrow{m}_i &= \eta^{ij} \partial_4 \jmath_j = \gamma_{12}^{ij} \jmath_j + \gamma_{22}^j J_1^{ij} \\
\partial_4 \vec{J}_1^i &= -\frac{\partial \mathcal{H}(\jmath_i, \vec{J}_1^i)}{\partial_i p^i} = \kappa_{11}^i \jmath_i + \kappa_{12}^i \vec{J}_1^i \\
\partial_4^2 \jmath_i &= -\eta_2^j \partial_4 J_1^{ij} = -\gamma_{11}^{ij} \jmath_j - \gamma_{12}^j J_1^{ij}
\end{aligned}$$

We assume the high- frequency modes of the vibration interact with the defect wave field therefore the problem is reduced to the solution the defect equations simultaneously with the 1, 2 equations

$$\nabla \times \vec{\alpha}^j - \frac{1}{c}\partial_4 \vec{J}_1^j = \vec{\sigma}_i^{fr} \tag{4}$$

$$\nabla \cdot \vec{J}_1^i - \jmath_i / c = 0 \text{ or } \nabla \cdot (\partial_4 \vec{J}_1^j + \overrightarrow{m}_i) = 0 \tag{5}$$

$$\nabla \times \vec{J}_1^i + \frac{1}{c}\partial_4 \vec{\alpha}^i = \vec{S}^i / c \quad \nabla \cdot \vec{\alpha}^i = -\theta^i$$

$$\partial_4^2 \jmath_i + \gamma_{11}^{ij} \jmath_j + \gamma_{12}^{ij} \vec{J}_1^j = 0 \quad \partial_4 \overrightarrow{m}_i = \gamma_{12}^{ij} \,\vec{\jmath}_j + \gamma_{22}^{ij} \vec{J}_1^j \tag{6}$$

We are looking for a one-dimensional solution in the case that the defect field has no disclinations.

$$\begin{aligned}
\alpha_2 &= \alpha_2(0) e^{i(\omega t - kz)};\, m_1 = i m_1(0) e^{i(\omega t - kz)} \\
\jmath_1 &= \jmath_1(0) e^{i(\omega t - kz)};\, J_1 = J_1(0) e^{i(\omega t - kz)}.
\end{aligned}$$

Then the differential equations become the system

$$\begin{aligned} ik\alpha_2 + i\omega/cJ_1 - im_1 &= 0; \qquad ikJ_1 - (i\omega/c)\,\alpha_y = 0 \\ (-\omega^2 + \gamma_{11}^{11})\jmath_1 + \gamma_{12}^{11}J_1 &= 0 \qquad - im_1 + \omega\gamma_{12}^{11}i\jmath_1 + \gamma_{22}^{11}\omega iJ_1 = 0. \end{aligned}$$

This system of equations has a nontrivial solution only if the determinant of coefficient of $J_1, \alpha_2, m_1, \jmath_1$ vanishes (for simplification upper indexes are omitted)

$$\begin{bmatrix} \omega/c & k & -1 & 0 \\ k & -\omega/c & 0 & 0 \\ \gamma_{12} & 0 & 0 & \omega^2 - \gamma_{11} \\ \gamma_{22}\omega & 0 & -1 & \gamma_{12}\omega \end{bmatrix} = 0$$

The characteristic polynomial has the form
$(1 - \gamma_{22}c)\,\omega^4 + (k^2c^2 + \gamma_{12}^2c + \gamma_{22}c\gamma_{11} - \gamma_{11})\,\omega^2 - k^2c^2\gamma_{11} = 0.$
with solutions (Fig1)

$$\begin{aligned} \omega_1^2 &= \frac{1}{2}\frac{k^2c^2 + \gamma_{12}^2c + \gamma_{22}c\gamma_{11} - \gamma_{11} - \sqrt{D}}{-1 + \gamma_{22}c} \\ \omega_2^2 &= \frac{1}{2}\frac{k^2c^2 + \gamma_{12}^2c + \gamma_{22}c\gamma_{11} - \gamma_{11} + \sqrt{D}}{-1 + \gamma_{22}c} \end{aligned}$$

where
$D = k^4c^4 + 2k^2c^3\gamma_{12}^2 - 2k^2c^3\gamma_{22}\gamma_{11} + 2k^2c^2\gamma_{11} + \gamma_{12}^4c^2 + 2\gamma_{12}^2c^2\gamma_{22}\gamma_{11} - 2\gamma_{12}^2c\gamma_{11} + \gamma_{22}^2c^2\gamma_{11}^2 - 2\gamma_{22}c\gamma_{11}^2 + \gamma_{11}^2.$
For $k \to 0$

$$\begin{aligned} \omega_1^2 &= 0 \\ \omega_2^2 &= \omega_l^2, \end{aligned}$$

where
$\omega_l^2 = \frac{\varepsilon_0}{\varepsilon_\infty}\omega_t^2,\ \omega_t^2 = \gamma_{11},\ \varepsilon_\infty = \gamma_{22}c - 1,\ \varepsilon_0 = (\gamma_{12}^2c/\gamma_{11} + \varepsilon_\infty)\,.$
For large k

$$\begin{aligned} \omega_2^2 &= \frac{c^2}{\varepsilon_\infty}k^2 \\ \omega_1^2 &= \omega_t^2. \end{aligned}$$

Consider the case $D = 0$ which gives values of two frequencies

$$\begin{aligned} \omega_1^* &= \frac{k_1^2c^2 + c\gamma_{22}\gamma_{11} + \gamma_{12}^2c - \gamma_{11}}{-2 + 2\gamma_{22}c} \\ \omega_2^* &= \frac{k_2^2c^2 + c\gamma_{22}\gamma_{11} + \gamma_{12}^2c - \gamma_{11}}{-2 + 2\gamma_{22}c} \end{aligned}$$

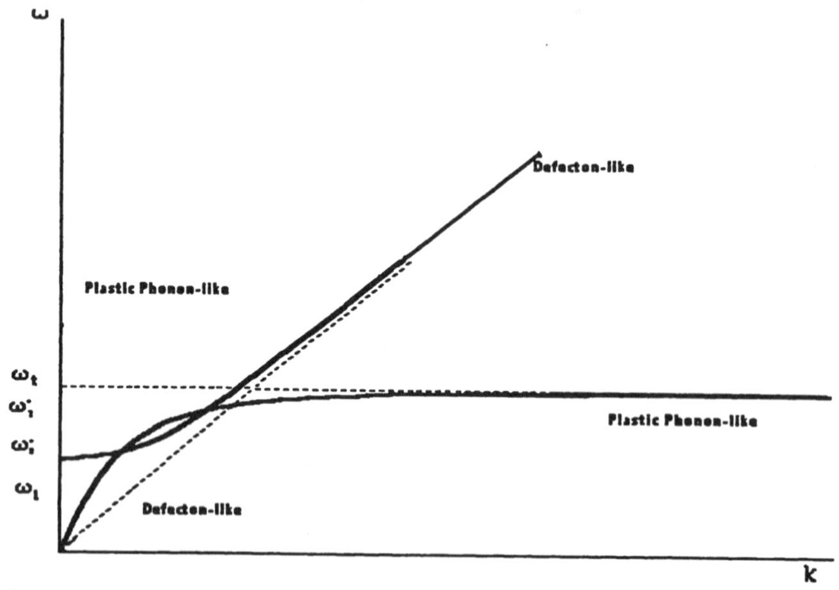

Figure 1: Couple modes of defect waves

where

$$k_1^2 = \frac{c\gamma_{22}\gamma_{11} - \gamma_{12}^2 c - \gamma_{11} + 2\sqrt{c\gamma_{12}^2\gamma_{11} - c^2\gamma_{12}^2\gamma_{22}\gamma_{11}}}{c^2}$$

$$k_2^2 = \frac{c\gamma_{22}\gamma_{11} - \gamma_{12}^2 c - \gamma_{11} - 2\sqrt{c\gamma_{12}^2\gamma_{11} - c^2\gamma_{12}^2\gamma_{22}\gamma_{11}}}{c^2}$$

In the case $\varepsilon_\infty = 0$ we obtain $\omega_1^* = \omega_2^*$

Thus interactions near the crossover where $\omega = \omega_2^*, \omega_1^*$ has the drastic effect of producing localized resonant vibrations. Thus we conclude that these local resonances are independent on macro-size of the specimen.

3 Conclusion

The use of this mesoscopic approach allows us to describe the behavior of dislocation tangles in terms of dislocation fields and to explain some resonances[4], detected at the macroscopic level.

ACKNOWLEDGMENT

This work was supported in part by Sandia National Laboratories under DOE Contract DE-AC04-98AL85000

References

[1] Aifantis E.C. 1993 *Proceedings of an International Symposium on Mechanics of Dislocations*, (edited by Aifantis E,C. and Hirch J.P.), Pensilvania: Choise Book.

[2] Kadić A. & Edelen D. G. B. 1983 *A Gauge Theory of Dislocations and Disclinations*. Heidelberg: Springer-Verlag.

[3] Eshelby J. D. 1951 *The Force on an Elastic Singularity,* Phil. Trans. Roy. Soc., Vol. A244, 87-112.

[4] Fitzgerald E. R. 1966 *Particle Waves and Deformation in Crystalline Solids.* New York: Interscience.

[5] Groma I. 1997 *Link between Microscopic and Mesoscopic length-scale description of the collective behavior of dislocations.* Physical review B, 56,5807-5813

[6] Gurtin M. E. 1995 *The nature of configuration forces.* Arch. Rational Mechanics. Anal. 131, 67-100

[7] Hirth J. P & Lothe J. 1968 *Theory of Dislocations,* New York: McGraw-Hill Book Company.

[8] Kröner E. 1983 *Field Theory of Defects in Solids: Present State, Merits and Open Questions.* Proceedings of the International Symposium on Mechanics of Dislocations, Houghton, Michigan.

[9] Maugin G.A. 1993 *The Role of Materials Forces in Inhomogeneous Elastic and Inelastic Solids,* in Anisotropy and Inhomogeneity in Elasticity and plasticity, Vol. 158, NY: AMD.

[10] Nabarro F. R. N. 1967 *Theory of Crystal Dislocations.* Oxford: University Press.

[11] Rojansky V. 1979 *Electromagnetic Fields and Waves,* New York: Dover.

[12] Truesdell C.A. & Nall W. 1965 *The Nonlinear Field Theories of Mechanics* in Handbuch der Physics, Bd. III/3, ed. S. Flugge, Berlin: Springer-Verlag.

Proceedings of Dynamic Systems and Applications 3 (2001) 577-584

Hopf Bifurcation Surfaces in Pioneer-Climax Competing Species Models

Suzanne Sumner
Mary Washington College, 1301 College Avenue, Fredericksburg, VA 22401-5358

ABSTRACT: Pioneer-climax competing species models using differential equations are studied where the per capita growth rates are functions of the total weighted densities of the populations. The per capita growth rate of the pioneer species is monotonically decreasing whereas the per capita growth rate of the climax species is a one-humped function of the total weighted density. Stocking and harvesting of either species are included in the model by constant rate forcing. The Hopf bifurcation surface is found for linear pioneer and one-humped climax fitnesses, thus indicating strategies for restoring the system to a stable equilibrium.

AMS (MOS) Subject Classification: 92B05, 92D25, 92D40.

1. INTRODUCTION

Competition for resources in an ecosystem of interacting, continuously reproducing species is both interspecific and intraspecific. To measure the effects of this competition between two different species of plants or animals, the total weighted density z_i of the populations is a linear combination of the population densities m_i of the individual species. Thus $z_i = c_{i1} m_1 + c_{i2} m_2$ where the constants $c_{ij} > 0$ represent the competitive effect of species j on species i. The per capita growth rate functions f_i, also called fitness functions, are taken to be functions of the total weighted density z_i. These systems are generalizations of Lotka-Volterra models, and other studies with this more general approach include Comins & Hassell [2], Selgrade & Namkoong [4], Franke & Yakubu [3], Buchanan & Selgrade [1], and Sumner [5].

The two types of species in question are pioneer and climax. The fitness function f_1 for the pioneer species m_1 is a monotonically decreasing function of the total density because it flourishes when the density is low, but crowding has a detrimental effect as density increases. However, the fitness function f_2 for the climax species m_2 is one-humped, i.e., it increases (because the climax experiences a beneficial effect from increasing density) until reaching a maximum and then decreases as a function of the total density (once overcrowding occurs). See Figure (1) for the form of the per capita growth rates. Examples of climax species can be found in Tonkyn's phytophagus insect study [6] as well as Selgrade & Namkoong's forestry models [4].

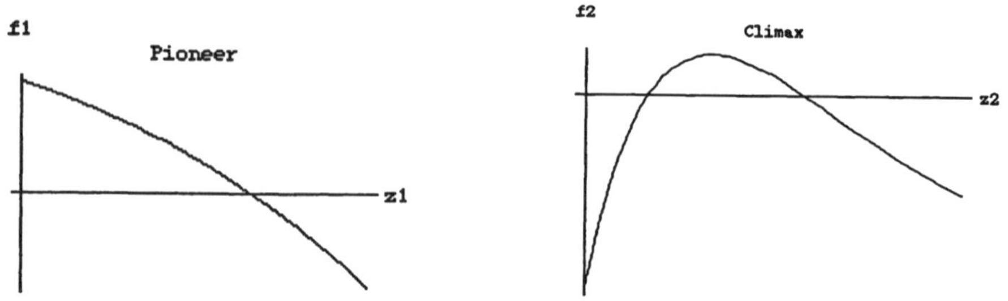

Figure (1): The per capita growth rates or fitness functions, f_1 and f_2, are functions of the total weighted densities of populations, z_1 and z_2, respectively. The pioneer fitness function f_1 is monotonically decreasing, and the climax fitness function f_2 increases to a maximum and then decreases.

Also included in the model are the effects of external constant rate forcing g_i on species i corresponding to stocking, if $g_i > 0$, or harvesting, if $g_i < 0$. For instance, a forest manager may use strategies of stocking or harvesting to assure stable coexistence of the pioneer and climax populations. Let $C = (c_{ij})$ be the matrix of competition coefficients c_{ij}. Note the z_i can be rescaled so that $c_{12} = c_{21} = 1$. The differential equation model becomes:

$$(1) \qquad \frac{dm_1}{dt} = m_1 f_1(z_1) + g_1, \qquad \frac{dm_2}{dt} = m_2 f_2(z_2) + g_2$$

Let $\vec{e} = (e_1, e_2)^T$ be an interior equilibrium of system (1), z_{ie} be z_i evaluated at $\vec{e}$, and F be the vector field of (1). Thus the Jacobian matrix of F, DF, evaluated at $\vec{e}$ has the form

$$(2) \quad DF(\vec{e}) = \begin{bmatrix} e_1 f_1'(z_{1e}) & 0 \\ 0 & e_2 f_2'(z_{2e}) \end{bmatrix} C + \begin{bmatrix} f_1'(z_{1e}) & 0 \\ 0 & f_2'(z_{2e}) \end{bmatrix} \quad \text{where}$$

$$(3) \quad \text{Trace}(DF(\vec{e})) = c_{11}\, e_1 f_1'(z_{1e}) - \frac{g_1}{e_1} + c_{22}\, e_2 f_2'(z_{2e}) - \frac{g_2}{e_2}$$

$$(4) \quad \text{Det}(DF(\vec{e})) = \text{Det}(C)\, e_1 e_2 f_1'(z_{1e}) f_2'(z_{2e}) - \frac{g_1 c_{22} e_2 f_2'(z_{2e})}{e_1} - \frac{g_2 c_{11} e_1 f_1'(z_{1e})}{e_2} + \frac{g_1 g_2}{e_1 e_2}.$$

Some information about the local stability of the system can be determined by Trace($DF(\vec{e})$) and Det($DF(\vec{e})$).

In particular, this paper studies the scenario where stable periodic behavior results from a Hopf bifurcation. A Hopf bifurcation is the emergence of a periodic cycle when an equilibrium changes stability as a parameter is varied. For a Hopf bifurcation to occur when the intraspecific competition parameter c_{11} is varied through the critical value c_{11}^*, the complex conjugate eigenvalues of $DF(\vec{e})$ must pass through the imaginary axis while the imaginary parts remain nonzero. Thus, a Hopf bifurcation occurs if Trace($DF(\vec{e})$) = 0, Det($DF(\vec{e})$) > 0 and $\mathrm{Re}'(\lambda_1(c_{11}^*)) \neq 0$. Selgrade & Namkoong [4] show that for the situation with no forcing, $g_1 = g_2 = 0$, a Hopf bifurcation occurs only if Det(C) < 0 and if the interior equilibrium $\vec{e}$ is specified by the smaller zero z_{2e}^* of the climax fitness so that $f_2'(z_{2e}^*) > 0$. Note for g_1 and g_2 very small, if Det(C) < 0, then Det($DF(\vec{e})$) > 0 from (4).

Buchanan & Selgrade [1] compute the Hopf bifurcation surface with respect to the intraspecific competition parameters c_{ii} and constant rate forcing parameters g_i for a specific linear pioneer fitness and quadratic climax fitness. Sumner [5] then generalizes this work to calculate the form of the Hopf bifurcation surface for any general linear pioneer and quadratic climax fitnesses. This work with general linear and quadratic fitnesses is reviewed in Section (2). In Section (3), a new bifurcation analysis is performed for any general linear pioneer and any one-humped climax to compute the corresponding Hopf bifurcation surface. Information about the Hopf bifurcation surface in terms of the intrinsic crowding parameters and constant rate forcing parameters yields tactics for returning a system with an unstable equilibrium to stability by either harvesting or stocking.

2. THE HOPF BIFURCATION SURFACE FOR THE LINEAR/QUADRATIC MODEL

This section reviews the analysis of the Hopf bifurcation surface for the general linear pioneer and general quadratic climax model found in Sumner [5] as a background for the work in Section (3). In that paper, Sumner [5] chooses $f_1(z_1) = \lambda - \mu z_1$ to be a general linear pioneer fitness function and $f_2(z_2) = \alpha(\gamma - z_2)(z_2 - \delta)$ to be a general quadratic climax fitness where the constants λ, μ, α, γ, and δ are all positive and $\gamma < \delta$.

To simplify the analysis of the linear/quadratic system (1) with the four parameters c_{11}, c_{22}, g_1, and g_2, Sumner [5] treats the case with forcing on only one population. For instance, the assumption that $g_2 = 0$ and g_1 is sufficiently small yields an interior equilibrium $\vec{e} = (e_1, e_2)^T$ and Det(C) < 0 if $0 < c_{11} < \dfrac{\lambda}{\mu\gamma}$ and $0 < c_{22} < \dfrac{\mu\gamma}{\lambda}$. Computing Trace($DF(\vec{e})$) = 0 from (3) yields the Hopf bifurcation surface $\{G = 0\}$ given below

(5) $$0 = G(c_{11}, c_{22}, g_1) = -2\, c_{11}\, \mu\, \gamma + \lambda + (2\, \mu\, c_{11} c_{22} + \alpha\, (\delta - \gamma)\, c_{22} - \mu)\, e_2$$

where e_2 is a function of c_{11}, c_{22}, and g_1. If the term $2\, \mu\, c_{11}\, c_{22} + \alpha\, (\delta - \gamma)\, c_{22} - \mu = 0$, then (5) is independent of g_1 so the vertical line $(c_{11}, c_{22}, g_1) = (\tilde{c}_{11}, \tilde{c}_{22}, g_1)$ where

(6) $$\tilde{c}_{11} = \frac{\lambda}{2\gamma\mu} \text{ and } \tilde{c}_{22} = \frac{\gamma\mu}{\lambda - \alpha\gamma^2 + \alpha\gamma\delta}$$

is contained on the surface $\{G = 0\}$.

Let C be the curve on the surface defined by C $= \{G = 0\} \cap \{g_1 = 0\}$. The point $(c_{11}, c_{22}) = (\tilde{c}_{11}, \tilde{c}_{22})$ is where the partial derivative $\frac{\partial G}{\partial g_1}$ vanishes on C. In addition, along C the partial derivatives $\frac{\partial g_1}{\partial c_{ii}}$ both change sign at $(c_{11}, c_{22}) = (\tilde{c}_{11}, \tilde{c}_{22})$. Therefore, the corresponding Hopf bifurcation surface has the form shown in Figure (2).

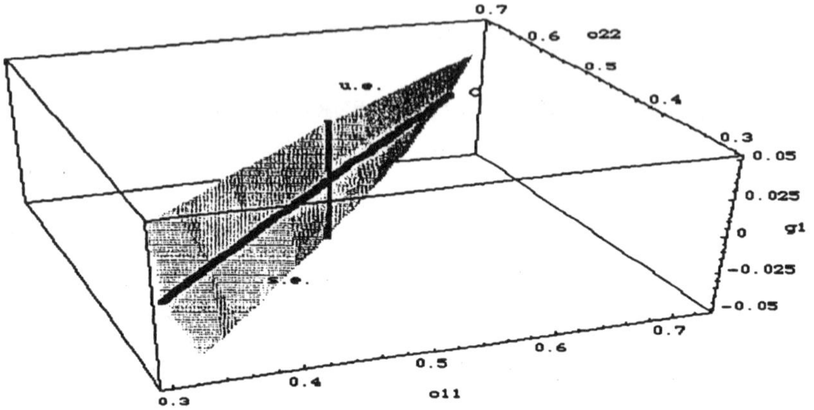

Figure (2): The graph of the Hopf bifurcation surface for a linear/quadratic model. The equilibrium is stable (s.e.) on one side of the surface and is unstable (u.e.) on the other side. If the competition parameters $c_{11} < \frac{1}{2}$ or $c_{22} < \frac{1}{2}$ and the equilibrium is unstable, then stocking is needed ($g_1 > 0$) to restore the system to a stable equilibrium as $(c_{11}, c_{22}, 0)$ lies below the Hopf bifurcation surface. Similarly, if $c_{11} > \frac{1}{2}$ or $c_{22} > \frac{1}{2}$ and the equilibrium is unstable, then harvesting ($g_1 < 0$) restores a stable equilibrium because $(c_{11}, c_{22}, 0)$ lies above the bifurcation surface.

For an unstable equilibrium where $g_1 = 0$ and either $c_{11} < \tilde{c}_{11}$ or $c_{22} < \tilde{c}_{22}$, then the point $(c_{11}, c_{22}, 0)$ lies below the bifurcation surface. Fixing c_{11} and c_{22} and increasing g_1 by stocking the pioneer species allows passage through the bifurcation surface to a stable equilibrium. Similarly, for an unstable equilibrium where $g_1 = 0$ and either $c_{11} > \tilde{c}_{11}$ or $c_{22} > \tilde{c}_{22}$, then the point $(c_{11}, c_{22}, 0)$ lies above the surface. Thus decreasing g_1 by harvesting the pioneer species restores the system to a stable equilibrium. Furthermore, for the case when $g_1 = 0$ and g_2 is sufficiently small, then the linear/quadratic Hopf bifurcation surface in terms of c_{11}, c_{22}, and g_2 is analogous to the one in Figure (2) so similar stocking or harvesting of the climax species can restabilize an unstable equilibrium.

3. THE HOPF BIFURCATION SURFACE FOR THE LINEAR/ONE-HUMPED MODEL

The research in this paper is motivated by the desire to generalize the fitness functions to produce a Hopf bifurcation surface that is qualitatively the same as in the linear/quadratic model. In other words, the goal is to determine the types of monotonically decreasing pioneer and one-humped climax fitnesses that yield a vertical line $(\tilde{c}_{11}, \tilde{c}_{22}, g_1)$ where the partial derivatives change sign. As a result, similar tactics of stocking and harvesting to restabilize an unstable equilibrium can be employed.

Assume the climax fitness f_2 is any one-humped function that is negative for low densities, increases until reaching a positive maximum, and then decreases as density increases. Suppose further that the monotonically decreasing pioneer fitness function f_1 has a power series representation in the variable z_1 so that $f_1(z_1) = a_0 + a_1 z_1 + a_2 z_1^2 + a_3 z_1^3 + a_4 z_1^4 + \cdots$. The Hopf bifurcation surface for either $g_1 = 0$ or $g_2 = 0$ will contain a vertical line $(\tilde{c}_{11}, \tilde{c}_{22}, g_1)$ or $(\tilde{c}_{11}, \tilde{c}_{22}, g_2)$ whenever $a_2 = a_3 = a_4 = \cdots = 0$, i.e., whenever f_1 is a linear pioneer fitness function.

This section outlines an analysis of the Hopf bifurcation surface for the general linear pioneer and general one-humped climax model and parallels the computation for the general linear pioneer and general quadratic climax model found in Sumner [5]. Let $f_1(z_1) = \lambda - \mu z_1$ be a linear pioneer fitness function where the constants λ, μ are positive and let $f_2(z_2)$ be any one-humped climax fitness function. Thus the corresponding system from (1) becomes

$$(7)\quad m_1' = m_1[\lambda - \mu(c_{11} m_1 + m_2)] + g_1, \qquad m_2' = m_2 f_2(m_1 + c_{22} m_2) + g_2 .$$

For the first scenario, assume $g_2 = 0$ and g_1 is sufficiently small. Let $z_{2e}^* = e_1 + c_{22}\, e_2$ be the smaller zero of f_2 so that $f_2'(z_{2e}^*) > 0$. Then the equilibrium $\vec{e} = (e_1, e_2)^T$ is interior and Det(C) < 0 if $0 < c_{11} < \dfrac{\lambda}{\mu z_{2e}^*}$ and $0 < c_{22} < \dfrac{\mu z_{2e}^*}{\lambda}$. Equating Trace($DF(\vec{e})$) = 0 using (3) produces the Hopf bifurcation surface $\{G = 0\}$ below.

$$(8) \qquad 0 = G(c_{11}, c_{22}, g_1) = -2\, c_{11}\, \mu\, z_{2e}^* + \lambda + (2\mu c_{11} c_{22} + c_{22} f_2'(z_{2e}^*) - \mu)\, e_2$$

where e_2 is a function of c_{11}, c_{22}, and g_1. For the term $2\, \mu\, c_{11}\, c_{22} + c_{22}\, f_2'(z_{2e}^*) - \mu = 0$, then (8) is independent of g_1 so the vertical line $(c_{11}, c_{22}, g_1) = (\tilde{c}_{11}, \tilde{c}_{22}, g_1)$ where

$$(9) \qquad \tilde{c}_{11} = \frac{\lambda}{2\mu z_{2e}^*} \text{ and } \tilde{c}_{22} = \frac{\mu z_{2e}^*}{\lambda + z_{2e}^* f_2'(z_{2e}^*)}$$

is contained on the surface $\{G = 0\}$.

As before, let C be the curve defined by C = $\{G = 0\} \cap \{g_1 = 0\}$. The point $(c_{11}, c_{22}) = (\tilde{c}_{11}, \tilde{c}_{22})$ is where the partial derivative $\dfrac{\partial G}{\partial g_1}$ again vanishes on C. As expected, along C the partial derivatives $\dfrac{\partial g_1}{\partial c_{ii}}$ both change sign at $(c_{11}, c_{22}) = (\tilde{c}_{11}, \tilde{c}_{22})$. Example (1) illustrates these previous computations with a specific example.

Example (1):
Let $f_1(z_1) = 1 - z_1$ be a linear pioneer fitness function, and let $f_2(z_2) = -1 + z_2 \exp(\frac{1}{2} - \frac{1}{2} z_2)$ be a climax fitness function having a linear times exponential form introduced by Selgrade & Namkoong [4]. Figure (3) below shows the corresponding Hopf bifurcation surface where $\tilde{c}_{11} = \frac{1}{2}$ and $\tilde{c}_{22} = \frac{2}{3}$.

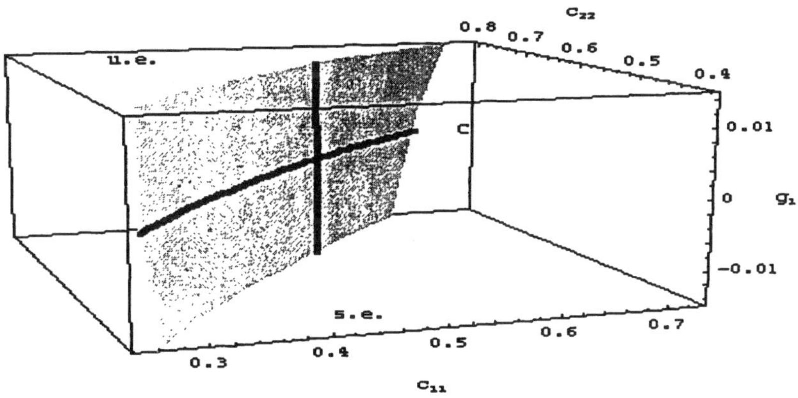

Figure (3): The graph of the Hopf bifurcation surface for a linear/exponential model. If the parameters $c_{11} < \frac{1}{2}$ or $c_{22} < \frac{2}{3}$ and the equilibrium is unstable, then stocking is needed ($g_1 > 0$) to restore the system to a stable equilibrium as $(c_{11}, c_{22}, 0)$ lies below the Hopf bifurcation surface. Similarly, if $c_{11} > \frac{1}{2}$ or $c_{22} > \frac{2}{3}$ and the equilibrium is unstable, then harvesting ($g_1 < 0$) restores a stable equilibrium because $(c_{11}, c_{22}, 0)$ lies above the bifurcation surface.

In the second scenario for $g_1 = 0$ and g_2 sufficiently small, then the following system of equations determines the equilibrium and Hopf bifurcation surface as functions of c_{11}, c_{22}, g_2 and m_2:

$$(10) \qquad 0 = G(c_{11}, c_{22}, g_2, m_2) = m_2 f_2\left(\frac{\lambda}{c_{11}\mu} - \frac{m_2}{c_{11}} + c_{22}m_2\right) + g_2$$

$$(11) \qquad 0 = H(c_{11}, c_{22}, g_2, m_2) = -\lambda + \mu\, m_2 + c_{22}\, m_2 f_2'\left(\frac{\lambda}{c_{11}\mu} - \frac{m_2}{c_{11}} + c_{22}m_2\right) - \frac{g_2}{m_2}$$

Employing the implicit function theorem at $g_2 = 0$ to solve (10) and (11) for m_2 and g_2 as functions of c_{11} and c_{22} requires that $\mathrm{Det}\left(\frac{\partial(G,H)}{\partial(m_2, g_2)}\right) \neq 0$. A long calculation shows that $\mathrm{Det}\left(\frac{\partial(G,H)}{\partial(m_2, g_2)}\right)$ is nonzero except at the point $(c_{11}, c_{22}) = (\hat{c}_{11}, \hat{c}_{22})$. If C is taken to be the curve $\{G = 0\} \cap \{H = 0\} \cap \{g_2 = 0\}$, then the partial derivatives $\frac{\partial g_2}{\partial c_{ii}}$ both change sign at $(c_{11}, c_{22}) = (\hat{c}_{11}, \hat{c}_{22})$. Thus the Hopf bifurcation surface is analogous to the case when $g_2 = 0$.

4. RESULTS

Stable behavior of the populations in the two-dimensional pioneer-climax system with constant rate forcing may be attained by passing through a Hopf bifurcation surface. With general linear/one-humped fitnesses, the Hopf bifurcation surface is computed for the scenario where forcing is applied to one species. A system with an unstable equilibrium may be restored to a stable equilibrium by stocking the pioneer (or climax) species if either intraspecific competition parameter is less than the given bounds. Likewise, harvesting the pioneer (or climax) species may restabilize an unstable equilibrium if either parameter is greater than the given bounds.

ACKNOWLEDGEMENTS

The author thanks Dr. W. A. Mangum for his helpful comments on this paper. The research reported in this paper was supported by a Faculty Development Grant from Mary Washington College, Fredericksburg, Virginia.

REFERENCES

1. Buchanan, J.R. & Selgrade, J.F. Discontinuous forcing of C^1 vector fields and applications to population interaction models. Canadian Appl. Math. Quarterly, Vol. 3 No. 2 (1995). 113-136.

2. Comins, H.N. & Hassell, M.P. Predation in multi-prey communities. Journal of Theoretical Biology, Vol. 62 (1976). 93-114.

3. Franke, J.E. & Yakubu, A. Mutual exclusion versus coexistence for discrete competitive systems. Journal of Mathematical Biology, Vol. 30 (1991). 161-168.

4. Selgrade, J.F. & Namkoong, G. Population interactions with growth rates dependent on weighted densities. Differential Equations Models in Biology, Epidemiology and Ecology (S. Busenberg, M. Martelli, eds.) Lecture Notes in Biomath. Vol. 92 (1991). Springer-Verlag, Berlin. 247-256.

5. Sumner, S. Hopf bifurcation in competing species models with linear and quadratic fitnesses. Dynamic Systems Appl. II, Conference Proceedings, (1996). 535-542.

6. Tonkyn, D.W. Predator-mediated mutualism: theory and tests in the homoptera. Journal of Theoretical Biology, Vol. 118 (1986). 15-31.

Proceedings of Dynamic Systems and Applications 3 (2001) 585-594

The Extension of the Quasilinearization Method for Nonlocal Reaction Diffusion Equations

A.S. Vatsala and Liwen Wang
Department of Mathematics
University of Louisiana at Lafayette
Lafayette, LA 70504

Abstract: In this paper we extend the method of quasilinearization to reaction diffusion equations with nonlocal conditions. The method yields monotone sequences which converge uniformly and monotonically to the unique solution of the nonlocal reaction diffusion equation. Furthermore, the rate of convergence is proved to be quadratic.
AMS (MOS) subject classification. 35K57, 35K60

1 Introduction

The method of quasilinearization of Bellman and Kalaba provides a numerical procedure for the computation of solutions of nonlinear problems. The method yields a monotone sequence of iterates which are solutions of linear problems. One can prove that the sequence converges uniformly and monotonically to the unique solution of a certain nonlinear problem in its interval of existence. The method also yields an increasing sequence of approximate solutions provided the forcing function is convex in the nonlinear terms. The main advantage of the method is that the iterates are solutions of linear equations and converge quadratically to the unique solution of the nonlinear problem. On the other hand if the forcing function is concave in u, one can obtain a dual result. However, the method of quasilinearization cannot be applied if the forcing function is the sum of a convex function and a concave function. Recently, the method of quasilinearization has been extended to ordinary differential systems [9] when the forcing function is split into the sum of a convex function and a concave function. In this chapter, we develop the extension of quasilinearization method to nonlocal reaction diffusion equations. Such problems can be formulated in terms of the theory of diffusion and heat conduction. For example, it can be applied to the description of the diffusion phenomenon of small amount of gas in a transparent tube. See [4, 5, 6] for details. For that purpose, consider the nonlocal reaction diffusion equation of the form

$$\begin{aligned} &\frac{\partial u}{\partial t} - Lu = f(t,x,u) + g(t,x,u) \qquad in\ Q_T \equiv (0,T] \times \Omega, \\ &u(t,x) = 0 \qquad on\ \Gamma \equiv (0,T] \times \partial\Omega \end{aligned} \tag{1.1}$$

$$u(0,x)+\sum_{i=1}^{N}\beta_i(x)u(x,T_i)=\psi(x) \qquad in \ \ \Omega,$$

where $T_i \in (0,T)$ $(i=1,2,\cdots,N)$, and L is a strictly uniformly elliptic operator defined by

$$L=\sum_{i,j=1}^{N} a_{i,j}(t,x)\frac{\partial^2}{\partial x_i \partial x_j}+\sum_{i=1}^{N} b_i(t,x)\frac{\partial}{\partial x_i}+c_0(t,x).$$

Here a_{ij}, b_i, c_0, ψ are bounded and sufficiently smooth functions. Furthermore, a_{ij} satisfies

$$d_0\left|\xi\right|^2 \leq \sum_{i,j=1}^{N} a_{ij}(t,x)\xi_i\xi_j \leq d_1\left|\xi\right|^2,$$

where d_0 and d_1 are independent of (t,x) and positive. Throughout this chapter, we also assume $\beta_i \leq 0$ and $-1 \leq \sum_{i=1}^{N} \beta_i(x) \leq 0$. $\psi(x)=0$ on $\partial\Omega$.

When one tries to extend the method of quasilinearization to (1.1), one can develop four main results as in [9]. However, in this chapter we develop the result corresponding to natural upper and lower solutions, only. The results corresponding to coupled upper and lower solutions of (1.1) are still open. For that purpose, appropriate comparison results have been developed. Two sequences are developed, starting from lower and upper solutions which will be proved to converge uniformly and monotonically to the unique solution of the nonlinear nonlocal reaction diffusion equation (1.1). Finally, it is proved that the convergence of the sequences to the unique solution of (1.1) is quadratic.

2 Preliminaries

In this section some existence and comparison results from [3] are needed to develop the main result.

Definition 2.1 A function $v_0 \in C^{1,2}[\overline{Q_T}, R]$ is called a lower solution of (1.1), if the inequalities

$$\frac{\partial v_0}{\partial t}-Lv_0 \leq f(t,x,v_0)+g(t,x,v_0),$$
$$v_0(t,x) \leq 0,$$
$$v_0(0,x)+\sum_{i=1}^{N}\beta_i(x)v_0(x,T_i) \leq \psi(x),$$

holds and an upper solution of (1.1) if the reversed inequalities hold.

The following maximum principle from [3] is needed to prove the monotonicity of iterates in the main result.

Theorem 2.1 If $u(t,x)$ satisfies

$$\frac{\partial u}{\partial t}-Lu+cu \geq 0 \qquad in \ \ \overline{Q_T}$$

$$u(t,x) \geq 0 \qquad on\ \Gamma,$$
$$u(0,x) + \sum_{i=1}^{N} \beta_i(x)u(x,T_i) \geq 0 \qquad in\ \Omega,$$

where $c_0(t,x) - c(t,x) \leq 0$. Then

$$u(t,x) \geq 0 \qquad in\ \overline{Q_T}.$$

Consider the linear nonlocal problem

$$\begin{aligned} &\frac{\partial u}{\partial t} - Lu = g(t,x) \qquad in\ \overline{Q_T}, \\ &u(t,x)|_{x\in\partial\Omega} = 0, \\ &u(0,x) + \sum_{i=1}^{N} \beta_i(x)u(x,T_i) = \psi(x) \qquad in\ \Omega. \end{aligned} \tag{2.1}$$

One can show that the solution of (2.1) exists on $\overline{Q_T}$.

Theorem 2.2 Assume that there exists v_0 and $w_0 \in C^{1,2}[\overline{Q_T}, R]$ which are lower and upper solutions of (2.1) such that $v_0(t,x) \leq w_0(t,x)$ on $\overline{Q_T}$. Furthermore, let $c_0(t,x) \leq 0$, $-1 \leq \sum_{i=1}^{N} \beta_i(x) \leq 0$ and $\beta_i \leq 0$. Then the nonlocal problem (2.1) has a unique solution belonging to $C^{1,2}[\overline{Q_T}, R]$ such that

$$v_0(t,x) \leq u(t,x) \leq w_0(t,x) \qquad \text{in } \overline{Q_T}.$$

Theorem 2.3 Assume the assumptions of Theorem 2.2 hold and $c_0(t,x) \leq -\mu$ in Ω, where μ is a positive constant. Then

$$|u(t,x)| \leq \frac{2}{\mu} e^{\frac{\mu T}{2}} \max_{Q_T}\{g(t,x)\} + (1 - e^{-\frac{\mu T}{2}})^{-1} \max_{Q_T}\{\psi(x)\}.$$

3 Extension of the Quasilinearization Method

In this section the extension of the quasilinearization method is developed. The method of extension of quasilinearization yields monotone sequences which converge uniformly and monotonically to the unique solution of (1.1). Furthermore, the convergence is quadratic.

Theorem 3.1 Assume that

(i) v_0 and w_0 satisfy

$$\frac{\partial v_0}{\partial t} - Lv_0 \leq f(t,x,v_0) + g(t,x,v_0), \text{ on } Q_T,$$
$$\frac{\partial w_0}{\partial t} - Lw_0 \geq f(t,x,w_0) + g(t,x,w_0), \text{ on } Q_T;$$

(ii) $f(t,x,u)$ and $g(t,x,u)$ are Hölder continuous such that $f_{uu} \geq 0$ and $g_{uu} \leq 0$;

(iii) $c_0(t,x) + f_u(t,x,w_0) + g_u(t,x,v_0) \leq 0$.

Then there exist monotone sequences $\{v_n\}$ and $\{w_n\}$ which converge uniformly and monotonically to the unique solution u of (1.1).

Proof. From (*ii*) one can see that $f(t,x,u)$ and $g(t,x,u)$ satisfy

$$f(t,x,u) \geq f(t,x,v) + f_u(t,x,v)(u-v) \tag{3.1}$$

and

$$g(t,x,u) \geq g(t,x,v) + g_u(t,x,u)(u-v) \tag{3.2}$$

for $u \geq v$.

Consider the nonlocal problem of the form

$$\begin{aligned} \frac{\partial v_1}{\partial t} - Lv_1 = &\ f(t,x,v_0) + g(t,x,v_0) \\ &+ [f_u(t,x,v_0) + g_u(t,x,w_0)](v_1 - v_0) \\ v_1(t,x) = 0 &\qquad \text{on } \Gamma, \\ v_1(0,x) + \sum_{i=1}^{N} &\beta_i(x) v_1(x,T_i) = \psi(x) \qquad in\ \ \Omega, \end{aligned} \tag{3.3}$$

and

$$\begin{aligned} \frac{\partial w_1}{\partial t} - Lw_1 = &\ f(t,x,w_0) + g(t,x,w_0) \\ &+ [f_u(t,x,w_0) + g_u(t,x,v_0)](w_1 - w_0) \\ w_1(t,x) = 0, &\qquad \text{on } \Gamma, \\ w_1(0,x) + \sum_{i=1}^{N} &\beta_i(x) w_1(x,T_i) = \psi(x) \qquad in\ \ \Omega. \end{aligned} \tag{3.4}$$

Initially, it will be proved that v_0 and w_0 are also lower and upper solutions of (3.3) and (3.4), respectively. Clearly v_0 is the lower solution of (3.3). Using (3.1) and (3.2), it follows that

$$\begin{aligned} \frac{\partial w_0}{\partial t} - Lw_0 &\geq f(t,x,w_0) + g(t,x,w_0) \\ &\geq f(t,x,v_0) + g(t,x,v_0) \\ &\quad + [f_u(t,x,v_0) + g_u(t,x,w_0)](w_0 - v_0), \text{ in } Q_T, \\ \text{and} \qquad w_0(t,x) &\geq 0. \qquad \text{on } \Gamma, \\ w_0(0,x) &+ \sum_{i=1}^{N} \beta_i(x) w_0(x,T_i) \geq \psi(x) \qquad \text{in } \Omega. \end{aligned}$$

This proves that w_0 is an upper solution of (3.3). Similarly one can prove v_0 and w_0 are the lower and upper solutions of (3.4). Using Theorem 2.2, the solutions of (3.3) and (3.4) exist and are unique and $v_0 \leq v_1$, $w_1 \leq w_0$.

Next, it will be established that $v_0 \leq v_1 \leq w_1 \leq w_0$. For this purpose, set $\gamma = v_1 - w_1$, then it follows that,

$$\begin{aligned}\frac{\partial \gamma}{\partial t} - L\gamma = \; & f(t,x,v_0) - f(t,x,w_0) \\ & + f_u(t,x,v_0)[(v_1 - w_1) + (w_0 - v_0)] \\ & + g(t,x,v_0) - g(t,x,w_0) \\ & + g_u(t,x,w_0)[(v_1 - w_1) + (w_0 - v_0)] \quad in\ Q_T\end{aligned}$$

Using the fact

$$\begin{aligned} f(t,x,v_0) - f(t,x,w_0) + f_u(t,x,v_0)(w_0 - v_0) \leq 0 \\ g(t,x,v_0) - g(t,x,w_0) + g_u(t,x,w_0)(w_0 - v_0) \leq 0, \end{aligned}$$

it follows that

$$\begin{aligned} & \frac{\partial \gamma}{\partial t} - L\gamma \leq [f_u(t,x,v_0) + g_u(t,x,w_0)]\gamma \qquad \text{on } Q_T, \\ & \gamma(0,x) + \sum_{i=1}^{N} \beta_i(x)\gamma(x,T_i) = 0 \qquad \text{in } \Omega, \end{aligned}$$

and $\gamma(0,x) \leq 0$ on Γ. Applying Theorem 2.1, it follows that $\gamma(t,x) \leq 0$. From this, one can conclude

$$v_0 \leq v_1 \leq w_1 \leq w_0 \, .$$

Now, assume that for some $k > 1$, the inequality

$$v_0 \leq v_1 \leq \cdots \leq v_i \leq \cdots \leq v_k \leq w_k \leq \cdots \leq w_i \leq \cdots \leq w_1 \leq w_0$$

holds, where v_i and w_i are solutions of following linear system

$$\begin{aligned} \frac{\partial v_i}{\partial t} - Lv_i = \; & f(t,x,v_{i-1}) + g(t,x,v_{i-1}) \\ & + [f_u(t,x,v_{i-1}) + g_u(t,x,w_{i-1})](v_i - v_{i-1}) \quad \text{in } Q_T, \\ v_i(t,x) = 0 \qquad & \text{on } \Gamma, \\ v_i(0,x) + \sum_{i=1}^{N} \beta_i(x) & v_i(x,T_i) = \psi(x) \qquad in\ \Omega, \end{aligned} \tag{3.5}$$

and it follows that,

$$\begin{aligned} \frac{\partial w_i}{\partial t} - Lw_i = \; & f(t,x,w_{i-1}) + g(t,x,w_{i-1}) \\ & + [f_u(t,x,w_{i-1}) + g_u(t,x,v_{i-1})](w_i - w_{i-1}), \quad \text{in } Q_T, \\ w_i(t,x) = 0 \qquad & \text{on } \Gamma, \\ w_i(0,x) + \sum_{i=1}^{N} \beta_i(x) & w_i(x,T_i) = \psi(x) \qquad in\ \Omega, \end{aligned} \tag{3.6}$$

for $i = 1, 2, \cdots k$.

Let v_{k+1} and w_{k+1} be solutions of the following linear equations

$$\begin{aligned} \frac{\partial v_{k+1}}{\partial t} - Lv_{k+1} = \; & f(t,x,v_k) + g(t,x,v_k) \\ & + [f_u(t,x,v_k) + g_u(t,x,w_k)](v_{k+1} - v_k) \\ v_{k+1}(t,x) = 0 \qquad & \text{on } \Gamma, \\ v_{k+1}(0,x) + \sum_{i=1}^{N} \beta_i(x) & v_{k+1}(x,T_i) = \psi(x) \qquad in\ \Omega, \end{aligned} \tag{3.7}$$

and

$$\begin{aligned} \frac{\partial w_{k+1}}{\partial t} - Lw_{k+1} = & \quad f(t,x,w_k) + g(t,x,w_k) \\ & +[f_u(t,x,w_k) + g_u(t,x,v_k)](w_{k+1} - w_k), \\ w_{k+1}(t,x) = 0 & \qquad \text{on } \Gamma, \\ w_{k+1}(0,x) + \sum_{i=1}^{N} \beta_i(x) w_{k+1}(x,T_i) = \psi(x) & \qquad in \ \Omega, \end{aligned} \tag{3.8}$$

respectively.

One will prove that $v_k \leq v_{k+1} \leq w_{k+1} \leq w_k$ on $\overline{Q_T}$ as follows.

Initially, one can show that v_k and w_k are lower and upper solutions of (3.7), respectively. Using (3.1) and (3.2), it follows that

$$\begin{aligned} \frac{\partial v_k}{\partial t} - Lv_k = & \quad f(t,x,v_{k-1}) + g(t,x,v_{k-1}) \\ & +[f_u(t,x,v_{k-1}) + g_u(t,x,w_{k-1})](v_k - v_{k-1}) \\ & \leq f(t,x,v_k) + g(t,x,v_k) \qquad \text{in } Q_T, \end{aligned}$$

which proves that v_k is the lower solution of (3.7). Similarly using (3.1) and (3.2), it follows that

$$\begin{aligned} \frac{\partial w_k}{\partial t} - Lw_k = & \quad f(t,x,w_{k-1}) + g(t,x,w_{k-1}) \\ & +[f_u(t,x,w_{k-1}) + g_u(t,x,v_{k-1})](w_k - w_{k-1}) \\ \geq & \quad f(t,x,w_k) + g(t,x,w_k) \\ & +f_u(t,x,v_k)(w_{k-1} - v_k) + g_u(t,x,w_{k-1})(w_{k-1} - v_k) \\ & +f_u(t,x,w_{k-1}) + g_u(t,x,v_{k-1})](w_k - w_{k-1}) \\ \geq & \quad f(t,x,w_k) + g(t,x,w_k) \\ & +f_u(t,x,v_k)(w_{k-1} - v_k) + g_u(t,x,w_k)(w_{k-1} - v_k) \\ & +[f_u(t,x,v_k) + g_u(t,x,w_k)](w_k - w_{k-1}) \\ \geq & \quad f(t,x,w_k) + g(t,x,w_k) \\ & +[f_u(t,x,v_k) + g_u(t,x,w_k)](w_k - v_k) \qquad \text{in } Q_T \end{aligned}$$

and

$$w_k(t,x) = 0 \qquad \text{on } \Gamma,$$

$$u(0,x) + \sum_{i=1}^{N} \beta_i(x) u(x,T_i) = \psi(x) \qquad in \ \Omega,$$

This proves that w_k is the upper solution of (3.7). Therefore, using Theorem 2.2, there exists a unique solution v_{k+1} such that $v_k \leq v_{k+1} \leq w_k$.

Similarly one can also prove that there exists a unique solution w_{k+1} such that $v_k \leq w_{k+1} \leq w_k$. In order to prove $v_{k+1} \leq w_{k+1}$, let $\gamma = v_{k+1} - w_{k+1}$. Then one can obtain

$$\frac{\partial \gamma}{\partial t} - L\gamma \leq [f_u(t,x,v_k) + g_u(t,x,w_k)]\gamma$$

$$\gamma(0,x) + \sum_{i=1}^{N} \beta_i(x) \gamma(x,T_i) = \psi(x) \qquad in \ \Omega,$$

and $\gamma(t,x) \leq 0$ on Γ. Using Theorem 2.1, it follows that $\gamma(t,x) \leq 0$. From this one can conclude

$$v_k \leq v_{k+1} \leq w_{k+1} \leq w_k.$$

Using mathematical induction, one can see that v_n and w_n satisfy

$$v_0 \leq v_1 \leq \cdots \leq v_n \leq w_n \leq \cdots \leq w_1 \leq w_0 \quad \text{for all } n.$$

One can show that v_n and w_n converge uniformly and monotonically such that $\lim_{n\to\infty} v_n = v(t,x) \leq \lim_{n\to\infty} w_n = w(t,x)$. Now, one can establish that $v(t,x)$, $w(t,x)$ are solutions of (1.1). Obviously v and w satisfy the boundary conditions and nonlocal initial conditions. Since a_{ij}, b_i, c_0, ψ are sufficiently smooth, $v(t,x)$ and $w(t,x)$ also satisfy

$$\begin{aligned} v(t,x) = {} & \int_\Omega G(t,x;y,0)v(0,y)dy && (3.9)\\ & - \int_0^t \int_\Omega G(t,x;\tau,y)(f(t,y,v(\tau,y)) + g(t,y,v(\tau,y))dyd\tau, \\ w(t,x) = {} & \int_\Omega G(t,x;y,0)w(0,y)dy && (3.10)\\ & - \int_0^t \int_\Omega G(t,x;\tau,y)(f(t,y,w(\tau,y)) + g(t,y,w(\tau,y))dyd\tau, \end{aligned}$$

where G is the Green's function for the operator $\frac{\partial}{\partial t} - L$. Using (3.9) and (3.10), one can verify that $v(t,x)$, $w(t,x)$ satisfy the equation (1.1). Hence $v(t,x)$, $w(t,x)$ are solutions of (1.1).

In order to prove $v(t,x) \equiv w(t,x)$ on $\overline{Q}_T$, set $\alpha(t,x) = v(t,x) - w(t,x)$. From this, it follows that

$$\begin{aligned} \frac{\partial \alpha}{\partial t} - L\alpha = {} & f(t,x,v) - f(t,x,w) \\ & + g(t,x,v) - g(t,x,w). \end{aligned}$$

Using the assumptions on f and g from the hypotheses, one can get

$$\begin{aligned} & f(t,x,v) \geq f(t,x,w) + f_u(t,x,w)(v-w), \\ & g(t,x,v) \geq g(t,x,w) + g_u(t,x,v)(v-w). \end{aligned}$$

From this, it follows that

$$\begin{aligned} & \frac{\partial \alpha}{\partial t} - L\alpha \geq [f_u(t,x,w) + g_u(t,x,v)]\alpha, \\ & \alpha(t,x)|_\Gamma = 0, \\ & \alpha(0,x) + \sum_{i=1}^N \beta_i(x)\alpha(x,T_i) = 0 \qquad in\ \Omega, \end{aligned}$$

Using Theorem 2.1, one can conclude $\alpha \geq 0$. This proves $v \equiv w = u$, where u is the unique solution of (1.1). ■

Theorem 3.2 Let the assumptions of Theorem 3.1 hold. Further let functions $|f_u|$, $|g_u|$, $|f_{uu}|$, $|g_{uu}|$ are bounded by M for $0 \leq t \leq T, x \in \Omega, v_0 \leq u \leq w_0$, where M is a positive constant. Assume further that, $c_0(t,x) + 2M \leq -\mu < 0$. Then the convergence of the sequences of Theorem 3.1 is quadratic.

Proof. In order to prove quadratic convergence, set $\alpha_{k+1} = u(t,x) - v_{k+1} \geq 0$, $\gamma_{k+1} = w_{k+1} - u \geq 0$. Then it follows that,

$$\alpha_{k+1}(0,x)+\sum_{i=1}^{N}\beta_i(x)\alpha_{k+1}(T_i,x)=0,$$
$$\gamma_{k+1}(0,x)+\sum_{i=1}^{N}\beta_i(x)\gamma_{k+1}(T_i,x)=0.$$

Hence

$$\begin{aligned}\frac{\partial\alpha_{k+1}}{\partial t}-L\alpha_{k+1}&=f(t,x,u)+g(t,x,u)-\{f(t,x,v_k)+g(t,x,v_k)\\&\quad+[f_u(t,x,v_k)+g_u(t,x,w_k)](v_{k+1}-v_k)\ \}\\&=f(t,x,u)+g(t,x,u)-f(t,x,v_k)-g(t,x,v_k)\\&\quad-[f_u(t,x,v_k)+g_u(t,x,w_k)](v_{k+1}-u+u-v_k)\\&\le[f_u(t,x,u)-f_u(t,x,v_k)]\alpha_k+[g_u(t,x,v_k)-g_u(t,x,w_k)]\alpha_k\\&\quad+[f_u(t,x,v_k)+g_u(t,x,w_k)]\alpha_{k+1}\\&\le M\alpha_k^2+M[\gamma_k+\alpha_k]\alpha_k+2M\alpha_{k+1}\\&=2M\alpha_k^2+2M\alpha_{k+1}+2M\alpha_k\gamma_k\\&\le 3M\alpha_k^2+2M\alpha_{k+1}+M\gamma_k^2,\quad\text{on } Q_T.\end{aligned}$$

Similarly, one can prove that

$$\frac{\partial\gamma_{k+1}}{\partial t}-L\gamma_{k+1}\le M\gamma_k^2+2M\gamma_{k+1}+M\alpha_k^2,\quad\text{on } Q_T.$$

Using the Comparison Theorem 2.3, it follows that

$$0\le\alpha_{k+1}(t,x)\le\tfrac{2}{\mu}e^{\frac{\mu T}{2}}\max_{Q_T}\{3M\alpha_k^2+M\gamma_k^2\}.$$

From this, one can conclude that

$$\max|u-v_{k+1}|\le\tfrac{2}{\mu}e^{\frac{\mu T}{2}}\max\{3M\,|u-v_k|_k^2+M\,|u-w_k|_k^2\}.$$

Similarly, one can also conclude that

$$\max|u-w_{k+1}|\le\tfrac{2}{\mu}e^{\frac{\mu T}{2}}\max\{M\,|u-w_k|_k^2+M\,|u-v_k|_k^2\}.$$

This completes the proof. ■

REFERENCES

[1] R. Bellman, *Methods of Nonlinear Analysis*, Vol. II, Academic Press, New York, 1973.

[2] R. Bellman and R. Kalaba, *Quasilinearization and Nonlinear Boundary Value Problems*, American Elsevier, New York 1965.

[3] J. Chabrowski, On nonlocal problems for parabolic equations, Nagoya Math. J., **93**, 109-131 (1984).

[4] W. A. Day, Extensions of a property of the heat equation to linear thermoelasticity and theories, Quart. Appl. Math., **40**, 319-330 (1982).

[5] W. A. Day, A decreasing property of solutions of parabolic equations with applications to thermoelasticity, Quart. Appl. Math., **40**, 468-475 (1983).

[6] K. Deng, Exponential decay of solutions of semilinear parabolic equations with nonlocal initial conditions, J. Math. Anal. Appl., **179**, No. 2, 630-637(1993).

[7] G.S. Ladde, V. Lakshmikantham and A.S. Vatsala, *Monotone Iterative Techniques for Nonlinear Differential Equations,* Pitman, Boston, 1985

[8] V. Lakshmikantham, An extension of the method of quasilinearization, J. Optimization, Theory and Applications, **82**, No. 2, 315-32(1994).

[9] V. Lakshmikantham and A.S. Vatsala, *Generalized Quasilinearization for Nonlinear Problem,* Kluwer Academic Publishers, 1998.

[10] C.V. Pao, *Nonlinear Parabolic and Elliptic Equations*, Plenum Press, New York 1992.

[11] A.S. Vatsala, Generalized quasilinearization and reaction diffusion equations, Nonlinear Times and Digest, **1**, 211-220 (1994).

[12] A.S. Vatsala, An extension of the method of quasilinearization for reaction diffusion equations, *Comparison methods and stability theory*, Marcel Dekker, Inc. 1994.

[13] A.S. Vatsala and L. Wang, The generalized quasi-linearization method for reaction diffusion equations on an unbounded domain, J. Math. Anal. Appl., **237**, 644-656 (1999).

Proceedings of Dynamic Systems and Applications 3 (2001) 595-602

Impulsive Reaction-Diffusion System and Stability by Vector Lyapunov Functions

A. S. Vatsala[1] and Xiaoquan Zhao[2]
[1]Department of Mathematics, The University of Louisiana, Lafayette, LA 70504
[2]The Center for Advanced Computer Studies, The University of Louisiana, Lafayette, LA 70504

ABSTRACT: In this paper, we study the stability properties of weekly coupled reaction-diffusion equations with fixed moment of impulses. We first prove a comparison result which will be used to prove our main theorem on stability of the above mentioned equations. Theorem 2 is a special case of Theorem 1 which assumes that the nonlinear term and the impulsive term of the equations should satisfy one side Lipschitz condition. Theorem 3 apply theorem 1 and theorem 2 to study the stability of the weekly coupled reaction-diffusion equations.

AMS(MOS) subject calssification. 34A37, 34D20

1 INTRODUCTION

It is known that using vector Lyapunov functions instead of a single Lyapunov function offers more flexibility [1,2,3], and vector Lyapunov function is a natural tool in the investigation of the stability properties of large scale dynamic systems [4]. Stability properties of weakly coupled reaction-diffusion systems by means of a vector Lyapunov function with initial conditions and nonlocal boundary conditions are presented in [1,5] respectively. Also it is known that the study of impulsive dynamic systems are very useful due to its application in population dynamics, chemical reactions, bursting rhythm models in biology, etc. In this paper, we investigate the stability properties of weakly coupled reaction-diffusion equations with fixed moment of impulses in terms of the stability properties of ordinary differential equations with impulses. This is achieved by developing appropriate comparison theorems, as in [6]. We present two examples to demonstrate the application of our main results.

2 PRELIMINARIES

We introduce some notations and definitions to develop our comparison and stability results. Let Ω be a smooth bounded domain in R^n with boundary $\partial\Omega \in C^{2+\alpha}$ and closure $\bar{\Omega}, Q_T = (0,T] \times \Omega$, $\bar{Q}_T = [0,T] \times \Omega$, $\Gamma_T = (0,T] \times \partial\Omega$, and $\bar{\Gamma}_T = [0,T] \times \partial\Omega$. For a given partition of interval $[0,T], 0 < \tau_1 < \tau_2 < \cdots < \tau_p < T$, we introduce the following notation:

$$M_i = \{(\tau_i, x) : \tau_i \in (0,T), x \in \Omega\}, M = \cup_{i=1}^{p} M_i$$
$$N_i = \{(\tau_i, x) : \tau_i \in (0,T), x \in \Omega\}, N = \cup_{i=1}^{p} N_i$$

If $t \in [0,\infty)$, then we denote Q_T as Q_∞ and Γ_T as Γ_∞. We define $C^{\tilde{}}(Q_T)$ as the set of all functions $u : [0,T] \times \Omega \to R^m$ satisfying the following conditions:

(i) $u(t,x)$ is of class $C^{1,1}_{t,x}$ for $(t,x) \in \bar{Q}_T \setminus (M \cup N), t \neq 0, T$;

(ii) $u_{xx}(t,x)$ exists and is continuous for $(t,x) \in Q_T \setminus M$;

(iii) for $v = (u, u_t, u_x, u_{xx})$, $\lim_{t \to \tau_k^-} v(t,x) = v(\tau_k, x)$, and $\lim_{t \to \tau_k^+} v(t,x)$ exists for $k = 1,2,3 \cdots p$, and $x \in \bar{\Omega}$, where $u_t = \partial u/\partial t, u_x = (\partial u/\partial x_1, \partial u/\partial x_2, \cdots \partial u/\partial x_n)$, and $u_{xx} = (\partial^2 u/\partial x_1^2, \partial^2 u/\partial x_2^2, \cdots \partial^2 u/\partial x_n^2)$.

Let $f \in C(Q_T \times R^m, R^m)$. (Here f represents a vector whose kth component is denoted by $f_k(t,x,u)$, and u_k is the kth component of $u, k = 1,2,3, \cdots m$). We say f is quasimonotone nondecreasing in u if for each i, $u \leq v, u_i = v_i$ implies $f_i(t,x,u) \leq f_i(t,x,v)$.

We consider the following system of impulsive parabolic equations:

$$\begin{cases} u_{k_t} = L_k u_k + f_k(t,x,u), & (t,x) \in Q_T \setminus M, \\ B_k u_k = \varphi_k(t,x), & (t,x) \in \Gamma_T \setminus N, \\ u_k(0,x) = \eta_{k_0}(x), & x \in \bar{\Omega}, \\ u_k(\tau_i^+, x) - u_k(\tau_i, x) = g_{k_i}(x, u_k(\tau_i, x)), & x \in \bar{\Omega}. \end{cases} \tag{1}$$

For convenience of use, we state the following assumptions for $k = 1,2,3 \cdots m$:

(A_0) (i) $L_k u_k = \sum_{i,j=1}^{n} a_{k_{ij}} u_{k_{x_i x_j}} + \sum_{j=1}^{n} b_{k_j} u_{k_{x_j}}$, L_k is a uniformly elliptic operator for each $k = 1,2,\cdots m$, i.e., there exist positive numbers ρ_1, ρ_2 such that $\rho_1 \mid \lambda \mid^2 \leq \sum_{i,j=1}^{n} a_{k_{ij}} \lambda_i \lambda_j \leq \rho_2 \mid \lambda \mid^2$, for any $\lambda \in R^n$.

(ii) f is quasimonotone nondecreasing in u and $\eta_{k_0} \in C(\bar{\Omega})$.

(iii)

$$B_k u_k = p_k(t,x)u_k(t,x) + q_k(t,x)\frac{\partial}{\partial v}u_k(t,x) \quad on\ \Gamma_T \backslash N,$$

where p_k and q_k are nonnegative continuous functions with $p_k + q_k > 0$ on Γ_T and $\varphi_k(t,x) \in C(\Gamma_T)$.

(iv) g_{k_i} is continuously differentiable in u and $(1 + g_{k_{iu}}) > 0$.

The function $u(t,x) \in C^{\cdot}(Q_T)$ is called an upper solution of the system (2.1) if it satisfies

$$\begin{cases} u_{k_t} \geq L_k u_k + f_k(t,x,u), & (t,x) \in Q_T \backslash M, \\ B_k u_k \geq \varphi_k(t,x), & (t,x) \in \Gamma_T \backslash N, \\ u_k(0,x) \geq \eta_{k_0}(x), & x \in \bar{\Omega}, \\ u_k(\tau_i^+,x) - u_k(\tau_i,x) \geq g_{k_i}(x, u_k(\tau_i,x)), & x \in \overline{\Omega}\,. \end{cases}$$

Lower solution can be defined analogously by reversing the above inequalities.

Consider the following system of impulsive differential equations:

$$\begin{aligned} &\frac{dx}{dt} = f(t,x), && t \neq \tau_i, i = 1, 2, \cdots, \\ &x(\tau_i^+) - x(\tau_i) = I_i(\tau_i, x(\tau_i)), \\ &x(t_0) = x_0, \end{aligned}$$

where $f : R^+ \times D \to R^n, R^+ = [0, \infty)$, D is a domain in R^n and $D \neq \emptyset$; $\{\tau_i\}$ is a sequence of impulsive moments, $t_0 < \tau_1 < \tau_2 < \cdots$; $I_i : R^+ \times D \to R^n$; $t_0 \in R^+$; $x_0 \in D$. We denote by $x(t; t_0, x_0)$ the solution of the above equations.

3 COMPARISON RESULTS

In this section, we develop comparison results relative to equation (1). These comparison results will be needed to develop our main results on stability.

Theorem 1. Assume that $v, w \in C^{\cdot}(Q_T)$ are given such that (A_0) holds. Further more,

(A_1) (i) v, w are lower and upper solutions of the system (1) ;

(ii) there exists a function $z(t,x)$ which is a C^1–function on $Q_T \backslash M \cup N$ such that $\inf\limits_{\bar{Q}_T \backslash M \cup N} z(t,x) > 0$, $\inf\limits_{\Gamma_T \backslash N} \dfrac{\partial z}{\partial \nu} > 0$ for sufficiently small $\varepsilon > 0$ and arbitrary given $u \in C^{\cdot}(Q_T)$;

(a) $\inf\limits_{(t,x)\in Q_T\setminus M} [\varepsilon z_{k_t} - \varepsilon L_k z_k - f_k(t,x,u+\varepsilon z) + f_k(t,x,u))] > 0;$

(b) $\varepsilon[z_k(\tau_i^+,x) - z_k(\tau_i,x)] \geq g_{k_i}(x, u_k(\tau_i,x) + \varepsilon z_k(\tau_i,x)) - g_{k_i}(x, u(\tau_i,x))$; $if\ x \in \bar{\Omega}$
then $v \leq u$ on Q_T.

In the next result, we show that if f_k and g_k satisfies the one-sided Lipschitz condition, then the assumption $(A_0)(ii)$ holds. Thus, we obtain the same conclusion as in Theorem 1 when f_k and g_k satisfy the Lipschitz condition.

Theorem 2.2. Assume condition (i) in Theorem 1 holds and $p > 0$. If there exist positive numbers L_1 and α such that

$$f(t,x,u) - f(t,x,v) \leq L_1(u-v), \quad u > v,$$

$$g_k(x,u_k) - g_k(x,v_k) \leq \alpha(u_k - v_k), \qquad u_k > v_k, k = 1,2,\cdots m,$$

then $v \leq w$ on Q_T.

Next, we present an example as an application of the above comparison results.

Example 1. Consider the system

$$\begin{cases} v_t = Av_{xx} - bv_x + f(t,v), & 0 \leq x \leq 1, 0 \leq t < \infty, t \neq \tau_i, \\ v(t,0) = v(t,1) = 0, & 0 \leq t < \infty, t \neq \tau_i, \\ v(0,x) = \eta_0(x) \geq 0, & 0 \leq x \leq 1, \\ v(\tau^+,x) - v(\tau,x) = l \cdot v(\tau,x), & 0 \leq x \leq 1, \end{cases} \tag{2}$$

where $A > 0$ is a diagonal matrix, $b > 0$ and $1 + l < 1$. Further more, f satisfies $f(t,0) = 0$,

$$f(t,u) - f(t,v) \leq L_1(u-v), L_1 > 0 \text{ whenever } 0 \leq v \leq u \leq Q, \text{ for some } Q > 0.$$

Define

$$R(t,x) = \begin{cases} k_1 e^{-\alpha_1 t + \beta(1-x)}\mathbf{e}, & (t,x) \in [0,\tau) \times [0,1] \\ k_1(1+l)e^{-\alpha_1 t + \beta(1-x)}\mathbf{e}, & (t,x) \in [\tau,\infty) \times [0,1], \end{cases}$$

where $\alpha_1, k_1 > 0$ and $\beta \in R$ to be chosen. Substituting R in the system (2), we get

$$\begin{aligned} R_t - AR_{xx} + bR_x - f(t,R) &\geq R(t,x)[-\alpha_1 - \beta^2\alpha - b\beta] - f \\ &\geq R[-\alpha_1 - \alpha\beta^2 - b\beta - L_1], \end{aligned}$$

where $\alpha = \max A_{ii}$. Let $\alpha_1 = L_0 - L_1 (> 0)$. Then,

$$R_t - AR_{xx} + bR_x - f \geq -R[L_0 + b\beta + \alpha\beta^2] = 0.$$

Let β be one of the roots of $\alpha\beta^2 + b\beta + L_0 = 0$. If $0 < \alpha < b^2/4L_0$, then we can choose $\beta < 0$. Suppose $| \eta_0(x) | < Q_1$. Let $k_1 = Q_1 e^{-\beta}$, then $R(0,x) \geq \eta_0(x)$ on $[0,1]$. We can see easily that $R(t,0) > 0, R(t,1) > 0$, and $R(\tau^+,x) - R(\tau,x) = lR(\tau,x)$. Now applying Theorem 2.2, we get

$$0 \leq v(t,x) \leq R(t,x) \quad on\ [0,\infty) \times [0,1].$$

4 STABILITY RESULTS

In this section we use the method of vector Lyapunov functions and the comparison results of the section 3 to discuss the stability of the trivial solution of weakly coupled reaction-diffusion systems. For that purpose, consider the following weakly coupled impulsive reaction-diffusion system:

$$\begin{cases} u_t = Lu + f(t,x,u), & (t,x) \in Q_\infty \backslash M, \\ u(0,x) = \eta_0(x) \geq 0, & x \in \bar{\Omega}, \\ u(t,x) = 0, & (t,x) \in \Gamma_\infty \backslash N, \\ u_k(\tau_i^+,x) - u_k(\tau_i,x) = g_k(u_k(\tau_i,x)), & x \in \bar{\Omega}, i = 1,2,\cdots p. \end{cases} \tag{3}$$

Here we assume existence of the trivial solutions of the above system in $C^{\cdot}(Q_\infty)$. We can now extend the method of vector Lyapunov functions of [7,8] to study the stability of the above system.

Theorem 3. Assume that

(i) $V \in C^{1,2}[R_+ \times s(\rho), R_+^m]$, where $S(\rho) = \{u \in R^m, | u | < \rho\}, V_u(t,u)Lu \leq LV(t,u)$, $V(t,0) = 0$ and

$$\begin{aligned} &V_t(t,u) + V_u(t,u)f(t,x,u) \leq F(t,V), \quad (t,x) \in Q_\infty \backslash M, u \in s(\rho) \\ &V_k(\tau_i^+,u) - V_k(\tau_i,u) \leq g_k(V_k(\tau_i,u)), \quad u \in s(\rho), i = 1,2,\cdots p, \end{aligned}$$

where $F(t,r) \in C[R_+ \times R_+^m, R_+^m]$ is quasi-monotone nondecreasing in r and satisfies the one-sided Lipschitzian in r.

(ii) $g_k(u_k)$ is a $C^1 - function$ with $1 + g_{k_{u_k}} > 0$ and satisfies $g_k(u_k) - g_k(v_k) \leq \alpha_k(u_k - v_k), \alpha_k > 0, u_k \geq v_k$.

(iii) $f(t,x,0) \equiv 0, F(t,0) \equiv 0$

(iv) on $R_+ \times s(\rho), b(\| u \|) \leq \sum_{i=1}^n V_i(t,u) \leq a(\| u \|)$, where $a, b \in k$ and $k = \{\varphi \in C[R_+, R_+], \varphi(0) = 0$, and $\varphi(s)$ is strictly increasing in $s\}$.

Then, the stability properties of the trivial solution of either

$$\begin{cases} y' = F(t,y), & t \geq 0, \\ y(0) = y_0(0) \geq V(0, \eta_0(x)), & \\ y_k(\tau_i^+) - y_k(\tau_i) = g_k(y_k(\tau_i)), & i = 1, 2, \cdots, p; k = 1, 2, \cdots m; \end{cases} \tag{4}$$

or

$$\begin{cases} w_t = Lu + F(t,w), & (t,x) \in Q_\infty \backslash M, \\ w(0,x) = w_0(x) \geq V(0, \eta_0(x)), & x \in \bar{\Omega}, \\ w(t,x) = 0, & (t,x) \in \Gamma_\infty \backslash N, \\ w_k(\tau_i^+, x) - w_k(\tau_i, x) = g_k(w_k(\tau_i, x)), & x \in \bar{\Omega}, i = 1, 2, 3, \cdots, p, \end{cases} \tag{5}$$

imply the corresponding stability properties of the trivial solution of the system (3).

Example 2. Next we present another example to illustrate the application of Theorem 3. Consider the reaction diffusion system:

$$\begin{cases} u_t = Au_{xx} - bu_x + f(t,x,u(t,x)), & t \neq \tau_i, 0 \leq t < \infty, 0 \leq x \leq 1, \\ u(0,x) = \eta_0(x) > 0, & 0, \leq x \leq 1, \\ u(t,0) = u(t,1) = 0, & t \neq \tau_i, \\ u(\tau_i^+, x) - u(\tau_i, x) = \alpha u(\tau_i, x), & 0 \leq x \leq 1, -1 < \alpha < 0. \end{cases} \tag{6}$$

where $A > 0$ is a diagonal matrix, $b > 0$, and

$$f = \begin{pmatrix} (\sin u_1 - u_2^3 - \cdots - u_n^3) \\ (\sin u_2 - u_3^3 - \cdots - u_n^3) \\ \cdots\cdots\cdots\cdots\cdots\cdots \\ \sin u_n \end{pmatrix}$$

Let $V(t,u) = (u_1^2/2, u_2^2/2, \cdots, u_n^2/2)$, where $u(t,x)$ is a nonnegative solution of (6). It is easy to see that $V_u(t,u)Lu \leq LV(t,u)$, and

$$V_t(t,x) + V_u(t,u)f(t,x,u) \leq 2[V(t,u)].$$

Also,

$$V(\tau_i^+, u) - V(\tau_i, u) = \alpha(1+\alpha)V(\tau_i, x) \leq \alpha V(\tau_i, x).$$

So, by Theorem 2.3, the stability properties of the trivial solution of the ordinary impulsive problem:

$$\begin{cases} y' = 2y(t), & t \neq \tau_i, \\ y(0) = y_0 \geq \frac{1}{2}(\eta_0(x))^2, & \\ y(\tau_i^+) - y(\tau_i) = \alpha y(\tau_i), & i = 1, 2, 3, \cdots, p, \end{cases}$$

implies the corresponding stability properties of the trivial solution of the system (6). It is easy to see that the above system is simple compared with the system (6).

REFERENCES

1. Lakshmikantham, V. & Leela, S., Reaction-Diffusion System and Vector Lyapunov Function, Differential and Integral Equations 1 (1988), 41-47.
2. Lakshmikantham, V., Matrosov, V. M. & Sivasumdram, S., Vector Lyapunov Functions and Stability Analysis of Nonlinear Systems, Kluwer Academic Publisher, Dordrechet 1991.
3. Matrosov, V., On the Theory of Stability of Motion, Prikl. Math. Mech. 26 (1962), 992-1002.
4. Matrosov, V., Comparison Method in Systems Dynamic, Differential Equation 9 (1974), 408-445.
5. Siljak, D.D., Large Scale Dynamical System, North-Holland, New York, 1978.
6. Vatsala, A. S., Stability Results of Nonlocal Reaction Diffusion System via Vector Lyapunov Functions, Proc. Dynamic System and Appl. 1 (1994), 371-376.
7. Lakshmikantham, V., Matrosov, V. M. & Sivasumdram, S., Vector Lyapunov Functions and Stability Analysis of Nonlinear Systems, Kluwer Academic Publisher, Dordrechet 1991.
8. Lakshmikantham, V., Bainov, D.D. & Simeonov, P.S., Theory of Impulsive Differential Equations, World Scientific, Singapore, 1989.
9. Lakshmikantham, V., Leela, S. & Kaul, S., Comparison Principle for Impulsive Differential Equations with Variable Times and Stability Theory, Nonlinear Anal. 22 (1994), 499-503.

Proceedings of Dynamic Systems and Applications 3 (2001) 603-610

Global Dynamics of SEIR Models with Vertical Transmission and Saturation Incidence

Liancheng Wang & Michael Y. Li
Mississippi State University, Mississippi State, MS 39762

ABSTRACT: Two SEIR epidemiological models with vertical transmission and a constant population size are considered. The incidences of infection are given by saturation forms which work for the situations that the host population is saturated with infectious individuals or susceptible individuals, respectively. The global dynamics are established by showing that the disease always dies out if a threshold number $\sigma \leq 1$, and the disease persists at a unique endemic equilibrium if it is initially present when $\sigma > 1$.
AMS(MOS) subject classification. 92D25, 34K20

1. INTRODUCTION

In [14], Li, Smith and Wang studied an SEIR epidemic model for the dynamics of an infectious disease that spreads in a population through both horizontal and vertical transmission. The model is described by the following system of differential equations

$$\begin{cases} S' = b - bS - \lambda IS - pbE - qbI \\ E' = \lambda IS + pbE + qbI - (\epsilon + b)E \\ I' = \epsilon E - (\gamma + b)I \\ R' = \gamma I - bR, \end{cases} \tag{1}$$

where $S(t), E(t), I(t)$ and $R(t)$ denote the fractions of the population that are susceptible, exposed (not yet infectious), infectious, and recovered individuals at time t, respectively. The horizontal transmission is assumed to occur through direct contact of hosts and the incidence is described by the standard mass action form λIS, where $\lambda > 0$ is the transmission coefficient. For the vertical transmission, it is assumed that a fraction p $(0 \leq p \leq 1)$ and a fraction q $(0 \leq q \leq 1)$ of the offspring from the exposed and infectious classes, respectively, are born into the exposed class. The birth rate and the death rate are assumed to be equal and denoted by b, consequently the total population is constant; $S + E + I + R = 1$. The parameter $\epsilon > 0$ is the rate at which exposed individuals become infectious, and $\gamma > 0$ is the rate that infectious individuals become recovered. The immunity is assumed to be permanent, so once recovered, an individual will not become susceptible again. In the limiting case when $\epsilon \to \infty$, the model (1) reduces to an SIR model with vertical transmission, see Busenberg and Cooke [1,2]. When $p = q = 0$, the model (1) becomes an SEIR model with no vertical transmission. For general SIR or SEIR models, the reader is refereed to [6,8,9,18].

It is proved in [14] that the dynamics of (1) is completely determined by the basic reproduction number

$$\sigma_{p,q} = \frac{\lambda \epsilon}{(\epsilon + b)(\gamma + b) - pb(\gamma + b) - qb\epsilon}. \tag{2}$$

If $\sigma_{p,q} \leq 1$, the disease always dies out, whereas when $\sigma_{p,q} > 1$, the disease persists at a unique endemic equilibrium if it is initially present.

Nonlinear incidence forms other than the standard bilinear incidence form λIS have been used in the literature. Busenberg and Cooke [1,2] use a saturation incidence form $\frac{\lambda IS}{1+aI}$, where $a \geq 0$, which describes the situation that the population is saturated with infectious individuals; May and Anderson [16,17] use another saturation incidence term $\frac{\lambda IS}{1+aS}$, when a large number of susceptible individuals are present in population.

In this paper, we study model (1) with the two different saturation incidence forms mentioned above. We find the basic reproduction numbers $\sigma_{p,q}$ and $\sigma_{a,p,q}$, respectively, and then establish that the disease-free equilibrium is globally stable in the feasible region Γ if $\sigma_{p,q}$ $(\sigma_{a,p,q}) \leq 1$, and the unique endemic equilibrium is globally stable in $\mathring{\Gamma}$ when $\sigma_{p,q}$ $(\sigma_{a,p,q}) > 1$. The proof uses a general approach to the global stability problem developed by Li and Muldowney [12]. Our results generalize the global stability results in [14] or [13] by setting $a = 0$ or $a = p = q = 0$, respectively, and global stability results in [1] on the corresponding SIR models with vertical transmission.

In the next section, we state definitions and results we need in establishing our main theorems. The two models with the above mentioned saturation incidence forms are studied in Sections 3 and 4, respectively.

2. PRELIMINARIES

Let A be a linear operator on $\mathbf{R}^n$ and also be its matrix representation with respect to the standard basis of $\mathbf{R}^n$. Let "$\wedge$" denote the exterior product in $\mathbf{R}^n$. With respect to the canonical basis in the second exterior product space $\wedge^2\mathbf{R}^n$, the *second additive compound matrix* $A^{[2]}$ of A is an $\binom{n}{2} \times \binom{n}{2}$ matrix representing a linear operator on $\wedge^2\mathbf{R}^n$ whose definition on a decomposable element $u_1 \wedge u_2$ is

$$A^{[2]}(u_1 \wedge u_2) = Au_1 \wedge u_2 + u_1 \wedge Au_2.$$

Definition over the whole $\wedge^2\mathbf{R}^n$ is done by linear extension. A detailed discussion on compound matrices and their properties can be found in [5,19]. The second additive compound matrix of a 3×3 matrix $A = (a_{ij})$ is

$$A^{[2]} = \begin{bmatrix} a_{11}+a_{22} & a_{23} & -a_{13} \\ a_{32} & a_{11}+a_{33} & a_{12} \\ -a_{31} & a_{21} & a_{22}+a_{33} \end{bmatrix}.$$

Let $|\cdot|$ denote a vector norm in $\mathbf{R}^n$ and the corresponding operator norm it induces. The *Lozinskiĭ measure* μ on operators with respect to $|\cdot|$ is defined by (see [4, p.41]),

$$\mu(A) = \lim_{h \to 0^+} \frac{|I + hA| - 1}{h}$$

for an operator A. The Lozinskiĭ measures of A with respect to the l_1 norm $|x|_1 = \sum_i |x_i|$ and l_∞ norm $|x|_\infty = \max_i |x_i|$ are, respectively (see [4]),

$$\mu_1(A) = \sup_k \left(\operatorname{Re} a_{kk} + \sum_{i, i \neq k} |a_{ik}| \right),$$
$$\mu_\infty(A) = \sup_i \left(\operatorname{Re} a_{ii} + \sum_{k, k \neq i} |a_{ik}| \right). \tag{3}$$

Let $\Omega \subset \mathbf{R}^n$ be an open subset and $f : \Omega \to \mathbf{R}^n$ be a C^1 function. Consider the differential equation

$$x' = f(x). \tag{4}$$

We say that $\bar{x} \in \Omega$ is an *equilibrium* of (4) if $f(\bar{x}) = 0$. The equilibrium $\bar{x}$ is said to be *globally stable* in Ω if it is locally stable and all trajectories in Ω converge to $\bar{x}$ (see [7]). Let $x(t, x_0)$ denote the solution of (4) satisfying $x(0, x_0) = x_0$ and $Df(x)$ the Jacobian matrix of f at x. We make the following two assumptions:
(H_1) System (4) has a compact absorbing set $K \subset \Omega$.
(H_2) System (4) has a unique equilibrium $\bar{x}$ in Ω.

Let $x \mapsto P(x)$ be an $\binom{n}{2} \times \binom{n}{2}$ matrix-valued function that is C^1 on Ω. Assume that $P^{-1}(x)$ exists and is continuous for $x \in K$. For a Lozinskiĭ measure μ, a quantity $\bar{q}_2$ is defined as

$$\bar{q}_2 = \limsup_{t\to\infty} \sup_{x_0 \in K} \frac{1}{t} \int_0^t \mu(B(x(s, x_0)))ds, \tag{5}$$

where

$$B = P_f P^{-1} + P(Df)^{[2]} P^{-1} \tag{6}$$

and the matrix P_f is obtained by replacing each entry p_{ij} in P by its directional derivative in the direction f, $\nabla p_{ij}{}^* f$.

The following criterion of global stability result for (4) is established in [12, Theorem 3.5].

Theorem 2.1. *Assume that Ω is simply connected and that the assumptions (H_1) and (H_2) hold. Then $\bar{x}$ is globally stable in Ω if there exists a function $P(x)$ and a Lozinskiĭ measure μ such that $\bar{q}_2$ defined in (5) satisfies $\bar{q}_2 < 0$.*

3. THE MODEL WITH SATURATION INCIDENCE $\frac{\lambda IS}{1+aI}$

Replacing the bilinear incidence term λIS in (1) by $\frac{\lambda IS}{1+aI}$, where $a \geq 0$, we arrive at the following SEIR epidemiological model

$$\begin{cases} S' = b - bS - \dfrac{\lambda IS}{1+aI} - pbE - qbI \\ E' = \dfrac{\lambda IS}{1+aI} + pbE + qbI - (\epsilon + b)E \\ I' = \epsilon E - (\gamma + b)I \\ R' = \gamma I - bR. \end{cases} \tag{7}$$

When $a = 0$, (7) reduces to (1).

Note that R does not appear in the first three equations, so we can investigate the following system

$$\begin{cases} S' = b - bS - \dfrac{\lambda IS}{1+aI} - pbE - qbI \\ E' = \dfrac{\lambda IS}{1+aI} + pbE + qbI - (\epsilon + b)E \\ I' = \epsilon E - (\gamma + b)I, \end{cases} \tag{8}$$

and then evaluate R from $R = 1 - S - E - I$.

The simplex

$$\Gamma = \{(S, E, I) \in \mathbf{R}^3_+ \mid S + E + I \leq 1\}$$

is positively invariant with respect to (8). Denote the interior of Γ by $\overset{\circ}{\Gamma}$. Set

$$\sigma_{p,q} = \frac{\lambda\epsilon}{(\epsilon+b)(\gamma+b) - pb(\gamma+b) - qb\epsilon}. \tag{9}$$

It can be verified that $\sigma_{p,q} > 0$. System (8) has only one equilibrium $P_0 = (1,0,0) \in \Gamma$ if $\sigma_{p,q} \le 1$, whereas when $\sigma_{p,q} > 1$, it has two equilibria P_0 and P^* such that $P^* = (S^*, E^*, I^*) \in \overset{\circ}{\Gamma}$ with

$$\begin{gathered} S^* = \frac{ab\epsilon + (\epsilon+b)(\gamma+b)}{ab\epsilon + \sigma_{p,q}(\epsilon+b)(\gamma+b)}, \\ E^* = \frac{b}{\epsilon+b}(1-S^*), \quad I^* = \frac{b\epsilon}{(\epsilon+b)(\gamma+b)}(1-S^*). \end{gathered} \tag{10}$$

It follows from (10) that $0 < S^* < 1$ if and only if $\sigma_{p,q} > 1$.

Assume that $\sigma_{p,q} \le 1$. The global stability of P_0 can be verified using the Lyapunov function

$$L = \epsilon E + (\epsilon + b - pb)I.$$

The derivative of L along a solution of (8) is

$$L' = \frac{I}{(\epsilon+b)(\gamma+b) - pb(\gamma+b) - qb\epsilon}\left(\frac{\sigma_{p,q}S}{1+aI} - 1\right) \le 0,$$

if $\sigma_{p,q} \le 1$. Furthermore, $L' = 0$ if and only if $I = 0$. Since $\{P_0\}$ is the largest compact invariant set in $\{(S,E,I) \mid L' = 0\}$, P_0 is globally stable by LaSalle's Invariance Principle [10]. We thus have the following result.

Theorem 3.1. (1) If $\sigma_{p,q} \le 1$, then P_0 is the only equilibrium and it is globally stable in Γ. (2) If $\sigma_{p,q} > 1$, then P_0 becomes unstable and there exists a unique endemic equilibrium P^* in $\overset{\circ}{\Gamma}$. Furthermore, all solutions starting in $\overset{\circ}{\Gamma}$ and sufficiently close to P_0 move away from P_0.

From the local behavior of (8) near P_0 when $\sigma_{p,q} > 1$ as stated in Theorem 3.1, and by a similar argument as in [11], we can show that the system (8) is uniformly persistent in $\overset{\circ}{\Gamma}$ when $\sigma_{p,q} > 1$, namely, there exists constant $c > 0$ such that

$$\liminf_{t\to\infty} S(t) \ge c, \quad \liminf_{t\to\infty} E(t) \ge c, \quad \text{and} \quad \liminf_{t\to\infty} I(t) \ge c,$$

provided $(S(0), E(0), I(0)) \in \overset{\circ}{\Gamma}$. The uniform persistence of (8) in $\overset{\circ}{\Gamma}$ is equivalent to the existence of a compact set $K \subset \overset{\circ}{\Gamma}$ that is absorbing for (8) (see [3]). In fact, the unique endemic equilibrium P^* is globally stable in $\overset{\circ}{\Gamma}$ when $\sigma_{p,q} > 1$.

Theorem 3.2. *Assume that $\sigma_{p,q} > 1$. Then the unique endemic equilibrium P^* is globally stable in $\overset{\circ}{\Gamma}$.*

Proof. We use Theorem 2.1 to prove this theorem. From the discussion above, we know that system (8) satisfies the assumptions (H_1) and (H_2). The Jacobian matrix of (8) along a solution $x(t) = (S(t), E(t), I(t))$ of (8) is

$$J = \begin{bmatrix} -b - \frac{\lambda I}{1+aI} & -pb & -qb - \frac{\lambda S}{1+aI} + \frac{\lambda a I S}{(1+aI)^2} \\ \frac{\lambda I}{1+aI} & pb - \epsilon - b & qb + \frac{\lambda S}{1+aI} - \frac{\lambda a I S}{(1+aI)^2} \\ 0 & \epsilon & -\gamma - b \end{bmatrix}$$

and its second additive compound matrix $J^{[2]}$ is

$$J^{[2]} = \begin{bmatrix} pb - 2b - \epsilon - \frac{\lambda I}{1+aI} & qb + \frac{\lambda S}{1+aI} - \frac{\lambda aIS}{(1+aI)^2} & qb + \frac{\lambda S}{1+aI} - \frac{\lambda aIS}{(1+aI)^2} \\ \epsilon & -2b - \gamma - \frac{\lambda I}{1+aI} & -pb \\ 0 & \frac{\lambda I}{1+aI} & pb - \epsilon - \gamma - 2b \end{bmatrix}.$$

Let the function $P(S, E, I)$ in Theorem 2.1 be

$$P(S, E, I) = \begin{bmatrix} a_1 & 0 & 0 \\ 0 & (1-a_2)\frac{E}{I} & 0 \\ 0 & a_2\frac{E}{I} & \frac{E}{I} \end{bmatrix},$$

where $a_1 \geq 1$ and $0 \leq a_2 < 1$ are constants to be determined later. Then

$$P_f P^{-1} = \begin{bmatrix} 0 & 0 & 0 \\ 0 & \frac{E'}{E} - \frac{I'}{I} & 0 \\ 0 & 0 & \frac{E'}{E} - \frac{I'}{I} \end{bmatrix}$$

and the matrix $PJ^{[2]}P^{-1}$ is given by

$$\begin{bmatrix} pb - 2b - \epsilon - \frac{\lambda I}{1+aI} & a_1\left(qb + \frac{\lambda I}{1+aI} - \frac{\lambda aIS}{(1+aI)^2}\right)\frac{I}{E} & a_1\left(qb + \frac{\lambda I}{1+aI} - \frac{\lambda aIS}{(1+aI)^2}\right)\frac{I}{E} \\ \frac{1-a_2}{a_1}\frac{\epsilon E}{I} & a_2 pb - 2b - \gamma - \frac{\lambda I}{1+aI} & -(1-a_2)pb \\ \frac{a_2}{a_1}\frac{\epsilon E}{I} & \frac{\lambda I}{1+aI} + \frac{a_2[\epsilon-(1-a_2)pb]}{1-a_2} & (1-a_2)pb - 2b - \epsilon - \gamma \end{bmatrix}.$$

Thus the matrix $B = P_f P^{-1} + PJ^{[2]}P^{-1}$ in (6) can be written in the block form $B = \begin{bmatrix} B_{11} & B_{12} \\ B_{21} & B_{22} \end{bmatrix}$ with

$$B_{11} = pb - 2b - \epsilon - \frac{\lambda I}{1+aI}, \qquad B_{21} = \begin{bmatrix} \frac{1-a_2}{a_1}\frac{\epsilon E}{I} \\ \frac{a_2}{a_1}\frac{\epsilon E}{I} \end{bmatrix},$$

$$B_{12} = \left[a_1\left(qb + \frac{\lambda I}{1+aI} - \frac{\lambda aIS}{(1+aI)^2}\right)\frac{I}{E},\ a_1\left(qb + \frac{\lambda I}{1+aI} - \frac{\lambda aIS}{(1+aI)^2}\right)\frac{I}{E}\right],$$

$$B_{22} = \begin{bmatrix} \frac{E'}{E} - \frac{I'}{I} + a_2 pb - 2b - \gamma - \frac{\lambda I}{1+aI} & -(1-a_2)pb \\ \frac{\lambda I}{1+aI} + \frac{a_2[\epsilon-(1-a_2)pb]}{1-a_2} & \frac{E'}{E} - \frac{I'}{I} + (1-a_2)pb - 2b - \epsilon - \gamma \end{bmatrix}.$$

Choose a_2 so that $a_2 = 0$ if $\epsilon \geq pb$ and $a_2 = 1 - \epsilon/pb$ if $\epsilon < pb$. Then $a_2[\epsilon - (1-a_2)pb] = 0$, and B_{22} can be rewritten as

$$B_{22} = \begin{bmatrix} \frac{E'}{E} - \frac{I'}{I} + a_2 pb - 2b - \gamma - \frac{\lambda I}{1+aI} & -(1-a_2)pb \\ \frac{\lambda I}{1+aI} & \frac{E'}{E} - \frac{I'}{I} + (1-a_2)pb - 2b - \epsilon - \gamma \end{bmatrix}.$$

Choose the vector norm $|\cdot|$ in $\mathbf{R}^3$ as $|(u, v, w)| = \max\{|u|, |v|+|w|\}$. Then the Lozinskiĭ measure with respect to $|\cdot|$ can be estimated as follows (see [15,20]),

$$\mu(B) \leq \sup\{g_1, g_2\},$$

where

$$g_1 = \mu_1(B_{11}) + |B_{12}|, \quad g_2 = |B_{21}| + \mu_1(B_{22}).$$

Since B_{11} is a scalar, its Lozinskiĭ measure with respect to any norm in $\mathbf{R}^1$ is B_{11}. $|B_{12}|$ and $|B_{21}|$ represent matrix norms with respect to the l_1 vector norm, $\mu_1(B_{22})$ is the Lozinskiĭ measure of with respect to the l_1 norm in $\mathbf{R}^2$ (see [15]). We thus have

$$\begin{aligned}
\mu_1(B_{11}) &= pb - 2b - \epsilon - \frac{\lambda I}{1+aI}, \\
|B_{12}| &= a_1\left[qb + \frac{\lambda I}{1+aI} - \frac{\lambda a I S}{(1+aI)^2}\right]\frac{I}{E}, \quad |B_{21}| = \frac{1}{a_1}\frac{\epsilon E}{I}, \\
\mu_1(B_{22}) &= \max\left\{\frac{E'}{E} - \frac{I'}{I} - \gamma - 2b + a_2 pb,\ \frac{E'}{E} - \frac{I'}{I} - \epsilon - \gamma - 2b + 2(1-a_2)pb\right\} \\
&= \frac{E'}{E} - \frac{I'}{I} - \gamma - 2b + \max\{a_2 pb,\ (1-a_2)pb - \epsilon + (1-a_2)pb\} \\
&\le \frac{E'}{E} - \frac{I'}{I} - \gamma - 2b + \max\{a_2 pb,\ (1-a_2)pb\} \le \frac{E'}{E} - \frac{I'}{I} - \gamma - 2b + pb,
\end{aligned}$$

since $(1-a_2)pb - \epsilon \le 0$ and $0 \le a_2 < 1$. Therefore

$$\begin{aligned}
g_1 &= pb - 2b - \epsilon - \frac{\lambda I}{1+aI} + a_1\left[qb + \frac{\lambda S}{1+aI} - \frac{\lambda a I S}{(1+aI)^2}\right]\frac{I}{E} \\
&= pb - 2b - \epsilon - \frac{\lambda I}{1+aI} + a_1\left[qb + \frac{\lambda S}{1+aI}\right]\frac{I}{E} - \frac{\lambda a_1 a I S}{(1+aI)^2}\frac{I}{E},
\end{aligned} \tag{11}$$

and

$$g_2 \le \frac{1}{a_1}\frac{\epsilon E}{I} + \frac{E'}{E} - \frac{I'}{I} - \gamma - 2b + pb. \tag{12}$$

Rewriting (8) as

$$\left(qb + \frac{\lambda S}{1+aI}\right)\frac{I}{E} = \frac{E'}{E} + \epsilon + b - pb \tag{13}$$

$$\frac{\epsilon E}{I} = \frac{I'}{I} + \gamma + b, \tag{14}$$

and substituting (13) into (11), we obtain

$$\begin{aligned}
g_1 &= a_1\frac{E'}{E} + (1-a_1)pb + (a_1-1)\epsilon + (a_1-2)b - \frac{\lambda I}{1+aI} - \frac{\lambda a_1 a I S}{(1+aI)^2}\frac{I}{E} \\
&\le a_1\frac{E'}{E} + (a_1-1)\epsilon + (a_1-2)b,
\end{aligned} \tag{15}$$

since $a_1 \ge 1$, $a \ge 0$, $p \ge 0$, and $\lambda, S, E, I > 0$. Since (15) holds for all $a_1 \ge 1$, we can choose $a_1 = 1+\delta$ and $\delta > 0$ sufficiently small that $(a_1-1)\epsilon + (a_1-2)b = \delta(\epsilon+b) - b < 0$.

Substituting (14) into (12), and using $a_1 = 1+\delta$, we obtain

$$g_2 \le \frac{E'}{E} - \frac{\delta}{1+\delta}\frac{I'}{I} - \frac{\delta}{1+\delta}(b+\gamma) + (p-1)b \le \frac{E'}{E} - \frac{\delta}{1+\delta}\frac{I'}{I} - \frac{\delta}{1+\delta}(b+\gamma) \tag{16}$$

since $0 \le p \le 1$. Therefore,

$$\mu(B) \le \sup\{g_1, g_2\} \le \max\left\{(1+\delta)\frac{E'}{E},\ \frac{E'}{E} - \frac{\delta}{1+\delta}\frac{I'}{I}\right\} - \bar{b} \tag{17}$$

by (15) and (16), where $\bar{b} = \min\{b - \delta(\epsilon + b),\ \frac{\delta}{1+\delta}(b+\gamma)\} > 0$. Therefore, along each solution $x(t, x_0)$ to (8) with $x_0 \in K$, we have

$$\frac{1}{t}\int_0^t \mu(B)ds \le \frac{1}{t}\max\left\{(1+\delta)\log\frac{E(t)}{E(0)},\ \log\frac{E(t)}{E(0)} - \frac{\delta}{1+\delta}\log\frac{I(t)}{I(0)}\right\} - \bar{b},$$

which implies $\bar{q}_2 < 0$ since $0 < c \le E(t), I(t) < 1$. Therefore, P^* is globally stable in $\mathring{\Gamma}$ by Theorem 2.1, completing the proof. □

4. THE MODEL WITH SATURATION INCIDENCE $\frac{\lambda \mathbf{IS}}{\mathbf{1+aS}}$

Replacing the bilinear incidence term λIS in (1) by $\frac{\lambda IS}{1+aS}$, we arrive at the following SEIR epidemiological model

$$\begin{cases} S' = b - bS - \dfrac{\lambda IS}{1+aS} - pbE - qbI \\ E' = \dfrac{\lambda IS}{1+aS} + pbE + qbI - (\epsilon + b)E \\ I' = \epsilon E - (\gamma + b)I \\ R' = \gamma I - bR. \end{cases} \tag{18}$$

When $a = 0$, (18) reduces to (1). We consider the following system without R

$$\begin{cases} S' = b - bS - \dfrac{\lambda IS}{1+aS} - pbE - qbI \\ E' = \dfrac{\lambda IS}{1+aS} + pbE + qbI - (\epsilon + b)E \\ I' = \epsilon E - (\gamma + b)I, \end{cases} \tag{19}$$

and then evaluate R from $R = 1 - S + E + I$. The simplex $\Delta = \{(S, E, I) \in \mathbf{R}^3_+ \mid S + E + I \le 1\}$ is positively invariant with respect to (19). Let

$$\sigma_{a,p,q} = \frac{\lambda\epsilon}{(\epsilon + b)(\gamma + b) - pb(\gamma + b) - qb\epsilon} - a. \tag{20}$$

System (19) has two possible equilibria $P_0 = (1, 0, 0)$ and $P^* = (S^*, E^*, I^*)$, where

$$S^* = \frac{1}{\sigma_{a,p,q}},\quad E^* = \frac{b}{\epsilon + b}(1 - S^*),\quad \text{and}\quad I^* = \frac{b\epsilon}{(\epsilon + b)(\gamma + b)}(1 - S^*). \tag{21}$$

It follows from (21) that $0 < S^* < 1$ if and only if $\sigma_{a,p,q} > 1$, and thus $P^* \in \mathring{\Delta}$ exists if and only if $\sigma_{a,p,q} > 1$. The following results can be proved exactly the same way as Theorems 3.1 and 3.2.

Theorem 4.1. (1) If $0 < \sigma_{a,p,q} \leq 1$, then P_0 is the only equilibrium and it is globally stable in Δ. (2) If $\sigma_{a,p,q} > 1$, then P_0 becomes unstable and there exists a unique endemic equilibrium P^* in $\mathring{\Delta}$. Furthermore, system (18) is uniformly persistent in $\mathring{\Delta}$ if $\sigma_{a,p,q} > 1$.

Theorem 4.2. *If $\sigma_{a,p,q} > 1$, then the unique endemic equilibrium P^* is globally stable in $\mathring{\Gamma}$.*

ACKNOWLEDGMENTS

LW acknowledges the support of a Graduate Research Fellowship from the Department of Mathematics and Statistics at Mississippi State University. The research of MYL is supported in part by National Science Foundation grant DMS-9626128 and by a Ralph E. Powe Junior Faculty Enhancement Award from Oak Ridge Associated Universities.

REFERENCES

1. S. Busenberg and K. Cooke, *Vertical Transmitted Diseases*, Biomathematics, vol 23, Springer-Verlag, Berlin, 1993.
2. ———, *The population dynamics of two vertical transmitted infections*, Theor. Popul. Biol. **33** (1988), 181-198.
3. G. J. Butler and P. Waltman, *Persistence in dynamical systems*, Proc. Amer. Math. Soc. **96** (1996), 425-430.
4. W. A. Coppel, *Stability and Asymptotic Behavior of Differential Equations*, Heath, Boston, 1965.
5. M. Fiedler, *Additive compound matrices and inequality for eigenvalues of stochastic matrices*, Czech Math. J. **99** (1974), 392-402.
6. D. Greenhalgh, *Holf bifurcation in epidemic models with a latent period and nonpermanent immunity*, Math. Comput. Modeling **25** (1997), 85-107.
7. J. K. Hale, *Ordinary Differential Equations*, New York: McGraw-Hill, 1969.
8. H. W. Hethcote and S. A. Levin, *Periodicity in epidemiology models, in Applied Mathematical Ecology (L.Gross and S. A. Levin, eds.)*, Springer, New York, 1989, pp. 193-211.
9. H. W. Hethcote, H. W. Stech and P. Van den Driessche, *Periodicity and stability in epidemic models: a survey, in Differential Equations and Applications in Ecology, Epidemics, and Population Problems (K. L. Cooke, ed.)*, Academic Press, New York, 1981, pp. 65-85.
10. J. P. LaSalle, *The Stability of Dynamical Systems*, SIAM, Philadelphia, 1976.
11. M. Li, J. R. Graef, L. Wang and J. Karsai, *Global dynamics of a SEIR model with a varying total population size*, Math. Biosci. **160** (1999), 191-213.
12. M. Y. Li and J. S. Muldowney, *A geometric approach to the global-stability problems*, SIAM J. Math. Anal. **27** (1996), 1016-1083.
13. ———, *Global stability for the SEIR model in epidemiology*, Math. Biosci. **125** (1995), 155-164.
14. M. Y. Li, H. L. Smith and L. Wang, *Global dynamics of a SEIR model with vertical transmission*, preprint.
15. R. H. Martin, Jr., *Logarithmic norms and projections applied to linear differential equations*, J. Math. Anal. Appl. **45** (1974), 432-454.
16. R. M. May and R. M. Anderson, *Regulation and stability of host-parasite population interactions. II. Destabilizing process*, J. Anim. Ecol. **47** (1978), 219-267.
17. ———, *Population biology of infectious diseases II*, Nature **280** (1979), 455-461.
18. J. Mena-Lorca and H. W. Hethcote, *Dynamic model of infectious diseases as regulators of population sizes*, J. Math. Biol. **30** (1992), 693-176.
19. J. S. Muldowney, *Compound matrices and ordinary differential equations*, Rocky Mountain J. Math. **20** (1990), 857-872.
20. ———, *Dichotomies and asymptotic behavior for linear differential systems*, Tran. Amer. Math. Soc. **283** (1984), 465-484.

Proceedings of Dynamic Systems and Applications 3 (2001) 611-618

EXISTENCE AND UNIQUENESS OF SOLUTIONS TO THE STATIONARY POWER-LAW NAVIER-STOKES PROBLEM IN BOUNDED CONVEX DOMAINS

Dongming Wei
Department of Mathematics, University of New Orleans
Lakefront, New Orleans, LA 70148

ABSTRACT: We investigate the existence and uniqueness of solutions to the steady state Navier-Stokes problem governed by the the power-law model for viscous incompressible non-Newtonian fluid flows in bounded convex domains in $\mathbb{R}^d$.

AMS(MOS) subject classification. 65J67, 35Q30, 76N10

1. INTRODUCTION

Throughout this paper, Ω denotes a bounded convex domain in the real Euclidean space $\mathbb{R}^d$, where $d \geq 2$ is a positive integer. The term *domain* is defined as a connected open set in $\mathbb{R}^d$. Bold face letters are used for vectors or vector valued functions and vector valued function spaces. We use C to denote a generic positive constant.

For steady state fluid flows in Ω, let $\mathbf{u}(\mathbf{x}) = (u_1(\mathbf{x}), ..., u_d(\mathbf{x}))$ denote the velocity of the fluid particle at $\mathbf{x} = (x_1, ..., x_d) \in \Omega$ and $p(\mathbf{x})$ the pressure of the fluid at $\mathbf{x}$. The rate of deformation tensor $D(\mathbf{u}) \in \mathbb{R}^d \times \mathbb{R}^d$ is defined componentwise as $D_{i,j}(\mathbf{u}) = \frac{1}{2}(\frac{\partial u_i}{\partial x_j} + \frac{\partial u_j}{\partial x_i})$, $1 \leq i, j \leq d$. The second invariant of D is $\Pi_D(\mathbf{u}) = \frac{1}{2} D : D = \frac{1}{2} \sum_{i,j=1}^d D_{i,j}(\mathbf{u})^2$. The stress tensor for the fluid is denoted by $\sigma \in \mathbb{R}^d \times \mathbb{R}^d$. For incompressible fluid flows, we have the continuity equation $\nabla \cdot \mathbf{u} = 0$. The momentum equations for the isothermal steady-state flow neglecting the time derivative term are

$$(\mathbf{u} \cdot \nabla)\mathbf{u} = \nabla \cdot \sigma + \mathbf{f} \quad \text{in} \quad \Omega, \tag{1.1}$$

where $\nabla = (\frac{\partial}{\partial x_1}, ..., \frac{\partial}{\partial x_d})$, $(\mathbf{u} \cdot \nabla)\mathbf{u} = \sum_{i=1}^d u_i \frac{\partial \mathbf{u}}{\partial x_i}$, and $\mathbf{f} = (f_1(\mathbf{x}), ..., f_d(\mathbf{x}))$ is the body force. For simplicity, we only consider the homogeneous boundary conditions of the Dirichlet type: $\mathbf{u} = 0$ on $\partial\Omega$. In the power-law model for non-Newtonian fluid flows, it is assumed that

$$\sigma = -pI + 2\eta(\Pi_D(\mathbf{u}))D(\mathbf{u}), \tag{1.2}$$

where I is the identity matrix in $\mathbb{R}^d \times \mathbb{R}^d$, $\eta(z) = \eta_0 |z|^{(n-1)/2}$ for $z \in \mathbb{R}^1$, $0 < n < +\infty$, and $\eta_0 > 0$. The parameter n is often referred to as the power-law index and (1.2) as the Ostwald-deWaele equation. The fluid is Newtonian if $n = 1$ and is non-Newtonian if $n \neq 1$. Such non-Newtonian fluids exhibit shear-thinning behavior if $0 < n < 1$ and shear-thickening behavior if $1 < n < +\infty$. The power-law model has long been very popular in chemical engineering (Wilkinson [14]), as well as in geophysics, for studying highly viscous fluids since the Newtonian model ($n = 2$)

fails to capture the shear-thinning or shear-thickening behavior of such flows. It has been used to model the shear thinning phenomena in designing extrusion dies (Liu et al [9]), and also used in modeling of lithosphere (Sonder & England [12]).

Substituting (1.2) into (1.1) and letting $r = 1+n$, we get the steady state power-law Navier-Stokes equations

$$-\nabla \cdot 2\eta((\Pi_D(\mathbf{u}))D(\mathbf{u})) + (\mathbf{u}\cdot\nabla)\mathbf{u} + \nabla p = \mathbf{f}. \tag{1.3}$$

If the dimensionless variables $\mathbf{X} = \frac{1}{L}\mathbf{x}, \mathbf{U} = \frac{1}{V_0}\mathbf{u}, P = \frac{1}{2\eta_0(V_0/L)^n}p$, and $\eta(z) = \eta_0\frac{|z|^{(n-1)/2}}{2\eta_0(V_0/L)^n}$ are used in (1.3), we obtain

$$-\nabla(\cdot 2\eta(\Pi_D(\mathbf{U}))D(\mathbf{U})) + Re\,(\mathbf{U}\cdot\nabla)\mathbf{U} + \nabla P = \mathbf{f}, \tag{1.4}$$

where the Reynolds number $Re = \frac{\rho L^{2-n}V_0^n}{2\eta_0}$, L and V_0 are characteristic length and bulk velocity. Eq(1.4) and the continuity equation together with the Dirichlet boundary condition form the following Navier-Stokes problem for the power-law flows

$$\begin{aligned} -\nabla\cdot(2\eta(\Pi_D(\mathbf{u}))D(\mathbf{u})) + Re\,(\mathbf{u}\cdot\nabla)\mathbf{u} + \nabla p = \mathbf{f} \quad &\text{in } \Omega, \\ \nabla\cdot\mathbf{u} = \mathbf{0} \quad &\text{in } \Omega, \\ \mathbf{u} = \mathbf{0} \quad &\text{on } \partial\Omega, \end{aligned} \tag{1.5}$$

where $\mathbf{U}$ and P are replaced by $\mathbf{u}$ and p for simplicity in notation. If the inertial term $(\mathbf{u}\cdot\nabla)\mathbf{u}$ is negleted and $r \neq 2$, then (1.5) reduces to the power-law Stokes problem which has been studied by several authors; see, e.g., Baranger & Najib [2] and Lefton & Wei [7] for more details. When $r = 2$, (1.5) is the Navier-Stokes problem for Newtonian flows which has been studied extensively; see Ladyzhenskaya [5], Lions [8], Girault & Raviart [11], and Teman [13].

The purpose of this paper is to extend some of the well-known results for the case $r = 2$ to the case when $r \neq 2$ for the Navier-Stokes problem (1.5). The main result is that if $\frac{3d}{d+2} \le r < d$ and Ω is a bounded convex domain in $\mathbb{R}^d$ with $d \ge 2$, then (1.5) has at least one solution $(\mathbf{u}, p) \in \mathbf{W}_0^{1,r}(\Omega) \times L_0^{r'}(\Omega)$ for each $\mathbf{f} \in \mathbf{W}^{-1,r'}(\Omega)$, where $\frac{1}{r} + \frac{1}{r'} = 1$, in the sense defined in §3. Moreover, when $1 < r \le 2$, there exists a positive constant $C(d,r)$ such that if $C||\mathbf{f}||_{-1,r'} < \mu^{\frac{2}{3-r}}$, where $\mu = \frac{1}{Re}$, then the solution is unique. Furthermore, for $1 < r < +\infty$, there exists a constant C depending only on $\mathbf{f}$ and Ω such that $||\mathbf{u}||_{1,r} \le C$ and $||p||_{0,r'} \le C$ for any such solutions.

The Navier-Stokes problem (1.5) does not have a variational structure similar to that of the Stokes problem due to the nonlinear term $(\mathbf{u}\cdot\nabla)\mathbf{u}$, and therefore a different method from the one used in [2] or [7] is adapted in proving the result. A Brouwer's fixed point theorem for nonlinear operators in finite dimensional Banach spaces and an idea of passing to the limit due to Minty & Brown are used to establish existence of the velocity field $\mathbf{u}$. And an L^r version of the LBB condition on bounded convex domains is used to obtain the pressure p. In §2, we formulate an abstract existence theory based on the Brouwer's theorem and a result of Girault & Raviart. The main result is proved in §3.

Note that $\mathbf{u} = (u(x_2, ..., x_d), 0, ..., 0)$ for unidirectional flows. In this special case, the operator on the left hand side of (1.5) becomes the well-known nonlinear Laplacian $-\nabla \cdot (|\nabla u|^{r-2}\nabla u)$, which is often written as $-\Delta_r u$, where $\nabla u = (\frac{\partial u}{\partial x_2}, ..., \frac{\partial u}{\partial x_d})$ and $\nabla \cdot \mathbf{v} = \sum_{i=2}^{d} \frac{\partial v_i}{\partial x_i}$ for $\mathbf{v} = (v_2(x_2, ..., x_d), ..., (v_d(x_2), ..., x_d))$.

2. SOME THEORY OF NONLINEAR OPERATORS

For completeness, we will need some well-known results from convex analysis. Let X and M be two real reflexive Banach spaces with norms $||\cdot||_X$ and $||\cdot||_M$, respectively. Let X' and M' be their corresponding dual spaces with norms $||\cdot||_{X'}$ and $||\cdot||_{M'}$, respectively. Denote by $\langle x', x\rangle$ the action of $x' \in X'$ on $x \in X$ as well as the action of $x' \in M'$ on $x \in M$. The standard notation $u_m \rightharpoonup u$ denotes that u_m converges weakly to u in X or M. Here we quote a Brouwer's fixed point theorem, (see, e.g., Browder [3] or Fučik & Kufner [4]) , which is a fundamental tool in studying existence of solutions to nonlinear partial differential equations:

Theorem 2.1. *Let F be a continuous mapping defined on the Banach space X of finite dimension with values in X'. Assume the there exists a real function $\phi(t)$, defined on the interval $(0, +\infty)$, such that $\lim_{t\to+\infty} \phi(t) = +\infty$ and such that $\langle F(u), u\rangle \geq \phi(||u||_X)||u||_X, \forall u \in X$. Then $F(X) = X'$, i.e., the equation $F(u) = f$ has at least one solution u in the space X for arbitrary $f \in X'$.*

Suppose that a mapping $a : X \times X \times X \to \mathbb{R}^1$ is such that for each $w \in X$ there exists an operator $A(w) : X \to X'$ satisfying $a(w; u, v) = \langle A(w)u, v\rangle \, \forall u, v \in X$. Also assume $b : X \times M \to \mathbb{R}^1$, as a bilinear and continuous mapping such that there is a linearoperator $B : X \to M'$ defined by $\langle Bv, \mu\rangle = b(v, \mu) \, \forall v \in X$ and $\forall \mu \in M$. Let $B' : M \to X'$ be the dual operator of B satisfying $\langle B'\mu, v\rangle = \langle Bv, \mu\rangle = b(v, \mu) \, \forall v \in X$ and $\forall \mu \in M$. Consider the following variational problem which is called *Problem (Q)*: Given $l \in X'$ and $\chi \in M'$, find a pair $(u, \lambda) \in X \times M$ such that

$$\begin{aligned} a(u; u, v) + b(v, \lambda) &= \langle l, v\rangle \quad \forall v \in X, \\ b(u, \mu) &= \langle \chi, \mu\rangle \quad \forall \mu \in M. \end{aligned} \tag{2.1}$$

Problem (Q) has the following equivalent form: Given $l \in X'$ and $\chi \in M'$, find a pair $(u, \lambda) \in X \times M$ such that

$$\begin{aligned} A(u)u + B'\lambda &= l \text{ in } X', \\ Bu &= \chi \text{ in } M'. \end{aligned} \tag{2.2}$$

Let $V(\chi) = \{v \in X : Bv = \chi\}$. In particular, let $V = \mathrm{Ker}(B) = V(0)$. We associate *Problem (Q)* with the following problem which is called *Problem (P)*: Given $l \in X'$ and $\chi \in M'$, find $u \in V(\chi)$ such that

$$a(u; u, v) = \langle l, v\rangle \, \forall v \in V. \tag{2.3}$$

The mapping $u \to a(u;u,v)$ is said to be *monotone* in V if $a(u;u,u-v)-a(v;v,u-v) \geq 0$, $\forall u,v \in V$, where equality holds only if $u=v$. It is weakly continuous in V if $u_m \rightharpoonup u$ in $V \Longrightarrow \lim_{m\to+\infty} a(u_m;u_m,v) = a(u;,u,v)\, \forall v \in V$.

Theorem 2.2. *Let X be a separable real Banach space, and suppose that there exists a strictly increasing real function $\phi(t)$ defined on the interval $(0,+\infty)$ such that $\lim_{t\to+\infty} \phi(t) = +\infty$ and $a(v;v,v) \geq \phi(||v||_X)||v||_X\, \forall v \in V$, and the form $u \to a(u;u,v)$ can be split into $a_1(u;u,v)+a_2(u;u,v)$ where $a_1(u;u,v)$ and $a_2(u;u,v)$ satisfy the following:*
(1) a_1 is monotone and for each $w \in X$, there exists an operator $A_1(w): X \to X'$ satisfying $a_1(w;u,v) = \langle A_1(w)u, v\rangle, \forall u,v \in X$; $\lim_{t\to 0} A_1(u+t\theta)(u+t\theta) = A_1(u)u$ in X', and $u_m \rightharpoonup u$ in $V \Longrightarrow \lim_{m\to+\infty}\langle A_1(u_m)u_m, v\rangle = \langle w^*, v\rangle\ \forall v \in V$ for some $w^* \in X'$;
(2) a_2 is weakly continuous in V, and $u_m \rightharpoonup u$ in $V \Longrightarrow \lim_{m\to+\infty} a_2(u_m;u_m,u_m) = a_2(u;u,u)$.
Then problem (P) has at least one solution u in V satisfying $||u||_X \leq \phi^{-1}(||l||_{X'})$.

Proof: Since X is separable, $V \subset X$ is also separable; we write $V = \cup_{m=1}^{+\infty} V_m$, where $V_m = \text{span}\{w_1,...,w_m\}$ and $\{w_k\}_{k=1}^{+\infty}$ is a basis of V. Let $\langle F(u),u\rangle = a(u;u,u)$ and apply Theorem 2.1 to each finite dimensional space V_m. A sequence $\{u_m\}_{m=1}^{+\infty}$ can be found to satisfy $a(u_m;u_m,v) = \langle l,v\rangle\, \forall v \in V$. By (1) and $\phi(||u_m||_X)||u_m||_X \leq a(u_m;u_m,u_m) = \langle l,u_m\rangle \leq ||l||_{X'}||u_m||_X$, we have $||u_m||_X \leq \phi^{-1}(||l||_{X'})$. Being a bounded sequence in a reflexive Banach space, a subsequence $\{u_{m_k}\}_{k=1}^{+\infty}$ of $\{u_m\}_{m=1}^{+\infty}$ can be extracted such that $u_{m_k} \rightharpoonup u$, as $k \to \infty$. For simplicity, $\{u_{m_k}\}_{k=1}^{+\infty}$ is still being denoted by $\{u_m\}_{m=1}^{+\infty}$. By monotonicity, $a_1(u_m;u_m,u_m-v) - a_1(v,v,u_m-v) \geq 0\ \forall v \in V$, which gives $\langle l, u_m - v\rangle - a_2(u_m;u_m,u_m-v) - a_1(v,v,u_m-v) \geq 0\ \forall v \in V$ for large enough m. Therefore, by taking the limit, $\langle l, u-v\rangle - a_2(u;u,u-v) - a_1(v,v,u-v) \geq 0 \forall v \in V$. This can be rewitten as $\lim_{m\to+\infty} a_1(u_m;u_m,u-v) - a_1(v,v,u-v) \geq 0 \forall v \in V$, which gives $\langle w^*, u-v\rangle - \langle A_1(v)v, u-v\rangle \geq 0\ \forall v \in V$ for some $w^* \in X'$. For $\epsilon > 0$ and $\theta \in V$ let $v = u - \epsilon\theta$, then $\langle w^*,\theta\rangle - \langle A_1(u-\epsilon\theta)(u-\epsilon\theta),\theta\rangle \geq 0 \forall\theta \in V$ and $\forall\epsilon > 0$. Let $\epsilon \to 0$, then $\langle w^*,\theta\rangle - \langle A_1(u)u,\theta\rangle \geq 0\ \forall\theta \in V$. By replacing θ by $-\theta$ in the above argument, we obtain $\langle w^*,\theta\rangle - \langle A_1(u)u,\theta\rangle \leq 0\ \forall\theta \in V$. Therefore $\langle w^*,\theta\rangle - \langle A_1(u)u,\theta\rangle = 0$ $\forall\theta \in V$ which gives $w^* = A_1(u)u$ and subsequently $a(u;,u,v) = \langle l,v\rangle\ \forall v \in V$. Finally, $||u_m||_X \leq \phi^{-1}(||l||_{X'})$ gives $||u||_X \leq \phi^{-1}(||l||_{X'})$.

The idea for the passage to the limit used in the above proof is due to Minty-Browder (in this connection, see [3], [6], and [10]).

If (u,λ) is a solution of *Problem (Q)*, then u is a solution of *Problem (P)* since $b(v,\lambda) = 0 \forall v \in V$. We now show that if *Problem (P)* has a solution, then *Problem (Q)* also has a solution. We will need the following lemma of Girault & Raviart [11]. Define $V^0 = \{g \in X' : \langle g,v\rangle = 0 \forall v \in V\}$.

Lemma 2.1. *The following are equivalent:*
(1) $\exists\, \beta > 0$ such that $\inf_{\mu\in M} \sup_{v\in X} \frac{b(v,\mu)}{||\mu||_M ||v||_X} \geq \beta > 0$.
(2) B' is an isomorphism from M onto V^0 and $||B'\mu||_{X'} \geq \beta||\mu||_M\, \forall\mu \in M$.

(3) B is an isomorphism from X/V onto M' and $||Bv||_{M'} \geq \beta||v||_X \, \forall v \in X$.

Theorem 2.3. *Suppose that one of the three conditions in Lemma 2.3 hold, and that u is a solution of Problem (P). Then there exists a unique $\lambda \in M$ such that (u, λ) is a solution of Problem (Q) and $||\lambda||_M \leq (||l||_{X'} + ||A(u)u||_{X'})$.*

Proof: Since $l - A(u)u \in V^0$, $\exists$ a unique $\lambda \in M$ such that $B'\lambda = l - A(u)u$ and $\exists \, \beta > 0$ such that $||B'\lambda||_{X'} \geq \beta||\lambda||_M$ by (2) of Lemma 2.1. Therefore, (u, λ) is a solution of *Problem (Q)* and $||\lambda||_M \leq (||l||_{X'} + ||A(u)u||_{X'})$.

3. WEAK FORMULATION FOR THE NAVIER-STOKES PROBLEM

Let $L^r(\Omega)$, $1 < r < +\infty$, be the space of real functions defined on Ω with the $r-$th power absolutely integrable for the Lebesgue measure $d\mathbf{x} = dx_1...dx_d$. This is a Banach space with the norm $||\mathbf{u}||_{0,r} = \left(\int_\Omega |\mathbf{u}(\mathbf{x})|^r d\mathbf{x}\right)^{1/r}$. The Sobolev space $W^{1,r}(\Omega)$ is the space of functions in $L^r(\Omega)$ with first order distributional derivatives in $L^r(\Omega)$. The norm for this space is $||\mathbf{u}||_{1,r} = \left(\int_\Omega \sum_{[j]\leq 1} |D^j\mathbf{u}(\mathbf{x})|^r d\mathbf{x}\right)^{1/r}$, where $D^j\mathbf{u}(\mathbf{x}) = \frac{\partial^{[j]}u}{\partial x_{j_1}\partial x_{j_2}...\partial x_{j_d}}$, $[j] = \sum_{i=1}^d j_i$ and $j_i \in Z_0^+$ which is the set of nonnegative integers. Let $\mathcal{D}(\Omega)$ be the set of C^∞ functions with compact support. The closure of $\mathcal{D}(\Omega)$ in $W^{1,r}(\Omega)$ is denoted by $W_0^{1,r}(\Omega)$. Let $\mathbf{W}_0^{1,r}(\Omega) = [W_0^{1,r}(\Omega)]^d$. The norms for these spaces are $||\mathbf{v}||_{1,r} = \left(\int_\Omega (\sum_{i=1}^d \sum_{[j]=1} |D^j v_i|^2)^{r/2} d\mathbf{x}\right)^{1/r}$ for $\mathbf{v} \in \mathbf{W}_0^{1,r}(\Omega)$, and $||\mathbf{v}||_{0,r} = \left(\int_\Omega (\sum_{i=1}^d |v_i|^2)^{r/2} d\mathbf{x}\right)^{1/r} d\mathbf{x}$ for $\mathbf{v} \in \mathbf{L}^r(\Omega)$, respectively.

The dual space of $\mathbf{W}_0^{1,r}(\Omega)$ is $\mathbf{W}^{-1,r'}(\Omega)$ with norm denoted by $||\mathbf{f}||_{-1,r'}$ for $\mathbf{f} \in \mathbf{W}^{-1,r'}(\Omega)$. Let $\langle \mathcal{A}(\mathbf{u}), \mathbf{v} \rangle = \int_\Omega 2\eta(\Pi_D(\mathbf{u}))D(\mathbf{u}) : D(\mathbf{v})d\mathbf{x}$. Since $\nabla \cdot \mathbf{u} = 0$, $\langle \mathcal{A}(\mathbf{u}), \mathbf{v} \rangle = \int_\Omega 2\eta(|\nabla \mathbf{u}|)\nabla \mathbf{u} : \nabla \mathbf{v} d\mathbf{x}$. It is well known that $||\mathbf{v}|| = \left(\int_\Omega |\Pi_D(\mathbf{v})|^{r/2} d\mathbf{x}\right)^{1/r}$ is a norm that is equivalent to the norm $||\mathbf{v}||_{1,r}$ on $\mathbf{W}_0^{1,r}(\Omega)$. It can be shown that for $1 < r \leq 2$:

$$\begin{aligned} C||\mathbf{u} - \mathbf{v}||_{1,r}^2 &\leq \langle \mathcal{A}(\mathbf{u}) - \mathcal{A}(\mathbf{v}), \mathbf{u} - \mathbf{v} \rangle (||\mathbf{u}||_{1,r} + ||\mathbf{v}||_{1,r})^{2-r}, \\ ||\mathcal{A}(\mathbf{u}) - \mathcal{A}(\mathbf{v})||_{-1,r'} &\leq C||\mathbf{u} - \mathbf{v}||_{1,r}^{r-1}; \end{aligned} \tag{3.1a}$$

and for $2 \leq r < \infty$:

$$\begin{aligned} C||\mathbf{u} - \mathbf{v}||_{1,r}^r &\leq \langle \mathcal{A}(\mathbf{u}) - \mathcal{A}(\mathbf{v}), \mathbf{u} - \mathbf{v} \rangle, \\ ||\mathcal{A}(\mathbf{u}) - \mathcal{A}(\mathbf{v})||_{-1,r'} &\leq C||\mathbf{u} - \mathbf{v}||_{1,r}(||\mathbf{u}||_{1,r} + ||\mathbf{v}||_{1,r})^{r-2}, \end{aligned} \tag{3.1b}$$

where $C > 0$ is independent of $\mathbf{u}$ and $\mathbf{v}$ in $\mathbf{W}_0^{1,r}(\Omega)$. See, e.g., Baranger & Najib [2] for a proof of these inequalities.

Let $c(\mathbf{u}; \mathbf{v}, \mathbf{w}) = \int_\Omega ((\mathbf{u} \cdot \nabla)\mathbf{v}) \cdot \mathbf{w} d\mathbf{x}$ and let $b(p, \mathbf{v}) = -\int_\Omega p\nabla \cdot \mathbf{v} d\mathbf{x}$ for $\mathbf{w}, \mathbf{u}, \mathbf{v} \in \mathbf{W}_0^{1,r}(\Omega)$ and $p \in L^{r'}(\Omega)$. The weak formulation for the stationary Navier-Stokes problem (1.5) is to solve fcr a pair $(\mathbf{u}, p) \in \mathbf{W}_0^{1,r}(\Omega) \times L^{r'}(\Omega)$ such that

$$\begin{aligned} \nu < \mathcal{A}(\mathbf{u}), \mathbf{v} > + c(\mathbf{u}; \mathbf{u}, \mathbf{v}) - b(p, \mathbf{v}) &= \langle \mathbf{f}, \mathbf{v} \rangle \, \forall \mathbf{v} \in X; \\ b(q, \mathbf{u}) &= 0 \, \forall q \in M, \end{aligned} \tag{3.2}$$

where $\nu = \frac{1}{Re}$, $X = \mathbf{W}_0^{1,r}(\Omega)$ and $M = L^{r'}(\Omega)$. Note that (3.2) corresponds to *Problem (P)* in §2. Let $\mathbf{V} = \{\mathbf{u} \in \mathbf{W}_0^{1,r}(\Omega) : \nabla \cdot \mathbf{u} = 0\}$. The associated *Problem (Q)* is to find a $\mathbf{u} \in \mathbf{V}$ such that

$$\nu\langle \mathcal{A}(\mathbf{u}), \mathbf{v}\rangle + c(\mathbf{u}; \mathbf{u}, \mathbf{v}) = \langle \mathbf{f}, \mathbf{v}\rangle \ \forall \mathbf{v} \in \mathbf{V}. \tag{3.3}$$

Lemma 3.1. *For $\frac{3d}{d+2} \le r < d$, the trilinear form c is well defined, continuous on $[\mathbf{W}_0^{1,r}(\Omega)]^3$, and there eists a positive constant $C(d,r,\Omega)$ such that*

$$|c(\mathbf{u}; \mathbf{v}, \mathbf{w})| \le C(d,r,\Omega)||\mathbf{u}||_{1,r}||\mathbf{v}||_{1,r}||\mathbf{w}||_{1,r} \ \forall \mathbf{w}, \mathbf{u}, \mathbf{v} \in \mathbf{W}_0^{1,r}(\Omega).$$

Proof: For any $\mathbf{w}, \mathbf{u}, \mathbf{v} \in \mathbf{W}_0^{1,r}(\Omega)$ and for fixed $1 \le k, i \le d$, by Hölder's inequality,

$$\left|\int_\Omega (u_k \frac{\partial v_i}{\partial x_k} w_i dx\right| \le \left|\left|\frac{\partial v_i}{\partial x_k}\right|\right|_{0,r} ||u_k w_i||_{0,r'} \le \left|\left|\frac{\partial v_i}{\partial x_k}\right|\right|_{0,r} ||u_k||_{0,\frac{rd}{d-r}} ||w_i||_{0,\frac{rd}{(d+1)r-2d}},$$

which gives

$$|c(\mathbf{u}; \mathbf{v}, \mathbf{w})| = \left|\int_\Omega (\mathbf{u} \cdot \nabla \mathbf{v}) \cdot \mathbf{w} dx\right| \le ||\mathbf{u}||_{0,\frac{rd}{d-r}} ||\nabla \mathbf{v}||_{0,r} ||\mathbf{w}||_{0,\frac{rd}{(d+1)r-2d}}. \tag{3.4}$$

For $\frac{3d}{d+2} \le r < d$, we get $\frac{rd}{(d+1)r-2d} \le \frac{rd}{d-r}$. By the Sobolev compact imbedding $\mathbf{W}_0^{1,r}(\Omega) \hookrightarrow \mathbf{L}^{\frac{rd}{d-r}}(\Omega)$ (see, e.g., [4]), we have

$$||\mathbf{v}||_{0,\frac{rd}{(d+1)r-2d}} \le C(d,r,\Omega)||\mathbf{v}||_{0,\frac{rd}{d-r}} \le C(d,r,\Omega)||\nabla \mathbf{v}||_{0,r} \ \forall \mathbf{v} \in \mathbf{W}_0^{1,r}(\Omega),$$

which together with (3.4) gives the result.

Lemma 3.2. *For $\frac{3d}{d+2} \le r < d$, $c(\mathbf{u}; \mathbf{v}, \mathbf{v}) = 0, \forall \mathbf{u}, \mathbf{v} \in V$ and $c(\mathbf{u}; \mathbf{v}, \mathbf{w}) = -c(\mathbf{u}; \mathbf{w}, \mathbf{v}) \ \forall \mathbf{u}, \mathbf{v}, \mathbf{w} \in \mathbf{V}$.*

The proof of Lemma 3.2 follows closely to the simple proof given by Temam [13] for the case when $r = 2$. It is omitted here.

Lemma 3.3. *The mapping $\mathbf{u} \to c(\mathbf{u}; \mathbf{u}, \mathbf{v})$ is weakly continuous in V, and $\mathbf{u}_m \rightharpoonup \mathbf{u}$ in $V \Longrightarrow \lim_{m\to\infty} c(\mathbf{u}_m; \mathbf{u}_m, \mathbf{u}_m) = c(\mathbf{u};, \mathbf{u}, \mathbf{u})$.*

Proof: Suppose that $\mathbf{u}_m \rightharpoonup \mathbf{u}$ *in* $\mathbf{V}$. Then by the same compact imbedding $\mathbf{W}_0^{1,r}(\Omega) \hookrightarrow \mathbf{L}^{\frac{rd}{d-r}}(\Omega)$, $\mathbf{u}_m \to \mathbf{u}$ strongly *in* $\mathbf{L}^{\frac{rd}{d-r}}(\Omega)$ and $||\mathbf{u}_\mathbf{m}||_{1,r}$ is bounded. Using Lemma 3.2 and (3.4), the following hold:

$$|c(\mathbf{u}_\mathbf{m}; \mathbf{u}_\mathbf{m}; \mathbf{v}) - c(\mathbf{u}; \mathbf{u}; \mathbf{v})| = |c(\mathbf{u}_\mathbf{m}; \mathbf{v}; \mathbf{u}_\mathbf{m}) - c(\mathbf{u}; \mathbf{v}; \mathbf{u})|$$
$$\le ||\mathbf{u}_\mathbf{m} - \mathbf{u}||_{0,\frac{rd}{d-r}}(||\mathbf{v}||_{1,r}||\mathbf{u}_\mathbf{m}||_{1,r} + ||\mathbf{v}||_{1,r}||\mathbf{u}||_{1,r});$$

and

$$|c(\mathbf{u_m};\mathbf{u_m};\mathbf{u_m}) - c(\mathbf{u};\mathbf{u};\mathbf{u})| =$$
$$|c(\mathbf{u_m};\mathbf{u_m};\mathbf{u_m}-\mathbf{u}) + c(\mathbf{u_m};\mathbf{u};\mathbf{u_m}-\mathbf{u}) + c(\mathbf{u_m}-\mathbf{u};\mathbf{u},\mathbf{u})$$
$$\leq ||\mathbf{u_m}-\mathbf{u}||_{0,\frac{rd}{d-r}}(||\mathbf{u_m}||_{1,r}^2 + ||\mathbf{u_m}||_{1,r}||\mathbf{u}||_{1,r} + ||\mathbf{u}||_{1,r}^2),$$

which lead to the result of the lemma.

Theorem 3.1. *Let Ω be a bounded convex domain in $\mathbb{R}^d, d \geq 2$. Then*

$$\inf_{\mu \in L_0^{r'}(\Omega)} \sup_{\mathbf{v}\in \mathbf{W}_0^{1,r}(\Omega)} \frac{\langle \mu, \nabla\cdot\mathbf{v}\rangle}{||\mu||_{L_0^{r'}(\Omega)}||\mathbf{v}||_{\mathbf{W}_0^{1,r}(\Omega)}} \geq \beta > 0,$$

where $r' = \frac{r}{r-1}, r > 1$.

This "inf – sup" condition is referred to as the L^r version of the LBB-condition or simply the $L^r - LBB$. For a proof of Theorem 3.1, see Amrouch &Girault [1].

The following result is a generalized version of the results for the case $r = 2$ given in Girault & Raviart [11] and Teman [13]. It is the main result of this paper.

Theorem 3.2. *Let Ω be a bounded convex domain in $\mathbb{R}^d$, $d \geq 2$. Let $\frac{3d}{d+2} \leq r < d$. Then (3.2) has at least one solution $(\mathbf{u},p) \in \mathbf{W}_0^{1,r}(\Omega) \times L^{r'}(\Omega)$. For $1 < r \leq 2$ there exists a positive constant $C(d,r)$ such that if $C(d,r)||\mathbf{f}||_{-1,r'} < \mu^{\frac{2}{3-r}}$, then the solution is unique. Furthermore, for $1 < r < +\infty$, there exists a positive constant C which depends only on Ω and $\mathbf{f}$ such that $||\mathbf{u}||_{1,r} \leq C$ and $||p||_{0,r'} \leq C$.*

Proof: Let $Bu = \nabla\cdot\mathbf{u}$, $l = \mathbf{f}$, and $a(\mathbf{w};\mathbf{u},\mathbf{v}) = a_1(\mathbf{w};\mathbf{u},\mathbf{v}) + a_2(\mathbf{w};\mathbf{u},\mathbf{v})$, where $a_1(\mathbf{w};\mathbf{u},\mathbf{v}) = \mu\langle\mathcal{A}(\mathbf{u}),\mathbf{v}\rangle$ and $a_2(\mathbf{w};\mathbf{u},\mathbf{v}) = c(\mathbf{w};\mathbf{u},\mathbf{v})$. Let $\phi(t) = C|t|^{r-2}t$, then by Lemma 3.2, (3.1a) & (3.1b) $a(u;u,u) = a_1(u;u,u) \geq \phi(||u||_{1,r})||u||_{1,r}$. Therefore, by Lemma 3.1, Lemma 3.3, and Theorem 3.1, one can apply Theorem 2.2 to $a(\mathbf{u};\mathbf{u},\mathbf{v}) = \langle f,\mathbf{v}\rangle, \forall\mathbf{v} \in V$ and have the existence of a solution $\mathbf{u}$ to (3.3) and then apply Theorem 2.3 to show the existence of a solution $(\mathbf{u},p)$ to (3.2). By the results of Theorem 2.2 and 2.3, we have $||\mathbf{u}||_{1,r} \leq \phi^{-1}(||\mathbf{f}||_{-1,r})$ and $||p||_{0,r'} \leq (||\mathbf{f}||_{-1,r'} + ||A(\mathbf{u})\mathbf{u}||_{-1,r'})$. By (3.1a), (3.1b), and Lemma 3.1, $||A(\mathbf{u})\mathbf{u}||_{-1,r'} \leq (\mu||\mathcal{A}(\mathbf{u})\mathbf{u}||_{-1,r'} + C||\mathbf{u}||_{1,r}^2)$. These give $||\mathbf{u}||_{1,r} \leq C$ and $||p||_{0,r'} \leq C$.
To prove uniqueness, let $\mathbf{u}^* = \mathbf{u}_1 - \mathbf{u}_2$, where $\mathbf{u}_1$ and $\mathbf{u}_2$ are two solutions to (3.3). Then

$$\mu\langle\mathcal{A}(\mathbf{u}_1),\mathbf{u}^*\rangle - \mu\langle\mathcal{A}(\mathbf{u}_2),\mathbf{u}^*\rangle + c(\mathbf{u}_1;\mathbf{u}_1,\mathbf{u}^*) - c(\mathbf{u}_2;\mathbf{u}_2,\mathbf{u}^*) = 0. \tag{3.5}$$

By Lemma 3.2, $c(\mathbf{u}_1;\mathbf{u}^*,\mathbf{u}^*) = 0$. Therefore $c(\mathbf{u}_1;\mathbf{u}_1,\mathbf{u}^*) - c(\mathbf{u}_2;\mathbf{u}_2,\mathbf{u}^*) = c(\mathbf{u}^*,\mathbf{u}_2,\mathbf{u}^*)$ and it yields

$$\mu\langle\mathcal{A}(\mathbf{u}_1),\mathbf{u}^*\rangle - \mu\langle\mathcal{A}(\mathbf{u}_2),\mathbf{u}^*\rangle = -c(\mathbf{u}^*,\mathbf{u}_2,\mathbf{u}^*). \tag{3.6}$$

Similarly, $\mu\langle\mathcal{A}(\mathbf{u}_i),\mathbf{u}_i)\rangle + c(\mathbf{u}_i;\mathbf{u}_i,\mathbf{u}_i) = \langle\mathbf{f},\mathbf{u}_i\rangle$ and $c(\mathbf{u}_i;\mathbf{u}_i,\mathbf{u}_i) = 0$ for $i = 1,2$ give

$$\mu\langle\mathcal{A}(\mathbf{u}_i),\mathbf{u}_i)\rangle = \langle\mathbf{f},\mathbf{u}_i\rangle. \tag{3.7}$$

By (3.1a), (3.1b), and (3.7),

$$\mu||\mathbf{u}_i||_{1,r}^{r-1} \leq ||\mathbf{f}||_{-1,r}, i = 1, 2. \tag{3.8}$$

By (3.1a), (3.1b), Lemma 3.1, (3.6), and (3.8), we have for $1 < r \leq 2$,

$$\begin{aligned} \mu||\mathbf{u}^*||_{1,r}^2 &\leq C||\mathbf{u}^*||_{1,r}^2||\mathbf{u}_2||_{1,r}(||\mathbf{u}_1||_{1,r} + ||\mathbf{u}_2||_{1,r})^{2-r} \\ &\leq C\mu^{\frac{r-3}{r-1}}||\mathbf{u}^*||_{1,r}^2||\mathbf{f}||_{-1,r}^{\frac{3-r}{r-1}}, \end{aligned} \tag{3.9}$$

which implies that $\mathbf{u}^* = \mathbf{0}$ if $C||\mathbf{f}||_{-1,r} < \mu^{\frac{2}{3-r}}$. This completes the proof.

Note: Similar to (3.9), for $2 < r < \infty$: $\mu||\mathbf{u}^*||_{1,r}^r \leq C||\mathbf{u}^*||_{1,r}^2||\mathbf{f}||_{-1,r}^{\frac{1}{r-1}}$, which, however, does not give $\mathbf{u}^* = \mathbf{0}$. Theorem 3.2 indicates that the shear-thinning power-law flows behave more like Newtonian flows in terms of existence and uniquenes of solutions in contrast to the shear-thickening flows.

REFERENCES

1. Amrouche, C., & Girault, V., Decomposition of vector spaces and application to the Stokes problem in arbitrary dimension, Vol 44 (1994), NO. 119, Czechoslovak Mathematical Journal, Praha, 109-140.
2. Baranger, J., & Najib, K., Analyse numerique des écoulements quasi-Newtoniens dont la viscosité obéit à la loi puissance ou loi de carreau, Vol 58 (1990), Numer. Math., 34-49.
3. Browder, F. E., Existence and uniqueness theorems for solutions of nonlinear boundary value problems, Vol 17 (1965), Pro. Sympos. Appl. Math., Amer. Math. Soc., Providence, R. I., 24-49.
4. Fučik, S. & Kufner, A., Nonlinear Differential Equations, Elsevier, New York, 1980.
5. Ladyzhenskaya, O. A., The mathematical theory of viscous incompressible flow, Revised 2nd ed., Gordon and Breach, New York, 1969.
6. Ladyzhenskaya, O. A., New equations for the description of the viscous incompressible fluids and solvability in the large of the boundary value problems for them, Boundary Value Problems of Mathematical Physics , Amer. Math. Soc., Providence, RI, 1970.
7. Lefton, L.,& Wei, D., Existence of solutions to the stationary power-law Stokes problem, preprint.
8. Lions, J. L., Quelques méthodes de résolution des problèmes aux limites nonlinéaires, Dunod, 1969.
9. Liu, T. J., Wen, S., & Tsou, J., Three-Dimensional Finite Element Analysis of Polymeric Fluid flow in an Extrusion Die. Part I: Entrance Effect, Vol 34 (1994), Polymer Engineering Science, No. 10, 827-834.
10. Minty, G. J., On a "monotonicity" method for the solution of nonlinear equations in Banach spaces, Vol 50 (1963), Proc. Nat. Acad. Sci. U.S.A., 1038-1041.
11. V. Girault, V., & Raviart, P. A., Finite Element Methods for Navier-Stokes Equations, Springer, Berlin, 1986.
12. Sonder, L. J., & England, P. C., Vertical averages of rheology of the continental lithosphere, Vol 77 (1986), Earth Planet. Sci. Lett., 81-90.
13. Temam, R., Navier-Stokes Equations, Theory and Numerical Analysis North-Holland, Amsterdam, 1977.
14. Wilkinson, W., Non-Newtonian Fluids, Pergamon Press, New York, 1960.

Proceedings of Dynamic Systems and Applications 3 (2001) 619-628

Persistence in Age - Structured, Discrete Predator - Prey Systems With Dispersion

Abdul-Aziz Yakubu
Department of Mathematics, Howard University, Washington, D.C. 20059

July 3, 2000

Abstract

We formulate and analyze a two-age class [juvenile and adult classes], discrete-time predator-prey system containing one predator, n competing species of prey with pre-growth dispersion and post-growth dispersion. If all the juvenile species of prey mature to adulthood and predation is very weak, then all the species of prey in the system persist [with the resulting extinction or existence of the predator species] whenever the interspecific prey competition is also very weak.

AMS (MOS) subject classification: 34C35, 39A10, 39A12, 92D25.

1. INTRODUCTION

The main model for this study is a two patch system that consists of n competing species of prey and one predator with two age-classes [juvenile and adult classes [3 - 9, 11, 12] and with two levels of dispersion [pre-growth dispersion and post-growth dispersion]. To produce an outbred and heterogeneous population, all the juvenile species and all the adult species are allowed to disperse between the two patches.

In the general two patch system, we obtain that if all the juvenile species of prey mature to adulthood and predation is very weak, then all the species in the system persist whenever the intraspecific prey competition is much greater than the interspecific competition and the predator has a positive searching efficiency for some species of prey [Theorem 1 and Corollary 1]. We illustrate, with an example in Section 5, that if predation and interspecific prey competition are very weak, then prey extinction could occur provided not all the juvenile species of prey survive to become adults.

2. MODEL DERIVATION

For each species of prey $i \in \{1, 2, ..., n\}$, in each patch $\lambda \in \{A, B\}$, $x_{\lambda i}(t)$ denotes the population of its larvae or juveniles at generation t, with the non-increasing positive continuous function $s_{\lambda i} : [0, \infty) \to (0, 1]$ being its survival function.

For each species of prey $i \in \{1, 2, ..., n\}$, let $\bar{x}_{Ai}(t) = x_{Ai}(t) - \mu_i[x_{Ai}(t) - x_{Bi}(t)]$ and $\bar{x}_{Bi}(t) = x_{Bi}(t) + \mu_i[x_{Ai}(t) - x_{Bi}(t)]$, where $\mu_i \in (0, \frac{1}{2}]$ is the dispersion coefficient for its pre-growth dispersion. At generation t, $\bar{x}_{\lambda i}(t)$ is the effective population density of the larvae or juveniles of species of prey i in patch λ.

Furthermore, for each species of prey i, in patch λ, $y_{\lambda i}(t)$ denotes the population density of its adults at generation t. To make each species of prey i a pioneer species and achieve the self-regulation of species in the main model, we assume throughout that in each patch λ the growth function of its adults, $g_{\lambda i} : [0, \infty) \to (0, \infty)$, is a strictly decreasing positive continuous function with $g_{\lambda i}(0) > 1$ and with $\lim_{y_{\lambda i} \to \infty} g_{\lambda i}(y_{\lambda i}) = 0$. As a result, density effects are deleterious.

In patch λ, we let $z_{\lambda x}(t)$ [respectively, $z_{\lambda y}(t)$] denote the population density of the juvenile [respectively, adult] predator, at generation t. Also, let $\bar{z}_{Ax}(t) = z_{Ax}(t) - \mu_{n+1}[z_{Ax}(t) - z_{Bx}(t)]$ and $\bar{z}_{Bx}(t) = z_{Bx}(t) + \mu_{n+1}[z_{Ax}(t) - z_{Bx}(t)]$, where $\mu_{n+1} \in (0, \frac{1}{2}]$ is the pre-growth dispersion coefficient for the juvenile predator. In patch λ, at generation t, $\bar{z}_{\lambda x}(t)$ is the effective population density of the juvenile predator. For each λ, the nonnegative constants $\alpha_{\lambda i}$ and $\beta_{\lambda i}$ respectively denote the searching efficiencies for the adults and juveniles of species of prey i by the adult predator.

Now, we introduce a notation for the vectors of prey population densities. For each $\lambda \in \{A, B\}$, let

$$\begin{aligned} x_\lambda(t) &= (x_{\lambda 1}(t), x_{\lambda 2}(t), ..., x_{\lambda n}(t)) \in \mathbb{R}^n_+, \bar{x}_\lambda(t) = (\bar{x}_{\lambda 1}(t), \bar{x}_{\lambda 2}(t), ..., \bar{x}_{\lambda n}(t)) \in \mathbb{R}^n_+ \text{ and} \\ y_\lambda(t) &= (y_{\lambda 1}(t), y_{\lambda 2}(t), ..., y_{\lambda n}(t)) \in \mathbb{R}^n_+, \text{ where } \mathbb{R}_+ = [0, \infty) \text{ and } \mathbb{R}^n_+ \text{ is the} \end{aligned}$$

nonnegative cone. In patch λ, $\bar{x}_\lambda(t)$ is the vector of juvenile prey population densities after the pre-growth dispersion at generation t. Also in patch λ, $y_\lambda(t)$ is the vector of adult prey population densities at generation t.

To allow the predator for this study to have a Holling's Type II functional response, we let $E_i : \mathbb{R}^n_+ \to [-1, \infty)$ denote a continuous function that accounts for the relative abundance of each species of prey $i \in \{1, 2, ..., n\}$ [17]. In their investigation of the effects of the addition of a single searching [polyphagous] predator to a system of n competing species of prey in the way envisaged by Nicholson and Bailey [2, 21], Comins and Hassell [2, 23] used $E_i(x_\lambda(t)) = \delta_i(x_{\lambda i}(t) - \frac{1}{n}\sum_{i=1}^{n} x_{\lambda i}(t))/\frac{1}{n}\sum_{i=1}^{n} x_{\lambda i}(t)$, where $\delta_i \in [0, 1]$ is the degree of searching for prey i.

In patch $\lambda \in \{A, B\}$, for each adult [respectively, juvenile] of species of prey i, we let the weighted density variable at generation t, $u_{\lambda i}(t) = \sum_{j=1}^{n}\{a^\lambda_{ij}\bar{x}_{\lambda j}(t) + b^\lambda_{ij}y_{\lambda j}(t)\} + [1 + E_i(\bar{x}_\lambda(t))]\alpha_{\lambda i}z_{\lambda y}(t)$ [respectively, $v_{\lambda i}(t) = \sum_{j=1}^{n}\{c^\lambda_{ij}\bar{x}_{\lambda j}(t) + d^\lambda_{ij}y_{\lambda j}(t)\} + [1 + E_i(y_\lambda(t))]\beta_{\lambda i}z_{\lambda y}(t)$] where $a^\lambda_{ij}, b^\lambda_{ij}, c^\lambda_{ij}$ and d^λ_{ij} are nonnegative constants. In patch λ, for the adults [respectively, juveniles] of species of prey i, a^λ_{ij} and b^λ_{ij} [respectively, c^λ_{ij} and d^λ_{ij}] are the interaction coefficients and model the effect of the $j-th$ species of prey on the corresponding $i-th$ species of prey. If $E_i \equiv 0$ in the main model, then the predator is a nonsearching predator with a constant searching efficiency. Otherwise, the predator is a searching predator with a nonconstant searching efficiency [23].

For each species of prey i and each patch λ, we let

$$M_{\lambda i}(t) = \bar{x}_{\lambda i}(t)[1 - \frac{s_{\lambda i}([1 + E_i(y_\lambda(t))]\beta_{\lambda i}z_{\lambda y}(t))g_{\lambda i}([1 + E_i(\bar{x}_\lambda(t+1))]\alpha_{\lambda i}\bar{z}_{\lambda x}(t))}{s_{\lambda i}(0)g_{\lambda i}(0)}],$$

$$N_{\lambda i}(t) = y_{\lambda i}(t)[1 - \frac{g_{\lambda i}([1 + E_i(\bar{x}_\lambda(t))]\alpha_{\lambda i}z_{\lambda y}(t))s_{\lambda i}([1 + E_i(y_\lambda(t+1))]\beta_{\lambda i}\bar{z}_{\lambda x}(t))}{s_{\lambda i}(0)g_{\lambda i}(0)}].$$

In patch λ, $M_{\lambda i}(t)$ [respectively, $N_{\lambda i}(t)$] is the population density of the juvenile [respectively, adult] species of prey i that do not escape predation. System (1), the main model for this study, consists of the following nonlinear difference equations:

$$\left.\begin{aligned}
x_{Ai}(t+1) &= y_{Ai}(t)g_{Ai}(u_{Ai}(t)) - D_i[y_{Ai}(t)g_{Ai}(u_{Ai}(t)) - y_{Bi}(t)g_{Bi}(u_{Bi}(t))],\\
x_{Bi}(t+1) &= y_{Bi}(t)g_{Bi}(u_{Bi}(t)) + D_i[y_{Ai}(t)g_{Ai}(u_{Ai}(t)) - y_{Bi}(t)g_{Bi}(u_{Bi}(t))],\\
y_{Ai}(t+1) &= \bar{x}_{Ai}(t)s_{Ai}(v_{Ai}(t)) - d_i[\bar{x}_{Ai}(t)s_{Ai}(v_{Ai}(t)) - \bar{x}_{Bi}(t)s_{Bi}(v_{Bi}(t))],\\
y_{Bi}(t+1) &= \bar{x}_{Bi}(t)s_{Bi}(v_{Bi}(t)) + d_i[\bar{x}_{Ai}(t)s_{Ai}(v_{Ai}(t)) - \bar{x}_{Bi}(t)s_{Bi}(v_{Bi}(t))],\\
z_{Ax}(t+1) &= \sum_{i=1}^{n}\{M_{Ai}(t)+N_{Ai}(t)\}\\
&\quad -D_{n+1}[\sum_{i=1}^{n}\{M_{Ai}(t)+N_{Ai}(t)\} - \sum_{i=1}^{n}\{M_{Bi}(t)+N_{Bi}(t)\}],\\
z_{Bx}(t+1) &= \sum_{i=1}^{n}\{M_{Bi}(t)+N_{Bi}(t)\}\\
&\quad +D_{n+1}[\sum_{i=1}^{n}\{M_{Ai}(t)+N_{Ai}(t)\} - \sum_{i=1}^{n}\{M_{Bi}(t)+N_{Bi}(t)\}],\\
z_{Ay}(t+1) &= \bar{z}_{Ax}(t) - d_{n+1}[\bar{z}_{Ax}(t) - \bar{z}_{Bx}(t)],\\
z_{By}(t+1) &= \bar{z}_{Bx}(t) + d_{n+1}[\bar{z}_{Ax}(t) - \bar{z}_{Bx}(t)],\ (i=1,2,...,n)
\end{aligned}\right\} \quad (1)$$

where each post-growth dispersion coefficient $d_i, d_{n+1}, D_i, D_{n+1} \in [0, \frac{1}{2}]$. Notice that in System (1), the post-growth dispersion coefficient of the juveniles of species of prey i [respectively, predator], d_i [respectively, d_{n+1}] is allowed to be different from that of its adult species, D_i [respectively, D_{n+1}].

The main model, System (1), is a spatially extended juvenile-adult ecological model that consists of n competing species of prey and a predator in both patch A and patch B, where all species are dispersive. In System (1), patch A and patch B are coupled by an exchange of a fraction of the corresponding difference of the local population densities in the two patches. If the predator is missing, then System (1) with post-growth dispersion and without pre-growth dispersion reduces to the two patch system of Franke and Yakubu in [11]. System (1) with no dispersion and predation reduces to an n-dimensional version of the age-structured model of Ebenman [7, 8], provided patch B is empty. Also, if both the juvenile predator and the adult predator in System (1) are confined to patch A and the local population of the predator in patch B is zero, then patch B is a prey refuge from predation. Yakubu [22] has studied the effects of a prey refuge in a multi-prey discrete predator-prey system with no age structure.

3. PREY PERSISTENCE IN TWO PATCH SYSTEMS

We obtain, in this section, persistence results for System (1) [1, 10, 16]. The condition $s_{\lambda n}(v_{\lambda n}) = 1$ for each $\lambda \in \{A, B\}$ of Theorem 1 and Corollary 1, is biologically restrictive. It implies that, in both patch A and patch B, all the juveniles of species of prey n survive to become adults in the next generation. If, in addition, the searching efficiency for species of prey n by the adult predator [in both patch A and patch B] is sufficiently small, then species of prey n persists provided the interspecific competition between species of prey n and all the other species of prey is very weak (Theorem 1).

Theorem 1. *In System (1), assume that for each* $\lambda \in \{A, B\}$, $s_{\lambda n}(v_{\lambda n}) = 1$ *and* $b^{\lambda}_{nn} = 1$. *If for each* $j \in \{1, 2, ..., n-1\}$ *and each* $\lambda \in \{A, B\}$, *the interspecific competition coefficients* a^{λ}_{nj}, b^{λ}_{nj} *and the searching efficiency* $\alpha_{\lambda n}$ *are small enough, then* F *is uniformly persistent with respect to* $C = \mathbb{R}^{n-1}_{+} \times \{0\} \times \mathbb{R}^{n-1}_{+} \times \{0\} \times \mathbb{R}^{n-1}_{+} \times \{0\} \times \mathbb{R}^{n-1}_{+} \times \{0\} \times \mathbb{R}^{2}_{+} \times \mathbb{R}^{2}_{+}$. *Hence, for each point* (x, y, z_x, z_y) *in the interior of* $\mathbb{R}^{2n}_{+} \times \mathbb{R}^{2n}_{+} \times \mathbb{R}^{2}_{+} \times \mathbb{R}^{2}_{+}$, $\omega((x, y, z_x, z_y)) \cap C = \emptyset$ *and species of prey* n *persists.*

The proof of Theorem 1 is similar to the proof of Theorem 5 in [10] and is omitted in this short article.

In System (1), if all the juvenile species of prey mature to adulthood and the predation is very weak, then all the species in the system persist whenever interspecific prey competition is very weak and is less than the intraspecific prey competition and the predator has a positive searching efficiency for some species of prey (Corollary 1).

Corollary 1. *In System (1), assume that* $s_{\lambda i}(v_{\lambda i}) = 1$ *and* $b^{\lambda}_{ii} = 1$ *for each* $i \in \{1, 2, ..., n\}$ *and each* $\lambda \in \{A, B\}$. *If each* $\alpha_{\lambda i}$ *is sufficiently small, then there exist intraspecific competition coefficients* a^{λ}_{ii} *and* b^{λ}_{ii} *greater than the interspecific competition coefficients* a^{λ}_{ij} *and* b^{λ}_{ij} *for each* $i \neq j$ *such that all the different species of prey persist. If, in addition,* $\alpha_{\lambda i}$ *is positive for some species of prey* i *in either patch* $\lambda = A$ *or patch* $\lambda = B$, *then the predator and all the different species of prey persist.*

To prove Corollary 1, choose $a^{\lambda}_{ii} = 1$ and $b^{\lambda}_{ii} = 1$. Then use Theorem 1 with sufficiently small $\alpha_{\lambda i}$, a^{λ}_{ij} and b^{λ}_{ij} that are sufficiently less than one for each $\lambda \in \{A, B\}$ and each $i \neq j$ in $\{1, 2, ..., n\}$. To complete the proof note that the predator population is positive whenever the predator has a positive searching efficiency for any of the persisting species of prey.

4. PERSISTENCE VERSUS EXTINCTION

In the persistence results of the previous section (Theorem 1 and Corollary 1), all the juvenile species of the persisting species of prey mature to become adults. We now illustrate, with an example, that if interspecific competition and predation are very weak, then exclusionary dynamics could occur in System (1) provided not all the juvenile species of prey mature to adulthood. For this illustration, we consider a reduced form of System (1) with two-age classes but with two competing species of prey and a predator and with Ricker's model as the per-capita growth rates and survival rates[2, 9 - 15, 18 - 23].

$$\left.\begin{array}{rcl}
x_{Ai}(t+1) & = & y_{Ai}(t)\exp(p_{Ai}-[u_{Ai}(t)]), \\
x_{Bi}(t+1) & = & y_{Bi}(t)\exp(p_{Bi}-[u_{Bi}(t)]), \\
y_{Ai}(t+1) & = & \overline{x}_{Ai}(t)\exp(-q_{Ai}[v_{Ai}(t)])- \\
 & & d_i\{\overline{x}_{Ai}(t)\exp(-q_{Ai}[v_{Ai}(t)])-\overline{x}_{Bi}(t)\exp(-q_{Bi}[v_{Bi}(t)])\}, \\
y_{Bi}(t+1) & = & \overline{x}_{Bi}(t)\exp(-q_{Bi}[v_{Bi}(t)])+ \\
 & & d_i\{\overline{x}_{Ai}(t)\exp(-q_{Ai}[v_{Ai}(t)])-\overline{x}_{Bi}(t)\exp(-q_{Bi}[v_{Bi}(t)])\}, \\
 & & (i=1,2) \\
z_{Ax}(t+1) & = & \sum_{i=1}^{2}\{M_{Ai}(t)+N_{Ai}(t)\}, \\
z_{Bx}(t+1) & = & \sum_{i=1}^{2}\{M_{Bi}(t)+N_{Bi}(t)\}, \\
z_{Ay}(t+1) & = & \overline{z}_{Ax}(t)-d_3\{\overline{z}_{Ax}(t)-\overline{z}_{Bx}(t)\}. \\
z_{By}(t+1) & = & \overline{z}_{Bx}(t)+d_3\{\overline{z}_{Ax}(t)-\overline{z}_{Bx}(t)\}.
\end{array}\right\} \quad (2)$$

Notice that if $n=2$ and for each $i\in\{1,2\}$ and each $\lambda\in\{A,B\}$, $g_{\lambda i}(u_{\lambda i}(t)) = \exp(p_{\lambda i}-u_{\lambda i}(t))$, $s_{\lambda i}(v_{\lambda i}(t)) = \exp(-q_{\lambda i}v_{\lambda i}(t))$ and $D_i=0$, then System (1) reduces to System (2). Thus, in System (2) all the juvenile species are dispersive but all the adult species are sedentary.

5. IllUSTRATIVE EXAMPLE

In System (2), let $E_i\equiv 0$. Now, set the following parameter values: For each $\lambda\in\{A,B\}$ and each $i,j\in\{1,2\}$, $a_{ii}^{\lambda}=b_{ii}^{\lambda}=1$, $a_{ij}^{\lambda}=b_{ij}^{\lambda}=0$ if $i\neq j$, $\alpha_{\lambda i}=$

$\beta_{\lambda i} = 0$, $p_{A1} = p_{B1} = 1$, $p_{A2} = p_{B2} = 1.1$, $\mu_1 = \mu_2 = 0.01$, $d_1 = d_2 = 0.001$ and $D_1 = D_2 = 0$.

With our choice of constants, the interspecific competition and predation are very weak in System (2). If in addition each $q_{\lambda i} = 0$, then all the juvenile species of prey mature to adulthood and by Corollary 1, the persistence of all the species of prey occurs. In fact, with our choice of constants, System (2) has a stable period 2 orbit that attracts positive population densities. The points in the period 2 orbit are $(0, 0, 0, 0, 1.0193,\ 1.1224,\ 1.0193,\ 1.1202,\ 0, 0, 0, 0) \rightarrow (0.9998,\ 1.0975,\ 2.7706,\ 3.3653,\ 0, 0, 0, 0, 0, 0, 0, 0)$. The period 2 orbit illustrates the coexistence of the two adult species of prey and the coexistence of the two juvenile species of prey in alternating generations.

To show that the extinction of a species of prey occurs in System (2) with weak predation and competition, and with intraspecific competition greater than interspecific competition, fix the constants q_{A2} and q_{B2} at $q_{A2} = q_{B2} = 0$. For each λ and each i, j set $c_{ij}^{\lambda} = d_{ij}^{\lambda} = 1$. Our numerical experiments show that as $q_{A1} = q_{B1}$ is increased from zero past 0.9, a dramatic bifurcation occurs. The system changes from the persistence of all the species of prey to the extinction of prey 1. In particular, at $q_{A1} = q_{B1} = 1$, System (2) has a stable period 2 orbit with species of prey 1 driven to extinction. The points in the period 2 orbit are $(0, 0, 0, 0, 0, 1.1224, 0, 1.1202, 0, 0, 0, 0) \rightarrow (0, 1.0975, 0, 3.3653, 0, 0, 0, 0, 0, 0, 0, 0)$.

In this example, not all the juveniles of species of prey 1 mature to adulthood [$q_{A1} > 0$ and $q_{B1} > 0$]. As a result, species of prey 2 [the species of prey with all of its juveniles maturing to adulthood] dominates the system and drives species of prey 1 to extinction.

6. CONCLUSION

In this short article, we have extended the classic Nicholson-Bailey's discrete-time predator-prey model to include age-structure in a two patch multi-prey predator-prey model with pre-growth dispersion and post-growth dispersion.

ACKNOWLEDGMENT

This research was partially supported by a grant from Howard University Faculty Excellence Award. The author thanks Prof. J.M. Cushing of University of Arizona for suggesting the inclusion of the biological concept of pre-growth dispersion in the main model, System(1).

REFERENCES

1. G.J. Butler and P. Waltman, Persistence in dynamical systems, *J. Differential Equations* 65(1986), 111-138.

2. H.N. Commins and M.P. Hassell, Predation in multi-prey communities, *J. Theor. Biol.* 62 (1976), 93-114.

3. J.M. Cushing, A discrete model for competing stage-structured species, *Theor. Popl. Biol.* 41(1992), 372-387.

4. J.M. Cushing, The allee effect in age-structured population dynamics, *Mathematical Ecology* (Trieste, 1986), World Sci. Publ., Teaneck, NJ (1988), 479-505.

5. J.M. Cushing and J. Li, Intra-specific competition and density dependent juvenile growth, *Bulletin of Math. Biol.* 54(1992), 503-519.

6. J.M. Cushing and J. Li, Juvenile versus adult competition, *J. Math. Biol.* 29 (1991), 457-473.

7. J.M. Cushing and J. Li, On Ebenman's model for the dynamics of a population with competing juveniles and adults, *Bulletin of Math. Biol.* 51(1989), 687-713.

8. B. Ebenman, Competition between age classes and population dynamics, *J. Theor. Biol.* 131(1988), 389-400.

9. J.E. Franke and A.-A. Yakubu, Dominance conditions in age-structured discrete competitive systems, *Proceedings of the First International Conference on Difference Equations and Applications* (eds. S.N. Elaydi, J.R. Graef, G. Ladas and A.C. Peterson), Gordon Breach Science Publication (1995) 197-211.

10. J.E. Franke and A.-A. Yakubu, Extinction and persistence of species in discrete competitive systems with diffusion, *J. Math. Anal. Appl.* 203 (1996), 746-761.

11. J.E. Franke and A.-A. Yakubu, Extinction of species in age-structured, discrete noncooperative systems, *J. Math. Biol.* 34 (1996), 442-454.

12. J.E. Franke and A.-A. Yakubu, Global asymptotic behavior and dispersion in age-structured, discrete competitive systems, *Canadian Applied Mathematics Quarterly,* 4 (1997), 375-395.

13. M.P. Hassell, "The dynamics of competition and predation," *Studies in Biology* 72(1976), Edward Arnold (Publishers) Ltd.

14. M.P. Hassell and R.M. May, Aggregation of predators and insect parasites and its effect on stability, J. Anim. Ecol. 43 (1974), 567-594.

15. A. Hastings, Complex interactions between dispersal and dynamics: Lessons from coupled logistic equation, *Ecology*, 74(1993), 1362-1372.

16. J. Hofbauer and J.W.-H. So., Uniform persistence and repellors for maps, *Proc. Amer. Math. Soc.* 107(1987), 1137-1142.

17. C. S. Holling, The components of predation as revealed by a study of small mammal predation of the European pine sawfly, *Can. Entomol.* 91(1959), 293-320.

18. S. Levin, "Population Biology," *Lecture Notes in Biomathematics*, Vol. 52 (eds: H.I. Freedman and C. Strobeck), Springer-Verlag, New York/Berlin (1982).

19. R.M. May, Simple mathematical models with very complicated dynamics, Nature 261(1976), 159-161.

20. R.M. May, "Theoretical Ecology", *Sinauer Associates Inc. (Publications)*, Sunderland/Massachusetts (1981).

21. A.J. Nicholson and V.A. Bailey, The balance of animal populations, *Proc. Zoological Soc. London* 551 (1935).

22. A.-A. Yakubu, Prey dominance in discrete predator-prey systems with a prey refuge, *Math. Biosc.* 144: 155-178(1997).

23. A.-A. Yakubu, Searching predator and prey dominance in discrete predator-prey systems with dispersion (SIAP, In press).

Proceedings of Dynamic Systems and Applications 3 (2001) 629-636

Empirical Bayes Estimators for Borel–Tanner Distribution

George P. Yanev[1] and Chris P. Tsokos
University of South Florida, PHY 114, Tampa, FL 33620
[1]On leave from the Bulgarian Academy of Sciences, Sofia, Bulgaria

ABSTRACT: The Borel-Tanner probability distribution was derived by Borel (1942) and Tanner (1953) to characterize the distribution behavior of the number of customers served in a queuing system with Poisson input and constant service time. Later this probability distribution was applied in some models for random trees and branching processes. In the latter case one of the parameters can be interpreted as the offspring mean in a Galton-Watson process with Poisson reproduction law. We propose empirical Bayes estimators for this parameter.
AMS subject classification. 62C10, 62F15, 60J80.

1. Introduction

Let us consider, following the remarks of Consul [4], a random model for the spread of a certain characteristic in a given population as follows. First, the characteristic of interest is transmitted to some members of a group from a source at an initial point in time. Then the individuals who have acquired the characteristic spread it, according to a probability distribution, to the members of other groups. The new "generation" spreads the same characteristic again and the process continues over and over until either it dies out or the entire population gets the characteristic. Let Z_0 be the number of individuals who first have acquired the particular characteristic, i.e., the zeroth generation of ancestors. All the individuals who have acquired the characteristic from each of the ancestors Z_0 will then form the first generation, and each of them will convey the characteristic to others, and so on. One objective of the model might be to find the probability distribution of all individuals possessing the characteristic under study up to the moment when it is no longer spread. Under certain assumptions (see Consul [4], p.40), the number of individuals possessing the characteristic in the zeroth, first, second, etc. generation, denoted by $Z_0, Z_1, Z_2, \ldots$ form a Bienaymé-Galton-Watson branching process with a Poisson offspring distribution. Two basic assumptions are: (i) the characteristic is spread independently by different individuals, and (ii) "the offspring", i.e., the number of individuals who have acquired the characteristic from one spreader follows a fixed with respect to the time probability distribution. Note that, if at a certain moment the offspring distribution changes, e.g., as result of a campaign for increasing or decreasing the spread of the characteristic, then the total number of individuals having the characteristic until that moment may be considered to belong to the zeroth generation. Thus, assumption (ii) will be still valid.

Let $g_\theta(s)$ be the probability generating function of the Poisson (θ) offspring distribution, given by

$$g_\theta(s) = e^{\theta(s-1)} \ .$$

Then it can be found that the total number X_θ of individuals having the particular characteristic generated by a single ancestor has probability distribution

$$B(x,\theta) = P(X_\theta = x | Z_0 = 1) = \frac{(\theta x)^{x-1}}{x!} e^{-\theta x} \ , \quad x = 1,2,\dots, \quad 0 < \theta < 1, \tag{1}$$

which was first studied by Borel [2]. If in (1) $Z_0 = r$ with probability 1, then

$$P(X_\theta = x | Z_0 = r) = \frac{r}{(x-r)!} x^{x-r-1} \theta^{x-r} e^{-\theta x} \ , \ x = r, r+1, \dots, \ 0 < \theta < 1, \tag{2}$$

which is known as the Borel–Tanner distribution, see e.g. Johnson & Kotz [7]. Note that, when Z_0 is a Poisson (λ) random variable it can be seen that

$$P(X_\theta = x) = \frac{\lambda(\lambda + \theta x)^{x-1}}{x!} e^{-\lambda - \theta x} \ ,$$

which is the generalized Poisson distribution.

Initially, (2) was derived as the probability distribution of the number of customers served in a queuing system with Poisson (θ) input and a constant service time, given that the length of the queue at the initial time is r. Later this distribution was applied in studying random trees, see e.g. Kolchin [8], and in branching processes as distribution of the total progeny assuming Poisson offspring, similarly to the interpretation we have described initially. More recently, the Borel-Tanner distribution appeared in a variety of applications, see e.g. Steutel & Hansen [15] and Aldous [1].

In this article we shall study an empirical Bayes procedure for estimating the parameter θ in (1). This approach has certain advantages to the ordinary Bayes one, because of the a priori assumptions that heed to be made. The case (2) can be treated similarly. Recall that the parameter θ can be thought of as the offspring mean in a branching process with Poisson reproduction law. Next for convinience, we give some important results which can be found in the literature concerning inferences for Borel-Tanner distribution. In Section 3 we propose an approximate simple empirical Bayes estimator for θ.

2. Inferences for Borel-Tanner Distribution

For the distribution (2) we have

$$EX_\theta = r(1-\theta)^{-1} \quad \text{and} \quad VarX_\theta = r\theta(1-\theta)^{-3} \ .$$

Note that, if $X_1, X_2, \ldots, X_n$ is a random sample taken from the Borel-Tanner distribution, then the sum $Y_n = X_1 + \ldots + X_n$ is a sufficient statistic for θ and it also has a Borel-Tanner distribution given by

$$P(Y_n = y) = \frac{rn}{(y - rn)!} y^{y-rn-1} \theta^{y-rn} e^{-y} .$$

Gupta in [6] has obtained the maximum likelihood estimate for θ from (2) to be

$$\hat{\theta}(x) = 1 - \frac{r}{x} .$$

In Charalambides [3] and Kumar & Consul [9] the minimum variance unbiased estimators, MVUE, for $\theta^k, k = 1, 2, \ldots$ and $P(X_\theta = x)$ were studied. In particular, the MVUE for θ is

$$\bar{\theta}(x) = 1 - \frac{r-1}{x} + \frac{x-2r}{xr} .$$

The MVUE for the variance of $\bar{\theta}(x)$ is

$$MVUE(V(\bar{\theta})) = \frac{1}{r^2} + \frac{1}{x} - \frac{1+r}{x^2} .$$

Goodness-of-fit tests for the Borel-Tanner distribution are proposed in Mirvaliev [11] and Nikulin & Voinov [13].

In the classical Bayesian setting, let us assume that the parameter θ is a realization of the random variable Θ with prior distribution $G(\theta)$ on $(0, 1)$. It is well-known that the Bayesian estimator $\theta_G(y)$ of θ w.r.t. $G(\theta)$ and assuming a squared-error loss is the posterior mean written as $E_{\theta|x}(\Theta)$. Suppose that $G(\theta)$ is the uniform distribution on $(0, 1)$. Then a simple computation gives

$$\begin{aligned} \theta_U(x) &= \frac{\int_0^1 \theta P(X_\theta = x | Z_0 = r) d\theta}{\int_0^1 P(X_\theta = x | Z_0 = r) d\theta} \\ &= 1 - \frac{r-1}{x} - \frac{x^{x-r}}{(x-r)!} \left(1 - e^{-x} \sum_{j=0}^{x-r} \frac{x^j}{j!} \right)^{-1} . \end{aligned}$$

3. Empirical Bayes Estimation of θ

Empirical Bayes methods rely on the existence of a prior distribution $G(\theta)$ which, however, is unknown, see Robbins [14]. Under this assumption the results of previous experiments can be used to estimate the prior $G(\theta)$ and/or the Bayes estimator $\theta_G(x)$. Let us consider the case where $X_1, \ldots, X_n$ is a sequence of independent random variables, independent of (X_θ, Θ) and with the same marginal distribution as X_θ. In the general case, in order to find an estimator of $\theta_G(x)$ based on $X_1, \ldots, X_n$, one has to construct an estimate, $\hat{G}$, of the prior $G(\theta)$ and then replace $G(\theta)$ by $\hat{G}$ in the formula for $\theta_G(x)$. In some special cases it turns out that $\theta_G(x)$ can be estimated without estimating the prior $G(\theta)$. All empirical Bayes estimates which are derived without explicit estimation of $G(\theta)$ are called simple empirical Bayes estimates, SEBEs. As it is pointed out in Maritz and Lwin [10], apparent advantages of using a SEBE are: it is distribution free w.r.t. $G(\theta)$; no assumption is needed about the class of distributions to which $G(\theta)$ might belong, or in which a good approximation of $G(\theta)$ might be found. A general rule is that a SEBE is obtainable if its corresponding Bayes estimator depends only on the marginal distribution of the observed variable.

Unfortunately, in the case of Borel–Tanner distribution, no SEBE for θ is obtainable. However, by an appropriate reparametrization, we shall derive an approximate SEBE for θ using a technique due to Cressie [5] (see also Nichols & Tsokos [12]).

From now on, for simplicity, we assume $r = 1$, i.e., we shall consider the Borel-Tanner distribution in the form (1).

Let us change the parameter θ in (1) to

$$\varphi = f(\theta) = \theta e^{-\theta} .$$

Then (1) can be written as

$$B(x,\varphi) = \frac{x^{x-1}}{x!}\varphi^{x}\left(f^{\leftarrow}(\varphi)\right)^{-1} , \tag{3}$$

where $g^{\leftarrow}$ stands for the inverse of a function g. Our next step will be to transform the problem of estimating $\theta_G(x)$ to the problem of estimating the Bayes estimator $\varphi_G^k(x)$, say, of φ^k for some integer k. The advantage of using the new parameter φ instead of θ is that for φ one can find a SEBE. Indeed, the Bayes estimator $\varphi_G^k(x)$ for $k = 1, 2, 3, \ldots$ is given by

$$\begin{aligned} \varphi_G^k(x) &= \frac{\int_0^1 \theta^k e^{-k\theta} B(x,\theta)dG(\theta)}{\int_0^1 B(x,\theta)dG(\theta)} \\ &= \frac{m(x+k)}{m(x)}\left(\frac{x}{x+k}\right)^{x-1}\prod_{i=1}^{k}\frac{x+i}{x+k} , \end{aligned} \tag{4}$$

where $m(x)$ is the marginal probability function of X_θ. Let $X_1, \ldots, X_n$ be a sequence of independent random variables, independent of (X_θ, Θ) and with the same marginal distribution as X. From (4) it is clear that an SEBE estimator $\varphi_n^k(x)$, say, for φ^k can be obtained by estimating $m(z)$ with $m_n(z)$, an estimate based on the observations $x_1, x_2, \ldots, x_n$, i.e., for $k = 1, 2, \ldots$

$$\varphi_n^k(x) = \frac{m_n(x+k)}{m_n(x)}\left(\frac{x}{x+k}\right)^{x-1}\prod_{i=1}^{k}\frac{x+i}{x+k} . \tag{5}$$

First, we shall derive some inequalities for the Bayes estimator $\theta_G(x)$.

Lemma *If* $0 < \Theta < 1$ *a.s., then*

$$\theta_G(x) \geq f^{\leftarrow}\left(\varphi_G(x)\right) . \tag{6}$$

If in addition, there exists $\varepsilon > 0$ such that $0 < \Theta < 1 - \varepsilon$ a.s., then

$$\theta_G(x) \le (f^k)^{\leftarrow}\left(\varphi_G^k(x)\right) , \tag{7}$$

where $k \ge 1/\varepsilon^2$.

Proof For $f^k(\theta) = \theta^k e^{-k\theta}$, we have for every $k > 0$,

$$\frac{d}{d\theta} f^k(\theta) = k\theta^{k-1}e^{-k\theta}(1-\theta) > 0 ,$$

and

$$\frac{d^2}{d\theta^2} f^k(\theta) = k^2\theta^{k-2}e^{-k\theta}\left((1-\theta)^2 - \frac{1}{k}\right) .$$

Thus, $f(\theta)$ is concave down, whereas $f^k(\theta)$ is concave up provided that $k \ge 1/\varepsilon^2$. Now, if $k \ge 1/\varepsilon^2$, using Jensen's inequality, we obtain

$$\begin{aligned} E_{\theta|x} f(\Theta) &\le f(E_{\theta|x}(\Theta)) , \\ E_{\theta|x} f^k(\Theta) &\ge f^k(E_{\theta|x}(\Theta)) . \end{aligned}$$

Taking into account that $f^k(\theta)$ is an increasing function w.r.t. θ for every k, we obtain

$$f^{\leftarrow}\left(E_{\theta|x}(\Phi)\right) \le E_{\theta|x}\Theta \le (f^k)^{\leftarrow}\left(E_{\theta|x}(\Phi^k)\right) , \tag{8}$$

where $\Phi = f(\Theta)$. Note that, (8) implies (6) and (7), which completes the proof.

Now we shall illustrate how tight the above inequalities are with an example. Assume that the prior is given by

$$G(\theta) = 1/\theta , \qquad 0 < \theta < 0.95 .$$

Some numerical results are given in Table 1.

x	30	50	70	90	200
$f^{\leftarrow}(\varphi_G(x))$	0.7776	0.8258	0.8497	0.8647	0.9000
$\theta_G(x)$	0.8208	0.8542	0.8716	0.8827	0.9094
$(f^{400})^{\leftarrow}(\varphi_G^{400}(x))$	0.8888	0.8964	0.9015	0.9052	0.9170

Table 1: Numerical bounds for $\theta_G(x)$

Thus, the calculations support the tightness of the derived inequalities, especially when the observed x is large.

Now, we are in a position to obtain an approximate SEBE estimator for θ.

Proposition *Assume that there exists $\varepsilon > 0$ such that $0 < \Theta < 1 - \varepsilon$ a.s., and let $k \geq 1/\varepsilon^2$. Then an approximate SEBE $\theta_n(x)$ for θ is given by*

$$\theta_n(x) = \frac{\theta_n^L(x) + \theta_n^R(x)}{2},$$

where

$$\theta_n^L(x) = \min\left\{ f^{\leftarrow}(\varphi_n(x)) + \frac{\sigma_1^2(x)}{2}\delta(\varphi_n(x)) \ , \ (f^k)^{\leftarrow}\left(\varphi_n^k(x)\right) \right\},$$

and

$$\theta_n^R(x) = \max\left\{ f^{\leftarrow}(\varphi_n(x)) \ , \ (f^k)^{\leftarrow}\left(\varphi_n^k(x)\right) + \frac{\sigma_k^2(x)}{2}\delta\left(\varphi_n^k(x)\right) \right\},$$

where $\delta(y^k)$, $\sigma_k^2(x)$, and $\varphi_n^k(x)$ are given in (13), (10), and (5), respectively.

The procedure of constructing an approximate empirical Bayes estimator for θ involves three steps: approximation of $\theta_G(x)$ in terms of the Bayes estimators for φ; in the obtained formulas replacing the Bayes estimators by their empirical Bayes counterparts, and constructing an estimator taking into account the inequalities from the Lemma.

Denote $\mu = E_{\theta|x}(f^k(\Theta))$ and $\sigma_k^2 = Var_{\theta|x}(f^k(\Theta))$. Let us carry out the Taylor series expansion of $(f^k)^{\leftarrow}$ about μ to the second order. Namely,

$$\begin{aligned} \Theta &= (f^k)^{\leftarrow}(f^k(\Theta)) \\ &\approx (f^k)^{\leftarrow}(\mu) + \left(f^k(\Theta) - \mu\right)\left((f^k)^{\leftarrow}(\mu)\right)' + \frac{1}{2}\left(f^k(\Theta) - \mu\right)^2\left((f^k)^{\leftarrow}(\mu)\right)''. \end{aligned}$$

To obtain an approximation of $\theta_G(x)$ we take expectation of both sides and replace μ by its Bayes estimate $\varphi_G^k(x)$

$$\theta_G(x) \approx (f^k)^{\leftarrow}(\varphi_G^k(x)) + \frac{\sigma_k^2}{2}\left((f^k)^{\leftarrow}(\varphi_G^k(x))\right)''. \tag{9}$$

Initially, we can estimate

$$\sigma_k^2 = E_{\theta|x} f^{2k}(\Theta) - (E_{\theta|x} f^k(\Theta))^2$$

by the corresponding Bayes estimators, i.e.,

$$\sigma_G^2(x) = \varphi_G^{2k}(x) - \left(\varphi_G^k(x)\right)^2.$$

Then applying (5) we obtain an empirical Bayes estimator for σ_k^2

$$\begin{aligned} \sigma_k^2(x) &= \varphi_n^{2k}(x) - \left(\varphi_n^k(x)\right)^2 \\ &= \frac{m_n(x+2k)}{m_n(x)}\left(\frac{x}{x+2k}\right)^{x-1}\prod_{i=1}^{2k}\frac{x+i}{x+2k} \\ &- \frac{m_n^2(x+k)}{m_n^2(x)}\left(\frac{x}{x+k}\right)^{2(x-1)}\prod_{i=1}^{k}\left(\frac{x+i}{x+k}\right)^2. \end{aligned} \tag{10}$$

Secondly, since $f'(\theta) = (1-\theta)e^{-\theta}$, we have for $k = 1, 2, \ldots$

$$\frac{d}{d\varphi}(f^k)^{\leftarrow}(\varphi) = \frac{\exp(f^{\leftarrow}(\varphi))}{k\varphi^{k-1}(1-f^{\leftarrow}(\varphi))} = \frac{\exp(f^{\leftarrow}(\varphi))}{k\varphi^{k-1}(1-\varphi\exp(f^{\leftarrow}(\varphi)))}.$$

For the second derivative of $(f^k)^{\leftarrow}(\varphi)$ one can obtain for $k = 1, 2, \ldots$

$$\frac{d^2}{d\varphi^2}(f^k)^{\leftarrow}(\varphi) = \frac{(f^{\leftarrow}(\varphi))^2(2-f^{\leftarrow}(\varphi)) - (k-1)f^{\leftarrow}(\varphi)(1-f^{\leftarrow}(\varphi))^2}{k\varphi^{k+1}(1-f^{\leftarrow}(\varphi))} \tag{11}$$

Now, replacing in (9) $\varphi_G^k(x)$ by the empirical Bayes estimator $\varphi_n^k(x)$, and taking into account (11) we obtain for $k = 1, 2, \ldots$

$$\theta_G(x) \approx (f^k)^{\leftarrow}\left(\varphi_n^k(x)\right) + \frac{\sigma_k^2(x)}{2}\delta\left(\varphi_n^k(x)\right), \tag{12}$$

where for $k = 1, 2, \ldots$

$$\delta(y^k) = \frac{(f^{\leftarrow}(y^k))^2(2-f^{\leftarrow}(y^k)) - (k-1)f^{\leftarrow}(y^k)\left(1-f^{\leftarrow}(y^k)\right)^2}{ky^{k+1}(1-f^{\leftarrow}(y^k))}, \tag{13}$$

and $\sigma_k^2(x)$ is given in (10). Note that, if $k = 1$ then $\delta(y) > 0$, whereas $\delta(y^k) < 0$ for sufficiently large k. Thus,

$$f^{\leftarrow}(\varphi_n(x)) < f^{\leftarrow}(\varphi_n(x)) + \frac{\sigma_1^2(x)}{2}\delta(\varphi_n(x)), \tag{14}$$

and for a sufficiently large k

$$(f^k)^{\leftarrow}\left(\varphi_n^k(x)\right) + \frac{\sigma_k^2(x)}{2}\delta\left(\varphi_n^k(x)\right) < (f^k)^{\leftarrow}\left(\varphi_n^k(x)\right). \tag{15}$$

On the other hand, recall that from the Lemma we have

$$f^{\leftarrow}(\varphi_G(x)) < \theta_G(x) < (f^k)^{\leftarrow}\left(\varphi_G^k(x)\right).$$

Replacing $\varphi_G^k(x)$ by the empirical Bayes estimate $\varphi_n^k(x)$, we have at least approximately

$$f^{\leftarrow}(\varphi_n(x)) < \theta_G(x) < (f^k)^{\leftarrow}\left(\varphi_n^k(x)\right). \tag{16}$$

Clearly, an approximate empirical Bayes estimator $\theta_n(x)$, say, for θ can be set to be equal to the right hand side in (12). On the other hand, $\theta_n(x)$ being close to the Bayes estimator $\theta_G(x)$ must satisfy (16). Therefore, taking into account the inequalities (14) and (15), $\theta_n(x)$ is to be between $\theta_n^L(x)$ and $\theta_n^R(x)$. As a reasonable choice we have picked the midpoint $\left(\theta_n^L(x) + \theta_n^R(x)\right)/2$. Thus, the estimator has been constructed.

As a final note it is worth mentioning that the reparametrization (3) of the Borel–Tanner distribution has another advantage: monotone likelihood ratio in x. Hence, any empirical Bayes estimator for φ should reflect this. Using the method in Van Houwelingen [16], starting with an estimator $\varphi(x)$, say, one can construct a monotonized version $\varphi^*(y)$. Such a construction will be subject of a future research.

Acknowledgments

The authors thank A. N. V. Rao for his helpful comments and suggestions. The first named author is partially supported by National Foundation of Scientific Investigations, Bulgaria, Grant MM-704/97.

References

[1] Aldous, D. J. (1999). Deterministic and stochastic models for coalescence (aggregation and coagulation): a review of the mean-field theory for probabilists. Bernoulli, Vol 5 (1999), 3-48.

[2] Borel, E. (1942). Sur l'emploi du théorème de Bernoulli pour faciliter le calcul d'un infinité de coefficients. Application au problème de l'attente à un quichet. C. R. Acad. Sci. Paris, Vol 214(1942), 452–456.

[3] Charalambides, Ch, (1974). The distribution of sufficient statistics of truncated generalized logarithmic series, Poisson, and negative binomial distributions. Canadian J. Statist., Vol 2 (1974), 261-275.

[4] Consul, P. C, Generalized Poisson Distributions. Marcel Dekker, New York, 1989.

[5] Cressie, N. (1982). A useful empirical Bayes identity. The Annals of Statistics, Vol 10 (1982), 625–629.

[6] Gupta, R. C, (1974). Modified power series distribution and some of its applications. Sankya, Ser. B, Vol 35 (1974), 288-298.

[7] Johnson, N. L. & Kotz, S, Discrete Distributions. Houghton Mifflin, Boston, 1969.

[8] Kolchin, V. F, Random Mappings. Translation Series in Mathematics, Optimization Software, Inc., Publication Division, New York, 1986.

[9] Kumar A. & Consul P. C, (1980). Minimum variance unbiased estimation for modified power series distribution. Commun. Statist. A - Theory and Methods, Vol 9 (1980), 1261-1275.

[10] Maritz, J. S. & Lwin, T, Empirical Bayes Methods. 2nd edn., Chapman and Hall, London, 1989.

[11] Mirvaliev M, (1989). Goodness-of-fit tests to the Borel-Tanner distribution. In: Ed. S. Kh. Sirazhdinov, Functionals of random processes and statistical inferences, 50-58, Fan, Tashkent, 1989, (In Russian).

[12] Nichols, W. G. & Tsokos C. P. (1972). Empirical Bayes point estimation in a family of probability distributions. Intern. Statist. Review, Vol 40, 147–151.

[13] Nikulin, M. S. & Voinov, V. G. (1994). A chi-squared goodness-of-fit test for modified power series distributions. Sankhya, Series B, Vol 56 (1994), 86-94.

[14] Robbins, H. (1955). An empirical Bayes approach to statistics. In: Eds. S. Kotz & N. L. Johnson, Breakthroughs in statistics, Vol 1, Foundations and basic theory. Springer, New York, 1992.

[15] Steutel, F. W. & Hansen, B. G. (1989). Haight's distribution and busy periods. Statistics & Probability Letters 7 (1989), 301-302.

[16] Van Houwelingen, J. C. (1977). Monotonizing empirical Bayes estimators for a class of discrete distributions with monotone likelihood ratio. Statist. Neerl., Vol 31 (1977), 95–104.

Proceedings of Dynamic Systems and Applications 3 (2001) 637-644

Stability of Delay Difference Systems

Shunian Zhang
Department of Applied Mathematics, Shanghai Jiaotong University
Shanghai, 200240, CHINA

ABSTRACT: In this paper, we deal with the stability of the zero solution for both delay difference systems and neutral (delay) difference systems. The criteria of uniform asymptotic stability are proposed in terms of the discrete Liapunov functionals as well as Liapunov functions with Razumikhin techniques.
AMS (MOS) subject classification: 39A10, 39A11

1. INTRODUCTION

In [1] we established stability criteria for the finite delay difference systems of the general form in terms of discrete Liapunov functionals as well as Liapunov functions. Furthermore, the results obtained in [1] have been further improved in [2]. However, to the best of our knowledge, there have only appeared several results on oscillations and asymptotic behaviors of some specific neutral (delay) difference equations, but we have not seen any stability results for neutral difference systems of the general form so far.

In this paper, we will introduce the relevant results on stability of delay difference systems. Then we will establish stability criteria for neutral difference systems of the general form by using discrete Liapunov functionals and functions.

2. DELAY DIFFERENCE SYSTEMS

Consider the finite delay difference systems of the general form

$$x(n+1) = f(n, x_n), \qquad n \in Z^+, \tag{1}$$

where Z^+ denotes the set of non-negative integers, $x \in R^k$ with some positive integer k. Let

$$C = \{\varphi : \{-r, -r+1, \ldots, -1, 0\} \to R^k\} \text{ with some } r \in Z^+,$$

and define the norm of $\varphi \in C$ as

$$\|\varphi\| = \max_{s=-r,-r+1,\ldots,0} |\varphi(s)| \text{ with } |\cdot| \text{ a norm in } R^k.$$

Let $C_H = \{\varphi \in C : \|\varphi\| < H\}$ for some constant $H > 0$, and $f : Z^+ \times C_H \to R^k$, while $x_n \in C$ is defined as

$$x_n(s) = x(n+s) \text{ for } s = -r, -r+1, \ldots, 0.$$

Suppose that for any given $n_0 \in Z^+$ and a given function $\varphi \in C_H$, there exists a unique solution of (1), denoted by $x(n_0, \varphi)(n)$, such that it satisfies (1) for all integer $n \geq n_0$ and

$$x(n_0, \varphi)(n_0 + s) = \varphi(s) \qquad \text{for } s = -r, -r+1, \ldots, 0.$$

To deal with stability, as usual, we assume that $f(n, 0) = 0$ so that (1) has

In the sequel, we always assume that the variables n, s, i, j, k take integer values and the corresponding inequalities as well as intervals are discrete ones.

Definition 2.1. The zero solution of (1) is said to be uniformly stable (U.S.) if for any given $\varepsilon > 0$, and any $n_0 \in Z^+$, there exists a $\delta = \delta(\varepsilon) > 0$ independent of n_0 such that if $\|\varphi\| < \delta$, then

$$\|x_n(n_0, \varphi)\| < \varepsilon \text{ (or } |x(n_0, \varphi)(n)| < \varepsilon) \text{ for all } n \geq n_0.$$

Definition 2.2. The zero solution of (1) is said to be uniformly asymptotically stable (U.A.S.) if it is U.S. and there is a $\delta_0 > 0$ such that for each $\gamma > 0$, there exists an integer $N(\gamma) > 0$ independent of n_0 such that if $n_0 \in Z^+$ and $\|\varphi\| < \delta_0$, then

$$\|x_n(n_0, \varphi)\| < \gamma \text{ (or } |x(n_0, \varphi)(n)| < \gamma) \text{ for all } n \geq n_0 + N(\gamma).$$

Definition 2.3. The class of functions, denoted by $\mathcal{K}$, is defined as

$$\mathcal{K} = \{a \in C[R^+, R^+] : a(u) \text{ is strictly increasing in } u \text{ and } a(0) = 0\}.$$

The following lemma (cf. [1]) is needed in the proof of the results in this section.

Lemma 2.1. Let $\{x(n)\}$ be a bounded sequence. Suppose that for $w_1 \in \mathcal{K}$ and $r \in Z^+$, $\sum_{j=n-r}^{n} w_1(|x(j)|) \geq \alpha$ with some $\alpha > 0$. Then for any other $w_2 \in \mathcal{K}$ there exists $\beta > 0$ such that

$$\sum_{j=n-r}^{n} w_2(|x(j)|) \geq \beta.$$

We are now in a position to state the main results in this section. The first one is in terms of the discrete Liapunov functionals.

Theorem 2.1. (cf. [1]) Suppose there exists a Liapunov functional $V : Z^+ \times C_H \to R^+$ with $R^+ = [0, \infty)$ such that

(i) $u(|\varphi(0)|) \leq V(n, \varphi) \leq v_1(|\varphi(0)|) + v_2[\sum_{j=-r}^{0} v_3(|\varphi(j)|)]$,

(ii) $\Delta V_{(1)}(n, x_n(n_0, \varphi)) \leq -w(|x(n_0, \varphi)(n)|)$, where u, $v_i (i = 1, 2, 3)$, $w \in \mathcal{K}$, and

$$\Delta V_{(1)}(n, x_n(n_0, \varphi)) \equiv V(n + 1, x_{n+1}(n_0, \varphi)) - V(n, x_n(n_0, \varphi))$$

with $x(n_0, \varphi)(n)$ being a solution of (1).

Then the zero solution of (1) is U.A.S..

The next important result is known as of Razumikhin type. Here we use the discrete Liapunov functions with Razumikhin techniques instead of using Liapunov functionals in Theorem 2.1.

Theorem 2.2. (cf. [1]) Suppose there exists a Liapunov function $V : Z^+ \times B_H \to R^+$ with $B_H = \{x \in R^k : |x| < H\}$ such that

(i) $u(|x|) \leq V(n, x) \leq v(|x|)$,

(ii) $\Delta V_{(1)}(n, x_n(n_0, \varphi)) \leq -w(|x(n_0, \varphi)(n)|)$ if

$$P(V(n + 1), x(n_0, \varphi)(n + 1)) > V(s, x(n_0, \varphi)(s)) \text{ for } n - r \leq s \leq n,$$

where $u,\ v,\ w \in \mathcal{K}$, $P : (0,\infty) \to (0,\infty)$ is a continuous function with $P(t) > t$ if $t > 0$, and

$$\Delta V_{(1)}(n, x_n(n_0,\varphi)) \equiv V(n+1, x(n_0,\varphi)(n+1)) - V(n, x(n_0,\varphi)(n))$$

with $x(n_0,\varphi)(n)$ being a solution of (1).

Then the zero solution of (1) is U.A.S..

We note that in Theorem 2.1, V is a functional and the estimate on ΔV is required to hold for all n and $x(n) = x(n_0,\varphi)(n)$; whereas in the Razumikhin-type Theorem 2.2, V is a function and the estimate on ΔV is required to hold only under certain circumstances. Actually, we can extend these theorems by using Razumikhin techniques with Liapunov functionals rather than Liapunov functions as indicated below so that the obtained result can be applied in a more general sense.

Theorem 2.3. (cf. [2]) Suppose there exists a $V : Z^+ \times C_H \to R^+$ such that

(i) $u(|\varphi(0)|) \le V(n,\varphi) \le v_1(|\varphi(0)|) + v_2[\sum_{j=-r}^{0} v_3(|\varphi(j)|)]$,

(ii) for any $x(n) = x(n_0,\varphi)(n)$ with $n_0 \in Z^+$ and $\varphi \in C_H$, if $V(n+1, x_{n+1}) \ge v_1(\|\varphi\|)$ $+\ v_2((r+1)v_3(\|\varphi\|))$ holds for $n_0 \le n \le n_0 + r - 1$ with $x_n = x_n(n_0,\varphi)$ then

$$\Delta V_{(1)}(n, x_n) \le 0;$$

whereas if $P(V(n+1, x_{n+1})) \ge V(s, x_s)$ holds for $n - r \le s \le n,\ n \ge n_0 + r$ then

$$\Delta V_{(1)}(n, x_n) \le -w(|x(n)|),$$

where P is defined as in Theorem 2.2.

Then the zero solution of (1) is U.A.S..

Proof. To prove the U.S., for any given $\varepsilon > 0$ $(\varepsilon < H)$, choose $\delta > 0$ such that $\delta < \varepsilon$, $v_1(\delta) < u(\varepsilon)/2$, and $v_2((r+1)v_3(\delta)) < u(\varepsilon)/2$. Let $n_0 \in Z^+$, $\varphi \in C_\delta$, and denote $x(n) = x(n_o,\varphi)(n)$, $V(n) = V(n, x_n(n_0,\varphi))$. We shall show that

$$V(n) < u(\varepsilon) \ \text{ for all } n \ge n_0. \tag{2}$$

Obviously,

$$V(n_0) \le v_1(\delta) + v_2((r+1)v_3(\delta)) < u(\varepsilon).$$

For any $n \in [n_0, n_0 + r]$, if $V(n) < v_1(\|\varphi\|) + v_2((r+1)v_3(\|\varphi\|)$, then $V(n) < u(\varepsilon)$. Now suppose that there exists some $n^* \in [n_0, n_0 + r - 1]$ such that

$$V(n) < v_1(\|\varphi\|) + v_2((r+1)v_3(\|\varphi\|) \ \text{ for } n_0 \le n \le n^*,$$

while

$$V(n^*+1) \ge v_1(\|\varphi\|) + v_2((r+1)v_3(\|\varphi\|).$$

Then by (ii) we have $\Delta V_{(1)}(n^*) \le 0$, and thus

$$V(n^*+1) \le V(n^*) < v_1(\|\varphi\|) + v_2((r+1)v_3(\|\varphi\|).$$

This leads to a contradiction. Therefore, if (2) fails, then there must be some $n_1 \ge n_0 + r$ such that $V(n) < u(\varepsilon)$ for $n_0 \le n \le n_1$, and

$$V(n_1 + 1) \ge u(\varepsilon).$$

Since now $P(V(n_1+1)) > V(n_1+1) \geq u(\varepsilon) > V(s)$ for $n_1 - r \leq s \leq n_1$, it follows from (ii) that $\Delta V_{(1)}(n_1) \leq 0$ and so

$$V(n_1+1) \leq V(n_1) < u(\varepsilon).$$

Again, a contradiction. Hence, (2) holds. It then follows from (i) that

$$|x(n)| < \varepsilon \quad \text{for all } n \geq n_0.$$

Since $\delta > 0$ is independent of n_0, this proves the U.S.

Next, we show the U.A.S.. For $\varepsilon = H$ find $\delta = \delta(H)$ of U.S. and let $\delta_0 = \delta(H)$. Let $\gamma > 0$ be given. We must find an integer $N(\gamma) > 0$ such that $[n_0 \in Z^+,\ \varphi \in C_{\delta_0},\ n \geq n_0 + N]$ imply that $|x(n)| < \gamma$.

From (i) we have

$$V(n) \leq v_1(H) + v_2((r+1)v_3(H)) \quad \text{for all } n \geq n_0.$$

Choose a constant $B > v_1(H) + v_2((r+1)v_3(H))$. For the given $\gamma > 0$ we may assume $u(\gamma) < B$. Let

$$d = \inf_{u(\gamma) \leq t \leq B} (P(t) - t) > 0,$$

and K be the positive integer satisfying

$$u(\gamma) + (K-1)d < B \leq u(\gamma) + Kd.$$

We can first claim by Lemma 1.1 that there must be some integer $N_1 \in [n_0 + r, n_0 + N^*]$ with N^* being independent of n_0 and such that

$$V(N_1) \leq u(\gamma) + (K-1)d.$$

Furthermore, we can show that

$$V(n) \leq u(\gamma) + (K-1)d \quad \text{for all } n \geq n_0 + N^*.$$

By induction, we can prove that there exist $N_i \in [n_0 + r + (i-1)N^*, n_0 + iN^*]$ such that

$$V(n) \leq u(\gamma) + (K-i)d \quad \text{for } n \geq n_0 + iN^*,\ i = 2, 3, \ldots, K.$$

Thus, $V(n) \leq u(\gamma)$ for all $n \geq n_0 + KN^*$, and by (i) we have

$$|x(n)| < \gamma \quad \text{for all } n \geq n_0 + N,$$

where $N = KN^*$ is independent of n_0. This completes the proof of the theorem. QED.

3. NEUTRAL DIFFERENCE SYSTEMS

Consider now the neutral difference systems of the general form

$$\Delta(D(n, x_n)) = f(n, x_n), \qquad n \in Z^+, \tag{3}$$

where Z^+ denotes the set of non-negative integers, $x \in R^k$ with some positive integer k, Δ is the forward difference operator, $D, f : Z^+ \times C_H \to R^k$, while $x_n \in C$ is defined as in Section 2. Suppose that for any given $n_0 \in Z^+$ and a given function $\varphi \in C_H$, there exists a unique solution of (3), denoted by $x(n_0, \varphi)(n)$, such that it satisfies (3) for all integer $n \geq n_0$ and

$$x(n_0, \varphi)(n_0 + s) = \varphi(s) \quad \text{for } s = -r, -r+1, \ldots, 0.$$

Also, we assume that $D(n,0) = f(n,0) = 0$ so that (3) has the zero solution.

The definitions of U.S. and U.A.S. for the zero solution of (3) are the same as in Definitions 2.1 and 2.2. In establishing the main results in this section, we need the following definition and lemma.

Definition 3.1. $D(n,\varphi)$ is said to be uniformly stable (U.S.) if there exist constants $K_1 > 0, K_2 > 0, A > 0$, and $a > 0$ such that for any $h : Z^+ \to R^k$, and any $\tau \in Z^+$, $\psi \in \mathcal{C}_A$ with $D(\tau,\psi) = h(\tau)$, the solution $x(\tau,\psi)(n)$ of

$$\begin{cases} D(n,x_n) = h(n), \\ x_\tau = \psi \end{cases}$$

satisfies the following estimate:

$$\|x_n\| \le K_1 e^{-a(n-\tau)}\|\psi\| + K_2 \max_{\tau\le i\le n} |h(i)|, \quad n \ge \tau. \tag{4}$$

Lemma 3.1. Let $D(n,\varphi)$ be U.S. and let $\alpha(t)$ be any continuous function with $\alpha(0) = 0$ and $\alpha(t) > 0$ for $t > 0$. Then for small $t > 0$, there is a $\mathcal{K}$ function $\beta(t) \ge K_1 + K_2\alpha(t)$ such that

(i) for each small $\delta > 0$ and each $\tau \in Z^+$, $[\|\psi\| < \delta,\ |D(n,x_n(\tau,\psi))| \le \alpha(\delta)$ for $\tau \le n \le \tau^*$ with some integer $\tau^* \le +\infty]$ imply that

$$\|x_n(\tau,\psi)\| \le \beta(\delta), \quad \tau \le n \le \tau^*; \tag{5}$$

(ii) for each small $\delta > 0$ and each small $\mu > 0$ there exists an integer $N = N(\delta,\mu) > 0$ such that $[\|\psi\| < \delta,\ |D(n,x_n(\tau,\psi))| \le \alpha(\mu)$ for $n \ge \tau]$ imply that

$$\|x_n(\tau,\psi)\| \le \beta(\mu), \quad n \ge \tau + N(\delta,\mu). \tag{6}$$

Proof. Let $\delta > 0$ be so small that $\delta < A$. Then it follows from (4) with $h(n) = D(n,x_n(\tau,\psi))$ that whenever $\|\psi\| \le \delta$ and $|D(n,x_n(\tau,\psi))| \le \alpha(\delta)$ for $\tau \le n \le \tau^*$ we have

$$\|x_n(\tau,\psi)\| \le K_1\delta + K_2\alpha(\delta) \quad \text{for } \tau \le n \le \tau^*.$$

Trivially, we can choose a $\mathcal{K}$ function $\beta(t) \ge K_1 t + K_2\alpha(t)$ so that (5) holds.

Also, for each $\mu > 0$, if $\|\psi\| \le \delta$ and $|D(n,x_n(\tau,\psi))| \le \alpha(\mu)$ for $n \ge \tau$, then by (4) it suffices to show that

$$K_1 e^{-\alpha(n-\tau)}\delta + K_2\alpha(\mu) \le K_1\mu + K_2\alpha(\mu),$$

which implies

$$\|x_n(\tau,\psi)\| \le \beta(\mu).$$

Therefore, if we pick an integer $N = \max\{0, [(\ln(\delta/\mu)/\alpha)] + 1\}$, where and in the sequel, $[\cdot]$ denotes the greatest integer function, then (6) holds. QED.

The first result in this section is established by means of the discrete Liapunov functionals with Razumikhin techniques.

Theorem 3.1. Let $D(n,\varphi)$ be U.S. Suppose there exists a $V : Z^+ \times \mathcal{C}_H \to R^+$ such that

(i) $u(|D(n,\varphi)|) \le V(n,\varphi) \le v(\|\varphi\|)$;

(ii) there exists an integer r_0 and a constant $\rho \ge 1$ such that when $n_0 \le n < n_0 + r_0$ and $V(n+1, x_{n+1}(n_0,\varphi)) \ge \rho v((\|\varphi\|)$ we have

$$\Delta V_{(3)}(n, x_n(n_0,\varphi)) \le 0,$$

and when $n \geq n_0 + r_0$ and $P(V(n+1, x_{n+1}(n_0, \varphi))) > V(s, x_s(n_0, \varphi))$ for $n - r_0 \leq s \leq n$ we have

$$\Delta V_{(3)}(n, x_n(n_0, \varphi)) \leq -w(|D(n, x_n(n_0, \varphi))|),$$

where $u, v, w \in \mathcal{K}$, P is the same as in Theorem 2.2., and

$$\Delta V_{(3)}(n, x_n(n_0, \varphi)) \equiv V(n+1, x_{n+1}(n_0, \varphi)) - V(n, x_n(n_0, \varphi))$$

with $x(n_0, \varphi)(n)$ being a solution of (3).

Then the zero solution of (3) is U.A.S..

Proof. (I) First, we show the U.S. It is possible to choose a continuous function $\alpha(t) \geq u^{-1}(\rho v(t))$ so that $\alpha(0) = 0$ and $\alpha(t) > u^{-1}(\rho v(t)) \geq 0$ for small $t > 0$. Thus, $\rho v(t) < u(\alpha(t))$ for small $t > 0$. For the above-chosen $\alpha(t)$, by Lemma 3.1 we can find a corresponding $\beta(t)$ with the desired properties.

Now for any given $\varepsilon > 0$ ($\varepsilon < H$), let $\delta < \min\{\varepsilon, A, \beta^{-1}(\varepsilon)\}$, and for any $n_0 \in Z^+$ and $\varphi \in C_\delta$, we denote $x(n) = x(n_0, \varphi)(n)$, $x_n = x_n(n_0, \varphi)$, $V(n) = V(n, x_n(n_0, \varphi))$, and $\Delta V(n) = \Delta V_{(3)}(n, x_n(n_0, \varphi))$. Then by (i) we have

$$u(|D(n_0, x_{n_0})|) \leq V(n_0) \leq v(\|x_{n_0}\|) = v(\|\varphi\|) < v(\delta) < u(\alpha(\delta)).$$

By reduction to absurdity, we can show that

$$u(|D(n, x_n)|) \leq V(n) \leq u(\alpha(\delta)) \quad \text{for all } n \geq n_0,$$

which implies that $|D(n, x_n)| \leq \alpha(\delta)$ for $n \geq n_0$, and thus by Lemma 3.1 we can obtain the required conclusion

$$\|x_n\| \leq \beta(\delta) < \varepsilon \quad \text{for } n \geq n_0.$$

(II) Furthermore, we show that the zero solution is U.A.S.. First of all, for $\varepsilon = H$, we can find the corresponding $\delta(H) > 0$ by the U.S. Let $\delta_0 = \delta(H)$. We want to show that for any given $\gamma > 0$ ($\gamma < \delta_0$), any $n_0 \in Z^+$ and any $\varphi \in C_{\delta_0}$, we can find an integer $N(\gamma) > 0$ such that

$$\|x_n(n_0, \varphi)\| < \gamma \quad \text{for all } n \geq n_0 + N(\gamma).$$

It is known from part (I) that for any $n_0 \in Z^+$ and any $\varphi \in C_{\delta_0}$ we have

$$V(n) \leq u(\alpha(\delta_0)), \quad |D(n, x_n)| \leq \alpha(\delta_0), \quad \text{and} \quad \|x_n\| < H \quad \text{for } n \geq n_0.$$

For the given $\gamma > 0$, we can choose $\lambda > 0$ such that $\lambda < \min\{\gamma, \beta^{-1}(\gamma)\}$. Let $d = \inf\{P(t) - t : u(\alpha(\lambda)) \leq t \leq u(\alpha(\delta_0))\}$, and K be the least positive integer with $u(\alpha(\lambda)) + Kd \geq u(\alpha(\delta_0))$.

By induction we can claim that

$$V(n) \leq u(\alpha(\lambda)) + (K - i)d \quad \text{for } n \geq n_i = n_0 + iN^*, \quad i = 0, 1, \ldots, K, \tag{7}$$

where N^* is to be determined later and which will be independent of n_0. In fact, (7) is trivially valid for $i = 0$. Now we assume that for some j with $0 \leq j \leq K - 1$, we have

$$V(n) \leq u(\alpha(\lambda)) + (K - j)d \quad \text{for } n \geq n_j.$$

We want to show that

$$V(n) \leq u(\alpha(\lambda)) + (K - j - 1)d \quad \text{for } n \geq n_{j+1}. \tag{8}$$

First, by reduction to absurdity and the estimate (4) in Definition 3.1, we can assert that there must exist some $n^* \in [n_j + r_0, n_j + r_0 + q\sigma]$ such that

$$V(n^*) \leq u(\alpha(\lambda)) + (K - j - 1)d,$$

where $q = [u(\alpha(\delta_0))/w(\lambda/(2K_2))] + 1$ and $\sigma = \max\{0, [(\ln(2K_1 H/\lambda)/a] + 1\}$ with constants $K_1,\ K_2,\ a$ in Definition 3.1.

Furthermore, we can assert that

$$V(n) \leq u(\alpha(\lambda)) + (K - j - 1)d \quad \text{for all } n \geq n^*.$$

Therefore, we arrive at (8) with $n_{j+1} = n_j + N^*$, where $N^* = r_0 + q\sigma$ is obviously independent of n_0. By induction, it follows from (7) and (i) that

$$u(|D(n, x_n)|) \leq V(n) \leq u(\alpha(\lambda)) \quad \text{for } n \geq n_K = n_0 + KN^*,$$

and thus,

$$|D(n, x_n)| \leq \alpha(\lambda) \quad \text{for } n \geq n_K = n_0 + KN^*.$$

By applying Lemma 3.1 with $\mu = \lambda,\ \delta = H,\ \tau = n_0 + KN^*$, and $\psi = x_\tau$ (Noting that $\|\psi\| = \|x_\tau\| < H$.) we conclude that

$$\|x_n(n_0, \varphi)\| = \|x_n(\tau, x_\tau)\| \leq \beta(\lambda) < \gamma \text{ for } n \geq n_0 + N(\gamma),$$

where $N(\gamma) = KN^* + N(H, \lambda)$ is also independent of n_0, while $N(H, \lambda)$ is the one from Lemma 3.1. This completes the proof. QED.

Instead of discrete Liapunov functionals in Theorem 3.1, if we adopt the discrete Liapunov functions together with Razumikhin technique, then we are able to establish another kind of stability criteria for (3) as follows.

Theorem 3.2. Let $D(n, \varphi)$ be U.S. and $|D(n, \varphi)| \leq v_0(\|\varphi\|)$ with $v_0 \in \mathcal{K},\ u, v, w \in \mathcal{K}$, and $\beta(t) \geq t$ is the corresponding $\mathcal{K}$ function to $\alpha(t) \geq u^{-1}(v(v_0(t)))$ by Lemma 3.1. Suppose there exist a $V : Z^+ \times R^k \to R^+$ and a $F \in C[R^+, R^+]$ which is nondecreasing and $F(v(v_0(t))) > v(\beta(t))$ for small $t > 0$ such that

(i) $u(|x|) \leq V(n, x) \leq v(|x|)$;

(ii) whenever $F(V(n+1, D(n+1, x_{n+1}))) > V(n+s, x(n+s))$ for $-r \leq s \leq 0$ we have

$$\Delta V_{(3)}(n, D(n, x_n)) \equiv V(n+1, D(n+1, x_{n+1})) - V(n, D(n, x_n)) \leq -w(|D(n, x_n)|).$$

Then the zero solution of (3) is U.A.S..

Proof. (I) First, we show that the zero solution of (1) is U.S. For any given $\varepsilon > 0$ ($\varepsilon < H$), pick $\delta > 0$ with $\delta < \min\{\varepsilon, A, \beta^{-1}(\varepsilon)\}$. For any $n_0 \in Z^+$ and $\varphi \in C_\delta$, we denote $x(n) = x(n_0, \varphi)(n)$, $V(n) = V(n, x(n))$, $\bar{V}(n) = V(n, D(n, x_n))$, and $\Delta\bar{V}(n) = \Delta V_{(1)}(n, D(n, x_n))$.

By reduction to absurdity and applying Lemma 3.1, we can claim that

$$u(|D(n, x_n)|) \leq \bar{V}(n) \leq v(v_0(\delta)) \text{ for all } n \geq n_0,$$

which implies that

$$|D(n, x_n)| \leq u^{-1}(v(v_0(\delta))) \leq \alpha(\delta) \quad \text{for } n \geq n_0.$$

As in Theorem 3.1, by applying Lemma 3.1 again we can conclude that

$$\|x_n\| \leq \beta(\delta) < \varepsilon \quad \text{for } n \geq n_0.$$

This shows the U.S. of the zero solution.

(II) Next, we assert that the zero solution is U.A.S.. For $\varepsilon = H$, by the U.S. we can find the corresponding $\delta(H) > 0$. Let $\delta_0 = \delta(H)$. Then for any $n_0 \in Z^+$ and $\varphi \in C_{\delta_0}$ we have

$$\bar{V}(n) \leq v(v_0(\delta_0)), \quad |D(n, x_n)| \leq \alpha(\delta_0), \quad \text{and} \quad \|x_n\| < H \quad \text{for all} \quad n \geq n_0. \tag{9}$$

Now for any given $\gamma > 0$ ($\gamma < \delta_0$) (we may assume $\gamma < 4\beta(\delta_0)$ and thus $\delta_0 > \beta^{-1}(\gamma/4)$), we will find $N(\gamma) > 0$ such that $\varphi \in C_{\delta_0}$ implies

$$\|x_n\| = \|x_n(n_0, \varphi)\| < \gamma \quad \text{for} \quad n \geq n_0 + N(\gamma).$$

For the given $\gamma > 0$, we choose $\lambda > 0$ so that $\gamma/2 \leq \beta(\lambda) < \gamma$. By the assumption, we can pick a constant $d > 0$ such that $v(v_0(\beta^{-1}(\gamma/2))) - d > v(v_0(\beta^{-1}(\gamma/4)))$ and

$$F(v(v_0(t)) - d) > v(\beta(t)) \quad \text{for} \quad \beta^{-1}(\gamma/4) \leq t \leq \delta_0.$$

It is easy to see that by the assumption on F and the uniform continuity we may choose $d > 0$ suitably small so that $F(v(v_0(t)) - d) > F(v(v_0(t))) - \rho \geq v(\beta(t))$ for $\beta^{-1}(\gamma/4) \leq t \leq \delta_0$. Let K be the least nonpositive integer such that $u(\alpha(\lambda)) + Kd \geq v(v_0(\delta_0))$.

First, we assume that $u(\alpha(\lambda)) < v(v_0(\delta_0))$, and thus $K \geq 1$. Then there must exist $\delta_i > 0$ such that

$$v(v_0(\delta_i)) = v(v_0(\delta_0)) - id, \qquad i = 0, 1, 2, \ldots, K.$$

The choices of λ, d, and K as well as the assumption $u(\alpha(t)) \geq v(v_0(t))$ imply that $\delta_0 > \delta_1 > \cdots > \delta_K > \beta^{-1}(\gamma/4) > 0$. Let

$$M = \inf_{v_0(\delta_K) \leq t \leq \alpha(\delta_0)} w(t) > 0.$$

By induction with Lemma 3.1, we can claim that

$$\bar{V}(n) \leq v(v_0(\delta_0)) - id \quad \text{for} \quad n \geq n_i = n_0 + iN^*, \quad i = 0, 1, \ldots, K,$$

where $N^* = \tilde{N} + [v(v_0(\delta_0))/M] + 2$, $\tilde{N} = \max_{0 \leq i \leq K} N(H, \delta_i)$, and $N(H, \delta_i)$ is the corresponding integer in Lemma 3.1. Therefore, we arrive at

$$u(|D(n, x_n)|) \leq \bar{V}(n) \leq v(v_0(\delta_0)) - Kd \leq u(\alpha(\lambda)) \quad \text{for} \quad n \geq n_0 + KN^*,$$

and thus

$$|D(n, x_n)| \leq \alpha(\lambda) \quad \text{for} \quad n \geq n_K = n_0 + KN^*. \tag{10}$$

On the other hand, if $u(\alpha(\lambda)) \geq v(v_0(\delta_0))$, then $K = 0$ and by (9) we know that (10) is still true.

Since $\|x_{n_K}\| < H$, by applying Lemma 3.1 again we obtain

$$\|x_n\| \leq \beta(\lambda) < \gamma \quad \text{for all} \quad n \geq n_K + N(h, \lambda) \equiv n_0 + N(\gamma),$$

where $N(\gamma) = KN^* + N(H, \lambda)$, which is obviously independent of n_0. Thus, the proof is now complete. QED.

ACKNOWLEDGMENT

This work was partially supported by the National Natural Science Foundation of China.

REFERENCES

1. Elaydi, S., & Zhang, Shunian, Stability and periodicity of difference equations with finite delay. *Funkcial. Ekvac.*, Vol 37 (1994), 401-413.
2. Zhang, Shunian, Razumikhin techniques in delay difference systems. *PanAmerican Math. J.*, Vol 3 (1993), 1-16.

EDITORS

G. S. Ladde, Department of Mathematics,
University of Texas at Arlington, Arlington, TX, 76019, USA,

N. G. Medhin, Department of Mathematics,
Clark Atlanta University, Atlanta, Atlanta, GA, 30314, USA., and

M. Sambandham, Department of Mathematics
Morehouse College, Atlanta, Atlanta, Atlanta, GA, 30314, USA.

Editorial Committee

R. P. Agarwal, Dept of Math, University of Singapore, Singapore
Shair Ahmad, Dept of Math, Univ. of Texas at San Antonio, San Antonio, TX 78285, USA
D. Bainov, P.O. Box 45, 1504 Sofia, BULGARIA
R. Bozeman, Dept of Math, Morehouse College, Atlanta, GA, 30314, USA.
C. Y. Chan, Dept of Math, University of Louisiana at Lafayette, Lafayette, LA 70504, USA
J. Graef, Dept of Math, The University of Tennessee, Chattanooga, TN 37403-2598
J. K. Hale, Dept of Math, Georgia Institute of Technology, Atlanta, GA 30332, USA
L. Hatvani, Dept of Math, Bolyai Institute, Aradi vertanuk Tere 1, Szeged, H-6720, Hungary
D. Kannan, Dept of Math, University of Georgia, Athens, GA 30602, USA
J. de Klerk, Dept of Math, Potchefstroom University, Potchefstroom 2520, South Africa
V. B. Kolmanovskii, Cybe. Dept, MIEM, Bolshoi Vuzovskii 3/12, Moscow 109028 USSR
V. Lakshmikantham, Dept of Appl.Math, Florida Inst. of Tech., Melbourne, FL 32901, USA
V. M. Matrosov, Nonl. Dyn. Res. Center, Dm. Ulyanova Str. Bld. 5, 11733 Moscow, Russia
R. E. Mickens, Dept of Physics, Clark Atlanta University, Atlanta, GA 30314
A. Msezane, Center for Theor. Studies, of Phy. Sys., Clark Atlanta Univ. Atlanta, GA 30314
M. Rama M. Rao, Dept of Math., Univ. of Texas at San Antonio, San Antonio, TX 78285, USA
S. Sathanantham, Dept. of Math, Tennessee State University, Nashville, TN, 37203-3401
R. Shivaji, Department of Math, Mississippi State University, Mississippi State, MS 39762
S. Sumner, Department of Math, Mary Washington College, Freferickburg, VA 22401-5358
C. P. Tsokos, Dept. of Math., University of South Florida, Tampa, FL, 33620
A. S. Vatsala, Dept of Math, University of Louisiana at Lafayette, Lafayette, LA 70504, USA
F, Zanolin, Dept di Math , Universita-via Zanon, 6-33100 Udine, ITALY

SPEAKERS

Ackleh, Azmy., University of Southwestern Louisiana, LA
Agarwal, R. P., National University of Singapore, Singapore
Ahlbrandt, Calvin., Utah State University, UT
Ahmad, Shair, University of Texas at San Antonio, TX
Akca, Haydar., King Fahd Univ. of Petroleum and Minerals, Saudi Arabia
Al-Dabass, David., The Nottingham Trent University, UK
Al-Said, Eisa A.,King Saud University, Saudi Arabia
Alexandre Zenchuk, Instituto de Fisica Teorica, Brazil
Alkahby, Hadi Y., Dillard University, LA
Anabtawi, Mahmoud., Tennessee State University, TN
Anderson, Douglas., Concordia College, MN
Atici, Ferhan Merdivenci., Ege University, Turkey
Avery, Richard I.,Dakota State University, SD
Bantsur, Nataliya Ins. Mat. Acac Scie. Ukraine
Bartusek, Miroslav.,Masaryk University, Czech Republic
Blackmore, Dennis., New Jersey Institute of Technology, NJ
Blayneh, Kbenesh., Florida A&M University, FL
Bogdanov, Andrew., Ulyanovsk State University, Russia
Bohner, Martin., University of Missouri Rolla, MS
Byszewski, Ludwik., Cracow University of Technology, Poland
Cahlon, Baruch., Oakland University, MI
Caldwell, S., Mississippi State University, MS
Chan, C.Y., University of Lusitania at Lafayette, LA
Chandramouli, R., Univ. of South Florida, FL
Chen, Chang., University of Texas at San Antonio, TX
Chen , Yuping., Wuhan University, P. R. China
Chen, Chao-Nien., National Changhua University of Education, Taiwan
Chen, Yong-Zhuo., University of Pittsburgh at Bradford, PA
Chhetri, Maya., Mississippi State University, MS
Christov, Christo I., University of Southwestern Louisiana, LA
Dads, Aid., Universite Cadi Ayyad, Morocco
Datta, Anjali., Tuskegee University, Alabama
Deng, Keng., University of Southwestern Louisiana, LA
Dobnikar, Andrej., Univerza v Ljubljani, Slovenia
Dosly, Ondrej., Masaryk University, Czech Republic
Drici, Zahia., Illinois Wesleyan University, IL
Edwards, Roderick., University of Victoria, Canada
Elbert, Arpad., Hungarian Academy of Sciences, Hungary
Erbe, Lynn H., University of Nebraska-Lincoln, NE
Escultura, E. E. ,University of Philippines, Phillippines
Ezzinbi, Khalil., Universite Cadi Ayyad, Morocco
Fedotov, Alexandr S., Ins. Mat. Acac Scie. Ukraine
Fedotov, Alexandr S., Ternopol Pedagogical University, Ukraine
Franceschini, V.,Universita ' di Modena, Italy

Galkin, V. A., Institute of Atomic Energetics, Russia
Garner, J. B. Mississippi State University, MS
George, Gravvanis., University of the Aegean, Greece
Gilberti, G., Universita ' di Modena, Italy
Golec, Janusz., Fordham University, NY
Gomatam, Jagnnathan., Glasgow Caledonian University, UK
Gomatam, Shanti., University of South Fl, FL
Gonzalez-Hernandez, Hugo G., Universidad La Salle, Mexico
Graef, John., University of Tennessee, TN
Guo, Jong-Shenq., National Taiwan Normal University, Taiwan
Gupta, Chaitan P., University of Nevada, NV
Guseinov, G.Sh.,Ege University, Turkey
Hai, D.D., Mississippi State University, MS
Hamaya, Yoshihiro., Okayama University of Science, Japan
Han, Maoan., Georgia Institute of Technology, GA
Handy, Carlos., Clark Atlanta University, GA
Harrison, Martin., Loughborough University, UK
Hatvani, L., Szeged University, Hungary
He, Ming., University of Wisconsin-La Crosse, WI
Heikkila, Seppo., University of Oulu, Finland
Henderson, Johnny., La Grange College, GA & Auburn Univ., AL
Hernandez, Jaime., Dillard University, LA
Hetzer, G. , Auburn University, AL
Hirsch, Morris, W., University of California, CA
Huynh, Quyen., Brown University, RI
Islam, Muhammad N.,University of Dayton, OH
Ivanov, Anatoli F., Pennsylvania State University, PA
Jalbout, Fouad N. , Dillard University, LA
Jitao, Sun., Shanghai Tiedao University, P. R. China
Kamenskii, G. A.,Russia
Kang , Lishan., Wuhan University, P. R. China
Kannan, D., University of Georgia, GA
Kaplan, Lance., Clark Atlanta University, GA
Karsai,Janos., Med. University, Hungary
Kaymakcalan, Billur., Middle East Technical University, Turkey
Kersner, Robert., Hungarian Academy of Sciences, Hungary
Kim, A., Seoul National University, Korea
Kim, Jong K., Kyungnam University, Korea
Klerk, Johan H de., Potchefstroom University, South Africa
Kolmanovskii, V., Moscow University of Electronics, Russia
Kolosov, Gennadii E. .,Moscow Univ. Info. Sciences and Control, Russia
Kolupaeva, Svetlana N. .,Tomsk State University, Russia
Konigsberg, Mordejai ., Mexico
Koralp, Erol., Middle East Technical University, Turkey
Kosareva Natalia, P., Russia
Kuang, Yang., Arizona State University, AZ
Kuzmina, Lyudmila K.,Kazan Aviation Institute, Russia
Ladde, G. S., University of Texas at Arlington, TA

Lakshmikantham, V., Fl. Institute of Technology, FL
Lawrence, Bonita A., University of South Carolina, SC
Lee, Keonhee., Chungnam National University, Korea
Liu, James.,James Madison University, VA
Lou, Yuan., Ohio State University, OH
Lu, Xin., University of North Carolina at Wilmington, NC
Ma, Xiaoyun., University of Wisconsin-La Crosse, WI
Mahesh, Nerurkar G., Rutgers University, NJ
Mao, Xuerong., University of Strathclyde, UK
Matrosov, V. M., Rissia
Medhin, Negash G.,Clark Atlanta University, GA
Medina, Rigoberto., Universidad de Los Lagos, Chile
Meehan, Maria., National University of Ireland, Ireland
Mickens, R. E., Clark Atlanta, University, GA
Migorski, Stanislaw., Jagiellonian University, Poland
Monika, Sobalovas., Masaryk University, Czech Republic
Montiel-Castellanos, Marcos., Universidad La Salle, Mexico
Mujica, Fernando., Georgia Institute of Technology, GA
Muldowney, James S., University of Alberta, Canada
Murenzi, Romain., Clark Atlanta University, GA
Muresan, Marian., Babes-Bolyai, University, Romania
Namuduri, Kameswara., Clark Atlanta University, GA
Navarro, Ramon., Universidad Simon Bolivar, Venezuela
Nu~nez, Carmen., Universidad de Valladolid, Spain
O'Regan, Donal., National University of Ireland, Ireland
Oh-Bok, Kwon., Kyugmoon Institute of Technology, Korea
Okonkwo, Zephyrinus C. ,Alabama State University, AL
Okunbor, Daniel., University of Maryland Eastern Shore, MD
Olmstead, W. E., Northwestern University, IL
Olusegun B. Oluyede, Morehouse College, GA
Onose, Hiroshi., Ibaraki University, Japan
Pederson, Steven M., Morehouse College, GA
Peterson, Allan., University of Nebraska-Lincoln, NE
Polotski, Vladimir., Ecole Polytechnique, Montreal
Raghavendra, V., Indian Institute of Technology, India
Rama Mohana Rao, M., University of Texas at San Antonia, TX
Ramachandran, K. M. ., University of South Fl, FL
Ramaswamy, Veeru., Clark Atlanta University, GA
Rammaha, Mohammad., University of Nebraska, NB
Ridenhour, Jerry., Utah State University, UT
Rivero, Jesus., Universidad de Ciencias, Venezuela
Rizzo, Rebecca L .,Florida Institute of Technology, FL
Roberts, Lila., Georgia Southern University, GA
Robinson, Stephen B.,Wake Forest University, NC
Rowell, Jonathan Tyrone., North Carolina State University, NC
Ruan, Shigui., Dalhousie University, Canada
Ryashko, Lev., Ural State University, Russia
Sanz, Ana., Universidad de Valladolid, Spain

Sathananthan, S.,Tennessee State University, TN
Schmidt, Darrell., Oakland University, MI
Shen, Wenxian., Auburn University, AL
Shilnikov, Andrey., Georgia Institute of Technology, GA
Shivaji, R., Mississippi State University, MS
Simpao, Valentino A., Mathematical Consultant Services, KY
Singh-Altmayer, Kumed., Southern University, GA
Smaoui, Nejib., Kuwait University, Kuwait
Smith, J. T.., GA Institute of Technology, GA
Snider, Arthur David., University of South Florida, FL
Sochacki, Jim., James Madison University, VA
Stapleton, David P. ,University of Central Oklahoma, OK
Steinberg, Lev G.,University of Puerto Rico, PRODUCTION
Stephens, Sonya A. F.,Florida A&M University, FL
Story, Troy., Morehouse College, GA
Sumner, Suzanne., Mary Washington College, VA
Swiniarski, Roman W., San Diego State University, CA
Szent-Gyorgyi, A., Med. University, Hungary
Takeuchi, Yasuhiro., Shizuoka University, Japan
Tineo, Antonio., Universidad de los andes, Merida, Venezuala
Travis, Betty., University of Texas at San Antonio, TX
Tsokos, C. P., University of South Florida, FL
Vanmaele, Michele., Southern University, GA
Vatsala, A.S., University of Southwestern Louisiana, LA
Vernia, Cecilia.,Universita ' di Modena, Italy
Vrinceanu, Daniel., Georgia Institute of Technology, GA
Wang, Liancheng., Mississippi State University, MS
Wang, Liwen., University of Southwestern Louisiana, LA
Wei, Dongming., University of New Orleans, LA
Wu, De Ting., Morehouse College, GA
Y.M, Maximov., Russia
Yakubu, Abdul-Aziz., Howard University, DC
Yanagiya, Akira., Waseda University, Japan
Yang, Bo., Mississippi State University, MS
Yin, George., Wayne State University, MI
Yin, William., La Grange College, GA & Auburn University, AL
Yinping, Zhang., Shanghai Tiedao University, P. R. China
Yuen, S. Irene., Lakeland Community College, OH
Zanolin, F., University of Udine, Italy
Zhai, Jian., Hokkaido University, Japan
Zhang, Shunian.,, Shanghai Jiao Tong University, P. R. China
Zhang, Weijiang., University of Wisconsin-La Crosse, WI
Zhao, Xiaquan., University of Southwestern Louisiana, LA
Zhu, Jianmin., Fort Valley State University, GA
Zvi Retchkiman., Mexico